SAUNDERS MAC LANE

SELECTED PAPERS

B. Gaye.

Saunders Mac Lane
Selected Papers

Edited by I. Kaplansky

Springer-Verlag
New York Heidelberg Berlin

Saunders Mac Lane

Department of Mathematics
University of Chicago
5734 University Avenue
Chicago, Illinois 60637
USA

Editor

I. Kaplansky

Department of Mathematics
University of Chicago
5734 University Avenue
Chicago, Illinois 60637
USA

AMS Subject Classification: 01 A 75

Library of Congress Cataloging in Publication Data

Mac Lane, Saunders,
Selected papers.

"Bibliography of Saunders Mac Lane": p.
1. Algebra—Collected works. I. Kaplansky, Irving, 1917–
QA155.M27 510 79-10105

Printed in the United States of America.

9 8 7 6 5 4 3 2 1

ISBN 0-387-90394-1 Springer-Verlag New York
ISBN 3-540-90394-1 Springer-Verlag Berlin Heidelberg

Table of Contents

*Numbers in brackets refer to Bibliography on pp. 545–553.

Preface

A preface usually ends with appropriate expressions of thanks to the people who have helped. I would like instead to begin in that way. Most important is my gratitude to Samuel Eilenberg, Roger Lyndon, and Max Kelly, who joined me in contributing essays, and to Alfred Putnam who wrote the biography that leads off the volume. Dorothy Mac Lane helped generously with her encouragement (and of course her special source of information). Freda Davidson pitched in gallantly in helping to proofread the selected papers for typographical slips that had somehow eluded the eagle eye of Saunders Mac Lane (there turned out to be mighty few). When tricky problems came up I could always rely on Herman Meyer to solve them. John Gray enriched the year by organizing a highly successful conference in Aspen on May 23–27, 1979. Lastly, Walter Kaufmann-Bühler of the Springer-Verlag brought to bear his immense know-how and met all my wishes (with a smile, I think).

Saunders Mac Lane will be seventy years old on August 4, 1979, and with a little luck, the date of publication will work out to be within epsilon of this target date. Happy birthday, Saunders! (My colleagues in Eckhart Hall who watch his close approximation to perpetual motion will find this hard to believe.)

As Max Kelly remarks in beginning his essay, Saunders has given us bits of his own overview of his scientific career in [93] and [115]. To this I would add [114], where there is a fascinating narrative of the long collaboration with Sammy (which we can now read about on both sides of the ledger), and [116], which offers a broad panorama of the way the cohomology of groups arose in various ways at the hands of various people.

In the February, 1975 issue of the American Mathematical Monthly there is a tribute by Ralph Boas, on the occasion of the selection of Saunders for the Association's Distinguished Service Award.

For both professional and personal reasons, helping to assemble this volume of selected papers has been an enjoyable task. One particularly gratifying aspect is that his Ph.D. thesis is at last appearing in the regular literature, with a candid second look accompanying it as a bonus. Up to now (except for fifty scattered copies) it was invisible except for being listed as number 504 in Church's bibliography of symbolic logic (Journal of Symbolic Logic, vol. 1, p. 214).

There is one final point to be mentioned. The erratum to [109] has not been reprinted, since the paper itself was slightly revised to take account of it.

Irving Kaplansky

A Biographical Note

Alfred Putnam

A biographical note constitutes a form of introduction. Yet to introduce anyone as widely known and universally recognized as Saunders Mac Lane poses a particular problem: Saunders has had his share of formal introductions but he is not a formal person. A solution would be to rehearse informally information available in standard references; that would omit the vitality of the man. An alternative solution would be to recount a selection of more personal and characteristic notes; that would omit the structure of a lifetime. Since my introduction may be serving to acquaint some with Saunders Mac Lane for the first time, I have tried to include both facts from a full life and more personal notes to enable others to gain a sense of this intensely human mathematician and intensely mathematical man. I hope that Saunders himself, from whom I have so often heard—when it especially counted—the admonition "Give me an example" will savor my solution of the problem he presents. I should offer my credentials: for close to forty years Saunders Mac Lane has been my mathematical mentor, colleague, friend.

Of one thing about Saunders Mac Lane I am as certain as when I began to work with him as a graduate student: anyone engaged with him is sure to learn something. What it is could be the significance for mathematics of a complex piece of research. It could be how, over the years, firmly held convictions as to the true nature of mathematics must shape the content and teaching of mathematics. Or it could be the right principles by which the contributions of a mathematician to mathematics ought to be judged. At the right moment, on a walk with him it could even be a simple theory to account for cyclical fluctuations in the level of Lake Michigan. In whatever way learning may be experienced with Saunders Mac Lane, it is certain to have the marks of intense concentration, directness of address, and strictest attention to fundamentals. Saunders Mac Lane is ever the teacher.

The life we are celebrating began in Taftville, Connecticut on August 4, 1909 when Saunders Mac Lane was born, the eldest child of Donald Bradford and Winifred (Saunders) MacLane. His New England descent traced to the Mayflower, his Scots family to the MacLeans of Mull. (Actually, it would be more nearly accurate to record that it was Saunders *MacLane* who was born on that day in 1909, for the Saunders *Mac Lane* we know came into being much later through a creative accident in typing his name.)

Saunders Mac Lane's father and grandfather were both Congregational ministers. His grandfather, who was then a Presbyterian, had espoused evolution in his church in Steubenville, Ohio and found therewith that the more open views of Congregationalism better matched his convictions. Saunders' father was a conscientious man with a strong sense of obligation. During World War I in Roxbury, Massachusetts, Saunders' father had given his support to pacifism and had felt obliged to relinquish his church there. He moved then to a small and struggling congregation in Wilbraham, Massachusetts. There Saunders attended a two-room school. It must have been with something of his grandfather's and father's spirit that Saunders, in his turn in the 1920's, could be found advocating in the high school paper in Leominster, Massachusetts the abolition of military drill and military spending.

Those who know Saunders Mac Lane at all know him as a man of utmost integrity, steadfast principles, and strong will. On occasion these can have meant pain for himself or others, when feelings and strict application of principles have been at odds, but he is made of stern stuff. In that, as in his gift for teaching, his didactic skill, and his standards of clear and direct expression, we see in Saunders the force of some of that same tradition which informed the lives of his father and grandfather.

When Saunders Mac Lane was fifteen his father died and the family moved to Leominster, where his grandfather had a church. Saunders graduated from high school in 1926 and was sent by an uncle to Yale. In his first year there he proceeded to win the Barge Prize in a freshman mathematics competition. He is reputed also to have achieved the distinction of having worked all the problems in the current redoubtable Yale calculus textbook. His subsequent mathematical development at Yale was powered by his own curiosity and drive, and was channeled by Ore, Wilson and Pierpont into areas represented in the strong but rather stiff department of that day. In another direction, his experience with E. J. Miles, whose love for teaching mathematics to undergraduates extended over a lifetime, must have begun to shape a feeling for teaching into what became later Saunders' own utterly distinctive and colorful style of forceful exposition.

In 1930 Saunders Mac Lane graduated from Yale with the degree of Ph.B. Under what seem now the rather quaint requirements of that day, and even a manifestation of "the two cultures," his was the degree reserved for those who arrived in New Haven with less Latin than required for candidates for the more usual A.B.

The year following graduation Saunders held a fellowship for $1000 at the University of Chicago. In 1929 he and Robert Maynard Hutchins, who had left Yale in that year as Law School dean to become President of the University of Chicago, each had received one of the Yale awards annually provided from Montclair, New Jersey. That departure from the regular custom of recognizing illustrious football performance with an award was a propitious one for the young mathematician in initiating an association with Hutchins and Chicago. When he arrived to take up his fellowship, however, even though Hutchins had personally invited him to Chicago, Saunders' first undertaking was to be admitted as a student, a formality he had completely overlooked.

His fellowship year at Chicago was a significant one. He came into contact with new mathematics and new ways of looking at mathematics. As was true for anyone at Chicago, he felt the influence of the powerful views of E. H. Moore. Then, in the spring of that fellowship year, he met a graduate student in economics who was to have a very important place in his future. Through the spring months Dorothy Jones, the daughter of a faculty family at the University of

Arkansas, and he found and shared together many common interests.

The following fall, with his degree of A. M. from Chicago, plans to continue his mathematics, and an admixture of wanderlust, Saunders Mac Lane was off to Göttingen. There was then at Göttingen an unparalleled group of productive mathematicians, leaders in creating exciting advances in new directions in mathematics. Saunders' studies took shape for him particularly in logic and he undertook his dissertation with Paul Bernays. Because of the ominous movement of events in Germany, however, he hastened to complete his dissertation in the early summer of 1933 and had his examination in mid-July with Hermann Weyl. That July was eventful for him in another way: only a few days after his successful examination he and Dorothy Jones were married in Göttingen. She had come on to Göttingen earlier and had typed his dissertation.

When Saunders and Dorothy returned to the United States it was to a Sterling Fellowship at Yale for 1933–34. His dissertation having been duly published in Germany, he received his D.Phil. from Göttingen in 1934. Then from Yale he moved to Harvard, where for the next two years he held a Benjamin Peirce instructorship.

By 1936 the economic depression which had gripped the United States since 1929 had eased, but jobs for beginning mathematicians—no matter how able—were still scarce. It was Saunders' good fortune that among these an instructorship for one year was open at Cornell for 1936–37. In the course of that year came an offer from the University of Chicago, to which he returned for 1937–38 as instructor. It was during that year at Chicago that the Mac Lanes' first daughter, Gretchen, was born and that Saunders was called back to Harvard as assistant professor.

I suppose it must have been when I arrived in Cambridge in the fall of 1938, fresh from Hamilton College and hopeful but untried, that I first met Saunders Mac Lane. During my graduate years I never had a formal course with him, but in courses of his which I attended I met a style of mathematics and mathematical thinking that captured my enthusiasm. I knew my bent was algebra, and when the time came Saunders Mac Lane accepted me as his student. Harvard was an exhilarating place to be for mathematics, and all of us there were fortunate to have had that experience before the anxieties preceding World War II could fully overtake us.

My meetings with Saunders to review my work were sometimes in his study in Eliot House, or on walks along the Charles River, or at the Mac Lanes' home at 7 Avon Street. These meetings were vigorous, stimulating, and productive. I had started out to look at some problems in algebraic geometry, following Saunders' suggestion, but had finally settled down with him to investigate the structure of certain domains complete in a valuation. At our meetings he invariably pressed me for genuine examples of the sometimes unrealistically abstract notions I proposed, and he saw to it that I worked these out as fully as I could. "Give me an example" was a challenge to discipline my speculations.

As Teaching Fellow in one of Harvard's freshman courses I experienced Saunders' style in another way. He visited my Math A class when we were doing some of the analytic geometry of lines. I had prepared diligently, but my exposition of the solution on one problem, though utterly transparently lucid to me, was obviously far less so to the students. At the end of the hour—which finally arrived—Saunders asked, "Do you have some time?" I agreed I had. "Then come across the hall and I'll show you how you should have taught that class." That next hour was as uncomfortable and as valuable as any in my whole career, and I emerged from it imbued with the makings of a teacher. Even today I find myself returning to some of those lessons. I cannot be sure of everything I learned, but to exhibit natural liveliness was one and to seek to make clear the essential lines of the plot of an argument was another.

In the spring of 1941, while Saunders was on leave at Yale, the Mac Lanes' second daughter, Cynthia, was born. That year also saw the publication of *A Survey of Modern Algebra*, fruit of the collaboration of Garrett Birkhoff and Saunders Mac Lane on his return to Harvard. Birkhoff and Mac Lane, with *A Survey of Modern Algebra*, opened to American undergraduates what had until then been largely reserved for mathematicians in van der Waerden's *Moderne Algebra*, published a decade earlier. The impact of Birkhoff and Mac Lane on the content and teaching of algebra in colleges and universities was immediate and long sustained. What we recognize in undergraduate courses in algebra today took much of its start with the abstract algebra which they made both accessible and attractive.

Pearl Harbor abruptly changed all our lives. I hurried ahead to complete my thesis, then moved to Yale and an undergraduate Navy Program. Saunders early engaged himself in the war effort in the Applied Mathematics Group–Columbia.

Saunders Mac Lane and I became colleagues at the University of Chicago in the fall of 1947. I had preceded him by two years in response to an invitation to join the College Mathematics Staff. This was a separate faculty group, then being organized under the leadership of E. P. Northrop, which was entrusted with the mathematics program in the undergraduate plan of general education in the newly structured College of the University of Chicago, that had been engineered by Robert Maynard Hutchins. Saunders came as professor to the Department of Mathematics from Harvard, where he had become associate professor in 1941 and professor in 1946. The Department of Mathematics, upon the initiative of Hutchins himself, had become by 1947, under the chairmanship of Marshall Stone, a research center unexcelled in the United States. In algebra there was assembled a most remarkable group: Adrian Albert, Irving Kaplansky, Saunders Mac Lane, Otto Schilling, André Weil.

Saunders Mac Lane succeeded Marshall Stone at Chicago in the chairmanship of the Department of Mathematics in 1952. The two three-year terms of his chairmanship were typically vigorous and personal. His advocacy of the research goals of the department was ardent and tireless. It was characteristic of Saunders that he should have made special efforts to know each graduate student and to follow the research of each. For the young mathematicians who held appointments at Chicago he never failed to seek ways to urge them on in their work and to provide them with what he judged to be the most stimulating environment.

In the spring of 1952 I had agreed to head the College Mathematics Staff. The divergences between the research interests centered in the department and the perceived commitment of the College Mathematics Staff to mathematics in general education generated tensions which sometimes ruffled the relationship between Saunders and me. He pressed his points forcefully. His disbelief in the appropriateness of the exposition of any formal logic in the mathematics course in the College was consistently maintained. His active belief in the centrality of the concept of function was unwavering. The concomitant exchange of views, which engaged members of the department and staff, opened the way for future constructive changes in the administrative structure of undergraduate mathematics at Chicago, culminating in the consolidation of the College Mathematics Staff into the Department of Mathematics in 1959.

The wider public aspects of Saunders Mac Lane's devotion to the cause of mathematics, and to science generally, have been many. He was President of the Mathematical Association of America from 1950 to 1952, and in 1975 received its Distinguished Service Award in recognition of his sustained and active concern for the advancement of undergraduate mathematics. In 1963 the University of Chicago named him Max Mason Distinguished Service Professor. He served as President of the American Mathematical Society from 1973 to 1975. The following year he headed a delegation of mathematicians for a formal visit to the Peoples Republic of China. In 1973 he became a Vice President of the National Academy of Sciences, to which body he had first been elected in 1949. He was appointed to the National Science Board in 1974.

Beginning as a graduate student with a brief exposure to group extensions, I have watched the development of Saunders Mac Lane's mathematics through homological algebra to category theory. It is a pleasure now to give him a special salute: Saunders Mac Lane belongs in a category by himself.

ABGEKÜRZTE BEWEISE IM LOGIKKALKUL

Inaugural-Dissertation

zur Erlangung der Doktorwürde
der Mathematisch-Naturwissenschaftlichen Fakultät
der Georg August-Universität zu Göttingen

vorgelegt von

Saunders MacLane
aus Norwalk, Connecticut, U.S.A.

GÖTTINGEN
Gedruckt bei Hubert & Co. G.m.b.H.
1934

Referent: Professor Dr. Hermann Weyl
Tag der mündlichen Prüfung: 19. Juli 1933

Inhaltsverzeichnis.

Einleitung.

§ 1. **Der Begriff eines Prozesses:** Symbolik ist vielleicht das Hauptthema der modernen mathematischen Logik. Die logische Symbolik ist jetzt im wesentlichen vollständig, denn mit den Symbolen der modernen Logik kann man jeden mathematischen Satz knapp und bequem ausdrücken. Das Grundproblem der logischen Deduktion bleibt aber durch dieses Studium der Symbolik im wesentlichen ungeändert, denn in jeder Symbolik muß man nichtsymbolische Regeln haben, die verschiedene Gruppen von Zeichen miteinander verknüpfen. Die Aufgabe der Logik ist letzten Endes die Ableitung von Schlüssen aus gegebenen Prämissen, und dazu muß man Methoden haben, die den Aufbau von Schlüssen aus Prämissen erlauben. Eine solche Konstruktionsmethode nennen wir einen Prozeß. Ein *Prozeß* hat also zwei Merkmale: erstens ist er eine Vorschrift, die eine neue Gruppe von Zeichen durch Transformation oder Kombination aus einer oder mehreren gegebenen Gruppen von Zeichen erzeugt; zweitens muß ein Prozeß als Deduktionsmethode anwendbar sein.

Schon die Logik von Aristoteles kann man als eine Logik der Prozesse auffassen: der Syllogismus ist ja ein Prozeß. Man kann z. B. mit der Prämisse (Untersatz) „Sokrates ist ein Mensch" anfangen, und dann, mittels des Bezugstheorems (Obersatz) „Alle Menschen sind sterblich", schließen: „Sokrates ist sterblich". In der Mathematik kommen andere Prozesse vor. Wenn wir z. B. mit der Gleichung

$$(1) \qquad (a + b)\,c = ac + bc$$

als Prämisse anfangen, dann können wir durch Bezugnahme auf das kommutative Gesetz den Schlußsatz (linkes distributives Gesetz)

$$(2) \qquad c(a + b) = ca + cb$$

ableiten. In beiden Fällen haben wir einen Prozeß, der mit einer Prämisse anfängt, und dann durch ein Bezugtheorem einen gewissen Schluß oder ein gewisses Ergebnis erreicht.

Jeden mathematischen Beweis kann man als eine Reihe solcher Prozesse auffassen; wir werden daher die Prozesse als die atomaren Bestandteile der Beweise betrachten. Der Zweck dieser Arbeit ist die Untersuchung solcher Beweisprozesse: die Aufstellung der wichtigsten Prozesse, die genaue Formulierung dieser Prozesse mittels der Prinzipien der symbolischen Logik, und endlich das Studium der Methoden, wodurch ein Beweis aus solchen Prozessen aufgebaut werden kann.

§ 2. **Prozesse als Abkürzungen:** Jeder Beweisprozeß hat eine wichtige Eindeutigkeitseigenschaft: das Ergebnis des Prozesses ist immer durch die Prämisse und das Bezugstheorem eindeutig bestimmt. Im Syllogismus ist der Schlußsatz vollständig eindeutig bestimmt, wenn nur der Obersatz und der Untersatz gegeben sind. Analog ist das Ergebnis (2) des algebraischen Prozesses im vorigen Abschnitt völlig durch die Vorschrift bestimmt, daß man mit dem kommutativen Gesetz überall in der Gleichung (1) substituieren sollte. Da das Ergebnis eines Prozesses in dieser Weise festgelegt ist, ist es keineswegs notwendig, das Ergebnis selbst tatsächlich anzugeben! Diese Bemerkung ist besonders nützlich für einen Beweis, der aus einer Reihe von Prozessen besteht, wo jeder Prozeß mit dem Ergebnis des vorigen Prozesses anfängt. Hier kann man sämtliche Zwischenergebnisse weglassen, denn man kann diese Ergebnisse aus den Prozessen mechanisch nachrechnen. Die bloße Reihe der Prozesse selbst ist also ein vollständiger, aber abgekürzter Beweis. Dieser abgekürzte Beweis wird die wesentlichen Prozesse und Operationen des Beweises hervorheben, ohne aber die technischen Einzelheiten ihrer Ausführung beizumischen, und wird so eine begriffliche Formulierung des Beweises ergeben. Unser Studium der Beweisprozesse wird also Methoden ergeben, die passende Abkürzungen von formalen Beweisen in der abstrakten Mathematik und in der Logik ermöglichen. Wir werden zeigen, daß solche Abkürzungen sowohl praktisch wie auch theoretisch wichtig sind.

Mittels des Studiums der Beweisprozesse kann man eine weitere und tiefere Abkürzungsmethode mittels des Begriffes, *Plan eines Beweises*, aufbauen. Unter einem Plan verstehen wir eine Regel oder Vorschrift, die die einzelnen Schritte eines Beweises zweckmäßig bestimmt. Solche Pläne, die sonst nur intuitiv faßbar sind, werden wir durch die Beweisprozesse für die einzelnen Schritte genau definieren können.

§ 3. **Prozesse als Manipulationen mit Symbolen:** Wenn ein Prozeß als eine Abkürzung dienen soll, dann muß dieser Prozeß genau und allgemein definiert werden. Das kann man in den einzelnen Fällen klar sehen: um den Syllogismus zu benutzen, müssen wir wissen, wie der Schluß in jedem Fall aus den Prämissen konstruiert werden kann. Um für alle Prozesse genaue Definitionen zu geben, müssen wir für jeden Prozeß die Prämisse, das Bezugstheorem und das Ergebnis in einer normalen Form ausdrücken. Eine solche normale Form für einen beliebigen mathematischen Satz ist durch die Symbolik der mathematischen Logik gegeben; denn bekanntlich kann man jedes Theorem mittels dieser Symbolik ausdrücken. Deshalb werden wir einen Prozeß als eine Operation an symbolischen Ausdrücken auffassen. Die Theorie der Prozesse beruht daher auf der symbolischen Logik und auf der Theorie der Operationen mit Symbolen.

Dasselbe wird sich herausstellen, wenn wir mit der Betrachtung der gewöhnlichen Beweise der symbolischen Logik anfangen. Ein streng symbolischer Beweis eines Satzes besteht ausschließlich aus Manipulationen mit Symbolen. Diese Manipulationen sind aber nicht planlos ausgeführt, sondern verfolgen einige wohlbestimmte Zwecke oder Grundideen. Daher werden dieselben Gruppen von Manipulationen wiederholt in Beweisen auftauchen. Eine solche Gruppe von Manipulationen kann man dann in einer Operation zusammenfassen. Die so entstandenen Operationen sind im wesentlichen unsere Beweisprozesse. Ein Prozeß ist daher nichts anderes als eine Abkürzung für eine organisch zusammenhängende Gruppe von symbolischen Manipulationen.

Durch diese Gedankengänge kommen wir zwangsläufig zur Einsicht, daß eine Untersuchung von Beweisprozessen mit einem Studium der Symbolik und der elementaren Operationen mit Symbolen anfangen muß. Dieses Studium werden wir im folgenden aufnehmen.

I. Kapitel.

Theorie der Symbolik.

§ 1. **Symbolische Ausdrücke:** Die Untersuchung der mathematischen Symbolik ist im wesentlichen die Analyse derjenigen Gruppen von Zeichen, die als symbolische Ausdrücke für mathematische Sätze und Definitionen auftreten können. Eine solche Gruppe von Zeichen nennen wir einen Ausdruck. Um einen festen Symbolismus zu haben, werden wir für alle Ausdrücke die klassische logische Symbolik der *Principia*[1] benutzen. Die Untersuchungen von Peano, Moore[2] und anderen haben *in extenso* gezeigt, daß jeder mathematische Satz und jeder mathematische Beweis in dieser Symbolik adäquat dargestellt werden kann. Der Terminus „Ausdruck" bedeutet also eine typographische Zusammenstellung[3] von Symbolen, die gemäß den Vorschriften dieses Systems der Logik aufgebaut ist. Hilbert[4] und von Neumann[5] haben genaue Definitionen dieses Terminus gegeben; im folgenden wollen wir auch eine Definition eines Ausdrucks geben, um dabei gewisse Hilfsbegriffe zu entwickeln.

Erstens bemerken wir, daß jede mathematische Theorie von gewissen *Kategorien* von Gegenständen spricht. Die Kategorie von Aussagen wird z. B. in jeder Theorie benutzt; andere Beispiele sind die Kategorien der ganzen Zahlen, der reellen Zahlen, der Elemente eines algebraischen Körpers usw. Jeder symbolische Ausdruck wird Elemente aus einer bestimmten Kategorie dar-

---

[1] *Principia* wie hier und im folgenden für Whitehead und Russell, *Principia Mathematica*, vol. I, 2nd edition, benutzt.

[2] E. H. Moore, *An Introduction to a Form of General Analysis*.

[3] Es wäre möglich, das Studium von Ausdrücken mit einer abstrakten Theorie solcher typographischen Zusammenstellungen (d. h. Symbolkomplexe) anzufangen.

[4] Hilbert und Ackermann, *Grundzüge der theoretischen Logik*, S. 52.

[5] von Neumann, *Zur Hilbertschen Beweistheorie*, Mathematische Zeitschrift, 26., S. 1 (1927). Statt „Ausdruck" wird hier das Wort „Formel" benutzt.

stellen; diese Kategorie wird kurz der *Typus* des Ausdrucks genannt. Wir wollen annehmen, daß keine Kategorie leer ist, denn mit dieser Annahme bleiben sämtliche formalen Gesetze der Logik gültig, insbesondere die Gesetze für „alle“ und „es gibt“.

Die einfachsten Ausdrücke bestehen aus einem einzelnen Zeichen — entweder einer Konstanten oder einer Variablen. Jede solche Konstante c wird als Element einer bestimmten Kategorie gedacht, die wir den *Typus* von c nennen wollen. Andererseits hat eine Variable x immer eine eindeutig bestimmte Kategorie K als *Spielraum* oder Variationsbereich; dieses K wird als der Typus von x genommen. Der Spielraum jeder Variable muß entweder explizit oder implizit angegeben werden. Die Variablen x und n werden z. B. meistens mit der Menge aller reellen bzw. aller natürlichen Zahlen als Spielräume gebraucht, und ähnlicherweise werden die Zeichen p und q oft als Variablen im Bereich aller Aussagen benutzt.

Weiter betrachten wir dann die kleinsten Ausdrücke, die mehr als ein Zeichen enthalten. Solche nennen wir *Grundformen.* Ein Beispiel ist der Ausdruck

$$p \supset q,$$

mit der logischen Verknüpfung $\supset$ und den zwei Aussagenvariablen p und q. In diesem Ausdruck nennen wir $\supset$ den *Operator* und p und q die 1-*Komponenten.* Da dieser Ausdruck immer eine Aussage darstellt, hat er sozusagen den Typus einer Aussage. Im allgemeinen werden wir diejenigen Grundformen betrachten, die als Operatoren entweder logische Verknüpfungen, Klammerzeichen, Funktionen oder Relationen haben, wie im folgenden dargestellt wird.

Die allgemeine Grundform besteht aus einem Operator λ und gewissen Variablen oder 1-Komponenten $w_1, \ldots, w_n$ in einer beliebigen aber festen symbolischen Anordnung, und hat also die schematische Gestalt[1]

$$\lambda(w_1, \ldots, w_n) \tag{1}$$

---

[1] Die Benutzung dieser Gestalt soll nicht etwa bedeuten, daß λ typographisch am Anfang der Grundform steht.

Mit jeder Grundform ist eine Kategorie als Typus gegeben. Wir betrachten nur die drei[1] folgenden Fälle:

Fall 1. λ ist eine feste mathematische Funktion der n Variablen w_i, wie z. B. *cos* oder $+$, und der Typus der Grundform enthält alle Funktionswerte. Dabei darf die Kategorie von Aussagen weder als Spielraum eines w_i noch als Typus der Grundform benutzt werden.

Fall 2. λ ist eine feste[2] Relation zwischen den n Variablen w_i, wie z. B. $>$, $=$, ε, usw. Der Typus der Grundform ist hier die Kategorie von Aussagen; diese Kategorie darf aber nicht als Spielraum eines w_i benutzt werden.

Fall 3 umfaßt die folgenden Grundformen, die alle vom Typus einer Aussage sind:

	Form	*Operator*	*1-Komponenten*
Logische Verknüpfungen	$p \supset q$	$\supset$ (folgt)	p, q
	$p \equiv q$	$\equiv$ (äquivalent)	p, q
	$p \,.\, q$	. (und)	p, q
	$p \vee q$	$\vee$ (oder)	p, q
	$\sim p$	$\sim$ (nicht)	p
Klammerzeichen:	$(\exists x) \,.\, p$	$(\exists)$ (es gibt)	x, p
	$(x) \,.\, p$	( ) (für alle)	x, p

Der Sinn der Grundformen liegt darin, daß alle Ausdrücke durch Kombination von Grundformen aufgebaut sind. Wir können also eine induktive Definition des Begriffes „Ausdruck" folgendermaßen angeben: ein Ausdruck A muß entweder ein einzelnes Zeichen sein, oder muß direkt aus einer Grundform

$$\lambda(w_1, \ldots, w_n)$$

durch Ersetzung der 1-Komponenten w_i entstehen. D. h. jedes w_i wird durch einen schon gebildeten Ausdruck A_i ersetzt, dessen

[1] Der Einfachheit halber werden in dieser Aufzählung gewisse Typen von Grundformen — z. B. das $\hat{x}\,(\varphi x)$ bei Russell oder das $\varepsilon\,(x)$ bei Hilbert — weggelassen. Unsere ganze Methode ist aber leicht auf solche Fälle anwendbar.

[2] Man kann veränderliche Funktionen und Relationen leicht auf festen zurückführen; vgl. von Neumann a. a. O. S. 18. Außerdem kann man Aussagefunktionen auf Relationen zurückführen.

Typus Unterkategorie des Spielraums der Variable w_i ist; der Ausdruck A hat also die Gestalt

(2) $$\lambda(A_1, \ldots, A_n).$$

Wenn insbesondere die Grundform ein Klammerzeichen $(\exists x)$ oder (x) als Operator hat, dann darf die Komponente x nur durch eine Variable y ersetzt werden; der Spielraum von y darf aber beliebig sein. In jedem Fall werden wir die Grundform (1) die *Hauptform* des Ausdrucks (2) nennen. Der Typus (2) wird gleich dem Typus der Hauptform gesetzt.

Für die so definierten Ausdrücke brauchen wir einige weitere Begriffe. Die *Komplexität* eines Ausdrucks wird durch die Anzahl der im Aufbau des Ausdrucks benutzten Grundformen gemessen, wobei mehrmals benutzte Grundformen mehrmals zu zählen sind. In (2) gilt also

$$\text{Komplexität}\ (A) = 1 + \Sigma\ \text{Komplexität}\ (A_i).$$

Viele Eigenschaften von Ausdrücken werden durch vollständige Induktion nach der Komplexität untersucht.

Die Ausdrücke A_i, die im Ausdruck (2) die 1-Komponenten der Hauptform ersetzen, heißen die 1-Komponenten von (2). Die 1-Komponenten eines Ausdrucks A haben oft selbst 1-Komponenten; diese heißen *2-Komponenten* in A. In dieser Weise fortschreitend, erhalten wir den allgemeinen Begriff einer *Komponente* in A. Die Aufteilung von A in Komponenten muß endlich auf *Endkomponenten* stoßen; d. h. auf Komponenten von der Komplexität Null. Diese Endkomponenten sind die in A enthaltenen Variablen und Konstanten. Der Operator der Hauptform von A heißt der *Hauptoperator* von A; die Hauptoperatoren der Komponenten in A heißen die *Operatoren von A*. Ein beliebiger Symbolkomplex ist in A *enthalten*, wenn dieser Komplex als Operator oder Komponente von A auftritt.

Jeder mathematische Satz, als symbolischer Ausdruck dargestellt, hat den Typus einer Aussage. Ausdrücke dieses Typus nennen wir kurz *satzartige* Ausdrücke. Alle anderen Typen (d. h. diejenigen mit Funktionen als Hauptoperatoren) sind dann *nicht-satzartig*.

Wenn zwei Ausdrücke als Symbolkomplexe in keiner Hinsicht voneinander verschieden sind, dann heißen diese Ausdrücke *iden-*

tisch. Diese Identität kann man durch Induktion streng definieren, und daraus zeigt man leicht, daß die Relation der Identität reflexiv, symmetrisch und transitiv ist. Der Unterschied zwischen der Identität und der Gleichheit zweier Ausdrücke ist wichtig, denn die Gleichheit verlangt nur die Identität der Werte der Ausdrücke, wie z. B.

$$a+b=b+a.$$

Kein satzartiger Ausdruck kann innerhalb (d. h. als Komponente) eines nicht-satzartigen Ausdrucks auftreten. Die Buchstaben L, M und N werden für satzartige Ausdrücke benutzt. Dagegen bedeuten C, D und E nicht-satzartige Ausdrücke, und A und B beliebige Ausdrücke.

Der Begriff eines symbolischen Ausdrucks, wie oben formuliert, dient als Grundlage für die Untersuchung von Beweisprozessen.

§ 2. **Gebundene Variablen:** In einem Ausdruck A heißt bekanntlich eine Variable y *gebunden*, wenn A ein Klammerzeichen $(\exists y)$ oder (y) enthält; sonst ist y frei. Die Bezeichnung der gebundenen Variablen kann eventuell Schwierigkeiten verursachen. Nach einem grundlegenden Prinzip des Symbolismus darf ein Zeichen nur eine Bedeutung innerhalb eines gegebenen Beweises haben. Ausdrücke der Form

$$(\exists x).\varphi x.\equiv.(\exists x).\psi x,$$

wo das Zeichen x für zwei verschiedene gebundene Variablen benutzt wird, kann man daher besser in der Gestalt

$$(\exists x).\varphi x.\equiv.(\exists y).\psi y$$

schreiben. Im allgemeinen werden wir die folgende *Übereinkunft für gebundene Variablen* treffen[1]: In jedem gegebenen oder aufgebauten Ausdruck A ist jede gebundene Variable, die kein besonderes Zeichen hat, sofort zu ersetzen. Genauer gesagt, wird jede gebundene Variable hinter einem Klammerzeichen, wie z. B. in

(1) $$(\exists x).L(x)$$

stehen; wenn das Zeichen x irgendwo in A außerhalb (1) auftritt, dann ist x überall in (1) durch eine bisher unbenutze Variable y,

[1] Diese Übereinkunft enthält die fünfte Bedingung der Hilbertschen Definition eines Ausdrucks; vgl. Hilbert und Ackermann a. a. O.)

die den Spielraum von x als Spielraum hat, zu ersetzen. Wenn mehrere gebundene Variablen in A auftreten, dann sind diejenigen Variablen erst zu ersetzen, deren Wirkungsbereiche[1] am kleinsten sind.

Durch diese Ersetzungen erhält man dann einen Ausdruck, wo jede gebundene Variable x in einer Komponente der Form $(\exists x) . L(x)$ oder $(x) . L(x)$ und sonst nirgendswo auftritt.

§ 3. **Freie Variablen:** Eine freie Variable ist im wesentlichen ein Zeichen, das ersetzt werden kann. Wenn man eine freie Variable x in einem Ausdruck A überall durch eine Konstante c ersetzt, und wenn c dem Spielraum von x angehört, dann wird das Resultat ein Ausdruck sein. Wenn jede freie Variable eines satzartigen Ausdrucks L in dieser Weise ersetzt wird, dann erhält man eine Aussage. Der Ausdruck L ist bekanntlich dann und nur dann *richtig*, wenn alle Aussagen, die in dieser Weise erhalten werden können, wahr sind[2].

Aus einem gegebenen Satz L kann man verschiedene ,,spezielle Fälle" von L aufbauen, entweder durch Änderung der Bezeichnung für die gebundenen Variablen, oder durch Substitution eines geeigneten Ausdrucks für eine freie Variable, oder durch Kombination solcher Operationen. Die allgemeinste solche Kombination lautet folgendermaßen: Gegeben ein Ausdruck A mit freien Variablen $x_1, \ldots, x_m$ und gebundenen Variablen $y_1, \ldots, y_n$. Ein *Fall* von A wird aus A erhalten, indem jedes x überall in A durch einen Ausdruck B_i und jedes y_j überall durch eine Variable z_j ersetzt wird. Dabei müssen die folgenden Bedingungen erfüllt sein.

1. Der Typus von B_i ist eine Unterkategorie des Spielraums von x_i.
2. Die z_j sind voneinander verschieden.
3. Die Spielräume von y_j und z_j sind identisch.
4. Kein z_j ist in einem B_i enthalten.

---

[1] Wirkungsbereich ist hier in dem gewöhnlichen Sinn zu verstehen; z. B. ist der Wirkungsbereich von x in (1) gleich $L(x)$.

[2] Diese Definition ist nur die gewöhnliche Übereinkunft, daß eine Behauptung

$$x + y = y + x$$

als eine Behauptung für alle Werte von x und y verstanden werden soll.

Durch diese Ersetzung entsteht aus A ein Symbolkomplex, der selbst ein Ausdruck ist.

Für diesen Begriff „Fall" hat man die zwei folgenden Sätze, die man mittels der Definitionen unter getrennter Behandlung von freien und gebundenen Variablen beweisen kann.

Satz 1. Jeder Fall eines richtigen Ausdrucks ist selbst richtig.

Satz 2. Die Relation „A' ist Fall von A" ist reflexiv und transitiv.

§ 4. **Der Vergleich von Ausdrücken:** Mit der oben gegebenen Definition eines Falles kann man aus einem gegebenen Ausdruck mehrere Fälle aufbauen, aber andererseits kann man nicht unmittelbar bestimmen, ob ein gegebener Ausdruck B ein Fall eines anderen Ausdrucks A ist. Die Definition ist also nicht konstruktiv, ein Mangel, den wir jetzt beseitigen wollen.

Spezielle Probleme der angeführten Art sind leicht zu behandeln. Wir können z. B. ohne weiteres feststellen, daß

(1) $$(3^2 + d) + d^{-1},$$

ein Fall des Ausdrucks

(2) $$(a + b) + b^{-1},$$

ist. Diese intuitive Feststellung entsteht aus einem induktiven Vergleich der Komponenten dieser zwei Ausdrücke. Ähnlicherweise nennen wir zwei beliebige Ausdrücke A und B *1-vergleichbar*, wenn sie dieselbe Hauptform haben. Sie haben dann die Gestalt (§ 1, (2))

$$A = \lambda(A_1, \ldots, A_n)$$
$$B = \lambda(B_1, \ldots, B_n)$$

und daher können wir eine 1-1-Zuordnung zwischen den 1-Komponenten aufstellen, wobei jedes A_i dem entsprechenden B_i zugeordnet ist. Wenn zwei zugeordnete Ausdrücke A_i und B_i selbst 1-vergleichbar sind, dann wird der Vergleich von A_i mit B_i eine Zuordnung zwischen gewissen 2-Komponenten von A und B stiften. Die Ausführung dieser Operation soweit wie möglich ergibt einen *Vergleich* von A mit B, der eine 1-1 Zuordnung zwischen einigen Komponenten (nicht notwendigerweise allen) von A und einigen Komponenten von B stiftet. Jede besondere Relation zwischen A und B wird besonderen Eigenschaften für diese Zuordnung entsprechen, und dadurch können wir das obige Problem der konstruktiven Definition des Begriffes „Fall“ behandeln.

Satz 1. Ein Ausdruck B ist dann und nur dann ein Fall eines Ausdrucks A, wenn der Vergleich von A mit B eine Zuordnung ergibt, die den folgenden Bedingungen genügt:

1. Jede Komponente in A entspricht einer Komponente in B; d. h., der Vergleich von A mit B ist bis auf die Endkomponenten von A ausführbar.

2. Wenn eine Endkomponente in A eine Konstante ist, dann ist die zugeordnete Komponente in B dieselbe Konstante.

3. Wenn eine Variable x mehr als einmal als Komponente auftritt, dann sind die zugeordneten Komponenten in B miteinander identisch.

4. Nach 1 und 3 wird jede freie Variable x in A einem Ausdruck B_i in B, und jede gebundene Variable y_j in A einer Variable z_j in B entsprechen. Für B_i und y_j müssen die Bedingungen § 3, (1) gelten.

Diesen Satz kann man durch Induktion nach der Komplexität von A beweisen, unter Benutzung der Definitionen für „Fall“ und „Vergleich“. Eine Anwendung dieses Satzes kann man oft durch die folgende Bemerkung vermeiden: Wenn B ein Fall von A ist, dann muß B alle Operatoren und Konstanten aus A enthalten, und muß außerdem mindestens so viele gebundene Variablen wie A besitzen.

Der folgende Satz ist eine Vorbereitung für das Studium der Substitution:

Satz 2. Wenn ein Ausdruck

$$L \equiv M \tag{3}$$

gegeben ist, und wenn L' ein gegebener Fall von L ist, dann gibt es einen Ausdruck M', so daß

$$L' \equiv M' \tag{4}$$

ein Fall von (3) ist.

Beweis. Im Aufbau von L' aus L ist jede freie Variable x in L durch einen Ausdruck B' ersetzt worden. Dieses B′ kann man nach Satz 1 durch Vergleich von L mit L' finden. Wir können dann ein M' mit der erwünschten Eigenschaft aus M konstruieren, wenn wir jedes x überall in M durch das entsprechende B' ersetzen, wobei die anderen Variablen in M ungeändert bleiben und diejenigen gebundenen Variablen in M, die auch in den Ausdrücken B' auftreten, durch neue voneinander verschiedene Variablen zu ersetzen sind. Das so aufgebaute M' nennen wir den Fall von M, der L' *entspricht*. Ähnliche „entsprechende" Fälle kann man mit $L \supset M$ oder $C = D$ statt (3) aufbauen.

§ 5. **Invariante Eigenschaften eines Ausdrucks:** Zwei Ausdrücke A und B, die nur in den für die Variablen benutzten Zeichen verschieden sind, haben im wesentlichen dieselbe Bedeutung und heißen daher *gleichbedeutend* oder *synonym*. Genauer, A und B sind gleichbedeutend, wenn B aus A dadurch erhalten werden kann, daß man jede Variable in A überall durch eine neue Variable ersetzt, wobei diese neuen Variablen voneinander verschieden sind. Daraus beweist man leicht, daß die Relation „gleichbedeutend" reflexiv, symmetrisch und transitiv ist. Mittels eines Vergleichs können wir auch folgendes konstruktive Kriterium für diese Relation erhalten:

Satz 1. Zwei Ausdrücke A und B sind dann und nur dann gleichbedeutend, wenn der Vergleich von A mit B eine Zuordnung mit folgenden Eigenschaften ergibt:

1. Jede Komponente von A entspricht einer Komponente von B und umgekehrt; d. h. der Vergleich von A mit B ist bis auf die Endkomponenten ausführbar;
2. Zugeordnete Endkomponenten haben denselben Typus;
3. Zwei Endkomponenten in A sind dann und nur dann identisch, wenn die zugeordneten Komponenten in B identisch sind.
4. Eine Konstante c in A muß derselben Konstante c in B entsprechen, und umgekehrt.

Satz 2. Wenn zwei logische Ausdrücke L und M gleichbedeutend sind, dann ist die Richtigkeit von L äquivalent der Richtigkeit von M. (Beweis: § 3, Satz 1.)

In jeder logischen Manipulation kann man gleichbedeutende Ausdrücke miteinander vertauschen. Alle Operationen oder Relationen für Ausdrücke müssen daher bezüglich der Relation „gleichbedeutend" invariant sein. Die Relation „Fall" ist z. B. invariant, denn wenn A ein Fall von B ist, und wenn A und A', B und B paarweise gleichbedeutend sind, dann ist A' ein Fall von B'. Derartige Invarianzforderungen sind durch Analogie mit ähnlichen Forderungen in der Geometrie zu verstehen. Dadurch kann man zwischen zufälligen und notwendigen Eigenschaften eines symbolischen Ausdrucks unterscheiden.

§ 6. **Die allgemeine Ersetzung:** Der Begriff eines Falles beruht auf der Ersetzung der Variablen eines Ausdrucks. Für die späteren Definitionen der Beweisprozesse brauchen wir einen allgemeineren Begriff der Ersetzung, wo nicht nur eine Variable, sondern auch eine beliebige Komponente eines Ausdrucks ersetzt werden kann. Offenbar ist eine 1-Komponente in einer Grundform und daher auch in einem beliebigen Ausdruck ersetzbar. Durch mathematische Induktion kann man die Ersetzung einer beliebigen Komponente eines Ausdrucks A in solcher Weise definieren, daß die so erhaltene abstrakte Operation völlig mit der intuitiven Vorstellung einer Ersetzung übereinstimmt. Eine solche Ersetzung in A wird aber meistens einen sinnlosen Symbolkomplex und nicht einen Ausdruck ergeben. Die Möglichkeit sinnvoller Ersetzung wird im folgenden Satz festgelegt.

Satz 1. Wenn die Komponente B eines Ausdrucks A durch einen Ausdruck B' ersetzt worden ist, dann wird das Ergebnis auch ein Ausdruck sein, wenn die folgenden Bedingungen erfüllt sind:

1. Der Typus von B' ist eine Unterkategorie des Typus von B.
2. Wenn B eine gebundene Variable von A ist, dann ist B' auch eine Variable.

Oft muß man mehrere Komponenten B_1, B_2, ... eines Ausdrucks A nacheinander ersetzen. Wenn die B_i getrennt sind (d. h. wenn kein B_i selbst Komponente eines anderen B_j ist), dann kann man die erwünschten Ersetzungen in irgendeiner Reihenfolge ausführen, ohne das Endresultat zu ändern, wenn nur die Bedingungen in Satz 1 in allen Fällen erfüllt bleiben. Eine derartige Ersetzung mehrerer Komponenten nennen wir *simultan.*

Wenn in einem Ausdruck A ein anderer Ausdruck B mehrmals als Komponente auftritt, und wenn alle diese Komponenten simul-

2

tan durch *einen* Ausdruck B' ersetzt sind, dann sagen wir, daß B durch B' *überall* in A ersetzt worden ist.

Diese allgemeine Theorie der Ersetzung enthält als speziellen Fall die früher benutzte Ersetzung von Variablen.

II. Kapitel.

Die einfachen Prozesse.

§ 1. **Die Struktur eines Prozesses:** In den nächsten drei Kapiteln werden wir mehrere spezifische Beweisprozesse aufstellen. Diese Prozesse sind alle durch induktive Analyse einfacher, aber typischer Beweise gefunden. Eine solche Analyse besteht erstens in einer Einteilung des Beweises in aufeinanderfolgende Schritte und dann in der Bestimmung des allgemeinen Charakters jedes Schrittes. Die so erhaltenen Prozesse sind dann für viele Beweise nützlich.

Die einfachsten Beweisprozesse bestehen in der Aufstellung einer Folgerung einer gegebenen Prämisse; d. h. man fängt mit einem gewissen Ausdruck L (die Prämisse) an, und konstruiert dann, mittels Bezugnahme auf ein schon bewiesenes Theorem M (das Bezugstheorem), einen neuen logischen Ausdruck N, das Ergebnis. Unter Voraussetzung von M muß die Richtigkeit dieses Ergebnisses eine Folgerung der Richtigkeit von L sein. Die Prämisse ist also *stärker* als das Ergebnis, und die Konstruktion von N wird daher einen *Schrumpfprozeß* genannt. Ein solcher Prozeß ist sozusagen ein Abstieg von Bekanntem zu Unbekanntem; das kann man aber umkehren, indem man von Unbekanntem zu Bekanntem aufsteigt. Genauer, man fängt mit einem Ausdruck L' an, und konstruiert daraus, mittels Bezugnahme auf ein schon bewiesenes Theorem M' (das Bezugstheorem), einen neuen logischen Ausdruck N', wobei aber dieser Endausdruck N' stärker als der Anfangsausdruck L' ist. Durch einen solchen *Schwellprozeß* wird das Problem, L' zu beweisen, auf das Problem, N' zu beweisen, zurückgeführt. Auch hier wird der Anfangsausdruck L' als die *Prämisse* und der Endausdruck N' als das *Ergebnis* bezeichnet. Noch ein dritter Fall ist möglich: ein Prozeß, der ein Ergebnis N'' aus einer Prämisse L'' mittels eines Bezugstheorems M'' konstruiert, wo die Richtigkeit der Prämisse und die Richtigkeit des Ergebnisses äquivalent sind. Die Bezeichnung *Stationärprozeß* wird hier benutzt.

Prozesse mit Prämisse und Ergebnis, aber ohne Bezugstheorem, sind auch möglich. Die drei Typen sind genau wie oben zu definieren.

Die Definition eines besonderen Prozesses hat mehrere Phasen. Erstens muß man festlegen, was für Ausdrücke als Prämisse und als Bezugstheorem in Betracht kommen. Zweitens muß man für jede Prämisse und jedes Bezugstheorem des festgelegten Typs das Ergebnis durch rein symbolische Manipulation aufbauen können. Endlich muß man angeben, ob der Prozeß ein Schwell-, Schrumpf-, oder Stationärprozeß ist, und in jedem Fall muß man die Gültigkeit beweisen; d. h. man muß zeigen, daß Prämisse und Ergebnis in der angegebenen Weise miteinander verbunden sind.

§2. **Die Schlußregel:** Die grundlegende Schlußregel der Logik besagt folgendes: Wenn L und $L \supset M$ richtige Ausdrücke sind, dann ist auch M richtig. In dieser Regel können wir entweder L, $L \supset M$, oder M als Prämisse nehmen; dadurch haben wir die drei folgenden Prozesse:

Inf schrumpf[1]: Prämisse, ein satzartiger Ausdruck L; Bezugstheorem, eine Implikation $L \supset M$, wo L die gegebene Prämisse ist; Ergebnis M. Ein Schrumpfprozeß.

Inf schwell: Prämisse M; Bezugstheorem $L \supset M$, Ergebnis L. Ein Schwellprozeß.

Hyp weg: Prämisse $L \supset M$; Bezugstheorem L, Ergebnis M. Ein Schrumpfprozeß. Er wird „Hyp weg" genannt, weil er die Hypothese L von der Prämisse wegnimmt.

Die Gültigkeit dieser Prozesse folgt sofort aus der Schlußregel (vgl. *Principia*, *1. 1, *1. 11, *9. 12).

§ 3. **Syllogismus**: Aus dem Prinzip des Syllogismus erhalten wir die folgenden Prozesse:

Syll unter schrumpf: Prämisse $L \supset M$; Bezugstheorem $M \supset N$; Ergebnis $L \supset N$.

[1] Inf steht für „inference" = Schlußschema.

2*

Syll ober schrumpf: Prämisse $M \supset N$;
Bezugstheorem $L \supset M$; Ergebnis $L \supset N$.
Syll unter schwell: Prämisse $L \supset N$;
Bezugstheorem $M \supset N$; Ergebnis $L \supset M$.
Syll ober schwell: Prämisse $L \supset N$;
Bezugstheorem $L \supset M$; Ergebnis $M \supset N$.

Die zwei ersten sind Schrumpfprozesse, die anderen Schwellprozesse. Die Bezeichnungen „ober" und „unter" bedeuten, daß der Ober- bzw. Untersatz des Syllogismus als Bezugstheorem dient. In allen Fällen kann man die Gültigkeit leicht beweisen.

§ 4. **Prozesse mit allgemeinen Bezugstheoremen**: Ein bewiesener Satz behauptet sowohl die Richtigkeit des Satzes selbst wie auch die Richtigkeit aller Fälle dieses Satzes (vgl. I, § 3, Satz 1). In einem Prozeß können wir daher nicht nur einen bewiesenen Satz, sondern auch einen beliebigen Fall eines solchen Satzes als Bezugstheorem benutzen. Dieser Fall muß sachgemäß gewählt sein. Wenn wir z. B. den Satz

$$M \supset N \tag{1}$$

bewiesen haben, und wenn wir den Prozeß „Inf schrumpf" auf die Prämisse L anwenden wollen, dann können wir einen Fall von (1) als Bezugstheorem benutzen, wenn die Prämisse L ein Fall von M ist (letzteres kann man mit I, § 4, Satz 1 prüfen). Dieses ergibt die folgende Verallgemeinerung des Prozesses:

Inf schrumpf: Prämisse L; Bezugstheorem, eine Implikation $M \supset N$, wobei L ein Fall von M ist; das Ergebnis ist der entsprechende Fall von N (Konstruktion nach I, § 4, Satz 2).

In genau derselben Weise kann man aus den anderen Prozessen in § 2 und § 3 *verallgemeinerte* Prozesse erzeugen. Die ursprünglichen Prozesse vor der Verallgemeinerung mögen dagegen *spezielle* Prozesse heißen. Für einen speziellen Prozeß und die entsprechende Verallgemeinerung wird derselbe Name benutzt; wenn das Gegenteil nicht ausdrücklich gesagt wird, ist immer der verallgemeinerte Prozeß gemeint.

Noch eine Bemerkung: im obigen allgemeinen Prozeß kann eine Variable sowohl in L wie auch in N, aber mit verschiedenen Bedeutungen auftreten. Um Verwechselungen zu vermeiden, können wir sämtliche Variable in Bezugstheorem durch neue voneinander verschiedene Variable ersetzen. Eine solche Vorbereitung wird vor der Anwendung jedes verallgemeinerten Prozesses durchgeführt: d. h. das Bezugstheorem L_1 wird durch einen synonymen Ausdruck L_2 (der nach I, § 5, Satz 2 mit L_1 äquivalent ist) ersetzt, der so gewählt ist, daß keine Variablen aus L_2 in der betrachteten Prämisse auftreten. Dieses nennen wir eine *synonyme Vorbereitung.*

§ 5. **Verkürzte Prozesse mit Fällen**: Da alle Fälle eines richtigen Ausdrucks richtig sind, haben wir den folgenden Prozeß:

Fall ($x \rightarrow A$, $y \rightarrow B$, ...): Prämisse, ein Ausdruck L mit reellen Variablen x, y, ...; kein Bezugstheorem; das Ergebnis ist derjenige Fall von L, den man erhält, indem man simultan x, y, ... überall durch A, B, ... ersetzt. Dies ist ein gültiger Schrumpfprozeß, wenn der Typus jedes A eine Teilmenge des Spielraums des entsprechenden x ist, und wenn A keine Variablen enthält, die in L gebunden sind.

Dieser Prozeß ist etwas kompliziert, denn er enthält die Angabe des Ersatzausdrucks für jede Variable aus L. Wenn ein solcher Prozeß in einem Beweis auftritt, dann hat er meistens einen bestimmten Zweck; etwa den, einen späteren Schritt des Beweises zu ermöglichen. Wenn z. B. eine Prämisse $L \supset M$ gegeben ist, dann wird oft ein geeigneter Fall $L' \supset M'$ dieser Prämisse aufgebaut, um nachher den speziellen Prozeß „Hyp weg" mit einem Bezugstheorem N anzuwenden. Dieser Prozeß wird nur dann anwendbar sein, wenn N Fall von L ist. Wenn diese Bedingung gilt, dann können wir für M' den entsprechenden Fall von M wählen. Wir haben also den für diesen Zweck geeigneten Fall konstruktiv bestimmt. Wir haben dadurch die Prozesse „Fall" und „Hyp weg" in den folgenden Prozeß vereinigt.

Fall + Hyp weg: Prämisse $L \supset M$; Bezugstheorem, ein Ausdruck N, der Fall von L ist; das Ergebnis ist dann der entsprechende Fall von M. Ein Schrumpfprozeß.

In derselben Weise können wir aus anderen Schrumpfprozessen die folgenden gültigen Kombinationen aufbauen: *Fall + inf schrumpf, Fall + syll unter schrumpf, Fall + syll ober schrumpf*

In allen solchen Prozessen werden wir vor der Ausführung eine synonyme Vorbereitung des Bezugstheorems vornehmen.

§ 6. **Die Grundprinzipien der Beweisabkürzung**: Die Anwendung der bisher erörterten Prozesse zur Abkürzung von Beweisen werden wir mit einem einfachen Beweis aus *Principia* *2 illustrieren. Wir wollen die Behauptung

*2. 08 $p \supset p$

beweisen. Wir fangen mit dem Axiom

*1. 2 $p \vee p . \supset . p$

an, und mit diesem Ausdruck als Prämisse wenden wir den „Inf schrumpf" Prozeß mit dem Bezugstheorem

*2. 05 $q \supset r . \supset : p \supset q . \supset . p \supset r$

an. Dazu müssen wir zunächst im letzten Ausdruck eine synonyme Vorbereitung ausführen, indem wir p durch s ersetzen; der geeignete Fall des Bezugstheorems ist dann

$$p \vee p . \supset . p : \supset :. s \supset . p \vee p : \supset . s \supset p.$$

Das Ergebnis des Prozesses ist daher[1]

$$s \supset . p \vee p : \supset . s \supset p$$

Mit diesem Ausdruck als Prämisse benutzen wir dann den Prozeß „Fall + Hyp weg", mit dem schon bewiesenen Satz

*2. 07 $p . \supset . p \vee p$

als Bezugstheorem. Das Ergebnis dieses Prozesses ist das erwünschte Theorem *2. 08, das dadurch aus *1 . 2 mit einer Reihe gültiger Schrumpfprozesse abgeleitet worden ist.

Der ganze Beweis stellt sich bequem in folgender Gestalt dar:

Anfang *1 . 2	$p \vee p . \supset . p$
Inf schrumpf *2 . 05	$s \supset . p \vee p : \supset . s \supset p$
Fall + hyp weg *2 . 07	$p \supset p.$

Auf der linken Seite stehen die Namen der Prozesse, mit den benutzten Bezugstheoremen; auf der rechten Seite die Ergebnisse. Die Prämisse des ersten Prozesses ist als „Anfang" bezeichnet; die Prämisse jedes anderen Prozesses ist das Ergebnis des vorhergehenden Prozesses. Das Endergebnis ist das erwünschte Theorem; eine Tatsache, die wir als „Ende Th." schreiben wollen.

[1] Wenn die synonyme Vorbereitung nicht durchgeführt worden wäre, dann wäre das Ergebnis nicht so allgemein.

Diese Darstellung eines Beweises ist im wesentlichen die Form, die normalerweise in der Mathenatik benutzt wird: erst die Beschreibung eines Schrittes, dann das Ergebnis dieses Schrittes. Wenn wir aber die Prozesse genau formuliert haben, dann können wir die Darstellung weiter vereinfachen; denn wenn Prozeß, Prämisse und Bezugstheorem bekannt sind, dann ist das Ergebnis vollständig bestimmt! Es ist unnötig, die Ergebnisse explizit anzugeben, denn sie sind mechanisch aus den Definitionen für die Prozesse konstruierbar. Man braucht nur die Beschreibung der Schritte niederzuschreiben. Für den obigen Beweis lautet diese Beschreibung folgendermaßen:

Anfang *1 . 2, Inf schrumpf *2 . 05,
Fall + Hyp weg *2 . 07, Ende Th.

Einen Beweis in dieser Form nennen wir einen *abgekürzten Beweis.* Wenn nur dieser abgekürzte Beweis gegeben ist, dann können wir mechanisch die angedeuteten Schritte ausführen, um die Einzelheiten des Beweises wieder aufzubauen. Dieser Aufbau ist rein mechanisch, denn er ist nur eine Anwendung der Definitionen der im Beweis benutzten Prozesse. Der abgekürzte Beweis ist also theoretisch vollständig, genau so wie

$$851 \div 37 = 23$$

eine theoretisch vollständige Beschreibung einer arithmetischen Division ist. Der abgekürzte Beweis gibt den entscheidenden Gedanken, der hinter jedem Schritt des Beweises steckt; in diesem Sinne vermeidet er detaillierte Manipulationen und stellt nur die wesentlichen Ideen dar. Solche Beweise vereinigen also die symbolische Genauigkeit der logischen Beweise der *Principia* mit der konzisen Klarheit der gewöhnlichen intuitiv abgekürzten mathematischen Beweise.

III. Kapitel.

Substitutionsprozesse.

§ 1. **Substitution vermittelst einer Gleichung**: Das bekannte elementare Verfahren der Substitution von Gleichem für Gleiches ist im wesentlichen ein Beweisprozeß, den wir „Sub Gl" oder „Sub" nennen wollen. Die genaue Formulierung ist folgende:

Sub Gl: Prämisse, ein satzartiger Ausdruck L; Bezugstheorem, eine Gleichung[1] $C = D$; das Ergebnis L' wird aus L erhalten, indem man C überall in L durch D ersetzt[2].

Wenn man z. B. die Prämisse

$$(a + b)\, c = ac + bc$$

und das Bezugstheorem

$$(1) \qquad ac = ca$$

hat, dann wird der Prozeß der Substitution das Ergebnis

$$(a + b)\, c = ca + bc$$

liefern.

Dieser Prozeß ist ein gültiger Stationärprozeß, wenn alle in L auftretende Funktionen, Aussagefunktionen und Relationen wohldefiniert[3] sind. Im Beweis dieser Gültigkeit können wir ohne Beschränkung der Allgemeinheit annehmen, daß nur eine Komponente C in L ersetzt worden ist. Wir bezeichnen mit M die kleinste satzartige Komponente in L, die C enthält, und mit E die größte nicht satzartige Komponente in L, die C enthält. Die Ersetzung von C durch D in M und E ergibt Ausdrücke M' bzw. E'. Wenn C eine n-Komponente in E ist, dann können wir durch Induktion nach n beweisen, daß E gleich E' ist. Daraus folgt sofort, daß M und M' formal äquivalent sind. Endlich, wenn M eine m-Komponente in L ist, dann können wir durch Induktion nach m beweisen, daß L und L' formal äquivalent sind, unter Benutzung der Eindeutigkeit[4] der logischen Verknüpfungen und der Klammerzeichen.

---

[1] Eine Gleichung — d. h. ein Ausdruck der Hauptform $x = y$ — muß immer die Gleichheit zweier nicht-satzartiger Ausdrücke behaupten, wie z. B.

$$(a + b) + c = a + (b + c).$$

[2] Wenn C nirgends in L als Komponente auftritt, dann hat der Prozeß keinen Einfluß.

[3] Eine Funktion $f(x, y)$ zweier Variablen ist dann und nur dann wohldefiniert, wenn

$$x = x' \,.\, y = y' \,.\supset.\, f(x, y) = f(x', y');$$

eine Relation $\varphi\,(x, y)$ ist dann und nur dann wohldefiniert, wenn

$$x = x' \,.\, y = y' \,.\supset.\, \varphi(x, y) \equiv \varphi(x', y').$$

Ähnliche Bedingungen gelten für andere Variablenanzahlen.

[4] Die Verbindung $\vee$ hat z. B. die Eindeutigkeitseigenschaft

$$p \equiv p' \,.\, q \equiv q' \,.\supset:\, p \vee q \,.\equiv.\, p' \vee q'.$$

Der Aufbau des Ergebnisses dieses Substitutionsprozesses ist besonders einfach, wenn das Bezugstheorem $C = D$ *balanciert* ist; d. h. wenn jede freie Variable in C auch in D auftritt und umgekehrt. Die Gleichung (1) oben ist z. B. balanciert, wie auch die meisten Gleichungen der Algebra. In einer unbalancierten Gleichung können die Variablen, die in C, aber nicht in D auftreten, keine Schwierigkeiten verursachen; der umgekehrte Fall muß aber etwas näher untersucht werden.

Wir bezeichnen die in D, aber nicht in C enthaltenen freien Variablen durch $x_1, \ldots, x_m$. Um vermittelst des Bezugstheorems $C = D$ in L zu substituieren, machen wir zunächst eine synonyme Vorbereitung, wodurch die Extravariablen durch neue Variablen $y_1, \ldots, y_m$ ersetzt werden, die nicht schon in L auftreten. Die oben definierte Substitution in L wird dann ausgeführt. Das Ergebnis L' wird alle freien Variablen aus L und auch die y_i enthalten. Wenn die Substitution in der Ersetzung mehrerer Komponenten von L besteht, dann werden verschiedene Extravariablen für jede solche Komponente benutzt. Die Gültigkeit des so definierten Prozesses ist wie oben zu beweisen.

Mit demselben unbalancierten Bezugstheorem kann man ein anderes Ergebnis bekommen. Betrachten wir den Fall, wo nur eine Extravariable x im Ergebnis $L'(x)$ der oben definierten Substitution auftritt. Dieses x kann man nachher durch einen beliebigen konstanten Wert aus dem Spielraum von x ersetzen, da man annehmen kann, daß dieser Wert im Bezugstheorem schon vor der Substitution festgelegt wurde. Die Prämisse L ist also dem Ausdruck

$$(\exists x) \,.\, L'(x)$$

formal äquivalent (im formalen Beweis benutzt man *Principia* *10 . 281). Wir können diesen Ausdruck statt L' selbst als Ergebnis eines neuen Stationärprozesses nehmen. Diesen Prozeß nennen wir *Sub Gl extra*. Wenn mehrere Extravariablen $y_1, \ldots, y_t$ in diesem Prozeß eingeführt sind, dann wird das Ergebnis die Gestalt

$$(\exists y_1, \ldots, y_t) \,.\, L'(y_1, \ldots, y_t)$$

haben, wobei wir der Eindeutigkeit halber die Reihenfolge der Variablen im Präfix folgendermaßen festsetzen: diejenige Variable

y_1, die im Symbolkomplex L' vor allen anderen y_i als Komponente auftritt, wird als erste Variable des Präfixes genommen, usw.

§ 2. **Verallgemeinerte Substitution**: Mit den Methoden aus II, § 4 können wir den Substitutionsprozeß verallgemeinern; statt vermittelst einer Gleichung selbst zu substituieren, können wir vermittelst jedes geeigneten Falles dieser Gleichung substituieren. Als Beispiel betrachten wir die Anwendung des distruibutiven Gesetzes

$$ac + bc = (a + b)c$$

auf die Prämisse

(1) $$(ad + bd) + cd = (a + b + c)d.$$

Wir können erst die Komponente $(ad + bd)$ ersetzen, mit dem Resultat

$$(a + b)d + cd = (a + b + c)d.$$

Hier ist die ganze linke Seite ein Fall der linken Seite des gegebenen Gesetzes, und eine Substitution vermittelst dieses Falles des Gesetzes ergibt eine Identität. Im ganzen haben wir zwei Fälle des distrubutiven Gesetzes benutzt — erst haben wir eine kleinere Komponente der Prämisse, und dann eine größere Komponente ersetzt.

Diese Reihenfolge — erst die kleinsten, dann die größeren Komponenten — wird immer benutzt, wenn Substitutionen mit mehreren Fällen einer Bezugsgleichung möglich sind. Ein solcher verallgemeinerter Substitutionsprozeß (Abkürzung: Sub Gl) ist für jede Bezugsgleichung $C = D$ und für jede Prämisse A möglich. Um die Reihenfolge der einzelnen Operationen festzulegen, müssen wir das Ergebnis dieses Prozesses durch vollständige Induktion nach der Komplexität von A festlegen.

1. *Schritt*: A hat die Komplexität Null. Wenn A ein Fall von C ist, dann wird der entsprechende Fall von D als Ergebnis genommen; sonst ist A selbst das Ergebnis.

2. *Schritt*: A hat die Komplexität n $(n > 0)$, mit

$$A = \lambda(A_1, \ldots, A_m).$$

Dann haben die Komponenten A_i eine kleinere Komplexität. In jedem A_i können wir also die schon definierte Substitution ausführen, mit einem Ergebnis B_i. Wir betrachten dann den Ausdruck

(2) $$\lambda (B_1, \ldots, B_n).$$

Wenn dieser Ausdruck selbst ein Fall von der linken Seite C der Bezugsgleichung ist, dann nehmen wir den entsprechenden Fall der rechten Seite D als Ergebnis; sonst ist der Ausdruck (2) selbst das Ergebnis. Diese Definition ergibt einen gültigen Stationärprozeß.

Es sind auch andere Übereinkünfte für die Reihenfolge der Substitutionen in solchen Prämissen möglich; wir werden aber unter Substitution mittels einer Gleichung immer den obigen verallgemeinerten Prozeß verstehen, wobei die kleinsten Komponenten in der obigen Weise zuerst zu ersetzen sind. Unbalancierte Bezugstheoreme sind genau wie im speziellen Prozeß zu behandeln.

§ 3. **Substitution vermittelst einer Äquivalenz:** Eine Äquivalenz ist ein Ausdruck der Form

$$M \equiv N,$$

mit zwei satzartigen 1-Komponenten. Vermittelst einer beliebigen Äquivalenz kann man eine Substitution ausführen. Die Definition dieses Substitutionsprozesses, sowohl im speziellen, wie auch im verallgemeinerten und im unbalancierten Fall, ist wörtlich dieselbe, wie für die Substitution vermittelst einer Gleichung; man muß nur in den vorigen Abschnitten $C = D$, C bzw. D durch $M \equiv N$, M bzw. N ersetzen. Der Beweis der Gültigkeit des Prozesses ist sogar einfacher; nur der letzte Teil des in § 1 gegebenen Beweises wird benutzt[1]. Bei der Konstruktion des Ergebnisses werden eventuell synonyme Transformationen nach II, § 4 nötig sein; wegen I, § 5, Satz 2 wird aber die Gültigkeit des Prozesses dadurch nicht beeinflußt.

Bei der Substitution vermittelst eines Bezugstheorems $M \equiv N$ kann man nicht nur die linke Seite durch die rechte, sondern auch die rechte Seite durch die linke ersetzen. Diesen Prozeß nennen wir

$$\text{Sub perm} \quad M \equiv N;$$

dieses bedeutet immer die gewöhnliche Substitution vermittelst des permutierten Bezugstheorems $N \equiv M$. Analoges gilt auch für Substitution vermittelst einer Gleichung.

---

[1] Vgl. Hilbert und Ackermann a. a. O. S. 61, Regel X.

§ 4. **Substitution vermittelst einer Implikation:** In einem Ausdruck können wir eine Komponente M nicht nur durch einen äquivalenten Ausdruck N_1, sondern auch durch einen schwächeren Ausdruck N ersetzen. Das ist im wesentlichen eine Substitution vermittelst des Bezugstheorems

(1) $$M \supset N.$$

Die Prozesse *Inf schrumpf* (II, § 2), *Syll unter schrumpf*, und *Syll ober schwell* (II, § 3) sind Beispiele solcher Substitutionen.

Bei einer solchen Substitution in einem Ausdruck L muß man beachten, ob die Substitution ein Schwell- oder ein Schrumpfprozeß ist. Das ergibt folgendes Problem (vgl. II, § 1): In einem Ausdruck L werde die Komponente M durch einen schwächeren Ausdruck N mit einem Ergebnis L' ersetzt. Unter welchen Bedingungen wird L' schwächer als L oder L schwächer als L' sein? Im ersten Fall, nennen wir M eine *positive* Komponente in L, im zweiten Fall eine *negative*. Um diesen Begriff des Vorzeichens einer Komponente zu untersuchen, werden wir zunächst die 1-Komponenten in L betrachten. Hier wird das Vorzeichen durch den Hauptoperator bestimmt. Man hat z. B. für die Hauptform $p \supset q$ die Sätze (aus *Principia* *2. 05. 06)

$$p \supset p' . \supset : p' \supset q . \supset . p \supset q$$
$$q \supset q' . \supset : p \supset q . \supset . p \supset q',$$

die zeigen, daß in $p \supset q$ die Komponente p negativ und die Komponente q positiv ist. Für die anderen Hauptoperatoren haben wir analoge Sätze: für $\sim$, *2. 16; für &, *3. 45; für $\vee$, *2. 38; für ($\exists$), *10. 28; für ( ), *10. 27. Dadurch ergeben sich die folgenden Definitionen für die *Vorzeichen* von 0-Komponenten und 1-Komponenten:

Hauptform	*Vorzeichen der Komponenten*
L	L ist +
$\sim L$	L ist −
L & M	L und M sind +
$L \vee M$	L und M sind +
$L \supset M$	L ist −, M ist +
$(\exists x) . L(x)$	L ist +
$(x) . L(x)$	L ist +

Dieser Begriff eines Vorzeichens wird von 1-Komponenten auf n-Komponenten durch die arithmetische Regel der Multiplikation von Vorzeichen übertragen: Eine positive bzw. negative 1-Komponente einer positiven n-Komponente ist positiv bzw. negativ; eine positive bzw. negative 1-Komponente einer negativen n-Komponente ist negativ bzw. positiv. Diese Regel ergibt für jede Komponente M eines Ausdrucks L eindeutig ein bestimmtes Vorzeichen (positiv oder negativ), vorausgesetzt, daß M satzartig ist und in keiner Komponente von L enthalten ist, die $\equiv$ als Hauptoperator hat[1]. Auf Grund dieser Definition können wir durch Induktion den folgenden Satz beweisen:

Satz 1: Ist M eine positive Komponente in L, und ergibt die Ersetzung von M durch N einen neuen satzartigen Ausdruck L', dann gilt: Wenn M stärker als N ist, dann ist auch L stärker als L', und wenn M schwächer als N ist, dann ist L schwächer als L'. Wenn andererseits M eine negative Komponente in L ist, dann gelten dieselben Behauptungen unter Vertauschung von L mit L'.

Nach dieser Vorbereitung können wir den Prozeß der Substitution vermittelst einer Implikation definieren:

> *Sub imp schrumpf*: Prämisse, ein satzartiger Ausdruck L; Bezugstheorem, eine Implikation $M \supset N$; das Ergebnis wird aus L aufgebaut, indem man jedes M, das als *positive* Komponente in L auftritt, durch den schwächeren Ausdruck N ersetzt. Die Gültigkeit folgt aus Satz 1.

In derselben Weise, nur unter Vertauschung von „positiv" mit „negativ", definieren wir den Prozeß *Sub imp schwell*. Beide kann man nach § 3 mit einem permutierten Bezugstheorem anwenden, wobei wir aber wieder „positiv" mit „negativ" vertauschen müssen. Beide Prozesse sind wie in § 2 zu verallgemeinern. Unbalancierte Bezugstheoreme werden nach § 1 behandelt.

Diese Substitutionen vermittelst einer Implikation sind in vielen Hinsichten Verallgemeinerungen der Prozesse für Syllogismus und Schlußschema. Diese sind aber nicht immer auf Sub-

[1] Diese Ausnahme ist notwendig, weil für die Grundform $p \equiv q$ kein Satz der oben angeführten Art möglich ist.

stitutionen zurückführbar. Man wird z. B. aus einer Prämisse

$$p . \supset . q \supset p$$

und einem Bezugstheorem

$$r \supset s . \supset . \sim s \supset \sim r$$

durch die Prozesse „Sub imp schrumpf" und „Inf schrumpf" zwei verschiedene Ergebnisse erhalten.

§ 5. **Substitution vermittelst Definitionen:** Das Wesen einer Definition ist die Einführung eines neuen Zeichens mit gewissen Regeln, wobei jeder Ausdruck, der das neue Zeichen enthält, auf einen eindeutig bestimmten nur aus alten Zeichen aufgebauten Ausdruck zurückgeführt werden kann[1]. Diese Regeln werden also vorschreiben, daß das neue Zeichen, oder aber gewisse Kombinationen, die das neue Zeichen enthalten, immer durch gewisse Kombinationen alter Zeichen zu ersetzen ist. Diese Ersetzung muß aber immer einen Symbolkomplex ergeben, der selbst ein Ausdruck ist. Die Ersatzkombination muß daher gewisse Eigenschaften haben, die durch den Charakter des neuen Zeichens bestimmt sind (man kann z. B. eine neue Konstante nur durch einen konstanten Ausdruck ersetzen). Aus diesen Forderungen können wir die Gestalt der vier möglichen Definitionstypen ableiten: die Definition einer Konstanten, einer Funktion, einer Relation oder einer Variablen.

Die Definition einer Konstanten (z. B. die Definition von 2 durch 1 + 1) muß die Form

$$b = C \quad \mathrm{Df}$$

haben, wo *C* ein nicht satzartiger Ausdruck ist, der keine freien Variablen enthält; d. h. ein Ausdruck, der nur aus Funktionen und bekannten Konstanten aufgebaut ist. Diese Definition ist im wesentlichen die Festsetzung eines Substitutionsprozesses; in jeden Ausdruck, der *b* enthält, kann *b* überall durch *C* ersetzt werden mit einem Ergebnis, das dem ursprünglichen Ausdruck definitionsgemäß äquivalent ist.

[1] Diese Analyse des Wesens der Definition ist für unsere Zwecke, nämlich die Untersuchung der Rolle der Definitionen in Beweisprozessen besonders geeignet. Vgl. *Principia*, S. 11; Dubislav, *Die Definition*, S. 30.

Die Definition einer neuen Funktion f [z. B. die Definition von *cot x* durch (cos x) / (sin x)] muß die Form

$$(1) \qquad f(x_1, \ldots, x_n) = C(x_1, \ldots, x_n) \text{ Df}$$

haben, wo C ein nicht-satzartiger Ausdruck ist, der nur $x_1, \ldots, x_n$ als freie Variablen hat, und der nur schon bekannte Funktionen enthält. Der linksstehende Ausdruck in (1) ist die Grundform (vgl. I, § 1) für die neue Funktion f. Der Substitutionsprozeß für diese Definition ist die gewöhnliche Substitution vermittelst der „Gleichung" (1) nach den Vorschriften in §§ 1 und 2.

Dieselben Bemerkungen gelten für die Definition einer neuen Relation, wobei aber statt C ein satzartiger Ausdruck L zu benutzen ist. Hier wird der Substitutionsprozeß einer Substitution vermittelst einer Äquivalenz entsprechen.

Alle drei Definitionstypen ergeben daher einen stationären Substitutionsprozeß, wodurch jede Prämisse mit dem neu definierten Zeichen auf ein Ergebnis, das nur alte Zeichen enthält, zurückgeführt wird. Diesen Prozeß nennen wir *Sub def*.

§ 6. **Aufbauende Definitionen:** Es ist noch die Definition einer neuen Variable zu behandeln. Der Spielraum einer Variablen ist die einzige invariante Eigenschaft (im Sinne von I, § 5) dieser Variablen; daher muß eine solche Definition im wesentlichen einen neuen Spielraum (d. h. eine neue Kategorie von Elementen) definieren. Das ist aber der Zweck einer aufbauenden Definition[1]. Die Struktur solcher Definitionen ist eine wichtige Frage der mathematischen Logik. Wir werden im folgenden eine Analyse dieser Struktur geben, die einen Substitutionsprozeß ergibt, der den anderen „Sub def"-Prozessen analog ist.

Durch die Untersuchung mehrerer Beispiele zeigt man leicht, daß eine aufbauende Definition die folgenden Momente enthält:

1. Aus gewissen gegebenen Gegenständen werden neue ideale Gegenstände (d. h. neue Variable) aufgebaut. Jeder neue Gegen-

---

[1] Die Wichtigkeit solcher Definitionen wurde von H. Weyl betont, vgl. H. Weyl, *Philosophie der Mathematik und Naturwissenschaft*, S. 8. Solche Definitionen erscheinen in *Principia* als „incomplete symbols" (*Principia*, Einleitung, III. Kapitel).

stand ξ ist also als eine Funktion der in seinem Aufbau benutzten alten Gegenstände darstellbar, in der Gestalt

(1) $$\xi = \Phi(x_1, \ldots, x_n) = \Phi(x_i),$$

wo der Ausdruck rechts alle formalen Eigenschaften einer Grundform mit der Funktion Φ hat. Insbesondere bauen die neuen Gegenstände ξ, η, ζ, ... einen neuen Spielraum auf.

2. Für die neuen Gegenstände werden gewisse Relationen und Funktionen definiert, aber immer mittels der schon bekannten Relationen und Funktionen für die alten Gegenstände: d. h. wir definieren Relationen zwischen den ξ indirekt als Relation zwischen den $\Phi(x_i)$. Die Form solcher Definitionen können wir durch zwei typische Fälle erläutern: den Fall einer Funktion zweier Variablen, $f(\xi, \eta)$, und den Fall einer Relation zwischen zwei Variablen, $r(\xi, \eta)$:

(2) $$f(\Phi(x_i), \Phi(y_i)) = C \quad \text{Df}$$

(3) $$r(\Phi(x_i), \Phi(y_i)) = L \quad \text{Df.}$$

Die Ausdrücke C und L dürfen nur schon bekannte Funktionen und Relationen enthalten, und können nur die x_i und y_i als freie Variablen haben. Der Typus der Grundform von f wird durch den Typus von C gegeben; wenn aber der Typus der Grundform von f mit dem neuen Spielraum der ξ identisch sein soll, dann muß C die Funktion Φ als Hauptoperator haben. In analoger Weise kann man die Definition von Funktionen und Relationen mit anderen Variablenanzahlen, oder von Relationen, die zwischen neuen Variablen ξ und alten z stattfinden, analysieren.

3. Die Gleichheit der neuen Gegenstände ist ausdrücklich zu definieren. Diese Definition muß die Form (3) haben:

$$\Phi(x_i) = \Phi(y_i) \,.\equiv.\, M \quad \text{Df.}$$

Hier muß M so beschaffen sein, daß die dadurch definierte Gleichheit reflexiv, symmetrisch und transitiv wird, und ferner so, daß alle in 2 definierten Relationen und Funktionen gegenüber dieser Gleichheit invariant sind. Das letztere bedeutet folgendes:

$$\xi = \xi' \,.\, \eta = \eta' \,.\supset.\, f(\xi, \eta) = f(\xi', \eta')$$
$$\xi = \xi' \,.\, \eta = \eta' \,.\supset.\, r(\xi, \eta) \equiv r(\xi', \eta').$$

Eine allgemeinere Konstruktion idealer Gegenstände, wobei die Gleichheit unmittelbar durch diese Invarianzbedingung definiert wird, ist auch möglich, wird aber hier nicht erörtert.

Die so konstruierten idealen Gegenstände sind in gewissem Sinne unvollständige Zeichen, denn sie haben eine Bedeutung nur im Zusammenhang mit Relationen oder Funktionen, wie z. B. r und f. Eine aufbauende Definition kann daher keine alleinstehenden idealen Gegenstände ergeben, sondern muß ein System definieren, das aus idealen Gegenständen und aus Relationen und Funktionen für diese Gegenstände besteht. Diese Sachlage betont die Wichtigkeit des Begriffes „System" für das Verständnis der mathematischen Logik.

Eine solche aufbauende Definition hat nicht die einfache Struktur der Definitionen des vorigen Abschnittes, die in einer Vorschrift der Form „A und B sind durch Definition identisch" bestehen, aber trotzdem können wir auch mit jenen Definitionen einen Substitutionsprozeß verbinden. Dieser Prozeß, den wir *Sub aufbau* nennen wollen, ist auf eine Prämisse L anwendbar, die neue Variablen ξ, η usw., und auch neue Funktionen oder Relationen, wie z. B. f oder r, enthält. Die neuen Variablen und Relationen werden auf folgende Weise in zwei Schritten eliminiert:

1. Die neuen Variablen ξ, η usw. sind durch die Ausdrücke $\Phi(x_i)$ bzw. $\Phi(y_i)$ usw. mit alten Variablen x_i und y_i zu ersetzen. Alle so benutzten x_i und y_i müssen voneinander verschieden sein. Ein Klammerzeichen $(\exists\xi)$ oder (η) mit einer neuen Variable ist folgendermaßen zu ersetzen:

$$(\exists\xi) \rightarrow (\exists x_1, \ldots, x_n).$$

2. Nachdem diese Ersetzungen ausgeführt worden sind, hat man die Funktionen und Relationen für die idealen Gegenstände vermittelst der Definitionen (2), (3) usw. zu eliminieren. Dabei hat man mit den kleinsten Komponenten von L zu beginnen und nach der in § 2 gegebenen Vorschrift schrittweise zu den größeren Komponenten überzugehen. Für jede Komponente ist die geeignete Definition zu benutzen.

Dieser Prozeß transformiert L in einen äquivalenten Ausdruck L', für den folgendes gilt: Wenn die betreffende aufbauende Definition ein neues System S_2 (mit neuen Variablen, Relationen und Funktionen) aus einem alten System S_1 erzeugt, und wenn die

3

Prämisse L nur Elemente aus S_2 enthält, dann enthält das Ergebnis L' nur Elemente aus dem alten System S_1.

Diese ganze Untersuchung kann man auf bedingte aufbauende Definitionen erweitern, die neuen Gegenstände

$$\Phi(x_1, \ldots, x_n)$$

nur aus solchen $x_1, \ldots, x_n$ konstruieren, welche einer festen Bedingung

$$L(x_1, \ldots, x_n) \tag{4}$$

genügen. Solche aufbauende Definitionen sind häufig; Beispiele sind: die Definition der reellen Zahlen als Dedekindsche Schnitte oder als konvergente Folgen oder die Definition der beschreibenden Funktionen in *Principia* usw. Der Substitutionsprozeß, der einer solchen bedingten aufbauenden Definition entspricht, geht aus der obigen Vorschrift durch passende Abänderungen hervor; insbesondere muß man bei der Ersetzung der gebundenen Variablen die feste Bedingung (4) berücksichtigen.

§ 7. **Beispiele:** In der Algebra wird das Divisionsgesetz

$$(b \,.\, a^{-1}) \,.\, a = b \tag{1}$$

aus den bekannten Gesetzen

$$a^{-1} \,.\, a = 1, \tag{2}$$

$$b \,.\, 1 = b, \tag{3}$$

$$(a \,.\, b) \,.\, c = a \,.\, (b \,.\, c) \tag{4}$$

bewiesen. Der gewöhnliche Beweis fängt mit den gegebenen Gleichungen an, und erreicht den Ausdruck (1) durch Kombination derselben. Der Gang des Beweises ist also durch das erwünschte Resultat bestimmt. Das kann man oft klarer darlegen, wenn man von dem erwünschten Satz ausgeht und diesen Satz schrittweise reduziert. In diesem Fall sind die Schritte

(5) Anfang Th., Sub (4), Sub (2), Ende (3).

Hier und im folgenden werden wir „Sub" für den verallgemeinerten „Sub Gl"-Prozeß benutzen; denn in einem Substitutionsprozeß ist es immer aus dem Bezugstheorem selbst klar, ob eine Substitution vermittelst einer Gleichung, einer Äquivalenz oder einer Implikation gemeint ist.

Ein abgekürzter Beweis der Form (5), der mit dem fraglichen Theorem (Abkürzung, Th) anfängt, nur Schwell- und Stationär-prozesse benutzt und mit einem bekannten Theorem endet, heißt ein *Reduktionsbeweis*. Oft wird der Ausdruck am Ende eines solchen Beweises nicht selbst ein bekanntes Theorem, sondern nur ein Fall eines solchen Theorems sein. Das ist eine Verallgemeinerung (im Sinne von § 2) des speziellen Endprozesses in (5). In allen solchen Fällen benutzen wir die Bezeichnung[1] „Ende Theorem so-und-so". Einen solchen Schwellprozeß, dessen Ergebnis ein bekanntes Theorem ist, nennen wir einen *Endprozeß*.

Die folgenden Beweise aus den *Principia* *22 lassen sich als Reduktionsbeweise darstellen und benutzen überdies viele andere Prozesse dieses Kapitels. In allen Fällen sind die hier gegebenen Beweise kürzer und klarer als dieselben Beweise in den *Principia* selbst[2].

*22. 39 Anfang Th, Sub *20. 33, Sub *22. 33, Sub *20. 3, Ende Identität.

*22. 481 Anfang Th, Sub perm *22. 41, Sub perm *22. 48, Ende *2. 08.

*22. 5 Anfang Th, Sub *20. 43, Sub *22. 33, Ende perm *4. 24.

*22. 68 Anfang Th, Sub *20. 43, Sub *22. 34, Sub *22. 33, Sub *22. 34, Ende Perm *4. 4.

*22. 91 Anfang Th, Sub *20. 43, Sub *22. 34, Sub *22. 05, Sub *22. 33, Sub *22. 35, Sub *20. 06, Sub *4. 3, Ende *5. 63.

Im zweiten Schritt des Beweises *22. 481 hat man eine Substitution vermittelst eines unbalancierten Bezugstheorems.

Die Konstruktion der rationalen Zahlen bietet ein Beispiel einer aufbauenden Definition und der Substitution vermittelst einer solchen Definition. Nehmen wir an, daß die ganzen Zahlen, ihre Addition, Multiplikation und die damit zusammenhängenden Eigenschaften gegeben sind. Dann kann man in bekannter Weise die rationalen Zahlen als Paare ganzer Zahlen einführen und dabei auch die Multiplikation, Addition und lineare Anordnung dieser

---

[1] Dieser Prozeß ist wohldefiniert, denn nach I, § 4, Satz 1 können wir feststellen, ob die Prämisse ein Fall des gegebenen Bezugstheorems sei.

[2] Für einige Beweise muß man unsere Theorie der symbolischen Ausdrücke in naheliegender Weise auf Symbole wie $\hat{x}(\varphi x)$ und $\equiv_x$ erweitern.

3*

Zahlen definieren. Diese Definitionen haben die in § 6 dargestellte Gestalt. Sie ermöglichen es, die Eigenschaften der rationalen Zahlen abzuleiten. Das assoziative Gesetz für die Multiplikation rationaler Zahlen wird z. B. in zwei Schritten bewiesen: erst eine Substitution vermittelst der aufbauenden Definition der Zahlen, dann eine Umformung der dadurch gegebenen Gleichung, um eine Identität zu erhalten. Der Beweis ist also

Anfang Th, Sub Aufbau, Ende algebraische Identität[1].

Genau derselbe abgekürzte Beweis gilt für jede richtige Gleichung, die nur die Addition und die Multiplikation rationaler Zahlen enthält! Wenn die aufbauende Definition einmal aufgestellt ist, dann ist die Bestätigung der Eigenschaften der so geschaffenen Zahlen rein automatisch; die oft gegebenen einzelnen Beweise sind nicht nötig. Schon die notwendigen Invarianzeigenschaften der neuen Gleichheit sind in derselben Weise zu verifizieren.

Dieselbe Analyse ist für die Definition der komplexen Zahlen als Zahlenpaare, der Quaternionen als Zahlenquadrupel oder der Vektoren als n-Tupel anwendbar. Alle diesbezüglichen Beweise können mit diesem Verfahren systematisch gewonnen werden.

Die bisher diskutierten Prozesse enthalten sämtliche Beweismethoden der Logik von Aristoteles. Die vier Formen eines Urteils sind folgendermaßen ausdrückbar:

$$\text{A}: \varphi x \supset \psi x; \qquad \text{E}: \varphi x \,.\supset.\, \sim \psi x$$

$$\text{I}: (\exists x) \,.\, \varphi x \,.\, \psi x; \qquad \text{O}: (\exists x) \,.\, \varphi x \,.\, \sim \psi x.$$

Die Umkehrung eines Urteils ist dann eine Substitutoin vermittelst einer der folgenden Äquivalenzen

$$p \supset q \,.\equiv.\, \sim p \supset \sim q$$

$$p \,.\, q \,.\equiv.\, q \,.\, p.$$

Die Modi des Syllogismus kann man als Substitutionen vermittelst einer Implikation ausdrücken, eventuell mit einer Umkehrung verbunden. Nur Darapti, Felapton, Bramantip und Fesapo kann man so nicht erhalten, da diese in der obigen Deutung nicht gültig sind[2].

[1] Diesen Prozeß kann man auch mit unseren Methoden definieren; vgl. VI, § 4.

[2] Hilbert und Ackermann a. a. O. S. 41.

IV. Kapitel.

Prozesse mit Klammerzeichen.

§ 1. **Normalform**: Jeden satzartigen Ausdruck L können wir in ein *Präfix* und eine *Matrix*, nach dem Schema

(1) $$(\exists x_1)\,(x_2)\,(x_3)\,(\exists x_4)\ldots(x_n)\,.\,M(x_1,\ldots,x_n),$$

aufspalten. Das Präfix besteht aus allen Klammerzeichen $(\exists x_1)$, ..., (x_n), deren Wirkungsbereiche der ganze Ausdruck mit Ausnahme gewisser anderer Klammerzeichen ist. Die Matrix M enthält keine weiteren solchen Klammerzeichen. Sie kann aber weitere gebundene Variable mit *kleinerem* Wirkungsbereiche enthalten; diese nennen wir *innere* gebundene Variablen. Doch kann man oft für einen Ausdruck eine *Normalform* finden, in der alle Klammerzeichen im Präfix stehen[1]. Die Herstellung dieser Normalform ist ein nützlicher Beweisprozeß, der die systematische Behandlung satzartiger Ausdrücke erleichtert.

Die einfachste Operation bei der Herstellung der Normalform ist das Herausziehen eines inneren Klammerzeichens. Nach *Principia* *10.34

(2) $$(x)\,.\,\varphi x\,.\supset.\,p\,:\equiv.\,(\exists x)\,.\,\varphi x\supset p$$

kann man z. B. das Zeichen (x) aus einer Implikation $\supset$ herausziehen. Andere ähnliche Sätze für $\sim$, $\supset$, & und $\vee$ werden in *10.2. 21.23.252.253.33.35.36.37 gegeben. Der Inhalt dieser Sätze wird am besten unter Verwendung des Begriffes des Vorzeichens einer Komponente (III, § 4) zusammengefaßt: In einem Ausdruck L, der $\sim$, $\supset$, $\vee$ oder & als Hauptoperator hat, und der eine 1-Komponente M' mit dem Hauptoperator $(\exists x)$ oder (x) hat, können wir dieses Klammerzeichen herausziehen (d.h. vor den ganzen Ausdruck stellen). Wenn M' eine positive Komponente ist, dann bleibt das herausgezogene Klammerzeichen ungeändert; wenn M' negativ ist, dann wird $(\exists x)$ in (x) verwandelt und umgekehrt.

Mittels vollständiger Induktion können wir dann das Herausziehen eines Klammerzeichens aus einer beliebigen Komponente untersuchen. Betrachten wir eine beliebige innere gebundene

[1] Hilbert und Ackermann a. a. O. S. 63.

Variable y, die in der Matrix M eines Ausdrucks entweder in der Form

$$(\exists y) . N(y)$$

oder in der Form

$$(y) . N(y)$$

auftreten muß. Diese Variable kann man oft schrittweise bis zum Präfix herausziehen, wenn keine anderen gebundenen Variablen im Wege stehen; d. h. wenn keine andere innere gebundene Variable einen Wirkungsbereich hat, der N als Komponente enthält. Wenn das der Fall ist, und wenn außerdem N entweder eine positive oder eine negative Komponente in M ist, dann heißt y *ausziehbar*. Diese Variable y in (1) *herausziehen* heißt dann, das Zeichen $(\exists y)$ oder (y) innerhalb M weglassen und am Ende des Präfixes einfügen. Wenn der Wirkungsbereich N eine positive Komponente in M ist, dann wird dieses Zeichen ungeändert eingefügt; wenn dagegen N negativ ist, dann wird $(\exists y)$ in (y) verwandelt, und umgekehrt. Dadurch ist der Stationärprozeß des Herausziehens einer ausziehbaren inneren gebundenen Variablen y definiert.

Die Normalform eines Ausdrucks ist meistens nicht eindeutig definiert; aber für einen Beweisprozeß müssen wir Eindeutigkeit haben. Eine solche Definition ist die folgende:

Norm: Prämisse, ein satzartiger Ausdruck L, wie z. B. (1). Der erste Schritt im Aufbau des Ergebnisses ist das Herausziehen aller in L ausziehbaren inneren gebundenen Variablen, und zwar in der folgenden Reihenfolge: erst die Variablen, die nach dem Herausziehen mit dem Allzeichen ( ) auftreten; und zwar in der Reihenfolge, in der sie in L (von links nach rechts gelesen) auftreten; dann alle andern in dem ursprünglichen Ausdruck L ausziehbaren Variablen, ebenfalls in der Reihenfolge, in der sie in L auftreten. Auf den so erhaltenen Ausdruck wird dasselbe Verfahren abermals angewendet usw., bis die Normalform erreicht ist.

Das ist ein Stationärprozeß. Die Definition ist so aufgestellt, daß das Ergebnis in formaler Hinsicht relativ schwach ist. Es ist aber zu bemerken, daß dieser Prozeß für Ausdrücke mit dem Operator $\equiv$ im allgemeinen keine vollständige Normalform ergibt.

Oft wird ein umgekehrter Prozeß angewandt, der auf dem

Begriff des Hineinziehens einer gebundenen Variablen beruht. Wenn z. B. ein Ausdruck

$$(\exists y) . N(y)$$

gegeben ist, dann kann man y unter den Hauptoperator von N *hineinziehen*, wenn der Hauptoperator $\sim$ ist, oder wenn der Hauptoperator $\supset$, & oder $\vee$ ist und wenn nur eine 1-Komponente in N die Variable y enthält. Das Vorzeichen dieser 1-Komponente spielt wie vorher eine Rolle. Der Prozeß *Umgekehrte Norm* für eine Prämisse L wird dann alle gebundenen Variablen so weit wie möglich hineinziehen, wobei die Variablen mit den kleinsten Wirkungsbereichen erst zu behandeln sind.

§ 2. **Prozesse für Fixierung**: Ein Existenzsatz wird oft durch Konstruktion bewiesen. Dieser Prozeß besteht in folgendem:

$x = C$ *Fixieren*: Prämisse, ein Ausdruck $(\exists x).L(x)$ mit Hauptoperator $\exists$. Wir wollen für die gebundene Variable x den Wert C einsetzen. Dazu muß C ein nicht-satzartiger Ausdruck sein, dessen Typus eine Untermenge des Spielraums von x ist, und dessen Variablen freie Variablen in der Prämisse sind. Das Ergebnis ist

(1) $$L(C);$$

d. h. das Ergebnis wird aus L erhalten, indem x überall in L durch C ersetzt wird. Nach *Principia* *10. 24 ist das ein gültiger Schwellprozeß.

Dieser Prozeß der Fixierung (oder Exemplifixierung) ist umständlich, denn man muß den Wert C ausdrücklich angeben. In vielen Fällen aber wird dieser Wert durch den späteren Gang des Beweises vollständig bestimmt. Insbesondere wird in einem Reduktionsbeweis das Ergebnis (1) eines Prozesses der Fixierung oft ein Fall eines bekannten Theorems sein. Dann können wir den Wert C durch Vergleich dieses bekannten Theorems mit der Prämisse des Prozesses der Fixierung bestimmen. Das ergibt den folgenden Prozeß:

Ende Fixieren: Prämisse, ein Ausdruck $(\exists x) . L(x)$ mit Hauptoperator $\exists$. Bezugstheorem, ein satzartiger Ausdruck M. Der Prozeß ist dann und nur dann ein gültiger Endprozeß für einen Reduktionsbeweis, wenn es ein solches C gibt, daß der Prozeß „$x = C$ Fixieren" anwendbar ist und ein Ergebnis $L(C)$ hat, das ein Fall von M ist. Um diesen Prozeß als Abkürzung zu benutzen, muß man außerdem eine Konstruktionsmethode für C haben. Das kann man oft durch Vergleich von $L(x)$ mit M nach den Vorschriften in I, § 4 folgendermaßen erreichen:

I. Wenn der Vergleich zeigt, daß $L(x)$ Fall von M ist (vgl. I, § 4, Satz 1), dann können wir für C eine beliebige Konstante im Spielraum von x wählen.

II. Eine Variable y in M kann im Vergleich mehreren Komponenten $D_1, D_2, \ldots, D_m$ in $L(x)$ entsprechen. Wenn C existiert, dann müssen sämtliche D_i nach Ersetzung von x durch C identisch sein. Um das zu prüfen, kann man D_i mit D_j vergleichen ($i = = j$). Wenn C existiert, dann muß jedes x in D_i einer Komponente in D_j entsprechen, und diese Komponente muß entweder x selbst sein oder ein Ausdruck C', der kein x enthält und dessen Typus Untermenge des Spielraums von x ist. In diesem Fall kann man $L(C')$ konstruieren und dann durch Vergleich bestimmen, ob $L(C')$ ein Fall von M ist. Wenn das zutrifft, dann hat C den Wert C'. Eine ähnliche Methode kann man für jedes Paar D_i und D_j von Komponenten aus $L(x)$, die zwei identischen Komponenten D_i' und D_j' in M entsprechen, anwenden.

III. Beim Vergleich von $L(x)$ mit M kann die Variable x eventuell einer Komponente D in M entsprechen. In diesem Fall kann man C aus D durch Ersetzung der Variablen in D durch passende Komponenten aus L konstruieren.

Die zweite Methode ist in den meisten praktischen Fällen die beste, aber keine der drei Methoden gibt in jedem Fall eine Konstruktionsmethode für C.

Wenn das Klammerzeichen $(\exists x)$ nicht (wie wir bisher angenommen haben) am Anfang, sondern irgendwo im Präfix eines Ausdrucks L auftritt, dann ist der Prozeß

$$x = C \text{ Fixieren}$$

anwendbar, wenn C einen passenden Typus hat und wenn die Variablen in C entweder freie Variablen in L sind oder gebundene Variablen, die vor x mit dem Allzeichen ( ) auftreten. Man kann mehrere gebundene Variablen in dieser Weise nacheinander fixieren (die Reihenfolge ist unwesentlich), und in vielen Fällen kann man die dazu notwendigen C durch Vergleich bestimmen.

§ 3. **Eine Halbnorm:** Der Satz

$$(x) . \varphi x : (y) . \psi y : \equiv . (z) . \varphi z . \psi z$$

aus *Principia* *10. 22 ergibt einen der Norm analogen Prozeß, denn nach diesem Satz können die zwei gebundenen Variablen (x) und (y) simultan vor die Verknüpfung & herausgezogen, und danach durch eine einzelne neue Variable z ersetzt werden. Für andere logische Verknüpfungen kann man analoge Resultate erhalten, wie durch die folgende Tabelle angedeutet wird.

Verknüpfung	*Prämisse*	*Ergebnis*	*Charakter*	*Principia*
&	$(\exists x) . \varphi x : (\exists y) . \psi x$	$(\exists z) . \varphi z . \psi z$	schwell	*10. 5
$\vee$	$(\exists x) . \varphi x . \vee . (\exists y) . \psi y$	$(\exists z) . \varphi z \vee \psi z$	stationär	*10. 42
$\vee$	$(x) . \varphi x . \vee . (y) . \psi y$	$(z) . \varphi z \vee \psi z$	schrumf	*10. 41
$\supset$	$(\exists x) . \varphi x . \supset . (\exists y) . \psi y$	$(z) . \varphi z \supset \psi z$	schwell	*10. 28
$\supset$	$(x) . \varphi x . \supset . (y) . \psi y$	$(z) . \varphi z \supset \psi z$	schwell	*10. 27
$\equiv$	Dasselbe wie für $\supset$			*10. 271. 281

Gemäß dieser Tabelle können wir jedes Klammerzeichenpaar, $(\exists x)$ und $(\exists y)$ oder (x) und (y) aus einer logischen Verknüpfung &, $\vee$, $\supset$ und $\equiv$ herausziehen. Dadurch erhalten wir den folgenden Prozeß.

Halbnorm schrumpf: Prämisse, ein satzartiger Ausdruck L; das Ergebnis wird durch wiederholte Anwendung der Prozesse der obigen Tabelle auf die Komponenten von L erhalten. Ein Prozeß der Tabelle ist auf eine Komponente M anwendbar, wenn M die Form der Prämisse dieses Prozesses hat, aber ein Schwell- bzw. Schrumpfprozeß wird nur auf eine negative bzw. positive Komponente angewandt. Die Reihenfolge der Benutzung dieser Prozesse ist unwesentlich.

Der Prozeß „Halbnorm schwell" wird in derselben Weise unter Vertauschung von „positiv" mit „negativ" definiert. Ein Prozeß „Halbnorm stationär", ist auch möglich.

Diese Prozesse werden nicht immer eine vollständige Normalform für die Prämisse ergeben, aber sie haben den Vorteil gegenüber dem einfachen Normprozeß, daß sie es ermöglichen, gebundene Variablen auch aus der Verknüpfung $\equiv$ herauszuziehen. Oft muß man diese Prozesse in Verbindung mit dem einfachen Normprozeß benutzen. Umgekehrte Halbnormprozesse sind auch möglich.

§ 4. **Gemischte Prozesse mit Klammerzeichen**: Für die Manipulation gebundener Variablen kann man einige weitere Prozesse aufstellen, wie in der folgenden Tabelle angedeutet wird.

Prozeß	*Prämisse*	*Ergebnis*	*Charakter*
Add (x)	$L(x)$	$(x) . L(x)$	stationär
Add $(\exists x)$	$L(x)$	$(\exists x) . L(x)$	schrumpf
Abnehmen	$(x) . L(x)$	$L(x)$	stationär
Abnehmen	$(\exists x) . L(x)$	$L(x)$	schwell
Vertauschen	$(x) : (y) . L(x, y)$	$(y) : (x) . L(x, y)$	stationär
Vertauschen	$(\exists x) : (\exists y) . L(x, y)$	$(\exists y) : (\exists x) . L(x, y)$	stationär
Vertauschen	$(x) : (\exists y) . L(x, y)$	$(\exists y) : (x) . L(x, y)$	schwell
Vertauschen	$(\exists x) : (y) . L(x, y)$	$(y) : (\exists x) . L(x, y)$	schrumpf

Die vier letzten Prozesse sind auch auf einen Ausdruck anwendbar, der eine Komponente N der Form der betreffenden Prämisse hat. Das Ergebnis wird in diesem Fall gebildet, indem man dieses N durch das entsprechende Ergebnis aus der Tabelle ersetzt. Der Charakter des Prozesses bleibt derselbe, wenn N eine positive Komponente ist; wenn aber N eine negative Komponente ist, dann wird ein Schrumpfprozeß aus der Tabelle in einen Schwellprozeß umgewandelt, und umgekehrt. Wenn N kein Vorzeichen hat, dann darf man nur die stationären Prozesse benutzen.

§ 5. **Beispiele**: Viele Beweise aus *Principia* ergeben Beispiele für die Prozesse dieses Kapitels. Wir werden einige Beweise aus *Principia* *22 (Theorie der Mengen) geben:

*22. 351: Anfang Th, Sub *13. 02, Sub *20. 43, Sub *22. 35, Sub *20. 06, Norm, Abnehmen, Sub *4. 21, Ende *5. 19.

*22. 4: Anfang Th, Sub *22. 01, Halbnorm, Abnehmen, Ende perm *4. 01.

*22. 45: Anfang Th, Sub *22. 01, Sub *22. 33, Halbnorm, Abnehmen, Ende *4. 76.

*22. 481: Anfang Th, Sub perm *22. 41, Sub extra *22. 48, Ende fixieren *2. 08.

*22. 59: Anfang Th, Sub *22. 01, Sub *22. 34, Halbnorm, Abnehmen, Ende *4. 77.

*22. 62: Anfang Th, Sub *13. 16, Sub *20. 43, Sub *22. 01, Sub *22. 34, Halbnorm, Ende *4. 72.

Der Beweis für *22. 481 benutzt eine Substitution vermittelst eines unbalancierten Bezugstheorems (vgl. III, § 1). Die zwei dadurch eingeführten Extravariablen werden dann durch Vergleich und Fixierung behandelt (vgl. § 2, Ende). In den Beweisen für *22. 59. 62. 4 tritt die Verknüpfung = auf, und deshalb wird „Halbnorm" statt „Norm" hier benutzt.

Man kann auch viele einfache mathematische Beweise mit diesen Prozessen bequem behandeln. Ein Beispiel wird durch den folgenden Beweis aus der Theorie der Stetigkeit geliefert. Nehmen wir an, man habe die Stetigkeit einer Funktion $f(y)$ im Punkte x definiert und die folgende Eigenschaft bereits bewiesen!

I. Wenn f und g im Punkte x stetig sind, dann ist auch $f + g$ im Punkte x stetig.

Wir brauchen ferner den Begriff der Stetigkeit einer Funktion für alle x-Werte.

II. Definition. Die Funktion f ist dann und nur dann stetig, wenn f für alle Punkte x stetig ist.

Auch für diese allgemeine Stetigkeit gilt ein zu I analoges Additionstheorem.

III. Wenn f und g stetig sind, dann ist $f + g$ stetig.

Wenn wir die drei Ausdrücke I, II und III in symbolische Form übersetzen, dann können wir für III den folgenden Beweis geben:

Anfang Th, Sub II, Halbnorm, Abnehmen, Ende I.

Der folgende Beweis für III ist ähnlich, aber etwas länger:

Anfang Th, Sub II, Norm, Abnehmen, Ende fixieren I.

Der Unterschied der zwei Beweise liegt darin, daß im ersten Beweis der Prozeß „Halbnorm" statt des Prozesses „Norm" mit einer späteren Fixierung benutzt wird.

Die in diesem Kapitel definierten Prozesse mit Klammerzeichen sind in dem Sinne vollständig, daß diese Prozesse überall als Ersatz für die gewöhnlichen Regeln für das Manipulieren mit gebundenen Variablen dienen können. Das bedeutet, daß jeder Schritt eines Beweises, der eine solche Regel (z. B. eine Regel aus *Principia* *10 oder *11) als Bezugstheorem hat, durch einen oder mehrere Prozesse dieses Kapitels ersetzt werden kann. Diese Behauptung können wir leicht durch Analyse aller Sätze aus *Principia* *10 und *11 bestätigen.

V. Kapitel.

Beweisstruktur.

Nachdem wir in den letzten drei Kapiteln verschiedene Beweisprozesse analysiert haben, werden wir in diesem Kapitel die Methoden für die Zusammenfügung solcher Prozesse in gültigen Beweisen studieren.

§ 1. **Beweisfiguren**: Ein Reduktionsbeweis (III, § 7) hat im wesentlichen die Struktur einer linearen Kette von Ausdrücken. Durch Untersuchung komplexerer Beweise kann man feststellen, daß auch diese Beweise aus verschiedenen solchen Ketten zusammengefügt sind. Daher kann eine Untersuchung über Beweisstruktur mit dem Begriff einer Kette anfangen. Eine Kette hat die Form

$$\begin{array}{l} L_1\ L_2, \ldots, L_{m-1}, L_m \\ P_1, P_2, \ldots, P_{m-1}, \end{array}$$

wo die L_i eine lineare Reihe satzartiger Ausdrücke bilden, und wo die P_i Prozesse mit Bezugstheoremen sind. Dabei muß der Prozeß P_i mit der Prämisse L_i und dem gegebenen Bezugstheorem immer das Ergebnis L_{i+1} haben. Drei Kettentypen sind wichtig:

1. **Die Schwellkette,** wo jeder Prozeß entweder ein Schwell- oder ein Stationärprozeß ist.

2. **Die Schrumpfkette,** wo jeder Prozeß entweder ein Schrumpf- oder ein Stationärprozeß ist.

3. **Die Stationärkette,** wo jeder Prozeß stationär ist.

In den drei Fällen wird der Endausdruck gegenüber dem Anfangsausdruck stärker bzw. schwächer bzw. äquivalent sein. Um eine solche Kette abzukürzen, muß man nur folgendes angeben: Anfangsausdruck, Endausdruck und die Kette der Prozesse mit den zugehörigen Bezugstheoremen. Wir verwenden folgende Bezeichnungen: $\leftarrow$ für eine Schwellkette, $\rightarrow$ für eine Schrumpfkette und $\leftrightarrow$ für eine Stationärkette. Das Zeichen

$$L \rightarrow M$$

bedeutet also eine Schrumpfkette, die mit dem Ausdruck L anfängt und mit M endet; andererseits bedeutet das Zeichen

$$L \leftarrow$$

eine Kette, die mit L anfängt und mit einem schon bewiesenen Satz endet. Wenn das Gegenteil nicht explizit gesagt wird, werden wir immer annehmen, daß alle Bezugstheoreme einer solchen Kette bekannte Sätze sind.

Mit dieser Bezeichnung können wir einen Reduktionsbeweis mit dem Diagramm

$$\text{(1)} \qquad \text{Th} \leftarrow$$

beschreiben, denn ein solcher Beweis fängt mit dem fraglichen Theorem an (Th) und erreicht dann eine bekannte Endformel durch eine Schwellkette. Ein Diagramm dieser Art nennen wir eine *Beweisfigur*. Die Beweisfigur muß immer die Form des Beweises angeben; d. h. die Art und Weise, wie die Ketten im Beweis zusammengesetzt sind. Man kann viele gültige Figuren angeben. Wenn z. B. der betrachtete Satz die Form $L . M$ hat, was wir durch die Bezeichnung

$$\text{Th} = L . M$$

ausdrücken wollen, dann kann man ihn beweisen, indem man L und M gesondert behandelt. Wenn insbesondere ein Reduktionsbeweis sowohl für L wie auch für M gegeben wird, dann haben wir die folgende Beweisfigur:

$$\text{(2)} \qquad \text{Th} = L . M: \qquad L \leftarrow; \; M \leftarrow.$$

Durch eine derartige Aufspaltung des gegebenen Theorems in Komponenten können wir viele andere gültige Beweisfiguren erhalten. Die wichtigsten sind die folgenden:

$$\text{(3)} \qquad \text{Th} = L \supset M: \qquad L \rightarrow M$$
$$\text{(4)} \qquad \text{Th} = L \equiv M: L \longleftrightarrow M$$
$$\text{(5)} \qquad \text{Th} = L \vee M . \supset . N: \quad N \leftarrow M, \; N \leftarrow L$$
$$\text{(6)} \qquad \text{Th} \leftarrow \; = L . M: \qquad L \leftarrow, \; M \leftarrow.$$

In der letzten Figur haben wir einen Beweis, der den fraglichen Satz mit Hilfe einer Schwellkette in einen Ausdruck der Form $L . M$ transformiert. Dieser Endausdruck wird dann in zwei Ketten nach der Figur (2) bewiesen. In derselben Weise kann man die

Figuren (3), (4) und (5) mit einer vorbereitenden Schwellkette verbinden.

Andere und verwickeltere Beweisfiguren sind auch möglich. Wenn z. B. ein Theorem der Form

$$L \supset M$$

zu beweisen ist, dann können wir oft eine einzelne Schwellkette benutzen, die mit der Behauptung M anfängt und die die Hypothese L einmal oder mehrmals als Bezugstheorem benutzt. Bei einer solchen Benutzung der Hypothese L ist aber zu beachten, daß jede freie Variable in L, die auch in M auftritt, in beiden Ausdrücken in derselben Weise zu behandeln ist. Wir müssen daher die Anwendung des Bezugstheorems L auf die *speziellen* Prozesse Schlußschema, Syllogismus und Substitution beschränken; d.h. auf Prozesse, die im Sinne von II, § 4 und III, § 2 nicht verallgemeinert sind. In einer derartigen Anwendung werden wir L ein *spezielles* Bezugstheorem nennen. Die Bezeichnung $\xleftarrow{L}$ wird für eine Kette benutzt, die bekannte Sätze und auch das spezielle Bezugstheorem L als Bezugstheoreme benutzt. Der oben diskutierte Beweis hat also die Figur

(7) Th $= L \supset M$: $M \xleftarrow{L}$

Andere ähnliche Beweisfiguren sind die folgenden:

(8) Th $= (L \,.\, M \supset N)$: $N \xleftarrow{L} M$

(9) Th $= (L \,.\, M \supset N)$: $N \xleftarrow{L,\, M}$.

In genau derselben Weise können wir ein Theorem der Form $L \,.\supset.\, M \supset N$ behandeln.

In (1) bis (9) haben wir einige einfache Beweisfiguren angegeben. Aus diesen Figuren können wir andere durch Umkehrung einer oder mehrerer Ketten erhalten; d. h. durch Ersetzung von $L \leftarrow$ durch $\rightarrow L$ und umgekehrt[1]. Diese Aufzählung von Beweisfiguren ist aber keineswegs vollständig.

Mit Hilfe dieser Beweisfiguren können wir unsere Abkürzungsmethoden wesentlich erweitern: Wenn wir einen Beweis abkürzen wollen, der eine gültige Beweisfigur hat, dann können wir erst diese Figur angeben und nachher die gewöhnlichen Abkürzungen für

[1] Diese Umkehrung darf aber nicht in der ersten Kette in (6) ausgeführt werden.

jede Kette der Figur hinschreiben. Durch die Angabe der Beweisfigur vor den Einzelheiten wird eine gute Übersicht über den ganzen Beweis geboten.

§ 2. **Beweise mit Hilfssätzen**: In diesem Abschnitt untersuchen wir noch eine Methode zur Erweiterung unserer Abkürzungsmethoden auf komplexe Beweise. Diese Methode stützt sich auf die Beweisanalyse, die in vielen abstrakten Theorien — z. B. in *Principia* und in der *General Analysis* von E. H. Moore — benutzt wird. Diese Analyse zerlegt den Beweis eines Satzes durch Auffindung aller irgendwie benutzten Hilfssätze, die dann unabhängig aufgestellt werden können. Dadurch erscheinen die schwierigen Momente einer Theorie nicht als wichtige Schritte eines Beweises, sondern als selbständige Sätze. Wenn dieses Programm durchgeführt wird, dann wird der Beweis jedes Hilfssatzes einfacher und unseren Methoden zugänglicher werden. Daher können wir oft einen abgekürzten Beweis in der folgenden Weise angeben: Erst die Formulierung der notwendigen Hilfssätze; dann eine Abkürzung der Beweise dieser Hilfssätze mit den schon behandelten Methoden (jeder solche Beweis darf aber die schon bewiesenen Hilfssätze als Bezugstheoreme benutzen); endlich ein Beweis des Satzes selbst, wobei alle Hilfssätze als Bezugstheoreme zulässig sind.

In vielen Fällen wird der Charakter des Satzes die Form der notwendigen Hilfssätze im wesentlichen bestimmen. Es ist wahrscheinlich möglich, allgemeine Prozesse aufzustellen, die eine derartige automatische Konstruktion der Hilfssätze systematisieren.

§ 3. **Beispiele:** Die Figur (3) in § 1 können wir durch einfache Beweise aus *Principia* *2 erläutern:

*2. 15 Th $= L \supset M, L \rightarrow M$. Anfang L, Syll unter schrumpf *2. 12, Inf schrumpf *2.03, Syll unter schrumpf *2. 14, Ende M.

In diesem Beweise bedeuten L und M die Hypothese bzw. die Behauptung des betrachteten Theorems. Eine analoge Konvention wird für alle Beweisfiguren gemacht.

*2. 16 Th $= L \supset M$: $L \rightarrow M$. Anfang L, Syll unter schrumpf *2. 12, Inf schrumpf *2. 03, Ende M.

*2. 17 Th $= L \supset M$: $L \rightarrow M$. Anfang L, Inf schrumpf *2. 03, Syll unter schrumpf *2. 14, Ende M.

Für 2. 18 und *2. 2 können wir analoge Beweise geben. Oft kann man die umgekehrte Figur benutzen, wie im folgenden Beweis:

*2. 17 Th $= L \supset$ M: $M \leftarrow L$. Anfang M, Syll unter schwell *2. 14, Inf schwell *2. 03, Ende L.

Für die Figur (2) in § 1 existieren viele Beispiele. Der Beweis einer notwendigen und hinreichenden Bedingung $L \equiv M$ in den zwei Schritten $L \supset M$ und $M \supset L$ ist eine Anwendung dieser Figur. Ein anderes Beispiel ist der folgende Beweis aus *Principia*:

*22. 6 Th $\rightarrow = (L \,.\, M)$: $L \leftarrow$, $M \leftarrow$

Anfang Th, Sub *4. 01, Ende $L \,.\, M$;

Anfang L, Sub *22. 01, Sub *22. 34, Halbnorm, Norm, Abnehmen, Sub *2. 04, Ende fixieren *3. 44

Anfang M, Norm, $\gamma = \alpha \supset \beta$ fixieren, Sub $p \supset q \,.\, \supset .\, q : \equiv p$, Ende *22. 58.

Beweisfiguren werden oft in verschiedenen Formen in gewöhnlichen mathematischen Beweisen benutzt. Die Vorschrift, „Beweis durch vollständige Induktion", ist eine solche Figur, die man vielleicht als

$$\text{Th} = \varphi(n) : \varphi(1) \leftarrow\!\!\text{———} ; \varphi(n+1) \xleftarrow{\varphi(n)}$$

darstellen kann.

VI. Kapitel.

Beweispläne.

In unseren bisherigen Betrachtungen haben wir die einzelnen Schritte eines Beweises untersucht und dadurch eine Reihe von Beweisprozessen gewonnen. Ein Beweis selbst wird durch Zusammenfügung solcher Prozesse konstruiert, wie wir im vorigen Kapitel gesehen haben. Die Wahl der Prozesse für einen Beweis ist aber keineswegs zufällig, sondern verfolgt in jedem Beweis einen bestimmten Plan. Auf Grund unserer Analyse der einzelnen Beweisschritte können wir im folgenden solche Beweispläne untersuchen.

§ 1. **Beweisregeln:** In der Mathematik wird oft von „ähnlichen" Beweisen geredet. Mit unseren Begriffen können wir eine scharfe Definition dieser Ähnlichkeit geben. In der Tat sind zwei

Beweise dann und nur dann *ähnlich*, wenn die entsprechenden abgekürzten Beweise identisch sind. Allgemeiner hat man auch *halbähnliche* Beweise, wo die abgekürzten Beweise sich nur durch Benutzung verschiedener (aber doch analoger) Bezugstheoreme unterschieden. Eine derartige Halbähnlichkeit bedeutet oft, daß die Beweise denselben Plan verfolgen.

Wenn in einer Theorie viele untereinander ähnliche Beweise auftreten, dann ist es zweckmäßig, diese Beweise unter eine allgemeine Beweisregel zusammenzufassen. Eine Beweisregel ist ein Abschnitt eines abgekürzten Beweises, der einen selbständigen Namen erhalten hat; diesen Namen kann man dann in vielen abgekürzten Beweisen statt des ganzen Abschnitts benutzen [1]. Die algebraische Regel für das Fortschaffen von Klammern ist eine Regel in diesem Sinne, denn sie besteht nur in wiederholter Substitution vermittelst des distributiven Gesetzes. Ähnlicherweise hat man Regeln für Differentiation und für Integration und Regeln für die Lösung einfacher Probleme. In allen Fällen funktioniert eine Regel in dem obigen allgemeinen Sinn als ein einfacher Beweisplan.

Mit unseren Methoden haben wir also eine adäquate Erklärung der Wichtigkeit von ähnlichen Beweisen und Beweisregeln in der Mathematik gegeben. Eine derartige Erklärung wird in der klassischen Logik überhaupt nicht versucht.

§ 2. **Substitution vermittelst Definitionen:** Der Beweis vieler einfachen Theoreme beruht nur auf den Definitionen der in den Theoremen enthaltenen Elemente. Dieser Beweisplan, den wir Substitution vermittelst Definitionen nennen können, ist in jeder Theorie, die mindestens eine Definition enthält, möglich.

Sub def: Prämisse, ein satzartiger Ausdruck L, der einige Elemente (Relationen, Funktionen, Konstanten usw.) enthält, die in der betreffenden Theorie durch Definition eingeführt worden

[1] Man kann viele Beweise durch Einschiebung unwirksamer Schritte unter eine Regel bringen; d. h. durch Schritte, die in der allgemeinen Regel angegeben werden, die aber in den Beweisen selbst unter Umständen keine Anwendung finden. Wir werden die Übereinkunft treffen, solche unwirksame Schritte in einem Beweis immer wegzulassen.

sind. Das Ergebnis wird aus L gebildet, indem man diese Elemente durch Substitution vermittelst ihrer Definitionen eliminiert; dabei wird die Substitution in passender Weise für jeden Definitionstypus ausgeführt (vgl. III, §§ 5 und 6). Die Reihenfolge der Eliminationen wird durch die Konvention festgesetzt, daß diejenigen Elemente zuerst zu eliminieren sind, die in der Theorie am letzten definiert worden sind[1]. Es handelt sich hier um einen Stationärprozeß. Unser Prozeß wird oft auf gewisse ausgewählte Definitionen beschränkt.

Es gibt zahlreiche Anwendungen dieses Prozesses, *Sub def*, besonders in den Beweisen der abstrakten Mathematik. Seine große Anwendbarkeit beruht auf der begrifflichen Struktur mathematischer Theorien. Jede abstrakte Theorie fängt mit einem gewissen Basissystem, das aus Funktionen und Relationen besteht, und mit gewissen Axiomen für dieses System an. Die Theorie definiert dann auf dieser Grundlage verschiedene neue Relationen, Funktionen und Systeme und stellt Sätze über diese neuen Elemente auf. Jeder solcher Satz wird durch den Prozeß *Sub def* auf einen Satz zurückgeführt, der nur Elemente aus dem Basissystem enthält. Dieser Prozeß wird also einen Beweis oft systematischer oder einfacher machen.

§ 3. **Die Auslassung von Bezugstheoremen:** Viele einfache Reduktionsbeweise werden mit einem Satz des Aussagekalküls (vgl. *Principia* *2—*5) beendet; d. h. der abgekürzte Beweis schließt mit einem Prozeß

(1) Ende N,

wo N ein Satz des erwähnten Charakters ist. Das können wir aber durch Auslassung von N vereinfachen, denn wir können eine notwendige und hinreichende Bedingung für die Existenz dieses N

---

[1] Die Reihenfolge der Elimination ist aber größtenteils unwesentlich. Nehmen wir als Beispiel an, daß zwei Funktionen f und g in der Theorie definiert worden sind, und daß die Definition für f die Funktion g nicht enthält und umgekehrt. Dann kann man f und g entweder in der Reihenfolge f, g oder in der Reihenfolge g, f eliminieren, ohne das Resultat zu beeinflussen. Eine analoge Behauptung gilt für die Definitionen von Relationen; beide Behauptungen kann man durch Induktion bestätigen.

geben, und, falls es existiert, eine Methode für seine Konstruktion festlegen, wie im folgenden genauer gezeigt wird.

Ende Aussage: Prämisse, ein satzartiger Ausdruck L, der keine Klammerzeichen enthält. Man bestimmt diejenigen Komponenten M_i in L, die Aussagenfunktionen und Relationen als Hauptoperatoren haben (diese Komponenten nennen wir die *kleinsten satzartigen Komponenten* in L). Jede solche Komponente M_i wird überall in L durch eine Aussagevariable p ersetzt, wobei verschiedene p für nicht identische M_i zu benutzen sind. Der Ausdruck L', der dadurch entsteht, wird nur Variablen p und logische Verknüpfungen enthalten. Dann bestimmt man, ob L' ein richtiger Satz im Aussagekalkül ist, entweder durch Bestimmung seiner Wahrheitswerte oder durch systematisches Nachschlagen in einem leidlich vollständigen Verzeichnis[1] der Sätze dieses Kalküls. Der Prozeß „Ende Aussage" ist dann und nur dann auf die Prämisse anwendbar, wenn dieses L' tatsächlich richtig ist, denn dann und nur dann ist ein Prozeß der Form (1) auf L anwendbar und zwar mit $N = L'$.

Dieser kurze Prozeß „Ende Aussage" ist typisch für viele einfache Beweispläne; der Plan des Beweises bestimmt die Bezugstheoreme, die im Aufbau des abgekürzten Beweises zu benutzen sind; dadurch ist es möglich, die explizite Angabe dieser Bezugstheoreme zu vermeiden! Man muß aber bemerken, daß eine derartige Auslassung das Vorhandensein eines Verzeichnisses oder eines Wörterbuchs voraussetzt, das die möglichen Bezugstheoreme systematisch klassifiziert.

§ 4. **Endprozesse:** Die letzten zwei oder drei Schritte eines Beweises werden oft unter einem gemeinsamen Gesichtspunkt gewählt. Ein Beispiel ist der Prozeß „Ende fixieren" (IV, § 2), der im wesentlichen eine planmäßige Kombination zweier Schritte, einer Festsetzung und eines Endprozesses, ist; die Angabe der Kombination durch ein Stichwort macht die Angabe der einzelnen Prozesse überflüssig. Viele andere ähnliche Pläne sind in derselben Weise als Endprozesse ausdrückbar. Einen solchen Plan können

---

[1] Das Nachschlagen wird erleichtert, wenn das Verzeichnis systematisch geordnet ist; z. B. nach den in den Sätzen enthaltenen Verknüpfungen.

4*

wir mit dem oben eingeführten Prozeß „Ende Aussage" aufstellen. Er ist aber, wie man weiß, nur dann direkt anwendbar, wenn die Prämisse keine gebundenen Variablen enthält. Sonst muß die Prämisse erst durch Elimination der gebundenen Variablen vorbereitet werden. Das kann man durch eine der folgenden drei Methoden durchführen:

(1) Halbnorm, Wiederholen abnehmen, Ende Aussage;
Norm, Wiederholen abnehmen, Ende Aussage;
Sub def $\equiv$, Norm, Wiederholen abnehmen, Ende Aussage.

In der letzten Methode wird die Phrase „Sub def $\equiv$" für eine Substitution vermittelst der Definition der Verknüpfung $\equiv$ durch & und $\supset$ verwendet. Jede dieser drei Methoden ergibt, wenn anwendbar, einen rein logischen Beweis der betrachteten Prämisse; daher werden sie *Ende Logik* genannt. Die Benutzung dieser Phrase am Ende einer Schwellkette bedeutet immer die Benutzung einer der drei Regeln in (1). Wenn es aus dem Zusammenhang nicht klar ist, welche der drei Methoden gemeint ist, dann kann man alle drei in der gegebenen Reihenfolge nacheinander versuchen. In jedem Fall wird eine notwendige und hinreichende Bedingung für die Anwendbarkeit durch die Theorie des vorigen Abschnitts gegeben. Die Nützlichkeit dieses Endprozesses wird durch Beispiele in den Beweisen für *22.351.4.45.59.62 in IV, § 5 erläutert.

Mit dem Begriff einer auswählenden Substitution können wir einen anderen nützlichen Endprozeß aufbauen. Nach unserer früheren Definition wird eine Substitution alle Komponenten einer Prämisse L gleichmäßig angreifen. Eine auswählende Substitution in einer Prämisse L wird dagegen nur gewisse willkürlich gewählte Komponenten ersetzen dürfen. Diese Auswahl findet meist in Bezug auf einen bestimmten Zweck statt, insbesondere in bezug auf eine Endformel, wie z. B.

(2) Auswahl so-und-so Sub M, end N.

Hier werden mit „so-und-so" die ausgewählten Komponenten angegeben; das Bezugstheorem M muß eine Gleichung, Äquivalenz oder Implikation sein. Wir werden die zwei Schritte in (2) in die Regel *Ende N durch Auswahl Sub M* zusammenfassen; denn für die Anwendbarkeit dieser Regel können wir ein konstruktives Kriterium geben: dazu brauchen wir nur alle möglichen ausgewählten Substitutionen (es gibt nur eine endliche Anzahl davon) auszuführen, um in jedem Fall zu bestimmen, ob das Ergebnis ein Fall von *N* ist. Auch für wiederholte Substitution oder für aufeinanderfolgende Substitutionen vermittelst mehrerer Bezugstheoreme ist ein solches Kriterium für Auswahl möglich.

Oft können wir ein bequemeres Kriterium für die Anwendbarkeit einer solchen Substitution geben. Die Prämisse *L* sei eine Gleichung, die nur Addition und Multiplikation umfaßt; eine auswählende Substitution vermittelst der kommutativen und assoziativen Gesetze für Addition und Multiplikation lasse sich so ausführen und werde so ausgeführt, daß eine Identität resultiert. Dafür, daß dies möglich ist, kann man Bedingungen angeben, die nur die Struktur der Prämisse *L* benutzen[1]. Wir können diese speziell algebraische Regel *Ende algebraische Identität* nennen.

§ 5. **Analyse der Beweise in Principia Mathematica:** In *Principia Mathematica* bewirken unsere Abkürzungsmethoden eine erhebliche Reduktion der Länge der Beweise. Diese Reduktion wird auch viele formale Manipulationen durch begriffliche Überlegungen ersetzen. Diese Vorteile sind alle ohne Verlust der Vollständigkeit oder der symbolischen Präzision der Beweise möglich.

Nach dem Plan der *Principia* werden alle Elemente (Funktionen, Relationen usw.) mittels der primitiven Begriffe der Logik definiert. Der Plan *Sub def* wird also jeden Satz in *Principia* auf einen Satz zurückführen, der nur die primitiven Begriffe der Logik enthält. Wenn dieses Resultat keine Existenzbehauptung ist, dann wird der Satz durch die Regel

Sub def, Ende Logik

beweisbar sein. Dieser „universelle" Beweisplan ist auf zahlreiche Sätze in *Principia* anwendbar, wird aber oft durch kürzere Pläne ersetzbar sein. Wenn dieser Plan nicht möglich ist, dann sind andere Pläne — insbesondere Pläne mit Prozessen der Fixierung —

[1] Wenn z. B. *L* nur Multiplikation enthält, dann muß jede Variable und Konstante so oft links wie rechts in *L* auftreten. Solche Bedingungen sind am besten mittels Induktion nach der Komplexität von *L* zu beweisen.

leicht möglich. Diese Analyse der *Principia* ist in den Einzelheiten in mehreren Kapiteln geprüft worden, u. a. in *22 (Klassenkalkül). Die so erhaltenen Beispiele abgekürzter Beweise zeigen, daß unsere Methoden tatsächlich rationelle Beweise liefern.

Dieser Gesichtspunkt verlangt aber eine Umarbeitung der Grundlagen der Logik gegenüber der Darstellung in *Principia*, denn man müßte in den Grundlagen eine Theorie der Symbolkomplexe einfügen. Eine solche Umarbeitung ist sowohl möglich wie wünschenswert, denn sie ist im wesentlichen nur eine Formulierung des Hilbertschen „finiten" Gesichtspunkts.

§ 6. **Substitutionsbeweise:** Wenn ein Satz L gegeben ist, und wenn wir wissen, daß der Beweis durch aufeinanderfolgende Substitutionsprozesse zu führen ist, dann können wir den vollständigen Beweis konstruieren, denn wir können aus der Struktur von L die notwendigen Bezugstheoreme finden. Das erste Bezugstheorem $M \equiv N$ (oder $M \supset N$ oder $C = D$) muß die Eigenschaft haben, daß wenigstens eine Komponente in L ein Fall von M ist, denn sonst hat die Substitution mit dem Bezugstheorem keine Wirkung. Da es nur eine endliche Anzahl bereits bewiesener Bezugstheoreme gibt, und da nur eine endliche Anzahl Komponenten in L vorhanden sind, so können wir systematisch alle Bezugstheoreme finden, die der obigen Bedingung genügen. Das ergibt eine endliche Anzahl erster Schritte für den Beweis. Für jeden solchen Schritt können wir in derselben Weise alle möglichen zweiten Schritte finden usw. In dieser Weise fortfahrend, können wir eine endliche Anzahl möglicher Beweise finden, worunter wenigstens einer tatsächlich ein Beweis sein muß. Der Plan „Beweis durch Substitution" wird also die Schritte des Beweises vollständig bestimmen. Dieser Plan ist, wie der Plan in § 3, als eine Auslassung der Bezugstheoreme zu betrachten.

Dieser Plan hat aber den Nachteil, daß er eine große Anzahl möglicher Beweise ergibt. Um das zu vermeiden, können wir dasselbe Verfahren unter der Annahme anwenden, daß nicht alle schon bewiesene Sätze, sondern nur gewisse ausgewählte Sätze als Bezugstheoreme in Betracht kommen. Ein derartiger Plan wird leichter anzuwenden, aber nicht so oft anwendbar sein. Ein solcher

Plan wird durch den Beweis des Satzes nahegelegt, daß das Komplement einer Menge α nie mit dieser Menge identisch sein kann; in Zeichen (*Principia* *22. 351)

$$-\alpha \neq \alpha.$$

Die allgemeine Form $\beta \neq \gamma$ dieses Satzes wird im Bezugstheorem

(1) $$\beta \neq \gamma \,.\equiv.\, \sim(\beta = \gamma)$$

wiedergefunden. Eine Substitution damit ergibt

$$\sim(-\alpha = \alpha).$$

Da eine Komponente der Form $\beta = \gamma$ hier auftritt, ist eine Substitution vermittelst der Äquivalenz

(2) $$\beta = \gamma \,.\equiv:\, (x) : x\varepsilon\beta \,.\equiv.\, x\varepsilon\gamma$$

angebracht, mit dem Resultat

(3) $$\sim((x) : x\varepsilon - \alpha \,.\equiv.\, x\varepsilon\alpha).$$

Wegendes Vorhandenseins der Komponente $x\varepsilon -\alpha$ werden wir mit dem Bezugstheorem

$$x\varepsilon - \alpha \,.\equiv.\, \sim(x\varepsilon\alpha)$$

substituieren. Der dadurch entstehende Ausdruck ist dann ohne weiteres mit dem Prozeß „Ende Logik" beweisbar.

Ziel dieses Beweises ist es, den gegebenen Ausdruck so zu reduzieren, daß der Prozeß „Ende Logik" anwendbar sein wird. Da dieser Endprozeß nur die logischen Verknüpfungen und Klammerzeichen eines Ausdrucks in Betracht zieht, muß die vorhergehende Reduktion ebenso viele logische Verknüpfungen wie möglich einführen. Mit dem ersten Bezugstheorem wird z. B. ein Ausdruck $\beta \neq \gamma$ ohne logische Verknüpfungen durch einen Ausdruck mit dem Operator $\sim$ ersetzt. Außerdem ist $\beta \neq \gamma$ der allgemeinste Ausdruck mit dem Hauptoperator $\neq$ und ist daher an jeden Ausdruck anwendbar, der diesen Operator enthält. Die Bezugstheoreme (2) und (3) haben einen ähnlichen Charakter.

Im allgemeinen Fall eines solchen Planes sind die benutzten Bezugstheoreme immer Äquivalenzen $M \equiv N$, die folgenden Bedingungen genügen:

1. M enthält keine logische Verknüpfungen oder Klammerzeichen; dagegen enthält N wenigstens einen solchen Operator.

2. M ist allgemein; d. h. M enthält keine Konstante, alle Endkomponenten in M sind voneinander verschieden, und keine

Variable ist durch eine Variable mit einem größeren Spielraum sinnvoll ersetzbar.

Der Beweisplan, den wir hier untersuchen, ist folgendermaßen zu beschreiben: in einer gegebenen Prämisse *L* werden alle möglichen Substitutionen vermittelst Bezugstheorem des obigen Typs ausgeführt. Von zwei solchen Bezugstheoremen wird immer dasjenige Theorem erst angewendet, dessen linke Seite *M* die kleinere Komplexität hat. Wenn solche Substitutionen nicht mehr möglich sind, dann wird der Prozeß „Ende Logik" angewandt. Dieser Plan wird die einzelnen Schritte des Beweises im wesentlichen eindeutig bestimmen, mit Ausnahme des Falles, wo zwei verschiedene Äquivalenzen des obigen Typs mit synonymen linken Seiten *M* und *M'* in Frage kommen. Diesen Plan nennen wir

(4) Sub für Ende Logik.

Wir können die Wirksamkeit dieses Plans durch seine Anwendung in *Principia* *22 erläutern. In diesem Abschnitt gibt es 72 Sätze; für 59 dieser Sätze wird der Plan (4) einen Beweis ergeben In 42 Fällen ist das Ergebnis im wesentlichen der in *Principia* selbst gegebene Beweis, in den anderen Fällen sind die Beweise aus *Principia* etwas kürzer. Unter den 13 Ausnahmen, wo der Plan (4) nicht anwendbar ist, gibt es 10 Sätze, wo analoge Pläne leicht aufgestellt werden können. Für die drei anderen (die Sätze *22. 6, *22. 94 und *22. 95) können wir mit unseren Methoden einzelne abgekürzte Beweise geben. Die Wirksamkeit des Plans (4) ist aber keineswegs auf diesen Abschnitt *22 beschränkt. Im Abschnitt *34 kann man z. B. alle acht Sätze von *34. 2 bis *34. 26 mit diesem Plan beweisen. Wir haben also gezeigt, daß viele Beweise in *Principia* und in ähnlichen Theorien durch allgemeine Beweispläne ersetzbar sind.

VII. Kapitel.

Übersicht.

1. **Zusammenfassung der Resultate:** Das Hauptergebnis dieser Arbeit ist eine Analyse der Struktur logischer Beweise und

der Nachweis, daß diese Analyse Methoden für Beweisabkürzung ergibt. Wir können unsere Resultate folgendermaßen kurz zusammenfassen: 1. Ein Beweis besteht aus Ketten, die in einer Beweisfigur gruppiert sind. 2. Eine Kette ist eine Reihe von Schritten, wobei jeder Schritt ein Fall eines allgemeinen Prozesses ist. 3. Für jeden solchen Prozeß können wir auf Grund einer finitkombinatorischen Theorie der symbolischen Ausdrücke eine universelle Definition geben. 4. Die einzelnen Schritte eines Beweises sind oft allgemeinen und systematischen Beweisplänen untergeordnet.

Die von uns aufgestellten Prozesse und Pläne sind die folgenden:

Prozesse des Schlußschemas (II, § 2): Inf schrumpf, Inf schwell, Hyp weg;

Prozesse des Syllogismus (II, § 3); Syll unter schrumpf, Syll ober schrumpf, Syll unter schwell, Syll ober schwell;

Prozesse im Präfix: Norm (IV, § 1), Halbnorm (IV, § 3), Fixieren (IV, § 2), Add (IV, § 4), Abnehmen (IV, § 4), Vertauschen (IV, § 4);

Prozesse der Substitution: Sub Gl (III, § 1), Sub Äq (III, § 3), Sub def (III, § 5), Sub aufbau (III, § 6), Sub Imp (III, § 4);

Verkürzte Prozesse (II, § 5), Fall + Hyp weg, Fall + Inf schrumpf, Fall + Syll unter schrumpf, Fall + syll ober schrumpf;

Endprozesse: Ende fixieren (IV, § 2), Ende Aussage (VI, § 3), Ende Logik (VI, § 4), Ende Sub auswahl (VI, § 4);

Pläne: Sub def (VI, § 2), Sub für Ende Logik (VI, § 6).

Durch unsere Abkürzungsmethoden haben wir die folgenden Resultate erzielt: 1. Eine weitgehende Vereinfachung und Verkürzung der Beweise in jeder logischen Abhandlung, wie z. B. *Principia.* 2. Eine theoretische Analyse der Struktur eines mathematischen Beweises, insbesondere eine adäquate logische Erklärung der Tatsache, daß wirkliche mathematische Beweise mit kurzen Beschreibungen gegeben werden können. 3. Eine gründliche Erklärung des Wesens ähnlicher Beweise und Beweisregeln. 4. Die Möglichkeit, in der Darstellung gewisser Gebiete der abstrakten

Mathematik abgekürzte Beweise anzuwenden. Diese Möglichkeit ist noch im einzelnen nachzuprüfen.

Unsere Abkürzungsmethoden sind keineswegs auf die besonderen Formen beschränkt, die wir für die Darstellung benutzt haben. Die klassische Logik der *Principia* diente als die Grundlage unserer Untersuchung, aber nur deshalb, weil diese Logik in einer bequemen und vollständigen Form vorhanden ist. Irgendeine andere symbolische Logik — z. B. die intuitionistische Logik von Heyting — wäre ebensogut gewesen. Die Grundbegriffe *Ausdruck* und *Prozeß* bleiben dabei ungeändert. Außerdem ist es möglich, symbolische Beweise je nach Bedarf in verschiedenen Formen zu schreiben. Man kann z. B. den Prozeß „Substitution vermittels einer Implikation" entweder in Worten oder in der Abkürzung „Sub imp" oder noch kürzer mit einem einzigen dazu gewählten Zeichen anschreiben. Man kann Abkürzungen für alle Beweise einer Theorie, oder aber nur für die Sätze, wo die Abkürzungen besonders einfach sind, benutzen. Wichtige oder schwierige Schritte eines abgekürzten Beweises kann man voll ausschreiben. Ferner ist die Benutzung einer Beweisabkürzung nicht davon abhängig, daß die Sätze in der Symbolik der mathematischen Logik gegeben sind. Wenn die Sätze in der gebräuchlichen mathematischen Form gegeben sind, dann ist es möglich, sie mechanisch in die logische Symbolik zu „übersetzen". Das ist nach allgemeinen Übersetzungsregeln ausführbar; für jede Sprache sind solche Regeln ohne weiteres aufzustellen. In dieser Hinsicht hat die Logik der Beweisprozesse eine weitaus größere Anpassungsfähigkeit als die gewöhnlichen Formen der mathematischen Logik.

§ 2. **Weitere Einzelprobleme:** Beim weiteren Ausbau unserer Methoden werden Probleme verschiedener Art entstehen. Die erste Art von Problemen betrifft die Aufstellung neuer Prozesse. In der Anwendung unserer Methoden auf einzelne Beweise können Schritte vorkommen, die mittels der bisher erörterten Prozesse nicht faßbar sind oder die vielleicht am besten einem neuen Prozeß unterzuordnen sind. In einem solchen Falle kann man durch induktive Analyse dieses Schrittes (und eventuell anderer ähnlicher Schritte) einen neuen Prozeß aufstellen. Dieses Verfahren der induktiven Analyse einer besonderen Operation, um die zugehörigen allgemeinen Prozesse zu finden, ist allgemein formulierbar, denn es ist nur ein Zweig der noch nicht entwickelten mathematischen Logik der Induktion. Für eine solche Aufstellung eines neuen Prozesses sind die folgenden Richtlinien nützlich:

1. Die Definition eines Beweisprozesses muß das Ergebnis eindeutig bestimmen; d. h. mit einem festen Prozeß, einem gegebenen Bezugstheorem und einer gegebenen Prämisse kann man nur ein Ergebnis erhalten.

2. Die Definition eines Beweisprozesses muß konstruktiv sein; d. h. wenn Prämisse und Bezugstheorem gegeben sind, dann muß das Ergebnis durch Ausführung gewisser vorgeschriebener symbolischer Operationen mit Prämisse und Bezugstheorem tatsächlich konstruierbar sein.

3. Der Prozeß für einen Beweisschritt muß so weit wie möglich die intuitiven Ideen oder den natürlichen Zweck dieses Schrittes wiedergeben.

Analoge Probleme entstehen bei der Aufstellung neuer Beweispläne. Wir haben nur sehr einfache Beweispläne analysiert; es ist anzunehmen, daß durch das Studium mathematischer Beweise viele andere solcher Pläne ausfindig gemacht werden können. Manche solcher Pläne werden, wie in VI, § 3 und VI, § 6, Methoden sein, die die Auslassung von Bezugstheoremen bewirken. Auch hier ist es wichtig, daß die Beweispläne den intuitiven Zwecken der Beweise entsprechen.

Es gibt auch weitere Anwendungen der Methoden der Beweisabkürzung. Man kann z. B. die Transformation von Beweisen untersuchen — d. h. wie ein Beweis aus einer Gestalt in eine andere Gestalt transformiert werden kann. Ein Beispiel wird durch die zwei Beweise für *2 . 17 in V, § 3 geliefert, wo der zweite Beweis aus dem ersten durch Transformation der Beweisfigur erhalten worden ist. Beim Studium solcher Transformationen kann man z. B. nach den notwendigen und hinreichenden Bedingungen suchen, daß zwei Substitutionen miteinander vertauscht werden können (vgl. die Untersuchung der Substitutionen vermittelst Definitionen in VI, § 2). Solche Untersuchungen der Beweistransformationen geben die Möglichkeit, die wesentlichen Momente eines Beweises von den zufälligen zu unterscheiden.

§ 3. **Analyse der Grundprinzipien:** Der Hauptbegriff unserer Untersuchung ist wohl die Idee eines Beweisprozesses, Aus diesem Begriff entspringt die Möglichkeit der Beweisabkürzung, denn ein Beweisprozeß ersetzt die einzelnen symbolischen Manipulationen eines Schrittes durch eine Idee, die diese Manipulationen beherrscht. Die Idee des Schrittes wird also direkt und begrifflich und nicht indirekt und symbolisch vorgeführt. Jeder Schritt eines Beweises gehorcht einer leitenden Idee, die als ein allgemeiner Beweisprozeß dargestellt werden kann.

Ein ähnliches Prinzip entspringt aus unserem Studium der Beweispläne. Ein Beweis ist nicht nur eine Reihe einzelner Schritte, sondern nur eine Gruppe von Schritten, die nach einem bestimmten Plan oder Zweck zusammengefügt sind. Dadurch ist der Begriff „Zweck" als eine grundlegende Kategorie der Logik erkannt. Wenn z. B. der Zweck eines Beweises der Nachweis des fraglichen Satzes durch Substitution ist, dann können wir aus diesem Zweck heraus die einzelnen Schritte des Beweises konstruieren. In diesem Sinne gibt es eine „Integrallogik", die Gruppen von Schritten als komplexe Einheiten betrachtet, als Ergänzung der gewöhnlichen „Differentiallogik", die nur den Charakter einzelner Schritte untersucht. Daher behaupten wir, daß jeder mathematische Beweis eine leitende Idee hat, die alle einzelnen Schritte des Beweises bestimmt, und die als ein Beweisplan dargestellt werden kann.

Dieser Gesichtspunkt wird durch die Betrachtung des Begriffes des Beweisstils erläutert. Für die Darstellung eines gegebenen Beweises kann man viele grundverschiedene Stile benutzen — den präzisen, symbolischen und detaillierten Stil, der in *Principia* und in vielen anderen Gebieten der Mathematik benutzt wird, wo auf Kosten der dahinterstehenden Ideen eine strenge Ausführung der Beweisschritte verlangt wird — und den intuitiven und begrifflichen Stil, der immer die Hauptideen und Methoden eines Beweises hervorhebt, um die einzelnen Manipulationen im Licht dieser Ideen zu verstehen. Dieser Stil ist in den Büchern und Vorlesungen von H. Weyl besonders ausgeprägt.

Unsere Untersuchung von Beweisplänen ist im wesentlichen der Nachweis, daß die Strenge eines detaillierten symbolischen Beweises auch in diesem intuitiven Stil möglich ist, denn man kann die intuitiven Ideen als spezifische Beweispläne deuten.

Sowohl für einzelne Schritte wie auch für Beweispläne haben wir ein Grundprinzip: Die Einzelheiten einer mathematischen Operation sind immer einer leitenden Idee untergeordnet, die genau und allgemein formuliert werden kann und die als Basis für die konstruktive Bestimmung der Einzelheiten genommen werden kann. Dies nennen wir das *Prinzip der leitenden Idee.* Wir vermuten, daß dieses Prinzip nicht nur für einzelne Beweisschritte und für Beweispläne, sondern auch für alle anderen mathematischen Verfahren anwendbar ist. Die Struktur einer ganzen mathematischen Theorie ist z. B. immer durch ein paar Hauptideen und Grundmethoden bestimmt, die einer präzisen logischen Analyse zugänglich sind, genau wie die Hauptideen in Beweisen und Beweisprozessen. Der Gang einer Theorie ist oft durch den Charakter der Probleme bestimmt, die die Theorie lösen muß. In jedem Fall deutet all dies auf ein umfangreiches und wichtiges Untersuchungsgebiet für die mathematische Logik hin: das Studium der Struktur der Bestandteile der Mathematik und die Bestimmung dieser Struktur durch leitende Ideen. Eine derartige Untersuchung muß den Zusammenhang zwischen der mathematischen Logik und der Mathematik verstärken, denn eine erfolgreiche Betrachtung der Struktur der Mathematik muß offenbar mit der Mathematik selbst anfangen. In dieser Hinsicht darf die mathematische Logik nicht ein abgesondertes und hochspezialisiertes Gebiet bleiben.

Die Untersuchung der abgekürzten Beweise ist also nur ein Teil der Strukturtheorie der Mathematik, die auf dem Prinzip der leitenden Idee gegründet ist, und die neue Probleme in der mathematischen Logik selbst und neue Methoden für die Koordination der mathematischen Wissenschaften ergibt.

Lebenslauf.

Ich, Saunders MacLane, wurde am 4. August 1909 in Norwich, Connecticut, Vereinigte Staaten von Amerika, geboren. Mein Vater ist Donald Bradford MacLane, meine Mutter Winifred MacLane (geb. Saunders). Ich bin amerikanischer (U. S. A.) Staatsangehöriger.

In Amerika habe ich die üblichen Schulen für zwölf Jahre besucht. Danach habe ich vier Jahre an der Yale-Universität (New Haven, Conn. U. S. A.) studiert und am Schluß dieses Studiums den Grad eines Ph. B. erworben. Es folgte ein Jahr (1930—31) an der Universität Chicago, wo ich den M. A. erlangte. Danach bin ich nach Deutschland gekommen, wo ich vier Semester (W.-S. 1931/32 bis S.-S. 1933) an der Universität Göttingen studierte.

An der Yale-Universität belegte ich mathematische Collegia bei den Dozenten Ore, Pierpont, E. W. Brown, Miles, Tracey und L. S. Hill. In Chicago habe ich bei den Professoren E. H. Moore, Bliß, Dickson, E. P. Lane und Barnard Vorlesungen gehört. Während meines Studiums in Göttingen habe ich Vorlesungen von Weyl, Herglotz, Geiger, Landau, H. Lewy, E. Noether, Born, Bernays und Hilbert beigewohnt.

Diese Gelegenheit möchte ich benutzen, um meinen Lehrern meinen Dank auszusprechen; dem vor kurzem verstorbenen Professor Moore (Chicago) für seine klare Auseinanderlegung der abstrakt-mathematischen Methoden; Professor Bernays für seine Kritik; und vor allem Professor Weyl für seine Ratschläge und für die Anregung seiner Vorlesungen.

A Late Return to a Thesis in Logic

Saunders Mac Lane

My Göttingen thesis for the D.Phil., "Abgekürzte Beweise im Logikkalkul" (Abbreviated Proofs in the Calculus of Logic), was written in the spring and summer of 1933. At that time in Göttingen there was a general rush to finish everything up before the Mathematical Institute collapsed under the pressures of the time and the anti-Semitic decrees of the new Hitler government in Germany. Moreover, I had long intended to finish writing a thesis in Mathematical Logic by that time.

Irving Kaplansky's efforts in preparing this volume of selected papers of mine have now brought me to read through this thesis again. I found it difficult to establish any real contact with my thesis ideas, over a gap of forty-five years and several considerable shifts of mathematical interest in the meantime.

At that time, I was much impressed with the need for mathematical rigor and the importance of making this rigor both clear and intuitively convincing. When I was an undergraduate student at Yale, the impact of the Weierstrassian emphasis on rigor via ϵ-δ was still apparent. I learned rigor from the chapter on Foundations in Edwin B. Wilson's *Advanced Calculus* (despite his typesetter, he favored both rigor and vigor). I also learned rigor from James Pierpont, the senior statesman of the Yale department; he had studied in Germany and had brought rigor to graduate students in this country through his careful books on real and complex variable theory. I also learned rigor from Pierpont's former student, Professor Wallace A. Wilson, who once regretfully told me that he understood rigor better than his master Pierpont, but that he had little mathematical substance, save metric and topological spaces, to which to apply this rigor. Undeterred by this example of form with little content, I went on to learn how rigor looked in logic from F. S. C. Northrop of the Philosophy Department at Yale. Through him, I came to admire *Principia Mathematica* by Whitehead and Russell, finding there a fine symbolic rigor almost untouched by the English language - though at the time I did not understand well why that touch of language was indeed still essential, nor did I press my curiosity beyond the first volume of *Principia*. As a Junior, I did ask Professor Wilson for a reading course in *Principia*, but he advised against it, and had me study Hausdorff's *Mengenlehre* instead - where I learned more rigor, as applied to sets, metric spaces, and ordinal numbers.

My first year of graduate study at Chicago was influenced most deeply by E. H. Moore. He and his disciple R. W. Barnard presented mathematical theorems in a formal and logistic notation (modeled on Peano), but gave proofs in a more informal fashion; I wondered how this could be effectively formalized, and wondered even more on the occasion when E. H. Moore had me give a seminar lecture on a paper by Ernst Zermelo–his proof that the axiom of choice implies that every set can be well-ordered (from Moore's very thorough critique after my lecture I learned a great deal about how to give a seminar lecture). I also listened to G. A. Bliss on the Calculus of Variations, wondering the while about all manner of minor inexactitudes in the construction of fields of extremals. One fine day I challenged Professor Bliss to produce the necessary ϵ's and δ's to make the proofs fully rigorous–and he did. I did take another course in (Aristotelian) Logic, this time with Hutchins' protegé Mortimer Adler; there I learned more about argument than about rigor. Finally I wrote a M.A. thesis about algebraic systems with 2, 3, or 4 binary operations; in retrospect, it seems to me now that I was trying to discover Universal Algebra–a search in which I did not succeed.

After a year I left Chicago to go to Göttingen where I hoped that the environment of Hilbert would give more encouragement to my study of logic and rigor. By then I must have vaguely begun to see that a good proof consisted of more than just rigorous detail, because there was also an important element of plan for the proof–the crucial ideas, which, over and above the careful detail, really make the proof function and get to the desired end. I clearly recall sitting in a vast lecture room listening to Edmund Landau lecture on Dirichlet series. As always, Landau's proofs were simply careful lists of one detail after another, but he gave this detail with such exemplary care that I could both copy down in my notebook all the needed detail and enter in the margin some overarching description of the plan of his proof (a plan which he never directly revealed).

Then I came gradually to the insight that proofs in mathematics combined rigorous detail and overall plan–and that overweening attention to the precision of detail could, as in the case of *Principia*, wholly hide the plan. There arose with me the notion that the necessary rigor could be codified and simplified, so as to be made almost automatic. If only the automatic could be properly described and organized, then the essential ideas of the proof would come through. This idea of organizing the *plan* of a proof in a formal way was evidently the germ of my thesis.

I had already started work on an earlier thesis idea, also in logic, early in the academic year 1932-33. I no longer know what was intended as the content of that thesis, but I do clearly recall that it did not find favor with either Professor Bernays or Professor Weyl when I explained it to them in February of 1933. For a brief period I toyed with the idea of going instead to Vienna, where I thought that Rudolph Carnap would be more sympathetic. Instead, I thought very hard in spurts about a thesis. A decisive spurt came on April 18-22, when I finally worked out a plan of the final thesis. In an exuberant letter of April 26th to my mother I wrote:

> "Perhaps I have time to tell you a bit about my new discovery. It's a new symbolic logic for *mathematical proofs*. It applies, as far as I can see now, to all

> proofs in all branches of mathematics (a rather big order!). It makes it possible to write down the proof of a theorem in a very much shorter space than by the usual methods, and at the same time it makes the proof very much clearer. In essence, it eliminates practically *all* the long mechanical manipulations necessary to prove a theorem. It is only necessary to give *leading ideas* of the *proof*. In fact, once these leading ideas are given–together with a few directions–then it becomes possible to compute *from* the leading ideas just what the *proof* of the theorem will be. In other words, once these leading ideas are given, all the rest is a purely mechanical sort of job. It is possible to define once and for all how the job is to be carried out (the general definition depends essentially upon the abstract methods I have been developing for the past year)."

Of somewhat later date is an exuberant first draft (in English) of the thesis: long-winded, full of rash philosophical assertions, and ending with a long table of things I still intended to develop.

The thesis itself (rewritten later, first in English and then translated into German) is more mathematical and businesslike. It observes that long stretches of formal proofs (written, say, in the style of *Principia*) are indeed trivial, and can be reconstructed by following well-recognized general rules. The thesis develops standard metamathematical terminology to describe formal expressions–as certain strings of symbols, suitably arranged. This is followed by a meticulous description of what it means to substitute y (or something more complex) for x in an expression. This description let me state exactly what it would mean to determine that one expression is a special case of another.

On this basis, I described exactly a number of the routine steps in a proof, giving each a label, as for example:

> *Inf schrumpf*: To prove a theorem $L \supset P$, search for a prior theorem of the form $M \supset N$, where L is a "special case" of M and P the corresponding special case of M.
>
> *Sub inf schrumpf*: Given a prior theorem $M \supset N$, one can conclude that $L \supset L'$, where L' is obtained from L by replacing every"positive" component of the form M by a new component N.
>
> *Sub Def*: Substitute the definitions.
>
> *Identität*: Use one of the standard identities of algebra (or of the propositional calculus).
>
> *Sub Theorem #20.43*: Use the cited theorem, in the (only) possible way.
>
> $x = C$ *fixieren*: Given a permise $(\exists x)L(x)$, assert $L(C)$ for some suitable "constant" C.
>
> *Halbnorm*: Move a quantifier $\exists x$ or $\forall x$ to the front of an expression.

All told, the thesis gives twenty or twenty-five of such rules (listed at the start of Chapter VII), and then observes that many proofs can be "abbreviated" by listing in order the rules to be applied. In this sense, the thesis gives a formal definition of a routine proof.

Chapter VI finally starts an analysis of *plans of proof*–a *plan* is a sequence of such standard steps. By describing such plans I hoped to define exactly what "similar proofs" would be. There are a number of examples of such plans, chiefly chosen from easier arguments in Algebra and from some early sections of *Principia* (and in those sections, this scheme worked well). For my present taste, the thesis does not give enough hard examples from the rest of mathematics.

In summary, the thesis observed that many proofs in mathematics are essentially *routine*–and that one can carefully write even a complete description of each type of routine step, so that the formal proof of the theorem, written in detail, can be replaced by the much shorter description of these steps. Moreover, since the *steps* are specified one can often summarize the directions of the proof by giving its *plan* (presumably the most crucial of the routine steps).

As a practical means of writing out proof, this scheme didn't (and couldn't) succeed. Mathematicians don't want *formal* proof; my proposed abbreviations were ugly, and analysis of such easy detail is best "left to the reader". Also, the thesis did not carry its ideas far enough into non-trivial proofs. Mathematicians do generally recognize that interesting proofs involve one or two "tricks" or "twists" added to a straightforward procedure. My ideas *might* have been carried to the point of giving a complete and formal description of all the straightforward procedures. They could then have been left aside–formalism of no great use–giving emphasis to the *real* tricks that make the proof work.

In an informal sense, this is what we still do in understanding harder proofs.

There are a few incidental points about the thesis. Initially, some of the English version of the thesis was translated for me by one of my fellow students. However, most of the translation is mine, and it looks "translated". I may have modified the presentation considerably, late in 1933 after I left Göttingen.

In the translation, E. H. Moore's favorite word, "range", has become "Spielraum", while the "scope" of a bound variable has become "Wirkungsbereich". There is an interesting anticipation of the Eilenberg–Mac Lane distinction between covariant and contravariant functors; for example, in the propositional calculus, in $p \supset q$ (for p implies q), p is contravariant and q is covariant (positive). The full description appears in Chapter II § 4 and was needed for sub inf schrumpf. It is now an easy and established item in logic, and may perhaps have appeared explicitly first in this thesis–though it is implicitly present in the propositional calculus even since *Principia* or perhaps since Boole.

My thesis was printed, in the requested number of copies, by one of the smaller printers in Göttingen. When I was done, I formally received the degree of D.Phil. (in 1934). For a couple of years, I continued to think about the matter, and wrote one paper (in the *Monist*) to summarize the ideas of my thesis. Some notes for lectures probably given in 1935 indicate that I did then apply the methods of my thesis to get at the structure of the standard proof that the limit of a uniformly convergent series of continuous functions is continuous. In a few years, however, my activities in algebra pushed aside those in logic.

There remains the real question of the actual structure of mathematical proofs and their strategy. It is a topic long given up by mathematical logicians, but one which still–properly handled–might give us some real insight.

November 1978

A CONSTRUCTION FOR ABSOLUTE VALUES IN POLYNOMIAL RINGS*

BY

SAUNDERS MacLANE

1. **Introduction.** An absolute value or "Bewertung" of a ring is a function $\|b\|$ which has some of the formal properties of the ordinary absolute value. More explicitly, for any b in the ring, $\|b\|$ must be a real number with the properties†

$$\|bc\| = \|b\| \cdot \|c\|, \qquad \|b + c\| \leqq \|b\| + \|c\|.$$

If the second inequality holds also in the stronger sense

$$\|b + c\| \leqq \max (\|b\|, \|c\|)$$

then the value $\|b\|$ is called non-archimedean (Ostrowski [17], p. 272). The thus delimited non-archimedean values are of considerable arithmetic interest. They are useful in questions of divisibility and irreducibility and in fact often correspond exactly to the prime ideals of the given ring. This paper is devoted to the explicit construction of non-archimedean values. More specifically, given all such values for the field R of rational numbers, we construct all possible values of the ring $R[x]$ of all polynomials in x with coefficients in R.

In treating a non-archimedean value it is convenient to replace $\|a\|$ by a related "exponential" value

$$Va = - \log \|a\|,$$

with corresponding forms (§2) of the formal properties of V (Krull [13], p. 531, and [14], p. 164). All possible non-archimedean values of the field of rational numbers have been determined by Ostrowski ([17], pp. 273–274). For every prime p there is a *p-adic* exponential value V_0 in which the value of any rational number is obtained by writing the number as $p^\alpha(u/v)$, where u and v are prime to p, and setting

$$V_0[p^\alpha(u/v)] = k\alpha, \tag{1}$$

where k is any fixed positive constant. This value we denote by the symbol $[V_0 p = k]$. The only other value V is a trivial one, in which Va is zero for $a \neq 0$.

* Presented to the Society, April 20, 1935, under the title, *Abstract absolute values and polygonal irreducibility criteria*; received by the editors June 28, 1935.

† See Kürschák [15]. (Numbers in brackets refer to the bibliography at the end of the paper.)

On this basis we can determine all possible values in the ring of polynomials with rational coefficients. Any such value W gives a p-adic or trivial value $V_0 a = Wa$ for the rational numbers and a value $\mu = Wx$ for the variable x. These facts alone give a first approximation V_1 to the value W, as follows:

$$
\begin{aligned}
(2)\qquad & V_1(a_n x^n + a_{n-1}x^{n-1} + \cdots + a_0) \\
& \qquad = \min\,[V_0 a_n + n\mu,\ V_0 a_{n-1} + (n-1)\mu,\ \cdots,\ V_0 a_0].
\end{aligned}
$$

This V_1 is actually a value and is never larger than W. If V_1 is not equal to W, we choose a $\phi(x)$ of smallest possible degree for which $W\phi(x) > V_1\phi(x)$. We then define a second approximation $V_2 f(x)$ by using* the true value for $\phi(x)$. In this manner we construct successive approximations $V_1, V_2, V_3, \cdots$ which in the limit will give the arbitrary value W (§8).

The succession of values $V_1, V_2, \cdots$ is defined in Part I for polynomials with coefficients in any field K. This requires a general method (§§4 and 5) of constructing a value V_k from a previously obtained value V_{k-1}. The value given by the limit of such a sequence needs a special study (§7). Here, as in §§8 and 16, we assume that every value of the field K is "discrete"; that is, that the real numbers used as values form an isolated point set, as in the case of p-adic values.

Part II investigates the structure of the values which have been constructed. The central problem is the construction of the "residue-class field" which arises when polynomials which differ by a polynomial of positive value are put into the same residue-class. For the absolute values constructed in Part I this field is determined by an inductive construction of the homomorphism of polynomials to residue-classes (§§10–14). This homomorphism also yields a more specific description of how our values can be built up (§§9, 13). Since a given value W can be represented in many ways by a sequence of approximations $V_1, V_2, V_3, \cdots$, we treat in §§15 and 16 the questions as to when two such sequences can give the same ultimate value W, and how such a sequence can be put in a normal form.

Among the applications of this construction of absolute values we mention the classification of irreducibility criteria of the Newton Polygon type. The theorem of Eisenstein [4] states that a polynomial

$$f(x) = x^n + a_{n-1}x^{n-1} + a_{n-2}x^{n-2} + \cdots + a_0$$

with integral coefficients a_i is irreducible if each coefficient a_i is divisible by some fixed prime p, while the last term a_0 is not divisible by p^2. In terms of the value V_1 of (2) with $\mu = 1/n$ these hypotheses on $f(x)$ become

* Similar "second-stage" values V_2 appear implicitly in the irreducibility investigations of Ore [7], Kürschák [6], and Rella [10].

$$V_1 f(x) = V_1 x^n = V_1 a_0 < V_1(a_i x^i) \qquad (i = 1, \cdots, n-1).$$

In this form a simple proof of the theorem can be given. The theorems of Königsberger [5], Dumas [3], and Ore [8] are likewise related to the values V_1. The second stage values V_2 can be similarly applied to interpret the irreducibility theorems of Schöneman [11], Bauer [1], Kürschák [6], and Ore [7]. By using the general value V_k one can obtain a still more extensive irreducibility criterion which includes all these previous theorems (MacLane [16]), and which asserts that certain polynomials $f(x)$ with irreducible homomorphic images of sufficiently high degree are themselves irreducible. Our construction for absolute values can also be applied to give a new and complete treatment of the problem of constructing the prime ideal factors of a given rational prime in a given algebraic field.*

I. The construction of non-archimedean values

2. Elementary properties of values in rings. A ring† S is said to have a *non-archimedean value* (for short, a value) V if to every element $a \neq 0$ in S there is assigned a unique real number Va with the properties

$$V(ab) = Va + Vb, \qquad V(a+b) \geqq \min(Va, Vb).$$

These we call the *product* and *triangle* laws respectively. We assume also that 0 is assigned the value $+\infty$, with the following conventions for any finite number γ:

$$\gamma < \infty, \qquad \infty + \gamma = \gamma + \infty = \infty + \infty = \infty.$$

Two simple consequences of the product law are

$$V(1) = V(-1) = 0, \qquad V(-a) = V(a). \tag{1}$$

More important is the strengthened form of the triangle law:

$$Va \neq Vb \text{ implies } V(a+b) = \min(Va, Vb). \tag{2}$$

For suppose instead that $Va > Vb$ and $V(a+b) > \min(Va, Vb)$. Then

$$Vb = V(a+b-a) \geqq \min(V(a+b), Va) > Vb,$$

a contradiction.

Since we are using a value analogous to the negative logarithm of the ordinary absolute value, a "small" absolute value will correspond to a "large"

* The application of the methods of this paper to the prime ideal construction is treated in another paper (Duke Mathematical Journal, vol. 3 (1936), pp. 492–510.) The general irreducibility criterion will be presented in a subsequent paper.

† Here and in the sequel "ring" means "commutative ring with unit element."

value V. Hence we say that two ring elements a and b are of the same order of magnitude or *equivalent* in* V—symbol $a \backsim b$(in V)—if and only if

$$V(a - b) > Va.$$

The product and strong triangle laws show that equivalent elements have the same value and that equivalence is a reflexive, symmetric, and transitive relation, provided the supplementary assumption† that $0 \backsim 0$ be made. Two equivalences

$$a \backsim b \quad \text{and} \quad c \backsim d$$

can be multiplied to give

$$ac \backsim bd. \tag{3}$$

An element b is *equivalence-divisible* in V by a if and only if there exists a c in S such that

$$b \backsim ca \quad (\text{in } V).$$

If this is true, it remains true when a or b is replaced by an equivalent element.

The product law implies that a ring S with a value V must be a domain of integrity. The value V may be extended to the quotient field of S by defining, in accord with the product law,

$$V\left(\frac{a}{b}\right) = Va - Vb \tag{4}$$

for any elements a and $b \neq 0$ in S. One then obtains the

THEOREM 2.1. *Let S be a domain of integrity with the quotient field K. If V is a value of S, then the function defined by* (4) *is a value of K. Conversely, every value of K can be obtained in this way from one and only one value of S.*

When $S = K$ is a field, the set of all real numbers Va for $a \neq 0$ in S is an additive group Γ, called the *value-group* of V. If the positive numbers of Γ have a positive minimum $\delta > 0$, then the value V is said to be *discrete.*‡ In this case the group Γ is cyclic and consists of all multiples of δ. If all elements not 0 have the value 0, then V is called *trivial*. Every ring has a trivial value, while the p-adic values for the field of rational numbers are examples of discrete values. Values of arithmetic interest are generally discrete.

* This term, used by Rella [10], is similar to Ore's "congruent modulo a polygon" ([8], p. 270) and Kürschák's "equipollence" ([6], p. 185).

† Here and subsequently the element 0 plays an exceptional role.

‡ Krull [14], p. 171, and Hasse-Schmidt [12], p. 31.

3. **The first stage values.** Our problem is this: Given all values of a field K; to construct all values for the ring $K[x]$ of all polynomials* in x with coefficients in K. By Theorem 2.1, this is equivalent to determining all values in the field $K(x)$ of rational functions of x with coefficients in K. No gain in generality would result were a ring S used instead of the field K.

As indicated in the introduction, the values for $K[x]$ will be constructed in stages. For the first step, take any value V_0 for the field K and any real number μ, and then define a corresponding *first stage value* V_1 for any polynomial by the equation (2) of §1. In particular, this gives

$$V_1 x = \mu, \qquad V_1 a = V_0 a \qquad (\text{any } a \in K).$$

Hence we use the symbol $[V_0, V_1 x=\mu]$ for the value V_1.

THEOREM 3.1. *If V_0 is a value of K and μ a real number, the function $V_1 = [V_0, V_1 x=\mu]$ defined above is a value of $K[x]$.*

This has been proved by Rella ([10], pp. 35–36) and by Ostrowski ([18], p. 363). The latter calls x an "invariant element." A particularly simple V_1 arises when $\mu=0$. On the other hand, if V_0 is trivial and $\mu<0$, then

$$V_1 a(x) = \mu \cdot \deg a(x).$$

The symbol deg $a(x)$ here and in the sequel denotes the *degree in x* of the polynomial $a(x)$.

4. **Augmented values.** Our construction now proceeds to build a second stage value on the basis of a first stage one; or, more generally, a kth stage value from one at the stage $k-1$. The process involved can be formulated once for all: Given a value W for $K[x]$; to construct an "augmented" value V by assigning larger values to a certain "key" polynomial $\phi(x)$ and to its equivalence-multiples. The *key polynomial* $\phi(x)$ must be suitably chosen.

DEFINITION 4.1. *A key polynomial $\phi(x)\neq 0$ over a value W of $K[x]$ is one which satisfies the following conditions:*

(i) *Irreducibility. If a product is equivalence-divisible in W by $\phi(x)$, then one of the factors is equivalence-divisible by $\phi(x)$.*

(ii) *Minimal degree. Any non-zero polynomial equivalence-divisible in W by $\phi(x)$ has a degree in x not less than the degree of $\phi(x)$.*

(iii) *The leading coefficient† of $\phi(x)$ is 1.*

This key polynomial is to be assigned a new value

$$V\phi(x) = \mu > W\phi(x). \tag{1}$$

* Henceforth all polynomials considered are to have coefficients in K, unless otherwise noted.

† This assumption, although unnecessary, will simplify the subsequent work.

To find the new values of other polynomials, we use *expansions in* ϕ; that is, expressions in powers of $\phi(x)$ of the form*

$$(2) \qquad f(x) = f_m(x)\phi^m + f_{m-1}(x)\phi^{m-1} + \cdots + f_0(x),$$

in which each coefficient polynomial $f_i(x)$ is either zero or of degree less than the degree of $\phi(x)$. Any polynomial has one and only one such expansion, which may be found by successive division by powers of ϕ. The new value $Vf(x)$ is computed from the expansion thus:

$$(3) \qquad V[f_m(x)\phi^m + f_{m-1}(x)\phi^{m-1} + \cdots + f_0(x)] = \min_i [Wf_i(x) + i\mu].$$

Here "min" with subscript i means the smallest quantity of the form $Wf_i(x)+i\mu$, for $i=0, 1, \cdots, m$.

THEOREM 4.2. *If W is a value of $K[x]$, $\phi(x)$ is a key polynomial over W and μ is a real number satisfying* (1), *then the function V defined in* (3) *is also a value of $K[x]$. V is called an augmented value, and is denoted by*

$$V = [W, V\phi = \mu].$$

Proof. The product and triangle laws for V must be verified.† We first prove the triangle law for a sum $f(x)+g(x)$. Let f and g have the expansions (2) and

$$(4) \qquad g(x) = g_n(x)\phi^n + g_{n-1}(x)\phi^{n-1} + \cdots + g_0(x)$$

respectively. By adjoining zero coefficients we can make $m=n$. Hence $f+g$ has the expansion

$$f(x) + g(x) = \sum_{i=0}^{n} [f_i(x) + g_i(x)]\phi^i.$$

By the definition of V and the triangle law for W,

$$\begin{aligned} V(f + g) &= \min_i [W(f_i + g_i) + i\mu] \geqq \min_i [\min (Wf_i, Wg_i) + i\mu] \\ &\geqq \min_i [Wf_i + i\mu, Wg_i + i\mu] \\ &= \min [\min_i (Wf_i + i\mu), \min_i (Wg_i + i\mu)], \\ V(f + g) &\geqq \min [Vf, Vg]. \end{aligned}$$

---

* We use ϕ as an abbreviation for $\phi(x)$, and similarly for other polynomials.

† The proof resembles one of Rella's ([10], pp. 36–37); the product-law proof is an extension of the Dumas-Kürschák-Ore proof for the Newton polygon of a product.

To prove the product law we will use the *quotient-remainder* expression for a polynomial $f(x)$,

$$f(x) = q(x)\phi + r(x), \tag{5}$$

where $r(x)$ is zero or of degree less than that of $\phi(x)$.

LEMMA 4.3. *If ϕ is a key polynomial over a value W of $K[x]$, and if $f(x)\neq 0$ has the quotient-remainder expression* (5), *then*

$$Wr(x) \geqq Wf(x), \tag{6}$$

$$W(q(x)\phi) \geqq Wf(x). \tag{7}$$

The inequality in (6) *holds if and only if $f(x)$ is equivalence-divisible by ϕ in W.*

Were the first conclusion (6) false, then $Wf(x) > Wr(x)$ in (5) and the definition of equivalence would give

$$r(x) \backsim (-q(x))\phi \qquad \text{(in } W\text{)}.$$

Hence $r(x)\neq 0$ is equivalence-divisible by ϕ, a contradiction to the minimal property of ϕ and the restricted degree of $r(x)$. The second conclusion (7) now follows from (6) by the triangle law.

The third conclusion gives a test for equivalence-divisibility in terms of ordinary division. When $Wr(x) > Wf(x)$, then (5) shows $f(x)$ equivalence-divisible by ϕ. Conversely, if $f(x)$ is equivalence-divisible in W by ϕ, then there exist polynomials $h(x)$ and $s(x)$ so that

$$f(x) = h(x)\phi + s(x), \qquad Ws(x) > Wf(x).$$

If now the equality sign in (6) should hold, we would have

$$r(x) = f(x) - q(x)\phi = (h(x) - q(x))\phi + s(x),$$

with

$$Ws(x) > Wf(x) = Wr(x),$$

making ϕ an equivalence-divisor of $r(x)$, again a contradiction. The lemma is proved.

Return to Theorem 4.2 and consider the product law first for a product of two monomial expansions $a(x)\phi^t$ and $b(x)\phi^u$. Because of the limited degrees of $a(x)$ and $b(x)$, the product $a(x)b(x)$ has an expansion with not more than two terms,

$$a(x)b(x) = c(x)\phi + d(x). \tag{8}$$

The product $a(x)b(x)$ is not equivalence-divisible in W by ϕ, for if it were, the equivalence-irreducibility of ϕ (Definition 4.1) would require that one of the factors be equivalence-divisible by ϕ, contrary to the minimal property. Lemma 4.3 and the triangle axiom yield then

$$W(c(x)\phi) \geqq W(a(x)b(x)) = Wd(x).$$

Since the new value of ϕ exceeds the old value,

$$Wc(x) + \mu > W(a(x)b(x)) = Wd(x). \tag{9}$$

The product under consideration has by (8) the expansion

$$(a(x)\phi^t)(b(x)\phi^u) = c(x)\phi^{t+u+1} + d(x)\phi^{t+u};$$

hence the definition of V and the conclusion (9) give

$$\begin{aligned} V[(a(x)\phi^t)(b(x)\phi^u)] &= \min [Wc(x) + \mu + (t+u)\mu, Wd(x) + (t+u)\mu] \\ &= Wd(x) + (t+u)\mu \\ &= Wa(x) + t\mu + Wb(x) + u\mu \\ &= V(a(x)\phi^t) + V(b(x)\phi^u). \end{aligned}$$

This is the product law for monomial expansions.

The product law for polynomials $f(x)$ and $g(x)$ with arbitrary expansions (2) and (4) respectively is an immediate consequence. The product $f(x)g(x)$ has an expansion obtained by adding expansions of monomial products; hence

$$V(f(x)g(x)) \geqq Vf(x) + Vg(x). \tag{10}$$

To show that the equality holds, choose t and u as the largest integers with

$$V(f_t(x)\phi^t) = Vf(x), \qquad V(g_u(x)\phi^u) = Vg(x)$$

respectively. The monomial case then shows* that the expansion of $f(x)g(x)$ has a term $r(x)\phi^{t+u}$ with the value $Vf+Vg$. The equality holds in (10), and Theorem 4.2 is established.

5. **Properties of augmented values.** An augmented value V is never less than the original value W. This characteristic property will now be established. As a consequence the method used to compute V can be extended (Theorem 5.2) in a way subsequently useful in §12.

THEOREM 5.1. (*Monotonicity.*) *The augmented value* $V=[W, V\phi=\mu]$ *makes*

$$Vf(x) \geqq Wf(x)$$

for all polynomials $f(x)\neq 0$. *The inequality sign holds if and only if* $f(x)$ *is equivalence-divisible in W by* ϕ. *In particular, the equality sign holds whenever the degree of* $\phi(x)$ *exceeds that of* $f(x)$.

* The details here omitted are given in Rella's proof.

The proof is by induction on the degree m of the expansion of $f(x)$ in ϕ (see §4, (2)). If $m=0$, the definition of V shows $Vf(x)$ and $Wf(x)$ equal. If $m>0$, the quotient-remainder expression

$$f(x) = q(x)\phi + r(x) \tag{1}$$

indicates that $q(x)$ has an expansion of degree $m-1$ in ϕ; hence the induction assumption will be

$$Vq(x) \geqq Wq(x).$$

The value of the first term on the right of (1), by §4, (1), and the quotient-remainder Lemma 4.3, is

$$V(q(x)\phi) \geqq Wq(x) + V\phi > W(q(x)\phi) \geqq Wf(x).$$

For the second term, the case $m=0$ and Lemma 4.3 imply

$$Vr(x) = Wr(x) \geqq Wf(x),$$

where the inequality holds if and only if $f(x)$ is equivalence-divisible by ϕ in W. The strong triangle law for V applied to (1) now gives the result (see §2, (2)).

THEOREM 5.2. *If in the expression*

$$a(x) = a_n(x)\phi^n + a_{n-1}(x)\phi^{n-1} + \cdots + a_0(x)$$

the degrees of the $a_i(x)$ are not limited, but if no $a_i(x)$ is equivalence-divisible in W by ϕ, then the augmented value $V=[W, V\phi=\mu]$ is

$$Va(x) = \min_i [Wa_i + i\mu] \qquad (i = 0, 1, \cdots, n).$$

Proof. A quotient-remainder expression for each coefficient polynomial gives

$$a_i(x) = q_i(x)\phi + r_i(x) \qquad (i = 0, 1, \cdots, n),$$

$$a(x) = \sum_{i=0}^{n} q_i(x)\phi^{i+1} + \sum_{i=0}^{n} r_i(x)\phi^i.$$

Lemma 4.3 shows that the second summation has a value

$$V\left(\sum_i r_i\phi^i\right) = \min_i [Wr_i + i\mu] = \min_i [Wa_i + i\mu]$$

and that the first summation has a larger value

$$V\left(\sum_i q_i\phi^{i+1}\right) \geqq \min_i [Vq_i + V\phi + i\mu] > \min_i [Wq_i + W\phi + i\mu],$$

because of the monotonicity. The strong triangle law for the sum of these two summations yields the desired conclusion.

6. **Inductive and limit-values.** This section classifies the values and value-groups obtained by successive augmented values.

DEFINITION 6.1. *A kth stage inductive value V_k is any value of $K[x]$ obtained by a sequence of values $V_1, V_2, \cdots, V_k$, where $V_1 = [V_0, V_1 x = \mu_1]$ is a first stage value* (§3) *and where each V_i is obtained by augmenting V_{i-1};*

$$V_i = [V_{i-1}, V_i\phi_i = \mu_i] \qquad (i = 2, 3, \cdots, k).$$

*Furthermore, for $i = 2, \cdots, k$, the key polynomials $\phi_i(x)$ must satisfy:**
(1) $\deg \phi_i(x) \geqq \deg \phi_{i-1}(x)$;
(2) *the equivalence $\phi_i(x) \backsim \phi_{i-1}(x)$ (in V_{i-1}) is false.*
Here the first key polynomial is understood to be $\phi_1(x) = x$.

This value V_k may be conveniently symbolized thus:

$$(3) \qquad V_k = [V_0, V_1 x = \mu_1, V_2\phi_2 = \mu_2, V_3\phi_3 = \mu_3, \cdots, V_k\phi_k = \mu_k].$$

Given an infinite sequence $V_1, V_2, \cdots, V_k, \cdots$ of such values, we set

$$(4) \qquad V_\infty f(x) = \lim_{k\to\infty} V_k f(x).$$

The monotonic character of V_k indicates that this limit, if not finite, is $+\infty$. V_∞ satisfies the product law for values, as can be shown by taking limits in the product law for V_k. As for the sum $f(x)+g(x)$, note that the triangle law in V_k indicates that one of the inequalities

$$V_k(f(x) + g(x)) \geqq V_k f(x), \qquad V_k(f(x) + g(x)) \geqq V_k g(x)$$

holds for infinitely many k. One of the conclusions

$$V_\infty(f(x) + g(x)) \geqq V_\infty f(x), \qquad V_\infty(f(x) + g(x)) \geqq V_\infty g(x)$$

then results, and thence follows the triangle law for V_∞. We have

THEOREM 6.2. *Let $\{\phi_k(x)\}$ and $\{\mu_k\}$ be fixed infinite sequences such that all the functions V_k indicated in* (3) *are inductive values. Then the function $V_\infty f(x)$ defined in* (4) *is a value of $K[x]$, provided some polynomials not zero be allowed to have the value $+\infty$.*

This function V_∞ will be called a *limit-value*. The case when several successive key polynomials have the same degree will often require separate treatment, based on

* These conditions involve no loss of generality, but simplify subsequent proofs (see Theorem 6.7 and the end of §9).

LEMMA 6.3. *If in the inductive value V_k in (3) the key polynomials $\phi_{t+1}(x)$, $\phi_{t+2}(x), \cdots, \phi_k(x)$ all have the same degree, for t with $0 \leqq t < k-1$, then*

(i) $V_t(\phi_{j+1}-\phi_j)=\mu_j$ $\qquad (j=t+1, t+2, \cdots, k-1)$,

(ii) $\mu_k > \mu_{k-1} > \cdots > \mu_{t+1}$,

(iii) $V_t\phi_k = V_t\phi_{k-1} = \cdots = V_t\phi_{t+1}$ $\qquad$ (*if* $t>0$).

Proof. Let j range from $t+1$ to $k-1$, and set

$$s_j(x) = \phi_{j+1}(x) - \phi_j(x). \tag{5}$$

Since both ϕ's have the first coefficient 1, the degree of $s_j(x)$ is less than that of $\phi_j(x)$. Therefore, by Theorem 5.1,

$$V_t s_j(x) = V_{t+1} s_j(x) = \cdots = V_k s_j(x).$$

If the first conclusion were false for some j, we would have

$$V_t s_j(x) = V_j(\phi_{j+1} - \phi_j) > \mu_j = V_j\phi_j,$$

for the other inequality is impossible by Lemma 4.3. This would give

$$\phi_{j+1} \backsim \phi_j \qquad (\text{in } V_j),$$

a contradiction of assumption (2). The conclusion (i) is thus established. Coupled with the monotonicity and the triangle axiom for (5), it gives the second conclusion, for

$$\mu_{j+1} = V_{j+1}\phi_{j+1} > V_j\phi_{j+1} \geqq \min\,[V_j\phi_j, V_t s_j] = \mu_j.$$

For similar reasons, assuming now that $t>0$,

$$V_t s_j(x) = \mu_j = V_j\phi_j > V_{j-1}\phi_j \geqq V_t\phi_j.$$

The strong triangle axiom for V_t in (5) then yields conclusion (iii),

$$V_t\phi_{j+1} = \min\,[V_t s_j(x), V_t\phi_j] = V_t\phi_j.$$

An interesting consequence of this lemma is the invariance of the values assigned to the key polynomials.

THEOREM 6.4. *If the ith stage of the inductive value V_k in (3) uses a key polynomial ϕ_i with an assigned value μ_i, then*

$$V_k\phi_i(x) = V_i\phi_i(x) = \mu_i.$$

For this conclusion follows directly from Theorem 5.1 if the degree of $\phi_{i+1}(x)$, and hence that of every subsequent key polynomial, exceeds the degree of $\phi_i(x)$. The only case remaining is that of Lemma 6.3, with $t=i-1$. But, by (5),

$$\phi_i = \phi_k - s_{k-1}(x) - s_{k-2}(x) - \cdots - s_i(x).$$

The terms on the right have by the preceding lemma the V_k values $\mu_k, \mu_{k-1}, \cdots, \mu_i$ respectively, so that the conclusion follows by the strong triangle law. Both this theorem and Lemma 6.3 hold equally well for limit-values.

The monotonic property of inductive values can be sharpened thus:

THEOREM 6.5. *Let a limit or inductive value be built up by the inductive values* $V_1, V_2, \cdots$. *Then, for any fixed polynomial* $f(x) \neq 0$, *either*

$$V_{k+1}f(x) > V_kf(x) \qquad (k = 1, 2, \cdots),$$

or else there is an $i \geqq 1$ *such that*

$$V_1f(x) < V_2f(x) < \cdots < V_{i-1}f(x) < V_if(x) = V_{i+1}f(x) = V_{i+2}f(x) = \cdots.$$

In the latter case there is an $r(x)$ *of degree less than that of* ϕ_{i+1} *with*

$$f(x) \sim r(x) \qquad (\text{in } V_k) \quad (k = i+1, i+2, \cdots).$$

Suppose, contrary to the first alternative, that for some i

$$V_{i+1}f(x) = V_if(x).$$

Then the quotient-remainder expression

$$f(x) = q(x)\phi_{i+1} + r(x)$$

must by Theorem 5.1 and Lemma 4.3 have $V_ir = V_if$. Hence, for any $k \geqq i+1$,

$$V_k(f - r) \geqq V_{i+1}(f - r) = V_{i+1}(q\phi_{i+1}) > V_i(q\phi_{i+1}) \geqq V_if = V_ir = V_kr.$$

Therefore $f(x)$ and $r(x)$ are equivalent in V_k and

$$V_kf = V_kr = V_ir = V_if,$$

so that $V_kf(x)$ is constant for $k \geqq i$, which is the second alternative.

An inductive value V_k of $K[x]$ gives by Theorem 2.1 a value for the field $K(x)$ of rational functions. This value has by §2 a value-group Γ_k, which we call the *value-group associated with* V_k. It may be determined in the following way:

THEOREM 6.6. *The value* V_k *in* (3) *has a value-group* Γ_k *consisting of all real numbers of the form*

$$\nu + m_1\mu_1 + m_2\mu_2 + \cdots + m_k\mu_k,$$

where the m_i *are integers and* ν *is an element of the value-group of the original value* V_0.

That every number of Γ_k must be of this form follows by induction from the definition of the augmented value V_k. Conversely, any number of this form is by Theorem 6.4 the value in V_k of the rational function

$$b x^{m_1} \phi_2^{m_2} \cdots \phi_k^{m_k},$$

where b is a constant in K with the value ν.

For a more precise description, designate a real number μ as *commensurable* with an additive group of numbers whenever some integral multiple of μ lies in the group (Ostrowski [18], p. 367). Then

THEOREM 6.7. *In an inductive value V_k from* (3) *every assigned value μ_i, except perhaps μ_k, is commensurable with the value-group Γ_{i-1} of the preceding value (the case $i=1$ included).*

For a proof, consider the expansion in ϕ_i of the next key,

$$\phi_{i+1}(x) = f_m(x)\phi_i^m + f_{m-1}(x)\phi_i^{m-1} + \cdots + f_0(x).$$

If μ_i is not commensurable with Γ_{i-1}, no two terms here can have the same value in V_i. Only one term, say the jth, has the minimum value, and

$$\phi_{i+1}(x) \backsim f_j(x)\phi_i^j \qquad \text{(in } V_i\text{)}.$$

By the irreducibility of ϕ_{i+1} at least one of the conditions:

(6) $f_j(x)$ is equivalence-divisible in V_i by ϕ_{i+1},

(7) $\phi_i(x)$ is equivalence-divisible in V_i by ϕ_{i+1},

must hold. Because of the minimal property of ϕ_{i+1} the first possibility (6) contradicts the assumption (1) of Definition 6.1. For the same reasons the second possibility (7) implies that ϕ_{i+1} and ϕ_i have the same degree, while

$$s(x) = \phi_{i+1}(x) - \phi_i(x)$$

has a smaller degree. Because of (7), Lemma 4.3 applied to V_i and the key polynomial ϕ_{i+1} shows $V_i s(x) > V_i\phi_i(x)$. Hence

$$\phi_i(x) \backsim \phi_{i+1}(x) \qquad \text{(in } V_i\text{)},$$

a contradiction of assumption (2). There can be no next key ϕ_{i+1}.

7. **Constant degree limit-values.** A limit-value V_∞ for polynomials does not give a value for all rational functions if some of the polynomials have the value $+\infty$. Hence the problem: When is V_∞ *finite*; that is, when is $V_\infty f(x)$ finite for all $f(x) \neq 0$? We obtain an answer in the discrete case.

If the key polynomials $\phi_k(x)$ increase indefinitely in degree, then $V_k f(x)$ is by Theorem 5.1 ultimately constant for fixed $f(x)$ and V_∞ is finite. A different situation arises if the degrees of $\phi_k(x)$ have an upper bound. By assumption (1) of Definition 6.1, the degrees of $\phi_k(x)$ are then all equal to some M for k sufficiently large. For an example of such a *constant degree limit-value,* start with the p-adic value $[V_0 3 = 1]$ for the rational field (see §1, (1)) and set

$$V_1 = [V_0, V_1 x = 1],$$
$$V_k = [V_{k-1}, V_k(x + 2p + p^2 + p^3 + \cdots + p^{k-1}) = k] \qquad (k = 2, 3, \cdots).$$

This gives a limit-value of constant degree 1. Since

$$\frac{p}{2} = 2p + p^2 + p^3 + \cdots + p^{k-1} - \frac{p^k}{2} \qquad (k > 1, p = 3)$$

holds by the usual methods for p-adic numbers, we find

$$V_\infty\left(x + \frac{p}{2}\right) = \lim_{k\to\infty} V_k\left[(x + 2p + p^2 + \cdots + p^{k-1}) - \frac{p^k}{2}\right] = \lim_{k\to\infty} k = \infty.$$

Hence this V_∞ is not finite.

This use of p-adic numbers suggests the general notion of a perfect ring. In any ring S with a value V, a sequence $\{a_n\}$ is a *Cauchy sequence* if $V(a_n - a_m)$ approaches ∞ with n and m. If every Cauchy sequence has a V-limit b such that $V(a_n - b)$ approaches ∞ with n, the ring S is said to be *perfect*. Any ring can be embedded in a perfect ring by the usual procedure of adjoining limits of Cauchy sequences (Kürschák [15] and Hasse-Schmidt [12], p. 24).

THEOREM 7.1. (*Finiteness criterion.*) *Let V_∞ be a limit-value with key polynomials $\phi_k(x)$ of constant degree M for $k > t > 0$. Extend the ring $K[x]$ with the value V_t to be a perfect ring S^*. Assume that all values of K are discrete. Then $\{\phi_k\}$ is a Cauchy sequence in V_t and has a limit ϕ in S^*. Furthermore V_∞ is finite if and only if there is no $g(x) \neq 0$ in $K[x]$ divisible in S^* by the limit ϕ.*

For V_∞ the symbolism of Theorem 6.2 may be used. Since $\phi_{t+1}, \phi_{t+2}, \cdots$ all have the same degree M, the conclusions of Lemma 6.3 on constant degree values are applicable. Each number μ_i is by Theorem 6.7 commensurable with the value-group Γ_{i-1} of V_{i-1}. Our assumption shows the original value-group Γ_0 of V_0 to be discrete, hence, by Theorem 6.6 and by induction, the group Γ_t is discrete. But Lemma 6.3 gives

$$\mu_i = V_t(\phi_{i+1} - \phi_i) \in \Gamma_t \qquad (i > t); \tag{1}$$

hence $\Gamma_i = \Gamma_t$ for $i > t$. This lemma also shows the sequence $\{\mu_i\}$ to be monotone increasing for $i > t$; it lies in the discrete set Γ_t, hence

$$\lim_{i\to\infty} \mu_i = \infty. \tag{2}$$

The strong triangle law combined with (1) then proves

$$V_t(\phi_{i+j} - \phi_i) = V_t\left[\sum_{k=i}^{i+j-1} (\phi_{k+1} - \phi_k)\right] = \min_k [\mu_k] = \mu_i.$$

Therefore, by (2), $\{\phi_i\}$ is a Cauchy sequence with a limit ϕ in S^*. This ϕ need not be a polynomial, but, by conclusion (iii) of Lemma 6.3, $\phi \neq 0$.

Now consider the necessary condition for finiteness. If $g(x) \neq 0$ is divisible by ϕ in S^*, then

$$g(x) = h \cdot \phi,$$

where h is the V_t-limit of a Cauchy sequence $\{h_i(x)\}$ from $K[x]$. The usual argument for the convergence of a product shows

$$\lim_{i\to\infty} V_t[g(x) - h_i(x)\cdot\phi_i(x)] = \infty. \tag{3}$$

By the triangle axiom and the monotonic property for $i > t$,

$$V_i g \geqq \min\,[V_i(h_i\phi_i),\ V_i(g - h_i\phi_i)] \geqq \min\,[V_t h_i + \mu_i,\ V_t(g - h_i\phi_i)]. \tag{4}$$

But $\{h_i(x)\}$ is a convergent sequence in V_t with a limit not zero, so that, as is well known ([12], p. 25), $V_t h_i$ is ultimately constant. Consequently (2), (3), and (4) prove

$$V_\infty g(x) = \lim_{i\to\infty} V_i g(x) = \infty, \tag{5}$$

so that V_∞ is not a finite limit-value.

Conversely, suppose that V_∞ is not finite. Then (5) holds for some $g(x) \neq 0$. If $g(x)$ has the quotient-remainder expressions $q_i(x)\phi_i + r_i(x)$, then, by Theorem 5.1 and by Lemma 4.3,

$$V_t(g - q_i\phi_i) = V_t r_i = V_{i-1} r_i \geqq V_{i-1} g(x) \to \infty \qquad (i > t).$$

Thus the sequence $\{q_i\phi_i\}$ converges in V_t to the limit $g(x) \neq 0$. Since $\{\phi_i\}$ already converges to the limit $\phi \neq 0$, the standard argument for the limit of a quotient $(q_i\phi_i)/\phi_i$ shows that $\{q_i\}$ must converge in V_t to some limit q in S^*, such that

$$f(x) = q\phi.$$

Hence ϕ is a factor of $f(x)$ in S^*, as asserted.

8. **Completeness.** We have the following theorem.

THEOREM 8.1. *If every value of the field K is discrete, then every non-archimedean value W of the ring $K[x]$ can be represented either as an inductive or as a limit-value.*

Given W, we shall construct by stages a corresponding inductive value V_k with the following three properties (notation as in §6, (3)):

$$Wf(x) \geqq V_k f(x) \qquad \text{(for all } f(x)), \tag{1}$$

$$\deg f(x) < \deg \phi_k \text{ implies } Wf(x) = V_k f(x), \tag{2}$$

$$W\phi_i(x) = V_k\phi_i(x) = \mu_i \qquad (i = 1, 2, \cdots, k). \tag{3}$$

The initial value V_1 is defined by

$$\mu_1 = Wx, \qquad V_0 a = Wa \qquad (\text{any } a \varepsilon K);$$

the triangle axiom for W and the definition of V_1 in §1, (2), then show that conditions (1), (2), and (3) hold for $k=1$.

Suppose now that an inductive value V_k with these three properties has already been constructed, and that the equality in (1) does not always hold. As a prospective key polynomial, choose a $\psi(x)$ of smallest possible degree with the property

$$W\psi(x) > V_k\psi(x). \tag{4}$$

Multiplication with some constant gives $\psi(x)$ the first coefficient 1. Furthermore the two statements,

$$Wf(x) > V_k f(x), \tag{5}$$

$$f(x) \text{ is equivalence-divisible in } V_k \text{ by } \psi(x), \tag{6}$$

are logically equivalent. For if (5) is given, and if $f(x)$ has the quotient-remainder expression $q(x)\psi + r(x)$, then

$$V_k(q\psi - f) = W(q\psi - f) \geqq \min [W(q\psi), Wf] > \min [V_k(q\psi), V_k f],$$

because of (2), the minimum degree choice of ψ and the induction assumption (1) for $q(x)$. Hence the strong triangle law shows $f \backsim q\psi$ in V_k, which is the conclusion (6). Conversely, if (6) holds there exist polynomials $h(x)$ and $s(x)$ with

$$f(x) = h(x)\psi + s(x), \qquad V_k s(x) > V_k f(x) = V_k(h(x)\psi).$$

Then, because of the induction assumption (1),

$$Wf \geqq \min [W(h\psi), Ws] \geqq \min [V_k h + W\psi, V_k s] > V_k h + V_k\psi = V_k f,$$

which gives conclusion (5). The equivalence of (5) and (6) is established.

From the equivalence one readily shows that $\psi(x)$ satisfies the Definition 4.1 of a key polynomial over the value V_k. Finally we can assign $\psi(x) = \phi_{k+1}$ the new value

$$\mu_{k+1} = W\psi > V_k\psi, \tag{7}$$

satisfying the proper inequality, and then construct the augmented value $V_{k+1} = [V_k, V_{k+1}\phi_{k+1} = \mu_{k+1}]$. This will be an inductive value if only conditions

(1) and (2) of Definition 6.1 hold. By the choice of $\phi_{k+1}=\psi$ and the induction assumption (2), $\phi_k(x)$ cannot exceed $\phi_{k+1}(x)$ in degree, therefore condition (1) of §6 is true. Condition (2) of §6 could only be false if ϕ_{k+1} and ϕ_k were equivalent in V_k; in other words, only if

$$V_k(\phi_k - \phi_{k+1}) > V_k\phi_k = V_k\phi_{k+1}.$$

By (2), (3), and the choice of ψ in (4) this would entail

$$W\phi_k \geqq \min\left[W\psi_{k+1}, W(\phi_k - \phi_{k+1})\right] > \min\left[V_k\phi_{k+1}, V_k\phi_k\right] = V_k\phi_k = W\phi_k,$$

a contradiction which establishes the desired condition.

The inductive value V_{k+1} thus constructed satisfies the analogues of the desired conditions (1), (2), and (3). The latter two are consequences of the definitions in (4) and (7), while (1) follows from the definition (see §4, (3)) of the augmented value V_{k+1} by the triangle axiom for W:

$$W\left(\sum_{i=0}^{m} f_i(x)\psi^i\right) \geqq \min_i\left[Wf_i(x) + i\mu_{k+1}\right] = V_{k+1}\left(\sum_{i=0}^{m} f_i(x)\psi^i\right).$$

The inductive construction of the value V_k associated with W is complete.

This process either will ultimately yield an inductive value V_k equal to W or will give an infinite sequence of inductive values with a limit-value V_∞ such that

$$Wf(x) \geqq V_\infty f(x) = \lim_{k\to\infty} V_k f(x) \qquad \text{(for all } f(x)).$$

In the discrete case the first inequality sign never occurs. For suppose instead that it did hold for some $f(x)$; then since $\{V_kf\}$ is monotone non-decreasing,

$$Wf(x) > V_kf(x) \qquad (k = 1, 2, \cdots).$$

The equivalence of (5) and (6) then implies that $f(x)$ is equivalence-divisible by $\phi'_{k+1}(x)$ in V_k. Hence the monotonicity Theorem 5.1 shows

$$V_{k+1}f(x) > V_kf(x) \qquad (k = 1, 2, \cdots).$$

This cannot hold if the degrees of the key polynomials $\phi_k(x)$ increase indefinitely, so that we have the case where $\phi_k(x)$ has the fixed degree M for $k>t$, as in Theorem 7.1. The monotonic increasing sequence $\{V_kf(x)\}$ consists of numbers all from the discrete group Γ_t, with the result

$$Wf(x) \geqq V_\infty f(x) = \lim_{k\to\infty} V_kf(x) = \infty.$$

This can occur only for $f(x)=0$, a trivial case. Accordingly, $W=V_\infty$, and the completeness theorem is established.

Ostrowski ([18], pp. 361–392) has developed another method for finding all non-archimedean values of a transcendental extension $K(x)$ of a field K. His method makes no discreteness assumptions, but requires that K first be extended to an algebraically closed field A. The values of $A(x)$ and thereby those of $K(x)$ are constructed by means of "pseudo-convergent sequences" analogous to limit-values with linear key polynomials. A higher degree would be impossible by the character of A. Note, however, that his method requires an elaborate construction to obtain values of A from those of K, and that the use of A precludes any application to irreducibility criteria or to the algebraic extensions of the residue-class field discussed in our Part II.

II. The structure of inductive values

9. **Properties of key polynomials.** To apply the preceding construction of values to any particular case it is necessary to know what polynomials can be used as key polynomials. This question is not constructively answered by the definition in §4. Part of this question will be answered at once (Theorem 9.4); the rest after the structure of the inductive values V_k has been more explicitly formulated. We first show that certain polynomials act like "equivalence-units":

LEMMA 9.1. *If V_k is an inductive value with $k>1$, then for every polynomial $b(x)$ with $V_kb(x)=V_{k-1}b(x)$ there is a polynomial $b'(x)$ with*

$$(1) \qquad b'(x)b(x) \backsim 1 \quad (\text{in } V_k), \qquad V_kb'(x) = V_{k-1}b'(x).$$

The hypothesis on $b(x)$ implies that $b(x)$ is not divisible by the last key polynomial $\phi_k(x)$. Since ϕ_k is certainly irreducible in the ordinary sense, there are polynomials $b'(x)$ and $c(x)$ with

$$b'(x)b(x) + c(x)\phi_k(x) = 1, \qquad \deg b'(x) < \deg \phi_k(x).$$

By Theorem 5.1, $V_kb'=V_{k-1}b'$. The transition from V_{k-1} to V_k increases the value of $c\phi_k$, but leaves unchanged the values of $b'b$ and 1 in this equation. Hence $b'b \backsim 1$, as in (1).

LEMMA 9.2. *In any inductive V_k, the last key polynomial ϕ_k is equivalence-irreducible in V_k; a polynomial $g(x)$ not equivalence-divisible by ϕ_k in V_k has a value V_kg in Γ_{k-1}.*

If a polynomial $f(x)$ has the expansion

$$(2) \qquad f(x) = f_n(x)\phi_k^n + f_{n-1}(x)\phi_k^{n-1} + \cdots + f_0(x), \qquad \deg f_i(x) < \deg \phi_k(x),$$

then $f(x)$ *is equivalence-divisible in V_k by ϕ_k if and only if $V_kf_0>V_kf$.* For if $V_kf_0>V_kf$, then $f-f_0$ is a polynomial equivalent to f with a factor ϕ_k. Conversely, if $f \sim h(x)\phi_k$, then the last term f_0 of the expansion for f is obtained from $f-h\phi_k$, where $V_k(f-h\phi_k)>Vf$, so that $V_kf_0>V_kf$. In particular, an f not equivalence-divisible by ϕ_k has $V_kf=V_kf_0=V_{k-1}f_0 \in \Gamma_{k-1}$, as asserted.

This criterion shows ϕ_k equivalence-irreducible in V_k. For suppose instead that $f(x)g(x)$ is equivalence-divisible by ϕ_k, although neither factor is so divisible. Then the criterion gives $V_kf_0=V_kf$, $V_kg_0=V_kg$, where $g_0(x)$ is the last term in the expansion for g. The last term in the expansion for fg is the remainder $r_0(x)$ obtained by dividing f_0g_0 by ϕ_k; but since f_0g_0 is not equivalence-divisible in V_{k-1} by ϕ_k, Lemma 4.3 shows

$$V_kr_0 = V_{k-1}r_0 = V_{k-1}(f_0g_0) = V_kf_0 + V_kg_0 = V_k(fg).$$

This means that fg is not equivalence-divisible by ϕ_k, a contradiction proving the lemma.

An inductive value V_k will be called *commensurable* if the value μ_k assigned the last key polynomial is commensurable with the previous value-group Γ_{k-1} (cf. Theorem 6.7). There is then a smallest positive integer τ_k such that $\tau_k\mu_k$ is in Γ_{k-1}. For each $t \leqq k$ there is a similar τ_t:

(3) τ_t is the smallest integer such that $\tau_t\mu_t \in \Gamma_{t-1}$.

We will subsequently need polynomials with any given values:

LEMMA 9.3. *If V_k is a commensurable inductive value, then for any real number λ in the value-group Γ_k of V_k there is a polynomial $R_\lambda=R_\lambda(x)$ with value λ in V_k and in every value V_{k+1} obtained by augmenting V_k.*

Proof. As in Theorem 6.6, λ has the form

$$\lambda = \nu + m_1\mu_1 + m_2\mu_2 + \cdots + m_k\mu_k, \qquad \nu \in \Gamma_0.$$

Each integer m_i may be made non-negative by adding to $m_i\mu_i$ and subtracting from ν a sufficiently large term $q\mu_i$, so chosen that $q\mu_i \in \Gamma_0$ (e.g., choose $q \equiv 0 \pmod{\tau_1, \tau_2, \cdots, \tau_i}$). If then a is a constant of value ν,

$$R_\lambda = R_\lambda(x) = ax^{m_1}\phi_2^{m_2}\phi_3^{m_3}\cdots\phi_k^{m_k}, \qquad V_kR_\lambda = \lambda, \qquad m_i \geqq 0,$$

is the required polynomial. In any augmented value V_{k+1}, R_λ has value λ by Theorem 6.4.

THEOREM 9.4. *A polynomial $f(x)$ is a key polynomial for an inductive value over V_k if and only if the following conditions hold*: (i) *the expansion* (2) *has a last term with $V_kf=V_kf_0$*; (ii) *the expansion has a first term $f_n(x)\phi_k^n$ with $f_n(x)=1$, $V_k\phi_k^n=V_kf$, and $n\equiv 0 \pmod{\tau_k}$*; (iii) *$f(x)$ is equivalence-irreducible in V_k.*

Proof. Condition (i) means, as in the proof of Lemma 9.2, that $f(x)$ is not equivalence-divisible in V_k by ϕ_k. Assume first that $f(x)$ is a key. Condition

(iii) is necessary by definition. Were (i) false, then $V_k f < V_k f_0$, so that $f \backsim (f-f_0)$, while (2) shows $f-f_0 = q(x)\phi_k$ for a $q(x)$ of degree less than $f(x)$. Thus $f \backsim q\phi_k$ in V_k. Since f is a key, this leads to a contradiction much as in the proof of Theorem 6.7. The assumption $V_k f_0 \neq V_k f$ is false.

Since $f(x)$ is minimal (Definition 4.1) it has no equivalence-multiples of degree less than itself. Hence $f_n(x)$ is a constant in K, for otherwise $k>1$, and Lemma 9.1 supplies a $b'(x)$ with $b'(x)f_n(x) \backsim 1$ in V_k. The product $b'(x)f(x)$ formed from (2) and modified by replacing the first coefficient by 1 and by reducing the other coefficients modulo ϕ_k is then an equivalence-multiple of $f(x)$. Its degree is $n \cdot \deg \phi_k$, and is less than that of $f(x)$ unless $f_n(x) \varepsilon K$. As the leading coefficient of f must be 1, $f_n(x) = 1$ follows, as in (ii). Certainly $V_k \phi_k^n = V_k f$ is necessary, for otherwise $f - \phi_k^n$ is an equivalent polynomial of smaller degree. Thus

$$V_k \phi_k^n = V_k f = V_k f_0 = V_{k-1} f_0 \ \varepsilon\ \Gamma_{k-1},$$

so that $n \equiv 0 \pmod{\tau_k}$ by (3). This establishes the necessity of (ii).

Conversely, if $f(x)$ satisfies (i), (ii), and (iii), it has first coefficient 1 and is minimal, because any equivalence-multiple of $f(x)$ must be of degree at least n in ϕ_k (cf. the proof of Theorem 4.2). The remaining restrictions of Definition 6.1 are readily verified, so that $f(x)$ is in fact a key polynomial.

10. **Residue-class fields.** The structure of a ring S with a value V involves the corresponding *value-ring* S^+, which consists of all elements a of S with $Va \geqq 0$ (these elements are the so-called "integers" of S). A *congruence* for integers can be defined thus

$$a \equiv b \pmod{V} \text{ if and only if } V(a-b) > 0. \tag{1}$$

All elements of S^+ congruent to a given b form a *residue-class*; these classes together yield as usual the *residue-class ring* of V in S. This ring can also be considered as the residue-class ring S^+/P, where P, the set of all elements of S^+ with positive value, is a prime ideal in S^+. If S is a field, then S^+/P is also a field, the *residue-class field* of V in S. The structure of V depends essentially* on this residue-class field. For the p-adic value V_0 of the rational numbers (see §1, (1)) this field is simply the field of integers modulo p. Our problem is the determination of the residue-class field for any discrete inductive value.

If the residue-class of each integer a be denoted by Ha, then H is a homomorphism of S^+ to the residue-class ring $\Delta = S^+/P$, so that H has the following properties:

I. H is a many-one correspondence between S^+ and Δ;

* Hasse-Schmidt [12], p. 7; Ostrowski [18], p. 321.

II. H leaves sums and products unchanged; i.e., for $Va \geqq 0$ and $Vb \geqq 0$,

$$H(a+b) = Ha + Hb; \qquad H(ab) = (Ha)(Hb). \tag{2}$$

III. If $Va \geqq 0$, then $Ha = 0$ if and only if $Va > 0$.
By II, the last condition means that H carries congruent elements and only congruent elements into the same residue-class.

For an inductive value V_k we denote the residue-class rings thus, for $t = 1, 2, \cdots, k$:

(3) Λ_t is the residue-class field of V_t in $K(x)$;

(4) H_t is the homomorphism from $K(x)^+$ to Λ_t;

(5) Δ_t is the residue-class ring of V_t in $K[x]$.

But $f(x)$ and $g(x)$ are congruent as polynomials (mod V_t) if and only if they are congruent as rational functions (mod V_t). Hence each residue-class of Δ_t is contained in a residue-class of Λ_t, and no two residue-classes of Δ_t are contained in the same class of Λ_t. Addition and multiplication of classes are defined as addition and multiplication on elements in the classes, and hence are the same in Δ_t as in Λ_t. Therefore Δ_t is isomorphic to a subring of Λ_t. Since isomorphism does not alter the structure of a ring, we will *replace* Δ_t *henceforth by the isomorphic subring of* Λ_t. Then the H_t of (4) is also the homomorphism from $K[x]^+$ to Δ_t.

The correspondence H_t for rational functions is usually determined by the H_t for polynomials. For if $f(x)/g(x) \neq 0$ is a rational function with non-negative value and if V_t is commensurable, there is by Lemma 9.3 a polynomial $R(x)$ with $V_t R = -V_t g$, and by (2)

$$H_t\left(\frac{f(x)}{g(x)}\right) = H_t\left(\frac{Rf}{Rg}\right) = \frac{H_t(Rf)}{H_t(Rg)}. \tag{6}$$

Both $H_t(Rf)$ and $H_t(Rg)$ are residue-classes of polynomials, while $H_t(Rg) \neq 0$ by Property III. We have proved

LEMMA 10.1. *For a commensurable* V_k, *the residue-class field* Λ_k *of* $K(x)$ *is the quotient-field of the residue-class ring* Δ_k *of* $K[x]$.

THEOREM 10.2. *For a commensurable first stage inductive value* $V_1 = [V_0, V_1 x = \mu_1]$, *the residue-class ring* Δ_1 *is isomorphic to the ring* $F_0[y]$ *of all polynomials in a variable* y *with coefficients in* F_0, *the residue-class field of the value* V_0 *for* K.

Proof. There is given a homomorphism H_0 from the ring K^+ of all V_0-integers b in K to the residue-class field F_0. Each residue-class $H_1 b$ of Δ_1 contains the residue-class $H_0 b$ of F_0, and this correspondence $H_1 b \longleftrightarrow H_0 b$ is an isomorphism between F_0 and the set of those classes of Δ_1 containing elements of K. We will identify F_0 with this isomorphic subfield of Δ_1; then Δ_1 is an extension of F_0 and $H_1b=H_0b$ for all b in K^+.

Any monomial bx^n of value zero has $V_0b=-nV_1x=-n\mu_1$, so that the exponent n is a multiple of the integer τ_1, defined in §9, (3). Any $f(x)$ with $V_1f\geqq 0$ thus has the form

$$f(x) \equiv b_m x^{m\tau_1} + b_{m-1}x^{(m-1)\tau_1} + \cdots + b_1x^{\tau_1} + b_0 \pmod{V_1}$$

after terms of positive value are omitted. If e is a constant in K of value $V_0e=\tau_1\mu_1$, each term $b_jx^{j\tau_1}$ may be rewritten as a product $(b_je^j)(e^{-j}x^{j\tau_1})$ of two factors of value 0. The application of the homomorphism H_1 then yields

$$(7) \qquad H_1f(x) = \sum_{j=0}^{m} H_0(b_je^j)y^j; \qquad y = H(e^{-1}x^{\tau_1}).$$

With y so defined, any H_1f in Δ_1 becomes a polynomial in y with coefficients $H_0(b_je^j)$ in F_0, so that the residue-class ring Δ_1 is contained in $F_0[y]$. Since $y \,\varepsilon\, \Delta_1$ and Δ_1 is a ring, $\Delta_1=F_0[y]$. The element y is transcendental (i.e., a variable) over F_0, for otherwise it would satisfy an algebraic relationship $\alpha(y)=0$, where

$$\alpha(y) = \alpha_m y^m + \alpha_{m-1}y^{m-1} + \cdots + \alpha_0; \qquad \alpha_m \neq 0, \qquad \alpha_j \,\varepsilon\, F_0.$$

Then the residue-class $\alpha(y)$ contains the polynomial

$$f(x) = a_m e^{-m}x^{m\tau_1} + a_{m-1}e^{-m+1}x^{(m-1)\tau_1} + \cdots + a_0,$$

where each a_j is a constant with $H_0a_j=\alpha_j$. Then $V_1f\geqq 0$ and $H_1f=\alpha(y)=0$, so that, by Property III of H_1, $V_1f>0$. But $\alpha_m\neq 0$, so that $V_0a_m=0$ and $V_1(a_me^{-m}x^{m\tau_1})=V_1f=0$, a contradiction. The theorem is established. We note also that (7) enables us to calculate the residue-class of any given $f(x)$.

11. **Conditions for equivalence-irreducibility.** A key polynomial ϕ_{k+1} over a value V_k is not equivalence-divisible by ϕ_k (Theorem 9.4, condition (i)). For any $f(x)$ with this property, questions of equivalence-divisibility can be handled as follows:

LEMMA 11.1. *In a commensurable V_k let $f(x)$ be a polynomial not equivalence-divisible by ϕ_k, and choose a polynomial $R(x)$ so that $V_{k-1}R=V_kR=-V_kf$. Then a polynomial $g(x)$ with $V_kg=0$ is equivalence-divisible by f in V_k if and only if H_kg is divisible by $H_k[Rf]$ in the residue-class ring Δ_k.*

By Lemma 9.2, V_kf is in Γ_{k-1}, so Lemma 9.3 yields the R desired, and $H_k[Rf]\neq 0$. Suppose first that H_kg (which is not 0 by Property III of H_k) is divisible by $H_k[Rf]$. Then $H_kg=\alpha\cdot H_k[Rf]$ for some residue-class $\alpha=H_kh(x)\neq 0$ in Δ_k, and

$$H_kg = \alpha\cdot H_k[Rf] = (H_kh)H_k[Rf] = H_k(hRf).$$

Thus g and hRf have the same residue-class, $V_k(g-hRf)>0=V_kg$, and $g\backsim hRf$ is equivalence-divisible by f, as asserted.

Conversely, if g is equivalence-divisible by f, then $g\backsim hf\backsim hR'Rf$, where $R'(x)$ is a polynomial chosen as in Lemma 9.1 so that $RR'\backsim 1$. But Rf, g, and hence hR' have value 0, so that

$$g \equiv hR'Rf \quad (\text{in } V_k); \qquad H_kg = H_k(hR')\cdot H_k[Rf],$$

which shows H_kg divisible by $H_k[Rf]$.

LEMMA 11.2. *For f and R as in Lemma 11.1, $f(x)$ is equivalence-irreducible in V_k if and only if every product in Δ_k divisible by $H_k[Rf]$ has* a factor divisible by $H_k[Rf]$ in Δ_k.*

Suppose first that f is equivalence-irreducible and that $(H_kg)(H_kh)=H_k(gh)$ is a multiple of $H_k[Rf]$. As we can assume $V_kg=V_kh=0$, the previous lemma shows the product gh equivalence-divisible by the equivalence-irreducible f, so that one of the factors is so divisible. By Lemma 11.1 this means that H_kg or H_kh is a multiple of $H_k[Rf]$, as asserted in the lemma.

Conversely, suppose that every product $(H_kg)(H_kh)$ divisible by $H_k[Rf]$ has a factor so divisible, and consider a product $a(x)b(x)$ equivalence-divisible by f, so that $ab\backsim c(x)f$ for some c. Write $a(x)\backsim g(x)\phi_k^d$ and $b(x)\backsim h(x)\phi_k^e$, where the powers d and e are chosen so large that g and h are not equivalence-divisible by ϕ_k in V_k. Then V_kg and V_kh are by Lemma 9.2 in Γ_{k-1}, so that there are polynomials $S(x)$ and $T(x)$ with $V_k(gS)=V_k(hT)=0$. Then

$$STab \backsim (Sg)(Th)\phi_k^{d+e} \backsim STcf \qquad (\text{in } V_k).$$

But f is not equivalence-divisible by ϕ_k while ϕ_k is equivalence-irreducible (Lemma 9.2), so STc is equivalence-divisible by ϕ_k^{d+e}. Removal of this factor makes $(Sg)(Th)$ equivalence-divisible by f, so that as in the previous lemma $H_k(Sg)H_k(Th)$ is divisible by $H_k[Rf]$. One of the factors, say $H_k(Sg)$, is then divisible by $H_k[Rf]$, and (Lemma 11.1) Sg is equivalence-divisible by f. But $a\backsim S'(Sg)\phi_k^d$, where S' is chosen so that $S'S\backsim 1$. Hence $a(x)$ is equivalence-divisible by f, and f is equivalence-irreducible.

12. **Residue-class rings for commensurable values.** We have

THEOREM 12.1. *If V_k is a commensurable inductive value of $K[x]$, given as in* §6, (3), *and if the original value V_0 of K has a residue-class field F_0, then there is a sequence of fields $F_1=F_0, F_2, F_3, \cdots, F_k$, each an algebraic extension of the preceding, such that for any $t=1,\cdots,k$ the V_t-residue-class ring of $K[x]$ is (isomorphic to) the ring $F_t[y]$ of polynomials in a variable y with coefficients in F_t. For $t>1$ the degree m_t of F_t is determined by* (cf. §9, (3))

* That is, the principal ideal $(H_k[Rf])$ is a prime ideal in Δ_k.

$$m_t\tau_{t-1}(\deg \phi_{t-1}) = \deg \phi_t; \qquad m_t = \deg [F_t : F_{t-1}].$$

By Lemma 10.1 we can then conclude at once

COROLLARY 12.2. *$F_t(y)$ is the V_t-residue-class field of $K(x)$.*

The case $t=1$ of this theorem is known (Theorem 10.2); hence we use induction, and assume the theorem true for V_t. It is convenient to omit the subscript $t+1$ (but not the subscript t) and to write V, ϕ, H, τ, etc., for V_{t+1}, ϕ_{t+1}, H_{t+1}, τ_{t+1}, etc. By the monotonic character of V (Theorem 5.1) polynomials $f(x)$ and $g(x)$ with $V_t f \geqq 0$ and $V_t g \geqq 0$ are congruent mod V_t only if they are congruent mod V. Each residue-class mod V_t is thus contained in a residue-class mod V, and this gives a homomorphism between $\Delta_t = F_t[y]$ and a subring F of the residue-class ring Δ (cf. §10, (5)), where $F = F_{t+1}$ *is composed of all residue-classes* mod V *containing an* $f(x)$ *with* $V_t f \geqq 0$. Polynomials f and g incongruent mod V_t become congruent mod V if and only if $f-g$ is equivalence-divisible by ϕ in V_t (Theorem 5.1). This means that $H_t f - H_t g$ is divisible by the polynomial

$$(1) \qquad \psi_{t+1}(y) = \psi(y) = H_t[R\phi]; \qquad (V_{t-1}R = V_t R = -V_t\phi,\ R = R_{t+1}(x)),$$

constructed as in Lemma 11.1. Since not all polynomials are equivalence-divisible by ϕ in V_t, $\psi(y)$ is not a constant in F_t, while Lemma 11.2 shows $\psi(y)$ an irreducible polynomial in $F_t[y]$. In the above homomorphism between $F_t[y]$ and F the multiples of $\psi(y)$ in $F_t[y]$ are the elements corresponding to 0, so that F is isomorphic to the ring of polynomials $F_t[y]$ modulo $\psi(y)$, or, alternatively, to the field obtained by adjoining to F_t a root θ of $\psi(y)$. We identify F with this isomorphic field:

$$(2) \qquad F = F_{t+1} = F_t(\theta); \qquad \psi(\theta) = 0 \qquad (\theta = \theta_{t+1}).$$

Then the residue-class $H_t f$, when reduced modulo $\psi(y)$, will be identical to the residue-class Hf; that is,

$$(3) \qquad V_t f(x) \geqq 0 \text{ implies } Hf = [H_t f]_{y=\theta}.$$

A monomial expansion $a(x)\phi^n$ of value 0 must have n a multiple of τ (cf. §9, (3)). Hence any $f(x)$ with $Vf \geqq 0$ has the form

$$(4) \quad f(x) \equiv f_m(x)\phi^{m\tau} + f_{m-1}(x)\phi^{(m-1)\tau} + \cdots + f_0(x) \pmod V, \ \deg f_i < \deg \phi.$$

Since $V\phi^\tau \varepsilon \Gamma_t$, there are by Lemmas 9.1 and 9.3 polynomials $Q_{t+1}(x)$ and $Q'_{t+1}(x)$ such that

$$(5) \qquad V_t Q = VQ = V\phi^\tau, \qquad QQ' \equiv 1 \pmod V, \qquad V_t Q' = VQ' = -V\phi^\tau.$$

The terms $f_i\phi^{i\tau}$ in the expansion (4) can be rewritten as products $(f_i Q^i)(\phi^{i\tau}Q^{-i})$, where $V_t(f_jQ^j)\geqq 0$, $V(\phi^{j\tau}Q^{-j})=0$, the former because $Vf\geqq 0$. The application of H, with (3), then proves

$$(6)\qquad Hf(x) = \sum_{j=0}^{m} [H_t(f_j(x)Q^j)]_{y=\theta}\cdot y_1^j; \qquad y_1 = H(\phi^\tau Q^{-1}).$$

This shows that every Hf in the residue-class ring Δ is also in $F[y_1]$, while by (5), $y_1=H(\phi^\tau Q^{-1})=H(\phi^\tau Q')$ is a residue-class of a polynomial, hence is in Δ. Consequently, $F[y_1]=\Delta$, as asserted in the theorem.

The element y_1 is transcendental over F; for suppose instead that y_1 satisfied an algebraic relation $\alpha(y_1)=0$, with

$$\alpha(y_1) = \alpha_m y^m + \alpha_{m-1}y^{m-1} + \cdots + \alpha_0; \qquad \alpha_m \neq 0, \qquad \alpha_j \,\varepsilon\, F.$$

By the original (italicized) definition of F each residue-class α_j of F contains a polynomial $b_j(x)$ with $V_tb_j\geqq 0$, so that $Hb_j=\alpha_j$. Then

$$f(x) = \sum_{j=0}^{m} b_j(x)Q'^j\phi^{j\tau} \equiv \sum_{j=0}^{m} b_j(x)Q^{-j}\phi^{j\tau} \qquad (\text{in } V)$$

is a polynomial of non-negative value which has $Hf=\alpha(y_1)=0$. By Property III of H, $Vf>0$. On the other hand Vf must equal 0, for $Hb_m=\alpha_m\neq 0$ gives $Vb_m=0$ and $Vf\leqq V(b_mQ^{-m}\phi^{m\tau})=0$, by Theorem 5.2. This contradiction shows y_1 a variable over F.

The formula (6) enables us to calculate $Hf(x)$ effectively for any $f(x)$ given in (4), provided only that $V_t(f_jQ^j)\geqq 0$ for all j.

It remains to determine the degree of the field F over F_t, which by (2) is the degree of $\psi(y)$. The key ϕ has by Theorem 9.4 an expansion of the form

$$(7)\qquad \phi = \phi_t^{m\tau_t} + \sum_{i=0}^{m\tau_t-1} a_i(x)\phi_t^i, \qquad V_t\phi = V_t\phi^{m\tau_t} = V_ta_0.$$

If $t>1$, $\psi=H_t[R\phi]$ can be computed by the analog of (6) for the preceding stage (with t in (6) replaced by $t-1$), for the coefficients Ra_i must by the choice of R have $V_{t-1}(Ra_i)\geqq 0$. This calculation shows $\psi(y)$ to be a polynomial in y with a first term $H_t(RQ_t^m)y^m$ arising from the first term of (7). But

$$V_t(RQ_t^m) = V_tR - V_tQ_t^m = V_t\phi - V_t\phi_t^{m\tau_t} = 0,$$

so that the coefficient of y^m is not 0. The polynomial ψ has degree m, and by (7)

$$m\tau_t \deg \phi_t = \deg \phi, \qquad m = \deg \psi = \deg\,[F:F_t],$$

as asserted* in Theorem 12.1. This theorem has now been demonstrated.

* The proof given holds for $t>1$, but may be simplified for the case $t=1$.

13. **Conditions for key polynomials.** In the criterion of Theorem 9.4 for a key polynomial the condition (iii) of equivalence-irreducibility can now be replaced by the condition of Lemma 11.2,

(iiia), $H_k[Rf]$ is an irreducible polynomial in $F_k[y]$.

This yields a final explicit description of key polynomials. A partial converse is possible:

THEOREM 13.1. *In a given V_k, let $\psi(y) \neq y$ be a polynomial* of degree $m>0$, irreducible in $F_k[y]$ and with first coefficient* 1. *Then there is one and, except for equivalent polynomials in V_k, only one $\phi(x)$ which is a key polynomial and which has $H_k[R\phi]=\psi(y)$, for a suitable R chosen as in* §11.

Proof. There is a polynomial $f(x)$ with the residue-class ψ, so that $H_kf=\psi$ and $V_kf=0$. If we multiply f by Q_k^m (chosen as in §12, (5)) and in the expansion of the resulting product drop all terms not of minimum value and then replace the leading coefficient of ϕ_k by 1, we obtain a polynomial $\phi(x)$ with the value $V_kQ_k^m$. For R we can then use $Q_k'^m$, so that

$$H_k[R\phi] = H_k[Q_k'^m Q_k^m f] = H_kf = \psi(y).$$

Furthermore ϕ can be shown to satisfy the remaining conditions (i) and (ii) of Theorem 9.4, hence ϕ is a key polynomial. The uniqueness is readily established.

Since $H_k[Rf]$ can be effectively constructed by §12, (6), the problem of testing whether a given $f(x)$ is a key polynomial is reduced to that of testing the image $H_k[Rf]$ of $f(x)$ for irreducibility in $F_k[y]$. If K is the field of rationals, then F_k is a finite field and the latter problem is completely solvable. This result can be used to construct examples for inductive values of any stage and for limit-values of both constant degree† and increasing degree types. The construction of constant degree values may be simplified by deducing from Theorem 9.4 the following partial converse of Lemma 6.3:

COROLLARY 13.2. *In V_k let $s(x)$ be a polynomial of degree less than that of $\phi_k(x)$ and with $V_ks(x)=V_k\phi_k$. Then $\phi_k(x)+s(x)$ is a key polynomial for an inductive value over V_k.*

14. **Special cases of homomorphism.** The residue-class fields can be similarly found for finite discrete limit-values and for inductive values where the value for K is trivial (§2) or where the last assigned value μ_k is incommensurable (§6).

* The assumption $\psi(y)\neq y$ is needed, for the condition $V_kf=V_kf_0$ in Theorem 9.4 implies $H_k[Rf]\neq y$.

† By using a transcendental p-adic number the finiteness condition of §7 can be satisfied.

THEOREM 14.1. *Let V_∞ be a limit-value constructed as in Theorem* 6.1 *from a sequence of values $V_0, V_1, V_2, \cdots$, and satisfying one of the conditions*:

(a) *the degrees of the keys ϕ_k are not bounded as $k \to \infty$*;

(b) *V_∞ is finite and discrete, and* $\deg \phi_k = M$ *for all $k > t$.*

The fields F_k of Theorem 12.1 *yield a (possibly infinite) extension field $F_\infty = F_0 + F_1 + F_2 + \cdots$ which is isomorphic to the V_∞ residue-class field both for $K[x]$ and for $K(x)$. In case* (b), $F_\infty = F_{t+1}$.

Let H_k be the homomorphism of $K[x]$ to the V_k residue-class ring $F_k[y]$ and H_∞ the homomorphism of $K[x]$ to the V_∞ residue-class ring Δ_∞. Then

$$(1) \qquad V_{k-1}f(x) \geqq 0 \text{ implies that } H_{k+j}f = H_k f \,\varepsilon F_k \qquad (j = 1, 2, \cdots).$$

For, according to §12, (3),

$$H_k f = [H_{k-1}f]_{y=\theta_k}, \qquad H_{k+1}f = [H_k f]_{y=\theta_{k+1}},$$

which indicates that $H_k f$ is a constant free of y in F_k, and that $H_{k+1}f$ must equal $H_k f$, and so on.

In case (a) there is for every $f(x)$ a k so large that $\deg \phi_k > \deg f$, so that $V_\infty f = V_{k-1}f$, as in Theorem 5.1. If $V_\infty f \geqq 0$, then, by (1), $H_{k+j}f = H_k f$ is a constant in F_k independent of j. The correspondence

$$(2) \qquad H_\infty f \longleftrightarrow H_k f, \text{ for } k \text{ with } H_k f = H_{k+1}f = H_{k+2}f = \cdots,$$

carries each element of Δ_∞ into an element of F_∞. Every element α of F_∞ is used, for, by the definition of F_∞, α is in some F_k so that α has the form $H_k f$, and $H_{k+j}f = H_k f$ as in (1), whence α corresponds in (2) to $H_k f$. The correspondence (2) is one-one, for elements are congruent mod V_∞ if and only if they are congruent modulo some V_k. Finally, (2) is an isomorphism, making $F_\infty \cong \Delta_\infty$ as asserted. The residue-class field of $K(x)$ is, by the argument of Lemmas 9.3 and 10.1, just the quotient field of F_∞, and must then be F_∞ itself.

In the case (b), the degrees of the extensions $F_{k+1}:F_k$ as determined in Theorem 12.1 are all 1 for $k > t$. Hence $F_k = F_{t+1}$. Because V_∞ is finite (§7) and discrete, Theorem 6.5 yields for any $f(x)$ with $V_\infty f \geqq 0$ an $i \geqq t$ so large that $V_\infty f = V_i f \geqq 0$. Then $H_k f$ is again ultimately constant, and (2) gives the isomorphism as before.

THEOREM 14.2. *For an incommensurable inductive value V_k of $K(x)$ the field F_k determined from F_{k-1} and ϕ_k exactly as in* §12 *is the V_k residue-class field of both $K[x]$ and $K(x)$.*

Proof. Since no non-zero multiple of $\mu_k = V_k\phi_k$ lies in Γ_{k-1}, no two terms in a ϕ_k-expansion can have the same value in V_k. Hence any polynomial is equivalent to a monomial expansion, and every rational function has by Lemma 9.1 the form

$$f(x)/g(x) \backsim c(x)\phi_k^m \quad (\text{in } V_k), \qquad V_k c = V_{k-1}c.$$

If f/g has value 0, then $m=0$, $V_{k-1}c \geqq 0$, and $H_k(f/g)=H_k c$. But F_k is defined in §12, italics (or, for $k=1$, in §10) as all residue-classes $H_k h$ with $V_{k-1}h \geqq 0$. In this case every residue-class has this form, so that F_k is as asserted the whole residue-class field, either for $K[x]$ or for $K(x)$.

In particular, over the trivial value V_0(§2) of K the only non-trivial inductive values are

$$V_1 = [V_0, V_1 x = \mu_1], \qquad \mu_1 \neq 0;$$
$$V_2 = [V_0, V_1 x = 0, V_2\phi = \mu_2], \qquad \mu_2 > 0, \phi(x) \text{ irreducible}.$$

Both are incommensurable (no multiple of μ_2 lies in the group Γ_0, which contains only 0). Furthermore, the residue-class field of K for the trivial V_0 is K itself. Hence the residue-class field for V_1 is K and for V_2 is $K(\theta)$, where θ is a root of $\phi(x)$.

15. **Equality conditions for values.** An inductive value is essentially a representation; the same value of $K[x]$ could easily have several such representations. This section and the next one will formulate necessary and sufficient conditions for the equality of two inductive or limit-values. In this connection two values V and W of a ring S will be called equal if and only if

$$Va = Wa \qquad (\text{all } a \in S). \tag{1}$$

In this section we consider the case when each key polynomial ϕ_k exceeds the preceding ϕ_{k-1} in degree—a case which can often be made to apply by omitting any ϕ_{k-1} without the above property.

LEMMA 15.1. *If an inductive value*

$$V_k = [V_{k-2}, V_{k-1}\phi_{k-1} = \mu_{k-1}, V_k\phi_k = \mu_k] \qquad (k \geqq 2)$$

has two key polynomials $\phi_{k-1}(x)$ and $\phi_k(x)$ of the same degree, then

$$W = [V_{k-2}, W\phi_k = \mu_k]$$

is an inductive value equal to V_k.

We first prove W an inductive value. Since ϕ_k exceeds $\phi_k - \phi_{k-1}$ in degree, the constant-degree Lemma 6.3 shows that

$$V_{k-2}(\phi_k - \phi_{k-1}) = V_{k-1}(\phi_k - \phi_{k-1}) = \mu_{k-1}, \mu_k > \mu_{k-1} > V_{k-2}\phi_{k-1}. \tag{2}$$

A combination of these two results proves

$$\phi_k \backsim \phi_{k-1} \qquad (\text{in } V_{k-2}). \tag{3}$$

Thus ϕ_k and ϕ_{k-1} have the same equivalence-divisibility properties in V_{k-2}, and so ϕ_k, like ϕ_{k-1}, is a key polynomial over V_{k-2}. By (2) and (3) the value $\mu_k > V_{k-2}\phi_k$ assigned to ϕ_k is sufficiently large. Therefore W is inductive, for conditions (1) and (2) of Definition 6.1 hold trivially. The definition of augmented values applied to the usual expansion (e.g., §9, (2)) of any $f(x)$ in powers of ϕ_k gives

$$V_k f(x) = \min_i \left[V_{k-1} f_i(x) + i\mu_k\right], \qquad Wf(x) = \min_i \left[V_{k-2} f_i(x) + i\mu_k\right].$$

The corresponding terms $V_{k-1}f_i$ and $V_{k-2}f_i$ are equal by Theorem 5.1, for each $f_i(x)$ has a degree less than that of ϕ_k or of ϕ_{k-1}. Therefore $V_k = W$. Successive applications of this lemma give

THEOREM 15.2. *Any inductive value is equal to an inductive value in which* $\{\deg \phi_k\}$ *is a monotone increasing sequence. A similar representation holds for any limit-value not of constant-degree type.*

For values in this particular form we can obtain necessary and sufficient conditions for equality.

THEOREM 15.3. *If the two inductive values*

$$(4) \qquad V_s = \left[V_0,\ V_1 x = \mu_1,\ V_2\phi_2 = \mu_2, \cdots, V_s\phi_s = \mu_s\right]$$

$$(5) \qquad W_t = \left[W_0, W_1 x = \nu_1,\ W_2\psi_2 = \nu_2, \cdots, W_t\psi_t = \nu_t\right]$$

both have a monotone character such that

$$1 < \deg \phi_2 < \cdots < \deg \phi_s, \qquad 1 < \deg \psi_2 < \cdots < \deg \psi_t,$$

then $V_s = W_t$ *holds if and only if*

(i) $V_0 = W_0, \quad s = t$;

(ii) $\deg \phi_k(x) = \deg \psi_k(x) \qquad (k = 1, \cdots, t)$;

(iii) $V_{k-1}(\psi_k - \phi_k) \geqq \mu_k = \nu_k \qquad (k = 1, \cdots, t)$.

The theorem is still true if either s or t is $+\infty$.

First prove the sufficiency of these conditions. Since (i) and (iii) make V_1 and W_1 identical, we can proceed by induction, assuming that $V_{k-1} = W_{k-1}$ is already established. Now compute the V_k value of ψ_k. Because key polynomials have the leading coefficient unity, (ii) shows that the degree of ϕ_k exceeds that of $\psi_k - \phi_k$, so that ψ_k has the expansion

$$(6) \qquad \psi_k = \phi_k + (\psi_k - \phi_k)$$

in powers of ϕ_k. The definition of V_k and (iii) prove

$$V_k\psi_k = \min\left[\mu_k, V_{k-1}(\psi_k - \phi_k)\right] = \mu_k = \nu_k.$$

The V_k value of any polynomial $f(x)$ can now be estimated from the expansion of $f(x)$ in ψ_k, for the triangle axiom for V_k gives

$$V_k\left[\sum_{j=0}^{n} f_j\psi_k^j\right] \geqq \min_j \left[V_{k-1}f_j + j\cdot V_k\psi_k\right] = \min_j \left[V_{k-1}f_j + j\cdot \nu_k\right] = W_k f,$$

because of the definition of W_k. Thus $V_kf \geqq W_kf$, while the inverse inequality is similarly proved. Hence $V_k = W_k$, and the induction is complete.

The necessity of the conditions depends chiefly on the invariance of the values assigned the key polynomials (Theorem 6.4). The assumption $V_s = W_t$ shows that $V_0 = W_0$ and $\mu_1 = \nu_1$. Hence (ii) and (iii) hold for $k=1$. We prove them by induction on k. If they hold through $k-1$, then the sufficiency proof shows $V_{k-1} = W_{k-1}$. By Theorem 5.1, deg ϕ_k can be characterized as the smallest degree of any polynomial $a(x)$ with the property that $V_sa > V_{k-1}a$. Since V_s and W_t are equal, deg ψ_k can be characterized by the same statement, so that

$$\text{deg } \phi_k(x) = \text{deg } \psi_k(x). \tag{7}$$

The monotonic assumption on $\{\text{deg } \phi_k\}$ then shows $V_s\psi_k = V_k\psi_k$. Hence, because of the invariance in W of the value assigned to ψ_k,

$$\nu_k = W_t\psi_k = V_s\psi_k = V_k\psi_k = V_k[\phi_k + (\psi_k - \phi_k)].$$

As before, (6) is an expansion in powers of ϕ_k, so that this equation becomes

$$\nu_k = \min\ [\mu_k,\ V_{k-1}(\psi_k - \phi_k)].$$

Combining this with the symmetric conclusion (using $V_{k-1} = W_{k-1}$)

$$\mu_k = \min\ [\nu_k,\ V_{k-1}(\psi_k - \phi_k)],$$

we obtain (iii) for index k. With (7) this completes the induction. The condition $s=t$ results, even in the case $s=t=\infty$.

16. **Normal forms for values.** The results of the previous section do not apply to constant degree limit values, nor do they yield unique normal forms. Both these goals can be reached in the discrete case by using key polynomials from which all unnecessary high-valued terms have been dropped.

In the expansion of any $f(x)$ in a value V_k, the coefficient $f_i(x)$ of any power of ϕ_k can itself be expanded in powers of ϕ_{k-1}. Since the degree of $f_i(x)$ is limited, the highest power of ϕ_{k-1} occurring is less than n_k/n_{k-1}, where n_i has the meaning

$$n_i = \text{deg } \phi_i(x) \qquad (i = 1, \cdots, k). \tag{1}$$

By an inductive process of this sort one can prove

THEOREM 16.1. *In any V_k every polynomial $f(x)$ can be expanded as a polynomial in the key polynomials with constant coefficients,*

$$f(x) = \sum_j a_j \phi_1^{m_{1j}} \phi_2^{m_{2j}} \cdots \phi_k^{m_{kj}}, \qquad (a_j \in K), \tag{2}$$

where the exponents m_{ij} are limited as follows

$$m_{ij} < n_{i+1}/n_i \quad (\text{all } j;\ i = 1, 2, \cdots, k-1). \tag{3}$$

The value of $f(x)$, when computed from the definition, is

$$V_k f(x) = \min_j V_k(a_j \phi_1^{m_{1j}} \cdots \phi_k^{m_{kj}}). \tag{4}$$

For a p-adic value, every number is equivalent to one of the numbers $c \cdot p^m$, $c=0, 1, \cdots, p-1$. For any value V_0 of a field K we can similarly (axiom of choice) pick from each class of equivalent elements a single representative element; in particular, we can make 1 one of the representatives. Given fixed representatives of this sort for each V_0, we say that a polynomial $f(x)$ is *homogeneous* in a value V_k derived from V_0 if in the expansion (2) of $f(x)$ all the coefficients a_j are representatives in V_0 and all the terms have the same minimum value $V_k f(x)$.

LEMMA 16.2. *Every polynomial $f(x)$ is equivalent in V_k to one and only one homogeneous polynomial $h(x)$. This $h(x)$ is called the "homogeneous part" of $f(x)$.*

Proof. Given $f(x)$, we find $h(x)$ by altering coefficients and dropping out terms in the expansion (2) for f. Were $f(x)$ also equivalent to a homogeneous $g(x)$, then all terms in the expansions of both $h(x)$ and $g(x)$ would have the same value $V_k h$, while $h-g$ would have a larger value. Thus corresponding coefficients are equivalent and therefore equal.

An inductive or limit value $V_k = [V_0, V_i\phi_i = \mu_i]$ may be called *homogeneous* if every key polynomial $\phi_i(x)$ is homogeneous in V_{i-1} $(i=2, \cdots, k)$. We will prove

THEOREM 16.3. *Any inductive or limit-value constructed from a discrete value V_0 of K is equal to a homogeneous inductive or limit-value.*

We have to prove that, if $U = [V_k, U\phi = \mu]$ is an augmented value over a homogeneous value V_k, then U itself is equal to a homogeneous inductive value. This is done by introducing successive homogeneous parts of ϕ as new keys. First use $\psi_1(x)$, the homogeneous part of ϕ in V_k. By Lemma 16.2

$$\psi_1(x) \sim \phi(x) \quad (\text{in } V_k), \qquad \deg \psi_1(x) = \deg \phi(x). \tag{5}$$

It follows that $\psi_1(x)$ is a key polynomial over V_k. Setting

$$\nu_1 = V_k[\phi(x) - \psi_1(x)] > V_k \psi_1(x),$$

we can construct a homogeneous value $W_1=[V_k, W_1\psi_1=\nu_1]$. If $\mu\leqq\nu_1$, then the sufficiency proof of Theorem 15.3 shows $U=W_1$. Otherwise $\mu>\nu_1$, and Corollary 13.2 proves $W'=[W_1, W'\phi=\mu]$ an inductive value, which by Lemma 15.1 is equal to U. We repeat the above argument, constructing a W_2 from $\psi_2(x)$, the principal part of ϕ in W_1. This gives a sequence of homogeneous inductive values,

$$W_t = [V_k, W_1\psi_1 = \nu_1, W_2\psi_2 = \nu_2, \cdots, W_t\psi_t = \nu_t] \qquad (t = 1, 2, \cdots).$$

The degrees of the $\psi_i(x)$ are all identical by (5), so that Lemma 6.3 proves that

$$\nu_1 < \nu_2 < \nu_3 < \cdots \tag{6}$$

and that each ν_i is in the value-group Γ_k of V_k. By hypothesis and Theorem 6.7, this Γ_k is discrete. Hence there is a smallest t with $\nu_t\geqq\mu$ in (6), and U is equal to the homogeneous value W_t. The advantage of so representing every value in a homogeneous form lies in the following uniqueness theorem:

THEOREM 16.4. *Two homogeneous inductive or limit-values which are equal must be identical.*

If the equal values are V_s and W_t as in §15, (4) and (5), then the asserted identity means simply that

$$V_0 = W_0, \qquad s = t, \tag{7}$$

$$\phi_k = \psi_k, \qquad \mu_k = \nu_k \qquad (k = 1, 2, \cdots, s). \tag{8}$$

The hypotheses readily give $V_0=W_0$ and (8) for $k=1$. Suppose (8) true up to $k-1$ inclusive. Then $V_{k-1}=W_{k-1}$. We can assume $s>k-1$, whence also $t>k-1$. Then ϕ_k has the following invariant properties which refer only to $V_{k-1}=W_{k-1}$ and $V_s=W_t$: ϕ_k is totally homogeneous in V_{k-1}, it has the first coefficient 1 and it has the minimum degree consistent with the property $V_s\phi_k>V_{k-1}\phi_k$. Furthermore ψ_k has the same properties. But these properties uniquely determine ϕ_k, for, since the difference $\phi_k-\psi_k$ is of degree less than ϕ_k, its value is

$$V_{k-1}(\phi_k - \psi_k) = V_s(\phi_k - \psi_k) \geqq \min [V_s\phi_k, V_s\psi_k] > \min [V_{k-1}\phi_k, V_{k-1}\psi_k].$$

Hence by the triangle law $\phi_k \backsim \psi_k$ in V_{k-1}, so that by Lemma 16.2, $\phi_k=\psi_k$. By Theorem 6.4, $\mu_k=\nu_k$, as in (8). The induction ends when k reaches $s=t$, and the identity $V_s\equiv W_t$ is proved. A simple consequence is

COROLLARY 16.5. *If every value of K is discrete, then no inductive value can ever equal a limit-value, and no limit-value of constant degree type* (§7) *can equal a limit-value not of this type.*

Bibliography

Papers on irreducibility:

1. M. Bauer, *Verallgemeinerung eines Satzes von Schönemann*, Journal für die Mathematik, vol. 128 (1905), pp. 87–89.
2. H. Blumberg, *On the factorization of expressions of various types*, these Transactions, vol. 17 (1916), pp. 517–544.
3. G. Dumas, *Sur quelques cas d'irreductibilité des polynômes à coëfficients rationnels*, Journal de Mathématiques, (6), vol. 2 (1906), pp. 191–258.
4. G. Eisenstein, *Ueber die Irreduzibilität und einige andere Eigenschaften der Gleichungen*, etc., Journal für die Mathematik, vol. 39 (1850), p. 166.
5. L. Königsberger, *Ueber den Eisensteinschen Satz von der Irreduzibilität algebraischer Gleichungen*, Journal für die Mathematik, vol. 115 (1895), pp. 53–78; especially (67) on p. 69.
6. J. Kürschák, *Irreduzible Formen*, Journal für die Mathematik, vol. 152 (1923), pp. 180–191.
7. O. Ore, *Zur Theorie der Irreduzibilitätskriterien*, Mathematische Zeitschrift, vol. 18 (1923), pp. 278–288.
8. O. Ore, *Zur Theorie der Eisensteinschen Gleichungen*, Mathematische Zeitschrift, vol. 20 (1924), pp. 267–279.
9. O. Perron, *Neue Kriterien für die Irreduzibilität algebraischer Gleichungen*, Journal für die Mathematik, vol. 132 (1907), p. 304.
10. T. Rella, *Ordnungsbestimmungen in Integritätsbereichen und Newtonsche Polygone*, Journal für die Mathematik, vol. 158 (1927), pp. 33–48.
11. Th. Schönemann, *Von denjenigen Moduln, welche Potenzen von Primzahlen sind*, Journal für die Mathematik, vol. 32 (1846), pp. 93–105, §61.

Papers on absolute values:

12. H. Hasse and F. K. Schmidt, *Die Struktur diskret bewerteter Körper*, Journal für die Mathematik, vol. 170 (1933), pp. 4–63.
13. W. Krull, *Idealtheorie in unendlichen algebraischen Zahlkörpern*, II, Mathematische Zeitschrift, vol. 31 (1930), pp. 527–557.
14. W. Krull, *Allgemeine Bewertungstheorie*, Journal für die Mathematik, vol. 167 (1932), pp. 161–196.
15. J. Kürschák, *Limesbildung und allgemeine Körpertheorie*, Journal für die Mathematik, vol. 142 (1912), pp. 211–253.
16. S. MacLane, *Abstract absolute values which give new irreducibility criteria*, Proceedings of the National Academy of Sciences, vol. 21 (1935), pp. 272–274.
17. A. Ostrowski, *Ueber einige Lösungen der Funktionalgleichung* $\phi(x)\cdot\phi(y)=\phi(xy)$, Acta Mathematica, vol. 41 (1917), pp. 271–284.
18. A. Ostrowski, *Untersuchungen zur arithmetischen Theorie der Körper*, Mathematische Zeitschrift, vol. 39 (1934), pp. 269–404.

Harvard University,
Cambridge, Mass.

A combinatorial condition for planar graphs[1]).

By

Saunders Mac Lane (Cambridge, Mass.).

1. *Introduction.* Kuratowski[2]) has proven that a topological graph is planar, i. e., that it can be mapped in a 1—1 continuous manner on the plane, if and only if it contains no subgraph having either of two specific forms. Whitney[3]) has shown that a graph is planar if and only if it has a combinatorial "dual". This paper establishes another combinatorial condition that a graph be planar. This condition may be stated in terms of ordinary combinatorial concepts:

Theorem I. *A combinatorial graph is planar if and only if the graph contains a complete set of circuits such that no arc appears in more than two of these circuits.*

A graph in the plane divides the plane into a number of regions, and each region is bounded by one or more circuits. These boundary circuits, with certain omissions, can be readily shown to form a complete set, and obviously no arc can appear on more than two of these boundaries. Hence the necessity of our condition is immediate (cf. § 5). The sufficiency proof is largely combinatorial in character. It is advantageous to first reduce the problem to the case of the non-separable graphs (cf. § 3) considered by Whitney. By removing a suitable arc (or arcs) from any non-separable graph we obtain a simpler non-separable graph. We first embed this simpler graph

---

[1]) Presented to the American Math. Society, April 11, 1936.

[2]) C. Kuratowski, Fund. Math. Vol. XV (1930), pp. 271—283. He considers a more general point set than a graph.

[3]) H. Whitney, *Non-separable and planar graphs,* Trans. Amer. Math. Soc. 34 (1932), pp. 339—362. We refer to this paper as Whitney I.

Reprinted from Fundamenta Math. 28 (1937), 22-32.

in the plane, then show from the assumed condition that the remaining arc (or arcs) can be added in the plane (§§ 4, 5).

By strictly combinatorial means the criterion of Theorem I will be shown equivalent (Theorem II, § 6) to the existence of a dual. This is in turn known [1]) to be equivalent (combinatorially) to the condition of Kuratowski.

2. *Definitions.* A *combinatorial graph* G consists of a finite set of elements $a, b, c, \ldots,$ called "arcs", and a finite set of "vertices" $p, q, r, \ldots,$ such that each arc b "joins" exactly two vertices p and q. Then p and q are the *ends* of b or are *on* b, while b may be denoted by pq. We assume that each vertex is on at least one arc [2]). Any set of arcs in G, together with all the vertices on these arcs, form themselves a *subgraph* of G. Each subgraph is determined by its arcs. If $m > 1$ and if $p_1, p_2, \ldots, p_m$ denote distinct vertices, then a subgraph C with arcs $p_1p_2, p_2p_3, \ldots, p_{m-1}p_m, p_mp_1$ is a *circuit*, a subgraph D with arcs $p_1p_2, p_2p_3, \ldots, p_{m-1}p_m$ is a *chain* with *ends* p_1 and p_m, and the chain D is *suspended* if p_1 and p_m are the only vertices of D on three or more arcs of G. If A and B are subgraphs, then $A \cap B$ is the subgraph containing those arcs in both A and B, $A+B$ contains those arcs in either A or B, and $G-A$ contains all arcs of G not in A.

If a graph G has $E(G)$ edges, $V(G)$ vertices, and $P(G)$ connected pieces, then

$$R(G) = V(G) - P(G), \quad N(G) = E(G) - V(G) + P(G) \tag{1}$$

are respectively the *rank* and *nullity* of G. A *sum modulo 2* of circuits $C_1 + C_2 + \ldots + C_m$ is the subgraph containing all arcs present in an odd number of the C_i's. The circuits $C_1, C_2, \ldots, C_n$ form a *complete set* in G if every circuit in G can be expressed uniquely as a sum mod 2 of certain of the C_i's. Every G contains at least one complete set of $n = N(G)$ circuits.

A planar *topological graph* H consists of a finite number of arcs (1—1 bicontinuous images of line segments) in the plane intersecting, if at all, only at their endpoints. These arcs and their endpoints, considered as elements and possibly renamed, form a combinatorial graph H'. Any such combinatorial graph H' is called *planar*, and H is a *map* of H'.

3. *Non-separable graphs.* A graph G is *separable* if it has two subgraphs F_1 and F_2 such that $F_1 + F_2 = G$, while F_1 and F_2 have no arcs and at most one vertex in common. If F_1 and F_2 have no common vertex, they are not connected; if they have one common vertex p, p is called a *cut vertex* of G. In either event G may be *separated* into F_1 and F_2. F_1 is either non-separable, or can itself be separated into F_3 and F_4, and likewise for F_2. Repetition of this finally yields subgraphs $G_1, G_2, \ldots, G_m$ which are no longer separable. These *non-separable components* of G are always the same, no matter

[1]) H. Whitney, *Planar Graphs*, Fund. Math. XXI (1933), pp. 73—84 We refer to this paper as Whitney II.

[2]) This exclusion of "isolated" vertices obviously does not affect Theorem I.

how the separation is carried out [1]). On the other hand, two arcs a and b in G are *cyclically connected* if $a=b$ or if there is a circuit in G containing both a and b. It can be proven that the relation "a is cyclically connected to b" is symmetric and transitive and that it holds if and only if [2]) a and b belong to the same non-separable component of G. This result gives an "invariant" definition of non-separable components, and so again proves their uniqueness.

Theorem 3.1. *If G is non-separable and has nullity greater than 1, and if R is a circuit in G, then there is* [3]) *in R a chain A which is suspended in G and whose removal leaves a non-separable graph $G—A$ of nullity $N(G)—1$.*

The proof proceeds by building up G from a sequence of non-separable subgraphs $H_1 \subset H_2 \subset H_3 \subset \ldots \subset G$. As $N(G)>1$, there is an arc a_1 not in R. Pick a circuit containing a_1 and an arc of R, and call this circuit H_1. If $H_{m-1} \neq G$ has been chosen, H_m is constructed thus: First pick an arc a_m in $G—H_{m-1}$, such that a_m is not in R unless $G—H_{m-1}$ contains only arcs of R. Since G is non-separable, there is a circuit D containing a_m and an arc of H_{m-1}. Denote by A_m the piece of D containing a_m and extending in each direction from a_m to the first vertex of H_{m-1}. Then choose H_m as $H_m=H_{m-1}+A_m$.

Each subgraph H_m is non-separable, as we now show by induction. The circuit H_1 must be non-separable. If H_{m-1} is non-separable, then the arcs of A_m are cyclically connected to the rest of H_m by the circuit consisting of A_m and a chain B in H_{m-1} joining the ends of A_m, so that H_m is non-separable.

We next prove that $R \subset H_m$ implies $H_m=G$. By the construction of H_1, $R \subset H_1$ is impossible. Hence let $m>1$ be the smallest integer for which $R \subset H_m$. Then R is not contained in H_{m-1}, and there is an arc b of R not in H_{m-1}. Denote by E the piece of R which contains b and extends along R in each direction from b up to the first vertex of H_{m-1}. This chain E is in R, while $R \subset H_m$, so

---

[1]) Whitney I, Theorem 12.

[2]) A similar proof in Whitney I, Theorem 7. Cf. also Kuratowski et Whyburn, *Sur les éléments cycliques et leurs applications*, Fund. Math. XVI (1930), pp. 305—331.

[3]) This includes the special case of this Theorem, which was proven by Whitney (I, Theorem 18), and which does not require $A \subset R$. Theorem 3.1 can also be proven by an induction from Whitney's Theorem.

that $E \subset H_m = H_{m-1} + A_m$. By construction, E has no arcs on H_{m-1}, so $E \subset A_m$. A_m was chosen to have its ends and no other vertices in common with H_{m-1}. E has the same property. Thus E is a sub-chain of A_m with the same ends as A_m, so that E must equal A_m. Now if $G - H_{m-1}$ were not contained in R, A_m would contain the arc a_n not in R, so that A_m cannot equal E, which is in R. Because of this contradiction, $G - H_{m-1}$ must be contained in R. But H_m contains all of H_{m-1} by construction and all of R by assumption, and so contains all of G. Hence $R \subset H_m$ implies $H_m = G$ and $A_m \subset R$.

Each A_m is a suspended chain in H_m, so that its addition to H_{m-1} increases the number of arcs by 1 more than the number of vertices and so increases the nullity by 1 (cf. (1)). Thus $N(H_m) = m$. The construction process finally stops with an $H_n = G$, and this n must be $N(G)$. The last added chain $A = A_n$ is a suspended chain in $H_n = G$ and is contained in R (see the paragraph above), while $G - A$ has nullity $n - 1$, just as required in the Theorem.

Separable graphs can also be built up in a standard fashion:

Theorem 3.2. *If G is separable, then there is a non-separable component H of G such that H and $G - H$ have at most one vertex in common.*

Proof: Pick out any component H_1 of G. If H_1 does not have the desired property, there is a component H_2 in $G - H_1$ with a vertex p_1 in common with H_1. If H_2 does not have the desired property, there is a component $H_3 \neq H_2$ containing a vertex $p_2 \neq p_1$ in common with H_2. Were $H_3 = H_1$, then chains of H_2 and H_1 joining p_1 to p_2 would form a circuit contained in no one component, an impossibility. Hence H_1, H_2 and H_3 are distinct. If H_3 does not have the desired property, we find a new component H_4, etc. The graph is finite, so the process must end with a component H_m with but one vertex in common with $G - H_m$, as required.

4. *The Induction Process.* Consider a *combinatorial graph G which satisfies the condition of Theorem I*. That is, assume that G contains a complete set of circuits

$$(2) \qquad C_1, C_2, \ldots, C_n; \qquad n = N(G),$$

which contain no arc more than twice. We call such a set (2) a *2-fold complete set*. The C_i are independent, so that the sum

$$(3) \qquad R = C_1 + C_2 + \ldots + C_n \qquad (\text{mod } 2)$$

is not zero. We call R the *rim* of G. It has the following property:

Lemma 4.1. *For a non-separable G, the rim* (3) *is a circuit.*

Suppose instead that R is not a circuit. It is then a cycle (i. e., each vertex is on an even number of arcs of R) and so contains a proper subcircuit D. The representation of D in terms of the complete set (2) may be written, if the C_i's are suitably renumbered, as

$$D = C_1 + C_2 + \ldots + C_m \qquad \text{(mod 2)}$$

As $D \neq R$, n exceeds m. Hence neither of the graphs

$$F_1 = C_1 + C_2 + \ldots + C_m, \qquad F_2 = C_{m+1} + C_{m+2} + \ldots + C_n$$

is void. We shall show that G separates into F_1 and F_2. An arc b of $F_1 \cap F_2$ must be in one of the first m C's and in one of the last $(n-m)$ C's. Since (2) is a 2-fold set, b is in no more of the C's. Thus b is in just one summand of D and in two summands of R, mod 2, so that b is in D and not in R, although $D \subset R$. This contradiction shows that F_1 and F_2 have no arcs in common.

As G is non-separable, there is a circuit E containing an arc of F_1 and one of F_2. The representation of E in the complete set is

$$E = \sum{}' C_i + \sum{}'' C_j \qquad \text{(mod 2)},$$

where the first sum runs over some of the indices from 1 to m and the second sum over some of the remaining indices. Since E contains edges of both F_1 and F_2, neither sum is void. The first sum $E' = \sum{}' C_i$ (mod 2) is thus not equal to E. But E' is contained in F_1, so that none of its arcs can be contained in the circuits $C_{m+1}, \ldots, C_n$ of F_2. Therefore $E' \subset E$. The circuit E has a proper subcycle E', an impossibility. Hence R is necessarily a circuit.

In a non-separable planar graph there must be a region boundary C_i which abuts on the outside boundary R only along a single chain. The corresponding combinatorial result can be stated thus:

Lemma 4.2. *If G is non-separable and has nullity greater than 1, then there is a suspended chain A in G such that*

(i) *$G - A$ is non-separable of nullity $N(G) - 1$;*

(ii) *A is contained in R and in one and only one C_i, say in C_n,*

(iii) *$C_1, C_2, \ldots, C_{n-1}$ form a 2-fold complete set for $G - A$,*

(iv) *The ends p and q of A are on R', where*

(4) $$R' = C_1 + C_2 + \ldots + C_{n-1} \qquad \text{(mod 2)},$$

(v) *R' consists of two chains $R - A$ and $C_n - A$ joining p to q.*

Proof: Pick a suspended chain $A \subset R$ as in Theorem 3.1. Then $G-A$ is non-separable, as in (i). Since G is non-separable, each arc of A, and hence all of A, is contained in some circuit. Any circuit is a sum, mod 2, of the C's, so that A is contained in some C_i. By renumbering we can make $A \subset C_n$. Since A is in R and at most two C's, A can be in no other $C_i \neq C_n$, as asserted in (ii). The circuits $C_1, C_2, \ldots, C_{n-1}$ are thus in $G-A$, and they are independent in $G-A$ as in G. Since there are at most $n-1 = N(G-A)$ independent circuits in $G-A$, these $n-1$ circuits form a 2-fold complete set in $G-A$, as stated in (iii).

We next show $A = C_n \cap R$. For suppose b were an arc in $C_n \cap R$ but not in A. Then b is in $G-A$, hence is in a circuit D of $G-A$. D is representable, by (iii), as a sum mod 2 of some of $C_1, C_2, \ldots, C_{n-1}$. Yet b is in C_n and R, hence can be in no other C_i, and so is not in the representation of D, contrary to $b \in D$. No such edge b is possible, so that $A = C_n \cap R$.

By (4), $R' = R + C_n$ (mod 2), so that R' consists of all arcs in R or in C_n but not in both. Those in both are in $R \cap C_n = A$, so that R' must consist of the arcs of $R-A$ and of $C_n - A$. As R is a circuit, $R-A$ (and also $C_n - A$) is a chain joining the ends p and q of A. $R-A$ is in R', so that its ends p and q are in R'. This gives the last conclusions (iv) and (v) of the Lemma.

5. *The Sufficiency Proof.*

Theorem 5.1. *A non-separable graph G with a 2-fold complete set* (2) *can be mapped on the plane in such a way that each C_i becomes the boundary of one of the finite regions into which G divides the plane, while R becomes the boundary of the exterior region.*

The proof will be by induction on the nullity $N(G)$. To avoid irrelevant topology, we shall show more explicitly that G can be embedded in such a way that *each arc of G becomes a broken line segment, while R becomes an equilateral triangle.* In the first case, if $N(G)=1$, G is non-separable and so is simply a circuit (cf. Whitney I, Theorem 10); hence it can be mapped on the plane. If $N(G) > 1$, then Lemma 4.2 yields a suspended chain A such that $G-A$ is non-separable, has a 2-fold complete set, and has a smaller nullity. By the induction assumption, $G-A$ can be mapped on the interior and boundary of a triangle, with R' on the boundary. By Lemma 4.2, (iv), the ends p and q of A are already on the boundary R', and by (v) the remainder C_n—A of C_n is one of the arcs of this boundary joining p to q. Hence we can add a new broken line segment A^* outside the triangle and with ends p and q in such a way that A^* and C_n—A together form the boundary of the new finite region. Then by (v) the boundary of the new exterior region is $R = A^* + (R - A)$. The induction proof will be complete if we show that R can be made an equilateral triangle.

To do this, let G—A be mapped on the triangle with vertices r, s, and t in the figure. Consider first the case when the arc

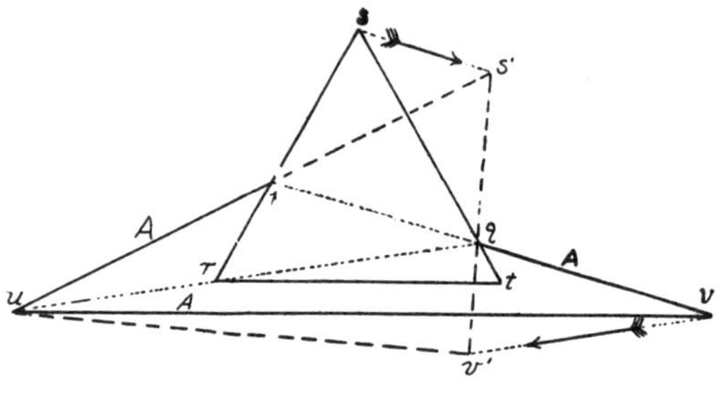

C_n—A on the edge of this triangle contains two vertices r and t of the triangle. Map A as the broken line $puvq$ in the figure. To make the rim $spuvqs$ a triangle, first shear[1]) the half plane which contains s and has the edge pq until the new position ps' of ps is a prolongation of up. Then shear the half plane with the edge uq until $s'qv$ is a straight line $s'qv'$. The new rim is now a triangle $us'v'$. It may be made isosceles by a shear with edge $v's'$, then equilateral by a compression toward this edge. As the shears used carry straight lines into broken lines, we do obtain a broken-line map of G on an equilateral triangle. The other cases, when C_n—A contains other vertices r, s, or t of the map G—A, may be similarly treated.

Theorem 5.2. *A graph G with a 2-fold complete set of circuits can be mapped on the plane in such a way that each arc becomes a broken line-segment.*

If G is separable, it can be reduced to non-separable components $H_1, H_2, \ldots, H_m$. Each component H_i which is not a single

[1]) A shear in a half-plane moves each point P parallel to the edge a distance which is a constant times the distance of P from that edge.

arc contains one or more of the given complete set of circuits (2), say the circuits $C_1, C_2, \ldots, C_k$. These circuits form a complete set for H_i: In the first place, they are independent (mod 2); secondly, any circuit D in H_i is expressible in terms of all the C's in the form $D=\sum' C_j+\sum'' C_j$ (mod 2), where the first sum runs over certain indices $j \leqq k$, the second sum over certain indices $j > k$. Then $D-\sum' C_j$ has arcs only in H_i and is equal to $\sum'' C_j$, which has arcs only in other components, and so must be void. $D=\sum' C_j$ (mod 2) is thus represented in terms of $C_1, C_2, \ldots, C_k$, and these circuits are a 2-fold complete set for H_i. Therefore H_i is planar, by Theorem 5.1. Furthermore it will suffice to prove the Theorem for a connected graph G.

The proof proceeds by induction on the number of components. Choose one component, say H_m, so that H_m and $G-H_m$ have but one vertex p in common (Theorem 3.2), and make the induction assumption that $G-H_m$, with components $H_1, H_2, \ldots, H_{m-1}$, has already been mapped on the plane as a graph F. If H_m has nullity zero, it is but a single arc and can be readily added to F in the plane. If $N(H_m)>0$, then the vertex p appears in at least one of the, circuits C_j of the remaining component H_m. If p is not already on the rim of H_m, we can make this the case by using a new 2-fold complete set for H_m like the old one except that one C_j containing p is replaced by the rim of H_m. Then p is on the new rim C_j. By Theorem 5.1, map H_m on the plane as the graph H_m^*, so that p becomes a point p_1^* on the outside boundary. But p also appears in the map of $G-H_m$ as point p_2^*; it remains to fit these maps together so that p_1^* and p_2^* will coalesce.

To do this, cut the plane of H_m^* into two halfplanes by a line through p_1^* and shear each half plane until the boundary of H_m^* makes at p_1^* an angle α smaller than the angle β between two of the adjacent (straight line) arcs which meet in the corresponding vertex p_2^* of the map of $G-H_m$. Then shrink H_m^* until it can fit inside the region which contains the angle β, place H_m^* in this region and let the points p_1^* and p_2^* coalesce, thus forming the desired cut vertex p^* in a planar map [1]) of $G=(G-H_m)+H_m$. All the transformations involved map the arcs of G into straight or broken line segments.

---

[1]) Another proof, using more topology, is indicated in Whitney I, Theorem 27.

This establishes the sufficiency of the criterion of Theorem I. It remains to verify its necessity. If each component of G has a 2-fold complete set of circuits, all these circuits together form a 2-fold complete set for G; hence it suffices to consider the non-separable case.

Theorem 5.3. *If G is a non-separable planar graph with $N(G) > 0$, then each finite region into which G divides the plane has as boundary a circuit of G, and these circuits together form a 2-fold complete set, while their sum, mod 2, is the boundary of the external region.*

For $N(G)=1$ this is the Jordan curve Theorem: in general, it follows by a simple induction on $N(G)$, using Theorem 3.1 and the fact that a cross-cut in the interior (exterior) of a Jordan curve cuts the interior (exterior) into two regions with suitable boundaries. This insures that each added arc is on the boundary of at most two regions.

6. *A Combinatorial Construction of Duals.* Whitney's condition for a planar graph, the existence of a combinatorial dual, must be equivalent to the existence of a 2-fold complete set of circuits. We shall establish this equivalence by combinatorial arguments, thus giving another proof for the criteria of Whitney and Kuratowski for planar graphs.

A graph G' is a *dual* of a graph G if there is a 1—1 correspondence between the arcs of G and those of G' such that, if H is any subgraph of G and H' the subgraph containing the corresponding arcs of G', then $R(G'-H') = R(G') - N(H)$; (cf. (1)). A set S of arcs in G is a *cut set* if $G-S$ has either more connected pieces or fewer vertices [1]) than G, and if neither of these results would hold were S replaced by a proper subset of S (Whitney II, p. 76). A graph has a dual if and only if each of its components has a dual (Whitney I, Theorems 23 and 25), and has a 2-fold complete set of circuits if and only if its components have such sets, by the arguments of the preceding section. Hence we restrict our attention to the non-separable case.

Theorem II. *A non-separable graph G has a combinatorial dual if and only if it has a 2-fold complete set of circuits.*

---

[1]) Practically, this second case means that $G-S$ is disconnected and has one piece which is an isolated vertex. We have excluded such vertices.

The trivial case when G is a single arc will be omitted. First, let G have the 2-fold complete set (2), with $n>0$, and denote the rim R in (3) by C_{n+1}. Since every arc is on at least one circuit, and on at most two circuits of the complete set, each arc is on exactly two of $C_1, \ldots, C_{n+1}$. Construct a new graph G' with vertices $p_1, \ldots, p_{n+1}$ corresponding to the circuits C_i and with arcs $b'_1, b'_2, \ldots, b'_t$ in 1—1 correspondence with the arcs of G, such that an arc b'_j has the ends p_i and p_k if the corresponding arc b_j in G is on the circuits C_i and C_k. We shall prove that this "circuit graph" G' is in fact a dual of G, by means of a combinatorial analog of the Jordan curve Theorem

Lemma 6.1. *If D is a circuit in G, then the corresponding subgraph D' in G' is a cut set in the circuit graph G'.*

Proof: D has a representation in terms of the complete set of C's. Renumber the C's so that the first k appear in this representation, and then use the definition (3) of C_{n+1} to obtain

$$D = C_1 + C_2 + \ldots + C_k = C_{k+1} + C_{k+2} + \ldots + C_{n+1} \pmod 2. \tag{5}$$

There is a corresponding subdivision of the vertices of G' into two sets $p_1, \ldots, p_k$ and $p_{k+1}, \ldots, p_{n+1}$. By the representation (5), D' consists of all the arcs of G' which have one end in the first set of vertices and the other end in the second. Thus in $G'-D'$ no vertex of the first set is connected to any vertex of the second. However, the vertices of the first set are all connected to each other in $G'-D'$. For were some of the vertices, say $p_1, p_2, \ldots, p_j$, connected by no arc of $G'-D'$ to the remaining vertices, then the sum $E = C_1 + + \ldots + C_j \pmod 2$ of the corresponding circuits of G would contain no arc also contained in $C_{j+1}, \ldots, C_k$ and so would contain only arcs in D. Thus E would be a proper subcycle of the circuit D, a contradiction. Similarly the vertices of the second set are all connected to each other in $G'-D'$. The addition of any single arc of D' connects the first set of vertices to the second, so that D' is a cut set, as asserted in the Lemma.

To show G' a dual of G it suffices [1] (Whitney II, Theorem 2) to show that a circuit in G always corresponds to a cut set of arcs in G', and conversely. The first half is the above Lemma. For the

[1] At this point we could alternatively parallel the argument of Whitney I, Theorem 29, that a planar graph has a dual, replacing the Jordan curve Theorem by the above Lemma.

converse, let S' be a cut set in G'. Since by the proof of the Lemma the cut set D' cuts G' into just two pieces, G' itself must be connected. Therefore S' cuts G' into two „pieces". Let the first contain the (re-numbered) vertices $p_1, \ldots, p_k$; the second, the vertices $p_{k+1}, \ldots, p_{n+1}$. Then the corresponding circuits in G give a cycle

$$D = C_1 + C_2 + \ldots + C_k = C_{k+1} + C_{k+2} + \ldots + C_{n+1} \qquad (\text{mod } 2)$$

Every arc on D belongs to one of the first k C's and to one of the remaining C's, and so corresponds to an arc in G' connecting the first set of vertices to the second set. This arc must belong to the cut set, so that $D \subset S$. But the cycle D must contain a subset D_1 which is a circuit. The corresponding set D_1' is, by Lemma 6.1, a cut set in G', although $D_1' \subset S'$. By definition a cut set has no proper subset which is a cut set, so that S must be identical with the circuit D_1, and cut sets do correspond to circuits. This shows G' to be a dual of G.

Conversely, let G have a dual G'. To find a 2-fold complete set of circuits, note first that the non-separability of G implies [1]) that of G'. By the definition of a dual, the nullity n of G is the same as the rank $V(G')-1$ of G' (cf. (1)), so that G' has $n+1$ vertices $p_1, p_2, \ldots, p_{n+1}$. The set D_i' of all arcs on any one vertex p_i is a cut set in G', for its removal deletes p_i. Consequently the corresponding set of arcs D_i in G must be a circuit, by the Theorem of Whitney quoted previously. These circuits $D_1, D_2, \ldots, D_{n+1}$ contain each arc of G exactly twice, so that their sum is zero (mod 2). No other relation mod 2 is possible. For suppose instead that

$$D_1 + D_2 + \ldots + D_m = 0 \quad (\text{mod } 2); \qquad m < n+1.$$

Then any arc of G on one of these D's is on another one, so that any arc of G' with a first end on one of $p_1, \ldots, p_m$ has a second end on another one of these vertices, and $p_1, \ldots, p_m$ are not connected to the remainder of G'. This contradicts the non-separability of G'. Thus $D_1, \ldots, D_n$ are independent (mod 2), are $n = N(G)$ in number, and so form a 2-fold complete set of circuits. Theorem II is established.

---

[1]) Whitney I, Theorem 26. The exclusion of isolated vertices is essential here.

PLANAR GRAPHS WHOSE HOMEOMORPHISMS CAN ALL BE EXTENDED FOR ANY MAPPING ON THE SPHERE.*

By V. W. Adkisson and Saunders MacLane.

Introduction. Certain conditions have been found under which a given homeomorphism of a given point set on a sphere can be extended to the sphere.[1]

It is the purpose of this paper to characterize the cyclicly connected planar graphs[2] G such that in every map of G on the sphere every homeomorphism of G into itself can be extended to the sphere.[3] The characterization will use only internal properties of the graph G, defined in G without reference to a map of G on any other space.

It is known that the extendability of a homeomorphism of a map of G depends on the behavior of the complementary domain boundaries of the map under the homeomorphism. Therefore the procedure for our problem is a study to determine which simple closed curves of a given graph can be boundaries of complementary domains of a map of a graph.

Definitions. The circuit[4] J of a cyclicly connected graph G is called a *bounding circuit* of G provided that for any two distinct maximal connected components, N_1 and N_2, of $G - J$ the sets $\bar{N}_1 \cdot J$ and $\bar{N}_2 \cdot J$ lie respectively on two distinct arcs AXB and AYB of J.[5]

* Presented to the Society, December 30, 1936. Received by the Editors February 6, 1937; revised April 8, 1937.

[1] H. M. Gehman, "On extending a homeomorphism between two subsets of spheres," *Bulletin of the American Mathematical Society*, vol. 42 (1936), pp. 79-81; V. W. Adkisson, "Cyclicly connected continuous curves whose complementary domain boundaries are homeomorphic," *Comptes Rendus des séances de la Société des Sciences et des Lettres de Varsovie*, vol. 23 (1930), Classe III, pp. 164-193. (Referred to in this paper as Thesis); V. W. Adkisson, "On extending a continuous (1 — 1) correspondence of continuous curves on a sphere," *ibid.*, vol. 27 (1934).

[2] For a discussion of cyclicly connected curves in general see G. T. Whyburn, "Cyclicly connected continuous curves," *Proceedings of the National Academy of Sciences*, vol. 13 (1927), pp. 31-38. A cyclicly connected graph is also non-separable. See H. Whitney, "Non-separable and planar graphs," *Transactions of the American Mathematical Society*, vol. 34 (1932), p. 339.

[3] This paper is the outgrowth of a problem originally suggested by Professor J. R. Kline.

[4] A circuit is a simple closed curve.

[5] "Bounding circuit" is equivalent to "*boundary curve*" as defined by S. Claytor,

Reprinted from the Amer. J. of Math. 59 (1937), 823-832 (with V.W. Adkinson), by permission of the Johns Hopkins University Press.

Note that each set $\bar{N}_i \cdot J$ consists of only a finite number of vertices of G. These vertices may be called "feet" of N_i.

A *split-circuit* of G is a bounding circuit J such that $G - J$ contains at least two components. The chief result of this paper may be stated in the following theorem.

THEOREM C. *Every homeomorphism of the cyclicly connected graph G into itself is extendable in every map of G on the sphere if, and only if, for each split circuit J and every homeomorphism σ of G such that $\sigma(G) = G$ and $\sigma(J) \neq J$, the sets $(G - J)$ and $J \cdot \sigma(J)$ each consists of two maximal connected components.*

The final theorem (Theorem D) characterizes more precisely the combinatorial structure of the graphs G satisfying the conditions of Theorem C. Specifically, such a graph either is a graph with just one map on the sphere or else is a graph with at most four essentially distinct homeomorphisms.

LEMMA 1. *Let the vertices p and q of a graph G lie on a circuit J of G, and in a map of G on a sphere S let R be one of the regions of S bounded by J. Then there is a complementary domain*[6] *boundary of G which contains both p and q and which lies in $\bar{R}$ if, and only if, there is no open arc of G in R which has ends on J and whose ends separate p and q on J.*[7]

THEOREM A. *If J is a circuit of G, then G can be mapped on a sphere so that J is a complementary domain boundary of G if, and only if, J is a bounding circuit.*[8]

This theorem can be proved by the following procedure. Take any map of G in which a certain component of $G - J$, say N, lies "outside" J.[9] One

---

"Topological immersion of peanian continua in a spherical surface," *Annals of Mathematics*, vol. 35 (1934), p. 809. A simple closed curve J of G is called a boundary curve of G provided that there do not exist in $G - J$ distinct components N_1 and N_2 such that (1) a point-pair of $\bar{N}_1 \cdot J$ separates a point pair of $\bar{N}_2 \cdot J$ on J, or (2) $\bar{N}_1 \cdot J = \bar{N}_2 \cdot J =$ three distinct points. Both definitions will be found useful in later proofs.

[6] A "complementary domain" will refer always to a domain complementary to some given map of G on a sphere. "Complementary domain boundary" will be abbreviated as "c. d. b."

[7] This lemma follows from a lemma by Kuratowski, "Sur les courbes gauches," *Fundamenta Mathematicae*, vol. 15 (1930), p. 274, Lemma III'; it can also be proved directly for graphs by more elementary combinatorial arguments.

[8] This is Proposition K of Claytor, *loc. cit.*, p. 828.

[9] It is sometimes convenient to refer to the two regions of a sphere bounded by a circuit J as "outside" and "inside" of J.

is able to show then, from the definition of a split-circuit, that the feet of N all lie on one c. d. b. of G "inside" and on J. Hence the given map can be altered by mapping N inside J instead of outside J. Proceeding in this manner we obtain a new map of G with all of G inside and on J.

COROLLARY 1. *If there is only one component of $G - J$ the circuit J is a c. d. b. of any map of G on a sphere.*

COROLLARY 2. *If J is a split circuit, G can be mapped so that J is not a c. d. b. of G.*

COROLLARY 3. *The c. d. b. J of a map of G will have a disconnected intersection with some other c. d. b. of the map of G if, and only if, J is a split circuit.*

Proof. The condition is sufficient. If J is a split circuit there are at least two components, N_1 and N_2, of $G - J$, while J is composed of two arcs AXB and AYB, such that $\bar{N}_1 \cdot J$ lies in AXB and $\bar{N}_2 \cdot J$ in AYB. Since G is cyclic, $\bar{N}_1 \cdot J$ is not a point, and A and B can be so chosen as to both belong to $\bar{N}_1 \cdot J$. Let R be the complementary domain of G bounded by J and set $R_1 = S - \bar{R}$. Now A and B both lie on a c. d. b. K of G with $K \subset \bar{R}_1$. For if not, by Lemma 1 there would exist a component N of $G - J$ such that two points of $\bar{N} \cdot J$ separate the points A and B on J and J would not be a split circuit. Since there is a component N_2 of $G - J$ such that $\bar{N}_2 \cdot J \subset AYB$, the arc AYB cannot be a subset of K, except possibly when $\bar{N}_2 \cdot J = A + B$. In this case a different choice of K readily gives the result. In like manner the arc AXB is not a subset of K. Therefore, since A and B are common to both K and J, $K \cdot J$ is disconnected.

The condition is necessary. Suppose G is mapped so that the c. d. b. J of G and the c. d. b. J_1 of G have a disconnected intersection F. Let F_1 and F_2 be two points in separate maximal connected subsets of this intersection. Then the removal of F_1 and F_2 disconnects G. For let the open arc $\langle F_1XF_2 \rangle$ be any arc lying in the domain complementary to G bounded by J and $\langle F_1YF_2 \rangle$ an arc lying in the domain complementary to G bounded by J_1. The simple closed curve $F_1XF_2YF_1$ has only the points F_1 and F_2 in common with G and disconnects G. Then F_1 and F_2 separate G into two components and neither component consists of a single arc, for otherwise F_1 and F_2 would belong to the same connected subset of F, contrary to assumption. Therefore there must exist two or more components of $G - J$, and since J is a c. d. b. of G, Theorem A shows J to be a bounding circuit. Then J is a split circuit.

THEOREM B. *In any mapping of G on a sphere S there are two complementary domain boundaries of G with a disconnected intersection if, and only if, G contains a split circuit.*

Proof. The condition is necessary from Corollary 3.

The condition is sufficient. Suppose G contains a split circuit J. Let N_1 and N_2 be two components of $G - J$ and divide J relative to N_1 and N_2 into two arcs AXB and AYB with A and B in $\bar{N}_1 \cdot J$, as in the proof of Corollary 3. Since J is a split circuit, every component M of $G - J$ satisfies either $\bar{M} \cdot J \subset AXB$ or $\bar{M} \cdot J \subset AYB$. Hence $G - (A + B)$ consists of two mutually separated sets; $H_1 = \langle AXB \rangle + N_1$ plus all components M, except N_1, with $\bar{M} \cdot J \subset AXB$, and $H_2 = \langle AYB \rangle + N_2$ plus all remaining components of $G - J$. Arcs of both sets H_1 and H_2 end on A, and so among the c. d. b.'s of G passing through A there will be at least one c. d. b. K containing arcs in both H_1 and H_2. Thus K must pass through A and B, the only points common to $\bar{H}_1$ and $\bar{H}_2$. Since A and B disconnect G into H_1 and H_2, neither one an arc, K also disconnects G, so that K is a split circuit. It is also a c. d. b. of G, so that by Corollary 3 it has a disconnected intersection with some other c. d. b., as asserted.

Proof of Theorem C. The conditions are necessary. First suppose $\sigma(J)$ does not contain any points of a component N_1 of $G - J$. Take a map of G in which J is a c. d. b. of G. As σ is extendable, $\sigma(J)$ must also be a c. d. b. of this map. We can map N_1 in the complementary domain of G bounded by J and leave $G - N_1$ fixed. But this gives a new map of G in which J is not a c. d. b. while $\sigma(J)$ remains a c. d. b. But then σ^{-1} cannot be extended to the sphere,[10] contrary to assumption. Hence $\sigma(J)$ contains points in every component of $G - J$.

Since there are at least two components N_1 and N_2 of $G - J$, $\sigma(J)$ to pass from N_1 to N_2 must contain points in J and also points not in J. Let $J \cdot \sigma(J)$ consist of k connected pieces, $g_1, g_2, \cdots, g_k$, each piece an arc or a point. Then $J - J \cdot \sigma(J)$ consists of k arcs $F_1, F_2, \cdots, F_k$, where the notation is chosen so that

$$J = F_1 + g_1 + F_2 + g_2 + \cdots + F_k + g_k \tag{1}$$

in the cyclic order shown. Because J and $\sigma(J)$ are both c. d. b.'s of G, it follows readily that the points of $g_1, \cdots, g_k$ are arranged in the same cyclic

[10] Thesis, Theorem 2. If M is a cyclicly connected continuous curve lying on a sphere S, and T a homeomorphism such that $T(M) = M$, a necessary and sufficient condition that T be extendable to S is that for every c. d. b. J of M, $T(J)$ be also a c. d. b.

order on $\sigma(J)$ as on J. Hence the arcs of $\sigma(J) - J \cdot \sigma(J)$ can be so denoted by $f_1, f_2 \cdots f_k$ that the cyclic order of arcs on $\sigma(J)$ is

$$\sigma(J) = f_1 + g_1 + f_2 + g_2 + \cdots + f_k + g_k. \tag{2}$$

The ends of f_i can then be so denoted by A_i and B_i that these points appear on J and on $\sigma(J)$ in the cyclic order $A_1B_1A_2B_2 \cdots A_kB_k$.

The circuits J and $\sigma(J)$ together divide the sphere into $k+2$ regions with the boundaries J, $\sigma(J)$, and $f_i + F_i$ $(i = 1, \cdots, k)$. As J and $\sigma(J)$ are c. d. b.'s of the map of G, the rest of G lies in the regions bounded by $f_i + F_i$. But each component of $G - J$ must contain some points of $\sigma(J)$ and hence some arc f_i. Therefore each closed region bounded by an $f_i + F_i$ contains exactly one component N_i of $G - J$, and there are k such components. Because σ carries J into $\sigma(J)$, $G - \sigma(J)$ must also have k components; namely, $N_i + F_i - (f_i + A_i + B_i)$.

Suppose now that there were three (or more) components of $G - J$. We can map N_1 in the complementary domain of G bounded by J, leaving the rest of G unaltered. This simply interchanges f_1 and F_1 in the boundaries J and $\sigma(J)$, so that there results a map of G with the c. d. b.,

$$K = f_1 + F_2 + \cdots + F_k + (J \cdot (\sigma J)), \quad L = F_1 + f_2 + \cdots + f_k + (J \cdot (\sigma J)).$$

These are the only c. d. b. containing $J \cdot \sigma(J)$, because every other c. d. b. is contained in one of $N_i + F_i$. Since σ must also be extendable in this map, it must carry K into K or L. But σ carries J into $\sigma(J)$, hence carries each F_i into some f_j, and carries K into a c. d. b. containing $\sigma(f_1)$, $k-1$ of the f_j, and $J \cdot (\sigma J)$. As this c. d. b. can be only L, and F_1 is in L, we have $\sigma(f_1) = F_1$. By the same argument, $\sigma(f_i) = F_i$ for every i.

But σ must leave fixed each end A_1 and B_1 of f_1, for otherwise $\sigma(A_1) = B_1$, $\sigma(B_1) = A_1$ which is impossible since $\sigma(A_i + B_i) = A_i + B_i$ $(i = 1, 2, \cdots, k)$ and $k \geqq 3$. This contradiction shows $\sigma A_1 = A_1$, $\sigma B_1 = B_1$.

Then also $\sigma f_1 = F_1$, $\sigma(F_1) = f_1$ and σ must carry $N_1 + F_1$ into itself. There is then another homeomorphism σ', equal to σ on N_1 and to the identity elsewhere. This σ' carries the c. d. b. J into $(J - F_1) + f_1$, which is not a c. d. b., so that σ' is not extendable in this map, contrary to hypothesis. The assumption of more than two components of $G - J$ is thus inconsistent. Therefore $G - J$ consists of two pieces N_1 and N_2, and we have already shown that $J \cdot \sigma(J)$ consists of two connected pieces, namely two arcs (or points) B_1A_2 and B_2A_1. Therefore the conditions of the theorem are necessary.

The conditions are sufficient. We need only show for each bounding circuit J that in any map in which J is a c. d. b. of G, $\sigma(J)$ must be a c. d. b. of G. (Thesis, Theorem 2). Consider a map in which J is a c. d. b. of G. If J is not a split circuit then $\sigma(J)$ is not a split circuit and must be a c. d. b. of G. (Corollary 1, Theorem A). If $\sigma(J) = J$ the case is trivial. Hence, assume $\sigma(J) \neq J$. Denote the two components of $J \cdot \sigma(J)$ by α and β. Then J consists of four parts F_1, α, F_2, β in that cyclic order, where F_1 and F_2 are arcs, α and β may be arcs or points. The circuit $\sigma(J)$ consists of four arcs f_1, α, f_2, β in that cyclic order, where f_i and F_i have end points in common. Hence $J + \sigma(J)$ divides the sphere into four regions, r_1, r_2, r_3, r_4, with the respective boundaries: J, $\sigma(J)$, $f_1 + F_1$, and $f_2 + F_2$.

Now the arcs F_1 and F_2 of J belong to distinct components of $G - \sigma(J)$. For otherwise there would be an arc f of $G - \sigma(J)$ joining F_1 to F_2, and it readily follows that part of this arc f must lie in the region r_1, contrary to the fact that r_1 is a complementary domain of G.

Suppose now that $\sigma(J)$ is not a c. d. b. of G. Then there are arcs of G in r_2 bounded by $\sigma(J)$, so that at least one component of $G - \sigma(J)$ is contained in this region. But there are at least two other components of $G - \sigma(J)$ (those containing F_1 and F_2) not in r_2. Hence there are at least three components of $G - \sigma(J)$. But this is impossible since there are only two components of $G - J$ by hypothesis, and σ carries components of $G - J$ into components of $G - \sigma(J)$. Therefore, $\sigma(J)$ is a c. d. b. of G and the theorem is proved.

We shall now obtain other necessary and sufficient conditions that every homeomorphism of G be extendable in every map. These new conditions will give in some detail a more complete characterization of the graphs G.

LEMMA 2. *If every homeomorphism of G into itself is extendable in every map on the sphere, and if for some homeomorphism σ, some split circuit J and some component N_1 of $G - J$, $\sigma(J) = J$ and $\sigma(N_1) = N_1$, then σ has two fixed points on J.*

Proof. Because J is a split circuit we can choose an arc $F = AYB$ of J such that F contains only A and B from $\bar{N}_1 \cdot J$, while F contains $\bar{N}_2 \cdot J$ for some component N_2 of $G - J$. Map G with N_1 inside J and all other components of $G - J$ outside J. There is then inside J a domain complementary to G whose boundary K contains F. Furthermore $K = F + f$, where f is an open arc of N_1.

Suppose first that $\sigma(K) = K$. Then $\sigma(J) = J$ implies that $\sigma(f) = f$, $\sigma(F) = F$ and hence that $\sigma(A + B) = A + B$. If A and B remain fixed we have the two desired fixed points. If A and B are interchanged by σ, there is a fixed point on[11] $F = AYB$ and also a fixed point on $J - F$.

Now assume $\sigma(K) \neq K$. The component N_2 is separated from $G - N_2$ by F and hence by $K = F + f$. Therefore K is a split circuit and N_2 a component of $G - K$. Theorem C applied to K shows that $\sigma(K)$ contains an arc of N_2. But this is impossible since $\sigma(K)$ consists of an arc of $\sigma(J) = J$ and an arc of $\sigma N_1 = N_1$, neither in N_2. Therefore $\sigma(K) = K$ and the lemma is proved.

Two homeomorphisms σ and τ are considered identical if $\sigma(G) = G$, $\tau(G) = G$ such that branch points that correspond under σ also correspond under τ, and edges that correspond under σ correspond under τ. (An edge of G means an arc of G joining two branch points, but containing no other branch points.)

THEOREM D. *Every homeomorphism of a cyclicly connected graph G into itself is extendable in every map of G on the sphere if, and only if, at least one of the following conditions holds:*

I. *Every split circuit of G is invariant under every homeomorphism σ. If there is a split circuit, then there is at most one $\sigma \neq 1$.*

II. *No split circuit is invariant under any $\sigma \neq 1$; there is only one homeomorphism $\sigma \neq 1$ and this σ is of order 2. For any split circuit J, $G - J$ consists of two connected pieces N_1 and N_2; $J \cdot \sigma(J)$ consists of just two connected pieces M_1 and M_2, and $\sigma(N_1) =$ a subset of N_1 plus an arc of J, $\sigma(N_2) =$ a subset of N_2 plus an arc of J, $\sigma(M_1) = M_2$, $\sigma(M_2) = M_1$.*

III. *There are in G only four split circuits, J_1, J_2, J_3, J_4 and only two topologically distinct maps of G on the sphere are possible. The complementary domain boundaries of each map include exactly two split circuits: J_1 and J_2 in one map, and J_3, J_4 in the other map. Any σ is of order 2 and either carries every split circuit into itself or else interchanges both J_1 and J_2, and J_3 and J_4. There are at most four distinct σ's, including the identity.*

Proof. Suppose every homeomorphism of G extendable.

Case 1. $\sigma(J) = J$ for every homeomorphism σ and every split circuit J. If some σ leaves some split circuit J pointwise fixed then σ is the identity.

[11] W. L. Ayres, "Concerning continuous curves and correspondences," *Annals of Mathematics*, vol. 28 (1927), p. 402.

For, map G so that J is a c. d. b. of G, and extend σ to be a periodic homeomorphism σ' of the sphere.[12] Since this σ' has a circuit of fixed points Eilenberg's results [13] concerning periodic transformations of a sphere show that the extension σ' is either the identity or is homeomorphic to a reflection in J In the latter case σ' would take $G - J$, on one side of J, into something or the other side of J; but there is nothing on the other side of J since J is a c. d. b. of G. Hence σ must be the identity, as asserted.

Consider now any split circuit J and any component N_1 of $G - J$. As in the proof of Lemma 2 choose a split circuit $K = f + F$, where $f \subset N_1$ and $F \subset J$. For any σ, the hypothesis of Case 1 shows that $\sigma(K) = K$, $\sigma(J) = J$ and hence that $\sigma(F) = F$. Let A and B be the ends of the arc F. If A and B are each invariant under σ, then every edge and branch point of J is fixed under σ, and σ is the identity. The only alternative is for σ to interchange A and B.

If there were two homeomorphisms σ and τ, neither the identity, then both σ and τ must interchange A and B. Since $\sigma \cdot \tau^{-1}$ leaves A and B fixed, it therefore will be the identity, so that $\sigma = \tau$. Hence there is, as asserted in I, only one homeomorphism not the identity.

Case 2. For some split circuit J, and some σ, $\sigma(J) \neq J$. By Theorem C there are then only two components N_1 and N_2 of $G - J$. Introduce the subarcs F_1 and F_2 of J and the subarcs f_1 and f_2 of $\sigma(J)$ as in (1) and (2) in the proof of Theorem C, and consider the four split circuits,

$$J,\ \sigma(J),\ F_1 + f_2 + J \cdot \sigma(J),\ f_1 + F_2 + J \cdot \sigma(J), \tag{3}$$

composed of J, f_1 and f_2. We shall show first that if G contains a split circuit K, other than one of (3), then K must lie entirely in one of the sets $H_i = N_i + F_i$ $(i = 1, 2)$.

Suppose G contains a split circuit K that is not one of (3) and has points in both H_1 and H_2. Let g_i be the arc of K that lies in H_i and A and B the ends of F_1. Then the circuits K and L $(= K - g_2 + F_2)$ each pass through A and B. Now one of the arcs, say g_1, must be different from F_1 and from f_1, and L is a split circuit; for any two components of $G - L$ with feet on the arc $L - F_2$ must have feet on distinct arcs (except for end points) of L since

[12] Since G is a graph (consisting of a finite number of vertices and edges) any transformation σ such that $\sigma(G) = G$ is necessarily periodic with a finite period k. Hence, if σ is extendable to the sphere, a transformation σ' of the sphere into itself can be so defined that $\sigma' = \sigma$ for points of G and that σ' is periodic of period k.

[13] S. Eilenberg, "Sur les transformations periodique de la surface de sphere," *Fundamenta Mathematicae* vol. 22 (1934), pp. 28-41.

they are on distinct arcs of the split circuit K, and there is only one component N_2 of $G - L$ with feet on F_2. If G is mapped so that J and $\sigma(J)$ are both c. d. b.'s of the map of G, the region R inside $f_1 + F_1$ (i. e. the region not containing F_2) is cut by the arc g_1. Since $\sigma(J) \neq J$ it follows easily that $\sigma(L) \neq L$, and from Theorem C, $G - L$ has only two components. But since the region R is cut by the arc g_1 of L there must be two components of $G - L$ in $\bar{R}$ and one component N_2 in H_2, making three in all. Therefore, no split circuit except those in (3) has arcs in both H_1 and H_2.

Suppose $\sigma(H_i) = H_i$. Then $\sigma(J) \neq J$, $\sigma(f_i) = F_i$ and $\sigma(F_i) = f_i$. Furthermore, if the end points of F_i remain invariant under σ it is possible to define a homeomorphism τ equal to σ on H_1 and to the identity on $G - H_1$. Then τ is not extendable (Theorem C) contrary to assumption. Hence the end points of F_i are interchanged by σ. Suppose now that H_1 contains a split circuit K such that $\sigma(K) = K$. Then K is in R_1, the "inside" of $f_1 + F_1$. Let R_2 be the region "outside" $f_1 + F_1$. Extend σ to be a periodic transformation σ' of the sphere. Since H_2 lies in R_2, $\sigma'(R_2) = R_2$. The periodic transformation σ' of R_2 must be homeomorphic to a rotation or to reflection [14] of $\bar{R}_2$. The latter is impossible as σ' has no fixed points on the boundary of R_2. The "rotation" σ' thus has a fixed point in R_2, and by Lemma 2 σ' has two fixed points in R_1. But $f_1 + F_1$ is fixed under σ' so that σ' must be homeomorphic to a rotation of the sphere, although a rotation of the sphere cannot have a fixed circuit and three fixed points not on this circuit (Eilenberg). Hence $\sigma(K) \neq K$ so that G contains no split circuit invariant under σ. Therefore, if there is no homeomorphism of G sending H_1 into H_2, σ is the only homeomorphism not the identity. This follows as in the last paragraph of Case 1. Then $\sigma^2 = 1$,[15] and we have condition II.

The only possibility remaining is that some σ has $\sigma(J) \neq J$ with $\sigma(H_1) = H_2$ and $\sigma(H_2) = H_1$. Suppose a split circuit K lies in H_1. Then $\sigma(K) \neq K$ and there are only two components of $G - K$. One of these components must contain H_2, and the other must lie in H_1. Now $\sigma(K)$ must contain an arc in each of these components, and hence must have an arc in H_1. But this is impossible since $\sigma(K)$ must lie in H_2. Therefore, G contains no split circuits other than (3).

Since c. d. b. must go into c. d. b. each of the following six σ's is possible. $\sigma_1(F_i) = f_i$, $\sigma_1(f_i) = F_i$ (as in Case II, interchanging A_i and B_i, where A_i and B_i are end points of F_i); $\sigma_2(F_1) = F_2$, $\sigma_2(F_2) = F_1$, $\sigma_2(f_1) = f_2$, $\sigma_2(f_2) = f_1$ (where A's go into B's and B's into A's); σ'_2 the same as σ_2, but

[14] Eilenberg, *loc. cit.*

[15] Since σ^2 leaves J pointwise fixed σ^2 is the identity.

A's go into A's and B's into B's; $\sigma_3(F_1) = f_2$, $\sigma_3(F_2) = f_1$, $\sigma_3(f_1) = F_2$, $\sigma_3(f_2) = F_1$ (A's into B's and B's into A's); σ'_3 same as σ_3 but A's go into A's and B's into B's; $\sigma_4(F_i) = F_i$, $\sigma_4(f_i) = f_i$ interchanging A_i and B_i. Now the following pairs cannot occur for the same mapping of G: (σ_1, σ_4), (σ_2, σ_3), (σ'_2, σ'_3). For suppose σ_1 and σ_4 both exist for a given G. Then $\sigma_4\sigma_1(F_i) = f_i$, $\sigma_4\sigma_1(f_i) = F_i$ leaving A_i and B_i fixed. But we have shown above (Case II) that there is no homeomorphism (other than the identity) leaving the end points of F_i fixed and carrying H_i into H_i. A similar contradiction may be obtained for the other two pairs of homeomorphisms. Therefore, there are at most three distinct homeomorphisms (excluding the identity) for any particular mapping of G.

The following relationships hold for the various σ's: $\sigma_2\sigma_1 = \sigma'_3$, $\sigma'_2\sigma_1 = \sigma_3$, $\sigma_2\sigma_4 = \sigma'_2$, $\sigma_3\sigma_4 = \sigma'_3$.

Furthermore, since each of the above σ's is such that σ^2 leaves at least one circuit of fixed points, it follows that $\sigma^2 = 1$ in each case (Eilenberg).

This establishes Case III, and the necessity of the given conditions.

The sufficiency of any one of the above conditions follows from Theorem C.

A further characterization of G when condition I is satisfied may be obtained by using Lemma 2. Let J be a split circuit with the invariant points X and Y. Then J can be expressed as the sum of two arcs, AXB and AYB, so that the feet of any component of $G - J$ lie entirely on AXB or entirely on AYB. Let N_1 be a component of $G - J$ with feet on AXB. Let $AXB = ApXqB$. Then for any point of $\bar{N}_1 \cdot J$ lying on $(ApX - X)$ there is a corresponding point of $\bar{N}_1 \cdot J$ lying on $(XqB - X)$.

In Case II, σ can be shown equivalent to a rotation of the sphere S through $180°$ about two invariant points, one lying in the region of S bounded by $f_1 + F_1$ that contains H_1, the other in the region of S bounded by $f_2 + F_2$ that contains H_2.

COROLLARY. *If G satisfies the conditions of Theorem D, then G either has no split circuits, or has four split circuits and at most four distinct homeomorphisms, or has at most two homeomorphisms.*

UNIVERSITY OF ARKANSAS,
AND
CORNELL UNIVERSITY.

Reprinted from DUKE MATHEMATICAL JOURNAL
Vol. 4, No. 3, September, 1938

A LATTICE FORMULATION FOR TRANSCENDENCE DEGREES AND p-BASES

BY SAUNDERS MAC LANE

1. **Introduction.** The transcendence degree of an extension of a field is the cardinal number of a maximal set of independent transcendents in the extension (Steinitz [11]);[1] in an Abelian group without elements of finite order the *rank* is the cardinal number of a maximal set of rationally independent group elements (Baer [1]). Both these cardinal numbers are invariants; the proofs of these two facts are similar, so that there should be an underlying theorem generalizing these proofs and stated in terms of the lattice of subfields or of subgroups, as the case may be.[2] This paper constructs, in §2, a type of lattice, called an "exchange" lattice, in which such a theorem can be proved (cf. §3). This lattice theorem includes also some investigations of Teichmüller ([12] and [13]) on fields of characteristic p; in particular, we establish the invariance of the cardinal number of a "relative p-basis" of an inseparable algebraic extension of such a field.

The crucial axiom for our lattices is an "exchange" axiom, related to the Steinitz exchange theorem. This axiom is equivalent to one of the axioms recently used by Menger [7] in investigating the algebra of affine geometry, and also to a certain covering property used by Birkhoff [2] in an analysis of the Jordan Theorem (cf. §4). This axiom can be viewed as a weakened form of the Dedekind or modular axiom for a lattice (Ore [9], or Birkhoff [4]). The Dedekind axiom itself could not apply to the lattices of fields with which we are concerned (see §5 and §6).

Unfortunately the exchange axiom is stated in terms of the "points" of the lattice, or alternately in terms of the covering relation. It thus applies only trivially to continuous geometries or to other infinite lattices having no points. In §7 we succeed in constructing two exchange axioms which in the presence of the other axioms are equivalent to the original exchange axiom, but which themselves do not involve points or coverings. These new axioms yield most of the usual properties of the dimension function in a finite lattice. Their title to be considered as substitutes for the modular law rests chiefly on their versatility: each of the new exchange axioms is seen in §8 to be equivalent to a natural assertion about the possibility of specified types of transpositions in any given chain of the lattice.

Received March 1, 1938; presented to the American Mathematical Society, April 9, 1938.

1 Numbers in brackets refer to the bibliography at the end of the paper.

2 This remark is due to R. Baer (in conversation).

2. **Exchange lattices**. The transcendence degree of a field $\mathfrak{L}$ over a subfield $\mathfrak{K}$ can be defined in terms of those subfields $\mathfrak{M}, \mathfrak{K} \subset \mathfrak{M} \subset \mathfrak{L}$, which are relatively algebraically closed in $\mathfrak{L}$, in the sense that every element of $\mathfrak{L}$ algebraic over $\mathfrak{M}$ is contained in $\mathfrak{M}$. With respect to field inclusion, these subfields $\mathfrak{M}$ form a continuous lattice.

A set L is a *continuous lattice*[3] (in the terminology of O. Ore, a *complete structure*) if a transitive and irreflexive relation $a < b$ is so defined for elements a and b of L that for every subset $A \subset L$ there exist in L an element $\Sigma(A)$, the *union*, and an element $\Pi(A)$, the *cross-cut*, such that $c \geqq \Sigma(A)$ holds if and only if $c \geqq a$ for every a in A, while $d \leqq \Pi(A)$ if and only if $d \leqq a$ for every a in A. If A consists of two elements a and b, we denote the union or join by $a + b$, and the cross-cut or meet by $a \cdot b$. The cross-cut $0 = \Pi(L)$ is the *zero* element, the union $1 = \Sigma(L)$ the *unit* element of L. If $a > b$ in L, but $a > c > b$ is impossible for c in L, then a is said to *cover* b or to be prime over b (Ore [9]). An element p prime over 0 is called a *point* of L (Menger [7]).

In the field case, each indeterminate x over $\mathfrak{K}$ generates a relatively algebraically closed field $\mathfrak{K}(x)'$ which is a point in the lattice of fields $\mathfrak{M}$. The crucial property of algebraic dependence is: if x is algebraic over $\mathfrak{M}(y)$, but not over $\mathfrak{M}$, then y is algebraic over $\mathfrak{M}(x)$. For lattices, we state a corresponding *exchange axiom*:

(E_1) If a is in L while p and q are points of L, then $a < a + p \leqq a + q$ implies $q \leqq a + p$.

We need also the existence of points and the "finiteness" of dependence:

(G_1) If $b < a$ in L, then there is in L a point p with $b < b + p \leqq a$.

(F_1) If Q is a set of points and p a point of L with $p \leqq \Sigma(Q)$, then there exists a finite set of points $q_1, q_2, \cdots, q_n$ of Q with $p \leqq q_1 + q_2 + \cdots + q_n$.

DEFINITION. *An exchange lattice is a continuous lattice satisfying* (E_1), (G_1), *and* (F_1).

Note that (E_1) is equivalent to the assertion that $a < a + p < a + q$ is impossible, while (G_1) is equivalent to the following formally stronger axiom:

(G_2) If $a \nleqq b$, then there is a point p with $p \leqq a$, $p \nleqq b$.

For $a \nleqq b$ implies $ab < a$, hence by (G_1) there is a p such that $p \leqq a$, $p \nleqq ab$. Thus $p \nleqq b$. In the presence of (G_1), (F_1) can be shown equivalent to the following point-free statement:

(F_2) If $b \leqq \Sigma(A)$, $A \subset L$, then there are in A elements $a_1, \cdots, a_n$ with $b(a_1 + \cdots + a_n) > 0$.

To prove (F_2), let Q be the set of all points $q \leqq$ some a of A, and choose any $p \leqq b$. Then $p \leqq \Sigma(Q) = \Sigma(A)$ by (G_2), whence $p \leqq q_1 + \cdots + q_n$ by (F_1). This gives the conclusion.

Our postulates are related to the properties of algebraic dependence as formulated by van der Waerden [14]. He considers a relation "b depends on S" for an element b and a subset S of a given set D, with the following properties:

[3] This definition is in von Neumann [8]. For further literature on lattices, cf. Ore [10] or Köthe [5].

1. each b depends on the set $\{b\}$;
2. if b depends on S and $S \subset T$, then b depends on T;
3. if b depends on S, then b depends on some finite subset of S;
4. if b depends on $S_0 = \{c_1, c_2, \cdots, c_n\}$ but on no proper subset of S_0, then c_n depends on $\{c_1, \cdots, c_{n-1}, b\}$;
5. if b depends on S and if every element of S depends on T, then b depends on T.

A relation "a depends on S" with these five properties we call for the moment a *dependence relation.* The exact connection with our axioms we state without proof as follows:

THEOREM 1. *If D is a set with a dependence relation, and if a subset $S \subset D$ is said to be closed whenever S contains every b of D dependent on S, then the set $\mathfrak{L}(D)$ of all closed subsets of D is an exchange lattice if $S < T$ means that S is a proper subset of T. Conversely, if L is any exchange lattice, and if $D = \mathfrak{D}(L)$ is the set of all points of L, then the relation $p \leqq \Sigma(S)$ for $S \subset D$ is a dependence relation of p to S. Furthermore, $\mathfrak{L}(\mathfrak{D}(L))$ is isomorphic to the given lattice L, while, for a given D, the set $D' = \mathfrak{D}(\mathfrak{L}(D))$ with its dependence relation is isomorphic to the set D^* obtained from D by identifying all pairs of mutually dependent elements of D. Here an isomorphism of D' to D^* means a one-to-one correspondence of D' to D^* which leaves unchanged the dependence relation.*

In the finite case, there is a similar connection to the matroids of Whitney [15], as expressed in terms of lattices by Birkhoff [3].

3. **The basis theorem.** A transcendence basis for a field is a maximal set of algebraically independent elements. Generally, a set P of points in any lattice L is *independent* if $(\Sigma(P'))\cdot(\Sigma(P'')) = 0$ for any two disjoint subsets P' and P'' of P (von Neumann [8], Chapter II). An independent set P of points with union $\Sigma(P) = a$ is called a *basis* of the element a of L. In the transcendence basis case, an independent set corresponds exactly to a set of indeterminates irreducible in the sense that no one indeterminate depends algebraically on the others. The same definition of independence holds in other cases:

THEOREM 2. *A set P of points in an exchange lattice is independent if and only if $p \leqq p_1 + p_2 + \cdots + p_n$ is impossible for distinct points $p, p_1, \cdots, p_n$ of P. Hence a set P is independent if and only if every finite subset of P is independent.*

Proof. The necessity of the condition is immediate. Suppose conversely that $p \leqq \Sigma p_i$ is impossible, but that P is dependent. There then are disjoint subsets P' and $P'' \subset P$ with $(\Sigma(P'))\cdot(\Sigma(P'')) > 0$. There then exists a point q such that $0 < q \leqq (\Sigma(P'))\cdot(\Sigma(P''))$. By the finiteness axiom, points p_i' in P' and p_j'' in P'' can be chosen so that

$$q \leqq p_1' + p_2' + \cdots + p_m', \qquad q \leqq p_1'' + p_2'' + \cdots + p_n''.$$

If $m > 0$ is so small that $q \nleqq p_1' + \cdots + p_{m-1}'$, the exchange axiom implies that

$$p_m' \leqq p_1' + \cdots + p_{m-1}' + q \leqq p_1' + \cdots + p_{m-1}' + p_1'' + \cdots + p_n'',$$

contrary to the assumed property of P. Hence P is independent. A similar use of the exchange axiom gives also the result (cf. Menger [7], p. 462):

COROLLARY. *If P is an independent set in the exchange lattice L, and if q is a point with $q(\Sigma(P)) = 0$, then the set obtained by adjoining q to P is independent.*

THEOREM 3. *If L is an exchange lattice and if P is any independent set of points of L with $\Sigma(P) \leqq b$, then there is a set $Q \supset P$ which is a basis of b. In particular, every element of L has a basis.*

Proof. When $b = 1$, we can construct Q from P by well ordering the points of L and by applying repeatedly Theorem 2 and its corollary. A basis for any $b \neq 1$ can then be found by applying the previous case to the quotient lattice $b/0$ (Ore [9], p. 425). For any $b \geqq c$ such a *quotient lattice* b/c consists of all elements d with $b \geqq d \geqq c$.

THEOREM 4. *If $b \geqq c$ are elements of the exchange lattice L, then the quotient lattice b/c is also an exchange lattice. Its points are all the elements of L of the form $c + p$, where p is a point of L such that $c < c + p \leqq b$.*

Proof. Any such $c + p$ is a point of b/c, for otherwise $c + p > d > c$, hence by (G_1) $c + p > c + q > c$, and this is impossible by the exchange axiom. Conversely, the existence axiom for points of L shows that any point of b/c must have the form $c + p$. The axioms for an exchange lattice can now be directly verified for b/c.

The comparison of different bases depends upon an exchange process (Steinitz [11], Theorem 7, p. 115).

THEOREM 5. *If P is an independent set of points in an exchange lattice, q a point with $q \leqq \Sigma(P)$, then there is a finite subset $P_0 \subset P$ such that* (i) $q \leqq \Sigma(P_0)$, *but this statement is false if P_0 is replaced by any of its proper subsets*; (ii) *if $R \subset P$ and $q \leqq \Sigma(R)$, then $P_0 \subset R$*; (iii) *if in P any point p_1 of P_0 be replaced by q, the resulting set P' is independent, and $\Sigma(P') = \Sigma(P)$.*

Proof. The finiteness axiom yields at once the set P_0 as in (i). If $q \leqq \Sigma(R)$, as in (ii), then $q \leqq \Sigma(R_0)$ for a similar minimal finite set R_0. If $P_0 \subset R$ is false, then $P_0 \neq R_0$. Since P_0 is minimal, R_0 contains at least one point r not in P_0. If R_0' is the set R_0 with r deleted, then $q \leqq \Sigma(R_0') + r$, so that the exchange axiom proves $r \leqq \Sigma(R_0') + q \leqq \Sigma(R_0') + \Sigma(P_0)$, contrary to the independence of P. Finally P' in conclusion (iii) is independent because of (ii) and the corollary to Theorem 2.

Repeated applications of this exchange process with a transfinite induction yield, exactly as in Steinitz [11] (Note 126 and correction thereto), a proof of the fundamental invariance theorem:

THEOREM 6. *If P and Q are two bases for an element b in an exchange lattice L, then the sets P and Q have the same cardinal number.*[4] *This number we call the rank of b.*

THEOREM 7. *Every exchange lattice L has relative complements*; *that is, given $b \geqq d \geqq c$ in L there is an element d' in L with $d + d' = b$, $dd' = c$.*

[4] An equivalent abstract basis theorem has been developed by Reinhold Baer, who also found several alternative postulational bases for abstract independence (all unpublished). (Added May 9, 1938.)

Proof. In the quotient lattice b/c there is by Theorems 4 and 3 a basis P_d for d. By Theorem 3 we can adjoin to P_d a set of points Q' of b/c such that P_d and Q' together form a basis for b. Then $d' = \Sigma(Q')$ is the desired complement, because the definition of independence insures that $(\Sigma(P_d))\cdot(\Sigma(Q')) = c$.

4. **Alternative exchange axioms and the Dedekind axiom.** The exchange axiom is a weakened form of the usual modular or Dedekind law (Ore [9], p. 412; Birkhoff [4]).

(D_1) DEDEKIND AXIOM. $a \geqq c$ implies $a(b+c) = ab + c$.

This axiom holds for the linear subspaces of a projective space, but not for an affine space. Menger showed that it could there be replaced in part by either of the two equivalent assertions, for all elements a, b and all points p of L, that

(E_2) $a \leqq b \leqq a + p$ implies $b = a$ or $b = a + p$; or that

(E_3) $p \nleqq a + b$ implies $ab = (a + p)b$

(see [7], Axiom 6^+). Birkhoff's investigations of the Jordan Theorem[5] involve the property that, for all a, b and d in L,

(E_4) if $a \geqq d$, $a + b > a$ and b covers d, then $a + b$ covers a.

The interrelations of these axioms are as follows.

THEOREM 8. *In any lattice L satisfying the point-existence axiom* (G_1), *any two of the conditions* (E_1), (E_2), (E_3), (E_4) *are equivalent. Each of them is a consequence of the Dedekind axiom, but there exist exchange lattices which satisfy* (E_1), (F_1), *and* (G_1) *but not the Dedekind axiom.*

In an arbitrary lattice the implications (E_4) $\rightarrow$ (E_2) $\rightarrow$ (E_1), (E_2) $\leftrightarrow$ (E_3) (proved by Menger) and (D_1) $\rightarrow$ (E_4) (cf. Birkhoff [2], Theorem 9.1) will hold. To prove (E_4) $\rightarrow$ (E_2), set $d = 0$ and $b = p$ in (E_4) to obtain the assertion "$a + p > a$ implies that $a + p$ covers a", an alternative statement for (E_2). Similarly (E_2) with $b = a + q$ will yield (E_1).

To establish (E_4) as a consequence of (E_1) and (G_1), suppose that in the conclusion of (E_4) $a + b$ fails to cover a. There is then a c, $a + b > c > a$, and by (G_1) there are points p and q such that $a < a + p \leqq c$, $d < d + q \leqq b$. Since b covers d, $d + q = b$ and

$$a < a + p \leqq c < a + b = a + d + q = a + q.$$

The exchange axiom then makes $a + q \leqq a + p < a + q$, a contradiction. Given (G_1) we now have (D_1) $\rightarrow$ (E_4) $\rightarrow$ (E_3) $\rightarrow$ (E_2) $\rightarrow$ (E_1) $\rightarrow$ (E_4), as asserted in the theorem.

The Dedekind axiom does not hold in every exchange lattice. For instance, let C be a finite class with m elements and n an integer, $1 \leqq n < m$. Consider the set L_{mn} of all subclasses $a, b, \cdots$ of C which have not more than n elements or exactly m elements, and let $a < b$ mean that a is a proper subclass of b.

[5] Lattices having this property (E_4) have been studied by F. Klein, who calls the property "axiom (γ'')" and who obtains several interesting equivalent axioms. Cf. F. Klein, *Birkhoffsche und harmonische Verbände*, Mathematische Zeitschrift, vol. 42 (1937), pp. 58–81. (Added May 9, 1938.)

Then L_{mn} is a lattice, in which the points are the subclasses of C with just one element, while the sum $a + b$ is the whole class C or the ordinary point-set union $a \cup b$ according as $a \cup b$ has more than n elements or not. The axioms for an exchange lattice are readily verified for L_{mn}. Whenever $m \geqq n + 2 \geqq 4$, and in no other cases, L_{mn} fails to be a Dedekind lattice. For instance, if $m = 4$, $n = 2$, L_{42} contains four points p, q, r and s. If we set $a = p + q$, $b = r + s$, $c = p$, then $a \geqq c$, but

$$a(b + c) = (p + q)(p + r + s) = (p + q)\cdot 1 = p + q,$$

$$ab + c = (p + q)(r + s) + p = p \text{ and } p + q \neq p,$$

contrary to the Dedekind law.

We turn to the Jordan Theorem. A *chain* of length n joining a to b, $a < b$, is a sequence of $n + 1$ elements x_i, $a = x_0 < x_1 < \cdots < x_n = b$. The chain is *principal* if each x_i covers x_{i-1}, for $i = 1, \cdots, n$. A lattice L is said to be of *finite dimensions* if any two elements $a < b$ can be joined by a principal chain.

THEOREM 9. *If two elements $a < b$ of an exchange lattice L are joined by a principal chain of length n, then no chain joining a to b has length greater than n, so that any principal chain joining a to b has the same length n. If the exchange lattice L is of finite dimensions, then the rank (or dimension) $\rho(a)$ defined in Theorem 6 is equal to the length of any principal chain joining 0 to a, and $\rho(a)$ satisfies the inequality*

$$\rho(ab) + \rho(a + b) \leqq \rho(a) + \rho(b). \tag{1}$$

The first assertion, which states in part that when a and b are joined by a principal chain they can be joined by no "infinite" chain, can be established by induction, with repeated applications of axiom (E_4). The rest of the theorem restates known results, for the exchange axiom (E_4) implies the following condition:

(ξ) If a and b cover d and $a \neq b$, then $a + b$ covers a and b. From this condition the Jordan Theorem and (1) can be established (Birkhoff [2], Theorems 8.2 and 9.2; Ore [9], Chapter II).

The following converse to Theorems 7 and 9 gives in effect an alternative definition for finite exchange lattices.

THEOREM 10. *If L is a lattice which has for every $a > d > 0$ a "relative complement" d' with $d + d' = a$, $d\cdot d' = 0$, and if to every element a of L an integer $\rho(a)$ can be so assigned that*

(i) *$a < b$ implies $\rho(a) < \rho(b)$;*

(ii) *if b covers a, then $\rho(b) = \rho(a) + 1$;*

(iii) *the inequality* (1) *holds for all a and b;*

then L is an exchange lattice.

Proof. The finiteness of each $\rho(a)$ and the condition (i) enforce ascending and descending chain conditions (Ore [9], p. 410) and thence the existence of infinite unions and cross-cuts, as well as the finiteness axiom (F_1). To establish the existence of points, let $a > d \geqq 0$ and pick a relative complement d' of d in a.

The descending chain condition yields a point p, $p \leqq d'$, with the requisite properties, $p \leqq a$ and $p \not\leqq d$.

If p is any point with $a < a + p$, then the inequality (1) and condition (ii) give

$$\rho(a + p) \leqq \rho(a) + \rho(p) - \rho(ap) = \rho(a) + \rho(p) - \rho(0) = \rho(a) + 1.$$

But $a + p > a$, so $\rho(a + p) > \rho(a)$ by (i), and we have $\rho(a + p) = \rho(a) + 1$. This means that $a + p$ covers a, and is in effect the exchange axiom (E_2). Therefore L is an exchange lattice.

A lattice L is *complemented* if for every a there is an a' with $a + a' = 1$, $aa' = 0$. If L is also modular, relative complements are known to exist, and in the finite case there is a dimension function. Therefore, we have the following

COROLLARY. *Any complemented Dedekind lattice of finite dimensions is an exchange lattice.*

The lattice of all linear subspaces of an affine space is also an exchange lattice, as can be established from Menger's axioms for affine geometry.

5. **Transcendence degrees of fields.** For fields $\mathfrak{L} \supset \mathfrak{K}$, the set of all relatively algebraically closed subfields $\mathfrak{M}$ between $\mathfrak{K}$ and $\mathfrak{L}$ forms an exchange lattice L. This is proved by noting that the relation "y depends algebraically on $\mathfrak{K}(S)$" is a dependence relation with the five properties used in Theorem 1 to construct an exchange lattice. The sets "closed" under this relation are exactly the subfields $\mathfrak{M}$, and the transcendence degree of $\mathfrak{M}$ over $\mathfrak{K}$ is the rank of $\mathfrak{M}$ in the lattice, and so is included in Theorem 6.

Such lattices of fields need not satisfy the Dedekind law. Consider over any field $\mathfrak{K}$ the field $\mathfrak{L} = \mathfrak{K}(x, y, z)$ of rational functions of three independent variables x, y and z. The subfields

$$\mathfrak{M} = \mathfrak{K}(x, y), \qquad \mathfrak{N} = \mathfrak{K}(z, x + yz), \qquad \mathfrak{R} = \mathfrak{K}(x) \tag{2}$$

are relatively algebraically closed in $\mathfrak{L}$, by Lüroth's theorem (cf. Steinitz [11], p. 126). The intersection of $\mathfrak{M}$ and $\mathfrak{N}$ is $\mathfrak{K}$. For let $\alpha \neq 0$ be an element of the intersection,

$$\alpha = f(x, y)/g(x, y) = r(t, z)/s(t, z), \qquad t = x + yz.$$

Then $g(x, y) \neq 0$, $s(t, z) \neq 0$, and we can assume that $r(t, z)$ and $s(t, z)$, as polynomials in t and z, have no factors in common except constants. Then $fs = gr$ is an identity in x, y and z; in it we set $z = 0$ to obtain

$$f(x, y)s(x, 0) = g(x, y)r(x, 0).$$

If $s(x, 0) = 0$, then, since $g(x, y) \neq 0$, $r(x, 0) = 0$. Since these are identities, $r(t, 0) = 0 = s(t, 0)$, which means that $r(t, z)$ and $s(t, z)$ have in common a factor z contrary to assumption. Hence $s(x, 0) \neq 0$ and

$$\alpha = f(x, y)/g(x, y) = r(x, 0)/s(x, 0),$$

so that α is in $\mathfrak{K}(x)$. If α were not in $\mathfrak{K}$, α would be transcendental over $\mathfrak{K}$ by Lüroth's theorem, so that x would be algebraic over $\mathfrak{K}(\alpha) \subset \mathfrak{N}$. Because $\mathfrak{N}$ is relatively algebraically closed, x is in $\mathfrak{N}$, and (2) shows that $\mathfrak{N}$ contains x, y, and z. This is a contradiction.

Because the intersection $\mathfrak{K} = \mathfrak{M} \cap \mathfrak{N}$ is the cross-cut $\mathfrak{M}\cdot\mathfrak{N}$ of $\mathfrak{M}$ and $\mathfrak{N}$ in the lattice L of relatively algebraically closed fields between $\mathfrak{K}$ and $\mathfrak{L}$, this lattice is not modular. For, by (2), $\mathfrak{M} > \mathfrak{R}$ and

$$\mathfrak{M}\cdot(\mathfrak{N} + \mathfrak{R}) = \mathfrak{M}\cdot\mathfrak{L} = \mathfrak{M}, \qquad \mathfrak{M}\cdot\mathfrak{N} + \mathfrak{R} = \mathfrak{K} + \mathfrak{R} = \mathfrak{R},$$

so that $\mathfrak{M}\cdot(\mathfrak{N} + \mathfrak{R}) \neq \mathfrak{M}\cdot\mathfrak{N} + \mathfrak{R}$, contrary to the Dedekind law.

This example can be extended in various ways. In any field $\mathfrak{K}(x_1, \cdots, x_n)$ with n independent variables one can construct two subfields $\mathfrak{M}$ and $\mathfrak{N}$, each of transcendence degree $n - 1$ over $\mathfrak{K}$, and with $\mathfrak{K}$ as intersection. In the field $\mathfrak{K}(x, y, z)$ one can also find a denumerable number of relatively algebraically closed subfields $\mathfrak{M}_i$, each of transcendence degree 2 over $\mathfrak{K}$, such that the intersection of any two of these subfields is $\mathfrak{K}$. These examples are typical of all fields, in the following sense:

THEOREM 11. *If $\mathfrak{K}$ is a relatively algebraically closed subfield of $\mathfrak{L}$, then the lattice of all relatively algebraically closed subfields $\mathfrak{M}$ with $\mathfrak{K} \subset \mathfrak{M} \subset \mathfrak{L}$ is a Dedekind lattice if and only if the transcendence degree of $\mathfrak{L}$ over $\mathfrak{K}$ is less than* 3.

Proof. When the transcendence degree is 1 or 2, the lattice has a simple form and the Dedekind law is trivially true. In the remaining cases, there are in $\mathfrak{L}$ three indeterminates x, y and z independent over $\mathfrak{K}$. Let $\mathfrak{M}$, $\mathfrak{N}$ and $\mathfrak{R}$ be the fields of (2) and $\mathfrak{M}'$, $\mathfrak{N}'$ and $\mathfrak{R}'$ be respectively their relative algebraic closures in $\mathfrak{L}$. Then $\mathfrak{M}' > \mathfrak{R}'$, and if we show that $\mathfrak{M}'$ and $\mathfrak{N}'$ intersect in $\mathfrak{K}$, the non-modularity follows as before.

Suppose then that w is an element common to $\mathfrak{M}'$ and $\mathfrak{N}'$, so that w is algebraic over both $\mathfrak{M}$ and $\mathfrak{N}$ and satisfies equations $f(u) = 0$ and $g(u) = 0$ irreducible over $\mathfrak{M}$ and $\mathfrak{N}$, respectively. Since $\mathfrak{K}(x, y, z)$ is a simple transcendental extension of $\mathfrak{M}$ and also of $\mathfrak{N}$, $f(u)$ and $g(u)$ must remain irreducible over the extension $\mathfrak{K}(x, y, z)$. If both f and g have the leading coefficient 1, then $f(u) \equiv g(u)$, so that w satisfies an equation $f(u) = 0$ whose coefficients are in $\mathfrak{M}\cdot\mathfrak{N}$; that is, in $\mathfrak{K}$. Therefore w, algebraic over $\mathfrak{K}$, is in $\mathfrak{K}$, and the intersection $\mathfrak{M}'\cdot\mathfrak{N}'$ is in fact $\mathfrak{K}$.

6. **Group ranks and p-bases.** Abelian groups furnish another example of exchange lattices. Let J be an additive Abelian group without elements of finite order (except 0). An element g of J is said to be dependent on a subset X of J if in X there are elements $x_1, \cdots, x_n$ and if there are integers $m, k_1, \cdots, k_n$, $m \neq 0$, with $mg = k_1x_1 + \cdots + k_nx_n$. This dependence relation has, as is readily verified, the five properties used in Theorem 1, so that the sets $H \subset J$ closed under this dependence form an exchange lattice. These closed subsets H are simply the subgroups $H \subset J$ for which the factor group J/H has no elements of finite order (except 0). A basis for J, in the lattice sense, is simply a

maximal independent set of elements of J, and the rank of J (Theorem 6) is a known invariant of J (Baer [1], Theorem 3.2). A simple proof shows also that the lattice of closed subgroups is a Dedekind lattice.

A field $\mathfrak{K}$ of characteristic p which is not perfect can be extended to a perfect field by the adjunction of p^n-th roots of a certain minimal set of elements of $\mathfrak{K}$. This minimal set is called a p-basis (Teichmüller [12], §3, and [13], p. 145, Hilfsatz 9). This notion can be generalized to any inseparable extension.[6] Let $\mathfrak{K}$ be a field of characteristic p, and $\mathfrak{L}$ a pure inseparable extension of exponent 1 over $\mathfrak{K}$; that is, an extension such that x^p is in $\mathfrak{K}$ for any x in $\mathfrak{L}$. An element y in $\mathfrak{L}$ will be called *p-dependent* on a subset X of $\mathfrak{L}$ if y is in the field $\mathfrak{K}(X)$. The properties 1, 2, 3 and 5 of a dependence relation are immediately verified. As for property 4, let y be in $\mathfrak{K}(x_1, \cdots, x_n)$, but not in $\mathfrak{K}(x_1, \cdots, x_{n-1})$. Since x_n^p is in $\mathfrak{K}$, $y = f(x_1, \cdots, x_n)$, where f is a polynomial with coefficients in $\mathfrak{K}$ and of degree less than p in x_n. Then x_n must actually occur in some term of f, so x_n is algebraic of degree $\leqq p - 1$ over $\mathfrak{K}_1 = \mathfrak{K}(y, x_1, \cdots, x_{n-1})$. But x_n also satisfies the inseparable equation $z^p - x_n^p = 0$ over $\mathfrak{K}_1$, so that x_n must be in $\mathfrak{K}_1$, and is p-dependent on $y, x_1, \cdots, x_{n-1}$, as asserted in property 4.

The exchange lattice corresponding to this dependence relation is simply the lattice of all subfields $\mathfrak{M}$ with $\mathfrak{K} \subset \mathfrak{M} \subset \mathfrak{L}$. The points of the lattice are the subfields of degree p over $\mathfrak{K}$, so that Theorem 6 becomes

THEOREM 12. *If $\mathfrak{L}$ is a pure inseparable extension of exponent* 1 *of a field $\mathfrak{K}$ of characteristic p, then there exists a set of subfields $\mathfrak{M}_\sigma$ of $\mathfrak{L}$, each of degree p over $\mathfrak{K}$, such that the adjunction of all $\mathfrak{M}_\sigma$ to $\mathfrak{K}$ gives $\mathfrak{L}$, while no $\mathfrak{M}_\sigma$ is a subfield of the field obtained by adjoining the remaining $\mathfrak{M}$'s to $\mathfrak{K}$. The cardinal number of subfields $\mathfrak{M}_\sigma$ is an invariant of $\mathfrak{L}/\mathfrak{K}$, called the relative degree of imperfection.*

Each subfield $\mathfrak{M}_\sigma$ has the form $\mathfrak{M}_\sigma = \mathfrak{K}(\sqrt[p]{x_\sigma})$ for some x_σ in $\mathfrak{K}$. The set of these x_σ's, one for each $\mathfrak{M}_\sigma$, can be called a *p-basis* of $\mathfrak{L}$ over $\mathfrak{K}$. This concept will apply to any inseparable algebraic extension, for such an extension $\mathfrak{F}$ of $\mathfrak{K}$ can be uniquely decomposed into $\mathfrak{K} \subset \mathfrak{L}_0 \subset \mathfrak{L}_1 \subset \mathfrak{L}_2 \subset \cdots \subset \mathfrak{F}$, where each $\mathfrak{L}_n$ contains all elements of $\mathfrak{F}$ of exponent not more than n over $\mathfrak{K}$, so that each $\mathfrak{L}_n$ is a pure inseparable extension of exponent 1 over $\mathfrak{L}_{n-1}$.

The lattice of subfields $\mathfrak{M}$ of $\mathfrak{L}$ is not always modular. Consider $\mathfrak{L} = \mathfrak{P}(x, y, z)$, $\mathfrak{K} = \mathfrak{P}(x^p, y^p, z^p)$, where x, y and z are independent indeterminates over the perfect field $\mathfrak{P}$ of characteristic p. $\mathfrak{L}$ over $\mathfrak{K}$ is pure inseparable of degree p^3, is obtained from $\mathfrak{K}$ by adjoining the independent p-th roots x, y and z, and so has the p-basis $\{x, y, z\}$ (cf. Teichmüller [12], Theorem 18). The subfields $\mathfrak{M} = \mathfrak{K}(x, y)$, $\mathfrak{N} = \mathfrak{K}(z, x + yz)$ are each of degree p^2 over $\mathfrak{K}$. To disprove the Dedekind law, it will suffice as in §5 to show that $\mathfrak{M}$ and $\mathfrak{N}$ intersect in $\mathfrak{K}$. Let them have in common the element

$$(3) \qquad \alpha = f(x, y) = g(z, t) \equiv \sum_{i,j=0}^{p-1} a_{ij} z^i t^j, \qquad t = x + yz,$$

[6] The possibility of such an extension is indicated by Teichmüller [12], §3.

where f and g are polynomials with coefficients in $\mathfrak{K}$ and of degree less than p in any one variable. The polynomial

$$(4) \qquad h(x, y, z) \equiv f(x, y) - g(z, x + yz)$$

must be zero in $\mathfrak{L}$ and is of degree at most $2p - 2$ in the variable z. If each term z^{p+s} be replaced by dz^s, with coefficient $d = z^p$ in $\mathfrak{K}$, then $h(x, y, z)$ is equal in $\mathfrak{L}$ to a new polynomial $h'(x, y, z)$ of degree less than p in any one variable. But the power products $x^q y^r z^s$ with exponents less than p form a basis for the algebraic extension $\mathfrak{L}/\mathfrak{K}$, so that $h'(x, y, z) = 0$ in $\mathfrak{L}$ implies that h' is identically zero. In h' terms in z^{p-1} never arise from a replacement $z^{p+s} \to dz^s$, but come only from terms $z^i t^j$ in g with $i + j \geqq p - 1$. These terms are, by expansion of (3),

$$\left(\sum_{j=0}^{p-1} \sum_{i=p-1-j}^{p-1} \binom{j}{p-1-i} a_{ij}\, x^{j+i-(p-1)} y^{p-1-i}\right) z^{p-1}.$$

These terms involve distinct power products $x^q y^s$ and have binomial coefficients not zero, so $h' \equiv 0$ implies $a_{ij} = 0$ for $i + j \geqq p - 1$. Thus (4) actually involves no term of degree p or more in x, in y or in z, and $h(x, y, z) \equiv 0$. But z arises only from $g(z, x + yz)$, so only the constant term of g can differ from 0, and the element $\alpha = g(0, 0)$ of $\mathfrak{M} \cdot \mathfrak{N}$ is in fact in $\mathfrak{K}$.

7. **Exchange axioms free of points.** A central feature of von Neumann's continuous geometry is the use of the modular law, which makes no reference to the points of the geometry. Similarly, Wilcox [16] has shown that affine geometry as developed algebraically by Menger can also be axiomatized without the use of points. His treatment depends on certain properties of a relation of modularity which do not hold in all exchange lattices. Nevertheless, our exchange axiom can be replaced by conditions which make no use of points or of covering relations.

To modify Menger's exchange axiom (E_3), which asserts that $p \nleqq a + c$ implies $(a + p)c = ac$, replace the point p by an arbitrary element b and the conclusion $(a + p)c = ac$ by the assertion that $(a + b_1)c = ac$ for some non-trivial part b_1 of b. No generality is lost if we require $a < c$; the hypothesis $p \nleqq a + c$ or $p(a + c) = 0$ might then become $bc < a$, and our modified statement is[7]

(E_5) $bc < a < c < b + c$ implies that there exists b_1 such that $bc < b_1 \leqq b$ and $(a + b_1)c = a$.

A similar elimination of points from the exchange axiom (E_1) leads eventually to the laws

(E_6) $bc < a < c < b + c$ implies that there exists b_1 such that $bc < b_1 \leqq b$ and $(a + b_1)c < c$;

(E_7) $bc < a < c < b + a$ implies the conclusion of (E_6).

The second assertion $(a + b_1)c < c$ of this conclusion is equivalent to $c \nleqq a + b_1$, which in turn is equivalent to $a + b_1 < c + b_1$.

[7] The statement that this law (or other similar laws) holds is to mean that it holds for all elements a, b, and c of the lattice.

These three point-free laws do not, like the exchange axiom (E_1), hold trivially in any lattice without points (as for instance in the lattice of all real numbers between 0 and 1, where $a < b$ has its usual meaning). The rôle of these laws can be stated thus:

THEOREM 13. *In any lattice,* (E_5) *implies* (E_6) *and* (E_6) *is equivalent to* (E_7).

THEOREM 14. *In a lattice satisfying the point-existence axiom* (G_1), *the exchange axiom* (E_1) *is equivalent to* (E_5) *and also to* (E_6). *Hence* (E_5) (*or* (E_6)) *can replace* (E_1) *in the definition of an exchange lattice.*

THEOREM 15. *In any lattice,* (E_7) (*and hence* (E_6) *or* (E_5)) *implies the covering law* (E_4). *Therefore a dimension function* $\rho(a)$ *can be defined in any complete lattice of finite dimensions satisfying* (E_7); *in other words, the conclusions of Theorem 9 hold for such a lattice.*

Proof. (E_5) $\rightarrow$ (E_6) $\rightarrow$ (E_7) is immediate. Conversely, to prove (E_7) $\rightarrow$ (E_6), let $bc < a < c < b + c$ as in (E_6). If $a + b > c$, then the hypothesis of (E_7) holds and yields the desired conclusion. If $a + b = c$, then $b \leqq c$, $b + c = c < b + c$, a contradiction. In the remaining case, $(a + b)c < c$, which states that the conclusion of (E_6) holds with $b_1 = b$. This gives Theorem 13.

For Theorem 15 we need only prove (E_7) $\rightarrow$ (E_4). Given the hypothesis of (E_4), the conclusion could be false only if $a + b > c > a$ for some c. Then $d \leqq bc \leqq b$, whence $bc = b$ or $bc = d$. In the former case, $b \leqq c$, $a + b \leqq a + c = c < a + b$, a contradiction. Therefore $bc = d$. Omitting the trivial case $d = a$, we have $bc < a < c < a + b$, as in (E_7), so that $(a + b_1)c < c$ for $bc < b_1 \leqq b$. Because b covers bc, $b_1 = b$ and $c = (a + b)c = (a + b_1)c < c$, a contradiction.

Since (E_7) $\rightarrow$ (E_4), Theorem 14 now needs only a proof that (E_5) holds whenever (E_1) and (G_1) do. Let $bc < a < c < b + c$. Since $b \nleqq c$, the property (G_2) of §2 furnishes a point p with $p \leqq b$, $p \nleqq c$. Hence $p \nleqq a + c = c$, so that $(a + p)c = ac$ by the exchange axiom (E_3), which is known to hold (Theorem 8). If we set $b_1 = bc + p$, then $bc < b_1 \leqq b$, while $(a + b_1)c = (a + bc + p)c = (a + p)c = ac = a$, as in the conclusion of (E_5).

Any one of these three point-free axioms can be viewed as a weaker form of the modular law (D_1), for in any lattice this modular law is equivalent to the following assertion, of a form similar to (E_5),

(D_2) $bc < a < c < b + c$ implies $(a + b)c = a$.

This is a direct consequence of the modular law. Conversely, the modular law requires that $c' = c(b + a)$ be equal to $a' = cb + a$ for any $c \geqq a$. In any event, $c' \geqq a'$ and $bc' \leqq a'$. If either $bc' = a'$ or $b + c' = c'$, the conclusion $c' = a'$ will follow readily. Hence the conclusion can be false only with $bc' < a' < c' < b + c'$, exactly as in the hypothesis of (D_2), whence we obtain the conclusion

$$cb + a = a' = (a' + b)c' = (cb + a + b)c(b + a)$$
$$= (a + b)c(b + a) = c(b + a),$$

which is the modular law.

The complexity of the axioms (E_5) to (E_7), attendant on the existence assertion for b_1, is unavoidable. Specifically, suppose that an axiom (E_M) could be found which, like (E_5), is equivalent to the exchange axiom (E_1) in the presence of the axioms (F_1) and (G_1), but which, unlike (E_5), contains no existence assertions. In other words, (E_M) is built up from certain elements $a_1, \cdots, a_n$ of the lattice L by any combinations using $<$, $=$, $+$, $\cdot$, "and", "implies" and "not", and is to be asserted for all choices of the elements a_i. Any such axiom (E_M) true in a lattice L is automatically true in any sublattice L' of L, for $<$, $+$, and $\cdot$ have the same meaning in L' as in L. But then (E_M) cannot be equivalent to (E_1). For the non-modular lattice L_{42} of §4 contains four points p, q, r and s and has a sublattice L' composed of 0, p, r, $p + r$, $q + s$ and 1. In L' axioms (F_1) and (G_1) hold, but axiom (E_1) fails, because in L' p and $q' = q + s$ are points, $p \leqq r + q'$, $p \nleqq r$, so that, according to (E_1), $q' = q + s \leqq r + p$ should hold, contrary to the construction of L_{mn}. But (E_M), true in L, is also true in L', and so is not equivalent to (E_1) in the presence of (F_1) and (G_1).

8. **Transposition axioms and modularity**. The point-free exchange axiom (E_5) states in part that the chain in its hypothesis can be subjected to the following successive transpositions,

$$bc < a < c < b + c, \tag{5}$$

$$bc < a < a + b_1 < b + c, \tag{6}$$

$$bc < b_1 < a + b_1 < b + c. \tag{7}$$

Note that the inclusions in (6) are all proper inclusions, for $a + b_1 = a$ would give $b_1 \leqq a$, $bc < b_1 \leqq ab \leqq cb$, a contradiction to ($E_5$), while $a + b_1 = b + c$ implies $(a + b_1)c = (b + c)c = c$, although $(a + b_1)c = a$ in (E_5). Similarly $b_1 < a + b_1$ in (7).

To investigate the relations of the exchange axiom to the possibility of such transpositions, consider any chain of length n,

$$C : a_0 < a_1 < a_2 < \cdots < a_n. \tag{8}$$

A *transposition* of C is the operation of replacing an element a_k by a new element a_k' between a_{k-1} and a_{k+1} ($a_{k-1} < a_k' < a_{k+1}$). We call the transposition *primary* if $a_k \cdot a_k' = a_{k-1}$, *proper* if $a_k \cdot a_k' < a_k$ and $a_k \cdot a_k' < a_k'$. A primary transposition is always proper, but not conversely. In these terms can be stated the following alternative modular and exchange axioms:

(T_5) If C is a chain of length $n \geqq 2$ and if $a_0 < b < a_n$ and $a_1 b = a_0$, then[8] one can find $n - 1$ or fewer successive primary transpositions of C which yield a chain $a_0 < b_1 < \cdots < a_n$ with second term $b_1 \leqq b$.

[8] The hypothesis $a_1 b = a_0$ is not highly restrictive, since in the contrary case $a_1 \geqq a_1 b > a_0$, so a simple insertion of $b_1 = a_1 b$ in C yields a new chain of the desired form. If in (T_5) the hypothesis $a_1 b = a_0$ were replaced by $a_{n-1} b = a_0$, and the conclusion "$n - 1$ or fewer" by "exactly $n - 1$", the resulting statement is equivalent to the original (T_5), by the arguments used for Theorem 16.

(T_6) Statement as in (T_5), with "primary" replaced by "proper".

(T_D) If C is a chain of length $n \geqq 2$ and if $a_0 < b < a_n$ and $a_{n-1}b = a_0$, then one can find $n - 1$ successive primary transpositions of C which yield a chain $a_0 < b < \cdots < a_n$ with second term b.

THEOREM 16. *In any lattice L, the Dedekind law is equivalent to* (T_D), *while the exchange axioms* (E_5) *and* (E_6) *are equivalent respectively to* (T_5) *and* (T_6).

We prove (E_5) $\to$ (T_5) by induction. If $n = 2$, C is $a_0 < a_1 < a_2$, and the transposition to $a_0 < b < a_2$ is primary by the hypothesis $a_1 b = a_0$. Let (T_5) be true for all chains in L of length $k < n$. If $a_{n-1}b > a_0$ in the given chain C, then $a_{n-1} > a_{n-1}b > a_0$, so we apply the induction assumption to $b' = a_{n-1}b$ and the chain $a_0 < \cdots < a_{n-1}$ of length $n - 1$. In the remaining case, $a_{n-1}b = a_0$, $b \nleqq a_{n-1}$, so the four-element chain $a_0 < a_{n-2} < a_{n-1} < a_{n-1} + b$ has the form of the hypothesis of (E_5). There exists a b_1, $a_0 < b_1 \leqq b$, with $(a_{n-2} + b_1) a_{n-1} = a_{n-2}$, which is to say that the transposition $a_{n-1} \to (a_{n-2} + b_1)$ in C is primary. The induction assumption applied to the chain $a_0 < a_1 < \cdots < a_{n-2} < a_{n-2} + b_1$ and to the element b_1 therefore gives $n - 2$ more transpositions leading to the desired type of chain.

As to the converse, (T_5) $\to$ (E_5), the chain (5) in the hypothesis of (E_5) is by (T_5) reducible by one or two primary transpositions to the form $bc < b_1 < c_1 < b + c$, with $b_1 \leqq b$. The first of these transpositions could not have introduced b_1, for then $b_1 < c$, which entails the contradiction $b_1 = bb_1 \leqq bc < b_1$. The first transposition is therefore $c \to c_1$; it is primary, so $cc_1 = a$. But $a < c_1$, $b_1 < c_1$, $a + b_1 \leqq c_1$ so $c(a + b_1) \leqq a$, and therefore $c(a + b_1) = a$, which is the conclusion of (E_5). This argument depends essentially on the circumstance that (T_5) allows only *two* transpositions on the chain (5).

An analogous proof that (T_6) $\leftrightarrow$ (E_6) is possible because (E_6) again gives two transpositions (5) $\to$ (6) and (6) $\to$ (7) in the chain (5) of length 3. The second transposition is primary, hence proper. The first transposition is proper because (E_6) asserts that $(a + b_1)c < c$ and because $(a + b_1)c < a + b_1$, for otherwise $a + b_1 \leqq c$ would give the paradox $bc \geqq b_1 > bc$.

To show that the Dedekind law implies (T_D), observe that the chain (8) can be subjected to the successive transpositions $a_k \to (a_{k-1} + b)$ $(k = n - 1, \cdots, 1)$. Each one is primary, for by the Dedekind law, $(a_{k-1} + b)a_k = a_{k-1} + ba_k \leqq a_{k-1} + ba_{n-1} = a_{k-1} + a_0 = a_{k-1}$. The final chain $a_0 < b < a_1 + b < \cdots < a_n$ has the specified form. The converse assertion that (T_D) $\to$ (D_1) can be readily checked as above by using the Dedekind law in the form (D_2) of §7.

BIBLIOGRAPHY

1. R. BAER, *The subgroup of elements of finite order of an Abelian group*, Annals of Mathematics, vol. 37(1936), p. 768.
2. G. BIRKHOFF, *On the combination of subalgebras*, Proceedings of the Cambridge Philosophical Society, vol. 29(1933), pp. 441–464.
3. G. BIRKHOFF, *Abstract linear dependence and lattices*, American Journal of Mathematics, vol. 57(1935), pp. 800–804.
4. G. BIRKHOFF, *Combinatorial relations in projective geometries*, Annals of Mathematics, (2), vol. 36(1935), pp. 743–748.

5. G. Köthe, *Die Theorie der Verbände*, Jahresbericht der deutschen Mathematiker-Vereinigung, vol. 47(1937), pp. 125–144.
6. S. Mac Lane, *Some interpretations of abstract linear dependence in terms of projective geometry*, American Journal of Mathematics, vol. 58(1936), pp. 236–240.
7. K. Menger, *New foundations of projective and affine geometry; algebra of geometry*, Annals of Mathematics, (2), vol. 37(1936), pp. 456–482.
8. J. von Neumann, *Continuous Geometry*, Notes of lectures at the Institute for Advanced Study, 1936.
9. O. Ore, *On the foundation of abstract algebra*. I, Annals of Mathematics, (2), vol. 36 (1935), pp. 406–437.
10. O. Ore, *On the decomposition theorems of algebra*, Comptes Rendus du Congrès International des Mathématiciens, 1936.
11. E. Steinitz, *Algebraische Theorie der Körper*, edited by R. Baer and H. Hasse, 1930.
12. O. Teichmüller, *p-Algebren*, Deutsche Mathematik, vol. 1(1936), pp. 362–388.
13. O. Teichmüller, *Diskret bewertete perfekte Körper mit unvollkommenem Restklassenkörper*, Journal für die Mathematik, vol. 176(1936), pp. 141–152.
14. B. L. van der Waerden, *Moderne Algebra*, vol. I, p. 204.
15. H. Whitney, *On the abstract properties of linear dependence*, American Journal of Mathematics, vol. 57(1935), pp. 509–533.
16. L. R. Wilcox, *Modularity in the theory of lattices*, Bulletin of the American Mathematical Society, vol. 44(1938), p. 50 (abstract 44-1-79).

University of Chicago.

Reprinted from DUKE MATHEMATICAL JOURNAL
Vol. 5, No. 2, June, 1939

MODULAR FIELDS. I

SEPARATING TRANSCENDENCE BASES

BY SAUNDERS MAC LANE

1. **Introduction**. Any extension K of a given field L has a transcendence basis T over L, that is, a set of elements $T = \{t_1, t_2, \cdots\}$ algebraically independent over L and such that all elements of K are algebraic over T. In other words, K can be considered as a (possibly infinite) field of algebraic functions of the variables $t_1, t_2, \cdots$. Many properties of algebraic equations must be restricted to separable equations, without multiple roots, so we enquire: When does a field K have over a subfield L a "separating" transcendence basis T such that all elements of K are roots of *separable* algebraic equations over T?

Forms of this question arise in the analysis of intersection multiplicities for general algebraic manifolds (B. L. van der Waerden [13][1]), in one method of discussing the structure of complete fields with valuations (Hasse and Schmidt [4], p. 16 and p. 46), and in the study of pure forms over function fields (Albert [2]). The properties of such separating transcendence bases may be also considered as one part of a systematic study of the algebraic structure of fields of characteristic p.

A first result, obtained independently by van der Waerden ([13], Lemma 1, p. 620) and by Albert and the author ([2], Theorem 3), is

THEOREM 1. *Any field K obtained by adjoining a finite number of elements to a perfect field P has a separating transcendence basis over P.*

A proof is given in §3 below. The fields treated in this theorem might also be described as finite algebraic function fields of n variables over P, for any integer n. A similar result for a more general ground field is (proof in §7, Theorem 14, Corollary)

THEOREM 2. *If L is a function field of one variable over a perfect coefficient field P, and if K is obtained by adjoining to L a finite number of elements in such a way that every element of K algebraic over L is in L, then K has a separating transcendence basis over L.*

The hypothesis that K is generated over L by a finite number of elements is essential to this theorem. For more general fields K there is a relation between the structure of K and that of its subfields over P.

Received January 13, 1939; presented to the American Mathematical Society, December 29, 1938.

[1] Numbers in brackets refer to the bibliography at the end of the paper.

THEOREM 3. *If a field K has a finite separating transcendence basis over a perfect subfield P, then any intermediate field M, such that $K \supset M \supset P$, also has a separating transcendence basis over P.*

The proof is given in §3.

Our chief purpose is to obtain further theorems of this type for an extension K/L in which the base field L is not restricted to be a perfect field P. Thus we obtain in Theorem 13 of §7 necessary and sufficient conditions that a given subset T be a separating transcendence basis for a given extension K/L, and also necessary and sufficient conditions that a given extension K/L have some finite separating transcendence basis (Theorem 14 in §7).

Extensions K/L with separating transcendence bases are special instances of extensions K/L which "preserve p-independence". The notion of the p-independence of subsets of L, due to Teichmüller ([10], §3), is formulated in §4. Those extensions K/L which "preserve" this p-independence, in the sense that p-independent subsets of L remain p-independent in K, can be characterized in a large number of different ways (Theorem 7 in §4, Theorem 10 of §5, Theorem 16 of §7). They arise naturally here and also in the study of the relative structure of discrete complete fields with valuations (Mac Lane [6]). Such extensions K/L which preserve p-independence seem to have most of the properties appertaining to arbitrary extensions of perfect base fields. An instance is the highly useful possibility of reducing certain equations involving only p-th powers of elements of K (see Theorem 10 of §5).

Extensions preserving p-independence provide a natural tool for investigating separating transcendence bases, largely because, given an extension K/L which preserves p-independence, any intermediate extension M/L with $M \subset K$ must automatically preserve p-independence. This analysis yields in §8 a theorem on separating transcendence bases for intermediate fields, which requires no mention of p-independence in its formulation, and which states that Theorem 3 above is valid without the hypothesis that the base field P be perfect.

In terms of p-independence we introduce for any extension K/L certain "relative p-bases" closely related to separating transcendence bases. In §9 we show that the notion of a p-basis can be used to generalize a theorem of A. A. Albert on pure forms, and we give another property of p-bases.

The earliest result on separating transcendence bases is one of F. K. Schmidt's [8] who proved that a function field of one variable over a perfect field P has a separating transcendence basis over P. His proof (§3 below) gives also the slightly more general statement:

THEOREM 4. *If a field K has transcendence degree 1 over its maximal perfect subfield P, then K has a separating transcendence basis over P.*

It would be tempting to conjecture that this theorem remains true without the restriction that the transcendence degree of K/P is 1. That this conjecture is false we show by a somewhat involved example given in §10. The difficulty of the example resides chiefly in the problem of explicitly computing the maximal perfect subfield of a field specifically given. The method used in this example (and also two corollaries to Theorem 19 of §8) throws some light on this general problem. Our example also provides an instance of a field whose "degree of imperfection", in the sense of Teichmüller, differs from its transcendence degree over its maximal perfect subfield.

2. **Preliminaries. Notation.** All fields to be considered will have characteristic a fixed prime p. If K is such a field, K^{p^n} denotes the field of all p^n-th powers of elements from K. Similarly, if T is any set of elements, T^p is the set of all p-th powers of elements of T, etc. If a field K and certain sets S, T, $\cdots$ are all contained in a larger field, then $K(S, T, \cdots)$ denotes the field obtained by adjoining to K all elements of S, of T, etc. If K contains a subfield L, properties of K relative to the ground field L will be called properties of the "extension" K/L. $K \cap M$ denotes the intersection of the fields K and M, $S \cup T$ the union of the sets S and T, $\{t\}$ the set whose only element is t.

DEFINITIONS.[2] Let K/L be a given extension. An element b of K is *separable* over L if b satisfies a polynomial equation $f(x) = 0$, with coefficients in L, having no multiple roots. If b in K has some power b^{p^e} in L, then b is *purely inseparable* over L. The extension K/L is *separable* (or, *purely inseparable*) if every b in K is separable (or, purely inseparable) over L. A set $T \subset K$ is *algebraically independent* over L if no element t of T is algebraic over the field $L(T - \{t\})$, where $T - \{t\}$ denotes the set T with t deleted. A *transcendence basis* T for K/L is a subset of K algebraically independent over L such that K is algebraic over $L(T)$. A *separating transcendence basis* for K/L is a transcendence basis T such that K is separable over $L(T)$. It can be shown (Mac Lane [5], Theorem 2.1) that K/L has a separating transcendence basis if and only if the elements of K can be so well ordered that each element is either transcendental or separable over the field obtained by adjoining to L all prior elements in the order. The *transcendence degree* for K/L is the (cardinal) number of elements in any transcendence basis for K/L.

A polynomial $f(x_1, \cdots, x_m)$ with coefficients in K may involve some variable x_1 only as powers of x_1^p; the polynomial f is then called *inseparable* in x_1. There is a largest integer e such that f can be written as a polynomial $g(x_1^{p^e}, x_2, \cdots, x_m)$ involving x_1 only as a p^e-th power; this largest p^e is called the *exponent* of x_1 in f. (If x_1 fails to appear in f, use $p^e = \infty$.) If b is inseparable over a field L, the irreducible equation $f(x) = 0$ for b over L is known to be inseparable (exponent $p^e > 1$) in x.

A field P is perfect if $P^p = P$, that is, if each element of P has a unique p-th root in P. The *maximal perfect subfield* of a given (imperfect) field K is the intersection K^{p^∞} of all the fields K^{p^n} ($n = 1, 2, 3, \cdots$).

A known result is (Teichmüller [10], Theorem 12; Mac Lane [5], §2)

LEMMA 1. *An element b which is both separable and purely inseparable over a field L lies in that field.*

[2] Cf. Albert [1], Chapter 7, or B. L. van der Waerden [12], Chapter 5.

3. **Bases over perfect ground fields.** Schmidt's Theorem for function fields of one variable over a perfect base field may be stated in a form including Theorem 4 of the introduction thus:

THEOREM 5. *If an imperfect field K has transcendence degree 1 over a perfect subfield P, then K has a separating transcendence basis over P.*

Proof. Let t be a transcendence basis for K/P. If every root $t^{p^{-e}}$ were in K, K would be algebraic over the perfect field $P(t, t^{p^{-1}}, t^{p^{-2}}, \cdots)$, so K itself would be perfect,[3] counter to assumption. Let e be the largest integer for which $u = t^{p^{-e}}$ is in K. Then $u^{1/p}$ is not in K, and u is a transcendence basis for K/P. It is also a separating transcendence basis for K/P unless K contains an x inseparable over $P(u)$. For such an x the irreducible equation $g(x^p, u) = 0$ over P involves only p-th powers of x, hence can be written as the p-th power of a polynomial over $P(u^{1/p})$. Thus the adjunction of $u^{1/p}$ to $P(u)$ reduces the degree of x over $P(u)$ by a factor p. This means that $u^{1/p}$ must be in $P(u, x) \subset K$, contrary to the choice of e.

A typical necessary and sufficient condition for the existence of a separating transcendence basis is

THEOREM 6. *A field K with a finite transcendence basis T over a perfect subfield P has a separating transcendence basis over P if and only if there is an integer e for which K^{p^e} is separable over $P(T)$.*

Proof. If K does have a separating transcendence basis X over P, each element x of this basis has some power $x^{p^{e'}}$ separable over $P(T)$. X, like T, has but a finite number[4] of elements, so we can choose e as the largest such exponent e' and have X^{p^e} separable over $P(T)$. As K is separable over $P(X)$, K^{p^e} is separable over $P(X^{p^e})$ and hence over $P(T)$, by the transitivity of separability.[5] The necessity of our condition is thereby established.

Conversely, let e be the smallest integer such that K^{p^e} is separable over $P(T)$. If $e = 0$, T itself is a separating transcendence basis, and we are done. Assume then that $e > 0$. We propose to replace $T = \{t_1, \cdots, t_n\}$ by a separably "equivalent" basis $\{u_1^p, u_2, \cdots, u_n\}$ in which one element is a p-th power u_1^p; for then the basis $u_1, \cdots, u_n$ with u_1 replacing u_1^p will "separate" more of K.

Since e was chosen as small as possible, there is in K an element y such that y^{p^e} but not $y^{p^{e-1}}$ is separable over $P(T)$. The irreducible equation $f(y, T) = 0$ for y over $P(T)$ then involves y only in the form y^{p^e}; i.e., has exponent p^e in y. We can also assume that $f(y, T)$ is irreducible in the ring of all polynomials in y and T with coefficients in P. All these coefficients are p-th powers in the perfect field P. If every variable t of T had exponent p or greater in f, f would contain only p-th powers, so could be written as a p-th power of another polynomial, in contradiction to its assumed irreducibility. If t is one of the variables

[3] Steinitz [9], §13, no. 4.

[4] The number of elements in a transcendence basis of K/P is an invariant of K/P. See Steinitz [9], §23.

[5] Steinitz [9], §13, no. 9.

of T which actually appears in f with exponent 1, we can consider $f(y, T) \equiv f(y, t, T - \{t\}) = 0$ as an equation for t over $P(T, y^{p^e})$. According to the Gauss Lemma, this equation is irreducible, while it has exponent 1 in t, hence is separable. Therefore t is separable[6] over $P(T - \{t\}, y^{p^e})$.

The set $T_1 = (T - \{t\}) \cup \{y\}$ obtained from T by replacing t by y is thus another transcendence basis for K/P, such that every element separable over $P(T)$ is separable over $P(T_1)$, by the transitivity of separability (see footnote 5). Furthermore y is not separable over $P(T)$ but is separable over $P(T_1)$. The subfield K_{T_1} of all elements b of K separable over $P(T_1)$ is thus larger than the subfield K_T of all b separable over $P(T)$. Because T is finite, the degree of K over this original subfield K_T is finite,[7] so a repetition of this transition from T to T_1 will finally yield a basis T_m over which all of K is separable.

Theorem 1 of the introduction is an immediate corollary of this theorem. Theorem 3 concerning intermediate fields is also a corollary. For let $K \supset L \supset P$, where K has a finite separating transcendence basis over P. Pick a transcendence basis X for L/P and a similar basis Y for K/L. Then the union $T = X \cup Y$ is a transcendence basis for K/P. The necessary condition of Theorem 6 applied to this basis T then shows that the sufficient condition of Theorem 6 must be satisfied by the original basis X. Hence the intermediate field L does in fact have a separating transcendence basis over P.

4. **Extensions preserving p-independence.** With any given extension K/L there is related in an invariant fashion a purely inseparable extension $K/L(K^p)$. The latter may be analyzed by the concepts of p-independence and p-basis introduced by Teichmüller in §3 of [10]. A subset X of K is *relatively p-independent* in K/L if $K^p(L, X')$ is a proper subfield of $K^p(L, X)$ whenever X' is a proper subset of X. A *relative p-basis* B for K/L is a relatively p-independent set such that $K = K^p(L, B)$. This notion of p-independence has the usual properties of an abstract dependence relation.[8] Therefore, every extension K/L has a relative p-basis and any two relative p-bases for the same extension have the same (cardinal) number of elements. Furthermore any set relatively p-independent in K/L can be embedded in a p-basis for K/L.

A subset X of K is (absolutely) *p-independent* if $X' < X$ implies $K^p(X') < K^p(X)$, where $<$ denotes proper inclusion. An (absolute) *p-basis* B for K is a p-independent subset for which $K = K^p(B)$. This is the special case of the above "relative" definitions obtained by assuming L perfect, for then $L = L^p$, $L \subset K^p$, and the field $K^p(L, X)$ used above becomes[9] $K^p(X)$. Ex-

[6] The argument underlying this exchange—from y^{p^e} separable over $P(T - \{t\}, t)$ to t separable over $P(T - \{t\}, y^{p^e})$—can be stated generally; cf. Lemmas I and II in Mac Lane [5].

[7] By Steinitz [9], Theorem 3 in §13 one proves $K_T \subset K \subset K_T(T^{p^{-e}})$, where $K_T(T^{p^{-e}})$ is certainly an extension of finite degree over K_T.

[8] Van der Waerden [12], p. 204, or Mac Lane [7], §6. The results stated above are given by Theorem 12 of [7] applied to the extension $K/K^p(L)$.

[9] Teichmüller's paper [10] considered chiefly this special case.

tensions which preserve the (absolute) p-independence of subsets of L will now be characterized in several different ways.

THEOREM 7. *Any two of the following properties of an extension K/L are equivalent*:

(i) *Every set $X \subset L$ p-independent in L is p-independent in K.*
(ii) *There is a p-basis B of L which is p-independent in K.*
(iii) $L^p(S) = L \cap K^p(S)$ *for every finite subset* $S \subset L$.
(iv) $L'(L^p) = L \cap L'(K^p)$ *for every subfield* $L' \subset L$.

Condition (iv) on the intersection $L \cap L'(K^p)$ states in effect that the adjunction to L' of p-th powers of elements of K yields no elements in L not obtainable by adjoining p-th powers from L.

DEFINITION. Any extension K/L with one of the equivalent properties (i) (ii), (iii), or (iv) will be said to *preserve p-independence.*

Proof. That (i) implies (ii) is trivial, so consider (ii) $\rightarrow$ (i). Were some p-independent subset X of L not p-independent in K, there would be an element x in X contained in $K^p(X')$, where X' is a finite subset of X not containing x. Since B of (ii) is a p-basis of L, there is a finite subset $B_1 \subset B$ such that x and X' are in $L^p(B_1)$. This means that $X' \cup \{x\}$ is p-dependent on B_1. In such circumstances one can exchange the elements x and $x' \in X'$ successively with suitable elements[10] of B_1, until all of B_1 is p-dependent on X', x and a remaining subset $B_2 \subset B_1$. This means that

$$L^p(B_1) = L^p(X', x, B_2) \tag{1}$$

and that the combined set $X' \cup \{x\} \cup B_2$ is p-independent in L. The two finite sets B_1 and $X' \cup \{x\} \cup B_2$ are mutually p-dependent over L and have the same number of elements, by construction. But the subset B_1 of B is assumed to remain p-independent in K, while the other subset $X' \cup \{x\} \cup B_2$ is p-dependent over K because we supposed x to be in $K^p(X')$. This is a contradiction since this makes $K^p(B_1)$ have a degree p^{β_1} over K^p, if β_1 is the number of elements in B_1, while the equal field $K^p(X', x, B_2) = K^p(X', B_2)$ has a smaller degree over K^p. Therefore (ii) $\rightarrow$ (i) in our theorem.

To demonstrate (iii) $\rightarrow$ (i), suppose counter to (i) that some p-independent set X of L becomes p-dependent in K, so that again some $x \in K^p(X')$. Therefore x is in the intersection $L \cap K^p(X') = L^p(X')$, by (iii). This result states that x is p-dependent on X' over L, counter to assumption.

The implication (iv) $\rightarrow$ (iii) may be obtained trivially by setting $L' = P(S)$, where P is used to denote some perfect subfield of L.

Finally, to prove that (i) $\rightarrow$ (iv), let L' be any subfield of L and pick a relative p-basis T for $L'(L^p)/L^p$. Then $L'(L^p) = L^p(T)$. If the conclusion $L \cap L'(K^p)$

[10] By the "exchange" property for dependence relations: If x depends on C and d, but not on C alone, then d depends on C and x. Cf. Teichmüller [10], §3 or Mac Lane [7], §6 and Axiom (E_1).

$= L'(L^p)$ of (iv) were false, there would be an element y of L in $L'(K^p) = K^p(L') = K^p(T)$ but not in $L'(L^p) = L^p(T)$. Thus y is not p-dependent on the set T in L, while T is by construction p-independent in L, so that the usual properties of dependence make[11] $T \cup \{y\}$ a p-independent subset of L. Condition (i) then insures that $T \cup \{y\}$ is p-independent in K, in conflict with the previous assertion that y is in $K^p(T)$. Hence (i) $\rightarrow$ (iv). The various implications (ii) $\leftrightarrow$ (i) $\rightarrow$ (iv) $\rightarrow$ (iii) $\rightarrow$ (i) completely establish the theorem.

Such an independence-preserving extension might be viewed as a generalization of the ordinary separable algebraic extensions, in the following sense:

THEOREM 8. *If K is an algebraic extension of L, then K/L preserves p-independence if and only if K/L is separable.*

Proof. If K/L is separable, each p-basis of L remains a p-basis of K (see footnote 14), and this insures that p-independence is preserved. Conversely, suppose that K/L preserves p-independence but is not separable, and denote by K_s the subfield consisting of those elements of K which are separable over L. Then $K \supset K_s$ and K/K_s is purely inseparable,[12] so K contains a c not in K_s with c^p in K_s. Any p-basis B of L is also a p-basis of K_s, so c^p is p-dependent on B, thus lies in $K_s^p(B)$. The exchange property (see footnote 10) of p-dependence provides for an exchange of c^p with some $b \in B$, with the result that $b \in K_s^p(B - \{b\}, c^p) \subset K^p(B - \{b\})$. This states that the set B is not p-independent in K, in violation of the assumption that the extension K/L preserves the p-independence of B. Hence the theorem is proved.

Other examples of extensions which preserve p-independence will now be cited (cf. Mac Lane [6], §6), but for our purposes it is especially important to note that any extension of a perfect field always preserves p-independence.

THEOREM 9. (a) *An extension K/L preserves p-independence if L is perfect or if K/L has a separating transcendence basis.*

(b) *If L is an extension of transcendence degree* 1 *over a perfect field P, then an extension K/L preserves p-independence if and only if no element of K is inseparable and algebraic over L. In particular, K/L preserves p-independence if L is relatively algebraically closed*[13] *in K.*

Proof. We first prove (a). If L is perfect, each p-basis of L is void, hence necessarily remains p-independent in any K. If K/L has a separating transcendence basis and if B is a p-basis of L, then B is part of a p-basis[14] of K, hence does remain p-independent in the extension K/L.

Part (b) refers especially to fields L which are function fields of one variable

[11] Mac Lane [7], corollary to Theorem 2.

[12] Steinitz [9], §14, Theorem 1.

[13] A subfield L of K is relatively algebraically closed in K if every element of K algebraic over L lies in L.

[14] By the following theorems of Teichmüller [10], §3: *If K is a separable algebraic extension of L, any p-basis of L is a p-basis of K. If $L(T)$ is a purely transcendental extension of L by algebraically independent elements T, then $B \cup T$ is a p-basis of $L(T)$ if B is a p-basis of L.* Both can be proved readily from the appropriate definitions.

over a perfect coefficient field P. Suppose first that K contains no element inseparable over L. If L were perfect, K/L would preserve p-independence by part (a), so we can assume that L is imperfect. By Theorem 5, L/P then has a separating transcendence basis of one element, t. This element is also (see footnote 14) a p-basis of L, so that if K/L were not to preserve p-independence, t would be in K^p. Then $t^{1/p}$ is in K, but is inseparable over L, contrary to the assumed character of L. Hence K/L preserves p-independence.

Conversely, suppose that K/L does preserve p-independence but that some b in K is inseparable over L. We can again suppose L imperfect and t a separating transcendence basis for L/P. The irreducible equation $f(y, t) = 0$ for $y = b$ over $P[t]$ then has an exponent $p^e > 1$ in y, but because of its irreducibility has exponent 1 in t. Viewed as an equation[15] for t over $P[y]$, it shows that t is separable over $P(b^{p^e}) \subset K^p$. But t is also purely inseparable over K^p, hence t is in K^p, in conflict with the hypothesis that K/L preserves the p-independence of the p-basis $\{t\}$. This completes the proof of part (b) of the theorem.

5. **Equations involving p-th powers.** The essential intrinsic property of extensions K/L which preserve p-independence is the possibility of reducing those algebraic equations between elements of K which involve only coefficients from L and p-th powers of elements from K. This includes a known, simple property of perfect fields L.

THEOREM 10. *A necessary and sufficient condition that an extension K/L preserve p-independence is that, for every finite subset Y of K, the linear dependence over L of the set Y^p implies the linear dependence over L of the set Y itself.*

Proof. Suppose first that K/L preserves p-independence, and that the elements $y_1, \cdots, y_m$ of some set Y have their p-th powers linearly dependent over L. Select a p-basis B of L, so that $L = L^p(B)$, and choose a finite subset $U \subset B$ with the property that $y_1^p, \cdots, y_m^p$ are linearly dependent over $L^p(U)$. If this is true for U the null set, $y_1, \cdots, y_m$ are certainly linearly dependent over L. Otherwise we can successively delete elements from U till we find a new U' and an element u such that $y_1^p, \cdots, y_m^p$ are linearly dependent over $L^p(U', u)$ but not over $L^p(U')$. Since u has degree p over $L^p(U')$, the given linear dependence relation may be expressed as

$$\sum_{i=1}^{m} \left(\sum_{j=0}^{p-1} b_{ij} u^j \right) y_i^p = 0, \qquad b_{ij} \in L^p(U'),$$

where not all $b_{ij} = 0$. Therefore

$$\sum_{j=0}^{p-1} \left(\sum_{i=1}^{m} b_{ij} y_i^p \right) u^j = 0. \tag{1}$$

If one of the coefficients $\sum b_{ij} y_i^p$ in this equation is not 0, (1) is a separable equation for u over $K^p(U')$, so that u is p-dependent on U' in K, in violation

[15] Compare the "exchange" argument for Theorem 6.

of the hypothesis that K/L preserves p-independence. If the coefficients of all powers u^j in (1) are zero, any one of these coefficients which involves a $b_{ij} \neq 0$ provides a linear dependence between $y_1^p, \cdots, y_m^p$ over $L^p(U')$, in contradiction to the choice of u. Hence $y_1, \cdots, y_m$ are linearly dependent over L.

Conversely, suppose that the linear dependence of a set Y^p always implies that of Y, but that K/L does not preserve p-independence. Then any p-basis B of L must become p-dependent in K, so that some b is in $K^p(b_1, \cdots, b_n)$, where $b, b_1, \cdots, b_n$ are distinct elements of B. The algebraic extension $K^p(b_1, \cdots, b_n)$ has over K^p a linear basis consisting of all power products $c = b_1^{e_1} \cdots b_n^{e_n}$ with exponents $e_k = 0, 1, \cdots, p-1$. If $c_1, \cdots, c_m$ are all such power products, $b \in K^p(b_1, \cdots, b_n)$ means that there are elements y_i not all zero in K for which

$$(2) \qquad b + y_1^p c_1 + \cdots + y_m^p c_m = 0.$$

Among the elements $1, y_1^p, \cdots, y_m^p$ of K^p pick a linearly independent basis, over L^p, consisting of $1, z_1^p, \cdots, z_t^p$, so that each y_i^p may be written as

$$y_i^p = d_{i0}^p + d_{i1}^p z_1^p + \cdots + d_{it}^p z_t^p, \qquad d_{ij} \text{ in } L.$$

If these expressions are substituted in (2) and the coefficients of each z_i^p collected, one finds

$$(3) \qquad \left(b + \sum_{i=1}^{m} d_{i0}^p c_i\right) + \left(\sum_{i=1}^{m} d_{i1}^p c_i\right) z_1^p + \cdots + \left(\sum_{i=1}^{m} d_{it}^p c_i\right) z_t^p = 0.$$

Here the first term $b + \sum d_{i0}^p c_i$ cannot be zero, because b is not p-dependent on $b_1, \cdots, b_n$ over the original field L. Therefore (3) asserts that $1, z_1^p, \cdots, z_t^p$ are linearly dependent over L. The hypothesis of the theorem then shows that $1, z_1, \cdots, z_t$ are linearly dependent over L. It therefore follows that $1, z_1^p, \cdots, z_t^p$ are linearly dependent over L^p and the construction of the z^p's as linearly independent over L^p is contradicted. This implies that the p-basis B of L remains p-independent, which is to say that the given extension does preserve p-independence.

LEMMA 2. *Let the extension K/L preserve p-independence, and let quantities $t_1, \cdots, t_n$ be algebraically independent over L, while t_0 in K is algebraically dependent on $t_1, \cdots, t_n$ according to a relation $f(t_0, t_1, \cdots, t_n) = 0$, with coefficients in L. If f is irreducible over L as a polynomial in the variables $t_0, \cdots, t_n$, then f necessarily has exponent 1 in at least one of these variables.*

Proof. The conclusion states in effect that such an irreducible f cannot be a polynomial in the p-th powers $t_0^p, \cdots, t_n^p$. If this were the case, f would be a linear relation between the p-th powers of a certain set of distinct power-products y_i

$$f(t_0, \cdots, t_n) \equiv \sum_{i=1}^{m} a_i y_i^p = 0, \qquad y_i = t_0^{e_{0i}} \cdots t_n^{e_{ni}}, \qquad (i = 1, \cdots, m),$$

where all the coefficients $a_i \neq 0$. This linear dependence of $y_1^p, \cdots, y_m^p$ over L implies by Theorem 10 a linear dependence of $y_1, \cdots, y_m$:

$$g(t_0, \cdots, t_n) \equiv \sum_{i=1}^{m} b_i y_i = 0, \qquad b_i \text{ in } L.$$

Not all $b_i = 0$, so some t_j, say t_n, must actually appear in g. The degree d of f in this quantity is then at least p times the degree of g in t_n. By the Gauss Lemma d is the degree of the element t_n of K over the field $L(t_0, \cdots, t_{n-1})$, although $g = 0$ provides an equation of smaller degree for t_n over that field. With this contradiction to the assumption that f contained only p-th powers the lemma is established.

6. **Relative p-bases.** Further preliminaries are necessary for the subsequent exposition of the close connection between relative p-bases of an extension K/L in the sense of §4 and separating transcendence bases for the same extension. A first result is the algebraic independence of p-bases, which was established by Teichmüller ([10], Theorem 15) for the case of absolute p-bases.

THEOREM 11. *The elements of any relative p-basis B of an extension K/L which preserves p-independence are algebraically independent over L.*

Proof. Were the elements of B algebraically dependent, one could find elements $t_0, t_1, \cdots, t_n$ in B algebraically dependent but with $t_1, \cdots, t_n$ algebraically independent over L. An irreducible polynomial relation $f(t_0, \cdots, t_n) = 0$ between these quantities must then as in Lemma 2 contain one variable, say t_n, of exponent 1. This equation provides an irreducible and separable equation for t_n over the field $L(t_0, \cdots, t_{n-1}) \subset K^p(L, t_0, \cdots, t_{n-1})$. Over the latter field $K^p(L, t_0, \cdots, t_{n-1})$, $t_n = (t_n^p)^{1/p}$ is also purely inseparable, so that t_n must lie in this field (Lemma 1, §2). This conclusion makes t_n relatively p-dependent on $t_0, \cdots, t_{n-1}$, contrary to the hypothesis on $B \supset \{t_0, \cdots, t_n\}$.

Explicit relative p-bases can be found from absolute p-bases in specific cases by the following process of composition and decomposition.

THEOREM 12. *If an extension K/L preserves p-independence and if B and C are disjoint subsets of K with $C \subset L$, then any two of the following statements imply the third:*
(i) *B is a relative p-basis of K/L;*
(ii) *C is a p-basis for*[16] *L;*
(iii) *$B \cup C$ is a p-basis for*[16] *K.*

Proof. We prove first that (i) & (ii) → (iii). The union $B \cup C$ might be p-dependent in two ways. In the first place, an element b of B might lie in $K^p(B - \{b\}, C) \subset K^p(L, B - \{b\})$, but this would violate the relative p-inde-

[16] From this theorem it is also possible to obtain a similar but more general theorem in which statements (ii) and (iii) concern relative p-bases for L/M and K/M respectively, where $K \supset L \supset M$ and the extension L/M is assumed to preserve p-independence (Theorem 12 is the case when M is perfect).

pendence of the set B. In the second place, an element c of C might lie in $K^p(B, C - \{c\})$. There then are distinct elements $b_1, \cdots, b_m$ in B such that

$$(1) \qquad c \in K^p(b_1, \cdots, b_m, C - \{c\})$$

and such that the statement (1) would be false were any b_i omitted. At least one b_i is present in (1) ($m \geqq 1$) because C is known to be p-independent in L and therefore in K. Over the smaller field $K^p(b_2, \cdots, b_m, C - \{c\})$ both c and b_1 have the degree p, so (1) yields an "exchanged" statement

$$b_1 \in K^p(c, b_2, \cdots, b_m, C - \{c\}) = K^p(b_2, \cdots, b_m, C).$$

This type of p-dependence has already been led to a contradiction, so $B \cup C$ is in fact p-independent in K. That it forms a p-basis for K is then readily shown.

The converse implication (ii) & (iii) $\rightarrow$ (i) is trivial, granted the hypothesis $B \cap C = 0$. As for the third implication, (i) & (iii) $\rightarrow$ (ii), the p-independence of the set C in L results at once from its assumed p-independence (hypothesis (iii)) in the larger field K. Were C not a p-basis of L, there would be in L an x not in $L^p(C)$. By (iii), x is in $K^p(B, C)$, so there are distinct elements $b_1, \cdots, b_m$ in B such that

$$(2) \qquad x \in K^p(b_1, \cdots, b_m, C),$$

but such that the statement (2) would be false were any b_i omitted. If $m > 0$, one deduces as in the previous argument from (1) that $b_1 \in K^p(x, b_2, \cdots, b_m, C) \subset K^p(L, b_2, \cdots, b_m)$, a violation of the relative p-independence of B in K/L (hypothesis (i)). Thus (2) is $x \in K^p(C)$, although we had assumed $x \in L^p(C)$ false. This states that the p-independent set (see footnote 11) $C \cup \{x\}$ of L has become p-dependent over K, contrary to the basic assumption on this extension K/L. The theorem is thereby completely proved.

We can now describe particular relative p-bases in two typical extended fields (involving transcendental, separable, and inseparable adjunctions).

Lemma 3. *A separating transcendence basis S for an extension K/L is always a relative p-basis for K/L.*

For, any p-basis B of L gives rise to a p-basis (see footnote 14) $B \cup S$ for K and this, by the decomposition Theorem 12, makes S a relative p-basis for K/L.

Lemma 4. *If $K \supset K_0 \supset L$, where K/K_0 is a finite purely inseparable extension and K/L preserves p-independence, then the (cardinal) number of elements in a relative p-basis for K/L is the same as the number of elements*[17] *in a relative p-basis for K_0/L.*

Proof. It suffices to consider the case when the degree $[K : K_0]$ is p, so that $K = K_0(a)$, where a^p is in K_0. We wish to construct for K/L a relative p-basis

[17] For the absolute case (L perfect) this lemma has been proved by M. Becker [3].

containing this element a; to this end we show first that a^p is relatively p-independent in K_0/L. Otherwise a^p would be in $K_0^p(L) = K_0^p(B)$, where B is a p-basis for L. Since a^p is not in K_0^p, a^p can here be exchanged with an element b of B, which means that $b \in K_0^p(a^p, B - \{b\}) \subset K^p(B - \{b\})$, contrary to the p-independence of B in L and K.

Since a^p is relatively p-independent in K_0/L, there is a relative p-basis C_0 for K_0/L containing a^p. The replacement of a^p by a in the set C_0 yields, as may be verified from the definitions, a p-basis C for K/L. C and C_0 have the same number of elements, so our lemma is proved.

7. **Criteria for separating transcendence bases.** The notions of p-independence will now be applied to obtain two types of theorems: first, necessary and sufficient conditions that a given set T be a separating transcendence basis for a given extension K/L; secondly, necessary and sufficient conditions that there exist some separating transcendence basis for a given extension.

THEOREM 13. *If the extension K/L preserves p-independence, then a subset $T \subset K$ is a separating transcendence basis for K/L if and only if T is both a transcendence basis for K/L and relatively p-independent in K/L.*

Proof. Lemma 3 insures that any separating transcendence basis has the two specified properties. Conversely, suppose that some T with these two properties is not a separating transcendence basis for K/L. Then some b of K is not separable over $L(T)$, hence satisfies for $y = b$ an irreducible polynomial equation $f(y, T) = 0$ with exponent at least p in y and with coefficients in L. At least one variable t of T must (by Lemma 2 of §5) have exponent 1 in this polynomial f. As in the "exchange" argument for Theorem 6, we can regard $f(y, T) = 0$ as an irreducible and separable equation for t over $L(T - \{t\}, b^p)$, for f involves $y = b$ only as y^p. Therefore t is separable over the larger field $K^p(L, T - \{t\})$. But t is also purely inseparable over this field, so, by Lemma 1, t must lie in the field $K^p(L, T - \{t\})$. This conclusion states that t is relatively p-dependent on $T - \{t\}$, contrary to the hypothesis that T is relatively p-independent. Hence T must be a separating basis.

THEOREM 14. *If the field K has a finite transcendence basis T over its subfield L, then K has a separating transcendence basis over L if and only if*

(i) *K/L preserves p-independence;*

(ii) *for some integer e, $L(K^{p^e})$ is separable over $L(T)$.*

Proof. That condition (i) is necessary was established in Theorem 9(a), while the necessity of (ii) results from the finiteness of the set T exactly as in Theorem 6 of §2.

Conversely, suppose that (i) and (ii) hold, and pick a relative p-basis B for K/L. According to (i) and Theorem 11 the elements of B are algebraically independent over L, so that we can[18] embed B in a transcendence basis $B \cup X$

[18] In any (abstract) dependence relation, an independent subset can be enlarged to form a maximal independent subset. Mac Lane [7], Theorem 3.

for K/L. From the assumed separability of $L(K^{p^e})$ over the original transcendence basis T one finds, since T is finite, a larger integer f such that $L(K^{p^f})$ is separable over $L(B, X)$. Thence we deduce successively the separability of each of the following extensions

$$(1)\quad L^p(K^{p^{f+1}})/L^p(B^p, X^p); \quad L(K^{p^{f+1}})/L(B^p, X^p); \quad K^{p^{f+1}}(L, B)/L(B, X^p).$$

But B is a relative p-basis for K/L, hence $K = K^p(L, B) = K^{p^2}(L, B) = \cdots = K^{p^{f+1}}(L, B)$. Thus (1) states that K is separable over the basis $B \cup X^p$, which is obviously a transcendence basis because $B \cup X$ is by construction a transcendence basis. Thus we have found, as required, a separating transcendence basis[19] $B \cup X^p$ for K/L.

COROLLARY. *If L is a field of transcendence degree* 1 *over a perfect subfield P, and if K is an extension of L of finite transcendence degree, containing no elements inseparable and algebraic over L, then K has a separating transcendence basis over L.*

This follows at once from Theorems 14 and 9(b); it includes Theorem 2 of the introduction as the special case when L is a function field of one variable over P. It is also possible to prove this corollary without using the notion of p-independence, by a suitable extension of the exchange process used for Theorem 6. The hypothesis that L is a function field of only one variable is essential to this theorem; for suppose instead that $L = P(x, y)$ is a rational function field of two independent variables x and y over a perfect field P, and consider the extension $K = L(z, u)$, where z is transcendent over L and u is a root of the inseparable equation $u^p = y + xz^p$. Any separating transcendence basis for K/L would consist of a single element t. Let $f(u, t) = 0$ and $g(z, t) = 0$ be respectively the separable irreducible polynomial equations for u and z over $L[t]$, of respective exponents p^α and p^β in t. Then u is separable over $L(t^{p^\alpha})$, z is separable over $L(t^{p^\beta})$, and t^{p^β} is separable over $L(z)$. If $\alpha \geqq \beta$, u is separable over $L(z)$, although the given equation $u^p = y + xz^p$ is inseparable. A similar contradiction arises if $\beta \geqq \alpha$. Hence this extension K of a function field L of two variables can have no separating transcendence basis t.

The close and natural relation between extensions preserving independence and extensions with separating transcendence bases will now be documented with a pair of theorems, one of which gives a necessary and sufficient condition for the existence of a separating basis, while the other gives conditions for the preservation of p-independence.

THEOREM 15. *Let K be a field obtained by adjoining a finite number of elements to L. Then K/L preserves p-independence if and only if K has a separating transcendence basis over L. Furthermore, if K/L does preserve p-independence, then a subset T of K is a separating transcendence basis for K/L if and only if it is a relative p-basis for K/L.*

[19] B alone is a transcendence basis, for X must be void. Any x in X is in K, hence by (1) is separable over $L(B, X^p)$, although x manifestly satisfies the inseparable equation $z^p - x^p = 0$ over $L(B, X^p)$, a contradiction to the assumption that X is not void.

Proof. We know that the presence of a separating transcendence basis S for K/L makes K/L preserve p-independence (Theorem 9(a)) and makes S a relative p-basis (Lemma 3 in §6). Thus we need only consider a relative p-basis T for an extension K/L which does preserve p-independence, and prove that T is a separating transcendence basis. If X is any transcendence basis for K/L, and if K_s is the subfield of all elements of K separable over $L(X)$, then X is a separating transcendence basis for K_s/L, hence a relative p-basis for K_s/L (Lemma 3 in §6). But K must be a purely inseparable extension of K_s, so that K/L has by Lemma 4 of §6 a relative p-basis Y consisting of exactly m elements, where m is the number of elements in X. The two p-bases Y and T both have the same number of elements,[20] so that we have in T a set of elements algebraically independent (Theorem 11) and equal in number to the number m of elements in a transcendence basis X. Therefore T is also a transcendence basis for K/L, so must be a separating basis by the sufficient condition of Theorem 13. Theorem 15 is established.

COROLLARY. *If an extension K/L has a finite separating transcendence basis, then any relative p-basis for K/L is a separating transcendence basis for K/L.*

Example. The hypothesis that the transcendence degree of K/L is finite is necessary for the validity of this corollary, even if we restrict the base field L to be perfect, as may be seen by the following example. Let P be a perfect field over which the elements $t_0, x_1, x_2, x_3, \cdots$ are algebraically independent, and introduce the quantities t_n successively as the roots of the (inseparable) equations

$$t_n^p = t_{n-1} + x_n \qquad (n = 1, 2, 3, \cdots).$$

K is the field $P(x_1, x_2, \cdots, t_0, t_1, t_2, \cdots)$ generated by all these quantities, and $T = \{t_0, t_1, t_2, \cdots\}$ is a separating transcendence basis for K/P. Furthermore, the set $X = \{x_1, x_2, \cdots\}$ can be shown to be a p-basis for K, hence a relative p-basis for K/P. Nevertheless, this p-basis is not even a transcendence basis, so is certainly not a separating transcendence basis.

THEOREM 16. *An extension K/L preserves p-independence if and only if $L(y_1, \cdots, y_n)$ has a separating transcendence basis over L for every finite set of elements $y_1, \cdots, y_n$ from K.*

Proof. If K/L does preserve p-independence, the extension from L to the subfield $K' = L(y_1, \cdots, y_n)$ must also preserve p-independence, so that Theorem 15 at once gives a separating transcendence basis for K'/L. Conversely, suppose that each K'/L has such a basis, but that the whole extension K/L does not preserve p-independence, so that some p-independent subset X of L becomes p-dependent in K. This means that some $x \in K^p(X')$, where $X' \subset X$ is a finite subset not containing x. Therefore $x \in L^p(y_1^p, \cdots, y_n^p, X')$ for suitable elements $y_1, \cdots, y_n$ in K. This result states that the original set X has

[20] This number is an invariant of the extension K/L. Mac Lane [7], Theorem 6.

already become p-dependent in the field $K' = L(y_1, \cdots, y_n)$, contrary to the hypothesis that every subfield with such a finite generation has a separating transcendence basis and hence preserves p-independence.

The distinction between extensions preserving p-independence and extensions with separating transcendence bases arises only for extensions with infinitely many generators, as for instance in the extension $K = L(x^{p^{-1}}, x^{p^{-2}}, \cdots)$ (where x is transcendental over L), which preserves p-independence but which, according to Theorem 14, does not have over L a separating transcendence basis.

8. **Intermediate fields.** The question next to be considered is this: If an extension K/M has a separating transcendence basis, and if L is a field between K and M, under what circumstances does L have a separating transcendence basis over M? Our answer, though dependent on the previous analyses of p-independence, can be stated independently of that notion.

THEOREM 17. *If the fields $K \supset L \supset M$ are such that K has a finite separating transcendence basis over M, then L also has a (finite) separating transcendence basis over M.*

Proof. Certainly K/M and thus L/M preserves p-independence (Theorem 9(a)). Pick transcendence bases X and Y respectively for K/L and L/M; $X \cup Y$ is then a transcendence basis for K/M. By the necessary condition (Theorem 14) for the existence of a separating basis, $M(K^{p^e})$ is separable over $M(X, Y)$ for some e. Therefore $M(L^{p^e})$ is separable over $M(X, Y)$. The adjunction of the indeterminates X cannot reduce any equations irreducible over $M(Y)$, so $M(L^{p^e})$ is also separable over $M(Y)$. This is the sufficient condition of Theorem 14 for the existence of a separating transcendence basis for L/M.

Example. This conclusion could not be asserted were the transcendence degree of K/M infinite, even if we restrict M to be a perfect field P. For let $T = \{t_0, t_1, \cdots\}$ be a set of elements algebraically independent over P, and define another set $Y = \{y_2, y_3, \cdots\}$ successively by the inseparable equations

$$y_n^p = t_{n-2} + t_{n-1}t_n^p \qquad (n = 2, 3, 4, \cdots). \tag{1}$$

Then the field $L = P(T, Y)$ does not have a separating transcendence basis[21] over P. Nevertheless L can be embedded in a larger field $K = L(T^{1/p}) = P(T, Y, T^{1/p}) = P(T^{1/p})$ which does have the separating transcendence basis $T^{1/p}$ over P. This counter-example depends essentially on the fact that the intermediate field L also has an infinite transcendence degree.

COROLLARY. *If the fields $K \supset L \supset M$ are such that K has a separating transcendence basis S over M and L has a finite transcendence degree over M, then L has a separating transcendence basis over M.*

[21] Proof given in Mac Lane [5], Lemma 8.5, where the present L appears as a field S_1, and where it is shown that L has a separating transcendence basis neither over P nor over $P(t_0)$.

Proof. Since L/M has a finite transcendence basis, all elements of this basis are algebraic over a finite subset S_0 of S. Hence L is contained in the field K_0 composed of those elements of K algebraic over $M(S_0)$. For this field, S_0 is a finite separating transcendence basis, so Theorem 17 applies to $K_0 \supset L \supset M$.

The question of a separating basis for K itself over the intermediate field L may now be considered.

THEOREM 18. *If $K \supset L \supset M$, where K has a finite separating transcendence basis over M, K has a separating transcendence basis over L if and only if K/L preserves p-independence.*

Proof. The necessity of this condition is known (Theorem 9(a)). Conversely, suppose that K/L does preserve p-independence, that T is any transcendence basis for K/L, and that S is the given separating basis for K/M. Because S is finite, there is an integer e such that S^{p^e} is separable over $L(T)$. Then $L(K^{p^e})$ is separable over $L(T)$, so that K/L must have a separating transcendence basis by the fundamental criterion of Theorem 14.

It is impossible that a field K have a separating transcendence basis over any subfield smaller than its maximal perfect subfield, as one sees by the following result.

THEOREM 19. *If a field K has a separating transcendence basis over a subfield L, then L contains the maximal perfect subfield of K.*

Proof. Let K^{p^∞} denote the maximal perfect subfield of K. If the conclusion were false, there would be a b in K^{p^∞} but not in L. All roots $b^{p^{-e}}$ lie in the perfect field $K^{p^\infty} \subset K$, so they all are separable over the extension $L(T)$ obtained from the given separating transcendence basis T for K/L. But $y = b$ is the root of some equation $g(y, T) = 0$ irreducible in the polynomial ring $L[y, T]$. Let t be a variable of T whose exponent p^e in g is as small as possible, so that no other t_i of T has exponent less than p^e, and let $T_0 = T - \{t\}$ be the set T with t deleted. Then $g(y, T) = g(y, t, T_0)$ is an irreducible separable equation for $y = b$ over $L(t^{p^e}, T_0^{p^e})$, with exponent 1 in the variable t^{p^e}. It follows (lemma below) that b is not separable over the smaller field $L(t^{p^{e+1}}, T_0^{p^e})$.

There is therefore an integer m and a corresponding set $S = T^{p^m}$ of transcendents such that all the roots $b^{p^{-n}}$ are separable over $L(S)$, while not all the roots $b^{p^{-n}}$ are separable over $L(S^p)$. Let $c = b^{p^{-d}}$ be one such inseparable root, so that c itself is not separable over $L(S^p)$, although $c^{p^{-1}}$ is separable over $L(S)$. If we apply the isomorphism $a \leftrightarrow a^p$, it follows that c is separable over $L^p(S^p)$, hence over the larger field $L(S^p)$, in contradiction to the choice of c. This contradiction establishes the theorem.

The lemma as to the inseparability of b used above is

LEMMA 5. *If a subset T of a field K is algebraically independent over a subfield L, and if an element b of K is a separable root $y = b$ of a polynomial $g(y, T)$ irreducible in the polynomial ring $L[y, T]$, then a variable t of T can appear in g with exponent 1 if and only if b is inseparable over $L(t^p, T - \{t\})$.*

The simple proof given in Mac Lane [5], Lemma II, §2, for the case when L is perfect, applies equally well for any field L. Easy consequences of the theorem are the following:

COROLLARY 1. *If a field K is obtained from a field L by successive transcendental and separable algebraic extensions, then the maximal perfect subfield of K is the maximal perfect subfield of L.*

COROLLARY 2. *If the elements of a set T are algebraically independent over a field L, then the intersection of all the fields $L(T^{p^e})$, for $e = 1, 2, \cdots$, is exactly the field L.*

9. **Pure forms.** The notion of a p-basis of a field can be used to show that Albert's results on pure null forms over a function field are in essence valid over an arbitrary coefficient field. We remark first that the *degree of imperfection* of a field K has been defined (Teichmüller [10]) to be the number of elements in a p-basis of K. A *pure form* f of degree q over K,

$$f(x_1, \cdots, x_m) = b_1 x_1^q + \cdots + b_m x_m^q \qquad (b_i \text{ in } K),$$

is a null form over K if $f(x_{10}, \cdots, x_{m0}) = 0$ for values $x_{10}, \cdots, x_{m0}$ not all zero in K.

THEOREM 20. *Every pure form $f(x_1, \cdots, x_m)$ of degree $q = p^e$ over a field K of characteristic p is a null form if the number of variables m exceeds q^r, where r is the degree of imperfection of K. However, there exist non-null forms over K for every $m \leqq q^r$.*

The proof is exactly similar to that given by Albert ([2], Theorem 6).

We also prove here a property of p-bases which we have used elsewhere without proof ([6], Lemma 3).

THEOREM 21. *If X and Y are disjoint subsets of a field K such that $X \cup Y$ is p-independent in K, and if $K(X^{p^{-\infty}})$ is obtained from K by adjoining all roots $x^{p^{-e}}$, for x in X and e an integer, then the set Y is p-independent in $K(X^{p^{-\infty}})$. Furthermore, if $X \cup Y$ is a p-basis of K, then Y is a p-basis of $K(X^{p^{-\infty}})$.*

Proof. If Y is not p-independent in $K(X^{p^{-\infty}})$, there is an element y in Y p-dependent on a finite subset Y_0 of Y not containing y. Hence for some integer e and some finite subset $X_0 \subset X$, $y \in K^p(X_0^{p^{-e}}, Y_0)$, so $y^{p^e} \in K^{p^{e+1}}(X_0, Y_0)$. This will lead to a contradiction on the degree of the field $L = K^{p^{e+1}}(X_0, Y_0, y)$ over $K^{p^{e+1}}$. For, on the one hand, the p-independence of $X \cup Y$ in K means that X_0, Y_0, and y together have degree $p^{\xi+\eta+1}$ over K^p, where ξ and η respectively denote the number of elements in X_0 and Y_0. Hence by an induction one finds

$$[L:K^{p^{e+1}}] = [K^{p^{e+1}}(X_0, Y_0, y):K^{p^{e+1}}] = p^{(\xi+\eta+1)(e+1)}.$$

On the other hand, $y^{p^e} \in K^{p^{e+1}}(X_0, Y_0)$, so that if we adjoin first all the elements of X_0 and Y_0, then finally the element y, we get for the same extension a degree not more than $p^{(\xi+\eta)(e+1)+e}$, a contradiction. This establishes the p-independence of Y in $K(X^{p^{-\infty}})$; that it becomes a p-basis is readily verified from the definition.

10. **Fields without separating transcendence bases.** We now give a counter-example to the possible extension of Theorem 4 of the introduction; in that we show:

(i) There is a field M which does not have a separating transcendence basis over its maximal perfect subfield P, but which does have a finite transcendence degree t over that field P. Here t may be any specified integer $t \geqq 2$.

This example will also show:

(ii) The number of elements in a p-basis of a field K (the so-called degree of imperfection of K) is not always equal to the transcendence degree of K over its maximal perfect subfield.

Teichmüller, in [10], has also proved (ii) by the example in which K is a field of formal power series, where the transcendence degree in question is infinite. Our example yields a field K in which both the transcendence degree and the degree of imperfection are finite.

Let P be a perfect field and $Z = \{z_1, z_2, \cdots\}$, a denumerable set of quantities algebraically independent over P. Denote by $P(Z^{p^{-\infty}})$ the perfect field

$$P(Z^{p^{-\infty}}) = P(Z, Z^{p^{-1}}, Z^{p^{-2}}, \cdots). \tag{1}$$

Let y and u_0 be algebraically independent over $P(Z^{p^{-\infty}})$, and define quantities u_n recursively by

$$u_n = y^{p^{n-1}} + z_n u_{n-1} \qquad (n = 1, 2, \cdots). \tag{2}$$

The field which we use as an example is then

$$M = P(Z^{p^{-\infty}}, y, u_0, u_1^{1/p}, u_2^{1/p^2}, \cdots, u_n^{1/p^n}, \cdots).$$

By (2), $u_{n-1}^{p^{1-n}}$ can be expressed in terms of $u_n^{p^{1-n}}$, so M is the union of a tower of fields $M_0 \subset M_1 \subset M_2 \subset \cdots$, where

$$M_n = P(Z^{p^{-\infty}}, y, u_n^{1/p^n}) \qquad (n = 0, 1, 2, \cdots).$$

From equations (2) we observe that

$$P(z_1, \cdots, z_n, y, u_0) = P(z_1, \cdots, z_n, y, u_n) \tag{3}$$

and hence that $z_1, \cdots, z_n, y, u_n$ are algebraically independent over P. Furthermore

$$M_0 = P(Z^{p^{-\infty}}, y, u_n), \qquad M_n = M_0(u_n^{p^{-n}}), \tag{4}$$

so that M_n is a purely inseparable extension of M_0 of degree p^n. Therefore the necessary condition of Theorem 6 applied to the transcendence basis $T = \{y, u_0\}$ for M proves the following result:

LEMMA 6. *The field M does not have a separating transcendence basis over $P(Z^{p^{-\infty}})$*

The extended field $M(y^{1/p})$ would by the equations (2) also contain each $u_{n-1}^{p^{-n}}$, so that we can assert

LEMMA 7. *The field M has a p-basis consisting of one element y and a transcendence basis* $\{y, u_0\}$ *over the perfect subfield* $P(Z^{p^{-\infty}})$.

Thus our example has the properties (i) and (ii) stated above, provided we can prove that $P(Z^{p^{-\infty}})$ is the maximal perfect subfield of M. This we now do.

LEMMA 8. *An element b of M is in* M_n *if and only if* b^{p^n} *is separable over* M_0.

Proof. By (4), all elements b of M_n certainly have the property stated. Conversely, suppose b^{p^n} separable over M_0, and take $k \geqq n$ so large that $b \in M_k = M_0(u_k^{p^{-k}})$. In such a field M_k it is known that any element b with b^{p^n} separable over the base field M_0 must be[22] in $M_0(u_k^{p^{-n}})$. If $k > n$, (2) yields $u_k^{p^{-n}} = y^{p^{k-n-1}} + z_k^{p^{-n}} u_{k-1}^{p^{-n}}$, hence $M_0(u_k^{p^{-n}}) = M_0(u_{k-1}^{p^{-n}})$. Therefore, by induction, $b \in M_0(u_n^{p^{-n}}) = M_n$, as asserted.

Consider now any element a in the maximal perfect subfield M^{p^∞} of M. The expansion for a in terms of the generators of M can involve but a finite number of u's, but a finite number of z's, and but a finite number of p^n-th roots of z's. Hence we can find a power $c = a^{p^n}$, also in M^{p^∞}, and an integer m such that $c \in P(Z_m, y, u_0)$, where Z_m is a finite subset

$$(5) \qquad Z_m = \{z_1, \cdots, z_m\}.$$

For each e, $c = c_e^{p^e}$ for some element c_e of M, while by Lemma 8, $c_e \in M_e$. Thus

$$(6) \qquad c \in P(Z_m, y, u_0), \qquad c \in M_e^{p^e} \qquad (e = 0, 1, 2, \cdots).$$

This situation will now be simplified by showing that only a finite number of z's need be used in these fields M_e, provided $e \geqq m$. Note that $M_e^{p^e} = P(Z^{p^{-\infty}}, y^{p^e}, u_e)$.

For each $e \geqq m$ pick the smallest n such that the first n z's suffice to make $c \in M_e^{p^e}$; i.e., such that $c \in P(Z_n^{p^{-\infty}}, y^{p^e}, u_e)$. If $n > e$, c is not in the field $R = P(Z_{n-1}^{p^{-\infty}}, y^{p^e}, u_e)$, but is in $R(z_n^{p^{-t}})$ for some t. As in the remark following (3), z_n is transcendental over $P(Z_{n-1}, y, u_e)$, hence over R. Thus c in the purely transcendental extension $R(z_n^{p^{-t}})$ must itself be transcendental over R. On the other hand, (6) and (4) state that

$$c \in P(Z_m, y, u_0) \subset P(Z_e, y, u_0) = P(Z_e, y, u_e) \subset P(Z_{n-1}, y, u_e),$$

and the last of these fields is algebraic over $R = P(Z_{n-1}^{p^{-\infty}}, y^{p^e}, u_e)$. Therefore c is algebraic over R, a contradiction from which we conclude that

$$c \in P(Z_e^{p^{-\infty}}, y^{p^e}, u_e) \qquad (e = m, m + 1, \cdots).$$

If the expression for c in terms of these generators actually involves some roots of Z_e, pick $s > 0$ so small that c is in $N^* = P(Z_e^{p^{-s}}, y^{p^e}, u_e)$ but not in $N =$

[22] Mac Lane [5], Lemma 6.1: *If u is transcendental over F, then the elements b of* $F(u^{p^{-\infty}})$ *with* b^{p^e} *separable over* $F(u)$ *all lie in* $F(u^{p^{-e}})$.

$P(Z_e^{p^{-s+1}}, y^{p^e}, u_e)$. The extension N^*/N is given by a tower $N \subset N_1 \subset N_2 \subset \cdots \subset N^*$, where

$$N_0 = N, \qquad N_i = N(z_1^{p^{-s}}, \cdots, z_i^{p^{-s}}) \qquad (i = 0, \cdots, e).$$

For some i, c is in N_i but not in N_{i-1}. Since $N^{*p} \subset N$, c is purely inseparable over N_{i-1}, and the two extensions $N_{i-1}(c)$ and $N_i = N_{i-1}(z_i^{p^{-s}})$, each of degree p, must be equal. Therefore $z_i^{p^{-s}} \in N_{i-1}(c)$. But now set $F = P([Z - \{z_i\}]^{p^{-\infty}}, y, u_0)$, so that $N_{i-1} \subset F(z_i^{p^{-s+1}})$, while by (6), $c \in P(Z_m, y, u_0) \subset F(z_i^{p^{-s+1}})$. Therefore $z_i^{p^{-s}} \in F(z_i^{p^{-s+1}})$. This conclusion is a contradiction because z_i is a quantity transcendental over F. Hence we have $s = 0$, and

$$c \in P(Z_e, y^{p^e}, u_e) \qquad (e = m, m+1, \cdots). \tag{7}$$

Combining (6) with (7), we next aim to prove that c is in each of the fields

$$D_{em} = P(Z_m, y^{p^e}, u_m^{p^{e-m}}) \qquad (e = m, m+1, \cdots). \tag{8}$$

Note that D_{ee} is the field $P(Z_e, y^{p^e}, u_e)$ of (7). A preliminary is

LEMMA 9. *Each D_{em} with $e \geqq m$ is relatively algebraically closed* (see footnote 13) *in D_{ee}.*

Proof. First simplify the notation of (8) thus:

$$D_{en} = P(Z_n, y', v) \qquad y' = y^{p^e}, \qquad v = u_n^{p^{e-n}}, \tag{9}$$

$$D_{en+1} = P(Z_n, z, y', u), \qquad z = z_{n+1}, \qquad u = u_{n+1}^{p^{e-n-1}} \tag{10}$$

In terms of these quantities u and v the defining equation (2) becomes

$$u^p = y' + z^{p^{e-n}} v. \tag{11}$$

Hence $D_{en} \subset D_{en+1}$, according to (9) and (10), and we can go from D_{en} to D_{ee} by a tower

$$D_{en} \subset D_{en+1} \subset D_{en+2} \subset \cdots \subset D_{ee}.$$

For the lemma it therefore suffices to prove D_{en} relatively algebraically closed in D_{en+1} for all $n < e$.

By (9) and (10) $D_{en+1} = D_{en}(z, u)$, where, as in (3), z is transcendental over D_{en}, while u is inseparable and algebraic over $D_{en}(z)$ in accord with (11). If D_{en} is not relatively algebraically closed in D_{en+1}, pick b in $D_{en+1} - D_{en}$ and algebraic over D_{en}. Then b^p is in $D_{en}(z)$ and is algebraic over D_{en}, so b^p is in D_{en}. Therefore b is purely inseparable over D_{en} and also over $D_{en}(z)$, and we must have $D_{en+1} = D_{en}(z, b)$. Therefore u is a rational function $g(z)/h(z)$ of z with coefficients in $D = D_{en}(b)$. We can assume that $g(0)$ and $h(0)$ are not both 0. This value of u in the defining equation (11) for u gives an identity

$$[h(z)]^p[y' + z^{p^{e-n}} v] = [g(z)]^p \tag{12}$$

in z over D. By setting $z = 0$, we find $y' = [g(0)/h(0)]^p$, with neither $h(0)$ nor $g(0)$ zero. This means that y' is in D^p, hence that $(y')^{1/p}$ is in D. A similar argument on the terms of highest degree in (12) proves that $v^{1/p} \in D$. However, in D_{en} of (9), Z_n, y' and v are algebraically independent over P, so that y', v are p-independent in D_{en}. This means that the extension $D_{en}((y')^{1/p}, v^{1/p})$ has degree p^2 over D_{en}, although we have just shown this extension to be contained in D, of degree p over D_{en}. This contradiction establishes the desired relative algebraic closure.

LEMMA 10. *For* $e \geqq m$, $c \in D_{em}$.

Proof. By (6), $c \in P(Z_m, y, u_0) = P(Z_m, y, u_m)$, a field which is certainly algebraic over D_{em} of (8). On the other hand $c \in D_{ee} = P(Z_e, y^{p^e}, u_e)$, by (7), and this field by the previous lemma contains no elements algebraic over D_{em} except the elements of D_{em} themselves. Hence we get the conclusion.[23]

If we put $L = P(Z_m)$, $\mu = e - m$, Lemma 10 states that c is in each of the fields $D_{em} = L((y^{p^m})^{p^\mu}, u_m^{p^\mu})$. Here y^{p^m} and u_m are algebraically independent over L, so the intersection of all these fields, for $\mu = 1, 2, \cdots$, is known by Corollary 2 to Theorem 19 of §8 to be L itself. Therefore $c \in P(Z_m)$, and the original element $a = c^{p^{-r}}$ of the maximal perfect subfield is therefore in $P(Z_m^{p^{-\infty}})$. We have thus completed our example by proving

LEMMA 11. *The field M has the maximal perfect subfield* $P(Z^{p^{-\infty}})$.

This field M thus has a transcendence degree 2 over its maximal perfect subfield. A field with analogous properties but with any desired transcendence degree $t \geqq 2$ over its maximal perfect subfield, is, by Corollary 1 of Theorem 19, the field $M^* = M(T)$, where T is a set of $t - 2$ elements algebraically independent over M.

BIBLIOGRAPHY

1. A. A. ALBERT, *Modern Higher Algebra*, Chicago, 1937.
2. A. A. ALBERT, *Quadratic null forms over a function field*, Annals of Mathematics, vol. 39(1938), pp. 494–505.
3. M. BECKER AND S. MAC LANE, *The minimum number of generators for inseparable algebraic extensions*. To be published.
4. H. HASSE AND F. K. SCHMIDT, *Die Struktur diskret bewerteter Körper*, Journal für die Mathematik, vol. 170(1934), pp. 4–63.
5. S. MAC LANE, *Steinitz field towers for modular fields*. Forthcoming in the Transactions of the American Mathematical Society.
6. S. MAC LANE, *Subfields and automorphism groups of p-adic fields*, Annals of Mathematics, vol. 40(1939), pp. 423–442.
7. S. MAC LANE, *A lattice formulation for transcendence degrees and p-bases*, this Journal, vol. 4(1938), pp. 455–468.
8. F. K. SCHMIDT, *Allgemeine Körper im Gebiet der höheren Kongruenzen*, Dissertation, Erlangen, 1925.
9. E. STEINITZ, *Algebraische Theorie der Körper*, edited by R. Baer and H. Hasse, Berlin, 1930.

[23] The arguments of Lemmas 8 and 9 are the crux of this example. As given, they depend essentially upon the algebraic independence of the z's. This is the inner reason for the complicated structure of the maximal perfect field $P(Z^{p^{-\infty}})$ used for this example.

10. O. TEICHMÜLLER, *p-Algebren,* Deutsche Mathematik, vol. 1(1936), pp. 362–388.
11. O. TEICHMÜLLER, *Diskret bewertete perfekte Körper mit unvollkommenen Restklassenkörper,* Journal für die Mathematik, vol. 176(1937), pp. 126–140.
12. B. L. VAN DER WAERDEN, *Moderne Algebra,* vol. I, first edition, Berlin, 1930.
13. B. L. VAN DER WAERDEN, *Zur algebraischen Geometrie* XIV, *Schnittpunktszahlen von algebraischen Mannigfaltigkeiten,* Mathematische Annalen, vol. 115(1938), pp. 619-644.

HARVARD UNIVERSITY.

SUBFIELDS AND AUTOMORPHISM GROUPS OF p-ADIC FIELDS[1]

By Saunders MacLane

(Received October 27, 1938)

1. Introduction

The fundamental theorem on the structure of perfect fields states that if a field K of given characteristic is perfect (or topologically complete) with respect to a discrete non-archimedean valuation $|a|$, with the properties,

$$(1) \qquad |ab| = |a|\,|b|, \qquad |a+b| \leqq \mathrm{Max}\,(|a|, |b|),$$

then K is uniquely determined by its field $\mathfrak{K}$ of residue classes, except in the case when K has characteristic 0, $\mathfrak{K}$ has characteristic p, and K is "ramified" over its rational subfield. The first proof of this structure theorem in the general case is due to Hasse and Schmidt [2], but for an imperfect residue class field $\mathfrak{K}$ their proof unfortunately uses an elaborate and unproven[2] lemma on the generation of $\mathfrak{K}$ by a Steinitz field tower. A second proof of the structure theorem has been given by Witt [13] and Teichmüller [10] and [11a]. For the case of an (algebraically) perfect residue class field $\mathfrak{K}$ Witt's proof uses a sophisticated "vector" analysis construction of the p-adic fields K, and Teichmüller's methods then treat the imperfect fields K on this basis.

The structure theorem involves two steps: first, the construction of a discrete complete field K with a given characteristic and a given residue class field $\mathfrak{K}$; second, the demonstration that the so constructed field is unique. The separation of these two steps shows that the previous constructions of K have been needlessly involved and can be replaced by an elementary stepwise construction (§3). This construction can then be so combined with the methods of Teichmüller for the imperfect fields $\mathfrak{K}$ as to give an elementary proof of the structure theorem (Corollary to Theorem 8 in §6). The cases when K and its residue class field $\mathfrak{K}$ have the same characteristic can be reduced simply by known methods (see Teichmüller [11a]), as K is then a power series field. Hence we consider only the so-called p-adic and $\mathfrak{p}$-adic fields; that is, discrete complete fields of characteristic zero with residue class fields of characteristic p.

Problems of the relative structure of complete fields can also be treated by these methods. One such question, which is suggested by investigations of valuations of higher rank (MacLane [4]) is this: Given two unramified p-adic extensions K and K' of a given p-adic field k with residue class field $\mathfrak{k}$, such that

[1] Presented to the American Mathematical Society, Sept. 10, 1938.

[2] Difficulties present in the proof of such a Lemma are discussed in another paper by the author, entitled "Steinitz field towers for modular fields," forthcoming in the Transactions of the Am. Math. Soc.

Reprinted from the Annals of Math. 40 (1939), 423-442, by permission of Princeton University Press.

K and K' have the same field $\mathfrak{K}$ of residue classes, are K and K' analytically equivalent over k? If $\mathfrak{k}$ and $\mathfrak{K}$ are finite fields, then k is an ordinary field of p-adic numbers, and the answer is known to be yes: K and K' are equivalent. For infinite fields $\mathfrak{K}/\mathfrak{k}$, any two extensions K and K' are still equivalent, as stated in Theorem 8 of §6, under certain restriction on the extension $\mathfrak{K}/\mathfrak{k}$. That some such restriction is necessary is indicated in §6 by an example.

Preliminary instances of this relative structure theory include the case of (transcendentally) separable extensions $\mathfrak{K}/\mathfrak{k}$, in §3, and the case of perfect fields $\mathfrak{K}$ and $\mathfrak{k}$, in §4. Two essential tools are the notions of multiplicative representatives (§2), and p-bases of imperfect fields, both due to Teichmüller.

Another question of relative structure of a p-adic field K concerns the existence of a p-adic subfield or intermediate field k with a specified residue class subfield $\mathfrak{k}$. The results on this problem given in §7 include the theorem that there is exactly one such field k if $\mathfrak{k}$ is perfect.

The group G of all automorphisms of a p-adic field K has an inertial subgroup G_1, consisting of automorphisms which leave all residue classes fixed, as well as certain ramification subgroups G_n. The quotient group G/G_1 is isomorphic (§8) to the automorphism group of the residue class field $\mathfrak{K}$, while each quotient group G_n/G_{n+1} is abelian and can be described explicitly in terms of $\mathfrak{K}$ (see §9). The results depend on certain constructions of automorphisms which are obtained from the relative structure theory of the preceding sections. The capstone of these investigations is the theorem of §10 which states that the subfield of elements left fixed by every automorphism of the inertial group G_1 is exactly that subfield k whose residue class field is the maximal perfect subfield of the whole residue class field. Strangely enough, the field of elements left fixed by the automorphisms of any ramification group G_n is the same field k!

2. Multiplicative representatives

A rank one "exponential" *valuation*[3] V of a field K is a real valued function $V(a)$ defined for $a \neq 0$ in K and having for all a and b the properties

(2-1) $$V(ab) = V(a) + V(b), \qquad V(a+b) \geqq \text{Min}\,(Va, Vb).$$

Associated with such a valuation there is the *value group* Γ, composed of all numbers $V(a)$, and the *residue class field* (res. field) $\mathfrak{K} = B/P$, obtained from the valuation ring B, which consists of all elements a with $Va \geqq 0$, and the prime ideal P, which contains all elements a with $Va > 0$. Denote by $H(a)$ the residue class of a in $\mathfrak{K}$, if a is in B, and set $H(a) = \infty$ if a is not in B. Then $H(a)$ is in the set $\{\mathfrak{K}, \infty\}$ composed of the elements of $\mathfrak{K}$ and the symbol ∞, while the many-one correspondence $a \rightarrow H(a)$ of K to $\{\mathfrak{K}, \infty\}$ has the properties that $Ha \neq \infty$ and $Hb \neq \infty$ imply

(2-2) $$H(a+b) = H(a) + H(b), \qquad H(ab) = (Ha)(Hb).$$

[3] The non-archimedean norm $|a|$ of (1) in the introduction generates an exponential valuation $V(a) = -\log |a|$.

The functions H and V are connected by the logical equivalence of the three statements

(2-3) $$H(a) = 0 \leftrightarrow H(1/a) = \infty \leftrightarrow Va > 0.$$

Such a homomorphism H reciprocally determines a corresponding V. By a *valuation* of K we henceforth mean a pair of functions $\{V, H\}$ consisting of a valuation function V and a homomorphism H of K on some $\{\mathfrak{K}, \infty\}$ with the properties (1), (2), and (3). Since V determines H and conversely, this simultaneous use of V and H is more symmetric; it also avoids conventions as to res. fields.[4] For instance if $K \subset L$, a valuation $\{W, H'\}$ of L is an *extension* of a valuation $\{V, H\}$ of K, if the functions W and H' are respectively extensions of V and H. Necessarily, then, the res. field $\mathfrak{L}$ of L contains that of K and the value group of L contains that of K. The extension L/K is *unramified* if these value groups are actually the same.

A valuation V is *discrete* if its value group Γ is isomorphic to the group of rational integers. A field K is *complete* in a valuation $\{V, H\}$, if it is complete in the topology introduced by the norm $|a| = e^{-Va}$. A discrete valuation of a field K is called $\mathfrak{p}$-*adic* if K is complete and of a characteristic 0 and if the res. field $\mathfrak{K}$ has as characteristic a prime p. A *prime element* π of K then is an element of minimum positive value in K. A $\mathfrak{p}$-adic valuation is called p-adic if the rational prime p is a prime element.

A *map* or *analytic isomorphism* T of K in K', where K and K' are fields with respective valuations $\{V, H\}$ and $\{V', H'\}$, is defined as an isomorphism $a \leftrightarrow a^T$ of K to a subset of K' which preserves valuations: $V'(a^T) = V(a)$. If the images a^T include all the elements of K, then T is an isomorphism of K *on* K'. Any isomorphism T generates a corresponding isomorphism τ of the res. field $\mathfrak{K}$ in a subfield of the res. field $\mathfrak{K}'$ of K', by the rule

(2-4) $$[H(a)]^\tau = H'(a^T).$$

The res. field $\mathfrak{K}$ may be replaced by this isomorphic subfield $\mathfrak{K}^\tau$ of $\mathfrak{K}'$ without changing the function V in the given valuation $\{V, H\}$. Hence, given such an isomorphism, we shall often assume that $\mathfrak{K} \subset \mathfrak{K}'$ and that T leaves the res. field $\mathfrak{K}$ elementwise fixed, that is

(2-5) $$H(a) = H'(a^T) \qquad (a \text{ in } K).$$

The congruence $a \equiv b \pmod{\pi^n}$, for π a prime element of a $\mathfrak{p}$-adic field K and for a and b in the valuation ring of K, means that $V(a - b) \geqq nV(\pi)$. In other words $a \equiv b \pmod{\pi^n}$ implies $a = b + d$ for some d with $V(d) \geqq V(\pi^n)$. Therefore

$$a^p = (b + d)^p = b^p + pb^{p-1}d + \cdots + pbd^{p-1} + d^p \equiv b^p \qquad (\text{mod } \pi^{n+1}),$$

hence the conclusion (Teichmüller [10], Lemma 8).

---

[4] Compare also the simpler proof obtained in Theorem 13 in §8.

LEMMA 1. *For elements a and b of a $\mathfrak{p}$-adic field, with $V(a) \geqq 0$, $V(b) \geqq 0$, $a \equiv b \pmod{\pi^n}$ implies $a^{p^m} \equiv b^{p^m} \pmod{\pi^{n+m}}$ for any $m \geqq 0$.*

For p-adic fields with perfect res. fields Teichmüller ([10], Lemma 5) has found a system R of "multiplicative representatives" of the residue classes. For any α in the res. field there is in R one *representative* a so that $H(a) = \alpha$.

LEMMA 2. *If $\mathfrak{k}$ is a perfect field of characteristic p, then any $\mathfrak{p}$-adic field K with residue class field $\mathfrak{K} \supset \mathfrak{k}$ contains a system R of representatives of the residue classes of $\mathfrak{k}$ which is uniquely determined by any one of the following three equivalent properties:*

(i) *If a_n of K is in the residue class $\alpha^{p^{-n}}$ for α in $\mathfrak{k}$, then $\lim (n \to \infty) a_n^{p^n}$ exists and is the R-representative of α;*

(ii) *If b_n is the R-representative of $\beta^{p^{-n}}$, for β in $\mathfrak{k}$, then $b_{n+1}^p = b_n$.*

(iii) *If a, b, and c are R-representatives, respectively, of elements α, β,* and *γ of $\mathfrak{k}$, then $\alpha\beta = \gamma$ implies $ab = c$.*

The proof of this Lemma is exactly like that given by Teichmüller for the special case $\mathfrak{K} = \mathfrak{k}$, and depends on using the description of (i) to define the system R.

I. RELATIVE STRUCTURE OF p-ADIC FIELDS

The Equivalence problem

Our primary problem is the relative equivalence of the extensions of complete fields with specified residue class extensions. Given a complete field k with a residue class field $\mathfrak{k}$ and two unramified complete extensions K and K' with the same residue class field $\mathfrak{K} = \mathfrak{K}'$; that is, given

$$(*) \qquad K \supset k, \qquad K' \supset k \text{ complete, unramified with } \mathfrak{K} = \mathfrak{K}' \supset \mathfrak{k};$$

when is K analytically equivalent to K'? Here an *analytic equivalence* of K to K' means an analytic isomorphism T of K on K' which is the identity on k and also on the common residue class field $\mathfrak{K}$. The "discrete," or "p-adic" equivalence problem will always refer to the problem (*), in which k is assumed discrete or p-adic, as the case may be. A gothic $\mathfrak{k}$ or $\mathfrak{L}$ will always denote the residue class field of the field represented by the corresponding roman letter k or L, etc.

3. Extension of maps

Though the construction of p-adic fields with specified res. fields has usually been accomplished by elaborate methods, an elementary procedure ([4], §5, especially Theorem 4) gives a step by step construction. For instance, if a given p-adic field k with valuation $\{V, H\}$ has the res. field $\mathfrak{k}$, and if $\mathfrak{K} = \mathfrak{k}(\xi)$ is a given transcendental extension of $\mathfrak{k}$, then a corresponding extension of the complete field k can be found by constructing a transcendental extension $k(x)$ and defining V' and H' for any polynomial $f(x) = a_n x^n + \cdots + a_0$ in $k[x]$ by

$$(3\text{-}1) \qquad V'\left(\sum_{i=0}^{n} a_i x^i\right) = \operatorname{Min}_i (Va_i), \qquad H'\left(\sum_{i=0}^{n} a_i x^i\right) = \sum_{i=0}^{n} (Ha_i)\xi^i.$$

The complete closure of $k(x)$ then yields a p-adic field with the res. field $\mathfrak{K} = \mathfrak{k}(\xi)$. Combining this with a similar treatment of algebraic extensions we have:

THEOREM 1. *Extension Theorem. Let k be a complete field with valuation $\{V, H\}$ and residue class field $\mathfrak{k}$, while $\mathfrak{K}$ is any extension of $\mathfrak{k}$. Then there exists an extension K of k, complete and unramified with respect to an extended valuation and having the residue class field $\mathfrak{K}$.*

For any prime res. field there is a corresponding field of p-adic numbers; hence the existence Theorem:

THEOREM 2. *Existence Theorem. There is a p-adic field K with any given residue class field $\mathfrak{K}$ of characteristic p.*

The extension Theorem 1 and subsequent studies of the uniqueness of this extension are closely related to the question of unramified division algebras with specified residue class division algebras, as treated by Witt ([12]) and Nakayama ([6]).

The problem of extending a given analytic isomorphism to parallel a specific residue class extension $\mathfrak{K}/\mathfrak{k}$ can be treated if $\mathfrak{K}$ has a separating transcendence basis[5] over $\mathfrak{k}$:

THEOREM 3. *Separable map extensions. Let two complete fields k and k' with the same residue class fields $\mathfrak{k}$ be analytically isomorphic under the map T which is the identity on $\mathfrak{k}$, and let K and K' be complete unramified extensions, $K \supset k$, $K' \supset k'$, with respective residue class fields $\mathfrak{K} \subset \mathfrak{K}'$. If the field $\mathfrak{K}$ has a separating transcendence basis $\mathfrak{Y}$ over $\mathfrak{k}$, the given map T can be extended to an analytic isomorphism T^* of K on a subfield of K', which is again the identity on $\mathfrak{K}$. There is only one such extension T^* if and only if $\mathfrak{K}$ is algebraic over $\mathfrak{k}$.*

The most natural instance of this Theorem arises when K and K' have the same res. field $\mathfrak{K} = \mathfrak{K}'$, so that T^* maps K on all K'. The proof of this instance is the same as for the general case, and the more general case is useful later in §7.

PROOF. Because $\mathfrak{Y}$ is a sep. trans. basis, $\mathfrak{K}/\mathfrak{k}$ can be built up by successive transcendental and sep. algebraic adjunctions. The required extension can then be found by transfinite induction if the case of a single transcendental or a separable algebraic extension can be treated (cf. MacLane [4], Lemma 2). *Transcendental case.* If $\mathfrak{K} = \mathfrak{k}(\xi)$ for ξ transcendental over $\mathfrak{k}$, and if x and x' are the elements from K and K' respectively representing ξ, then x and x' are transcendental over k and k', and both $k(x)$ and $k'(x')$ have the res. field $\mathfrak{k}(\xi)$. T can be extended in one and only one way to a value-preserving isomorphism T^* of $k(x)$ onto $k'(x')$ such that $x^{T^*} = x'$. Specifically, if $f(x) = a_n x^n + \cdots + a_0$ is a polynomial in $k[x]$, then $V(f)$ and $H(f)$ are given by (3-1) and $[f(x)]^{T^*} = \sum a_i^T x'^i$ is the desired extension. This T^*, extended in the natural way[6] to the complete closure of $k(x)$, is still the identity on $\mathfrak{K}$. *Separable algebraic case.* If $\mathfrak{K} = \mathfrak{k}(\beta)$, where β satisfies the irreducible, separable, and monic equation

[5] A subset $\mathfrak{Y}$ of $\mathfrak{K}$ is a separating transcendence basis (sep. trans. basis) for $\mathfrak{K}$ over $\mathfrak{k}$ if the elements of $\mathfrak{Y}$ are algebraically independent over $\mathfrak{k}$ and if $\mathfrak{K}$ is separable and algebraic over $\mathfrak{k}$ $(\mathfrak{Y})$.

[6] Hasse-Schmidt, [2], Fortsetzbarkeitsatz p. 26 and Satz 2 on p. 28.

$g(z) = 0$ over $\mathfrak{k}$, choose in $k[z]$ any monic polynomial $G(z)$ with the same degree as $g(z)$ and with coefficients representative of the coefficients of $g(z)$. Denote by $G'(z)$ the corresponding polynomial in K'. By the Hensel-Rychlik reducibility Theorem,[7] K contains exactly one root b of $G(z) = 0$ with residue class β. Furthermore $k(b)$ has res. field $\mathfrak{K} = \mathfrak{k}(\beta)$, and the map T can be extended to $k(b)$ by taking as b^T the correspondingly unique root b' of $G'(z) = 0$ with residue class β. This is clearly the only extension of T to $K = k(b)$, so that the Theorem is proven.

In the special case when $\mathfrak{K} = \mathfrak{K}'$ the extension T^* found by combining the above methods maps K on all K'. For, the map K^{T^*} is a subfield of K' with the same value group and the same res. field as K', hence is necessarily equal to K' (MacLane [4], Theorem 8). Hence a first equivalence Theorem, which applies without any assumption of discreteness:

Theorem 4. *Separable Equivalence. In an unramified equivalence problem,*

$$(*) \qquad K \supset k, \qquad K' \supset k, \qquad \text{with } \mathfrak{K} = \mathfrak{K}' \supset \mathfrak{k},$$

the complete fields K and K' are analytically equivalent over k if the residue class field $\mathfrak{K}$ has a separating transcendence basis over the original residue class field $\mathfrak{k}$.

4. Complete fields with perfect residue class fields

Theorem 5. *Given a discrete and unramified equivalence problem (*) with perfect residue class fields $\mathfrak{K}$ and $\mathfrak{k}$, the extensions K and K' are always analytically equivalent over k. More generally, if the given unramified extensions K and K' have distinct residue class fields $\mathfrak{K}$ and $\mathfrak{K}'$ with $\mathfrak{K} \subset \mathfrak{K}'$ and if $\mathfrak{K}$ and $\mathfrak{k}$ are perfect, there is an analytic isomorphism T of K in K' which leaves elements of k and residue classes of $\mathfrak{K}$ fixed.*

This Theorem, when proven, will have the following consequences:

Corollary 1. *If K and K' are p-adic fields with respective residue class fields $\mathfrak{K} \subset \mathfrak{K}'$, where $\mathfrak{K}$ is perfect, there is an analytic isomorphism of K into a subfield of K' which leaves fixed the residue classes of $\mathfrak{K}$.*

Proof. Apply Theorem 5, using for k the minimal complete subfield (of p-adic numbers) contained in both K and K'. A special case of this corollary is the usual "structure theorem" for a perfect residue class field:

Corollary 2. *Any two p-adic fields with the same perfect residue class field are analytically isomorphic under a map which leaves each residue class fixed.*

Theorem 5 may be proven directly from the separable map extension of Theorem 3 if the res. field $\mathfrak{K}$ has characteristic 0. Suppose then that $\mathfrak{K}$ has characteristic p, and let the set $\mathfrak{Y}$ of elements η form a transcendence basis for $\mathfrak{K}/\mathfrak{k}$. The perfect field $\mathfrak{K}$ must contain all roots $\eta^{p^{-e}}$ and hence also the subfield $\mathfrak{K}_0 = \mathfrak{k}(\mathfrak{Y})^{p^{-\infty}})$ consisting of all rational functions of all quantities $\eta^{p^{-e}}$, for which η is in $\mathfrak{Y}$ and e is an integer. For each such η Lemma 2 provides multi-

[7] Albert, [1], Lemma on p. 296: Hasse-Schmidt, [2], p. 31. The proof given by Albert for the discrete case holds equally well for a non-discrete valuation.

plicative representatives y and y' in K and K' respectively. The set Y of all the representatives y in K is a set of independent transcendents over k because the homomorphic images $\mathfrak{Y}$ are algebraically independent over $\mathfrak{K}$. The multiplicative representative $y^{p^{-e}}$ of $\eta^{p^{-e}}$ is by Lemma 2, (ii) a uniquely determined p^{eth} root of y. Therefore K contains the ring $R_0 = k[Y^{p^{-\infty}}]$, which consists of all polynomials f in the quantities $y^{p^{-e}}$, of the form

$$(4\text{-}1) \qquad f(y_1, \cdots, y_m) = a_1(y_1^{q_{11}} \cdots y_m^{q_{1m}}) + \cdots + a_t(y_1^{q_{t1}} \cdots y_m^{q_{tm}}),$$

in which $y_1, \cdots, y_m$ are any elements of Y, each coefficient a_i is in k and each exponent q_{ij} is a rational fraction with denominator a power of p. The valuation $\{V, H\}$ of K has $Vy = 0$, $Hy = \eta$. The algebraic independence of the η's and the definition (2-2) of a homomorphism give, by computation in (4-1),

$$(4\text{-}2) \qquad V(f) = \text{Min}(Va_1, \cdots, Va_t)$$

$$(4\text{-}3) \qquad H(f) = (Ha_1)(\eta_1^{q_{11}} \cdots \eta_m^{q_{1m}}) + \cdots + (Ha_t)(\eta_1^{q_{t1}} \cdots \eta_m^{q_{tm}}).$$

In other words, $H[f(y_1, \cdots, y_m)] = f^H(\eta_1, \cdots, \eta_m)$, where f^H denotes the polynomial obtained by applying H to each coefficient of f.

The set Y' of all multiplicative representatives y' of elements η in K' generates a similar ring $R_0' = k[Y'^{p^{-\infty}}] \subset K'$, for which the valuation and homomorphism are given by similar formulas. This means that the correspondence

$$T_0 : f(y_1, \cdots, y_m) \leftrightarrow f(y_1', \cdots, y_m')$$

is an analytic isomorphism of R_0 to R_0' which leaves fixed elements of k and of the residue-class ring $\mathfrak{k}[\mathfrak{Y}^{p^{-\infty}}]$. In natural fashion T_0 can be extended to the quotient field of R_0 and thence to the complete closure K_0 of this quotient field in K. The res. field of K_0 is $\mathfrak{K}_0 = \mathfrak{k}(\mathfrak{Y}^{p^{-\infty}})$; since $\mathfrak{Y}$ was chosen a trans. basis for $\mathfrak{K}/\mathfrak{k}$, the whole res. field is algebraic over the smaller field $\mathfrak{K}_0$. Furthermore $\mathfrak{K}_0$, like $\mathfrak{k}$, is perfect, so that $\mathfrak{K}$ must be separable over $\mathfrak{K}_0$. The map of K_0 in K' can therefore be extended by Theorem 3 to a map of K in K'. By construction, this map leaves fixed all elements of k and all res. classes.

5. Teichmüller's embedding process

Imperfect residue class fields may be analysed by p-bases. In any field $\mathfrak{K}$ of characteristic p consider the generation of $\mathfrak{K}$ over the subfield $\mathfrak{K}^p$ consisting of the pth powers of all its elements. A subset $\mathfrak{M}$ of $\mathfrak{K}$ is *p-independent* if $\mathfrak{K}^p(\mathfrak{M}')$ is a proper subfield of $\mathfrak{K}^p(\mathfrak{M})$ whenever $\mathfrak{M}'$ is a proper subset of $\mathfrak{M}$. A subset $\mathfrak{M}$ is a *p-basis* of $\mathfrak{K}$ if $\mathfrak{M}$ is p-independent and if $\mathfrak{K}^p(\mathfrak{M}) = \mathfrak{K}$. Teichmüller ([11], §3), who gave definitions equivalent to these, showed that every $\mathfrak{K}$ has a p-basis and can be obtained from the subfield $\mathfrak{K}^{p^n}$ of the $p^{n\text{th}}$ powers ($n \geqq 0$) of all its elements by the adjunction of any p-basis $\mathfrak{M}$;

$$(5\text{-}1) \qquad \mathfrak{K} = \mathfrak{K}^{p^n}(\mathfrak{M}).$$

A p-adic field with imperfect res. field can be embedded in a larger field with perfect res. field by successive extensions like those of Theorem 1. Teichmüller's exposition of this process[8] may be summarized thus:

THEOREM 6. (Teichmüller.) *If the* $\mathfrak{p}$*-adic field* K *has the residue class field* $\mathfrak{K}$ *containing a p-independent subset* $\mathfrak{M}$*, and if* $M \subset K$ *is a fixed set of representatives in* K *of the residue classes of* $\mathfrak{M}$*, then there exists a complete unramified extension* L *of* K *with a residue class field* $\mathfrak{L} = \mathfrak{K}(\mathfrak{M}^{p^{-\infty}})$*, whose maximal perfect subfield contains* $\mathfrak{M}$*, such that the multiplicative representatives in* L *of each element of* $\mathfrak{M}$ *are exactly the corresponding elements of* M*. If* $\mathfrak{M}$ *is a p-basis of* $\mathfrak{K}$*,* $\mathfrak{L}$ *is the minimal perfect extension* $\mathfrak{K}^{p^{-\infty}}$ *of* $\mathfrak{K}$*.*

In this theorem $\mathfrak{K}(\mathfrak{M}^{p^{-\infty}})$ denotes the field obtained from $\mathfrak{K}$ by adjoining all elements $\mu^{p^{-e}}$, for e an integer and μ in $\mathfrak{M}$. If $\mathfrak{M}$ is not a p-basis, a p-basis for $\mathfrak{L}$ may be found by the following subsequently useful process (proof omitted).

LEMMA 3. *If the disjoint subsets* $\mathfrak{M}$ *and* $\mathfrak{N}$ *of a field* $\mathfrak{K}$ *together constitutes a p-basis of* $\mathfrak{K}$*, then* $\mathfrak{N}$ *is a p-basis of* $\mathfrak{K}(\mathfrak{M}^{p^{-\infty}})$*.*

The existence and uniqueness of certain subfields $K \subset L$ in Theorem 6 were also established by Teichmüller for the p-adic case:

THEOREM 7. *If the p-adic field* L *has a residue class field* $\mathfrak{L}$ *with the maximal perfect subfield* $\mathfrak{L}^{p^\infty}$*, if* $\mathfrak{M}$ *is a p-basis of a subfield* $\mathfrak{K} \subset \mathfrak{L}^{p^\infty}$*, and if* M *is the set of multiplicative representatives of* $\mathfrak{M}$ *in* L*, then there is one and only one p-adic subfield* K *of* L *which contains* M *and has the residue class field* $\mathfrak{K}$*.*

PROOF. The existence of K, proven by Teichmüller by a suitable appeal to the Witt vector addition, can be obtained in elementary fashion. First construct any p-adic field K' with res. field $\mathfrak{K}$ (Theorem 1), and let M' be any set of representatives of $\mathfrak{M}$ in K'. By Theorem 6 embed K' in a p-adic field L_0 with res. field $\mathfrak{L}_0 = \mathfrak{K}^{p^{-\infty}} \subset \mathfrak{L}$ in such a way that the elements of M' become multiplicative representatives. By Corollary 1 of Theorem 5 there is a map T of L_0 in the given L. The definition of multiplicative representatives in Lemma 2 is invariant and T leaves residue classes fixed, so that T carries the set M' into the multiplicative representatives M. Therefore the image of K' under T is the required subfield K of L.

The uniqueness of K as asserted in Theorem 7 will be established in a more general form, useful subsequently in Theorem 12 and Lemma 6:

LEMMA 4. *In a* $\mathfrak{p}$*-adic field* L *any two complete subfields* K *and* K' *with the same residue class field* $\mathfrak{K}$ *are identical* ($K = K'$) *if* K *and* K' *contain in common some prime element* π *of* L *and a set* M *of representatives for some p-basis* $\mathfrak{M}$ *of the residue class field* $\mathfrak{K}$*.*

Note that $\mathfrak{L}$ need not be perfect nor M multiplicative, while if L is p-adic, the prime element p of L is necessarily present in every K, so the hypothesis on π is true.

Our proof arises from a simplification of certain res. class formulas due to

[8] Teichmüller, [10], pp. 146–147, Lemmas 14 to 18. This part of the paper uses no vector calculus, and proves Theorem 6 below only for $\mathfrak{M}$ a p-basis. Our more general statement is established by exactly the same methods.

Teichmüller ([10], Lemma 17). For any given b' of non-negative value in K' we need only find elements b_n in K such that $b' \equiv b_n \pmod{\pi^n}$ for each n. For then $b' = \lim b_n$ is in the complete field K, $K' \subset K$, and hence $K' = K$.

For $n = 1$, $b_1 \equiv b'$ exists because K and K' have the same res. field. In general, b_{n+1} is to be constructed by induction from b_n. As in (5-1), the res. class β of b' can be expressed as a polynomial $f(\mathfrak{M})$ in elements of the p-basis $\mathfrak{M}$ with coefficients α^{p^n} in $\mathfrak{K}^{p^n}$. If each coefficient α is replaced by a representative a' in K' and if each variable ξ of $\mathfrak{M}$ is replaced by its representative x in the set M, there results a polynomial $F'(M)$ lying in the residue class $f(\mathfrak{M})$, so that $F'(M) \equiv b' \pmod{\pi}$. Therefore $b' = F'(M) + c'\pi$ for some c' in K' with $Vc' \geqq 0$. Construct $F(M)$ from $F'(M)$ by replacing each coefficient a'^{p^n} by a^{p^n}, where a in K is chosen so that $a \equiv a' \pmod{\pi}$. Then by Lemma 1, $F'(M) \equiv F(M) \pmod{\pi^{n+1}}$, while the induction assumption yields a c_n in K with $c' \equiv c_n \pmod{\pi^n}$. Consequently $b' = F'(M) + c'\pi \equiv F(M) + c_n\pi \pmod{\pi^{n+1}}$, so that $F(M) + c_n\pi$ is the required element b_{n+1} of K.

An analogous proof yields a related Lemma concerned with maps, to be used in Lemma 6 of §8:

LEMMA 5. *Let K/k be an unramified $\mathfrak{p}$-adic extension with residue class fields $\mathfrak{K}/\mathfrak{k}$, and let $m \subset k$ be a system of representatives for a p-basis $\mathfrak{m}$ of $\mathfrak{k}$. Then any analytic automorphism T of K which carries m into itself and leaves $\mathfrak{k}$ and a prime element π of k elementwise fixed must leave k elementwise fixed.*

This includes the useful case when $K = k$, while if K is p-adic the hypothesis on π is superfluous. To prove the Lemma, we need only show $b^T \equiv b \pmod{\pi^n}$ for each n and each b in k with $Vb \geqq 0$. For $n = 1$ this must hold, since by hypothesis T leaves the res. class of b fixed. In the induction proof, $b = F(m) + c\pi$, as before, so that $b^T = F^T(m) + c^T\pi$, where $c^T \equiv c \pmod{\pi^n}$ and F^T is obtained from F by replacing each coefficient a^{p^n} by $(a^T)^{p^n}$. But

$$a \equiv a^T \pmod{\pi}, \qquad a^{p^n} \equiv (a^T)^{p^n} \qquad \pmod{\pi^{n+1}},$$

as in Lemma 1, so that we have the required conclusion $b^T \equiv b \pmod{\pi^{n+1}}$.

6. Equivalence of unramified extensions

A more general result than the equivalence Theorems 4 and 5 is given by

THEOREM 8. *Given a p-adic unramified equivalence problem (*) of Theorem 4, the extensions K and K' are analytically equivalent over k if some p-basis $\mathfrak{m}$ of the original residue class field $\mathfrak{k}$ is p-independent in the extended field $\mathfrak{K}$. If k is not p-adic but $\mathfrak{p}$-adic, the theorem is still true, provided k is unramified over some $\mathfrak{p}$-adic subfield k_0 whose residue class field is perfect.*

First some remarks: Any two p-adic fields K and K' with the same res. field $\mathfrak{K}$ both contain the (prime) field k of p-adic numbers. The res. field $\mathfrak{k}$ of k is finite and perfect, hence has a void p-basis, so that Theorem 8 includes the usual structure theorem for arbitrary res. fields:

COROLLARY 1. *Any two p-adic fields with the same residue class field are analytically isomorphic under a map which preserves residue classes.*

The essential hypothesis of Theorem 8 on $\mathfrak{K}/\mathfrak{k}$ is the requirement that some p-basis of $\mathfrak{k}$ is a subset of a p-basis of $\mathfrak{K}$. It can be shown that this will be the case if and only if every subset of $\mathfrak{k}$ p-independent in $\mathfrak{k}$ remains p-independent in $\mathfrak{K}$. An extension $\mathfrak{K}/\mathfrak{k}$ with this property is said to *preserve p-independence.* An algebraic extension $\mathfrak{K}/\mathfrak{k}$ preserves p-independence if and only if it is separable. The force of this hypothesis may also be indicated by the following examples of extensions $\mathfrak{K}/\mathfrak{k}$ preserving p-independence:

(i) Any extension $\mathfrak{K}$ of a perfect field $\mathfrak{k}$;

(ii) Any separable algebraic extension $\mathfrak{K}$ of any field $\mathfrak{k}$;

(iii) Any purely transcendental extension $\mathfrak{K}$ of any field $\mathfrak{k}$;

(iv) Any extension $\mathfrak{K}$ with a separating transcendence basis over $\mathfrak{k}$ (see §3).

The p-bases necessary to prove that these extensions preserve independence may be found directly or by the constructions for p-bases given by Teichmüller in [11]. These examples do not exhaust the independence-preserving extensions; for instance, over $\mathfrak{k} = \mathfrak{P}(x)$, $\mathfrak{K} = \mathfrak{P}(x, t, t^{p^{-1}}, t^{p^{-2}}, \cdots)$ preserves p-independence if $\mathfrak{P}$ is perfect and x and t are algebraically independent over $\mathfrak{P}$.

Proof of Theorem 8. The given p-basis $\mathfrak{m}$ of $\mathfrak{k}$ is p-independent in $\mathfrak{K}$ and hence can be enlarged to a p-basis $\mathfrak{M}$ of $\mathfrak{K}$ by known abstract properties of dependence relations. Choose a representative system m for $\mathfrak{m}$ in k and two representative systems $M \supset m$ and $M' \supset m$ for $\mathfrak{M}$ in K and K' respectively. Denote by $\mathfrak{L}$ the minimal perfect field $\mathfrak{K}^{p^{-\infty}}$ containing $\mathfrak{K}$ and use the Teichmüller embedding process to obtain unramified complete fields $L \supset K$, $L' \supset K'$ both with the res. field $\mathfrak{L}$ and containing M and M' respectively as multiplicative representatives. Both L and L' are unramified over the $\mathfrak{p}$-adic field k_0 with perfect res. field given in the hypothesis (if k is p-adic, let k_0 be the subfield of p-adic numbers). By the equivalence theorem for perfect residue class extensions (Theorem 5) there is a map T of L on L' leaving res. classes fixed and hence carrying the multiplicative representatives M of $\mathfrak{M}$ into M'. Furthermore K^T is a subfield of L' with res. field $\mathfrak{K}$ and containing $M' = M^T \subset K^T$, so that K^T and K' have both M' and a prime element π of k_0 in common. By Lemma 4, $K^T = K'$, so that T is actually a map of K on K'. By a similar argument, $k^T = k$, and T is an automorphism of k leaving the residue classes, the prime element π of k_0 and the multiplicative representatives m of the p-basis $\mathfrak{m}$ fixed. Hence T leaves k elementwise fixed, as in Lemma 5, and T does yield the required equivalence of K to K' over k.

Though we do not have a converse proof that preservation of p-independence is a necessary condition for the equivalence of unramified extensions K and K', some restrictive condition on $\mathfrak{K}/\mathfrak{k}$ is needed to establish this equivalence, as is indicated by the following partial converse.

Theorem 9. *Let the discrete field k have a residue class field $\mathfrak{k}$ with an algebraic extension $\mathfrak{K}$. All unramified complete extensions of k with the fixed residue class field $\mathfrak{K}$ are equivalent over k if and only if $\mathfrak{K}$ is separable over $\mathfrak{k}$.*

The proof will be applicable not only to a discrete field k, but also to any field k with a value group Γ such that $\Gamma \neq p\Gamma$, where $p\Gamma$ denotes the set of all multiples $p\gamma$ of elements γ in Γ. That separability suffices for equivalence was shown in Theorem 4. Assume conversely that $\mathfrak{K}/\mathfrak{k}$ is not separable, let $\mathfrak{k}_s$ be the maximal subfield of $\mathfrak{K}$ separable over $\mathfrak{k}$ and construct by Theorem 1 an unramified extension k_s of k with the extended res. field $\mathfrak{k}_s$. Pick an α in $\mathfrak{k}_s$ with $\alpha^{1/p}$ in $\mathfrak{K}$ but not in $\mathfrak{k}_s$, select any representative a of α in k_s, and form the algebraic extensions $L_1 = k_s(b_1)$, $L_2 = k_s(b_2)$ in which $b_1^p = a$, $b_2^p = a(1 + \pi)$. Since $\Gamma \neq p\Gamma$, π can be chosen in k so that

$$(6\text{-}1) \qquad 0 < V(\pi) \leqq 1, \qquad V(\pi) \text{ not in } p\Gamma.$$

The valuation of k_s has a unique extension to each field L_i, and in each case the res. field of L_i is $\mathfrak{k}_s(\alpha^{1/p})$. Unramified extensions $K_i \supset L_i$ then exist with the extended residue class field $\mathfrak{K} \supset \mathfrak{k}_s(\alpha^{1/p})$.

Were all unramified extensions corresponding to $\mathfrak{K}$ equivalent, there would be a map T of K_1 on K_2 leaving $\mathfrak{K}$ and k and hence (end of Theorem 3) leaving k_s elementwise fixed. The construction of b_i makes $(b_2/b_1^T)^p = 1 + \pi$ in K_2, wherefore $\pi = -1 + (1 + u)^p$, with $u = 1 - b_2/b_1^T$ in K_2. The binomial formula and the assumption $V(p) = 1$ entail

$$\pi = -1 + 1 + pu + \cdots + pu^{p-1} + u^p,$$

$$V(\pi) = \text{Min}\,[1 + V(u), \cdots, 1 + (p-1)V(u), pV(u)];$$

consequently $V(u) > 0$ and either $V(\pi) = 1 + V(u) > 1$ or else $V(\pi) = pV(u)$, both counter to (6-1). The non-uniqueness so established is the central reason for the non-uniqueness of certain maximally perfect extensions of valuations of higher rank (Mac Lane [4], §7, Example II).

7. Subfields of complete fields

When does a given p-adic field have a p-adic subfield with a specified res. field? A first answer is

THEOREM 10. *For any subfield $\mathfrak{k}$ of the residue class field $\mathfrak{K}$ of a p-adic field K there is a p-adic subfield $k \subset K$ with residue class field $\mathfrak{k}$. If $\mathfrak{k}$ is perfect, there is only one such field. More generally, if $\mathfrak{K}/\mathfrak{k}$ preserves p-independence, any two p-adic subfields k and k' with residue class field $\mathfrak{k}$ are conjugate under K.*

The statement that k and k' are *conjugate* means that there is an analytic automorphism of K leaving residue classes fixed and carrying k into k'. To prove the existence of k, construct by Theorems 1 and 2 fields $k' \subset K'$ with the res. fields $\mathfrak{k} \subset \mathfrak{K}$ and map K' on K by the structure Theorem; the map of k' is the required subfield. If k and k' are two p-adic subfields with the same perfect res. field $\mathfrak{k}$, then k and k' contain multiplicative representative systems R and R' respectively. These systems can be characterized uniquely in K by Lemma 2, hence $R = R'$. Any element in k or k' is then the limit of a series $\sum a_i p^i$ with coefficients a_i in R, so that[9] $k = k'$.

Finally, if $\mathfrak{K}/\mathfrak{k}$ preserves p-independence, the field k has a map T on k', by

[9] Examples of $k \neq k'$ with $\mathfrak{k}$ imperfect are discussed at the end of §8.

the structure Theorem, and the map may be extended to a map of $K \supset k$ on a new field $K' \supset k'$. Then K'/k' and K/k' are equivalent, by Theorem 8, under some map S leaving k' elementwise fixed. The map TS is an automorphism of K carrying k into k', so that k and k' are conjugate, as in the Theorem.

The subfield-res. field correspondence $k \rightarrow \mathfrak{k}$ has further properties related to the fact that the set S of all p-adic subfields L of K, under the ordinary inclusion relation, forms a lattice.[10]

THEOREM 11. *The p-adic subfields $k \subset K$ with perfect residue class fields form a sublattice S_p of the lattice S, while all perfect subfields $\mathfrak{k}$ of the residue class field $\mathfrak{K}$ constitute a sublattice $\mathfrak{S}_p$ of the lattice $\mathfrak{S}$ of all subfields $\mathfrak{L}$ of $\mathfrak{K}$. The correspondence $L \rightarrow \mathfrak{L}$ in which each $L \subset K$ corresponds to its residue class field $\mathfrak{L}$ is a lattice isomorphism of S_p to $\mathfrak{S}_p$, because it preserves union and intersection.*

PROOF. $\mathfrak{S}_p$ is a sublattice of $\mathfrak{S}$ (Ore [7]) because in $\mathfrak{S}$ unions and intersections of perfect fields are perfect. For the intersection this is clear; while the union $\mathfrak{k}_1 \cup \mathfrak{k}_2$ contains a maximal perfect subfield $\mathfrak{k}^*$ which then must in turn contain the perfect subfields $\mathfrak{k}_1$ and $\mathfrak{k}_2$, so $\mathfrak{k}^* = \mathfrak{k}_1 \cup \mathfrak{k}_2$ is perfect.

The res. field correspondence $L \rightarrow \mathfrak{L}$ preserves certain inclusions, for if $L \rightarrow \mathfrak{L}$ and $k \rightarrow \mathfrak{k}$, then

$$\text{(7-1)} \qquad \mathfrak{k} \subset \mathfrak{L} \text{ and } \mathfrak{k} \text{ perfect imply } k \subset L.$$

For, L must contain some subfield k' with the res. field $\mathfrak{k}$ of k, and by Theorem 10, k and k' must be identical.

To prove S_p a sublattice of S we must prove that unions and intersections in S of fields in S_p lie in S_p. If $k_1 \rightarrow \mathfrak{k}_1$, $k_2 \rightarrow \mathfrak{k}_2$ with $\mathfrak{k}_1$ and $\mathfrak{k}_2$ perfect, then the unique p-adic subfield k of K with res. field $\mathfrak{k}_1 \cup \mathfrak{k}_2$ is the union of k_1 and k_2 in S. For, $k \supset k_1$ and $k \supset k_2$ by (7-1), while any p-adic subfield $k' \supset k_1 \cup k_2$ must have a res. field $\mathfrak{k}' \supset \mathfrak{k}_1 \cup \mathfrak{k}_2$, so that $k' \supset k$, whence $k = k_1 \cup k_2$. On the other hand, let $k_1 \cap k_2$ be the intersection in S of the fields k_1 and k_2, while k_0 is the unique p-adic subfield of K with the perfect res. field $\mathfrak{k}_1 \cap \mathfrak{k}_2$. By (7-1), both k_1 and k_2 contain k_0, so $k_1 \cap k_2 \supset k_0$. However, the res. field of $k_1 \cap k_2$ is contained in that of k_0, so the unramified extension $k_1 \cap k_2 \supset k_0$ is necessarily k_0, and the intersection $k_1 \cap k_2 = k_0$ has the perfect res. field $\mathfrak{k}_0 = \mathfrak{k}_1 \cap \mathfrak{k}_2$, as asserted.

Finally, the correspondence $k \rightarrow \mathfrak{k}$ is one-to-one from S_p to $\mathfrak{S}_p$ by Theorem 10, and preserves the inclusion relation by (7-1), so that it is indeed a lattice isomorphism.

Fields intermediate to given unramified extensions suggest analogous questions:

THEOREM 12. *Let $\mathfrak{p}$-adic fields $k \subset K$, K unramified over k, have residue class fields $\mathfrak{k} \subset \mathfrak{K}$ with an intermediate field $\mathfrak{L}$, $\mathfrak{k} \subset \mathfrak{L} \subset \mathfrak{K}$. A complete field L with residue class field $\mathfrak{L}$ and with $k \subset L \subset K$ will exist if $\mathfrak{L}$ has a separating trans-*

[10] In S the union $L_1 \cup L_2$ is the smallest complete field containing L_1 and L_2, and not necessarily the smallest subfield of K containing L_1 and L_2.

cendence basis over $\mathfrak{k}$ (*cf.* §3) *or if* $\mathfrak{K}/\mathfrak{k}$ *preserves p-independence and k is p-adic* (see[11] §6). *If* $\mathfrak{L}^p(\mathfrak{k}) = \mathfrak{L}$ *at most one such field L exists.*

Proof. If $\mathfrak{L}$ has a sep. trans. basis over $\mathfrak{k}$, construct an unramified extension L' of k with res. field $\mathfrak{L}$, and map L' in K by Theorem 3; this gives the desired field L as the map of L'. If $\mathfrak{K}/\mathfrak{k}$ preserves p-independence, construct p-adic extensions $k \subset L' \subset K'$ with the res. class extensions $\mathfrak{k} \subset \mathfrak{L} \subset \mathfrak{K}$, and again map K' into K by the equivalence Theorem 8 of §6. Finally the hypothesis $\mathfrak{L}^p(\mathfrak{k}) = \mathfrak{L}$ is equivalent to the assumption that $\mathfrak{k}$ contains[12] a p-basis $\mathfrak{M}$ of $\mathfrak{L}$, which implies that k contains a representative system M of $\mathfrak{M}$. Any two intermediate fields L and L' with res. field $\mathfrak{L}$ must then contain k and M, so that Lemma 4 of §5 implies $L = L'$.

Such intermediate subfields L may not exist in cases when the hypotheses of Theorem 12 fail.

Example I. Let $\mathfrak{K}/\mathfrak{k}$ be a finite inseparable algebraic extension with an infinite number of intermediate fields, and let k be p-adic with res. field $\mathfrak{k}$. A finite, algebraic, and unramified extension K/k with res. field $\mathfrak{K}$ can be constructed as in Theorem 1. This extension K is complete[13] and hence p-adic. The finite separable extension K/k contains only a finite number of intermediate fields L, which cannot correspond to the infinite set of intermediate fields $\mathfrak{k} \subset \mathfrak{L} \subset \mathfrak{K}$.

Example II. Even when there are but a finite number of fields $\mathfrak{L}$ there may not exist corresponding intermediate fields L. We now construct such an extension K/k with res. fields of relative degrees $[\mathfrak{K} : \mathfrak{L}] = [\mathfrak{L} : \mathfrak{k}] = 2$, but where no intermediate p-adic field corresponding to $\mathfrak{L}$ can exist. Let $\mathfrak{P}$ be a perfect field of characteristic 2, $\mathfrak{k} = \mathfrak{P}(\tau)$ a simple transcendental extension of $\mathfrak{P}$, k a p-adic field with res. field $\mathfrak{k}$, and t a representative of τ in k. Set $K = k(\theta)$, where θ is a root of

$$f(x) = x^4 + 2x^3 + 2x^2 + t,$$

which is irreducible (mod 2) and hence irreducible over k. In the extension of the valuation V to K, θ has the residue class $\sqrt[4]{\tau}$ and K is complete and unramified with res. field $\mathfrak{K} = \mathfrak{k}(\sqrt[4]{\tau})$. Then $\mathfrak{L} = \mathfrak{k}(\sqrt{\tau})$ is an inseparable quadratic intermediate field which cannot be the res. field of an intermediate L simply because there is no quadratic field between K and k. This is the case because the resolvent cubic[14]

$$g(y) = y^3 - 2y^2 - 4ty + 4t$$

of the defining quartic equation $f(x) = 0$ is irreducible. The irreducibility of g

[11] If K/k is unramified and $\mathfrak{p}$-adic but not p-adic, the result still holds under the added hypothesis in the last sentence of Theorem 8.

[12] This will be the case, for instance, if $\mathfrak{L}/\mathfrak{k}$ is separable algebraic or if $\mathfrak{L}$ is obtained by adjoining to $\mathfrak{k}$ a transcendent x and all roots $x^{p^{-e}}$, etc.

[13] Cf. for instance Albert, [1], ch. XI, Theorem 11.

[14] Albert [1], ex. 9 and 17 on pp. 178–9.

may be established by the usual Eisenstein criterion,[15] for the Newton polygon of g relative to V is a straight line of slope 2/3.

This example shows that in Theorem 12 it would be impossible to state conditions on $\mathfrak{k} \subset \mathfrak{L} \subset \mathfrak{K}$ alone which would be necessary and sufficient for the existence of a corresponding intermediate p-adic field L, for it is possible to construct successively other p-adic fields $k \subset L' \subset K'$ with exactly the res. fields of this example.

II. Automorphism Groups

8. The inertial group

$G = G(K)$ denotes the group of all analytic automorphisms of a given p-adic field K and $G_1(K)$ the subgroup of "inertial" automorphisms T; that is, of those T which, as in (2-4), generate the identity automorphism on the res. field. Since K is complete in a discrete valuation it is not algebraically closed and hence not complete in any other (non-equivalent) valuation.[16] Therefore any automorphism of K is necessarily analytic, and $G(K)$ is the group of all automorphisms of K.

$G_1(K)$ is a normal subgroup of $G(K)$, exactly as for the Hilbert inertial group in an algebraic number field. We now seek the structure of the quotient group G/G_1 in the more general case when G consists only of automorphisms "relative" to a subfield k (i.e., leaving k elementwise fixed).

Theorem 13. *If $K \supset k$ are p-adic fields*[11] *with residue class fields $\mathfrak{K}/\mathfrak{k}$ preserving p-independence* (cf. §6), *and if $G(K/k)$ is the group of relative automorphisms of K over k and $G_1(K/k)$ its inertial subgroup, then the quotient group $G(K/k)/G_1(K/k)$ is isomorphic to the group $\mathfrak{G}(\mathfrak{K}/\mathfrak{k})$ of relative automorphisms of the residue extension $\mathfrak{K}$ over $\mathfrak{k}$. The isomorphism is given by the correspondence $T \to \tau$ in which τ is the residue class automorphism generated by T.*

Proof. The correspondence $T \to \tau$, where τ is given as in (2-4) by

$$\tau[H(a)] = H(a^T) \qquad (V(a) \geqq 0) \tag{8-1}$$

is certainly a homomorphism of G on a subgroup of $\mathfrak{G}$, and under this homomorphism the subgroup of G mapped on the identity is exactly G_1. It remains to construct a T for any given τ. If $\{V, H\}$ is the given valuation of K, then $\{V, \tau^{-1}H\}$ is another valuation of K with a different residue class homomorphism $H' = \tau^{-1}H$ but with the same res. field $\mathfrak{K}$. The two complete fields K with $\{V, H\}$ and K with $\{V, \tau^{-1}H\}$ are extensions of k with the same res. extension, so must be equivalent, by Theorem 8, under a map T of K on itself which is the identity on $\mathfrak{K}$. Hence, as in (2-5)

$$Ha = (\tau^{-1}H)(a^T)$$

for any a with $V(a) \geqq 0$. In other words, (8-1) holds, as desired.

---

[15] Mac Lane [5], or papers of Ore there mentioned.

[16] F. K. Schmidt, [9]. Our definition of equivalence of valuations and of analytic isomorphisms can be shown equivalent to those used by Schmidt; cf. Mac Lane, [4], footnote 15.

A special case, with k = the minimal complete subfield, is:

COROLLARY. *If G and G_1 are respectively the automorphism and inertial groups of a p-adic field, then G/G_1 is isomorphic to the automorphism group of the residue class field.*

Note that Theorem 13 includes also the known result that the Galois group of an unramified algebraic extension of a p-adic number field (with finite res. field) is isomorphic to the Galois group of the res. extension.

For further investigations we use the following systematic construction of automorphisms T of $G_1(K/k)$, where K is an unramified extension[11] of k with res. fields $\mathfrak{K}/\mathfrak{k}$.

LEMMA 6. *Relative inertial automorphs. If $\mathfrak{m}$ is a p-basis of $\mathfrak{k}$ contained in a p-basis $\mathfrak{m} + \mathfrak{M}$ of $\mathfrak{K}$, with $\mathfrak{m}$ and $\mathfrak{M}$ disjoint, and if M and M' are any two systems of representatives of $\mathfrak{M}$ in K, then there exists one and only one automorphism T of $G_1(K/k)$ which carries M into M'. All automorphisms T of $G_1(K/k)$ can be obtained thus, for fixed M and variable M'.*

If K is p-adic, this Lemma also gives the whole group $G(K)$.

PROOF. By Theorem 6 embed K in complete unramified extensions L and L' with res. field $\mathfrak{L} = \mathfrak{K}(\mathfrak{M}^{p^{-\infty}})$ so that M and M' respectively are multiplicative representatives of $\mathfrak{M}$. The res. field $\mathfrak{L}$ has as p-basis the p-basis $\mathfrak{m}$ of $\mathfrak{k}$, by Lemma 3 of §5, so L/k preserves p-independence and the fundamental equivalence Theorem 8 applies. It provides a map T of L on L' which leaves elements of $\mathfrak{L}$ and k fixed and which carries the multiplicative representatives M into M'. Pick a system m of representatives of $\mathfrak{m}$ in k. The subfield K^T of L' contains k and M', hence contains the representative system $m + M'$ of the p-basis $\mathfrak{m} + \mathfrak{M}$ of $\mathfrak{K}$, while the subfield K originally embedded in L' contains the same set $m + M'$. Any prime element π of k is contained in both K and K^T, hence $K^T = K$ by Lemma 4 and T is an automorphism of K. By construction, T leaves all residue classes fixed, hence T is in $G_1(K/k)$.

Any relative inertial automorphism T manifestly maps M on some other representative system M', and M' uniquely determines T by Lemma 5, as asserted. A special case of this uniqueness is

THEOREM 14. *A p-adic field with a perfect residue class field $\mathfrak{K}$ has except for the identity no automorphism which leaves every residue class fixed.*

The lemma may also be applied to show that the subfield $k \subset K$ with specified res. field $\mathfrak{k}$ constructed in Theorem 10 need not be unique in all cases when $\mathfrak{k}$ is not perfect. Specifically, if the p-adic field K has a res. field $\mathfrak{K}$ with a proper subfield $\mathfrak{k}$ containing some element ξ not in $\mathfrak{K}^p$, then K contains distinct p-adic subfields with res. field $\mathfrak{k}$. For there is a p-basis $\mathfrak{M}$ of $\mathfrak{K}$ containing the element ξ not in $\mathfrak{K}^p$, by known properties of p-independence. Let $k \subset K$ be a p-adic subfield with res. field $\mathfrak{k}$ and let M be any system of representatives for $\mathfrak{M}$ in K containing a representative x of ξ. Since $k < K$ there is a b in K but not in k with $Vb > 0$, so that $y = x + b$ is another representative of ξ in K. The replacement of x in M by this new representative y yields a new system M' of representatives of $\mathfrak{M}$. By Lemma 6 there is a map T of $G_1(K)$ with $M^T =$

M'. Under T the map k^T of the subfield k is another p-adic subfield with res. field $\mathfrak{k}$, and $k^T \neq k$ because k^T contains the element y in M' but not in k.

9. The series of ramification groups

If K/k is unramified and p-adic, denote by $G_n(K/k)$ the group of all those relative automorphisms T of K/k for which b in K and $V(b) \geqq 0$ imply $b^T \equiv b \pmod{\pi^n}$. For $n = 1$, G_n is the inertial group already considered. If $n > 1$, $G_n(K/k)$ is a *pseudo-ramification group* because its formal definition is like that for the usual Hilbert ramification group, although it cannot correspond to any (non-existent!) ramification of K over k. As in the classical case, one proves that each $G_n(K/k)$ is a normal subgroup of $G_1(K/k)$. The successive quotient groups are abelian, as for the ordinary ramification groups, and have the following explicit structure:

THEOREM 15. *Let K/k be p-adic*[11] *with residue class fields $\mathfrak{K}/\mathfrak{k}$ preserving p-independence and with pseudo-ramification groups $G_n = G_n(K/k)$, $n \geqq 1$. Then each quotient group G_n/G_{n+1} is abelian and is isomorphic to the direct product $\prod \mathfrak{K}^+$ of $\mathfrak{c}$ abelian groups, where each factor $\mathfrak{K}^+$ is isomorphic to the additive group of the residue field $\mathfrak{K}$, while the cardinal number $\mathfrak{c}$ of factors is the cardinal number of any set*[17] *$\mathfrak{N}$ which has the property that there is a p-basis $\mathfrak{m}$ of $\mathfrak{k}$ such that $\mathfrak{N}$ and $\mathfrak{m}$ are disjoint and $\mathfrak{N} + \mathfrak{m}$ is a p-basis of $\mathfrak{K}$.*

COROLLARY. *A p-adic field K with pseudo-ramification groups G_n has each G_n/G_{n+1} abelian and isomorphic to the direct product of $\mathfrak{c}$ factors $\mathfrak{K}^+$, where $\mathfrak{c}$ is the degree of imperfection of the residue class field $\mathfrak{K}$.*

The *degree of imperfection* of $\mathfrak{K}$ is by definition the cardinal number of elements in a p-basis of $\mathfrak{K}$; it is independent of the choice of the p-basis, as shown by Teichmüller [11].

PROOF OF THEOREM 15. For each residue class λ of the set $\mathfrak{N}$ pick a representative d_λ in K. By definition, G_n consists of maps T for which

$$d_\lambda^T = d_\lambda + \pi^n t_\lambda \tag{9-1}$$

for some elements t_λ in the valuation ring of K. The correspondence $T \to \{H(t_\lambda)\}$ carries G_n into the ordered set $\{H(t_\lambda)\}$ of $\mathfrak{c}$ residue classes $\alpha_\lambda = H(t_\lambda)$, where λ ranges over all $\mathfrak{c}$ elements of $\mathfrak{N}$. If $\mathfrak{K}_\lambda^+$ is the additive group of all residues α_λ, this correspondence is a homomorphism $G_n(K/k) \sim \mathfrak{K}^+$ because $S \to \{H(s_\lambda)\}$, $T \to \{H(t_\lambda)\}$ yields by definition (9-1) $ST \to \{H(t_\lambda + s_\lambda^T)\} = \{H(t_\lambda + s_\lambda)\}$. The correspondence carries T's in G_{n+1}, and only these (see Lemma 7, below) into the identity. Finally, every possible element $\{\alpha_\lambda\}$ of the direct product is the correspondent of some T, for by Lemma 6 there is an automorphism T with $d_\lambda^T = d_\lambda + \pi^n a_\lambda$, where a_λ is any representative of α_λ. That this T must lie in G_n is a result of the following Lemma, which can be readily established by a decomposition proof resembling that of Lemma 5 in §5:

LEMMA 7. *In a p-adic*[11] *field $K \supset k$, an inertial automorphism T of $G_1(K/k)$*

[17] There always is such a set $\mathfrak{N}$, because $\mathfrak{K}/\mathfrak{k}$ preserves p-independence. It can be shown that the cardinal number $\mathfrak{c}$ of $\mathfrak{N}$ is independent of the choice of $\mathfrak{N}$ in $\mathfrak{K}/\mathfrak{k}$.

is in the pseudo-ramification group $G_n(K/k)$ *if and only if* $d^T \equiv d \pmod{p^n}$ *holds for all* d *of a fixed system* M *of representatives of a p-basis* $\mathfrak{M}$ *of* $\mathfrak{K}$.

10. The inertial subfield

Each ramification group G_n of a p-adic field determines a corresponding subfield consisting of all elements of the field invariant under G_n. Unlike the classical case, these pseudo-ramification fields all coincide.

THEOREM 16. *In a p-adic field* K *let* k *be the unique p-adic subfield (cf. Theorem 10 in §7) whose residue class field* $\mathfrak{k}$ *is the maximal perfect subfield of the residue class field* $\mathfrak{K}$ *of* K, *and let* G_m *be the* m^{th} *pseudo-ramification group of* K, *for* $m \geqq 1$. *Then* k *consists of all those and only those elements of* K *invariant under all the automorphisms of* G_m.

That any T in G_m leaves k elementwise fixed is immediate, for the unique subfield k can only go into itself under T, so that T must leave the elements of k fixed as in Theorem 14. The crux of the theorem is the converse statement, that for any element b in $K - k$ there is a T of G_m with $b^T \neq b$. This will be established by explicit construction of a T as in Lemma 6. After some reductions, we can assume that the given b in the valuation $\{V, H\}$ of K has $V(b) = 0$ and $H(b) = \beta$ for some β not in $\mathfrak{k}$. Hence the query: given b, is there a T in G_m such that

$$Vb = 0,\ Hb = \beta \text{ in } K - k_0 \text{ imply } b^T \neq b? \tag{10-1}$$

First, a p-basis on which to build T! Since β is not in $\mathfrak{k}$, the maximal perfect subfield obtainable as the intersection of all fields[18] $\mathfrak{K}^{p^n}$, there is some index u for which β is in $\mathfrak{K}^{p^u}$, but not in $\mathfrak{K}^{p^{u+1}}$. For some γ,

$$\beta = \gamma^{p^u}, \qquad \gamma \text{ in } \mathfrak{K} \text{ but not in } \mathfrak{K}^p. \tag{10-2}$$

There is then a p-basis $\{\gamma\} + \mathfrak{M}$ of $\mathfrak{K}$ containing this element γ. Fix on a representative c of γ in K and a representative system M for $\mathfrak{M}$ in K, and construct by Lemma 6 of §8 the relative automorphism T of K/k with

$$c^T = c(1 + p^m), \qquad d^T = d \qquad (\text{all } d \text{ in } M). \tag{10-3}$$

By Lemma 7 this T is in $G_m(K)$. We propose to show T effective in (10-1).

T leaves fixed the representatives of M and hence all elements of the subfield L which is constructed as the complete closure in K of the field $k(M)$. The res. field $\mathfrak{L}$ of L is $\mathfrak{L} = \mathfrak{k}(\mathfrak{M})$. For, the elements of the p-independent set $\mathfrak{M}$ are algebraically independent over $\mathfrak{k}$ (Teichmüller [11]), so that the representing elements of M are necessarily algebraically independent over k. Furthermore the residue-class homomorphism H of $k(M)$ on $\mathfrak{L}$ is given by a formula like (4-1) and (4-3), so $\mathfrak{L} = \mathfrak{k}(\mathfrak{M})$, as asserted.

$\mathfrak{L}$ has certain extensions $\mathfrak{K}_n = \mathfrak{K}^{p^n}(\mathfrak{M})$ such that (Teichmüller [10], Lemma 11)

$$\mathfrak{K}_n = \mathfrak{L}(\mathfrak{K}^{p^n}), \qquad \mathfrak{K} = \mathfrak{K}_n(\gamma), \qquad [\mathfrak{K} : \mathfrak{K}_n] = p^n. \tag{10-4}$$

---

[18] Here $\mathfrak{K}^{p^n}$ denotes the field of all p^n powers ξ^{p^n}, for ξ in $\mathfrak{K}$.

Thus γ functions as a "relative p-basis" of $\mathfrak{K}/\mathfrak{L}$. We obtain representatives of all residue classes of $\mathfrak{K}_n$ in terms of the valuation rings B and B_L of K and L, respectively and of the set B^{p^n} of all p^{nth} powers of elements of B, as follows.

LEMMA 8. *Any element a of the ring $R_n = B_L[B^{p^n}]$ has $a^T \equiv a \pmod{p^{n+m}}$ and $H(a)$ in $\mathfrak{K}_n$. Conversely, any element of $\mathfrak{K}_n$ has a representative in R_n.*

PROOF. By the definition (10-4), $\mathfrak{K}_n = \mathfrak{K}^{p^n}(\mathfrak{L}) = \mathfrak{K}^{p^n}[\mathfrak{L}]$ has a set of representatives in the correspondingly constructed ring $R_n = B_L[B^{p^n}]$. On the other hand, any a is a polynomial $f(B^{p^n})$ with coefficients in B_L, hence $H(a)$ is in $\mathfrak{K}_n$ and $a^T \equiv a \pmod{p^{n+m}}$ by Lemma 1.

The original element b will subsequently be represented by polynomials in the quantity c of (10-3) with coefficients in R_n. Hence the following computation of the effect of T on such a polynomial:

LEMMA 9. *If the polynomial $F(y)$ is in $R_n[y]$, where $n \geqq i + 1 \geqq 1$, then*

$$[F(c^{p^i})]^T - F(c^{p^i}) \equiv c^{p^i}p^{i+m}F'(c^{p^i}) \pmod{p^{i+m+1}}, \tag{10-5}$$

where $F'(y)$ is the derivative of $F(y)$. In particular, if $i \geqq 0$,

$$(c^{p^i})^T - c^{p^i} \equiv c^{p^i}p^{i+m} \pmod{p^{i+m+1}}. \tag{10-6}$$

PROOF. The relation (10-6) results from the construction $c^T = c(1 + p^m)$ of T in (10-3) and from the binomial expansion

$$(1 + p^m)^{p^i} \equiv 1 + p^{m+i} \pmod{p^{m+i+1}}.$$

This congruence, together with Lemma 9, fails when $p = 2$ and $m = 1$, but this exception is avoided by assuming $m \geqq 2$; for the construction of a T in G_2 certainly provides a T in the larger group G_1! The congruence (10-5) is derived from the Taylor expansion of $[F(c^{p^i})]^T \equiv F(c^{p^i} + p^{i+m}c^{p^i})$, using (10-6) and Lemma 8 to find the effect of T on coefficients of F.

The elements a left fixed by T will next be sought. Suppose a in B and $a^T \equiv a \pmod{p^{m+s}}$ for some integer $s \geqq 1$. The res. $H(a) = \alpha$ is by (10-4) in $\mathfrak{K}_s(\gamma)$, so there is a polynomial $g(y)$ in $\mathfrak{K}_s[y]$ of degree less than p^s with $\alpha = g(\gamma)$. Then $a \equiv G(c) \pmod{p}$, where $G(y)$ is a polynomial whose coefficients are in R_s and represent corresponding coefficients of $g(y)$, as in Lemma 8. Therefore $a - G(c) = pa_1$, for some a_1 in B, and

$$0 \equiv a^T - a \equiv [G(c)]^T - G(c) + p(a_1^T - a_1) \pmod{p^{m+1}}.$$

But $a_1^T \equiv a_1 \pmod{p^m}$ because T is in G_m, so (10-5) for $i = 0$ gives

$$0 \equiv a^T - a \equiv cp^mG'(c) \pmod{p^{m+1}}.$$

Consequently $cG'(c) \equiv 0 \pmod{p}$ and $\gamma g'(\gamma) = 0$. This gives an equation $yg'(y) = 0$ of degree less than p^s for γ over $\mathfrak{K}_s$, which contradicts (10-4) unless $g'(y) = 0$ identically. In this event, g has the form $g(y) = f(y^p)$ and as a result $G(y) = F(y^p)$, provided G was originally so chosen that each zero coefficient of g corresponds to a zero coefficient in G. Thus $a \equiv F(c^p) \pmod{p}$, where F has degree less than p^{s-1} and coefficients in R_s. Repeating the process for higher moduli, we finally obtain

LEMMA 10. *If the element a in B has $a^T \equiv a \pmod{p^{m+s}}$, and if $0 < n \leqq s$, then a has an expansion*

$$a \equiv F_n(c^{p^n}) + F_{n-1}(c^{p^{n-1}})p + \cdots + F_1(c^p)p^{n-1} \pmod{p^n}, \tag{10-7}$$

where each $F_i(y)$ is a polynomial in $R_{s-n+i}[y]$ of degree less than p^{s-n}, and where any coefficient of F_i congruent to $0 \pmod p$ is equal to 0.

The proof is by induction on n for fixed s. The case $n = 1$ has just been established. Assume next that (10-7) holds for some n and that $s \geqq n + 1$. The next term in this expansion could be written as $a_n p^n$, for some a_n in B. We can again write $a_n \equiv F_0(c)$ where $F_0(y)$ is in $R_{s-n}[y]$ and of degree less than p^{s-n}. The expansion now is

$$a = F_n(c^{p^n}) + \cdots + F_1(c^p)p^{n-1} + F_0(c)p^n + a_{n+1}p^{n+1}, \tag{10-8}$$

for some element a_{n+1} in B, so that $a_{n+1}^T \equiv a_{n+1} \pmod{p^m}$. The difference $a^T - a$ is $\equiv 0 \pmod{p^{m+n+1}}$ by hypothesis and may be calculated by applying Lemma 9 to (10-8). The congruence $\sum c^{p^i}F_i'(c^{p^i}) \equiv 0 \pmod p$ results, where the sum runs from $i = 0$ to n. The homomorphic image of this congruence becomes

$$\gamma^{p^n}f_n'(\gamma^{p^n}) + \cdots + \gamma^p f_1'(\gamma^p) + \gamma f_0'(\gamma) = 0, \tag{10-9}$$

where $\gamma = H(c)$ and $f_k(y)$ denotes the H-image of $F_k(y)$.

An induction on j from this equation shows that $f_j'(y) = 0$. The initial case $j = 1$ can be subsumed in the induction by starting with the general assumption that the polynomial $h_j(z)$ determined by

$$z^{p^j}h_j(z) = z^{p^n}f_n'(z^{p^n}) + \cdots + z^{p^j}f_j'(z^{p^j}) \tag{10-10}$$

has a root $z = \gamma$; for when $j = 0$ this assumption is (10-9). Each $f_i(z)$, as image of $F_i(z)$, has coefficients in $\mathfrak{R}_{s-n+i}$, so that $h_j(z)$ has coefficients in $\mathfrak{R}_r$, where $r = s - n + j > 0$. Over this field $\mathfrak{R}_r$, γ must by (10-4) satisfy the irreducible equation $z^{p^r} - \gamma^{p^r} = 0$, for which reason the other equation $h_j(z) = 0$ for γ is a multiple

$$h_j(z) = q(z)(z^{p^r} - \gamma^{p^r}). \tag{10-11}$$

In (10-11) the terms on the left are of two sorts: first, the terms $z^{p^i - p^j}f_i(z^{p^i})$ from (10-10) with $i > j$, which are sums of terms of the form dz^e of degree $e = p^i - p^j + e_0 p^i \equiv - p^j \pmod{p^{j+1}}$; secondly, terms from $f_j'(z^{p^j})$ in which no terms z^e with $e \equiv - p^j \pmod{p^{j+1}}$ can occur, because f_j' is a derivative in a field of characteristic p. The terms of this second type have a degree $e' = e_0'p^j < p^{s-n}p^j = p^r$, less than the minimum degree of terms in $q(z)z^{p^r}$. One argues from this circumstance that $q(z)$ can contain only terms z^ν with exponents $\nu \equiv - p^j \pmod{p^{j+1}}$ and therefore that $f_j'(z^{p^j})$, which is in (10-11) the only source of terms not satisfying this congruence condition, must be identically zero. This conclusion, $f_j'(y) = 0$, completes the induction.

This identity $f_i'(y) = 0$ implies that $f_i(y) = g_{i+1}(y^p)$ for some polynomial $g_i(z)$ in $\mathfrak{R}_{s-n+i}[z]$ of degree less than p^{s-n-1}. The hypothesis concerning zero coefficients for the F's, inclusing of a proper such choice for the F_0 constructed at the start of the proof, thereby insures that $F_i(y) = G_{i+1}(y^p)$, for G_{i+1} some polynomial with coefficients in R_{s-n+i}. With these G's, (10-8) can be rewritten as the desired expansion (10-7) of $a \pmod{p^{n+1}}$, so that the induction on n for the Lemma is proven.

Return to the query (10-1) and suppose that the element b there had $b^T = b$. Apply Lemma 10 to b with $n = s = u + 1$, for u chosen as in (10-2). The Lemma states that $b \equiv F_n(c^{p^n}) \pmod{p}$, where F_n is of degree less than $p^0 = 1$, so that b must be in $R_n = R_{u+1}$ with a residue class $\beta = \gamma^{p^u}$ in $\mathfrak{R}_n = \mathfrak{R}_{u+1}$, contrary to (10-4). The second half of Theorem 16 is thus completed.

HARVARD UNIVERSITY.

BIBLIOGRAPHY

(1) A. A. ALBERT, *Modern Higher Algebra*, Chicago, 1937.

(2) H. HASSE AND F. K. SCHMIDT, *Die Struktur diskret bewerteter Körper*, Journal für die Mathematik, 170 (1934), pp. 4–63.

(3) S. MACLANE, *A Construction for Absolute Values in Polynomial Rings*, Trans. Am. Math. Soc., 40 (1936), pp. 363–395.

(4) S. MACLANE, *The Uniqueness of the Power Series Representation of certain Fields with Valuations*, Annals of Math., 39 (1938), pp. 370–382.

(5) S. MACLANE, *The Schönemann-Eisenstein Irreducibility Criteria in Terms of Prime Ideals*, Trans. Am. Math. Soc., 43 (1938), pp. 226–239.

(6) T. NAKAYAMA, *Divisionsalgebren über diskret bewerteten perfekten Körpern*, J. f. Math., 178 (1937), pp. 11–13.

(7) O. ORE, *On the Foundations of Abstract Algebra* I, Annals of Math., 36 (1935), pp. 406–437.

(8) A. OSTROWSKI, *Untersuchungen zur arithmetischen Theorie dcr Körper*, Math. Zeit., 39 (1934), pp. 269–404.

(9) F. K. SCHMIDT, *Mehrfach perfekte Körper*, Math. Ann., 108 (1933), pp. 1–25.

(10) O. TEICHMÜLLER, *Diskret bewertete perfekte Körper mit unvollkommenem Restklassenkörper*, J. f. Math., 176 (1937), pp. 141–152.

(11) O. TEICHMÜLLER, p-*Algebren*, *Deutsche Mathematik*, 1 (1936), pp. 362–388.

(11a) O. TEICHMÜLLER, *Über die Struktur diskret bewerteter Körper*, Göttinger Nachrichten, 1936, Math. Phys. Klasse, Fachgruppe I, vol. 1, Nr. 10.

(12) E. WITT, *Schiefkörper über diskret bewerteten Körpern*, J. f. Math., 176 (1936), pp. 153–156.

(13) E. WITT, *Zylische Körper und Algebren der Charakteristik* p *von Grade* p^n, J. f. Math., 176 (1936), pp. 126–140.

ANNALS OF MATHEMATICS
Vol. 41, No. 4, October, 1940

NOTE ON THE RELATIVE STRUCTURE OF p-ADIC FIELDS

BY SAUNDERS MAC LANE

(Received February 3, 1940)

This paper contains the solution of a problem stated and partially solved in a previous paper[1] on fields with discrete valuations. It also contains a correction of certain results given previously for non-discrete valuations.

1. The Problem of Relative Structure

A *p-adic field* K is a field of characteristic zero, complete under a valuation function $V(a)$ with integral values, such that the rational prime p has the value $V(p) = 1$. The structure theorem states that such a field K is uniquely determined (up to an analytic isomorphism) by the field $\mathfrak{K}$ of its residue classes (mod p). The original proof of Hasse and Schmidt for this structure theorem involved certain difficulties[2] with Steinitz towers of fields, but recently a different analysis of these towers has been found which re-establishes[3] a modified form of the Hasse-Schmidt proof.

Let K and K' be two p-adic extensions of a given p-adic field k, both with the same residue class field $\mathfrak{K}$. (Then $\mathfrak{K}$ is an extension of the residue class field $\mathfrak{k}$ of the original k). K and K' are *analytically equivalent* over $\mathfrak{K}$ and k if there is an analytic isomorphism which maps K on K' and leaves each element of k and each residue class of $\mathfrak{K}$ fixed. The relative structure problem is that of finding conditions on the residue class extension $\mathfrak{K}/\mathfrak{k}$ under which any two extensions K and K' are analytically equivalent. It suffices (*loc. cit.*, Theorem 8) to assume that $\mathfrak{K}$ "preserves p-independence" in $\mathfrak{k}$, in the sense defined later. This condition is also necessary.

THEOREM. *Let $\mathfrak{K}$ be any extension of the residue class field $\mathfrak{k}$ of a given p-adic field k. A necessary and sufficient condition that any two p-adic extensions of k with residue class field $\mathfrak{K}$ be analytically equivalent over $\mathfrak{K}$ and k is the requirement that $\mathfrak{K}/\mathfrak{k}$ preserve p-independence.*

Our proof that this requirement is necessary is an extension of a previously given example[1] of two non-equivalent extensions for the special case when $\mathfrak{K}$ is an inseparable algebraic extension of $\mathfrak{k}$. The employment of this type of example requires that we first (Lemma 1) break up an arbitrary extension $\mathfrak{K}/\mathfrak{k}$ which does not preserve p-independence into parts, one of which is an inseparable extension.

---

[1] S. MacLane, *Subfields and Automorphism groups of p-adic fields*, these Annals, vol. 40(1939), pp. 423–442. Subsequently referred to as "subauto."

[2] S. MacLane, *Steinitz field towers for modular fields*, Transactions of the American Mathematical Society, vol. 46(1939), pp. 23–45.

[3] See a forthcoming paper *Über inseparable Körper*, by F. K. Schmidt and S. MacLane, in Mathematische Zeitschrift.

Reprinted from the Annals of Math. 41 (1940), 751-753, by permission of Princeton University Press.

2. Extensions Preserving p-Independence

All residue class fields of p-adic fields necessarily have characteristic p. In a filed $\mathfrak{K}$ of this characteristic a subset X is said to be *p-independent* if no element of X lies in the field obtained by adjoining all the other elements of X to $\mathfrak{K}^p$, the field of all p^{th} powers of elements in $\mathfrak{K}$. An extension $\mathfrak{K}/\mathfrak{k}$ is said to *preserve p-independence*[4] if every subset of $\mathfrak{k}$ p-independent in $\mathfrak{k}$ remains p-independent in $\mathfrak{K}$.

Lemma 1. *An extension $\mathfrak{K}$ of a field $\mathfrak{k}$ of characteristic p fails to preserve p-independence if and only if $\mathfrak{K}$ contains a subfield $\mathfrak{L} \supset \mathfrak{k}$ and an element η not in $\mathfrak{L}$ such that η^p lies in a field $\mathfrak{L}^p(\gamma_1, \cdots, \gamma_n)$ generated over $\mathfrak{L}^p$ by elements γ_i from $\mathfrak{k}$.*

Given such an $\mathfrak{L}$ and η, one readily verifies that certain elements γ p-independent in $\mathfrak{k}$ must become p-dependent in $\mathfrak{K}$. Conversely, if $\mathfrak{K}/\mathfrak{k}$ does not preserve p-independence we can find a finite set $\Gamma = [\gamma_1, \cdots, \gamma_n]$ of elements of $\mathfrak{k}$ p-independent in $\mathfrak{k}$ but not in $\mathfrak{K}$. Γ is then also p-dependent in some subfield of $\mathfrak{K}$ generated over $\mathfrak{k}$ by a finite number of elements. Let us add these elements to $\mathfrak{k}$ one at a time, taking care to make each such adjunction either a transcendental extension, a separable algebraic extension, or the addition of a pth root. There is then some one of these elements η, the adjunction of which to the previously obtained field $\mathfrak{L}$ first brings about the p-dependence of Γ. This element η can be neither transcendental nor separable algebraic over $\mathfrak{L}$, for an extension of either of these types is known[1] to preserve p-independence. Therefore η is a pth root of some element of $\mathfrak{L}$.

By definition the p-dependence of Γ in $\mathfrak{L}(\eta)$ means that some one element of Γ, say γ_1, can be written as a polynomial

$$\gamma_1 = f(\gamma_2, \cdots, \gamma_n, \eta^p) \tag{1}$$

with coefficients in $\mathfrak{L}^p$, and with degree at most $p - 1$ in η^p or in any γ_i. Since Γ was not p-dependent in the field $\mathfrak{L}$, this polynomial actually involves η^p. Therefore η^p satisfies the separable equation (1) over $\mathfrak{L}^p(\gamma_1, \cdots, \gamma_n)$. But η^p also satifies over $\mathfrak{L}^p$ a purely inseparable equation, for $(\eta^p)^p$ is in $\mathfrak{L}^p$. Therefore η^p lies in $\mathfrak{L}^p(\gamma_1, \cdots, \gamma_n)$, as asserted.

Suppose again that $\mathfrak{K}/\mathfrak{k}$ does not preserve p-independence, and return to the proof of the Theorem. The conclusion of the Lemma states that η can be written as

$$\eta^p = \alpha_1^p\beta_1 + \alpha_2^p\beta_2 + \cdots + \alpha_m^p\beta_m, \qquad \alpha_i \text{ in } \mathfrak{L}, \tag{2}$$

where each β_i is a power product of elements γ_j, and hence is in $\mathfrak{k}$. Extend the given p-adic field k to obtain[5] two p-adic fields L and L', both with the residue class field $\mathfrak{L}$. For each residue class β_i of (2) choose in k some representative b_i, and for each α_i choose representatives a_i and a_i' in L and L', respectively. Since the residue class equation (2) is an irreducible equation of degree p in η, we know that corresponding equations over L and L' must be irreducible. Let y and y' be defined by two such equations,

[4] A discussion of such extensions appears in S. MacLane, *Modular fields. I. Separating transcendence bases.* Duke Mathematical Journal, vol. 5(1939), pp. 372–393.

[5] Such an extension of a p-adic field is possible, by Theorem 1 of "subauto."

(3) $$y^p = \sum_i a_i^p b_i \qquad y'^p = (1+p)(\sum_i a_i'^p b_i).$$

These elements generate p-adic fields $L(y)$ and $L'(y')$. Since y and y' both have the residue class η defined by (2), these fields both have the residue class field $\mathfrak{L}(\eta)$. Again extend[5] these fields $L(y)$ and $L'(y')$ to the p-adic fields K and K', respectively, with the extended residue class field $\mathfrak{K}$.

We assert that these two p-adic extensions K and K' are not analytically equivalent over $\mathfrak{K}$ and k. For suppose instead that K and K' were equivalent under some isomorphism $a \leftrightarrow a^T$. Since T is to leave k and residue classes of $\mathfrak{K}$ fixed, we know that

$$b_i^T = b_i, \qquad a_i^T \equiv a_i' (\mathrm{mod}\ p), \qquad y^T \equiv y' (\mathrm{mod}\ p).$$

According to Lemma 1 of "subauto," this implies

$$(a_i^p)^T \equiv a_i'^p (\mathrm{mod}\ p^2), \qquad (y^p)^T \equiv y'^p (\mathrm{mod}\ p^2).$$

Substitution of these in the first equation of (3) and subtraction of the second gives

$$p(\sum_i a_i'^p b_i) \equiv 0\ (\mathrm{mod}\ p^2), \qquad \sum_i a_i'^p b_i \equiv 0\ (\mathrm{mod}\ p),$$

a contradiction, for the residue class η of $\sum a_i'^p b_i$ is not 0. The existence of the two non-equivalent fields completes the solution of the relative structure problem, as stated in the Theorem.

3. A Correction

I. Kaplansky has called to my attention the fact that Theorems 3 and 4 of "subauto" cannot be correct without restriction in the case of non-discrete valuations. The difficulty arises because a non-discrete complete field L can have a proper complete extension L which has the same value group and residue class field as does L. Then K is an *immediate* extension of L in the sense of W. Krull,[6] who observed this phenomenon in connection with power series fields.

In the proof of Theorem 4, this difficulty appears in the argument that $K^{T*} = K'$ (p. 428 of "subauto"). The same trouble arises in Theorem 3. Both Theorems remain valid for discrete valuations, so the subsequent Theorems of the paper, which treat only this case, are not affected. The argument given for Theorem 3 in the non-discrete case does in fact prove a modified Theorem 3* which asserts, not that T^* is an analytic isomorphism of K on a subfield of K', but that T^* is an analytic isomorphism of some *subfield* K_0 of K on a subfield of K', where K_0 and K have the common residue class field $\mathfrak{K}$.

HARVARD UNIVERSITY

[6] W. Krull, *Allgemeine Bewertungstheorie*, Journal für die Mathematik, vol. 167(1932), pp. 160–196.

ANNALS OF MATHEMATICS
Vol. 43, No. 4, October, 1942

GROUP EXTENSIONS AND HOMOLOGY*

BY SAMUEL EILENBERG AND SAUNDERS MACLANE

(Received May 21, 1942)

CONTENTS

---

* Presented to the American Mathematical Society, September 4 and December 31, 1941. Part of the results was published in a preliminary report [5] and also in an appendix to Lefschetz [7]. The numbers in brackets refer to the bibliography at the end of the paper.

Reprinted from the Annals of Math. 43 (1942). 757-831, by permission of Princeton University Press.

INTRODUCTION

In 1937 the following problem was formulated by Borsuk and Eilenberg: Given a solenoid[1] Σ in the three sphere S^3, how many homotopy classes of continuous mappings $f(S^3 - \Sigma) \subset S^2$ are there? In 1939 Eilenberg proved ([4], p. 251) that the homotopy classes in question are in a 1-1-correspondence with the elements of the one-dimensional homology group $H^1(K, I) = Z^1(K, I)/B^1(K, I)$, where K is any representation of $S^3 - \Sigma$ as a complex, $Z^1(K, I)$ is the group of infinite 1-cycles in K with the additive group I of integers as coefficients and $B^1(K, I)$ is the subgroup of bounding cycles. This homology group is generally much "larger" than the conventional homology group $H_t^1(K, I) = Z^1/\bar{B}^1$ where $\bar{B}^1(K, I)$ is the group of cycles that bound on every finite portion of K; with an appropriate topology in the group Z^1, $\bar{B}^1$ turns out to be exactly the closure of B^1.

At this point the investigation was taken up by Steenrod [10]. By using "regular cycles" he computed the groups $H^1(S^3 - \Sigma)$ for the various solenoids Σ. The groups are uncountable and of a rather complicated nature.[2]

This paper originated from an accidental observation that the groups obtained by Steenrod were identical with some groups that occur in the purely algebraic theory of *extensions of groups*. An abelian group E is called an ex-

[1] For the definition see Appendix B below.

[2] A popular exposition of Steenrod's results can be found in his article in Lectures in Topology, Ann Arbor, University of Michigan Press, 1941, pp. 43–55.

tension of the group G by the group H if $G \subset E$ and $H = E/G$. With a proper definition of equivalence and addition, the extensions of G by H themselves form an abelian group Ext $\{G, H\}$. It turns out that $H^1(S^3 - \Sigma, I)$ is isomorphic with Ext $\{I, \Sigma^*\}$ where Σ^* is a properly chosen subgroup of the group of rational numbers.[3]

The thesis of this paper is that the theory of group extensions forms a natural and powerful tool in the study of homologies in infinite complexes and topological spaces. Even in the simple and familiar case of finite complexes the results obtained are finer than the existing ones.

Our fundamental theorem concerns the homology groups of a star finite complex K. Let $H^q(G)$ denote the homology group of infinite cycles with coefficients in an arbitrary topological group G. We obtain an explicit expression for $H^q(G)$ in terms of G and the cohomology groups $\mathcal{H}_q$ of *finite* cocycles with integral coefficients. ($\mathcal{H}_q$ is the factor group $\mathcal{Z}_q/\mathcal{B}_q$ of cocycles modulo coboundaries). This expression is

$$H^q(G) = \text{Hom}\,\{\mathcal{H}_q, G\} \times \text{Hom}\,\{\mathcal{B}_{q+1}, G\}/\text{Hom}\,\{\mathcal{Z}_{q+1} \mid \mathcal{B}_{q+1}, G\}.$$

Here Hom $\{H, G\}$ stands for the (topological) group of all homomorphisms of H into G, while Hom $\{\mathcal{Z}_{q+1} \mid \mathcal{B}_{q+1}, G\}$ denotes the group of those homomorphisms of $\mathcal{B}_{q+1}$ into G which can be extended to homomorphisms of $\mathcal{Z}_{q+1}$ into G. The factor group on the right in this expression appears to depend on the groups $\mathcal{B}_{q+1}$ and $\mathcal{Z}_{q+1}$, but actually depends only on the cohomology group $\mathcal{H}_{q+1} = \mathcal{Z}_{q+1}/\mathcal{B}_{q+1}$. In fact this factor group can best be interpreted as the group "Ext" of group extensions of G by $\mathcal{H}_{q+1}$. The fundamental theorem then has the form

$$H^q(G) = \text{Hom}\,\{\mathcal{H}_q, G\} \times \text{Ext}\,\{G, \mathcal{H}_{q+1}\}.$$

The paper is self contained as far as possible, both in algebraic and topological respects. The first four chapters below develop the requisite group-theoretical notions. Chapter I discusses the groups of homomorphisms involved in the above formula, while Chapter II introduces the group of group extensions, and proves the fundamental theorem relating this group to groups of homomorphisms. This fundamental theorem is essentially a formulation of the known fact that a group extension of G by H can be described either by generators of H (and hence by homomorphisms) or by certain "factor sets." Chapter III analyzes the group Ext $\{G, H\}$ for some special cases of G. Chapter IV introduces some additional groups, closely related to Ext, which arise as inverse limit groups in the treatment of homologies of topological spaces.

The last two chapters analyze homology groups. Chapter V treats the case of a complex, and proves the fundamental theorem quoted above, as well as parallel theorems for some of the other homology groups of a complex. Chapter

---

[3] More precisely Σ^* is the character group of Σ. The detailed treatment appears in Appendix B below.

VI obtains analogous theorems for the Čech homology groups of a topological space.

Appendix A discusses the case when G is a group with operators. Appendix B contains a computation of the group Ext $\{I, \Sigma^*\}$ mentioned above.

Each chapter is preceded by a brief outline. The chapters are related as in the following diagram:

$$\begin{array}{ccccccc} \mathrm{I} & \rightarrow & \mathrm{II} & \rightarrow & \mathrm{III} & \rightarrow & \mathrm{V} \\ & & & & & \searrow & & \searrow \\ & & & & & & \mathrm{IV} & \rightarrow & \mathrm{VI} \end{array}$$

Almost all of V can be read directly after I and II, and a major portion after I alone.

Chapters V and VI are strongly influenced by S. Lefschetz's recent book "Algebraic Topology" [7], that the authors had the privilege of reading in manuscript.

CHAPTER I. TOPOLOGICAL GROUPS AND HOMOMORPHISMS

After a certain preliminary definitions, this chapter introduces the basic group Hom $\{R, G\}$ of homomorphisms. In the case when R is a subgroup of a free group, we require two subgroups of "extendable" homomorphisms. The topology of these subgroups is investigated when the "coefficient group" G is itself topological.

1. Topological spaces

A set X is called a *space* if there is given a family of subsets of X, called *open sets*, such that

(1.1) *X and the void set are open,*

(1.2) *the union of any number of open sets is open,*

(1.3) *the intersection of two open sets is open.*

Complements of open sets are called *closed*. X is called a *Hausdorff space* if in addition

(1.4) *every two distinct points are contained respectively in two disjoint open sets.*

X is called a *compact* (= bicompact) space if

(1.5) *every covering of X by open sets contains a finite subcovering.*

A space X is *discrete* if every set in X is open.

The intersection of an open set of a space X with a subset A of X will be called *open in A*. With this convention A becomes a space.

Let X and Y be spaces and $x \rightarrow f(x) = y$ a mapping of X into a subset of Y. The mapping f is *continuous* if for every open set $U \subset Y$ the set $f^{-1}(U)$ is open (in X). The mapping f is *open* if for every open set $U \subset X$ the set $f(U)$ is open (in Y). A well known result is

LEMMA 1.1. *If f is a continuous mapping of a compact space X into a Hausdorff space Y, then $f(X)$ is closed in Y.*

A *product space* $\prod_\alpha X_\alpha$ of a given collection $\{X_\alpha\}$ of spaces X_α is defined as the space whose points are all collections $\{x_\alpha\}$, $x_\alpha \in X_\alpha$ and in which open sets are unions of sets of the form $\prod_\alpha U_\alpha$, where U_α is an open subset of X_α and $U_\alpha = X_\alpha$ except for a finite number of indices α.[4] It is known that $\prod X_\alpha$ is a Hausdorff or compact space if and only if for every α the space X_α is a Hausdorff or compact space.[5]

Let Λ be a set of elements and X be a space. We consider the set X^Λ of all functions with arguments in Λ and values in X. The set X^Λ is clearly in a 1-1 correspondence with the product $\prod X_\lambda$ where $\lambda \in \Lambda$ and $X_\lambda = X$. Hence we may consider X^Λ as a space.

2. Topological groups

Only abelian groups (written additively) will be considered.

A group G will be called a *generalized topological group* if G is a space in which the group composition (as a mapping $G \times G \to G$) and the group inverse (as a mapping $G \to G$) are continuous.

If G, considered as a space, is a Hausdorff space, then G will be called a *topological group*.[6] Similarly, if G is compact as a space we shall say that G is a *compact group*.

A subgroup of a (generalized) topological group is a (generalized) topological group. A closed subgroup of a compact group is compact.

LEMMA 2.1. *In a generalized topological group G the following properties are equivalent*:

(a) *every point of G is a closed set,*

(b) *the zero element of G is a closed set,*

(c) *G is a topological group.*[7]

The *factor group* $H = G/G_1$ of a generalized topological group G modulo a subgroup G_1 is the group of all cosets $g + G_1$ of G_1 in G. The correspondence $\varphi(G) = H$ carrying each $g \in G$ into its coset $\varphi g = g + G_1$ in H is the "natural" mapping of G on H. We introduce a topology in H by calling a set $U \subset H$ open if and only if $\varphi^{-1}(U)$ is open in G. It can be shown that this topology is the only one under which φ will be both open and continuous.

LEMMA 2.2. *If G is a generalized topological group and G_1 is an arbitrary subgroup of G, then the factor group $H = G/G_1$ is a generalized topological group; it is a topological group if and only if G_1 is a closed subgroup of G. If G is compact, then G/G_1 is compact.*

LEMMA 2.3. *The closure $\bar{0}$ of the zero element of a generalized topological group is a closed subgroup of G. Its factor group $G/\bar{0}$ is the "largest" factor group of G which is a topological group.*

The preceding two statements show the utility of the study of generalized

---

[4] If $\{\alpha\} = 1, 2, \cdots, n$ we also use the symbol $X_1 \times X_2 \times \cdots \times X_n$ for the product space.

[5] See C. Chevalley and O. Frink, Bulletin Amer. Math. Soc. 47 (1941), pp. 612–614.

[6] G is then a topological group in the sense of Pontrjagin [8].

[7] To prove that a) implies c) one first proves that each neighborhood of g contains the closure of a neighborhood of g, as in Pontrjagin [8], p. 43, proposition F.

topological groups. Several times in the sequel we need to consider an isomorphism

(2.1) $$G_1/H_1 \cong G_2/H_2$$

where the G_i are topological groups, while the H_i are not closed, so that G_i/H_i are only generalized topological groups. However, if we are able to prove that the isomorphism (2.1) is continuous in both directions in the "generalized" topology of the groups G_i/H_i, we obtain as a corollary the bicontinuous isomorphism of the topological groups $G_i/\bar{H}_i$.

If $\{G_\alpha\}$ is a collection of generalized topological groups the direct product $\prod_\alpha G_\alpha$ is a generalized topological group, provided we define the sum $\{g_\alpha\} = \{g'_\alpha\} + \{g''_\alpha\}$ by setting $g_\alpha = g'_\alpha + g''_\alpha$ for every α. Similarly, if Λ is any set and G is a generalized topological group, then the set G^Λ of all mappings of Λ into G is a generalized topological group. It follows from the results quoted in §1 that $\prod_\alpha G_\alpha$ and G^Λ are topological or compact groups if and only if the groups G_α and G are all topological or compact, respectively.

3. The group of homomorphisms

Let G and H be generalized topological groups. A *homomorphism* θ of H into G is a continuous function $\theta(h)$ defined for all $h \in H$ with values in G, such that $\theta(h_1 + h_2) = \theta(h_1) + \theta(h_2)$. For instance, the natural mapping of a group into one of its factor groups is a homomorphism. If θ_1 and θ_2 are two homomorphisms their sum $\theta_1 + \theta_2$, defined by

$$(\theta_1 + \theta_2)(h) = \theta_1(h) + \theta_2(h), \qquad \text{(all } h \text{ in } H)$$

is also a homomorphism. Under this addition, the set of all homomorphisms θ of H into G constitutes a group, which we denote by Hom $\{H, G\}$:

(3.1) $$\text{Hom } \{H, G\} = [\text{all homomorphisms } \theta \text{ of } H \text{ into } G].$$

To introduce a (generalized) topology in Hom $\{H, G\}$, take any compact subset X of H and any open subset V of G with $0 \in V$ and consider the set $U(X, V)$ of all θ with $\theta(X) \subset V$. In the usual sense ([8], p. 55) these sets $U(X, V)$ constitute a complete set of neighborhoods of 0 in Hom $\{H, G\}$, and are used to define the topology of Hom $\{H, G\}$.[8]

If H is discrete, the compact subsets X of H are just the finite ones. In this case Hom $\{H, G\}$ is a subgroup of the group G^H with the topology as defined in §2.

LEMMA 3.1. *If G is a topological group and H is discrete, then Hom $\{H, G\}$ is a closed subgroup of the group G^H of all mappings of H into G.*

PROOF. Let $\phi_0 \in G^H$ be a mapping of H into G that is not a homomorphism. There are then elements h_1, h_2, h_3 in H such that $h_1 + h_2 = h_3$ and $\phi_0(h_1) + \phi_0(h_2) \neq \phi_0(h_3)$. Since G is a Hausdorff space and the group composi-

[8] This is the general definition stated by Weil [11], p. 99, and Lefschetz [7], Ch. II.

tion is continuous there are in G three open sets U_1, U_2, U_3 containing $\phi_0(h_1)$, $\phi_0(h_2)$, and $\phi_0(h_3)$, respectively, such that[9] $(U_1 + U_2) \cap U_3 = 0$. Consequently the open subset U of G^H consisting of the mappings ϕ such that $\phi(h_1) \in U_1$, $\phi(h_2) \in U_2$, and $\phi(h_3) \in U_3$ has no elements in common with Hom $\{H, G\}$. Hence Hom $\{H, G\}$ is closed.

COROLLARY 3.2. *If H is discrete and G is a topological (and compact) group, then* Hom $\{H, G\}$ *is a topological (and compact) group.*

Note that the topology of Hom $\{H, G\}$ may not be discrete even though H and G both have discrete topologies. Observe also that if H is discrete, an alteration in the topology of G may alter the topology of Hom $\{H, G\}$ but not its algebraic structure. However, if H carries a non-discrete topology, an alteration in the topology of either H or G may alter the algebraic structure of Hom $\{H, G\}$, in that continuous homomorphisms may cease to be continuous, or vice versa.

If H is compact, we can take H itself to be the compact set X used in the definition of the topology in Hom $\{H, G\}$. Consequently, given any open set V in G containing 0, the homomorphisms θ, such that $\theta(H) \subset V$, constitute an open set. Hence if V can be picked so as not to contain any subgroups but 0, we see that Hom $\{H, G\}$ is discrete.

Subgroups and factor groups of H will correspond respectively to factor groups and subgroups of Hom $\{H, G\}$, as stated in the following lemmas.

LEMMA 3.3. *If H/H_1 is a factor group of the discrete group H, then* Hom $\{H/H_1, G\}$ *is (bicontinuously) isomorphic to that subgroup of* Hom $\{H, G\}$ *which consists of the homomorphisms θ mapping every element of H_1 into zero.*

The proof is readily given by observing that each homomorphism θ with $\theta(H_1) = 0$ maps each coset of H_1 into a single element of G, so induces a homomorphism θ' of H/H_1. The continuity of the isomorphism $\theta \to \theta'$ can be established, as always for isomorphisms between groups, by showing continuity at $\theta = 0$. ([8], p. 63).

LEMMA 3.4. *If L is a subgroup of H, then each homomorphism θ of H into G induces a homomorphism $\theta' = \theta \mid L$ of L into G. The correspondence $\theta \to \theta'$ is a (continuous) homomorphism of* Hom $\{H, G\}$ *into* Hom $\{L, G\}$. *If L is a direct factor of H, this correspondence maps* Hom $\{H, G\}$ *onto* Hom $\{L, G\}$.

4. Free groups and their factor groups

The homology groups will be interpreted later as certain groups of homomorphisms of "free" groups, which we now define. If the elements z_α of a discrete group F are such that every element of F can be represented uniquely as a finite sum $\sum n_\alpha z_\alpha$ with integral coefficients n_α, F is said to be a *free abelian group* with generators (or basis elements) $\{z_\alpha\}$. The number of generators may be infinite. A free group can be constructed with any assigned set of symbols as basis elements.

[9] $U_1 + U_2$ is the set of all sums $g_1 + g_2$, with $g_i \in U_i$. The symbol $\cap$ stands for the set-theoretic intersection.

LEMMA 4.1. *Every proper subgroup of a free group is free.*

For the denumerable case, this is proved by Čech [3]; a general proof is given in Lefschetz [7] (II, (10.1)).

Any discrete group H can be represented as a homomorphic image of a free group. Specifically, if we choose any set of elements t_α in H which together generate all of H, and if we then construct a free group F with generators z_α in 1-1 correspondence $z_\alpha \leftrightarrow t_\alpha$ with the given t's, the correspondence $\sum n_\alpha z_\alpha \rightarrow \sum n_\alpha t_\alpha$ will map the free group F homomorphically onto the given group H. If the kernel of this homomorphism[10] is R, H may be represented as the factor group $H = F/R$. R is essentially the group of "relations" on the generators t_α of H.

Given $R \subset F$, each homomorphism ϕ of F into G induces a homomorphism $\theta = \phi \mid R$ of the subgroup R into G, and the homomorphisms so induced form a subgroup of Hom $\{R, G\}$, denoted as

$$(4.1) \qquad \text{Hom}\ \{F \mid R, G\} = [\text{all } \theta = \phi \mid R, \text{ for } \phi \in \text{Hom}\ \{F, G\}].$$

Alternatively, the elements of this subgroup can be described as those homomorphisms θ of R into G which can be extended (in at least one way) to homomorphisms of F into G.

A similar, but lighter, restriction may be imposed as follows: Given $\theta \in \text{Hom}\ \{R, G\}$, require that for every subgroup $F_0 \supset R$ of F for which F_0/R is finite there exist an extension of θ to a homomorphism of F_0 into G. The θ's meeting this requirement also constitute a subgroup,

$$(4.2) \qquad \text{Hom}_f\ \{R, G; F\} = [\text{all } \theta \in \text{Hom}\ \{F_0 \mid R, G\} \text{ for every finite } F_0/R].$$

These two subgroups,

$$\text{Hom}\ \{F \mid R, G\} \subset \text{Hom}_f\ \{R, G; F\} \subset \text{Hom}\ \{R, G\},$$

are important because the corresponding factor groups in Hom $\{R, G\}$ are invariants of the group $H = F/R$, in that they do not depend on the particular free group F chosen to represent H. This fact may be stated as follows.

THEOREM 4.2. *If H is isomorphic to two factor groups F/R and F'/R' of free groups F and F', then*

$$(4.3) \qquad \text{Hom}\ \{R, G\}/\text{Hom}\ \{F \mid R, G\} \cong \text{Hom}\ \{R', G\}/\text{Hom}\ \{F' \mid R', G\},$$

the isomorphism being both algebraic and topological. The same result holds for the factor groups

$$(4.4) \qquad \text{Hom}\ \{R, G\}/\text{Hom}_f\ \{R, G; F\}, \qquad \text{Hom}_f\ \{R, G; F\}/\text{Hom}\ \{F \mid R, G\}.$$

This theorem is a corollary of a result to be established in Chapter II, as Theorem 10.1. It can also be proved directly, by appeal to the following lemma, which we state without proof.

---

[10] The *kernel* of a homomorphism θ of a group H is the set of all elements $h \in H$ with $\theta(h) = 0$.

LEMMA 4.3. *Let $F/R = E/G$, where $F \supset R$ is a free group and $E \supset G$ is any other group. There exists a homomorphism ϕ of F into E such that, in the given identification of cosets of G with cosets of R,*

$$\phi(x) + G = x + R, \qquad \textit{for all } x \in F. \tag{4.5}$$

Any other $\phi^* \in$ Hom $\{F, E\}$ with this property (4.5) has the form $\phi^* = \phi + \beta$, for some $\beta \in$ Hom $\{F, G\}$. Conversely, given ϕ with the property (4.5) any such $\phi^* = \phi + \beta$ has the same property.

Although a given group H can be represented in many ways as a factor group $H = F/R$ of a gree group, there is a "natural" such representation, in which F is the additive group F_H of the (integral) group ring of H. Specifically, given H, we choose for each $h \in H$ a symbol z_h and construct a free group F_H generated by the symbols z_h. The correspondence $z_h \to h$ induces a homomorphism. of F_H on H. Let R_H denote the kernel of this homomorphism. The factor group (4.3) of the Theorem can then be described invariantly in terms of H and G as the group

$$\text{Hom}\ \{R_H, G\}/\text{Hom}\ \{F_H \mid R_H, G\}.$$

The same remark applies to the factor groups of (4.4). It would be possible to use the groups so described as substitutes for the group of group extensions to be introduced in Chapter II.

5. Closures and extendable homomorphisms

If G is topological, we wish to examine the closures of the groups Hom $\{F \mid R, G\}$ and Hom$_f$ in the topological group Hom $\{R, G\}$. A preliminary is a characterization of the subgroup Hom$_f$

LEMMA 5.1. *A homomorphism θ of* Hom $\{R, G\}$ *lies in* Hom$_f$ $\{R, G; F\}$ *if and only if for each element t in F with a multiple mt in R there exists $h \in G$ with $\theta(mt) = mh$.*

PROOF. Let F_t be the subgroup of F generated by t and R. If $mt \in R$ for $m \neq 0$, F_t/R is finite and cyclic, so that $\theta \in$ Hom$_f$ is extendable to F_t. Hence the condition stated on $\theta(mt)$ is necessary. Conversely, for any given group $F_0 \subset F$ with F_0/R finite we can write F_0/R as a direct product of cyclic groups. By applying the given condition on θ to each of these cyclic groups, we find an extension of θ to F_0, as required.

Another characterization of Hom$_f$ can be found; the proof is similar:

LEMMA 5.2. *A homomorphism θ of* Hom $\{R, G\}$ *lies in* Hom$_f$ $\{R, G; F\}$ *if and only if θ can be extended to a homomorphism (into G) of each subgroup F_0 of F which contains R and for which the factor group F_0/R has a finite number of generators.*

We now consider the topology on Hom $\{R, G\}$.

LEMMA 5.3. *If G and hence* Hom $\{R, G\}$ *are generalized topological groups,* Hom$_f$ $\{R, G; F\}$ *is contained in the closure of* Hom $\{F \mid R, G\}$, or

$$\text{Hom}\,\{F \mid R, G\} \subset \text{Hom}_f\,\{R, G; F\} \subset \overline{\text{Hom}}\,\{F \mid R, G\} \subset \text{Hom}\,\{R, G\}.$$

PROOF. Let θ_0 be in $\text{Hom}_f\,\{R, G; F\}$, while U is any open set of Hom $\{R, G\}$ containing θ_0. Since F is discrete, the definition of the topology in Hom $\{R, G\}$ implies that there is a finite set of elements $r_1, \cdots, r_n$ of R such that U contains all θ for which each $\theta(r_i) = \theta_0(r_i)$. The elements r_i are all contained in a subgroup F_0 of F generated by a finite number of the given independent generators of the free group F. Since $\theta_0 \in \text{Hom}_f$, θ_0 has an extension θ' to the group generated by F_0 and R (Lemma 5.2). Introduce a new homomorphism θ^* of F by setting $\theta^*(z_\alpha) = \theta'(z_\alpha)$ for each generator z_α of F_0, $\theta^*(z_\alpha) = 0$ otherwise. This θ^* induces a homomorphism θ of R, which agrees with θ_0 on the original elements $r_1, \cdots, r_n$ and which is by construction an element of Hom $\{F \mid R, G\}$. In other words, the arbitrary neighborhood U of θ_0 does contain a homomorphism $\theta \in \text{Hom}\,\{F \mid R, G\}$. This proves the lemma.

LEMMA 5.4. *If G is a compact topological group,* Hom $\{F \mid R, G\}$ *is a closed sub-group of* Hom $\{R, G\}$, *and hence* Hom $\{F \mid R, G\} = \text{Hom}_f\,\{R, G; F\}$.

PROOF. By Corollary 3.2, both the groups Hom $\{R, G\}$ and Hom $\{F, G\}$ are compact and topological. The second of these groups is mapped homomorphically onto Hom $\{F \mid R, G\}$ by the continuous correspondence $\theta \to \theta \mid R$ of Lemma 3.4. Therefore, by Lemma 1.1, the image Hom $\{F \mid R, G\}$ is closed.

For any integer m, let mG be the subgroup of all elements of the form mg, with g in G. A condition for the closure of Hom_f may be stated in terms of these subgroups.

LEMMA 5.5. *If G is a generalized topological group, then* $\text{Hom}_f\,\{R, G; F\}$ *is closed in* Hom $\{R, G\}$ *whenever every subgroup mG of G is closed in G, for* $m = 2, 3, \cdots$.[11]

PROOF. Let θ be a homomorphism in the closure of $\text{Hom}_f\,\{R, G; F\}$. Consider an arbitrary t in F such that $mt \in R$. By Lemma 5.1 and the given condition on G it will suffice to prove that $\theta(mt) \in \overline{mG}$. Let V be any open set containing 0 in G. By the definition of the topology in Hom $\{R, G\}$, there exists for θ in the closure of Hom_f an element θ' in Hom_f itself, such that $\theta'(mt) - \theta(mt) \in V$. But $\theta'(mt)$ is in mG, so that the arbitrary open set $V + \theta(mt)$ does contain an element of mG. This proves $\theta(mt)$ in $\overline{mG}$, as required.

An examination of this proof shows that the given condition on G can be somewhat weakened. It suffices to require that the subgroup mG be closed in G for every integer m which is the order of an element of F/R. The same remark will apply in various subsequent cases when this condition on G is used.

CHAPTER II. GROUP EXTENSIONS

This chapter introduces the basic group Ext $\{G, H\}$ of all group extensions of G by H, and its subgroup $\text{Ext}_f\,\{G, H\}$ of all extensions which are "finitely trivial"

[11] If every subgroup mG is closed in G, Steenrod [9] and Lefschetz [7] say that G has the "division closure property."

(§8). Each individual group extension can be described either by a suitable "factor set" (§7) or by a certain homomorphism. The equivalence of these two representations is the fundamental theorem of this chapter (Theorem 10.1); it gives an expression of Ext $\{G, H\}$ as one of the factor-homomorphism groups already considered in Chapter I. This fundamental theorem, which is implicit in previous algebraic work on group extensions, is of independent algebraic interest. The chapter closes with a proof that the representation of Ext $\{G, H\}$ by homomorphisms is a "natural" one (§12). This conclusion is needed for the subsequent limiting process, which is used in defining the Čech homology groups.

6. Definition of extensions

A group E having G as subgroup and $H = E/G$ as the corresponding factor group is said to be an "extension" of G by H. More explicitly, if the groups G and H are given, a *group extension* of G by H is a pair (E, β), where E is a group containing G and β is a homomorphism of E onto H under which exactly the elements of G are mapped into $0 \in H$.[12] Such a β induces an isomorphism of E/G to H. For given G and H, two extensions (E_1, β_1) and (E_2, β_2) are regarded as *equivalent* if and only if there is an isomorphism ω of E_1 to E_2 which leaves elements of G and cosets of H fixed. In other words, the isomorphism ω of E_1 to E_2 must have $\omega g = g$ for $g \in G$ and $\beta_2 \omega x = \beta_1 x$ for $x \in E_1$. We regard equivalent extensions as identical, and so study the equivalence classes of extensions of G by H. It will appear that these equivalence classes are themselves the elements of a group.

For given G and H, the direct product $G \times H$ has the "natural" homomorphism $(g, h) \to h$ onto H, and so can be regarded as an extension of G by H. Any extension (E, β) equivalent to this direct product (with its natural homomorphism) is said to be a *trivial* extension of G by H.

7. Factor sets for extensions

A given extension (E, β) of G by H can be described in terms of representatives for elements of H. To each h in H select in E a representative $u(h)$, such that $\beta(u(h)) = h$. Every element of E lies in some coset h, so has the form $g + u(h)$ for g in G. The sum of any two representatives $u(h)$ and $u(k)$ will lie in the same coset, modulo G, as does the representative of the sum $h + k$. Hence there is an addition table of the form

$$u(h) + u(k) = u(h + k) + f(h, k), \tag{7.1}$$

where $f(h, k)$ lies in G for each pair of elements h, k in H. The commutative and associative laws in the group E imply two corresponding identities for f,

$$f(h, k) = f(k, h), \tag{7.2}$$

[12] Group extensions are discussed by Baer [2], Hall [6], Turing [11], Zassenhaus [15], and elsewhere. Much of the discussion in the literature treats the more general case in which G but not H is assumed to be abelian and in which G is not necessarily in the center of H.

$$f(h, k) + f(h + k, l) = f(h, k + l) + f(k, l). \tag{7.3}$$

The sum of any two elements $g_1 + u(h)$ and $g_2 + u(k)$ of E is determined by the addition table (7.1) and the addition given within G and H.

The extension E does not uniquely determine the corresponding function f. An arbitrary set of representatives $u'(h)$ for the elements of H can be expressed in terms of the given representatives as

$$u'(h) = u(h) + g(h), \qquad \text{each } g(h) \in G;$$

they will have an addition table like that of (7.1) with a function f' given by

$$f'(h, k) = f(h, k) + [g(h) + g(k) - g(h + k)]. \tag{7.4}$$

Conversely, a *factor set* of H in G is any function $f(h, k)$, with values in G for h, k in H which satisfies the "commutative" and "associative" conditions (7.2) and (7.3) for all h, k, and l in H. A *transformation set* is any function of h and k like the term in brackets in (7.4); thus for any function $g(h)$ defined for each $h \in H$ and taking on values in G, the function

$$t(h, k) = g(h) + g(k) - g(h + k) \tag{7.5}$$

is a transformation set. Such a set automatically satisfies the conditions (7.2) and (7.3), hence is always a factor set. Two factor sets f and f' are said to be *associate* if their difference is, as in (7.4), a transformation set. The correspondence between group extensions and factor sets may now be formulated as follows.

THEOREM 7.1. *For given groups G and H, there is a many-one correspondence $f \to (E, \beta)$ between the factor sets f of H in G and the group extensions (E, β) of G by H, where $f \to (E, \beta)$ holds if and only if f is the factor set which appears in one of the possible "addition tables" (7.1) for E. Two factor sets f and f' of H in G determine equivalent group extensions of G by H if and only if they are associate. In particular, the group extension determined by f is trivial if and only if f is a transformation set.*

PROOF. As a preliminary, observe that the associative relations (7.3) for f show (with $k = l = 0$, $h = k = 0$) that $f(0, 0) = f(h, 0) = f(0, l)$. Now, given f, we construct E_f as the group of all pairs (g, h) with addition given by the rule

$$(g_1, h) + (g_2, k) = (g_1 + g_2 + f(h, k), h + k),$$

and the homomorphism β_f defined by $\beta_f(g, h) = h$. Since $f(0, 0) = f(0, l)$, each element $(g, 0)$ may be identified with the corresponding element $g + f(0, 0)$ in G; the pair (E_f, β_f) is then indeed an extension of G by H. As a representative of h in E_f, we may choose $u(h) = (0, h)$; the addition table (7.1) then involves exactly the original factor set f. If E is an arbitrary group extension of G by H in which f appears as the factor set of E, the correspondence $g + u(h) \leftrightarrow (g, h)$ shows that the extension E is in fact equivalent to the extension E_f just constructed. Therefore $f \rightarrow (E_f, \beta_f)$ is a many-one correspondence with the defining property stated in the theorem.

If f and f' are associate, as in (7.4), the correspondence

$$(g, h) \rightarrow (g - g(h), h)'$$

shows that the corresponding extensions E_f and $E_{f'}$ are equivalent. Conversely, the argument leading to (7.4) shows in effect that E_f is equivalent to $E_{f'}$ only if f is associate to f'.

We turn now to two special applications of transformation sets. In the first place, the representative for the zero element of H may always be chosen as the zero in E. This means that $u'(0) = 0$, $u'(0) + u'(h) = u'(h)$, so that

$$f'(0, h) = f'(h, 0) = 0 \qquad (\text{all } h \in H). \tag{7.6}$$

A factor set f' with the property (7.6) may be called *normalized*; we have proved that every factor set f is associate to a normalized factor set.

Free groups may be characterized in terms of group extensions as follows:

THEOREM 7.2. *A group with more than one element H is free if and only if every extension of any group by H is the trivial extension.*

PROOF. Suppose first that H satisfies the condition that every extension of every G is trivial. Represent H as F/R, where F is free. Then F is a trivial extension of R by H, hence is a direct sum of R and H. Therefore H, as a subgroup of the free group F, is itself free. The other half of the theorem is stated in more detail in the following Lemma.

LEMMA 7.3. *Every factor set f' of a free group F in a group G is a transformation set, so that*

$$f'(x, y) = \phi(x + y) - \phi(x) - \phi(y), \qquad \phi(x) \in G, \tag{7.7}$$

holds for all $x, y \in F$. If F has generators z_α, the function ϕ may be chosen so that $\phi(0) = -f'(0, 0)$, $\phi(z_\alpha) = 0$ for each generator z_α.

PROOF. In the extension $E_{f'}$ of G by F we have an addition table

$$u'(x) + u'(y) = u'(x + y) + f'(x, y) \qquad (x, y \in F).$$

In E we introduce a new set of representatives $u(\sum e_\alpha z_\alpha) = \sum e_\alpha u'(z_\alpha)$ for the elements $\sum e_\alpha z_\alpha$ of F. These are related to the original representatives by an equation $u(z) = u'(z) + \phi(z)$, where $\phi(z)$ has values in G. Because F is a free group, $z \rightarrow u(z)$ as defined is a homomorphism of F into E, so that $u(x + y) = u(x) + u(y)$, and the factor set belonging to u is identically zero. But the given f' is associate to this zero factor set, as in (7.4). Setting $f = 0$, $\phi = -g$ in (7.4) gives (7.7), as desired. By construction, $u(z_\alpha) = u'(z_\alpha)$, so $\phi(z_\alpha) = 0$. Also $u'(0) + u'(0) = u'(0) + f'(0, 0)$, so that $u'(0) = f'(0, 0)$, $u(0) = 0$, and therefore $\phi(0) = -f'(0, 0)$. This completes the proof.

8. The group of extensions

For fixed H and G the sum of two factor sets f_1 and f_2 is a third factor set, defined as

$$(f_1 + f_2)(h, k) = f_1(h, k) + f_2(h, k) \qquad (h, k \in H).$$

Under this addition, the factor sets and the transformation sets form groups, denoted respectively by

(8.1) $\quad$ Fact $\{G, H\}$ = group of all factor sets of H in G,

(8.2) $\quad$ Trans $\{G, H\}$ = group of all transformation sets of H in G.

The factor sets belonging to a given group extension E constitute a coset of the subgroup Trans $\{G, H\}$, as in (7.4). Hence the correspondence of factor sets to extensions is a one-one correspondence between cosets of Fact/Trans and equivalence classes of extensions. This correspondence carries the addition of factor sets into an addition of group extensions. We are thus led to define the *group of group extensions* of G by H as[13]

(8.3) $\quad$ Ext $\{G, H\}$ = Fact $\{G, H\}$/Trans $\{G, H\}$.

If H is discrete while G is a (generalized) topological group, there will be a corresponding induced topology on Ext $\{G, H\}$. For each factor set f is a function on $H \times H$ with values in G, so that Fact $\{G, H\}$ is a subgroup of the generalized topological group $G^{H \times H}$ of all such functions. The subgroup "Trans" and the factor group "Ext" also carry topologies. Much as in §3 one can prove that if H is discrete and G topological, then Fact $\{G, H\}$ is a closed subgroup of $G^{H\times H}$. This proves

LEMMA 8.1. *If H is discrete and G is a topological* (*and compact*) *group, then* Fact $\{G, H\}$ *is a topological* (*and compact*) *group.*

In general, however, Trans $\{G, H\}$ will not be closed in Fact $\{G, H\}$, even when G is topological. In such cases Ext $\{G, H\}$ is necessarily a generalized topological group.

If (E, β) is an extension of G by H, each subgroup $S \subset H$ determines a corresponding subgroup $E_S \subset E$, consisting of all $e \in E$ with $\beta(e) \in S$. Since $E_S \supset G$, we may thus say that E "induces" an extension (E_S, β) of G by S. We call an extension E *finitely trivial* if E_S is trivial for every finite subgroup $S \subset H$.

Similarly, any factor set f of H in G determines for each subgroup $S \subset H$ a factor set f_S of S in G, where $f_S(h, k) = f(h, k)$ for h, k in S (i.e., f_S is obtained by "cutting off" f at S). The correspondence between factor sets and group extensions readily gives

LEMMA 8.2. *A factor set f of H in G determines a finitely trivial extension of*

---

[13] It is possible to define the sum of two group extensions directly, without using the factor sets (see Baer [2] p. 394); it also is possible to give an analogous definition of the topology introduced below in Ext $\{G, H\}$.

G by H if and only if, for every finite subgroup $S \subset H$, *the factor set* f_S *"cut off" at S is a transformation set of S in G. Hence the finitely trivial extensions of G by H constitute a subgroup* $\text{Ext}_f\{G, H\}$ *of* $\text{Ext}\{G, H\}$.

9. Group extensions and generators

A group extension can be described not only by factor sets, but also by certain homomorphisms related to the generators of the extending group H. For let (E, β) be a given extension of G by H, and $H = F/R$ a representation of H as a factor group of a free group F. Let F have the generators z_α, as in §4; the corresponding elements (or cosets) t_α of H will then be a set of generators of H. For each generator t_α choose a corresponding representative u_α in the given group extension E, so that $\beta u_\alpha = t_\alpha$. Then $\beta(\sum e_\alpha u_\alpha) = \sum e_\alpha t_\alpha$, so that any element $\sum e_\alpha t_\alpha \in H$ has a representative of the form $\sum e_\alpha u_\alpha$. This means that each element of E can be written in the form

$$x = g + \sum e_\alpha u_\alpha, \qquad g \in G, \qquad e_\alpha \text{ integers.}$$

From this representation one can at once determine how to add the elements of E. However, this representation is not in general unique, for $(\sum e_\alpha u_\alpha) \in G$ is equivalent to $\sum e_\alpha t_\alpha = 0'$, which in turn is equivalent to $(\sum e_\alpha z_\alpha) \in R$. Thus to each $r = \sum e_\alpha z_\alpha$ in the group R of "relations" there is assigned an element $\theta(r) \in G$, defined as

$$\theta(r) = \theta(\sum e_\alpha z_\alpha) = \sum e_\alpha u_\alpha$$

These assignments $\theta(r)$ completely determine the extension E.

The function θ hereby defined[14] is a homomorphism of R into G. Conversely every such homomorphism θ can be used to construct a corresponding group extension of G by $H = F/R$; it suffices to construct E by reducing the direct product $F \times G$ modulo the subgroup of all elements of the form $(r, \theta(r))$, for $r \in R$. There is thus a correspondence between homomorphisms of R into G and extensions of G by $H = F/R$.[15]

10. The connection between homomorphisms and factor sets

Given G and $H = F/R$, an extension E of G by H may be given either by a factor set or by a homomorphism of R into G. There must therefore be a relation between factor sets and homomorphisms of this type. We now propose to establish this relation directly, without using extensions explicitly. (Actually, the correspondence which we obtain is identical with that obtained by going from a homomorphism first to the corresponding group extension and then to its factor set.)

THEOREM 10.1. *If* $H = F/R$ *is a factor group of a free group F, while G is any other group, then*

---

[14] Actually θ may be obtained by "cutting off" one of the homomorphisms ϕ as described in Lemma 4.3.

[15] This correspondence has been stated by Baer ([2], p. 395) and used by Hall [6].

$$\text{(10.1)} \qquad \text{Ext}\,\{G, H\} \cong \text{Hom}\,\{R, G\}/\text{Hom}\,\{F \mid R, G\}.$$

Under the correspondence which gives this isomorphism

$$\text{(10.2)} \qquad \text{Ext}_f\,\{G, H\} \cong \text{Hom}_f\,\{R, G; F\}/\text{Hom}\,\{F \mid R, G\},$$

$$\text{(10.3)} \qquad \text{Ext}\,\{G, H\}/\text{Ext}_f\,\{G, H\} \cong \text{Hom}\,\{R, G\}/\text{Hom}_f\,\{R, G; F\}.$$

If G is a generalized topological group while F and H are discrete, all these isomorphisms are bicontinuous.

PROOF. As a preliminary, observe that the representation $H = F/R$ means that the free group F is a group extension of R by H. In this extension choose a representative $u_0(h)$ in F for each $h \in H$. F is then described, as in (7.1), by an addition table

$$\text{(10.4)} \qquad u_0(h) + u_0(k) = u_0(h + k) + f_0(h, k),$$

where f_0 is a factor set of H in R. This factor set will be fixed throughout the proof.

Since Ext $\{G, H\}$ is defined as Fact/Trans, the required isomorphism (10.1) could be established by a suitable correspondence of homomorphisms to factor sets. Let $\theta \in$ Hom $\{R, G\}$ be given, and define f_θ by

$$\text{(10.5)} \qquad f_\theta(h, k) = \theta[f_0(h, k)] \qquad (h, k \in H).$$

The requisite commutative and associative laws (7.2) and (7.3) for f_θ follow from those for f_0, and the correspondence $\theta \to f_\theta$ is a homomorphism of Hom $\{R, G\}$ into Fact $\{G, H\}$, and therefore into Ext $\{G, H\}$.

Suppose next that θ can be extended to a homomorphism θ^* of F into G. This homomorphism applied to (10.4) gives

$$\theta^*[f_0(h, k)] = \theta^*[u_0(h)] + \theta^*[u_0(k)] - \theta^*[u_0(h + k)].$$

If we set $g(h) = \theta^*[u_0(h)]$, the result asserts that $\theta^* f_0 = \theta f_0 = f_\theta$ is a transformation set.

Conversely, suppose that f_θ is a transformation set, so that $f_\theta(h, k) = g(h) + g(k) - g(h + k)$ for some function g. Now any element in F can be written, in only one way, in the form $r + u_0(h)$, with r in R, h in H. We define $\theta^*(r + u_0(h))$ as $\theta(r) + g(h)$. Clearly θ^* is an extension of θ; a straightforward computation with (10.4) shows that θ^* is actually a homomorphism. In this case, then, θ is extendable to F.

We know now that the correspondence $\theta \to f_\theta$ is an isomorphism of Hom $\{R, G\}$/Hom $\{F \mid R, G\}$ into a subgroup of Ext $\{G, H\}$. It remains to prove that it is a homomorphism onto. At this juncture we use for the first time the assumption that F is a free group. Let E be a given extension of G by H, with a factor set f which we can assume is normalized, as in (7.6). Let β_0 be the given homomorphism of F on H. Use f to define a factor set f' of F in G by the equation

$$f'(x, y) = f(\beta_0 x, \beta_0 y), \qquad x, y \in F. \tag{10.6}$$

Since F is free, f' is a transformation set, so we can find, as in Lemma 7.3, a function $\phi(z)$ on F to G with

$$\phi(x + y) = \phi(x) + \phi(y) + f'(x, y). \tag{10.7}$$

In particular, if x and y lie in R, $\beta_0 x = \beta_0 y = 0$, and $f'(x, y) = f(0, 0) = 0$, because f is normalized. Thus ϕ, restricted to R, is a homomorphism $\theta = \phi \mid R$ of R into G. Furthermore, if ϕ is applied to the addition table (10.4) for F, the property (10.7) gives

$$\phi[u_0(h) + u_0(k)] = \phi[u_0(h + k)] + \phi[f_0(h, k)],$$

where a term $f'(u_0(h + k), f_0(h, k))$, which would have entered by (10.7), is zero because f is normalized, $f_0(h, k) \in R$, and $\beta_0 f_0(h, k) = 0$. Now compute $f(h, k)$ for h, k in H. By (10.6),

$$\begin{aligned} f(h, k) &= f'(u_0(h), u_0(k)) \\ &= \phi[u_0(h) + u_0(k)] - \phi[u_0(h)] - \phi[u_0(k)] \\ &= \phi[u_0(h + k)] - \phi[u_0(h)] - \phi[u_0(k)] + \phi[f_0(h, k)], \end{aligned}$$

in virtue of the equation displayed just above. This equation asserts that f is associate to the factor set $\phi f_0 = \theta f_0$. In other words, given the normalized factor set f, we have constructed a homomorphism θ for which f is essentially θf_0. This completes the proof of (10.1).

It is desirable to find a more explicit expression for this dependence of θ on f. A simple induction applied to (10.7) will show that, for z_i in F,

$$\phi\left(\sum_{i=1}^{n} z_i\right) = \sum_{i=1}^{n} \phi(z_i) + \sum_{k=1}^{n-1} f'\left(\sum_{i=1}^{k} z_i, z_{k+1}\right).$$

If z_i is one of the generators z_α of F, then $\phi(z_i) = 0$, by Lemma 7.2. If $z_i = -z_\alpha$ is the negative of a generator, then by (10.7)

$$\phi(0) = \phi(z_\alpha + (-z_\alpha)) = \phi(z_\alpha) + \phi(-z_\alpha) + f'(z_\alpha, -z_\alpha),$$

so that $\phi(-z_\alpha) = -f'(z_\alpha, -z_\alpha)$. Now any element of F can be written as a finite linear combinations of generators and hence as a sum $\sum x_i$, where each x_i is either a generator or the negative of a generator z_α, and where any given generator may appear several times in this sum. In particular, for any element $r = \sum x_i$ in the subgroup R, the previous formula for ϕ becomes a formula for $\theta = \phi \mid R$,

$$\theta\left(\sum_{i=1}^{n} x_i\right) = -\sum{}' f(\beta_0 x_i, -\beta_0 x_i) + \sum_{k=1}^{n-1} f\left(\sum_{i=1}^{k} \beta_0 x_i, \beta_0 x_{k+1}\right), \tag{10.8}$$

where β_0 is the given homomorphism of F into H, and where the sum $\sum'$ is taken over those elements x_i which are the negatives of generators. The essential feature of this formula is the fact that it expresses $\theta(r)$ for $r \in R$ as a sum of a finite number of values of the given factor set f of H in G.

Now consider the continuity of the correspondence $\theta \to f_\theta$ used to establish (10.1). It suffices to establish the continuity at 0. If U is any open set, containing zero, in Hom $\{R, G\}$/Hom $\{F \mid R, G\}$, there will be an open set V containing 0 in G and a finite set of elements $r_1, \cdots, r_s \in R$ such that U contains the cosets of all homomorphisms θ with $\theta(r_i) \in V$, $i = 1, \cdots, s$.

For a given f, the expressions $\theta(r_i)$ of (10.8) for these elements r will involve but a finite number of elements of the factor set f. Because of the continuity of addition in G, we can construct an open set U' in Fact $\{G, H\}$ such that each $\theta(r_i)$ does in fact lie in the given V. This establishes the continuity of the correspondence $f \to \theta$. The continuity of the inverse correspondence is obtained by a similar argument on the definition (10.5) of this correspondence.

It remains only to consider the formulas (10.2) and (10.3) on finitely trivial extensions. Let θ and its correspondent f_θ be given, and let $F_0 \supset R$ be any subgroup of F for which F_0/R is finite. A previous argument, applied to F_0 instead of F, shows that θ can be extended to a homomorphism of F_0 into G if and only if f_θ, regarded as a factor set for F_0/R in G, is a transformation set. But the subgroup $\text{Hom}_f \{R, G; F\}$ by definition consists of all those θ which are extendable to every such F_0, while Ext_f by Lemma 8.2 is obtained from those factor sets which are transformation sets on every such subgroup F_0. $\text{Hom}_f \{R, G; F\}$/Hom $\{F/R, G\}$ is the subgroup corresponding to $\text{Ext}_f \{G, H\}$ under $\theta \to f_\theta$. This proves (10.2) and with it (10.3). The continuity of the isomorphisms in this case follows from the continuity of the isomorphism (10.1).

For subsequent purposes we observe that the correspondence $\theta \to f_\theta$ obtained in this proof is essentially independent of the choice of the fixed factor set f_0 for H in R. Specifically, if f_0 is replaced by an associate factor set f_0', f_θ will be replaced also by an associate factor set, so that the corresponding element of Ext $\{G, H\}$ is not altered.

11. Applications

The representation of Ext $\{G, H\}$ as Hom $\{R, G\}$/Hom $\{F \mid R, G\}$ gives an immediate proof of the invariance of the latter group, as stated in Theorem 4.2 of Chapter I. There are a number of other simple corollaries.

COROLLARY 11.1. *For a direct product $H \times H'$,*

$$\text{Ext}\,\{G, H \times H'\} \cong \text{Ext}\,\{G, H\} \times \text{Ext}\,\{G, H'\}. \tag{11.1}$$

If G is a generalized topological group, the isomorphism is bicontinuous.

PROOF. If $H = F/R$ and $H' = F'/R'$, we may write $H \times H' = (F \times F')/(R \times R')$, where $F \times F'$, like F and F', is free. Each homomorphism of $R \times R'$ into G determines homomorphisms θ and θ' of the subgroups R and R' into G, and this correspondence yields a (bicontinuous) isomorphism

$$\text{Hom}\,\{R \times R', G\} \cong \text{Hom}\,\{R, G\} \times \text{Hom}\,\{R', G\}.$$

Furthermore, under the same correspondence

$$\text{Hom}\,\{(F \times F') \mid (R \times R'), G\} \cong \text{Hom}\,\{F \mid R, G\} \times \text{Hom}\,\{F' \mid R', G\}.$$

These two relations yield a corresponding isomorphism between the respective factor groups such as Hom $\{R, G\}$/Hom $\{F \mid R, G\}$. By the fundamental theorem, the latter isomorphism is the one asserted in (11.1).

This conclusion can also be established without using homomorphisms, by a direct argument like that of Lemma 7.2. (Choose new representatives in E for elements of $H \times H'$ by setting $u'(hh') = u(h)u(h')$). Another simple argument directly with the factor sets will give a companion "direct product" representation,

$$\text{(11.2)} \qquad \text{Ext}\,\{G \times G', H\} \cong \text{Ext}\,\{G, H\} \times \text{Ext}\,\{G', H\};$$

this isomorphism is also bicontinuous.

COROLLARY 11.2. *If H is a cyclic group of order m, then*

$$\text{(11.3)} \qquad \text{Ext}\,\{G, H\} \cong G/mG, \qquad (mG = \textit{all } mg, \textit{for } g \in G).$$

This isomorphism is also bicontinuous.

This is a well known result, which can be derived directly from our main theorem. The cyclic group H can be written as $H = F/R$, where F is an infinite cyclic group with generator z, R the subgroup generated by mz. Then any $\theta \in$ Hom $\{R, G\}$ is uniquely determined by the image $\theta(mz) = h$ of the generator mz. This correspondence $\theta \to h$ (mod mG) gives the isomorphism (11.3).

A similar representation can be found for any finite abelian group H, simply by representing H as a direct product of cyclic groups of orders m_i, $i = 1, \cdots, t$. By Corollary 11.1, Ext $\{G, H\}$ is then isomorphic to the direct product of the groups G/m_iG. A similar decomposition applies if the abelian group H has a finite number of generators. The result may be stated as follows.

COROLLARY 11.3. *If H has a finite number of generators, and T is the subgroup of all elements of finite order in H, then* Ext $\{G, H\} \cong$ Ext $\{G, T\}$, *algebraically and topologically. The latter group is a direct product of groups of the form G/mG.*

Theorem 7.2 (extensions by a free group are trivial) has an analogue for infinitely divisible groups. Recall that G is *infinitely divisible* if for each $g \in G$ and each integer $m \neq 0$ the equation $mx = g$ has a solution $x \in G$.

COROLLARY 11.4. *A group G is infinitely divisible if and only if every extension of G by any group is the trivial extension.*

PROOF. If G is not infinitely divisible, some $G/mG \neq 0$, so that there will be a non-trivial extension of G by a cyclic group, as in Corollary 11.2. Conversely, suppose G is infinitely divisible. If $R \subset F$ are groups, a transfinite induction will show that every homomorphism of R into G can be extended to a homomorphism into G of the larger group F. Therefore the subgroup Hom $\{F \mid R, G\}$ exhausts the group Hom $\{R, G\}$, and Ext $\{G, F/R\} = 0$.

COROLLARY 11.5. *If T is the subgroup of all elements of finite order in H, then*

(11.4) $\quad \operatorname{Ext}\{G, H\}/\operatorname{Ext}_f\{G, H\} \cong \operatorname{Ext}\{G, T\}/\operatorname{Ext}_f\{G, T\}.$

This isomorphism is bicontinuous (if G is a generalized topological group).

PROOF. In the representation $H = F/R$, let F_T denote the set of all elements of F of finite order modulo R. The group T then has the representation $T = F_T/R$, while F_T, as a subgroup of a free group, is itself free. Now the group $\operatorname{Hom}_f\{R, G; F\}$ by definition consists of all homomorphisms extendable to subgroups of F finite over R; as these subgroups are all contained in F_T, the group Hom_f is identical with $\operatorname{Hom}_f\{R, G; F_T\}$. If both factor groups in (11.4) are now represented by groups of homomorphisms, as in (10.3), the result is immediate.

Observe that when T has only elements of finite order, the group $\operatorname{Ext}_f\{G, T\}$, though it consists of extensions E of G by T trivial on every finite subgroup of T, can contain non-trivial extensions. This is illustrated by the following example. Let p be a prime, and G a group with generators $g, h_1, h_2, \cdots$ and relations $p^i h_i = g$, for $i = 1, 2, \cdots$. In this group G the intersection of all the subgroups p^iG is the group generated by g alone. Let T be the group of all rational numbers of the form a/p^i, reduced modulo 1. Then all elements of T have finite order, and T may be written as $T = F/R$, where F is a free group with generators $z_1, z_2, \cdots$, and R the free subgroup generated by pz_1, $pz_2 - z_1, pz_3 - z_2, \cdots$. (The homomorphism $F \to T$ maps z_i into $1/p^i$.)

To prove $\operatorname{Ext}_f\{G, T\} \neq 0$ it suffices to find a $\theta \in \operatorname{Hom}_f\{R, G; F\}$ which is not in $\operatorname{Hom}\{F \mid R, G\}$. Such a θ is determined by setting $\theta(pz_1) = g$, $\theta(pz_{i+1} - z_i) = 0$, $i = 1, 2, \cdots$. The definition $\theta^*(z_{n-i}) = p^i h_n$ will provide an extension θ^* of θ to the finite subgroup of F generated by $z_1, \cdots, z_n$. However, suppose that θ had an extension ϕ to F. Then $\phi(pz_{i+1}) = \phi(z_i)$, so that $\phi(z_1) = p^n\phi(z_{n+1})$ for every n. This means that $\phi(z_1)$ is in every subgroup p^nG, hence has the form eg for an integer e. But then $g = \theta(pz_1) = p\phi(z_1) = epg$ gives a contradiction. Therefore $\operatorname{Ext}_f\{G, T\} \neq 0$ in this case. However, if G has no elements of finite order, one can prove easily that $\operatorname{Ext}_f\{G, T\} = 0$, using Lemma 5.1 (see §17 below).

For several types of topological groups G, §5 gives information on the topology of the various relevant subgroups of $\operatorname{Hom}\{R, G\}$. By the main theorem, the conclusions of Lemmas 5.3, 5.4, and 5.5 can now be rewritten as conclusions about the topology of $\operatorname{Ext}\{G, H\}$, as follows.

COROLLARY 11.6. *If H is discrete and G a generalized topological group, the closure of the zero element in the generalized topological group* $\operatorname{Ext}\{G, H\}$ *contains* $\operatorname{Ext}_f\{G, H\}$. *If, in addition, every subgroup mG is closed in G, for $m = 2, 3, \cdots$, then* $\operatorname{Ext}_f\{G, H\}$ *is closed in* $\operatorname{Ext}\{G, H\}$.

In particular, if H has no elements of finite order, then every extension of G by H is trivial on (the non-existent) finite subgroups of H, consequently $\operatorname{Ext}_f\{G, H\} = \operatorname{Ext}\{G, H\}$ and the closure of 0 is the whole group $\operatorname{Ext}\{G, H\}$. This means that $\operatorname{Ext}\{G, H\}$ carries the "trivial" (generalized) topology in which the only open sets are the whole group and the empty set.

COROLLARY 11.7. *If H is discrete and G compact and topological, then* $\mathrm{Ext}_f \{G, H\} = 0$ *and* Ext $\{G, H\}$ *is itself a compact topological group.*

This conclusion is obtained from Lemma 8.1 and from Lemma 5.4.

12. Natural homomorphisms

The basic homomorphism $\eta(\theta) = f_\theta$ mapping elements θ of Hom $\{R, G\}$ into factor sets f, as in Theorem 10.1, is a "natural" one. Specifically, this means that the application of η "commutes" with the application of any homomorphism T to the free group F and its subgroup R. To state this more precisely, we need to consider first the homomorphisms which T induces on the groups Hom $\{R, G\}$ and Ext $\{G, H\}$.

Let F' be a free group with subgroup R', T a homomorphism $z' \to Tz'$ of F' into the free group F such that $T(R') \subset R$. T induces a homomorphism of $H' = F'/R'$ into $H = F/R$. This induced homomorphism will be written with the same letter T, so that $T(g + R') = Tg + R$, for any coset $g + R'$.

Now consider $\theta \in$ Hom $\{R, G\}$. Clearly the product $\theta' = \theta T$ is an element of Hom $\{R', G\}$, and the correspondence $\theta \to \theta'$ is a homomorphism T_h^* of Hom $\{R, G\}$ into Hom $\{R', G\}$. Furthermore $\theta \in$ Hom $\{F \mid R, G\}$ implies $\theta T \in$ Hom $\{F' \mid R', G\}$, so that T_h^* also induces a homomorphism T_h^*,

(12.1) T_h^* : Hom $\{R, G\}$/Hom $\{F \mid R, G\} \to$ Hom $\{R', G\}$/Hom $\{F' \mid R', G\}$.

Similarly, consider $f \in$ Fact $\{G, H\}$. The function f' defined by

$$f'(h', k') = f(Th', Tk') \qquad (h', k' \in H')$$

is a factor set of H' in G, and the correspondence $f \to f'$ is a homomorphism T_e^* of Fact $\{G, H\}$ into Fact $\{G, H'\}$. Furthermore, $f \in$ Trans $\{G, H\}$ implies $f' \in$ Trans $\{G, H'\}$, so that T_e^* also induces a homomorphism T_e^* for the corresponding factor groups Ext = Fact/Trans,

(12.2) $$T_e^* : \mathrm{Ext}\,\{G, H\} \to \mathrm{Ext}\,\{G, H'\}.$$

By the (dual) homomorphisms induced on Ext or Hom by T we always mean these homomorphisms T_h^* and T_e^*.

THEOREM 12.1. *Let T be a homomorphism of F' into F with $T(R') \subset R$, where $F \supset R$ and $F' \supset R'$ are free groups, while η (or η') is the homomorphism of* Hom $\{R, G\}$ *onto* Ext $\{G, F/R\}$ *established in the proof of Theorem* 10.1. *Then*

(12.3) $$\eta' T_h^* = T_e^* \eta,$$

where T_h^*, T_e^* are the appropriate homomorphisms induced by T on Hom *and* Ext, *respectively.*

PROOF. The figure involved is

$$\begin{array}{ccc}
\mathrm{Hom}\,\{R, G\} & \xrightarrow[\eta]{} & \mathrm{Ext}\,\{G, F/R\} \\
\downarrow T_h^* & & \downarrow T_e^* \\
\mathrm{Hom}\,\{R', G\} & \xrightarrow[\eta']{} & \mathrm{Ext}\,\{G, F'/R'\}
\end{array}$$

The correspondence η was constructed from a factor set f_0 for F as an extension of R; similarly, η' is based on a factor set f_0' for H' in R', such that

$$u_0'(h') + u_0'(k') = u_0'(h' + k') + f_0'(h', k'), \tag{12.4}$$

where $u_0'(h')$ is a representative of $h' \in H'$ in F'. First we determine the relation between f_0 and f_0'. The given homomorphism T carries F' into F, H' into H and thus $u_0'(h')$ into $Tu_0'(h')$, a representative in F of Th' in H. This representative will differ from the given representative $u_0(Th')$ by an element of R, so that

$$Tu_0'(h') = u_0(Th') + \rho(h') \qquad \text{(all } h' \text{ in } H').$$

where each $\rho(h')$ lies in R. Now the representatives $Tu_0'(h')$ will add with a factor set $Tf_0'(h', k')$, as may be seen by applying T to both sides of (12.4). This factor set in associate (in the group TH') to the originally given factor set f_0 of $H \supset TH'$; explicitly we have, by the argument leading to (7.4), that

$$Tf_0'(h', k') = f_0(Th', Tk') + [\rho(h') + \rho(k') - \rho(h' + k')].$$

Suppose now that $\theta \in \text{Hom}\{R, G\}$ is given. Application of η and then T_e^* will give, by the definitions of these correspondences, a factor set f', with

$$\begin{aligned} f'(h', k') &= \theta[f_0(Th', Tk')] \\ &= \theta T[f_0'(h', k')] + [\theta\rho(h' + k') - \theta\rho(h') - \theta\rho(k')]. \end{aligned}$$

On the other hand, application of T_h^* and then η' will give, again by the appropriate definitions, a factor set f^* with

$$f^*(h', k') = \theta'[f_0'(h', k')] = \theta T[f_0'(h', k')].$$

Since $\theta\rho(h')$ is an element in G for each $h' \in H'$, these two equations show that f^* and f' are associate, hence that $f' = T_e^*\eta\theta$ and $f^* = \eta' T_h^*\theta$ do determine the same element of Ext $\{G, H\}$, as asserted in the theorem.

CHAPTER III. EXTENSIONS OF SPECIAL GROUPS

In this chapter we shall determine Ext $\{G, H\}$ more explicitly for various special groups G and H. We begin with a brief review of the theory of characters, which will be used extensively in this chapter and also in Chapters V and VI.

13. Characters[16]

Let G, H, and J be three generalized topological groups. G and H are said to be *paired* to J if a continuous function[17] $\phi(g, h)$ with values in J is given

[16] The character theory was discovered by Pontrjagin (see [8]), generalized by van Kampen (see Weil [12], Ch. VI and Lefschetz [7], Ch. II).

[17] As a mapping $G \times H \to J$; for discussion of pairing, cf. [8], [14].

such that for any fixed g_0, $\phi(g_0, h)$ is a homomorphism of H into J and for any fixed h_0, $\phi(g, h_0)$ is a homomorphism of G into J.

Each subset $A \subset G$ determines a corresponding subset Annih $A \subset H$, called the *annihilator* of A, such that $h \in$ Annih A if and only if $\phi(g, h) = 0$ for all $g \in A$. Annihilators of subsets of H are defined similarly. It is clear that the annihilators are subgroups.

LEMMA 13.1. *If G and H are paired to a topological group J, then for each $A \subset G$, Annih A is a closed subgroup of H.*

This is an immediate consequence of the continuity of ϕ for fixed g.

G and H are said to be *dually paired* to J if they are so paired that

$$\text{Annih } G = 0 \quad \text{and} \quad \text{Annih } H = 0.$$

LEMMA 13.2. *If G and H are paired to J then $G/$Annih H and $H/$Annih G are dually paired to J.*

The most important group pairings arise when $J = P$ is the additive group of reals reduced modulo 1. A homomorphism of a group G into P will be called a *character* and the group Hom $\{G, P\}$ will be written as Char G. Since P has no "arbitrarily small" subgroups, it follows from a remark in §3 that if G is compact, Char G is discrete. Vice versa, by Corollary 3.2, if G is discrete, Char G is compact and topological.

The basic lemma of the theory of characters is

LEMMA 13.3. *Let G be a discrete or compact topological group and let $g \neq 0$ be an element of G. There is then a character $\theta \in$ Char G such that $\theta(g) \neq 0$.*

In the case of discrete G the lemma follows easily from the proof of Corollary 11.4, since P is infinitely divisible. In the compact case the proof is much less elementary and uses the theory of invariant integration in compact groups.

The lemma can be equivalently formulated as follows:

LEMMA 13.4. *Let G be a discrete or compact topological group. G and Char G are dually paired to P with the multiplitation*

$$\phi(g, \theta) = \theta(g), \qquad g \in G, \theta \in \text{Char } G.$$

Now let G and H be paired to P with $\phi(g, h)$ as multiplication. Since, for a fixed g, $\phi(g, h)$ is a character of H and, for fixed h, $\phi(g, h)$ is a character of G, we obtain induced mappings

$$G \to \text{Char } H, \qquad H \to \text{Char } G. \tag{13.1}$$

A basic result of the character theory is

THEOREM 13.5. *Let the compact topological group G and the discrete group H be paired to P. The pairing is dual if and only if the induced mappings* (13.1) *are isomorphisms:*

$$G \cong \text{Char } H \quad \textit{and} \quad H \cong \text{Char } G.$$

The following two theorems are consequences of the previous results:

THEOREM 13.6. *If G is a discrete or a compact topological group, then*

$$\text{Char Char } G \cong G.$$

THEOREM 13.7. *If the compact topological group G and the discrete group H are dually paired to P, then for every closed subgroup G_1 of G and every subgroup H_1 of H we have*

$$\text{Annih [Annih } G_1] = G_1, \qquad \text{Annih [Annih } H_1] = H_1.$$

14. Modular traces

To study Ext $\{G, H\}$ for compact G we need a certain modification of the "trace" of an endomorphism of a free group. The simplest case of this modification refers to a correspondence which is not a homomorphism, but is a homomorphism, modulo m-folds of elements. It may be stated as follows.

LEMMA 14.1. *Let m be an integer, and let $r \to S(r)$ be a correspondence carrying the free group R into a finite subset of itself in such manner that*

$$S(r_1 + r_2) \equiv S(r_1) + S(r_2) \qquad (\textit{mod } mR), \tag{14.1}$$

for all $r_1, r_2 \in R$. Let the elements y_α be any independent basis for R, and write $S(y_\alpha) = \sum_\beta c_{\alpha\beta} y_\beta$, with integral coefficients $c_{\alpha\beta}$. Then the "trace"

$$t_m(S) \equiv \textstyle\sum_\alpha c_{\alpha\alpha} \qquad (\textit{mod } m) \tag{14.2}$$

is a well defined finite integer, modulo m, independent of the choice of the basis y_α for R.

The proof is exactly parallel to the standard one (e.g. [1], p. 569) for an actual homomorphism of R to itself, using the "modular" homomorphism condition (14.1) at the appropriate junctures in place of the full homomorphism condition. A similar analogue of a special case of the "additivity" of traces will give the following conclusion.

LEMMA 14.2. *If in Lemma* 14.1 *the elements $w_1, \cdots, w_t$ are any independent elements of R such that S(R) lies in the group generated by $w_1, \cdots, w_t$, and if $S(w_i) = \sum_j d_{ij} w_j$, then $t_m(S) \equiv \sum_i d_{ii}$ (mod m).*

Now let R be a subgroup of the free group F, σ a homomorphism of R into a finite subgroup of F/R. There will then be at least one integer m for which $m\sigma(R) = 0$. Choose for each coset u of F/R a representative $\rho(u)$ in F; then $\rho(u + v) \equiv \rho(u) + \rho(v)$ (mod R). For each $r \in R$, $m(\rho\sigma r)$ is also an element of R, and $S(r) = m(\rho\sigma r)$, where

$$R \xrightarrow[\sigma]{} F/R \xrightarrow[\rho]{} F \xrightarrow[m]{} R,$$

is a correspondence of R to R with the modular homomorphism property (14.1).[18] The trace of the original homomorphism σ is now defined as

$$t(\sigma) \equiv t_m(S)/m \equiv t_m(m\rho\sigma)/m \qquad (\text{mod } 1). \tag{14.3}$$

[18] S could also be described in terms of m and σ as follows: S is the essentially unique correspondence of R to a finite subset of $mF \cap R$ with property (14.1) and such that each $\sigma(r)$ is the coset modulo R of $S(r)/m$.

THEOREM 14.3. *If $R \subset F$, F a free group, and if σ is any homomorphism of R into a finite subgroup of F/R, then the trace $t(\sigma)$ defined by (14.3) is a unique real number, modulo 1, independent of the choices of m and ρ made in its definition. If σ_1 and σ_2 are two such homomorphisms of R to F/R,*

$$t(\sigma_1 + \sigma_2) \equiv t(\sigma_1) + t(\sigma_2) \quad (mod\ 1). \tag{14.4}$$

In particular, $t(_0) \equiv 0$ (mod 1). Furthermore, if T_0 is a fixed finite subgroup of F/R, the correspondence $\sigma \to t(\sigma)$ is a continuous homomorphism of Hom $\{R, T_0\}$ *into the reals modulo* 1.

We are to prove the invariance of the definition of t. First, hold ρ fixed and replace m by a proper multiple $m' = km$. Then S and $t_m(S)$ are both multiplied by k, hence $t'(\sigma) \equiv t_{km}(kS)/km \equiv kt_m(S)/km \equiv t(\sigma)$ is unaltered, mod 1. Now hold m fixed and let ρ' be any second set of representatives $\rho'(u)$ for cosets $u \in F/R$. Then $\rho'(u) \equiv \rho(u)$ (mod R), so $S'(r) \equiv S(r)$ (mod mR), which implies that $t_m(S') \equiv t_m(S)$ (mod m). This shows that the trace is independent of ρ and m.

The additive property (14.4) is readily established; it is only necessary to choose a single integer in such a way that both $m\sigma_1 R$ and $m\sigma_2 R$ are zero.

Before establishing the continuity of $t(\sigma)$, we propose a more explicit representation of the finiteness of $t(\sigma)$. Let T_0 be a fixed finite subgroup of F/R, and choose a direct summand F_0 of F with a finite number of generators such that $F_0/(F_0 \cap R)$ contains T_0. We can choose simultaneously ([1], p. 566) a basis $z_1, \cdots, z_n$ for F_0 and a basis $y_1, \cdots, y_s$ for $F_0 \cap R$ so that $y_i = d_i z_i$, for integers d_i, $i = 1, \cdots, s \leqq n$. Furthermore, one can prove $F_0 \cap R$ a direct summand of R; there is then a (not necessarily denumberable) basis for R of the form $y_1, \cdots, y_s, y_\alpha, y_\beta, \cdots$. In particular, if $\sigma(R) \subset T_0$, we may choose $\rho(0) = 0$, $\rho(T_0) \subset F_0$, hence $S(R) = m\rho\sigma(R) \subset F_0 \cap R$. The equations for S and its trace then take the form

$$S(y_\gamma) = \sum_{i=1}^{s} c_{\gamma i} y_i, \qquad t_m(S) \equiv \sum_{i=1}^{s} c_{ii} \quad (\text{mod } m), \tag{14.5}$$

where $\gamma = 1, 2, \cdots, s, \alpha, \beta, \cdots$.

To prove $t(\sigma)$ continuous it suffices to establish the continuity at $\sigma = 0$, and hence to prove that $t(\sigma) \equiv 0$ for σ in a suitable neighborhood U of 0 in Hom $\{R, T_0\}$. Let U be the open set in Hom $\{R, T_0\}$ consisting of all σ with $\sigma(y_1) = \cdots = \sigma(y_s) = 0$, where y_i is the special basis constructed from F_0 above. Then, because $\rho(0) = 0$, we have $S(y_i) = 0$, $t_m(S) \equiv 0$ (mod m), and therefore $t(\sigma) \equiv 0$ (mod 1) for σ in U.

We next consider circumstances under which the traces will vanish.

LEMMA 14.4. *If $\sigma \in$ Hom $\{R, F/R\}$ has an extension σ^* which carries F homomorphically into a finite subgroup T_0 of F/R, then $t(\sigma) \equiv 0$ (mod 1).*

PROOF. For T_0 we choose $y_i = d_i z_i$ as above, and then select ρ with $\rho(T_0) \subset F_0$ and m with $mT_0 = 0$ and each $d_i \equiv 0$ (mod m). Then, for suitable integers e_{ij},

$$\rho\sigma^*(z_i) = \sum_{j=1}^{s} e_{ij} z_j, \qquad i = 1, \cdots, n;$$

furthermore $\rho\sigma^*(kz_i) \equiv k\rho\sigma^*(z_i)$ (mod R_0), for any integer k. But $S(y_i) = m\rho\sigma(y_i) = m\rho\sigma^*(d_iz_i) \equiv md_i\rho\sigma^*(z_i)$ (mod mR_0). Then computing $t_m(S)$ by (14.5) and using the fact that $m \equiv 0$ (mod d_j) for each j, we find that $t_m(S) \equiv m \sum e_{ii} \equiv 0$ (mod m), as asserted.

Conversely, we can find certain circumstances in which the trace will assuredly not vanish.

LEMMA 14.5. *If $z \in F$ has order n, modulo R, and if σ is a homomorphism of R into the subgroup of F/R generated by the coset of z, then $\sigma(nz) \neq 0$ implies $t(\sigma) \not\equiv 0$ (mod 1).*

PROOF. Let u denote the coset of z, modulo R. Choose the system of representatives so that $\rho(iu) = iz$, for $i = 0, \cdots, n-1$, and use n as the integer m in the definition of the trace. Then $S = m\rho\sigma$ carries R into the cyclic subgroup generated by mz. Since $\sigma(nz) = ku$, where $k \not\equiv 0$ (mod m), $S(nz) \equiv knz$, and the trace, as computed by Lemma 14.2, is $t_m(S) \equiv k \not\equiv 0$ (mod m), as asserted.

15. Extensions of compact groups

The group of extensions of a compact topological group G can be expressed as an appropriate character group.

THEOREM 15.1. *If G is compact and topological, H discrete, then* $\text{Ext}_f\,\{G, H\} = 0$ *and there is a (bicontinuous) isomorphism:*

$$\text{(15.1)} \qquad \text{Ext}\,\{G, H\} \cong \text{Char Hom}\,\{G, H\}.$$

If G_0 is the component of 0 in G and T the subgroup of all elements of finite order in H, then also

$$\text{Ext}\,\{G, H\} \cong \text{Char Hom}\,\{G, T\} \cong \text{Char Hom}\,\{G/G_0, T\}.$$

The last conclusion follows at once from the first, for Hom $\{G, H\}$ includes only continuous homomorphisms ϕ of the compact group G; every such homomorphism must map the connected subgroup G_0 into 0. Furthermore each ϕ carries G into a finite subgroup of the discrete group H, hence into a subgroup of T. Observe also that H is discrete, hence has no arbitrarily small subgroups; therefore (cf. §3) Hom $\{G, H\}$ is discrete, as should be the case for a character group of the compact group Ext $\{G, H\}$.

It remains to prove (15.1). Represent H as F/R; then, according to the fundamental theorem of Chapter II, (15.1) is equivalent to

$$\text{(15.2)} \qquad \text{Hom}\,\{R, G\}/\text{Hom}\,\{F \mid R, G\} \cong \text{Char Hom}\,\{G, F/R\}.$$

According to Theorem 13.5 it will thus suffice to provide a suitable pairing of the compact group Hom $\{R, G\}$ and the discrete group Hom $\{G, F/R\}$ to the reals modulo 1. To this end, take any $\theta \in$ Hom $\{R, G\}$ and $\phi \in$ Hom $\{G, F/R\}$. As just above, $\phi(G)$ is a finite subgroup of F/R. Therefore $\sigma = \phi\theta$ is a homomorphism of R into a finite subgroup of F/R, so that the trace introduced in the previous section can be used to define

$$t(\theta, \phi) \equiv t(\phi\theta) \pmod{1}. \tag{15.3}$$

We propose to show that this is the requisite pairing.

In the first place, this product is additive, for

$$t(\theta + \theta', \phi) \equiv t(\theta, \phi) + t(\theta', \phi) \pmod{1},$$

$$t(\theta, \phi + \phi') \equiv t(\theta, \phi) + t(\theta, \phi') \pmod{1}$$

follow from the corresponding property (14.4) for $\sigma = \phi\theta$. Secondly, if ϕ is fixed, the product $t(\theta, \phi)$ is continuous in θ. For when ϕ is fixed, $\sigma = \phi\theta$ maps R into a fixed finite subgroup of F/R. Since $\theta \to \phi\theta = \sigma$ is continuous, and since $\sigma \to t(\sigma)$ is continuous, by Theorem 14.3, the continuity of $t(\theta, \phi)$ follows.

As to the annihilators under this pairing, we assert that

$$\text{Annih Hom } \{G, F/R\} = \text{Hom } \{F \mid R, G\}. \tag{15.4}$$

For suppose first that $\theta \in \text{Hom } \{F \mid R, G\}$ and let θ^* be an extension of θ to F. Then $\sigma^* = \phi\theta^*$ is an extension of $\sigma = \phi\theta$ to F, and σ^* still carries F into (the same) finite subgroup of F/R. Therefore, by Lemma 14.4, $t(\theta, \phi) \equiv t(\sigma) \equiv 0 \pmod{1}$. Hence θ is in the annihilator in question.

Conversely, let θ be fixed, and suppose that $t(\theta, \phi) \equiv 0 \pmod{1}$ for every ϕ; then $\theta \in \text{Hom } \{F \mid R, G\}$. Since G is compact and topological, it will suffice by Lemma 5.4 to prove that $\theta \in \text{Hom}_f \{R, G; F\}$. If this were not the case, there would be in F an element z of some order n, modulo R, such that $\theta(nz) = g_0$ is not an element of nG. But nG is a continuous image (under $g \to ng$) of the compact group G, hence (Lemma 1.1) is a closed subgroup of G; therefore G/nG is compact and topological. By Lemma 13.3 there is then character χ of G/nG with $\chi(g_0') \neq 0$, where g_0' is the coset of g_0 modulo nG. Since every coset of G/nG has as order some divisor of n, this character χ carries G/nG into the group generated by the fraction $1/n$, modulo 1. This is a cyclic group of order n, and so can be replaced by the isomorphic cyclic group of order n generated by the coset z' of z in F/R. The so-modified character X of G/nG then induces a continuous homomorphism ϕ of G into F/R, where

$$\phi(g_0) \neq 0, \qquad \phi(G) \subset [0, z', z'^2, \cdots, z'^{n-1}].$$

For this particular ϕ, the homomorphism $\sigma = \phi\theta$ carries nz into $\phi\theta(nz) = \phi(g_0) \neq 0$. Lemma 14.5 of the previous section then shows that $t(\sigma) \equiv t(\theta, \phi) \not\equiv 0 \pmod{1}$, contrary to the assumption $t(\theta, \phi) \equiv 0$ for every ϕ. Therefore θ does lie in Hom $\{F \mid R, G\}$, and 15.4 is proved.

Finally, we assert that, under the pairing t,

$$\text{Annih Hom } \{R, G\} = 0. \tag{15.5}$$

For suppose instead that $t(\theta, \phi) \equiv 0 \pmod{1}$ for all θ and for some $\phi \neq 0$. Then for some $g_0 \in G$, $\phi(g_0) = u \neq 0$. The element u of F/R is the coset of some element w of F; as before, ϕ maps G into a finite subgroup of F/R, so that w

has a finite order m, modulo R. It is then possible to select in the free group F an independent basis with a first element z_0 such that $w = kz_0$ for some integer k. If z_0 has order n, modulo R, there is then a corresponding basis for R of elements y_α, with $y_0 = nz_0$. Now construct $\theta \in \text{Hom}\{R, G\}$ by setting

$$\theta(y_0) = g_0, \qquad \theta(y_\alpha) = 0, \qquad y_\alpha \neq y_0.$$

This particular homomorphism carries R into the subgroup of G generated by g_0, so that the product $\sigma = \phi\theta$ carries R into the finite subgroup of F/R generated by $\phi(g_0) = u$. Since u is the coset of $w = kz_0$, this is contained in the subgroup of F/R generated by the coset of z_0. Furthermore $\sigma(nz_0) = u \neq 0$. Lemma 14.5 again applies to show that $t(\sigma) \equiv t(\theta, \phi) \not\equiv 0 \pmod 1$, counter to assumption.

Given the assertions (15.4) and (15.5) as to annihilators, it follows from Lemma 13.2 that the groups $\text{Hom}\{R, G\}/\text{Hom}\{F \mid R, G\}$ and $\text{Hom}\{G, F/R\}$ are dually paired. Formula (15.2) is then a consequence of Theorem 13.5.

16. Two lemmas on homomorphisms

A generalized topological group G is said to have no arbitrarily small subgroups if there is in G an open set V containing 0 but containing no subgroups other than the group consisting of 0 alone.

LEMMA 16.1. *If the discrete group T has no elements of infinite order and the generalized topological group G has no arbitrarily small subgroups, while G_0 is the same group with the discrete topology, then* $\text{Hom}\{T, G\}$ *and* $\text{Hom}\{T, G_0\}$ *have the same topology.*

PROOF. $\text{Hom}\{T, G\}$ and $\text{Hom}\{T, G_0\}$ are algebraically identical. The hypotheses on T insure that every finite set of elements of T generates a finite subgroup of T. A complete set of neighborhoods U of 0 in $\text{Hom}\{T, G\}$ may therefore be found thus: take a finite subgroup $T_0 \subset T$ and an open set V_0 in G containing 0, and let U consist of all homomorphisms θ with $\theta(T_0) \subset V_0$. In particular, if V_0 is contained in the special open set V of G which contains no proper subgroups, the subgroup $\theta(T_0)$ is zero, so that U consists of all θ with $\theta(T_0) = 0$. The special sets U so described also form a complete set of neighborhoods of 0 in $\text{Hom}\{T, G_0\}$. Therefore the two topologies on the group are equivalent.

LEMMA 16.2. *Let $F \supset R$ be a free (discrete) group, $G' \supset G$ a discrete group, while* $\text{Hom}\{F, G'; R, G\}$ *denotes the set of all homomorphisms $\phi \in$* $\text{Hom}\{F, G'\}$ *with $\phi(R) \subset G$. Then*

$$\text{Hom}\{F, G'; R, G\}/\text{Hom}\{F, G\} \cong \text{Hom}\{F/R, G'/G\}. \tag{16.1}$$

PROOF. Any homomorphism of F/R into G'/G may be regarded as a homomorphism of F into G'/G which carries R into zero (Lemma 3.3), so that (16.1) becomes

$$\text{Hom}\{F, G'; R, G\}/\text{Hom}\{F, G\} \cong \text{Hom}\{F, G'/G; R, 0\}. \tag{16.2}$$

For each $\phi \in \text{Hom}\ \{F, G'\}$ let ϕ^* be the corresponding homomorphism reduced modulo G, so that for $x \in F$, $\phi^*(x)$ is the coset of $\phi(x)$, modulo G. The correspondence $\phi \to \phi^*$ is a homomorphism mapping Hom $\{F, G'; R, G\}$ into Hom $\{F, G'/G; R, 0\}$. Furthermore $\phi^* = 0$ if and only if $\phi(F) \subset G$, or $\phi \in \text{Hom}\ \{F, G\}$. Therefore $\phi \to \phi^*$ provides an (algebraic) isomorphism of the left hand group in (16.2) to a subgroup of the right hand group.

Conversely, select a fixed basis z_α for the free group F, and for each coset $b \in G'/G$ pick a fixed representative element $\rho(b)$ in G'. For given $\sigma \in \text{Hom}\ \{F, G'/G; R, 0\}$, define a corresponding homomorphism $\phi = \phi(\sigma)$, for any $x = \sum k_\alpha z_\alpha \in F$, as

$$\phi(\sum_\alpha k_\alpha z_\alpha) = \sum_\alpha k_\alpha \rho[\sigma z_\alpha].$$

This is a homomorphism of F into G'. By construction, $\rho[\sigma z_\alpha]$ modulo G is just σz_α, hence $\phi(x)$, modulo G, is $\sigma(x)$, or $\phi^* = \sigma$. This implies that $\phi(R) \subset G$, and so that each σ is the correspondent of some ϕ in the homomorphism $\phi \to \phi^*$.

To show (16.2) bicontinuous, we first analyze the topology in the groups involved. By the definition of the topology in a factor group, we have to consider only open sets in Hom $\{F, G'; R, G\}$ which are unions of cosets of Hom $\{F, G\}$. If $z_1, \cdots, z_n$ is any finite selection from the fixed set of generators for F, the set $U(z_1, \cdots, z_n)$ consisting of all ϕ with $\phi(z_1) \equiv \cdots \equiv \phi(z_n) \equiv 0 \pmod G$ is such an open set, and contains $\phi_0 = 0$. We assert that any open set V containing 0 which is a union of cosets contains one of these sets U. For, given V, there will be elements $x_1, \cdots, x_m$ in F such that V contains all ϕ with $\phi x_i = 0$. Select generators $z_1, \cdots, z_n$ such that each x_i can be expressed in terms of $z_1, \cdots, z_n$; then V contains all ϕ with $\phi z_i = 0$. Moreover, if $\phi z_i \equiv 0 \pmod G$, there is a homomorphism ϕ_1 of F into G with $\phi z_i = \phi_1 z_i$; since $\phi - \phi_1 \in V$, since V is a union of cosets of Hom $\{F, G\}$, and since $\phi_1 \in \text{Hom}\ \{F, G\}$, we conclude that $\phi \in V$. Thus $V \supset U(z_1, \cdots, z_n)$.

A similar but simpler argument for Hom $\{F, G'/G; R, 0\}$ will show that every open set containing zero in this group contains all σ with $\sigma z_1 = \cdots = \sigma z_n = 0$, for a suitable set of the generators of F. The mapping $\sigma \to \phi$ carries open sets of this special type into the open sets $U(z_1, \cdots, z_n)$ described above, and conversely. This shows that the correspondence $\phi \to \phi^*$ is continuous at 0, and hence everywhere.

17. Extensions of integers

Next we consider the case in which every element of H has finite order; we then write T instead of H for this group. The group of extensions of the integers by such a group T can be written as a group of characters. In case T is finite, the result is a generalization of Corollary 11.2, for in this case Char $T \cong T$.

THEOREM 17.1. *If T has only elements of finite order, and if I is the (additive) group of integers,*

$$\text{(17.1)} \qquad \text{Ext}_f\ \{I, T\} = 0,$$

$$\text{Ext}\ \{I, T\} \cong \text{Char}\ T. \tag{17.2}$$

The methods used to establish this result apply with equal force if I is replaced by any discrete group G which has no elements of finite order. The group Char T of homomorphisms of T into the group of reals modulo 1 must then be replaced by a group of homomorphisms of T into another group suitably constructed from G. In fact, any G with no elements of finite order can be embedded in an essentially unique discrete group G_∞ with the following properties:[19]

(i) G_∞ has no elements of finite order,

(ii) G_∞/G has only elements of finite order,

(iii) G_∞ is infinitely divisible.

For any $g \in G_\infty$ and any integer m there is then a unique $h = g/m$ in G_0 with $mh = g$. The (discrete) factor group G_∞/G is the analogue of the topological group P' of rationals modulo 1. Specifically, if $G = I$, $G_\infty = I_\infty$ is the group of rational numbers, and G_∞/G is the group P', but with a discrete topology. Since T has only elements of finite order, Char T is Hom $\{T, P'\}$. But P' clearly has no arbitrarily small subgroups, so that the latter group, by Lemma 16.1, is identical (algebraically and topologically) with Hom $\{T, I_\infty/I\}$. The exact generalization of Theorem 17.1 is thus

THEOREM 17.2. *If T has only elements of finite order, while G is discrete and has no elements of finite order, and G_∞ is defined as above,*

$$\text{Ext}_f\ \{G, T\} = 0, \tag{17.3}$$

$$\text{Ext}\ \{G, T\} \cong \text{Hom}\ \{T, G_\infty/G\}. \tag{17.4}$$

The isomorphism is bicontinuous if G and G_∞/G are both discrete.

PROOF. If T is represented in the form $T = F/R$, for F free, the conclusions of this theorem can be reformulated, according to the fundamental theorem of Chapter II, as

$$\text{Hom}_f\ \{R, G; F\} = \text{Hom}\ \{F \mid R, G\}, \tag{17.3a}$$

$$\text{Hom}\ \{R, G\}/\text{Hom}\ \{F \mid R, G\} \cong \text{Hom}\ \{F/R, G_\infty/G\}. \tag{17.4a}$$

Observe first that any homomorphism $\theta \in \text{Hom}\ \{R, G\}$ can be extended in a unique way to a homomorphism $\theta^* \in \text{Hom}\ \{F, G_\infty\}$. For, since every element of $T = F/R$ has finite order, every $z \in F$ has a finite order modulo R. For each such z pick an integer m such that $mz \in R$, and define

$$\theta^*(z) = (1/m)\theta(mz), \qquad z \in F,\ mz \in R. \tag{17.5}$$

This definition of θ^* is independent of the choice of m, and does yield a homomorphism of F into G_∞. Clearly it is the only such homomorphism extending the given θ.

Suppose now that $\theta \in \text{Hom}_f\ \{R, G; F\}$. Each element $z \in F$ then generates a

---

[19] G_∞ could also be described as a tensor product; see §18.

finite subgroup of F/R, so θ can be extended to a homomorphism mapping z and R into G. This extension of θ must agree with the unique extension θ^*. This shows that $\theta^*(z) \in G$ for each z, so that θ^* is in fact a homomorphism of F into $G \subset G_\infty$, and $\theta \in \text{Hom}\{F \mid R, G\}$. This proves (17.3a).

As in §16, let $\text{Hom}\{F, G_\infty ; R, G\}$ denote the group of all homomorphisms $\phi \in \text{Hom}\{F, G_\infty\}$ with $\phi(R) \subset G$. This is a topological group, under the usual specification (§1) that any open set in $\text{Hom}\{F, G_\infty ; R, G\}$ is the intersection of this group with an open set in the topological group $\text{Hom}\{F, G_\infty\}$.

The correspondence $\phi \to \phi \mid R$ provides a bicontinuous isomorphism

$$(17.6) \qquad \text{Hom}\{F, G_\infty ; R, G\} \cong \text{Hom}\{R, G\}.$$

For, by Lemma 3.4, $\phi \to \phi \mid R$ is a continuous homomorphism. It is an isomorphism because each $\theta \in \text{Hom}\{R, G\}$ has a unique extension $\theta^* = \phi \in \text{Hom}\{F, G_\infty ; R, G\}$, by (17.5). This inverse correspondence is also continuous; for if U is the open set consisting of all ϕ with $\phi z_i = g_i$, for given $z_i \in F$ and $g_i \in G_\infty$, $i = 1, \cdots, n$, there is an open set U_m in $\text{Hom}\{R, G\}$ consisting of all θ with $\theta(mz_i) = mg_i$, where m is chosen so that each $mz_i \in R$ and each $mg_i \in G$. The correspondence $\theta \to \theta^*$ of (17.5) carries U_m into U. This proves (17.6).

The correspondence $\phi \to \phi \mid R$ maps the subgroup $\text{Hom}\{F, G\}$ of $\text{Hom}\{F, G_\infty ; R, G\}$ onto $\text{Hom}\{F \mid R, G\}$. Hence (17.6) also yields an isomorphism

$$\text{Hom}\{F, G_\infty ; R, G\}/\text{Hom}\{F, G\} \cong \text{Hom}\{R, G\}/\text{Hom}\{F \mid R, G\}.$$

On the other hand, Lemma 16.2 provides an isomorphism

$$\text{Hom}\{F, G_\infty ; R, G\}/\text{Hom}\{F, G\} \cong \text{Hom}\{F/R, G_\infty/G\}.$$

These two combine to give the required isomorphism (17.4a).

It should be remarked that the results of this section can also be obtained by arguments directly on factor sets, without the interposition of the fundamental theorem of Chapter II. Specifically, to prove Theorem 17.2, one could consider an extension E of G by T, determined by a factor set $f(s, t)$ for $s, t \in T$. If $t \in T$ has order m, let $\phi_E(t) \equiv (1/m)\sum_i f(it, t) \pmod G$, where $i = 0, 1, \cdots, m-1$. In this fashion E determines a homomorphism $\phi_E \in \text{Hom}\{T, G_\infty/G\}$. Conversely, given such a homomorphism ϕ, one may select for each $\phi(t) \in G_\infty/G$ a representative element $\phi'(t) \in G_\infty$ and construct the corresponding factor set as $f(s, t) = \phi'(s) + \phi'(t) - \phi'(s + t)$. These correspondences will establish (17.4). The device of constructing ϕ_E by summation over the terms of the factor set is an application of the so-called "Japanese homomorphism," as commonly used for (multiplicative) factor sets.

18. Tensor products

Some of our formulas can be expressed more easily by means of the tensor products introduced by Whitney [13]. If A and B are given discrete abelian groups the *tensor product* $A \circ B$ is a set whose elements are finite formal sums

$\sum a_i b_i$ of formal products $a_i b_i$, with each $a_i \epsilon A$, $b_i \epsilon B$. Two such elements are added simply by combining the two formal sums into a single sum. Two such elements are equal if and only if the second can be obtained from the first by a finite number of replacements of the forms $(a + a')b \leftrightarrow ab + a'b$, $a(b + b') \leftrightarrow ab + ab'$. The tensor product $A \circ B$ so defined is a discrete abelian group, and the multiplication $a \cdot b$ is a pairing of A and B to $A \circ B$.

In the special case when $B = G$ is a group containing no elements of finite order, and $A = R_0$ is the additive group of rational numbers, any sum $\sum a_i b_i$ can, by the distributive law, be rewritten as a single term $(r/s)b$, where s is a common denominator for the rational numbers a_i. This representation is essentially unique. Therefore $R_0 \circ G$ is simply the group G_∞ used in §17 above, and G_∞/G is $(R_0 \circ G)/G$ (for details, cf. Whitney [13], pp. 507–508).

The tensor product can equivalently be defined in terms of characters, in the following fashion:

THEOREM 18.1. *If A and B are (discrete) abelian groups,*

$$A \circ B \cong \text{Char Hom } \{B, \text{Char } A\}. \tag{18.1}$$

PROOF. This conclusion can also be written in the form

$$\text{Char } (A \circ B) \cong \text{Hom } \{B, \text{Char } A\}. \tag{18.2}$$

Since the group of characters is the group of homomorphisms into the group P of reals modulo 1, this conclusion is a special case (with $C = P$) of the following

LEMMA 18.2. *If A and B are discrete abelian groups, C any generalized (topological) abelian group, then there is a bicontinuous isomorphism*

$$\text{Hom } \{A \circ B, C\} \cong \text{Hom } \{B, \text{Hom } (A, C)\}. \tag{18.3}$$

PROOF. Let $\theta \epsilon \text{Hom } \{A \circ B, C\}$ be given. For each $b \epsilon B$, let $\phi_b(a) = \theta(ab)$. Then $\phi_b \epsilon \text{Hom } (A, C)$. Let $\omega_\theta(b) = \phi_b$. Then $\omega_\theta \epsilon \text{Hom } \{B, \text{Hom } (A, C)\}$, and the correspondence $\theta \rightarrow \omega_\theta$ is a homomorphism of Hom $\{A \circ B, C\}$ into Hom $\{B, \text{Hom } (A, C)\}$. One verifies readily that it is an (algebraic) isomorphism ($w_\theta = 0$ only if $\theta = 0$). Furthermore, it is an isomorphism onto the whole group Hom $\{B, \text{Hom } (A, C)\}$. For let any ω in the latter group be given, with $\omega(b) = \phi'_b \epsilon \text{Hom } (A, C)$ for each $b \epsilon B$. Then define

$$\theta_\omega(\sum a_i b_i) = \sum \phi'_{b_i}(a_i), \qquad a_i \epsilon A, \ b_i \epsilon B.$$

One verifies that θ_ω is uniquely defined, under the identifications $(a + a')b \rightarrow ab + a'b$, $a(b + b') \rightarrow ab + ab'$ used in the definition of $A \circ B$. Furthermore, $\theta_\omega \epsilon \text{Hom } \{A \circ B, C\}$, and $\theta_\omega \rightarrow \omega$ in the previously given correspondence. Therefore $\theta \rightarrow \omega_\theta$, $\omega \rightarrow \vartheta_\omega$ does yield the indicated isomorphism (18.3). The continuity of the isomorphism in both directions is readily established from these explicit formulas and the appropriate definitions of open sets in the given topologies of the groups concerned.

Chapter IV. Direct and Inverse Systems

The Čech homology groups for a space are defined as limits of certain "direct" and "inverse" systems of homology groups for finite coverings of the space (Chap. VI). In view of our representation of homology groups in terms of groups of homomorphisms and groups of group extensions we are led to consider limits of groups of this sort. We shall show that the limit of a group of homomorphisms is itself a group of homomorphisms (§21) and that the corresponding proposition holds in certain special cases for groups of group extensions (§22). In the general case, however, we must introduce a new group to represent the limit of a group of group extensions. This group can also be introduced as a limit of tensor products (§25).

19. Direct systems of groups

A directed set J is a partially ordered set of elements $\alpha, \beta, \gamma, \cdots$ such that for any two elements α and β there always exists an element γ with $\alpha < \gamma$, $\beta < \gamma$. For each index α in a directed set J let H_α be a (discrete) group, and for each pair $\alpha < \beta$, let $\phi_{\beta\alpha}$ be a homomorphism of H_α into H_β. If $\phi_{\gamma\alpha} = \phi_{\gamma\beta}\phi_{\beta\alpha}$ whenever $\alpha < \beta < \gamma$, the groups H_α are said to form a *direct system* with the *projections* $\phi_{\beta\alpha}$.[20]

Any direct system determines a unique (discrete) limit group $H = \underrightarrow{\text{Lim}}\, H_\alpha$ as follows. Every element h_α of one of the groups H_α is regarded as an element h_α^* of the limit H, and two elements h_α^*, h_β^* are equal if and only if there is an index γ, $\alpha < \gamma$, $\beta < \gamma$, with $\phi_{\gamma\alpha}h_\alpha = \phi_{\gamma\beta}h_\beta$. Two elements h_α^* and h_β^* in H are added by finding some γ with $\alpha < \gamma$, $\beta < \gamma$; the sum is then the element $h_\gamma^* = (\phi_{\gamma\alpha}h_\alpha + \phi_{\gamma\beta}h_\beta)^*$. Under this addition and equality, the elements h_α^* form a group $H = \underrightarrow{\text{Lim}}\, H_\alpha$. Each of the given groups H_α has a homomorphism $\phi_\alpha(h_\alpha) = h_\alpha^*$ into the limit group, and $\phi_\beta\phi_{\beta\alpha} = \phi_\alpha$, for $\alpha < \beta$.

In case each given projection $\phi_{\beta\alpha}$ is an isomorphism (of H_α into H_β), the limit group can be regarded as a "union" of the given groups: each group H_α has an isomorphic replica $\phi_\alpha H_\alpha$ within H, and H is simply the union of these subgroups.

A subset J' of the set J of indices α is said to be *cofinal* in J if for each index α there is in J' an α' with $\alpha < \alpha'$. The limit $\underrightarrow{\text{Lim}}\, H_{\alpha'}$, taken over any such cofinal subset, is isomorphic to the original limit H.

20. Inverse systems of groups

For each index α in a directed set let A_α be a (generalized topological) group, and for each $\alpha < \beta$ let $\psi_{\alpha\beta}$ be a (continuous) homomorphism of A_β in A_α. If $\psi_{\alpha\beta}\psi_{\beta\gamma} = \psi_{\alpha\gamma}$ whenever $\alpha < \beta < \gamma$, the groups A_α are said to form an *inverse system* relative to the *projections* $\psi_{\alpha\beta}$. Each inverse system determines a limit group $A = \underleftarrow{\text{Lim}}\, A_\alpha$. An element of this limit group is a set $\{a_\alpha\}$ of elements $a_\alpha \in A_\alpha$ which "match" in the sense that $\psi_{\alpha\beta}a_\beta = a_\alpha$ for each $\alpha < \beta$. The sum

[20] Direct (and inverse) systems were discussed in Steenrod [9], Lefschetz [7], Chap. I and II, and in Weil [12], Ch. I.

of two such sets is $\{a_\alpha\} + \{b_\alpha\} = \{a_\alpha + b_\alpha\}$; since the ψ's are homomorphisms, this sum is again an element of the group. This limit group A is a subgroup of the direct product of the groups A_α. The topology of the direct product $\prod A_\alpha$ thus induces (§1) a topology in $\varprojlim A_\alpha$; an open set in the latter group is the intersection with $\varprojlim A_\alpha$ of an open set of $\prod A_\alpha$. This makes $\varprojlim A_\alpha$ a generalized topological group. As before, a cofinal subset of the indices gives an isomorphic limit group.

Let each B_α be a subgroup of the corresponding group A_α of an inverse system, and assume, for $\alpha < \beta$, that $\psi_{\alpha\beta} B_\beta \subset B_\alpha$. Then the system B_α is an inverse system under the same projections $\psi_{\alpha\beta}$, and the limit $\varprojlim B_\alpha$ is, in natural fashion, a subgroup of $\varprojlim A_\alpha$. On the other hand, $\psi_{\alpha\beta}$ induces a homomorphism $\psi'_{\alpha\beta}$ of the (generalized topological) group A_β/B_β into A_α/B_α. Relative to these projections, the factor groups themselves form an inverse system A_α/B_α. The limit group of the latter system contains a homomorphic image of $\varprojlim A_\alpha$; if each a_α in A_α determines a coset a'_α in A_α/B_α, the map $\{a_\alpha\} \to \{a'_\alpha\}$ is a homomorphism of $\varprojlim A_\alpha$ into $\varprojlim (A_\alpha/B_\alpha)$ in which exactly the elements of $\varprojlim B_\alpha$ are mapped on zero. Thus we have

$$\varprojlim A_\alpha / \varprojlim B_\alpha \subset \varprojlim (A_\alpha/B_\alpha). \tag{20.1}$$

For compact topological subgroups this is an isomorphism:

LEMMA 20.1. *If the A_α form an inverse system relative to the $\psi_{\alpha\beta}$, and if each B_α is a compact topological subgroup of A_α with $\psi_{\alpha\beta} B_\beta \subset B_\alpha$, then*

$$\varprojlim A_\alpha / \varprojlim B_\alpha \cong \varprojlim (A_\alpha/B_\alpha). \tag{20.2}$$

PROOF. Consider any $c = \{c_\alpha\}$ in $\varprojlim (A_\alpha/B_\alpha)$, where $\psi'_{\alpha\beta} c_\beta = c_\alpha$ for each $\alpha < \beta$. Each $c_\alpha \in A_\alpha/B_\alpha$ is a coset of the compact topological subgroup B_α, hence itself is a compact Hausdorff subspace of the space A_α. Furthermore $\psi_{\alpha\beta}$ is a continuous mapping of the set c_β into c_α, for each $\alpha < \beta$. Since $\psi_{\alpha\gamma} = \psi_{\alpha\beta}\psi_{\beta\gamma}$, the sets c_α form an inverse system of compact non-empty Hausdorff spaces. Their limit space is therefore[21] non-vacuous. This means that there is a set of elements $a_\alpha \in c_\alpha$ with $\psi_{\alpha\beta} a_\beta = a_\alpha$ for $\alpha < \beta$. The element $\{a_\alpha\}$ in the group $\varprojlim A_\alpha$ is therefore an element which maps onto the given element $\{c_\alpha\}$ in the homomorphism $\{a_\alpha\} \to \{a'_\alpha\}$ used to establish (20.1). The continuity of (20.2), in both directions, follows readily.

There is also an "isomorphism" theorem for inverse systems.

LEMMA 20.2. *If the groups A_α form an inverse system relative to the projections $\psi_{\alpha\beta}$, while C_α form an inverse system (with the same set of indices) relative to projections $\phi_{\alpha\beta}$, and if σ_α are (bicontinuous) isomorphisms of A_α to C_α, for every α, such that the "naturality" condition $\sigma_\alpha \psi_{\alpha\beta} = \phi_{\alpha\beta}\sigma_\beta$ holds, then the groups $\varprojlim A_\alpha$ and $\varprojlim C_\alpha$ are bicontinuously isomorphic.*

---

[21] See Lefschetz [7], Theorem 39.1 or Steenrod [9], p. 666. Observe, however, that the latter proof is incomplete, because of the gap in lines 10–11 on p. 666.

21. Inverse systems of homomorphisms

Consider the group of all homomorphisms of H into G. As in Chap. II, §12, each projection $\phi_{\beta\alpha}$ of a direct system of groups H_α will induce a "dual" homomorphism $\phi^*_{\alpha\beta}$ of Hom $\{H_\beta, G\}$ into Hom $\{H_\alpha, G\}$. Furthermore $\phi^*_{\alpha\beta}\phi^*_{\beta\gamma} = \phi^*_{\alpha\gamma}$ for all $\alpha < \beta < \gamma$, so that the groups Hom $\{H_\alpha, G\}$ form an inverse system relative to these dual projections.

THEOREM 21.1. *If the (discrete) groups H_α form a direct system, then*

$$\text{Hom}\,\{\underrightarrow{\text{Lim}}\, H_\alpha, G\} \cong \underleftarrow{\text{Lim}}\,\text{Hom}\,\{H_\alpha, G\}. \tag{21.1}$$

PROOF. Consider any element $\omega = \{\theta_\alpha\}$ in $\underleftarrow{\text{Lim}}$ Hom $\{H_\alpha, G\}$. To define a corresponding homomorphism θ_ω on $H = \underrightarrow{\text{Lim}}\, H_\alpha$, represent each element $h \in H$ as a projection $h = \phi_\alpha h_\alpha$ of some element $h_\alpha \in H_\alpha$, and set

$$\theta_\omega(h) = \theta_\omega(\phi_\alpha h_\alpha) = \theta_\alpha(h_\alpha), \qquad h = \phi_\alpha h_\alpha. \tag{21.2}$$

The "matching" requirement that $\theta_\alpha = \phi^*_{\alpha\beta}\theta_\beta$ for $\alpha < \beta$ readily shows that $\theta_\omega(h)$ has a unique value, independent of the representation $h = \phi_\alpha h_\alpha$ chosen. Furthermore, $\theta_\omega \in$ Hom $\{H, G\}$, and the correspondence $\omega \to \theta_\omega$ is an isomorphism.

Conversely, let any $\theta \in$ Hom $\{H, G\}$ be given, and define

$$\theta_\alpha(h_\alpha) = \theta(\phi_\alpha h_\alpha), \qquad h_\alpha \in H_\alpha. \tag{21.3}$$

If $\alpha < \beta$, $\phi^*_{\alpha\beta}\theta_\beta(h_\alpha) = \theta_\beta[\phi_{\beta\alpha}h_\alpha] = \theta[\phi_\beta\phi_{\beta\alpha}h_\alpha] = \theta(\phi_\alpha h_\alpha) = \theta_\alpha h_\alpha$; so $\phi^*_{\alpha\beta}\theta_\beta = \theta_\alpha$, and these θ's match. Therefore $\omega = \{\theta_\alpha\}$ is an element of the inverse limit group $\underleftarrow{\text{Lim}}$ Hom $\{H_\alpha, G\}$, and clearly θ_ω is the original homomorphism θ. The correspondence $\omega \to \theta_\omega$ therefore does establish the desired isomorphism (21.1). The continuity in both directions follows directly from the formulas (21.2) and (21.3) and the appropriate definition of neighborhoods of zero in the groups concerned.

22. Inverse systems of group extensions

Consider a direct system of discrete groups H_α. As in Chap. II, §12, each projection $\phi_{\beta\alpha}$ of H_α into H_β will induce a homomorphism $\phi^*_{\alpha\beta}$ of Ext $\{G, H_\beta\}$ into Ext $\{G, H_\alpha\}$. Furthermore $\phi^*_{\alpha\beta}\phi^*_{\beta\gamma} = \phi^*_{\alpha\gamma}$ for all $\alpha < \beta < \gamma$, so that the groups Ext $\{G, H_\alpha\}$ form an inverse system. Contrary to the situation in the previous section, the limit group $\underleftarrow{\text{Lim}}$ Ext $\{G, H_\alpha\}$ may not be isomorphic to Ext $\{G, \underrightarrow{\text{Lim}}\, H_\alpha\}$. An example to this effect will be given below. However, there are two important cases when "Lim" and "Ext" are interchangeable.

THEOREM 22.1. *If G is compact and topological, while the (discrete) groups H_α form a direct system, then*

$$\text{Ext}\,\{G, \underrightarrow{\text{Lim}}\, H_\alpha\} \cong \underleftarrow{\text{Lim}}\,\text{Ext}\,\{G, H_\alpha\}. \tag{22.1}$$

This is proved by repeated applications of Lemma 20.1 to the representation

(22.2) $$\text{Ext}\,\{G, H\} = \text{Fact}\,\{G, H\}/\text{Trans}\,\{G, H\},$$

where $H = \underrightarrow{\text{Lim}}\, H_\alpha$. Recall that any $f \in \text{Trans}\,\{G, H\}$ has the form

$$f(h, k) = g(h) + g(k) - g(h + k), \qquad h, k \in H.$$

Here $g \in G^H$ is any mapping of H into G. Clearly $f = 0$ if and only if $g \in \text{Hom}\,\{H, G\}$, so

(22.3) $$\text{Trans}\,\{G, H\} \cong G^H/\text{Hom}\,\{H, G\}.$$

The correspondence $g \to f$ is clearly continuous; since the isomorphism (22.3) is one-one and since the groups G^H and $\text{Hom}\,\{H, G\}$ are compact, by Lemma 3.1, the bicontinuity of (22.3) follows. Furthermore, this isomorphism is a "natural" one relative to homomorphisms, so that the isomorphism theorem for inverse systems (Lemma 20.2) gives

$$\underleftarrow{\text{Lim}}\, \text{Trans}\,\{G, H_\alpha\} \cong \underleftarrow{\text{Lim}}\, [G^{H_\alpha}/\text{Hom}\,\{H_\alpha, G\}].$$

In this representation the groups G^{H_α} and $\text{Hom}\,\{H_\alpha, G\}$ with the "dual" projections $\phi^*_{\alpha\beta}$ form inverse systems with the respective limits G^H and $\text{Hom}\,\{H, G\}$. Furthermore each group $\text{Hom}\,\{H_\alpha, G\}$ is compact and topological so Lemma 20.1 gives

(22.4) $$\begin{aligned}\underleftarrow{\text{Lim}}\, \text{Trans}\,\{G, H_\alpha\} &\cong \underleftarrow{\text{Lim}}\, G^{H_\alpha}/\underleftarrow{\text{Lim}}\, \text{Hom}\,\{H_\alpha, G\}\\ &= G^H/\text{Hom}\,\{H, G\} \cong \text{Trans}\,\{G, H\}.\end{aligned}$$

On the other hand one may show exactly as in the proof of Theorem 21.1 on homomorphisms that there is a bicontinuous isomorphism

(22.5) $$\underleftarrow{\text{Lim}}\, \text{Fact}\,\{G, H_\alpha\} \cong \text{Fact}\,\{G, H\}.$$

Furthermore, each of the groups $\text{Trans}\,\{G, H_\alpha\}$ is compact and topological, so that Lemma 20.1 applies again to prove

$$\underleftarrow{\text{Lim}}\, [\text{Fact}/\text{Trans}] \cong \underleftarrow{\text{Lim}}\, \text{Fact}/\underleftarrow{\text{Lim}}\, \text{Trans}.$$

This, with (22.4) and (22.5), gives the desired conclusion.[22]

THEOREM 22.2. *If G is discrete and has no elements of finite order, while T_α is a direct systems of discrete groups with only elements of finite order, then*

(22.6) $$\text{Ext}\,\{G, \underrightarrow{\text{Lim}}\, T_\alpha\} \cong \underleftarrow{\text{Lim}}\, \text{Ext}\,\{G, T_\alpha\}.$$

The proof appeals directly to the result found in Theorem 17.2 of Chapter III, to the effect that

(22.7) $$\text{Ext}\,\{G, T_\alpha\} \cong \text{Hom}\,\{T_\alpha, G_\infty/G\}.$$

The groups $\text{Hom}\,\{T_\alpha, G_\infty/G\}$ will form an inverse system under the dual projections $\phi^*_{\alpha\beta}$; as in Theorem 21.1 we then have

$$\text{Hom}\,\{\underrightarrow{\text{Lim}}\, T_\alpha, G_\infty/G\} \cong \underleftarrow{\text{Lim}}\, \text{Hom}\,\{T_\alpha, G_\infty/G\}.$$

[22] Theorem 22.1 can also be proved by representing Ext by means of Char Hom $\{G, H\}$ as in Theorem 15.1. This argument, however, requires a tedious proof that the isomorphism established in the latter theorem is "natural," in the sense of §12.

But the group on the left is simply Ext $\{G, \varinjlim T_\alpha\}$, by another application of Theorem 17.2. The desired result should then follow by taking (inverse) limits on both sides in (22.7).

To carry out this argument, it is necessary to have the naturality condition which gives the isomorphism theorem (Lemma 20.2) for inverse systems. This naturality condition requires that the isomorphism (22.7) permute with the projections of the inverse systems. This is just a statement of the fact that the isomorphism (22.7) established in Theorem 17.2 is "natural" in the sense envisaged in §12. The proof of this naturality is straightforward, so details will be omitted.

COROLLARY 22.3. *If the discrete group G has only a finite number of generators, while T_α is a direct system of discrete groups with only elements of finite order, then*

$$\text{Ext}\ \{G, \varinjlim T_\alpha\} \cong \varprojlim \text{Ext}\ \{G, T_\alpha\}.$$

PROOF. Write G as $F \times L$ where F is free, L is finite (and thus compact). By (11.2) there is a "natural" isomorphism

$$\text{Ext}\ \{G, \varinjlim T_\alpha\} \cong \text{Ext}\ \{F, \varinjlim T_\alpha\} \times \text{Ext}\ \{L, \varinjlim T_\alpha\}.$$

The asserted result now follows by applying Theorem 22.2 to the first factor on the right, and Theorem 22.1 to the second, using Lemma 20.2.

We now show by an example that "Ext" and "Lim" do not necessarily commute. Let p be a fixed prime number, H the additive group of all rationals with denominator a power of p, and H_n the subgroup consisting of all multiples of $1/p^n$. Then $\varinjlim H_n = H$, since H is the union of the groups H_n. Furthermore H_n is a free group, so Ext $\{I, H_n\} = 0$, where I is the group of integers. On the other hand, Ext $\{I, \varinjlim H_n\} =$ Ext $\{I, H\}$ is a group computed in appendix B; it is decidedly not zero, in fact it is not even denumerable.

23. Contracted extensions

Before further consideration of the inverse limits of groups of extensions, we make a comparison of the group of extensions of a group G by a group H with the group of extensions by a subgroup H_0 of H. The identity mapping I of H_0 into H is a homomorphism, hence, as in §12, will give dual homorphisms

$$I^*: \text{Fact}\ \{G, H\} \rightarrow \text{Fact}\ \{G, H_0\}, \tag{23.1}$$

$$I^*: \text{Trans}\ \{G, H\} \rightarrow \text{Trans}\ \{G, H_0\}. \tag{23.2}$$

Specifically, I^* is the operation of "cutting off" a factor set $f \in$ Fact $\{G, H\}$ to give a factor set $f_0 = I^*f \in$ Fact $\{G, H_0\}$; $f_0(h, k)$ is defined only for $h, k \in H_0$, and always equals $f(h, k)$. Clearly I^* carries transformation sets into transformation sets, as in (23.2). Thus I^* also induces a dual homomorphism

$$I^*: \text{Ext}\ \{G, H\} \rightarrow \text{Ext}\ \{G, H_0\}. \tag{23.3}$$

This homomorphism may be visualized as follows: given E such that $G \subset E$

and $E/G = H$, there is an $E_0 \subset E$ such that $G \subset E_0$ and $E_0/G = H_0$. Then $I^*(E) = E_0$.

LEMMA 23.1. *If H_0 is a subgroup of the group H then for any group G the homomorphism I^* of* (23.3) *maps the group* Ext $\{G, H\}$ *onto* Ext $\{G, H_0\}$.

PROOF.[23] Represent H as F/R, where F is free. There is then a subgroup F_0 of F such that $R \subset F_0$ and $F_0/R = H_0$. By the fundamental theorem we have isomorphisms

$$\text{Ext } \{G, H\} \cong \text{Hom } \{R, G\}/\text{Hom } \{F \mid R, G\},$$

$$\text{Ext } \{G, H_0\} \cong \text{Hom } \{R, G\}/\text{Hom } \{F_0 \mid R, G\},$$

where Hom $\{F \mid R, G\} \subset$ Hom $\{F_0 \mid R, G\}$. According to the "naturality" theorem of §12 the homomorphism I^* between the groups on the left can be represented on the right as that correspondence which carries each coset of Hom $\{F \mid R, G\}$ into the coset of Hom $\{F_0 \mid R, G\}$ in which it is contained. This makes it obvious that the homomorphism is a mapping "onto."

LEMMA 23.2. *If $H_0 \subset H$, then the dual homomorphisms I^* of factor and transformation sets, as in* (23.1) *and* (23.2), *are mappings "onto."*

PROOF. Any element in Trans $\{G, H_0\}$ has the form

$$f(h, k) = g(h) + g(k) - g(h + k),$$

where g is an arbitrary function on H_0 to G. Let g^* be an arbitrary extension of g to H, and

$$f^*(h, k) = g^*(h) + g^*(k) - g^*(h + k).$$

Then f^* is a transformation set with $I^*f^* = f$. This proves that (23.2) is a mapping onto. Since (23.3) and (23.2) are mappings onto, the same holds for (23.1).

24. The group Ext*

Since limits do not always permute with groups of extensions, we now introduce a new group which is the limit of an inverse system of groups of group extensions.

Consider a discrete group T with only elements of finite order. The set $\{S_\alpha\}$ of all finite subgroups of T is a direct system, if $\alpha < \beta$ means that $S_\alpha \subset S_\beta$, and that the projection $I_{\beta\alpha}$ of S_α into S_β is simply the identity. The direct limit of $\{S_\alpha\}$ is the group T.

Let G be any generalized topological group. Since $\{S_\alpha\}$ is a direct system, it follows from a previous section that the groups Ext $\{G, S_\alpha\}$ form an inverse system with projections $I^*_{\alpha\beta}$. We define our new group as the limit of this system

[23] The lemma can also be proved directly in terms of the group extensions E, E_0, using a suitable transfinite induction.

$$\text{Ext}^* \{G, T\} = \underleftarrow{\text{Lim}}\ \text{Ext}\ \{G, S_\alpha\}. \tag{24.1}$$

The two theorems of §22 as to cases in which "Ext" and "Lim" commute give at once

COROLLARY 24.1. *If G is compact and topological, or is discrete without elements of finite order, then*

$$\text{Ext}^* \{G, T\} \cong \text{Ext}\ \{G, T\}.$$

In the definition of Ext* we used the approximation of T by its finite subgroups S_α. However, any approximation by finite groups will give the same result:

THEOREM 24.2. *If T_α is any direct system of finite groups, the corresponding inverse system of groups* Ext $\{G, T_\alpha\}$ *has a limit*

$$\underleftarrow{\text{Lim}}\ \text{Ext}\ \{G, T_\alpha\} \cong \text{Ext}^* \{G, \underrightarrow{\text{Lim}}\ T_\alpha\}. \tag{24.2}$$

PROOF. In case T_α is the system of all finite subgroups of the limit $T = \underrightarrow{\text{Lim}}\ T_\alpha$, this equation is simply the definition of Ext*. In general, $T = \underrightarrow{\text{Lim}}\ T_\alpha$ is a group in which every element has finite order. Each T_α has a homomorphic projection $T'_\alpha = \phi_\alpha T_\alpha$ into the limit T, and T is simply the union of these subgroups T'_α. The set of these subgroups T'_α is therefore cofinal in the set of all finite subgroups of T. The inverse system of the groups Ext $\{G, T'_\alpha\}$, relative to the "identity" projections $I^*_{\alpha\beta}$, is cofinal in the inverse system used to define Ext*, hence gives the same limit group,

$$\text{Ext}^* \{G, T\} \cong \underleftarrow{\text{Lim}}\ \text{Ext}\ \{G, T'_\alpha\}. \tag{24.3}$$

An element f^* in this limit group can be represented (but not uniquely) as a set $\{f_\alpha\}$ of factor sets $f_\alpha \in$ Fact $\{G, T'_\alpha\}$ which "match" modulo transformation sets. This means that for each $\beta > \alpha$ there is a transformation set $t_{\alpha\beta} \in$ Trans $\{G, T'_\alpha\}$ such that

$$f_\alpha(h', k') = f_\beta(h', k') + t_{\alpha\beta}(h', k'), \qquad h', k' \in T'_\alpha .$$

Now each homomorphism ϕ_α of T_α into T'_α determines, as in §12, a dual homomorphism ϕ^*_α of Fact $\{G, T'_\alpha\}$ into Fact $\{G, T_\alpha\}$, defined so that $e_\alpha = \phi^*_\alpha f_\alpha$ is the factor set given by the equations

$$e_\alpha(h, k) = f_\alpha(\phi_\alpha h, \phi_\alpha k), \qquad h, k \in T_\alpha . \tag{24.4}$$

If the f_α match, one readily proves that the corresponding e_α also match, modulo transformation sets. If the representation of f^* by $\{f_\alpha\}$ is changed by adding to each f_α a transformation set, the e_α's are changed accordingly by transformation sets. Therefore the correspondence

$$f^* = \{f_\alpha\} \to e^* = \{\phi^*_\alpha f_\alpha\} = \omega f^* \tag{24.5}$$

carries each element f^* in $\underleftarrow{\text{Lim}}$ Ext $\{G, T'_\alpha\}$ into a well defined element e^* in $\underleftarrow{\text{Lim}}$ Ext $\{G, T_\alpha\}$. One verifies at once that this correspondence is a homomorphism.

Now we use the assumption that each T_α is finite. If $\phi_\alpha h_\alpha = 0$ for some $h_\alpha \in T_\alpha$, the definition of equality in a direct system shows that $\phi_{\beta\alpha} h_\alpha = 0$ for some $\beta > \alpha$. Since the whole group T_α is finite, we can select a single $\beta = \beta_0(\alpha) > \alpha$ which will do this for all h_α, so that

$$\phi_\alpha h_\alpha = 0 \quad \text{implies} \quad \phi_{\beta\alpha} h_\alpha = 0, \qquad \beta = \beta_0(\alpha).$$

Since $\phi_\beta \phi_{\beta\alpha} = \phi_\alpha$, ϕ_β is now an *isomorphism* of $\phi_{\beta\alpha} T_\alpha$ onto T'_α. Let ϕ_β^{-1} denote the inverse correspondence.

Next we show that ω, as defined by (24.5), is an isomorphism. For suppose $\omega f^* = 0$; every $\overset{*}{\phi}_\alpha f_\alpha$ is then a transformation set t_α. Using (24.4) and $\beta = \beta_0(\alpha)$, we then have, for any $h', k' \in T'_\alpha$,

$$f_\alpha(h', k') \equiv f_\beta(h', k') = e_\beta(\phi_\beta^{-1} h', \phi_\beta^{-1} k') = t_\beta(\phi_\beta^{-1} h', \phi_\beta^{-1} k').$$

This shows that f_α is a transformation set, hence that $f^* = \{f_\alpha\} = 0$ in Ext* $\{G, T\}$.

To construct a correspondence inverse to ω, let $e^* = \{e_\alpha\}$ be a given element in $\varprojlim$ Ext $\{G, T_\alpha\}$, where each $e_\alpha \in$ Fact $\{G, T_\alpha\}$. Define

$$f_\alpha(h', k') = e_\beta(\phi_\beta^{-1} h', \phi_\beta^{-1} k'), \qquad \beta = \beta_0(\alpha) \tag{24.6}$$

for each $h', k' \in T'_\alpha$. Since the e_α's are known to match, we may verify that the replacement of β by any larger index γ in this definition will only alter f_α by a transformation set. To show that f_α and f_γ match properly for $\alpha < \gamma$, one then chooses $\beta > \beta_0(\alpha)$, $\beta > \beta_0(\gamma)$ in (24.6) and uses the given matching of the e_α's (modulo transformation sets). Finally, one verifies easily that the correspondence $\{e_\alpha\} \to \{f_\alpha\}$ of (24.6) is the inverse of the given correspondence ω of (24.5). This establishes the isomorphism (24.2) required in the theorem. The continuity, in both directions, follows from the formulae (24.5) and (24.6).

THEOREM 24.3. *If every element of T has finite order, the group* Ext* $\{G, T\}$ *contains an everywhere dense subgroup isomorphic to* Ext $\{G, T\}/\text{Ext}_f\ \{G, T\}$.

This will be established by constructing a "natural" homomorphism of Ext $\{G, T\}$ into Ext* $\{G, T\}$. To this end, let E be any extension of G by T determined by a factor set f. As in §23, f may be "cut off" to give a factor set f_α for any given finite subgroup $S_\alpha \subset T$. These factor sets match properly, so $\{f_\alpha\}$ determines a definite element in the inverse limit group Ext* $\{G, T\}$. Alteration of f by a transformation set alters each f_α by the correspondingly "cut off" transformation set, hence does not alter the element $\{f_\alpha\} = f^*$ of Ext*. Therefore $f \to \{f_\alpha\}$ is a well defined homomorphism of Ext into Ext*. In case f lies in $\text{Ext}_f\ \{G, T\}$, each f_α is a transformation set, by the very definition of Ext_f, so that $\{f_\alpha\} = 0$. Conversely, if each f_α is a transformation set, $f \in \text{Ext}_f$. We thus have a (bicontinuous) isomorphism of Ext/Ext_f onto a subgroup of Ext*.

To show this subgroup everywhere dense in Ext* it will suffice, whatever the topology in G, to show the following: Given an element $f^* = \{f'_\alpha\}$ in Ext* $\{G, T\}$ and a finite set J_0 of indices, there exists a factor set f in Fact $\{G, T\}$ such that

$f_\alpha - f'_\alpha$ is a transformation set for every index $\alpha \in J_0$. To prove this, choose a finite subgroup S_γ which contains all the groups S_α, for $\alpha \in J_0$. By Lemma 23.2, the given factor set f'_γ can be obtained by "cutting off" a suitable factor set f, so that $f_\gamma - f'_\gamma$ is the transformation set 0. The matching condition for the f'_α then shows that each difference $f_\alpha - f'_\alpha$ is also a transformation set, for $\alpha \in J_0$. This proves the property stated above, and with it, the theorem.

In many cases the subgroup considered in Theorem 24.3 is the whole group Ext*. It follows from previous considerations that this is the case when G is compact or when G is discrete and has no elements of finite order. Another important case is that when T is countable:

THEOREM 24.4. *If T is countable then*

$$\text{Ext}^* \{G, T\} \cong \text{Ext}\, \{G, T\}/\text{Ext}_f\, \{G, T\}. \tag{24.7}$$

PROOF. Since T is countable, the system of all finite subgroups of T used to define Ext* $\{G, T\}$ may be replaced by a cofinal sequence of finite subgroups T_n with $T_1 \subset T_2 \subset \cdots \subset T_n \subset \cdots \subset T$, with the identity projections $I_n : T_n \to T_{n+1}$. Therefore $\text{Ext}^* \{G, T\} = \underleftarrow{\text{Lim}}\ \text{Ext}\, \{G, T_n\}$. An element e^* of this group can then be represented as a sequence $\{f_n\}$ of factor sets $f_n \in \text{Fact}\, \{G, T_n\}$ which match, in the sense that, for some g_n,

$$f_{n+1}(h, k) = f_n(h, k) + [g_n(h) + g_n(k) - g_n(h + k)] \tag{24.8}$$

for all $h, k \in T_n$. The transformation set shown in brackets may be extended to all of T by extending g_n to a function g_n^* on T, as in Lemma 23.2. We introduce a new function $s_n(h) = g_1^*(h) + \cdots + g_{n-1}^*(h)$, for all $h \in T$, and a new family of factor sets

$$f'_n(h, k) = f_n(h, k) - [s_n(h) + s_n(k) - s_n(h + k)],$$

for $h, k \in T_n$.[24] Since f'_n differs from f_n by a transformation set, the given element e^* of Ext* has both representations $\{f_n\}$ and $\{f'_n\}$. But (24.8) also shows that f'_{n+1}, cut off at T_n, is exactly f'_n. Therefore these factor sets match exactly, and provide a composite factor set f of T in G. This factor set f is one which corresponds to the given element e^* of Ext* in the "natural" homomorphism of Ext into Ext* as constructed in Theorem 24.3, so this homomorphism maps Ext on all of Ext*, as asserted in (24.7).

25. Relation to tensor products

The group Ext* introduced in this chapter is closely related to the tensor product. Since an early form ([5]) of our results was formulated in terms of tensor products, we shall briefly state the connection. Let G be any group, A a compact zero-dimensional group, $\{A_\alpha\}$ the family of all open and closed subgroups of A. Then the groups A/A_α and $G \circ (A/A_\alpha)$ both form inverse

[24] This construction is an exact group theoretic analog of a similar matching process for chains, as devised by Steenrod ([9], p. 692).

systems. The modified tensor product $G \bullet A$ is defined as the limit of the groups $G \circ (A/A_\alpha)$.

Now let the group T with all elements of finite order be represented in terms of a free group F as $T = F/R$. Each finite subgroup S_α then has a representation F_α/R, and the fundamental theorem of Chapter II asserts that

$$(25.1) \qquad \text{Ext } \{G, S_\alpha\} \cong \text{Hom } \{R, G\}/\text{Hom } \{F_\alpha \mid R, G\}.$$

The groups on both sides here form inverse systems, relative to the identity as projections. Furthermore, the isomorphism of (25.1) permutes with these projections, so that the limits of the two direct systems in (25.1) are also isomorphic. In view of the definition of Ext*, this gives

$$(25.2) \qquad \text{Ext* } \{G, T\} \cong \underleftarrow{\text{Lim}} \, [\text{Hom } \{R, G\}/\text{Hom } \{F_\alpha \mid R, G\}].$$

Now if I is the group of integers, any element $\sigma = \sum g_i \phi_i$ in the tensor product $G \circ \text{Hom } \{R, I\}$ determines in natural fashion the homomorphism $\theta \in \text{Hom } \{R, G\}$ with $\theta(r) = \sum \phi_i(r) g_i$. By a somewhat lengthy argument, this correspondence $\sigma \to \theta$ can be used to "factor out" the G in (25.2) to give

$$(25.3) \qquad \text{Ext* } \{G, T\} \cong \underleftarrow{\text{Lim}} \, G \circ [\text{Hom } \{R, I\}/\text{Hom } \{F_\alpha \mid R, I\}].$$

The group in brackets here is Ext $\{I, F_\alpha/R\}$, by the fundamental theorem on group extensions. According to Theorem 17.1 it can be expressed as Char S_α. Therefore (25.3) is[25]

$$(25.4) \qquad \text{Ext* } \{G, T\} \cong \underleftarrow{\text{Lim}} \, (G \circ \text{Char } S_\alpha).$$

But the group Char S_α can, by the theory of characters (Lemma 13.2, Theorem 13.5), be rewritten as a factor group Char T/Annih S_α, where the subgroups of the form Annih S_α in Char T are exactly the open and closed subgroups in the zero-dimensional group Char T. Thus (25.4) may be restated in terms of the modified tensor product, as

$$(25.5) \qquad \text{Ext* } \{G, T\} \cong G \bullet \text{Char } T.$$

The use of the "modified" tensor product is therefore equivalent to the use of the group Ext*.

Chapter V. Abstract Complexes

Turning now to the topological applications, we will establish the fundamental theorem on the decomposition of the homology groups of an infinite complex in terms of the integral cohomology groups of the complex. This theorem will be obtained in several closely related forms (Theorems 32.1, 32.2 and 34.2) for three different types of homology groups. The largest (or "longest") homology group is that consisting of infinite cycles, with coefficients in G, reduced modulo

[25] This argument requires an application of the isomorphism theorem for inverse systems, and hence rests on the fact that the isomorphism of Theorem 17.1 is "natural" in the sense of §12.

the subgroup of actual boundaries. Since the latter subgroup is not in general closed, this homology group will be only a generalized topological group. This suggests the introduction of the shorter "weak" homology group, which consists of cycles modulo "weak" boundaries; i.e. those cycles which can be regarded as boundaries in any finite portion of the complex. The fundamental theorem for this type of homology group uses the group Ext_f which has been already analyzed. Finally, the group of cycles modulo the closure of the group of boundaries gives (following Lefschetz) a homology group which is always topological; for this we derive a corresponding form of the fundamental theorem. Furthermore, the standard duality between homology and cohomology groups enables us to deduce a corresponding theorem (Theorem 33.1) for the cohomology groups with coefficients in an arbitrary discrete group G.

The fundamental theorem expresses a homology group by means of a group of homomorphisms and a group of group extensions; the latter can also be represented by groups of homomorphisms, as in the basic theorem of Chapter II. The requisite connection between cycles of the homology group and homomorphisms is provided by the Kronecker index (§29).

26. Complexes

The complexes considered here will be abstract cell complexes[26] satisfying a star finiteness condition. More precisely, we consider a collection K of abstract elements σ^q called *cells*. With each cell there is associated an integer q called the dimension of σ^q. (There is no restriction requiring the dimension to be non-negative.) To any two cells σ_i^{q+1}, σ_j^q there corresponds an integer $[\sigma_i^{q+1}:\sigma_j^q]$, called the *incidence number*. K will be called a *star finite complex* provided the incidence numbers satisfy the following two conditions:

(26.1) *Given* σ_j^q, $[\sigma_i^{q+1}:\sigma_j^q] \neq 0$ *only for a finite number of indices* i;

(26.2) *Given* σ_j^{q+1} *and* σ_k^{q-1}, $\sum_i [\sigma_j^{q+1}:\sigma_i^q][\sigma_i^q:\sigma_k^{q-1}] = 0$.

Condition (26.1) is the star finiteness condition. It insures that the summation in (26.2) is finite.

If we consider the "incidence" matrices of integers

$$A^q = \| [\sigma_j^{q+1}:\sigma_i^q] \|$$

we can rewrite the two conditions as follows:

(26.1′) A^q *is column finite*;

(26.2′) $A^q A^{q-1} = 0.$

Actually we could have defined a complex as a collection of matrices $\{A^q\}$, $q = 0, \pm 1, \pm 2, \cdots$, such that (26.1′) and (26.2′) hold; we must assume then that the columns of A^q have the same set of labels as do the rows of A^{q-1}, in

[26] Essentially like those introduced by A. W. Tucker, for the case of finite complexes. Homology and cohomology are treated as in Whitney [14].

order to form the product A^qA^{q-1}. A q-cell will be then either a column of A^q or the corresponding row of A^{q-1}.

A subset L of the cells of K is called an *open subcomplex* if L contains with each q-cell all incident $(q + 1)$-cells; that is, if σ_i^q in L and $[\sigma_j^{q+1}:\sigma_i^q] \neq 0$ imply $\sigma_j^{q+1} \in L$. The incidence matrix A_L^q of L is then the submatrix obtained from A^q by deleting all rows and all columns belonging to cells not in L. Conditions (26.1) and (26.2) automatically hold in L, the latter because of the requirement that L be "open."

A subset $L \subset K$ *is a closed subcomplex* if L contains with each q-cell all incident $(q - 1)$-cells; that is, if $\sigma_j^q \in L$ and $[\sigma_j^q:\sigma_k^{q-1}] \neq 0$ imply $\sigma_k^{q-1} \in L$. The incidence matrix of L is obtained as before, and the conditions (26.1) and (26.2) again hold in L. Whenever L is a closed subcomplex, its complement $K - L$ is an open one, and vice versa.

A subset L of K will be called *q-finite* if L contains only a finite number of q-cells. Because K is star-finite, every $(q - 1)$-cell is contained in a q-finite open subcomplex of K.

27. Homology and cohomology groups

Let G be an abelian group. A q-dimensional *chain* c^q in K with coefficients in G is a function which associates to every q-cell σ_i^q in K an element g_i of G. We write c^q as a formal infinite sum

$$c^q = \sum_i g_i \sigma_i^q.$$

The sum of two chains $\sum g_i\sigma_i^q$ and $\sum h_i\sigma_i^q$ is the chain $\sum (g_i + h_i)\sigma_i^q$, and the chains form a group denoted by $C^q(K, G)$. If $g_i \neq 0$ for only a finite number of indices i then the chain c^q is *finite*. The finite chains form a subgroup $\mathcal{C}_q(K, G)$ of C^q.

The *coboundary* δc^q of a finite chain $c^q = \sum g_j\sigma_j^q$ is defined as

$$\delta c^q = \sum_i \Big(\sum_j [\sigma_i^{q+1}:\sigma_j^q] g_j\Big) \sigma_i^{q+1}.$$

Because of (26.1) δc^q is a finite $(q + 1)$-chain, while, because of (26.2), $\delta\delta c^q = 0$. The operation δ is a homomorphic mapping of $\mathcal{C}_q$ into $\mathcal{C}_{q+1}$. The kernel of this homomorphism is a subgroup $\mathcal{Z}_q(K, G)$ of $\mathcal{C}_q$. The chains of $\mathcal{Z}_q$ are called (finite) *cocycles*:

$$\mathcal{Z}_q(K, G) = [\text{all finite } q\text{-chains } c^q \text{ with } \delta c^q = 0].$$

A *coboundary* is a q-chain of the form δd^{q-1} for some $d^{q-1} \in \mathcal{C}_{q-1}$; these coboundaries form a subgroup

$$\mathcal{B}_q(K, G) = [\text{all finite chains } \delta d^{q-1}].$$

From the relation $\delta\delta = 0$ it follows that $\mathcal{B}_q \subset \mathcal{Z}_q$. The corresponding factor group

$$\mathcal{H}_q(K, G) = \mathcal{Z}_q(K, G)/\mathcal{B}_q(K, G)$$

is called the qth *cohomology* group of finite cocycles of K with coefficients in G. We also define the *co-torsion* group $\mathcal{T}_q(K, G)$ as the subgroup of all elements of finite order in $\mathcal{H}_q(K, G)$.

For a chain $c^q = \sum g_i \sigma_i^q$ of $C^q(K, G)$ we also define the *boundary*

$$\partial c^q = \sum_j \left(\sum_i [\sigma_i^q : \sigma_j^{q-1}] g_i\right) \sigma_j^{q-1}.$$

It again follows from (26.1) that ∂c^q is a well defined chain of $C^{q-1}(K, G)$ and from (26.2) that $\partial\partial c^q = 0$. The operation ∂ is a homomorphic mapping of C^q into C^{q-1}. The kernel of this homomorphism is a subgroup $Z^q(K, G)$ of C^q. The chains of Z^q are called *cycles*:

$$Z^q(K, G) = [\text{all chains } c^q \text{ with } \partial c^q = 0].$$

The chains of the form ∂d^{q+1} where $d^{q+1} \in C^{q+1}$ are the *boundaries*. They form a subgroup

$$B^q(K, G) = [\text{all chains } c^q = \partial d^{q+1}].$$

Because $\partial\partial = 0$ it follows that $B^q \subset Z^q$. The group

$$H^q(K, G) = Z^q(K, G)/B^q(K, G)$$

is called the qth *homology* group of K with coefficients in G.

Let L be a (closed or open) subcomplex of K. Each chain c^q in K, considered as a function on the q-cells, defines a corresponding chain c_L^q in L. If $c^q = \sum g_i \sigma_i^q$, $c_L^q = \sum' g_i \sigma_i^q$ is the sum found by deleting all terms $g_i \sigma_i^q$ for which σ_i^q is not in L. If L is open, then $\partial_L(c_L^q) = (\partial c^q)_L$, so that one can establish the following facts.

LEMMA 27.1. *$c^q \in Z^q(K, G)$ if and only if $c_L^q \in Z^q(L, G)$ for every q-finite open subcomplex L of K.*

LEMMA 27.2. *If $c^q \in B^q(K, G)$ then $c_L^q \in B^q(L, G)$, provided L is an open subcomplex of K.*

A statement analogous to Lemma 27.1 concerning B^q is not generally true. In this connection we define the group $B_w^q(K, G)$ of the *weak boundaries* as follows: $c^q \in B_w^q(K, G)$ provided $c_L^q \in B^q(L, G)$ for every q-finite open subcomplex L of K. For each such open subcomplex L we can construct a subcomplex L' consisting of all q-cells of L, all those $(q + 1)$-cells of L which lie on coboundaries of q-cells of L, and all $(q + i)$-cells of K, for $i > 1$. This subcomplex L' is open, is both q and $(q + 1)$-finite, and has $B^q(L, G) = B^q(L', G)$. Hence we conclude that $c^q \in B_w^q(K, G)$ if and only if $c_L^q \in B^q(L, G)$ for every open subcomplex L of K which is both q- and $(q + 1)$-finite. Clearly $B^q = B_w^q$ when K itself is q-finite.

It follows from Lemmas 27.1 and 27.2 that

$$B^q(K, G) \subset B_w^q(K, G) \subset Z^q(K, G).$$

The factor group

$$H^q_w(K, G) = Z^q(K, G)/B^q_w(K, G)$$

will be called the *weak q^{th} homology group* of K with coefficients in G. Clearly $H^q = H^q_w$ when K is q-finite

LEMMA 27.3. *$c^q \in B^q_w(K, G)$ if and only if for each finite subset M of K there is a chain c^q_{I} in $K - M$ such that $c^q - c^q_{\text{I}} \in B^q(K, G)$.*

PROOF. Suppose that $c^q \in B^q_w$. Given the finite set M there is a q-finite open subcomplex L containing M. Since $c^q_L \in B^q(L, G)$ there is a d^{q+1} in L such that $(\partial d^{q+1})_L = c^q_L$. Set $c^q_{\text{I}} = c^q - \partial d^{q+1}$. Clearly $c^q - c^q_{\text{I}} \in B^q$ and $(c^q_{\text{I}})_L = c^q_L - (\partial d^{q+1})_L = 0$, hence $c^q_{\text{I}} \subset K - L \subset K - M$.

Suppose now that c^q satisfies the condition of Lemma 27.3. Given a q-finite open subcomplex L of K there is a c^q_{I} in $K - L$ such that $c^q - c^q_{\text{I}} \in B^q(K, G)$. There is then a d^{q+1} such that $\partial d^{q+1} = c^q - c^q_{\text{I}}$. Since L is open we have

$$\partial_L(d^{q+1}_L) = (\partial d^{q+1})_L = c^q_L - (c^q_{\text{I}})_L = c^q_L ;$$

therefore $c^q_L \in B^q(L, G)$.

28. Topology in the homology groups

The group of q-chains $C^q(K, G)$ is isomorphic with $\prod_i G_i$, where $G = G_i$ and the set of indices i is in a 1-1 correspondence with the set of q-cells σ^q_i. Hence, if G is a generalized topological group, we can consider $C^q(K, G)$ as a generalized topological group, under the direct product topology, as defined in §1. If G is topological or compact, then $C^q(K, G)$ is also topological or compact, as the case may be.

The boundary operator ∂, regarded as a homomorphism of C^q into C^{q-1}, is continuous. Since Z^q is the group mapped into 0 by ∂, we obtain

LEMMA 28.1. *If G is topological then $Z^q(K, G)$ is a closed subgroup of $C^q(K, G)$.*

From Lemma 27.3 we deduce

LEMMA 28.2. $B^q(K, G) \subset B^q_w(K, G) \subset \bar{B}^q(K, G)$.

The homology groups $H^q = Z^q/B^q$ and $H^q_w = Z^q/B^q_w$ as factor groups of generalized topological groups are generalized topological groups; this is the way they will be considered in the rest of this paper. Even in the case when G and consequently Z^q is topological the groups H^q and H^q_w may be only generalized topological groups, for B^q and B^q_w need not be closed subgroups of Z^q.

If G is compact and topological, then $Z^q(K, G)$ and $C^{q+1}(K, G)$ are compact; since $B^q(K, G)$ is a continuous image of C^{q+1} (under the operation ∂), $B^q(K, G)$ is compact and therefore closed (see Lemma 1.1). Consequently we obtain

LEMMA 28.3. *If G is compact and topological, then $B^q(K, G) = B^q_w(K, G) = \bar{B}^q(K, G)$, $H^q(K, G) = H^q_w(K, G)$ and the groups are all compact and topological.*

Despite the fact that $\mathfrak{C}_q$ is a subgroup of the generalized topological group C^q we consider $\mathfrak{C}_q$ discrete and consequently the cohomology groups $H_q(K, G)$ are taken discrete.

29. The Kronecker index

Let G be a generalized topological group, H a discrete group and assume that a product $\phi(g, h) \in J$ is given pairing G and H to a group J (see §13).

Given two chains

$$c^q \in C^q(K, G), \qquad d^q \in \mathcal{C}_q(K, H),$$

we define the Kronecker index as

$$c^q \cdot d^q = \sum_i \phi(g_i, h_i) \in J;$$

the summation is finite since d^q is a finite chain. We verify at once that in this way the groups $C^q(K, G)$ and $\mathcal{C}_q(K, H)$ are paired to J.

Given $c^{q+1} \in C^{q+1}(K, G)$ and $d^q \in \mathcal{C}_q(K, H)$ we have

$$(29.1) \qquad (\partial c^{q+1}) \cdot d^q = c^{q+1} \cdot (\delta d^q).$$

This is a restatement of the associative law for matrix multiplication, since the operator ∂ is essentially a postmultiplication by the incidence matrix, while the coboundary operator δ is a premultiplication by the same matrix.

We now examine the annihilators relative to the Kronecker index.

$$(29.2) \qquad \mathcal{Z}_q(K, H) \subset \text{Annih } B^q_w(K, G) \subset \text{Annih } B^q(K, G).$$

$$(29.3) \qquad Z^q(K, G) \subset \text{Annih } \mathcal{B}_q(K, H).$$

PROOF. Let $z^q \in B^q_w$ and $w^q \in \mathcal{Z}_q$. Since w^q is finite there is a finite subset M of K such that $w^q \subset M$. In view of Lemma 27.3 there is a cycle $z^q_1 \subset K - M$ and a chain $c^{q+1} \in C^{q+1}(K, G)$ such that $\partial c^{q+1} = z^q - z^q_1$. Consequently

$$z^q \cdot w^q = (z^q - z^q_1) \cdot w^q = (\partial c^{q+1}) \cdot w^q = c^{q+1} \cdot \delta w^q = c^{q+1} \cdot 0 = 0.$$

Therefore $\mathcal{Z}_q \subset \text{Annih } B^q_w$. The proof of (29.3) is analogous.

It follows from (29.2) and (29.3) that

(29.4) *$H^q(K, G)$ and $\mathcal{H}_q(K, H)$ are paired to J,*

(29.5) *$H^q_w(K, G)$ and $\mathcal{H}_q(K, H)$ are paired to J.*

LEMMA 29.1. *If G and H are dually paired to J then, relative to the Kronecker index,*

(29.6) *$C^q(K, G)$ and $\mathcal{C}_q(K, H)$ are dually paired to J,*

$$(29.7) \qquad \mathcal{Z}_q(K, H) = \text{Annih } B^q_w(K, G) = \text{Annih } B^q(K, G),$$

$$(29.8) \qquad Z^q(K, G) = \text{Annih } \mathcal{B}_q(K, H).$$

PROOF. Given $c^q = \sum g_i \sigma^q_i \neq 0$ in C^q, we have $g_{i_0} \neq 0$ for some i_0. Select $h \in H$ so that $\phi(g_{i_0}, h) \neq 0$. Consider the chain $d^q = h\sigma^q_{i_0}$. Then $c^q \cdot d^q = \phi(g_{i_0}, h) \neq 0$. This proves that $\text{Annih } \mathcal{C}_q(K, H) = 0$. Similarly we prove that $\text{Annih } C^q(K, G) = 0$. This establishes (29.6).

Let $d^q \in \text{Annih } B^q(K, G)$. Hence $c^{q+1}\cdot(\delta d^q) = (\partial c^{q+1})\cdot d^q = 0$ for every c^{q+1}, and therefore $\delta d^q = 0$, in view of (29.6). This shows that $\text{Annih } B^q \subset \mathfrak{Z}_q$, which, together with (29.2), gives (29.7).

The proof of (29.8) is analogous to the previous one.

We remark that even when the pairing of the coefficient groups G and H is dual, the pairing (29.4) or (29.5) of the homology and cohomology groups need not be dual, as observed by Whitney ([14], p. 42).

We shall be especially interested in the pairing of G with the group I of integers to G by means of the product $\phi(g, m) = mg$. This pairing has the property that $\text{Annih } I = 0$. This is half of the definition of a dual pairing; the other half ($\text{Annih } G = 0$) may fail in case the order of every element in G divides a fixed integer m. Nevertheless the argument for Lemma 29.1 shows in this case that

(29.6′) $$\text{Annih } \mathfrak{C}_q(K, I) = 0,$$

(29.8′) $$Z^q(K, G) = \text{Annih } \mathfrak{B}_q(K, I).$$

We now introduce a subgroup of the group of cycles by the following definition:

(29.9) $$A^q(K, G) = \text{Annih } \mathfrak{Z}_q(K, I);$$

in other words, $c^q \in A^q(K, G)$ if and only if $c^q\cdot w^q = 0$ for every finite integral cocycle w^q. The position of this group A^q may be described as follows:

(29.10) $$B^q_w(K, G) \subset A^q(K, G) \subset Z^q(K, G).$$

By (29.2) we have $\mathfrak{Z}_q \subset \text{Annih } B^q_w$; consequently $B^q_w \subset \text{Annih } \mathfrak{Z}_q = A^q$. Since $\mathfrak{B}_q \subset \mathfrak{Z}_q$, we have $A^q = \text{Annih } \mathfrak{Z}_q \subset \text{Annih } \mathfrak{B}_q = Z^q$ by (29.8′).

LEMMA 29.2. *If G is a topological group, $A^q(K, G)$ is closed.*

This follows immediately from the continuity of the Kronecker index.

In case G is topological, the various subgroups of cycles of $C^q(K, G)$ are therefore related as follows:

$$B^q \subset B^q_w \subset \bar{B}^q \subset A^q = \bar{A}^q \subset Z^q = \bar{Z}^q \subset C^q.$$

30. Construction of homomorphisms

The essential device of this chapter is that of using the Kronecker index to generate homomorphisms. For a given chain $c^q \in C^q(K, G)$ define θ_{c^q} by

(30.1) $$\theta_{c^q}(d^q) = c^q\cdot d^q, \qquad d^q \in \mathfrak{C}_q(K, I).$$

LEMMA 30.1. *The correspondence $c^q \to \theta_{c^q}$ establishes an isomorphism*

$$C^q(K, G) \cong \text{Hom } \{\mathfrak{C}_q(K, I), G\}.$$

PROOF. It is clear that $\theta_{c^q} \in \text{Hom } \{\mathfrak{C}_q, G\}$, and that the correspondence $c^q \to \theta_{d^q}$ preserves sums. Also, if $\theta_{c^q} = 0$ then $c^q\cdot d^q = 0$ for all $d^q \in \mathfrak{C}_q$ and consequently $c^q = 0$. Conversely, given $\theta \in \text{Hom } \{\mathfrak{C}_q, G\}$, define

(30.2) $$c^q = \sum_i \theta(\sigma^q_i)\sigma^q_i.$$

Clearly $c^q \in C^q(K, G)$, while, for any given $d^q = \sum h_i\sigma_i^q \in \mathcal{C}_q$, we have

$$\theta_{c^q}(d^q) = c^q \cdot d^q = \sum h_i\theta(\sigma_i^q) = \theta(\sum h_i\sigma_i^q) = \theta(d^q).$$

This establishes the algebraic part of the Lemma.

We now recall that

$$C^q(K, G) \cong \prod_i G_i$$

where $G_i = G$ and the subscripts i are in a 1-1 correspondence with the q-cells σ_i^q. On the other hand, since the $\{\sigma_i^q\}$ constitute a set of generators for $\mathcal{C}_q(K, I)$, we have

$$\text{Hom } \{\mathcal{C}_q(K, I), G\} \cong \prod_i G_i.$$

Both these isomorphisms are bicontinuous, hence the combined isomorphism, which is precisely the isomorphism $c^q \leftrightarrow \theta_{c^q}$, is also bicontinuous.

LEMMA 30.2. *$\mathcal{Z}_q(K, I)$ is a direct factor of $\mathcal{C}_q(K, I)$.*

PROOF. The coboundary operator δ maps $\mathcal{C}_q$ onto $\mathcal{B}_{q+1}$ and the kernel is $\mathcal{Z}_q$. Hence $\mathcal{C}_q$ is a group extension of $\mathcal{Z}_q$ by $\mathcal{B}_{q+1}$. As a subgroup of the free group $\mathcal{C}_{q+1}$ the group $\mathcal{B}_{q+1}$ is free (Lemma 4.1) and therefore the group extension is trivial (Theorem 7.2). Hence $\mathcal{C}_q$ is the direct product of $\mathcal{Z}_q$ and a subgroup isomorphic with $\mathcal{B}_{q+1}$.

THEOREM 30.3. *$A^q(K, G)$ is a direct factor of $Z^q(K, G)$ and of $C^q(K, G)$.*

PROOF. Since $A^q \subset Z^q$ it will be sufficient to show that A^q is a direct factor of C^q. In the group Hom $\{\mathcal{C}_q(K, I), G\}$ consider the subgroup A of those homomorphisms that annihilate Z_q. Since $\mathcal{Z}_q$ is a direct factor of $\mathcal{C}_q$, A is a direct factor of Hom $\{\mathcal{C}_q, G\}$. However, under the isomorphism $\theta_{c^q} \to c^q$ of Lemma 30.1 the group A is mapped onto $A^q(K, G) = \text{Annih } \mathcal{Z}_q$, hence the conclusion. This proof also shows (Lemma 3.3) that

$$A^q(K, G) \cong \text{Hom } \{\mathcal{C}_q(K, I)/\mathcal{Z}_q(K, I), G\}. \tag{30.3}$$

Theorem 30.3 leads to the following direct product decompositions of the homology groups:

$$H^q(K, G) \cong (Z^q/A^q) \times (A^q/B^q), \tag{30.4}$$

$$H_w^q(K, G) \cong (Z^q/A^q) \times (A^q/B_w^q). \tag{30.5}$$

We proceed with the study of the first factor, Z^q/A^q.

THEOREM 30.4. *The correspondence $c^q \to \theta_{c^q}$ establishes an isomorphism*

$$Z^q(K, G)/A^q(K, G) \cong \text{Hom } \{\mathcal{H}_q(K, I), G\}.$$

PROOF. Since $Z^q = \text{Annih } \mathcal{B}_q$, by (29.8′), it follows that under the isomorphism $c^q \to \theta_{c^q}$ the group Z^q is mapped onto the subgroup of Hom $\{\mathcal{C}_q, G\}$ consisting of those homomorphisms annihilating $\mathcal{B}_q$. By Lemma 3.3 the latter subgroup can be identified with Hom $\{\mathcal{C}_q/\mathcal{B}_q, G\}$, so $Z^q \cong \text{Hom } \{\mathcal{C}_q/\mathcal{B}_q, G\}$. On the other hand, $\mathcal{Z}_q/\mathcal{B}_q$ is a direct factor of $\mathcal{C}_q/\mathcal{B}_q$, so that Lemma 3.4 shows that Hom $\{\mathfrak{Z}_q/\mathfrak{B}_q, G\}$ is a factor group of Hom $\{\mathfrak{C}_q/\mathfrak{B}_q, G\}$, corresponding to the subgroup consisting of homomorphisms annihilating $\mathfrak{Z}_q/\mathfrak{B}_q$. This subgroup in turn corresponds to the subgroup A^q of Z^q, hence

$$Z^q/A^q \cong \text{Hom }\{\mathfrak{Z}_q/\mathfrak{B}_q, G\}.$$

This is the desired conclusion.

31. Study of A^q

The correspondence $c^q \to \theta_{c^q}$ of Lemma 30.1 maps the group A^q of annihilators of cocycles onto the group of those homomorphisms of $\mathfrak{C}_q$ into G which carry $\mathfrak{Z}_q$ into zero. As observed in Lemma 3.3, the latter group is isomorphic to Hom $\{\mathfrak{C}_q/\mathfrak{Z}_q, G\}$. Since $\mathfrak{C}_q/\mathfrak{Z}_q \cong \mathfrak{B}_{q+1}$, this gives the isomorphism

$$A^q(K, G) \cong \text{Hom }\{\mathfrak{B}_{q+1}(K, I), G\}. \tag{31.1}$$

An examination of this construction shows that the homomorphism corresponding to a given $z^q \in A^q$ is determined as follows. For each $d^{q+1} \in \mathfrak{B}_{q+1}$ choose a $d^q \in \mathfrak{C}_q(K, I)$ for which $\delta d^q = d^{q+1}$, and define[27]

$$\phi_{z^q}(d^{q+1}) = z^q \cdot d^q.$$

Because z^q is in A^q, this result is independent of the choice of d^q for given d^{q+1}. Furthermore ϕ_{z^q} is a homomorphism of $\mathfrak{B}_{q+1}$ into G, and it is obtained from θ_{z^q} by the process indicated above, for one has

$$\phi_{z^q}(\delta d^q) = \theta_{z^q}(d^q).$$

We therefore have the following result.

LEMMA 31.1. *The correspondence* $z^q \to \phi_{z^q}$ *establishes the (bicontinuous) isomorphism* (31.1).

The properties of this isomorphism can be collected in the following

THEOREM 31.2. *The isomorphism* $z^q \to \phi_{z^q}$ *induces the isomorphisms*

$$A^q(K, G)/B^q(K, G) \cong \text{Hom }\{\mathfrak{B}_{q+1}, G\}/\text{Hom }\{\mathfrak{Z}_{q+1} \mid \mathfrak{B}_{q+1}, G\},$$

$$B^q_w(K, G)/B^q(K, G) \cong \text{Hom}_f\,\{\mathfrak{B}_{q+1}, G; \mathfrak{Z}_{q+1}\}/\text{Hom }\{\mathfrak{Z}_{q+1} \mid \mathfrak{B}_{q+1}, G\},$$

$$A^q(K, G)/B^q_w(K, G) \cong \text{Hom }\{\mathfrak{B}_{q+1}, G\}/\text{Hom}_f\,\{\mathfrak{B}_{q+1}, G; \mathfrak{Z}_{q+1}\},$$

where $\mathfrak{B}_{q+1} = \mathfrak{B}_{q+1}(K, I)$ *and* $\mathfrak{Z}_{q+1} = \mathfrak{Z}_{q+1}(K, I)$.

PROOF. We shall show that the groups $B^q(K, G)$ and $B^q_w(K, G)$ are mapped onto Hom $\{\mathfrak{Z}_{q+1} \mid \mathfrak{B}_{q+1}, G\}$ and Hom$_f$ $\{\mathfrak{B}_{q+1}, G; \mathfrak{Z}_{q+1}\}$, respectively.

Assume that $z^q \in B^q(K, G)$; then $\partial z^{q+1} = z^q$ for some $z^{q+1} \in C^{q+1}(K, G)$. Define

$$\phi^*(d^{q+1}) = z^{q+1} \cdot d^{q+1}; \qquad d^{q+1} \in \mathfrak{C}_{q+1}.$$

---

[27] Notice the analogy with the definition of the so-called "linking coefficient" (cf. Lefschetz [7], Ch. III).

Clearly $\phi^* \in \text{Hom}\,\{\mathcal{C}_{q+1}, G\}$. If $d^{q+1} = \delta d^q$ then

$$\phi^*(d^{q+1}) = z^{q+1}\cdot d^{q+1} = z^{q+1}\cdot \delta d^q = \partial z^{q+1}\cdot d^q = z^q\cdot d^q = \phi_{z^q}(d^{q+1}).$$

Hence ϕ^* is an extension of ϕ_{z^q} to $\mathcal{C}_{q+1}$ and in particular also to $\mathcal{Z}_{q+1}$.

Suppose conversely that ϕ_{z^q} can be extended to $\mathcal{Z}_{q+1}$. Since $\mathcal{Z}_{q+1}$ is a direct factor of $\mathcal{C}_{q+1}$ (Lemma 30.2) we may then find an extension ϕ^* of ϕ_{z^q} to $\mathcal{C}_{q+1}$. Define

$$z^{q+1} = \sum_i \phi^*(\sigma_i^{q+1})\sigma_i^{q+1}.$$

Clearly $z^{q+1} \in C^{q+1}(K, G)$ and $z^{q+1}\cdot\sigma_j^{q+1} = \phi^*(\sigma_j^{q+1})$ and hence $z^{q+1}\cdot d^{q+1} = \phi^*(d^{q+1})$ for all $d^{q+1} \in \mathcal{C}_{q+1}$. Consequently

$$\partial z^{q+1}\cdot\sigma_j^q = z^{q+1}\cdot\delta\sigma_j^q = \phi^*(\delta\sigma_j^q) = \phi_{z^q}(\delta\sigma_j^q) = z^q\cdot\sigma_j^q.$$

Since this holds for every σ_j^q we have $\partial z^{q+1} = z^q \in B^q(K, G)$.

Suppose $z^q \in B_w^q(K, G)$. In view of Lemma 5.1 it is sufficient to prove that if the cocycle $md^{q+1} \in \mathcal{B}_{q+1}(K)$ then $\phi_{z^q}(md^{q+1})$ is divisible by m. Let $\delta d^q = md^{q+1}$ and let M be a finite subset of K such that $d^q \subset M$. In view of Lemma 27.3 there is a chain $z_1^q \subset K - M$ such that $z^q - z_1^q = z_2^q \in B^q(K, G)$. It follows that $z_2^q \in A^q$ and so that $z_1^q \in A^q$, hence $\phi_{z_1^q}$ and $\phi_{z_2^q}$ are defined and $\phi_{z^q} = \phi_{z_1^q} + \phi_{z_2^q}$. Since $z_2^q \in B^q(K, G)$, then, as we just proved, $\phi_{z_2^q}$ can be extended to Z_{q+1} and therefore $\phi_{z_2^q}(md^{q+1})$ must be divisible by m. Since $d^q \subset M$ and $z_1^q \subset K - M$ we have $\phi_{z_1^q}(md^{q+1}) = z_1^q\cdot d^q = 0$. Hence $\phi_{z^q}(md^{q+1})$ is divisible by m.

Suppose conversely that ϕ_{z^q} can be extended to every subgroup of $\mathcal{Z}_{q+1}(K, I)$ of finite order over $\mathcal{B}_{q+1}(K, I)$. Then, as in Lemma 5.2, ϕ_{z^q} can also be extended to every subgroup $\mathcal{D}$ of $\mathcal{Z}_{q+1}(K, I)$ such that $\mathcal{D}/\mathcal{B}_{q+1}$ has a finite number of generators. Now let L be any open subcomplex of K which is both q and $(q + 1)$-finite; there is then an extension of ϕ_{z^q} to the group $\mathcal{D}_L$ generated by $\mathcal{B}_{q+1}(K, I)$ and $\mathcal{Z}_{q+1}(L, I)$. But in the complex L the homomorphism ϕ_{y^q} induced by $y^q = z_L^q$ agrees on $\mathcal{B}_{q+1}(L, I)$ with the homomorphism ϕ_{z^q}. Therefore $\phi_{y^q} \in \text{Hom}\,\{\mathcal{B}_{q+1}(L, I), G\}$ has an extension to $\mathcal{Z}_{q+1}(L, I)$. In view of what we proved before, we therefore have $y^q = z_L^q \in B^q(L, G)$. Since this holds for each L considered, $z^q \in B_w^q(K, G)$. This concludes the proof of Theorem 31.2.

In this theorem the factor homomorphism groups on the right can be reinterpreted as groups of group extensions, in accord with the results of Chapter II.

THEOREM 31.3. *The isomorphism $z^q \leftrightarrow \phi_{z^q}$ combined with the isomorphisms establishing relations between group extensions and homomorphisms lead to the following isomorphisms:*

(31.2) $$A^q(K, G)/B^q(K, G) \cong \text{Ext}\,\{G, \mathcal{H}_{q+1}\},$$

(31.3) $$B_w^q(K, G)/B^q(K, G) \cong \text{Ext}_f\,\{G, \mathcal{H}_{q+1}\},$$

(31.4) $$A^q(K, G)/B_w^q(K, G) \cong \text{Ext}\,\{G, \mathcal{T}_{q+1}\}/\text{Ext}_f\,\{G, \mathcal{T}_{q+1}\}$$

where $\mathcal{H}_{q+1} = \mathcal{H}_{q+1}(K, I)$ and $\mathcal{T}_{q+1} = \mathcal{T}_{q+1}(K, I)$ is the corresponding co-torsion group.

The isomorphisms established so far have all been bicontinuous.

32. Computation of the homology groups

As we have shown in §29, the Kronecker index establishes a pairing of the group $H^q(K, G)$ or $H^q_w(K, G)$ with the group $\mathcal{H}_q(K, I)$, the values of the products being in the group G. Accordingly we define the following subhomology groups:

(32.1) $$Q^q(K, G) = \text{Annih } \mathcal{H}^q(K, I) \text{ in } H^q(K, G),$$

(32.2) $$Q^q_w(K, G) = \text{Annih } \mathcal{H}_q(K, I) \text{ in } H^q_w(K, G).$$

We verify at once that $Q^q = A^q/B^q$ and $Q^q_w = A^q/B^q_w$. Consequently the results of the last two sections furnish the following two basic theorems:

THEOREM 32.1. *For a star finite complex K the homology group $H^q(K, G)$ of infinite cycles with coefficients in a generalized topological group G can be expressed in terms of the integral cohomology groups $\mathcal{H}_q = \mathcal{H}_q(K, I)$ and $\mathcal{H}_{q+1} = \mathcal{H}_{q+1}(K, I)$ of finite cocycles. The explicit relation is*

(32.3) $$H^q(K, G) \cong \text{Hom } \{\mathcal{H}_q, G\} \times \text{Ext } \{G, \mathcal{H}_{q+1}\}.$$

More explicitly, H^q has a subgroup Q^q, defined by (32.1), where

(32.4) $$Q^q(K, G) \text{ is a direct factor of } H^q(K, G),$$

(32.5) $$Q^q(K, G) \cong \text{Ext } \{G, \mathcal{H}_{q+1}\},$$

(32.6) $$H^q(K, G)/Q^q(K, G) \cong \text{Hom } \{\mathcal{H}_q, G\}.$$

THEOREM 32.2. *For a star finite complex K the weak homology group $H^q_w(K, G)$ of infinite cycles with coefficients in a generalized topological group G can be expressed in terms of the integral cohomology group $\mathcal{H}_q = \mathcal{H}_q(K, I)$ and the integral co-torsion group $\mathcal{T}_{q+1} = \mathcal{T}_{q+1}(K, I)$ of finite cocycles. The explicit relation is*

(32.7) $$H^q_w(K, G) \cong \text{Hom } \{\mathcal{H}_q, G\} \times (\text{Ext } \{G, \mathcal{T}_{q+1}\}/\text{Ext}_f \{G, \mathcal{T}_{q+1}\}).$$

More explicitly, H^q_w has a subgroup Q^q_w, defined by (32.2), where

(32.8) $$Q^q_w(K, G) \text{ is a direct factor of } H^q_w(K, G),$$

(32.9) $$Q^q_w(K, G) \cong \text{Ext } \{G, \mathcal{T}_{q+1}\}/\text{Ext}_f \{G, \mathcal{T}_{q+1}\},$$

(32.10) $$H^q_w(K, G)/Q^q_w(K, G) \cong \text{Hom } \{\mathcal{H}_q, G\}.$$

Both factors in (32.3) and (32.7) are generalized topological groups and the isomorphisms are bicontinuous.

If G is topological then by Corollary 3.2 the group Hom $\{\mathcal{H}_q, G\}$ is topological. If we also assume that mG is a closed subgroup of G for $m = 2, 3, \cdots$ then Corollary 11.6 shows that $\text{Ext}_f \{G, \mathcal{T}_{q+1}\}$ is a closed subgroup of Ext $\{G, \mathcal{T}_{q+1}\}$. Consequently we obtain

THEOREM 32.3. (Steenrod [9]). *If G is a topological group and mG is a closed subgroup of G for $m = 2, 3, \cdots$ then $H^q_w(K, G)$ is topological.*

The expressions for Q^q and Q^q_w can be simplified if additional information concerning the group G is available. If G is infinitely divisible then, by Corollary 11.4, Ext $\{G, H\} = 0$ for all H and therefore

COROLLARY 32.4. *If G is infinitely divisible then* $Q^q(K, G) = Q^q_w(K, G) = 0$ *and*

$$H^q(K, G) = H^q_w(K, G) \cong \text{Hom}\,\{\mathcal{H}_q, G\}.$$

From Theorem 17.2 we deduce

COROLLARY 32.5. *If G has no elements of finite order then*

$$Q^q_w(K, G) \cong \text{Ext}\,\{G, \mathcal{T}_{q+1}\}.$$

If, in addition, G is discrete then

$$Q^q_w(K, G) \cong \text{Hom}\,\{\mathcal{T}_{q+1}, G_\infty/G\}.$$

In particular, if $G = I$ then, by Theorem 17.1, $Q^q_w(K, I) \cong \text{Char}\,\mathcal{T}_{q+1}$ and therefore

$$H^q_w(K, I) \cong \text{Hom}\,\{\mathcal{H}_q, I\} \times \text{Char}\,\mathcal{T}_{q+1}. \tag{32.11}$$

THEOREM 32.6. *If G is compact and topological then* $H^q(K, G) = H^q_w(K, G)$ *is compact and topological and*

$$Q^q(K, G) = Q^q_w(K, G) \cong \text{Ext}\,\{G, \mathcal{T}_{q+1}\} \cong \text{Char Hom}\,\{G, \mathcal{T}_{q+1}\}. \tag{32.12}$$

This is a consequence of Corollary 11.7 and Theorem 15.1. Since G is compact, $\mathcal{T}_{q+1}$ discrete, and only continuous homomorphisms are taken in Hom $\{G, \mathcal{T}_{q+1}\}$, it follows that in the formula (32.12) for $Q^q(K, G)$ we may replace G by G/G_0 where G_0 is the component of 0 in G.

COROLLARY 32.7. *If* $\mathcal{H}_{q+1}(K, I)$ *has a finite number of generators then* $B^q(K, G) = B^q_w(K, G)$ *and*

$$H^q(K, G) = H^q_w(K, G) \cong \text{Hom}\,\{\mathcal{H}_q, G\} \times \text{Ext}\,\{G, \mathcal{T}_{q+1}\}. \tag{32.13}$$

In fact, since $\text{Ext}_f\,\{G, \mathcal{H}_{q+1}\} = 0$ (Corollary 11.3) it follows from (31.3) that $B^q = B^q_w$. Since also $\text{Ext}_f\,\{G, \mathcal{T}_{q+1}\} = 0$, formula (32.13) follows from Theorem 32.2.

In particular, Corollary 32.7 applies if K is a finite complex (cf. Alexandroff-Hopf [1], Ch. V and Steenrod [9], p. 675).

33. Computation of the cohomology groups

We start out with a brief review of the duality between homology and cohomology. Let G be a discrete group and $\hat{G} = \text{Char}\,G$ compact and topological. Since $\hat{G}$ and G are dually paired to the group P of reals mod 1 (see §13) the Kronecker index $c^q \cdot d^q \in P$ is defined as in §29 for $c^q \in C^q(K, \hat{G})$ and $d^q \in \mathcal{C}_q(K, G)$. Since the pairing of $\hat{G}$ and G is dual (Theorem 13.5) we have by Lemma 29.1

(33.1) $C^q(K, \hat{G})$ *and* $\mathcal{C}_q(K, G)$ *are dually paired to* P,

(33.2) $\mathcal{Z}_q(K, G) = \text{Annih}\, B^q(K, \hat{G})$; $\quad Z^q(K, \hat{G}) = \text{Annih}\, \mathcal{B}_q(K, G)$.

These formulas, Theorem 13.7, Lemma 13.2 and Theorem 13.5 imply that the Kronecker index defines a dual pairing of $\mathcal{H}_q(K, G)$ and $H^q(K, \hat{G})$ to P and that

$$\mathcal{H}_q(K, G) \cong \text{Char}\, H^q(K, \text{Char}\, G). \tag{33.3}$$

Using this result and the formulas established in the previous section for $H^q(K, \text{Char } G)$ we could write down a formula expressing $\mathcal{H}_q(K, G)$. For convenience we first define a subcohomology group

(33.4) $$\mathcal{P}_q(K, G) = \text{Annih } Q^q(K, \text{Char } G) \text{ in } \mathcal{H}_q(K, G),$$

in order to get a more detailed form for our result.

THEOREM 33.1. *For a star finite complex K the cohomology group $\mathcal{H}_q(K, G)$ of finite cocycles with coefficients in a discrete group G can be expressed in terms of the cohomology group $\mathcal{H}_q = \mathcal{H}_q(K, I)$ and the integral co-torsion group $\mathcal{T}_{q+1} = \mathcal{T}_{q+1}(K, I)$. The explicit relation is*

(33.5) $$\mathcal{H}_q(K, G) \cong (G \circ \mathcal{H}_q) \times \text{Hom } \{\text{Char } G, \mathcal{T}_{q+1}\}.$$

More explicitly, $\mathcal{H}_q(K, G)$ has a subgroup $\mathcal{P}_q(K, G)$, defined by (33.4), *where*

(33.6) $\mathcal{P}_q(K, G)$ *is a direct factor of* $\mathcal{H}_q(K, G)$,

(33.7) $$\mathcal{P}_q(K, G) \cong G \circ \mathcal{H}_q$$

(33.8) $$\mathcal{H}_q(K, G)/\mathcal{P}_q(K, G) \cong \text{Hom } \{\text{Char } G, \mathcal{T}_{q+1}\}.$$

PROOF. Since Q^q is a direct factor of H^q it follows from the character theory that $\mathcal{P}_q = \text{Annih } Q^q$ is a direct factor of $\mathcal{H}_q(K, G) = \text{Char } H^q$. It also follows that

$$\mathcal{P}_q \cong \text{Char } (H^q/Q^q), \qquad \mathcal{H}_q(K, G)/\mathcal{P}_q(K, G) \cong \text{Char } Q^q.$$

The first formula and (32.6) imply

$$\mathcal{P}_q(K, G) \cong \text{Char Hom } \{\mathcal{H}_q, \text{Char } G\},$$

which in view of Theorem 18.1 gives (33.7). The second formula combined with (32.12) proves (33.8).

If G has no elements of finite order, then Char G is connected and therefore Hom $\{\text{Char } G, \mathcal{T}_{q+1}\} = 0$. From (33.7) and (33.8) we therefore obtain

COROLLARY 33.2. *If G has no elements of finite order then*

$$\mathcal{H}_q(K, G) = \mathcal{P}_q(K, G) \cong G \circ \mathcal{H}_q(K, I).$$

We now proceed to give an intrinsic characterization of the subgroup $\mathcal{P}_q$ of $\mathcal{H}_q(K, G)$. A cocycle $w^q \in \mathcal{Z}_q(K, G)$ will be called *pure* if it is a linear combination of integral cocycles, as

$$w_q = \sum_{i=1}^{k} g_i w_i^q, \qquad g_i \in G, \qquad w_i^q \in \mathcal{Z}_q(K, I).$$

LEMMA 33.3. *The group $\mathcal{P}_q(K, G)$ is the subgroup of $\mathcal{H}_q(K, G)$ determined by the pure cocycles.*

PROOF. Let $\mathcal{S}$ be the subgroup of $\mathcal{Z}_q(K, G)$ consisting of all the pure cocycles. It may be shown that $\mathcal{B}_q(K, G) \subset \mathcal{S}$. In order to prove that $\mathcal{S}/\mathcal{B}_q(K, G) = \mathcal{P}_q(K, G)$ we must prove that $\mathcal{S}/\mathcal{B}_q(K, G) = \text{Annih } Q^q(K, \hat{G})$ where $\hat{G} = \text{Char } G$.

This is equivalent to proving that $Q^q(K, \hat{G}) = \text{Annih}\ (\mathfrak{S}/\mathfrak{B}_q(K, G))$, which reduces to the formula

$$A^q(K, \hat{G}) = \text{Annih}\ \mathfrak{S},$$

that we now propose to establish.

Let $z^q \in A^q(K, \hat{G})$ and let $w^q \in \mathfrak{S}$. Since $w^q = \sum g_i w_i^q$, where $w_i^q \in \mathfrak{Z}_q(K, I)$ and since $z^q \cdot w_i^q = 0$ by the definition of A^q, it follows that $z^q \cdot w^q = 0$.

Suppose now that c^q lies in $C^q(K, \hat{G})$ but not in $A^q(K, \hat{G})$. There is then a $w_i^q \in \mathfrak{Z}_q(K, I)$ such that $c^q \cdot w_i^q = \hat{g} \neq 0$ where $\hat{g} \in \hat{G}$. Pick $g \in G$ so that $\hat{g}(g) \neq 0$ and define $w^q = gw_i^q$. Clearly $w^q \in \mathfrak{S}$ is a pure cocycle and $c^q \cdot w^q = \hat{g}(g) \neq 0$, hence c^q is not in Annih $\mathfrak{S}$. This concludes the proof of the Lemma.

Using the description of $\mathfrak{P}_q(K, G)$ given in the Lemma we could easily establish the isomorphism $\mathfrak{P}_q \cong G \circ \mathfrak{H}_q(K, I)$ directly, using the definition of the tensor product. This was the procedure adopted by Čech [3] who essentially has proved all the results of this section. Our main improvement is that our isomorphisms are given explicitly and invariantly, while Čech used generators and relations throughout.

34. The groups H_t^q

The fact that the groups H^q and H_w^q may not be topological groups even though the coefficient group G is chosen to be topological induced Lefschetz and others to introduce the following group, for a topological coefficient group G,

$$H_t^q(K, G) = Z^q(K, G)/\bar{B}^q(K, G)$$

as a standard homology group for K.

The relation of this group to the groups previously considered is immediate:

$$(34.1) \qquad H_t^q \cong H^q/\bar{0} \cong H_w^q/\bar{0}.$$

Theorem 32.3 can now be reformulated as follows.

THEOREM 34.1 (Steenrod [9]) *If G is topological and mG is closed for $m = 2, 3, \cdots$ then $H_w^q(K, G) = H_t^q(K, G)$.*

Since G is a topological group, $A^q(K, G)$ is a closed subgroup of $Z^q(K, G)$ (Lemma 29.2) and consequently $\bar{B}^q \subset A^q$. It follows that the Kronecker index can be defined for elements of $H_t^q(K, G)$ and $\mathfrak{H}_q(K, I)$. We define a subhomology group

$$(34.2) \qquad Q_t^q(K, G) = \text{Annih}\ \mathfrak{H}_q(K, I) \text{ in } H_t^q(K, G).$$

THEOREM 34.2. *For a star finite complex K the topological homology group $H_t^q(K, G)$ of infinite cycles with coefficients in a topological group G can be expressed in terms of the integral cohomology group $\mathfrak{H}_q = \mathfrak{H}_q(K, I)$ and the integral co-torsion group $\mathfrak{T}_{q+1} = \mathfrak{T}_{q+1}(K, I)$ of finite cocycles. The explicit relation is*

$$(34.3) \qquad H_t^q(K, G) \cong \text{Hom}\ \{\mathfrak{H}_q, G\} \times (\text{Ext}\ \{G, \mathfrak{T}_{q+1}\}/\bar{0}).$$

More explicitly, H_t^q *has a subgroup* Q_t^q, *defined by* (34.2), *where*

(34.3) $Q_t^q(K, G)$ *is a direct factor of* $H_t^q(K, G)$,

(34.5) $Q_t^q(K, G) \cong \text{Ext}\,\{G, \mathcal{T}_{q+1}\}/\bar{0}$,

(34.6) $H_t^q(K, G)/Q_t^q(K, G) \cong \text{Hom}\,\{\mathcal{H}_q, G\}$.

PROOF. From the direct product decomposition (30.5) we obtain

$$H_t^q \cong (Z^q/A^q) \times [(A^q/B_w^q)/\bar{0}].$$

Consequently $Q_t^q = Q_w^q/\bar{0}$ is a direct factor. Since $Q_w^q \cong \text{Ext}\,\{G, \mathcal{T}_{q+1}\}/\text{Ext}_f\,\{G, \mathcal{T}_{q+1}\}$ and since, by Corollary 11.6, $\overline{\text{Ext}_f}\,\{G, \mathcal{T}_{q+1}\} = \bar{0}$, we obtain (34.5). Formula (34.6) follows from Theorem 30.4.

It might be interesting to notice that, while the groups $H^q(K, G)$ and $H_w^q(K, G)$ were algebraically independent of the choice of the topology in G, the group $H_t^q(K, G)$ depends both algebraically and topologically upon the topology chosen in G.

35. Universal coefficients

The results of the previous three sections can be summarized in the following fashion.

UNIVERSAL COEFFICIENT THEOREM. *In a star finite complex* K *the integral cohomology groups of finite cocycles determine all the homology and cohomology groups that were defined for a star finite complex, specifically:*

The groups G, $\mathcal{H}_q(K, I)$ *and* $\mathcal{H}_{q+1}(K, I)$ *determine the generalized topological homology group* $H^q(K, G)$ *of infinite cycles with coefficients in a generalized topological group* G.

The groups G, $\mathcal{H}_q(K, I)$ *and* $\mathcal{T}_{q+1}(K, I)$ *determine*:

(a) *the generalized topological weak homology group* $H_w^q(K, G)$ *of infinite cycles with coefficients in a generalized topological group* G;

(b) *the topological homology group* $H_t^q(K, G)$ *of infinite cycles with coefficients in a topological group* G;

(c) *the discrete cohomology group* $\mathcal{H}_q(K, G)$ *of finite cocycles with coefficients in a discrete group* G.

This shows that the group I of integers is a universal coefficient group for the homology theory of the complex K. Since the group P of reals mod 1 is the group of characters of I we have in view of (33.3) the fact that $\mathcal{H}_q(K, I) \cong \text{Char}\, H^q(K, P)$; therefore all the groups can be expressed in terms of $H^q(K, P)$ and $H^{q+1}(K, P)$, so that P is also universal.

Given a closed subcomplex L of K one often has to consider the relative groups of K mod L. However, the complexes used here are so general that $K - L$ is also a complex and the usual groups of K mod L coincide with the groups of $K - L$ as we have defined them. Consequently all our formulas remain valid in the relative theory.

36. Closure finite complexes

Closure finite complexes are obtained by replacing condition (26.1) in the definition of a complex by the following

(36.1) *Given* σ_i^q, $[\sigma_i^q : \sigma_k^{q-1}] \neq 0$ *for only a finite number of indices* k.

Simplicial complexes are all closure finite.

In a closure finite complex we consider finite cycles and infinite cocycles and obtain the discrete homology groups $\mathcal{H}^q(K, G)$ and the topologized cohomology groups $H_q(K, G)$, $H_q^w(K, G)$ and $H_q^t(K, G)$. All our development can be repeated with the modification of interchanging homology and cohomology groups and replacing $q + 1$ by $q - 1$. For instance formula (32.3) will take the form:

$$H_q(K, G) \cong \text{Hom}\,\{\mathcal{H}^q(K, I), G\} \times \text{Ext}\,\{G, \mathcal{H}^{q-1}(K, I)\}.$$

Instead of repeating the arguments for closure finite complexes we can use the previous results for star finite complexes and apply them to closure finite complexes by means of the concept of the dual complex. If the complex K is described by the incidence matrices A^q, the dual complex K^* will be defined by the transposed matrices

$$B^q = (A^{-q})'$$

The dual of a star finite complex is closure finite and vice versa. Also $(K^*)^* = K$. Moreover by passing from a complex to its dual, the boundary operation becomes the coboundary, and vice versa. Hence the homology and cohomology group are interchanged, and our formulas apply.

A locally finite (i.e. both closure and star finite) complex carries therefore two homology theories, namely, the theory of a star finite complex and the theory of a closure finite one. In the case of a manifold the Poincaré duality establishes a relation between the two theories. In general the theories are unrelated and in any specific problem we only use one at a time. We will quote two examples to this effect.

A) In the following chapter we define for every compact metric space a complex called the fundamental complex. This complex is locally finite, but its closure finite theory is trivial, while its star finite theory is extremely useful for the study of the underlying space.

B) Let us consider two infinite polyhedra represented as two locally finite complexes K and K'. Given a continuous mapping f of K into K' it is well known that f induces homomorphisms: 1°) of the groups of finite cycles of K into the corresponding groups of K', 2°) of the groups of infinite cocycles of K' into the corresponding groups of K. This explains why in problems connected with continuous mappings (like Hopf's mapping theorem and its generalizations; see [4]) we use only finite cycles and infinite cocycles, or in other words we use only the closure finite theory of K and K'.

Chapter VI. Topological Spaces

Here we formulate our results for the homology groups of a space. In the case of a compact metric space, Steenrod has shown that the homology groups can all be expressed as corresponding homology groups of the fundamental complex of the space, so that the results of Chapter V apply directly (§44). For a general space, the Čech homology groups are obtained as (direct or inverse) limits, so that the decomposition of the homology group is obtained as a limit of the known decompositions for the homology groups of finite complexes, and here the techniques developed in Chapter IV apply. The results obtained for a general space are not as complete as those for complexes, partly because the limit of a set of direct sums apparently need not be a direct sum, and partly because "Lim" and "Ext" do not permute, so that the group Ext* discussed in Chap. IV is requisite. We also discuss (§45) Steenrod's homology groups of "regular" cycles.

37. Chain transformations

Let $K = \{\sigma_i^q\}$ and $K' = \{\tau_j^q\}$ be two star finite complexes. Suppose also that for every integer q there is given a matrix of integers,

$$B^q = \| b_{ij}^q \|$$

with rows indexed by the q-cells of K, columns by the q-cells of K', and with only a finite number of non-zero entries in each column.

Given a q-chain $c^q = \sum g_i \sigma_i^q \in C^q(K, G)$ in K, define

$$Tc^q = \sum_j \left(\sum_i g_i b_{ij}^q\right) \tau_j^q .$$

The column finiteness condition implies that the summation $\sum_i g_i b_{ij}^q$ is finite and therefore that Tc^q is a well defined element of $C^q(K', G)$. We thus obtain homomorphisms (one for each q and G)

$$T\colon\ C^q(K, G) \to C^q(K', G).$$

Given a finite q-chain $d^q = \sum g_j \tau_j^q \in \mathfrak{C}_q(K', G)$ in K', define

$$T^* d^q = \sum_i \left(\sum_j g_j b_{ij}^q\right) \sigma_i^q$$

This time the column finiteness of B^q implies that T^*d^q is finite; hence we obtain homomorphisms

$$T^*\colon\ \mathfrak{C}_q(K', G) \to \mathfrak{C}_q(K, G).$$

T^* is called the *dual* of T.

It can be verified at once that if c^q is a chain in K and d^q is a finite chain in K' then

$$(Tc^q)\cdot d^q = c^q \cdot (T^* d^q), \tag{37.1}$$

whenever the coefficients are such that the Kronecker index has a meaning (§29).

T is called a *chain transformation* of K into K' if $\partial Tc^q = T(\partial c^q)$ for every q chain; that is, if

$$\partial T = T\partial. \tag{37.2}$$

It can be shown that this condition is equivalent to the requirement that

$$\delta T^* = T^*\delta. \tag{37.3}$$

It follows that a chain transformation T maps the groups Z^q, A^q, B^q_w and B^q of K homomorphically into the corresponding groups of K'. Similarly T^* maps the groups of K' into the corresponding groups of K. In particular a chain transformation induces homomorphisms of the homology groups

$$T\colon\ H^q(K, G) \to H^q(K', G), \tag{37.4}$$

$$T^*\colon\ \mathcal{H}_q(K', G) \to \mathcal{H}_q(K, G), \tag{37.5}$$

and of the corresponding subgroups defined by (32.1) and (33.4)

$$T\colon\ Q^q(K, G) \to Q^q(K', G), \tag{37.6}$$

$$T^*\colon\ \mathcal{P}_q(K', G) \to \mathcal{P}_q(K, G). \tag{37.7}$$

38. Naturality

We are now in a position to give a precise meaning to the fact that the isomorphisms established in Chapter V are all "natural."

THEOREM 38.1. *If T is a chain transformation of a complex K into K', then T permutes with the isomorphisms established in Theorems* 30.4 *and* 31.2, *provided the application of T in any group is taken to mean the application of the appropriate transformation induced by T on that group.*

PROOF. If the homomorphism established in Theorem 30.4 be denoted by μ (or by μ', for K'), then we have the homomorphisms

$$\begin{array}{ccc} Z^q(K) & \xrightarrow{\mu} & \text{Hom}\,\{\mathcal{H}_q, G\} \\ \downarrow T & & \downarrow T_h^{**} \\ Z^q(K') & \xrightarrow{\mu'} & \text{Hom}\,\{\mathcal{H}'_q, G\}, \end{array}$$

where T_h^{**} is the homomorphism of Hom $\{\mathcal{H}_q(K, I), G\}$ into Hom $\{\mathcal{H}_q(K', I), G\}$, induced as in §12 by the dual chain transformation T^*. The theorem then asserts that

$$\mu' T = T_h^{**}\mu.$$

To show this, take $c^q \in Z^q(K, G)$. The corresponding homomorphism $\theta = \mu c^q$ is then defined, for each cocycle d^q in $\mathcal{Z}_q(K)$, by $\theta(d^q) = c^q \cdot d^q$ (cf. §30). Then $\theta' = T_h^{**}\theta$ is, according to the definition of T_h, simply $\theta'(d'^q) = \theta(T^*d'^q)$. Hence, for any cocycle d'^q,

$$\theta'(d'^q) = \theta(T^*d'^q) = c^q \cdot (T^*d'^q) = (Tc^q)\cdot d'^q.$$

In the other direction, Tc^q maps under μ' into the homomorphism ϕ', defined for $d'^q \in Z_q(K')$ by

$$\phi'(d'^q) = (Tc^q)\cdot d'^q.$$

The formulas show that $\phi' = \mu' Tc^q$ and $\theta' = T_h^{**}\mu c^q$ are in fact identical, as required by Theorem 38.1.

To treat Theorem 31.2, let τ (or τ') denote the homomorphism of $A^q(K, G)$ onto Hom $\{\mathfrak{B}_{q+1}(K, I), G\}$ given in that theorem, while η is the map of the latter group onto Ext $\{G, \mathfrak{H}_{q+1}\}$. The figure is

$$\begin{array}{ccccc} A^q & \xrightarrow[\tau]{} & \text{Hom}\,\{\mathfrak{B}_{q+1}, G\} & \xrightarrow[\eta]{} & \text{Ext}\,\{G, \mathfrak{H}_{q+1}\} \\ \downarrow T & & \downarrow T_h^{**} & & \downarrow T_e^{**} \\ A'^q & \xrightarrow[\tau']{} & \text{Hom}\,\{\mathfrak{B}'_{q+1}, G\} & \xrightarrow[\eta']{} & \text{Ext}\,\{G, \mathfrak{H}'_{q+1}\} \end{array}$$

where T_h^{**}, T_e^{**} are again the induced homomorphisms. If $z^q \in A^q(K, G)$ is given, $\phi = \tau z^q$ is defined on each coboundary δd^q as $\phi(\delta d^q) = z^q\cdot d^q$, while $\phi' = T_h^{**}\phi$ is defined in turn as

$$\phi'(\delta d'^q) = \phi(T^*\delta d'^q) = \phi(\delta T^* d'^q) = z^q\cdot(T^* d'^q).$$

On the other hand, $\chi = \tau'(Tz^q)$ is defined on a coboundary $\delta d'^q$ of K' as

$$\chi(\delta d'^q) = (Tz^q)\cdot d'^q = z^q\cdot(T^* d'^q).$$

The results are identical, so $T_h^{**}\tau = \tau' T$. Now the "naturality" theorem for group extensions showed that T permutes with η, as in $T_e^{**}\eta = \eta' T_h^{**}$. Combination of these results gives

$$(\eta'\tau')T = T_e^{**}(\eta\tau).$$

This is the required commutativity condition, for $\eta\tau$ is the isomorphism envisaged in Theorem 31.3.

39. Čech's homology groups

We now briefly outline Čech's method of defining the homology and cohomology groups for a space X. Let U_α be a finite open covering of X and N_α the nerve of U_α. If U_β is a refinement of U_α we write $\alpha < \beta$. For $\alpha < \beta$ we have a chain transformation $T_{\alpha\beta}$: $N_\beta \to N_\alpha$ defined as follows: for each open set of the covering U_β select a set of U_α containing it; this maps the vertices of N_β into the vertices of N_α and leads to a simplicial mapping $T_{\alpha\beta}$. This chain transformation is not defined uniquely, but the induced homomorphisms

$$T_{\alpha\beta}: \quad H^q(N_\beta, G) \to H^q(N_\alpha, G),$$

$$T^*_{\beta\alpha}: \quad \mathfrak{H}_q(N_\alpha, G) \to \mathfrak{H}_q(N_\beta, G)$$

are unique. Using the directed system of all the finite open coverings of X we define[28]

$$\mathcal{H}^q(X, G) = \varprojlim H^q(N_\alpha, G) \tag{39.1}$$

$$\mathcal{H}_q(X, G) = \varinjlim \mathcal{H}_q(N_\alpha, G). \tag{39.2}$$

In (39.2) the groups are all discrete. In (39.1) G can be any generalized topological group and $\mathcal{H}^q(X, G)$, as an inverse limit of generalized topological groups, also is a generalized topological group. If G has the property that each of its subgroups mG ($m = 2, 3 \cdots$) is closed in G, the finiteness of each N_α implies that $H^q(N_\alpha, G)$ and hence $\mathcal{H}^q(X, G)$ is topological. If G does not have this property, it would still be possible to consider the group

$$\varprojlim H^q_t(N_\alpha, G) = \varprojlim [H^q(N_\alpha, G)/\bar{0}].$$

This group is always topological but its relation to the other groups is rather obscure.

In view of (37.6) the subgroups $Q^q(N_\alpha, G)$ of $H^q(N_\alpha, G)$ form an inverse system. We define

$$\mathcal{Q}^q(X, G) = \varprojlim Q^q(N_\alpha, G). \tag{39.3}$$

Clearly $\mathcal{Q}^q$ is a subgroup of $\mathcal{H}^q(X, G)$.

Similarly, in view of (37.7), the subgroups $\mathcal{P}_q(N_\alpha, G)$ of $\mathcal{H}_q(N_\alpha, G)$ form a direct system so we define

$$\mathcal{P}_q(X, G) = \varinjlim \mathcal{P}_q(N_\alpha, G). \tag{39.4}$$

$\mathcal{P}_q$ is a subgroup of $\mathcal{H}_q(X, G)$.

LEMMA 39.1. *The Kronecker index establishes a pairing of* $\mathcal{H}^q(X, G)$ *and* $\mathcal{H}_q(X, I)$ *with values in* G; *under this pairing*

$$\mathcal{Q}^q(X, G) = \text{Annih } \mathcal{H}_q(X, I).$$

LEMMA 39.2. *Let G be discrete and* $\hat{G} = \text{Char } G$. *The Kronecker index establishes a dual pairing of* $\mathcal{H}^q(X, \hat{G})$ *and* $\mathcal{H}_q(X, G)$ *with values in the group P of reals* mod 1; *under this pairing*

$$\mathcal{H}_q(X, G) \cong \text{Char } \mathcal{H}^q(X, \hat{G})$$

$$\mathcal{P}_q(X, G) = \text{Annih } \mathcal{Q}^q(X, \hat{G}).$$

Both lemmas have been established for each of the complexes N_α. The passage to the limit is possible in view of formula (37.1.)

In $\mathcal{H}_q(X, G)$ we also consider the subgroup $\mathcal{T}_q(X, G)$ of all elements of finite

[28] For more detail see Lefschetz [7]. Although the definition of the homology and cohomology groups given here is valid for any space X, it is well known that its interest is restricted to compact spaces only. This is due to the fact that only in compact spaces is the family of finite open coverings cofinal with the family of all open coverings.

order. Since each approximating group $\mathcal{H}_q(N_\alpha, G)$ has a finite set of generators, one can show, by arguments resembling those of §24, that

$$\mathcal{T}_q(X, G) = \underrightarrow{\text{Lim}}\ \mathcal{T}_q(N_\alpha, G).$$

40. Formulas for a general space

Using the formulas for complexes and applying a straightforward passage to the limit we obtain here some relations for $\mathcal{H}^q(X, G)$ and $\mathcal{H}_q(X, G)$ in terms of the groups $\mathcal{H}_q(X, I)$ and $\mathcal{T}_{q+1}(X, I)$. The results are not as complete as in the case of a complex.

THEOREM 40.1. *For a space X and a generalized topological coefficient group G the subgroup $\mathcal{Q}^q$ of the Čech homology group is expressible, in terms of a co-torsion group, as*

$$\mathcal{Q}^q(X, G) \cong \text{Ext}^* \{G, \mathcal{T}_{q+1}(X, I)\}, \tag{40.1}$$

while the corresponding factor group $\mathcal{H}^q(X, G)/\mathcal{Q}^q(X, G)$ is isomorphic to a subgroup of Hom $\{\mathcal{H}_q(X, I), G\}$.

If G/mG is compact and topological for $m = 2, 3, \cdots$ then

$$\mathcal{H}^q(X, G)/\mathcal{Q}^q(X, G) \cong \text{Hom } \{\mathcal{H}_q(X, I), G\}. \tag{40.2}$$

PROOF. For each nerve N_α we have (Theorem 32.1)

$$Q^q(N_\alpha, G) \cong \text{Ext } \{G, \mathcal{T}_{q+1}(N_\alpha, I)\}$$

The groups on either side form inverse systems and it follows from Theorem 38.1 and Lemma 20.2 that the limits of these systems are isomorphic,

$$\mathcal{Q}^q(X, G) \cong \underleftarrow{\text{Lim}}\ \text{Ext } \{G, \mathcal{T}_{q+1}(N_\alpha, I)\}.$$

However since $\mathcal{T}_{q+1}(X, I) = \underrightarrow{\text{Lim}}\ \mathcal{T}_{q+1}(N_\alpha, I)$ and the groups $\mathcal{T}_{q+1}(N_\alpha, I)$ are finite it follows from Theorem 24.2 that the limit on the right is Ext* $\{G, \mathcal{T}_{q+1}\}$. This proves formula (40.1).

From Theorem 32.1 we also have

$$H^q(N_\alpha, G)/Q^q(N_\alpha, G) \cong \text{Hom } \{\mathcal{H}_q(N_\alpha, I), G\},$$

and again the limits of the two inverse systems are isomorphic in view of Theorem 38.1. Consequently from Theorem 21.1 we get

$$\underleftarrow{\text{Lim}}\ [H^q(N_\alpha, G)/Q^q(N_\alpha, G)] \cong \text{Hom } \{\mathcal{H}_q(X, I), G\}.$$

Now it follows from (20.1) (Chap. IV) that the group

$$\mathcal{H}^q(X, G)/\mathcal{Q}^q(X, C) = \underleftarrow{\text{Lim}}\ H^q_\alpha/\underleftarrow{\text{Lim}}\ Q^q_\alpha$$

is isomorphic with a subgroup of the group $\underleftarrow{\text{Lim}}\ (H^q_\alpha/Q^q_\alpha)$. This proves the second assertion of the theorem. The subgroup will turn out to be the whole group whenever we are able to prove that $Q^q(N_\alpha, G)$ are compact topological groups.

Suppose now that G/mG is compact and topological for $m = 2, 3, \cdots$.

Given a cyclic group T of order $m \geqq 2$ we have Ext $\{G, T\} \cong G/mG$ (Corollary 11.2) and consequently Ext $\{G, T\}$ is compact and topological. It follows that Ext $\{G, T\}$ is compact and topological for every finite group T. In particular the groups

$$Q^q(N_\alpha, G) \cong \text{Ext } \{G, \mathcal{T}_{q+1}(N_\alpha, I)\}$$

are all compact and topological.

This completes the proof of the theorem. Notice that if G/mG is compact and topological for $m = 2, 3, \cdots$ then the group $\mathcal{Q}^q(X, G)$, as a limit of compact topological groups, is compact and topological.

If G is discrete and has no elements of finite order, or if $\mathcal{T}_{q+1}$ is countable, then by Theorem 24.4 and Corollary 24.1, the group Ext* in (40.1) may be replaced by Ext/Ext$_f$. In particular if $G = I$ then by Theorems 17.1 and 40.1,

(40.3) $$\mathcal{Q}^q(X, I) \cong \text{Char } \mathcal{T}_{q+1}(X, I),$$

(40.4) $$\mathcal{H}^q(X, I)/\mathcal{Q}^q(X, I) \cong \text{Hom } \{\mathcal{H}_q(X, I), I\}.$$

THEOREM 40.2. *The Čech homology group $\mathcal{H}^q(X, G)$ of a space X over a compact topological group G has a subgroup $\mathcal{Q}^q$, with factor group $\mathcal{H}^q/\mathcal{Q}^q$, both expressible in terms of integral cohomology groups of X as*

(40.5) $$\mathcal{Q}^q(X, G) \cong \text{Char Hom } \{G, \mathcal{T}_{q+1}(X, I)\},$$

(40.6) $$\mathcal{H}^q(X, G)/\mathcal{Q}^q(X, G) \cong \text{Hom } \{\mathcal{H}_q(X, I), G\}.$$

PROOF. From Theorem 40.1 we have $\mathcal{Q}^q \cong$ Ext* $\{G, \mathcal{T}_{q+1}\}$. However since G is compact topological we have Ext* $\{G, \mathcal{T}_{q+1}\} \cong$ Ext $\{G, \mathcal{T}_{q+1}\}$ (Corollary 24.1) and Ext $\{G, \mathcal{T}_{q+1}\} \cong$ Char Hom $\{G, \mathcal{T}_{q+1}\}$ (Theorem 15.1). This proves formula (40.5). We recall here that only continuous homomorphisms are considered. Formula (40.6) is a consequence of (40.2).

THEOREM 40.3. *The Čech cohomology groups $\mathcal{H}_q \supset \mathcal{P}_q$ of a space X over a discrete coefficient group G can be expressed, in part, in terms of the integral cohomology groups as*

(40.7) $$\mathcal{P}_q(X, G) \cong G \circ \mathcal{H}_q(X, I),$$

(40.8) $$\mathcal{H}_q(X, G)/\mathcal{P}_q(X, G) \cong \text{Hom } \{\text{Char } G, \mathcal{T}_{q+1}(X, I)\}.$$

PROOF. Let $\hat{G} = \text{Char } G$. Since $\mathcal{H}_q(X, G) \cong \text{Char } \mathcal{H}^q(X, \hat{G})$ and $\mathcal{P}_q =$ Annih $\mathcal{Q}^q(X, \hat{G})$ we have

$$\mathcal{P}_q(X, G) \cong \text{Char } [\mathcal{H}^q(X, \hat{G})/\mathcal{Q}^q(X, \hat{G})],$$

and using Theorems 40.2 and 18.1 we get

$$\mathcal{P}_q(X, G) \cong \text{Char Hom } \{\mathcal{H}_q(X, I), \text{Char } G\} \cong G \circ \mathcal{H}_q(X, I).$$

This formula could have been proved directly, passing to the limit with $\mathcal{P}_q(N_\alpha, G) \cong G \circ \mathcal{H}_q(N_\alpha, I)$. Since also $\mathcal{H}_q/\mathcal{P}_q \cong \text{Char } \mathcal{Q}^q(X, \text{Char } G)$, formula (40.8) is a consequence of Theorem 40.2.

The theorems and proofs carry over without change to the homology theory of X modulo a closed subset. Another generalization can be obtained by replacing the space X by a net of complexes, as defined by Lefschetz ([7] Ch. VI).

We are unable to answer the question whether $\mathcal{Q}^q(X, G)$ and $\mathcal{P}_q(X, G)$ are direct factors of $\mathcal{H}^q(X, G)$ and $\mathcal{H}_q(X, G)$. This is why we do not obtain expressions for $\mathcal{H}^q(X, G)$ and $\mathcal{H}_q(X, G)$ in terms of $\mathcal{H}_q(X, I)$ and $\mathcal{T}_{q+1}(X, I)$. The best we achieve in the case of a general space X is a description of the subgroups $\mathcal{Q}^q$ and $\mathcal{P}_q$ and of the corresponding factor groups, leaving the direct product proposition undecided.[29]

In the following sections of this chapter we shall discuss the case when X is a compact metric space, using the method of the fundamental complex. In this case we are able to obtain complete results, including the direct product decomposition.

41. The case $q = 0$

Before we proceed with the treatment of compact metric spaces we will discuss some details connected with the definition of the homology and cohomology groups for the dimension zero.

Let K be a finite simplicial complex. If we assume that there are no cells of dimension less than zero then every 0-chain will be a 0-cycle and the groups $H^0(K, G)$ and $\mathcal{H}_0(K, G)$ will be isomorphic to the product of G by itself n times, n being the number of components of K.

An alternate procedure is to consider K "augmented" by a single (-1)-cell σ^{-1} such that $[\sigma_i^0:\sigma^{-1}] = 1$ for all σ_i^0. In this case, given a 0-chain $c^0 = \sum g_i\sigma_i^0$, we have $\partial c^0 = (\sum g_i)\sigma^{-1}$ and consequently c^0 is a cycle if and only if $\sum g_i = 0$. The cohomology group gets affected also because the cocycle $\sum \sigma_i^0$ that was not a coboundary in the first approach is a coboundary in the augmented complex, since $\delta\sigma^{-1} = \sum \sigma_i^0$. It turns out that $H^0(K, G)$ and $\mathcal{H}_0(K, G)$ are isomorphic to the product of G by itself $n - 1$ times.

In defining the groups $\mathcal{H}^0(X, G)$ and $\mathcal{H}_0(X, G)$ for a space we again have two alternatives according as the nerves N_α are augmented or not.

Both the augmented and unaugmented complexes are abstract complexes in the sense of Ch. V and therefore all our previous results hold for either definition of $\mathcal{H}^0$ and $\mathcal{H}_0$. However in the discussion of compact metric spaces that follows there is an advantage in considering the nerves as augmented complexes, so as to have $\mathcal{H}^0(X, G) = \mathcal{H}_0(X, G) = 0$ if X is a connected space.

42. Fundamental complexes

Let X be a compact metric space. There is then a sequence U_n $(n = 0, 1, \cdots)$ of finite open coverings of X such that U_n is a refinement of U_{n-1} and every finite

[29] Steenrod [9] §10 brings an argument, which if correct would settle the question positively. Unfortunately an error occurs on p. 681, line 5. The error was noticed by C. Chevalley, who has also constructed an example showing that the argument could not be corrected in the general case. If X is metric compact, Steenrod's argument can be corrected to give the desired direct product decomposition (see §44 below).

open covering of X has some U_n as a refinement. This last property asserts that in the directed family of all the finite open coverings of X the sequence $\{U_n\}$ constitutes a cofinal subfamily and therefore the Čech homology and cohomology group can be equivalently defined using only the sequence of coverings U_n. We shall assume that U_0 is a covering consisting of only one set, namely X itself, so that the nerve N_0 of U_0 is a vertex. For each n we select a projection $T_n: N_n \rightarrow N_{n-1}$ of the nerve of U_n into the nerve of U_{n-1}. The projections $N_n \rightarrow N_{n-k}$ we define by transitivity.

We now define the fundamental complex K of X as follows. The complexes N_n for $n = 0, 1, \cdots$ shall be disjoint subcomplexes of K. For each $n = 1, 2, \cdots$ and each simplex σ^q of N_n we introduce a new $(q + 1)$-cell $\mathcal{P}\sigma^q$ whose boundary is $T_n\sigma^q - \sigma^q - \mathcal{P}\partial\sigma^q$. This formula gives a recursive definition of the incidence numbers.

In order to give a more intuitive picture of K we may consider each of the nerves N_n as a geometric simplicial complex, the projection T_n can then be regarded as a continuous simplicial transformation; that is, as linear on every simplex σ^q of N_n, while $\mathcal{P}\sigma^q$ can be visualized as a deformation prism consisting of intervals joining each point of σ^q with its image under T_n. With this interpretation K becomes a geometric complex and the cells $\mathcal{P}\sigma^q$ can be subdivided so as to furnish a simplicial subdivision of K. It is clear from this picture that K can be contracted to a point, namely by moving every point up its projection lines towards the vertex N_0.

The complex K is countable and is locally finite; i.e., both closure and star finite. Viewing K as a closure finite complex, we can define finite cycles and infinite cocycles. However, since K is contractible all the homology group with finite cycles will vanish. Using the results of Ch. V we conclude that the cohomology groups with infinite cocycles also will vanish. Consequently, regarded as a closure finite complex, the structure of K is trivial. If we approach K as a star finite complex we obtain cohomology groups with finite cocycles and homology groups with infinite cycles. Regarded this way the complex K furnishes a true picture of the combinatorial structure of the space X.

43. Relations between a space and its fundamental complex

THEOREM 43.1. *The compact metric space X and its fundamental complex K are linked by isomorphisms*

$$\mathcal{H}^q(X, G) \cong H_w^{q+1}(K, G), \tag{43.1}$$

$$\mathcal{H}_q(X, G) \cong \mathcal{H}_{q+1}(K, G). \tag{43.2}$$

We shall restrict ourselves here to indicate the definitions of the isomorphisms without going into the complete proof, which involves lengthy but straightforward calculations.[30]

Let $\mathbf{z}^q$ be an element of $\mathcal{H}^q(X, G)$. Then $\mathbf{z}^q$ can be represented by a sequence

[30] This proof is closely related to one given by Steenrod; see [10], §4.

of cycles $z_n^q \in Z^q(N_n, G)$ such that $z_{n-1}^q - T_n z_n^q \in B^q(N_{n-1}, G)$. For each $n = 1, 2, \cdots$ select a chain c_{n-1}^{q+1} in N_{n-1} such that

$$\partial c_{n-1}^{q+1} = z_{n-1}^q - T_n z_n^q,$$

and consider the chain

$$z^{q+1} = \sum_{n=1}^{\infty} c_{n-1}^{q+1} + \sum_{n=1}^{\infty} \mathfrak{D} z_n^q.$$

We verify that

$$\partial z^{q+1} = \sum_{n=1}^{\infty} (z_{n-1}^q - T_n z_n^q) + \sum_{n=1}^{\infty} (T_n z_n^q - z_n^q - \mathfrak{D}\partial z_n^q)$$

$$= z_0^q - \sum_{n=1}^{\infty} \mathfrak{D}\partial z_n^q = 0,$$

since $\partial z_n^q = 0$, while $z_0^q = 0$ for $q \geqq 0$, $z_0^0 = 0$ by §41. Consequently z^{q+1} is a cycle of K. If instead of $\{c_n^{q+1}\}$ we use a sequence $\{\bar{c}_n^{q+1}\}$ to define a cycle $\bar{z}^{q+1}$, then

$$z^{q+1} - \bar{z}^{q+1} = \sum_{n=1}^{\infty} (c_{n-1}^{q+1} - \bar{c}_{n-1}^{q+1})$$

Each term $c_{n-1}^{q+1} - \bar{c}_{n-1}^{q+1}$ is a finite cycle and therefore bounds in K, therefore $z^{q+1} - \bar{z}^{q+1}$ is a weakly bounding cycle and z^{q+1} determines uniquely an element $\mathbf{z}^{q+1} \in H_w^{q+1}(K, G)$. We define

$$\phi(\mathbf{z}^q) = \mathbf{z}^{q+1}.$$

Now let $\mathbf{w}^q \in \mathcal{H}_q(X, G)$. The element $\mathbf{w}^q$ can be represented for suitable n by a single cocycle $w^q \in \mathcal{Z}_q(N_n, G)$. We verify that $\mathfrak{D}w^q$ is then a $(q + 1)$-cocycle of K. Using the formula

$$\delta w^q = \mathfrak{D} T_n^* w^q - \mathfrak{D} w^q \text{ in } K,$$

and the fact that $\mathfrak{D}$ and δ commute we show that $\mathfrak{D}w^q$ determines uniquely an element $\mathbf{w}^{q+1}$ of $\mathcal{H}_q(K, G)$. We define

$$\psi(\mathbf{w}^q) = \mathbf{w}^{q+1}.$$

We also notice that the pair of isomorphisms ϕ, ψ preserves the Kronecker index

(43.3) $$\phi(\mathbf{z}^q) \cdot \psi(\mathbf{w}^q) = \mathbf{z}^q \cdot \mathbf{w}^q.$$

If X_0 is a closed subset of X then every covering U_n of X determines a covering of X_0 whose nerve L_n is a subcomplex of the nerve N_n of U_n. The subcomplex

$$L = \sum_{n=1}^{\infty} L_n + \sum_{n=1}^{\infty} \mathfrak{D} L_n$$

of K is then a fundamental complex of X_0. The isomorphisms (43.1) and (43.2) of Theorem 43.1 can be generalized as follows

$$\mathcal{H}^q(X \bmod X_0, G) \cong H_w^{q+1}(K \bmod L, G) \tag{43.1'}$$

$$\mathcal{H}_q(X - X_0, G) \cong \mathcal{H}_{q+1}(K - L, G). \tag{43.2'}$$

44. Formulas for a compact metric space

Using the fundamental complex and the results of Ch. V we shall now establish theorems for a compact metric space quite analogous to the ones proved for a complex in Ch. V.

THEOREM 44.1. *The Čech homology groups of a compact metric space X over a generalized topological coefficient group G can be expressed in terms of the integral cohomology groups $\mathcal{H}_q = \mathcal{H}_q(X, I)$, $\mathcal{T}_{q+1} = \mathcal{T}_{q+1}(X, I)$ as*

$$\mathcal{H}^q(X, G) \cong \operatorname{Hom}\{\mathcal{H}_q, G\} \times (\operatorname{Ext}\{G, \mathcal{T}_{q+1}\}/\operatorname{Ext}_f\{G, \mathcal{T}_{q+1}\}).$$

More precisely, in terms of the subhomology group $\mathcal{Q}^q$ of (39.3) *we have*

$$\mathcal{Q}^q(X, G) \textit{ is a direct factor of } \mathcal{H}^q(X, G), \tag{44.1}$$

$$\mathcal{Q}^q(X, G) \cong \operatorname{Ext}\{G, \mathcal{T}_{q+1}\}/\operatorname{Ext}_f\{G, \mathcal{T}_{q+1}\}, \tag{44.2}$$

$$\mathcal{H}^q(X, G)/\mathcal{Q}^q(X, G) \cong \operatorname{Hom}\{\mathcal{H}_q, G\}. \tag{44.3}$$

To prove the theorem we use the fact that the Kronecker intersection is preserved under the pair of isomorphisms ϕ, ψ of the previous section. Consequently, since

$$\mathcal{Q}^q(X, G) = \operatorname{Annih} \mathcal{H}_q(X, I) \text{ in } \mathcal{H}^q(X, G),$$
$$Q_w^{q+1}(K, G) = \operatorname{Annih} \mathcal{H}_{q+1}(K, I) \text{ in } H_w^{q+1}(K, G),$$

we have

$$\phi[\mathcal{Q}^q(X, G)] = Q_w^{q+1}(K, G),$$

and the theorem becomes a consequence of Theorems 43.1 and 32.2.

THEOREM 44.2. *The Čech cohomology groups of a compact metric space X with coefficients in a discrete group G can be expressed in terms of the integral cohomology groups $\mathcal{H}_q = \mathcal{H}_q(X, I)$, $\mathcal{T}_{q+1} = \mathcal{T}_{q+1}(X, I)$ as*

$$\mathcal{H}_q(X, G) \cong (G \circ \mathcal{H}_q) \times \operatorname{Hom}\{\operatorname{Char} G, \mathcal{T}_{q+1}\}.$$

More precisely, in terms of the subgroup $\mathcal{P}_q$ of (39.4), *we have*

$$\mathcal{P}_q(X, G) \textit{ is a direct factor of } \mathcal{H}_q(X, G), \tag{44.4}$$

$$\mathcal{P}_q(X, G) \cong G \circ \mathcal{H}_q, \tag{44.5}$$

$$\mathcal{H}_q(X, G)/\mathcal{P}_q(X, G) \cong \operatorname{Hom}\{\operatorname{Char} G, \mathcal{T}_{q+1}\}. \tag{44.6}$$

To prove the theorem we notice that

$$\mathcal{P}_q(X, G) = \text{Annih } \mathcal{Q}^q(X, \text{Char } G) \text{ in } \mathcal{H}_q(X, G),$$

$$\mathcal{P}_{q+1}(K, G) = \text{Annih } Q_w^{q+1}(K, \text{Char } G) \text{ in } \mathcal{H}_{q+1}(K, G),$$

and therefore

$$\psi[\mathcal{P}_q(X, G)] = \mathcal{P}_{q+1}(K, G)$$

and the theorem becomes a consequence of Theorems 43.1 and 33.1.

All these results remain valid for the homologies of X modulo a closed subset.

We now proceed to compare the results obtained here for the metric compact case with the results of §40 concerning general spaces.

Statements (44.1) and (44.4) contain a positive solution for the direct product problem which is still unsolved for the general space. Formula (44.3) was proved in (40.2) for general spaces only under the additional condition that G/mG be compact and topological for $m = 2, 3, \cdots$. Formula (44.2) was proved for general spaces under the form

$$\mathcal{Q}^q(X, G) \cong \text{Ext}^* \{G, \mathcal{T}_{q+1}(X, I)\}$$

which is equivalent to (44.2) because

$$\text{Ext}^* \{G, T\} \cong \text{Ext } \{G, T\}/\text{Ext}_f \{G, T\}$$

for countable groups T with only elements of finite order (Theorem 24.4) and the group $\mathcal{T}_{q+1}(X, I) \cong \mathcal{T}_{q+2}(K, I)$ is countable for a compact metric X, since K is countable.

Formulas (44.5) and (44.6) coincide with the ones proved in Theorem 40.3 for a general space.

45. Regular cycles

Using the concept of a "regular cycle" Steenrod ([10]) has defined a new homology group $H^q(X, G)$ of "regular" cycles, for a compact metric space X. This group is useful especially in the case when X is a subset of the n-sphere S^n, because it provides information about the structure of the open set $S^n - X$.

Steenrod ([10], Theorem 7) has proved that if K denotes a fundamental complex of X then

(45.1) $$H^q(X, G) \cong H^q(K, G).$$

From this, using Theorems 43.1 and 32.1 we derive the formula

(45.2) $$H^q(X, G) \cong \text{Hom } \{\mathcal{H}_{q-1}(X, I), G\} \times \text{Ext } \{G, \mathcal{H}_q(X, I)\},$$

for $q > 0$. This formula expresses $H^q(X, G)$ in terms of $\mathcal{H}_{q-1}(X, I)$ and $\mathcal{H}_q(X, I)$ and hence shows that, essentially, $H^q(X, G)$ is no *new* invariant.

Let us specialize formula (45.2), assuming that $q = 1$, and that X is connected. We have then $\mathcal{H}_0(X, I) = 0$ and therefore

(45.3) $$H^1(X, G) \cong \text{Ext } \{G, \mathcal{H}_1(X, I)\}.$$

Let us further assume $G = I$ and that X is one of the solenoids Σ. Since Σ is a connected, compact abelian group we have $H^1(\Sigma, P) \cong \Sigma$ (Steenrod [9], Theorem 15) where P (Steenrod's $\mathfrak{X}$) is the group of reals mod 1. Further, since Char $I \cong P$ we have $H_1(\Sigma, I) \cong \text{Char } H^1(\Sigma, P) \cong \text{Char } \Sigma$. Hence finally

(45.4) $$H^1(\Sigma, I) \cong \text{Ext } \{I, \text{Char } \Sigma\}.$$

This group will be explicitly computed in Appendix B; it was the starting point of this investigation (see introduction).

Steenrod has defined a subgroup $\tilde{H}^q(X, G)$ of $H^q(X, G)$ by considering regular cycles that are sums of finite cycles. He has also proved that under the isomorphism (45.1) this group is mapped onto the subgroup $B^q_w(K, G)/B^q(K, G)$ of $H^q(K, G)$.

We shall now show that, for $q > 1$,

(45.5) $$\tilde{H}^q(X, G) \cong \text{Ext}_f \{G, \mathcal{H}_q(X, I)\}.$$

(45.6) $$H^q(X, G)/\tilde{H}^q(X, G) \cong \mathcal{H}^{q-1}(X, G).$$

In fact, from Theorems 31.3 and 43.1 we deduce that $B^q_w(K, G)/B^q(K, G) \cong \text{Ext}_f \{G, \mathcal{H}_{q+1}(K, I)\} \cong \text{Ext}_f \{G, \mathcal{H}_q(X, I)\}$. This proves (45.5). In order to prove (45.6) notice that $H^q(X, G)/\tilde{H}^q(X, G) \cong H^q(K, G)/[B^q_w(K, G)/B^q(K, G)] \cong H^q_w(K, G) \cong \mathcal{H}^{q-1}(X, G)$.

Formulas (45.5) and (45.6) provide a splitting of $H^q(X, G)$ different from the one used in (45.2). The isomorphism (45.6) was established by Steenrod [10], who has also shown that $\tilde{H}^q$ can be computed using G and $\mathcal{H}_q(X, I)$, without however getting the explicit formula (45.5).

From (45.5) we immediately deduce the theorem of Steenrod that $\tilde{H}^q(X, G) = 0$ and $H^q(X, G) \cong \mathcal{H}^{q-1}(X, G)$ whenever $\mathcal{H}_q(X, I)$ has a finite number of generators.

Appendix A. Coefficient Groups with Operators

In many topological investigations it is convenient to construct homology groups $H^q(K, G)$ in cases when G is not just a group, but a ring or even a field. More generally, G can be allowed to be a group with operators. We show here that our results extend unchanged to such cases, and in particular, that the resulting homology groups are still completely determined by the integral cohomology groups.

G is called a *group with operators* Ω if G is a generalized topological group, Ω a space, and if to each element $\omega \in \Omega$ and each $g \in G$ there is assigned an element $\omega g \in G$ (the result of operating on g with ω), in such wise that

(i) ωg is a continuous function of the pair (ω, g),

(ii) $\omega(g_1 + g_2) = \omega g_1 + \omega g_2 \qquad (g_1, g_2 \in G)$.

It then follows that each element ω determines a (continuous) homomorphism $g \to \omega g$ of G into G; however, distinct elements of Ω need not determine distinct homomorphisms. The set Ω may have a discrete topology, or may even consist of just one operator ω.

If both G_1 and G_2 have operators Ω, a homomorphism (or isomorphism) ϕ of G_1 into G_2 is said to be *Ω-allowable* if $\phi[\omega g_1] = \omega[\phi g_1]$ for all $g_1 \in G_1$, $\omega \in \Omega$.

If G has operators Ω, a subgroup $S \subset G$ is said to be *allowable* if $\omega(S) \subset S$ for all $\omega \in \Omega$. The operators Ω may then be applied in natural fashion to the factor group G/S, by setting $\omega(g + S) = \omega g + S$. Then G/S is a group with operators Ω, and the natural homomorphism of G on G/S is allowable.[31]

If G is a group with operators Ω, the various groups introduced as functions of G in Chapters I–IV are also groups with operators. Specifically, let H be a discrete group, and for each $\theta \in \text{Hom}\,\{H, G\}$ define $\omega\theta$ as $[\omega\theta](h) = \omega[\theta(h)]$. Then $\omega\theta \in \text{Hom}\,\{H, G\}$, and

(A.1) $\text{Hom}\,\{H, G\}$ has operators Ω.

Furthermore, if $H = F/R$, where F is free, the groups $\text{Hom}\,\{F \mid R, G\}$ and $\text{Hom}_f\,\{R, G; F\}$ are allowable subgroups of $\text{Hom}\,\{R, G\}$, so

(A.2) $\text{Hom}\,\{R, G\}/\text{Hom}\,\{F \mid R, G\}$ has operators Ω.

Again, let f be a factor set of H in G, and define another factor set ωf by taking $[\omega f](h, k)$ as $\omega[f(h, k)]$. Then Ω becomes a space of operators for the group $\text{Fact}\,\{G, H\}$. Furthermore $\text{Trans}\,\{G, H\}$ is an allowable subgroup; therefore

(A.3) $\text{Ext}\,\{G, H\}$ has operators Ω.

In similar fashion one concludes that $\text{Ext}_f\,\{G, H\}$ and Ext/Ext_f have operators Ω.

As another case, take $\phi \in \text{Hom}\,\{G, H\}$ and define a homomorphism $\omega\phi \in \text{Hom}\,\{G, H\}$ by setting $[\omega\phi](g) = \phi[\omega(g)]$ for each $g \in G$. If G is compact or discrete, one may show that $\omega\phi$ is a continuous function of ω and ϕ. In this case, and for any generalized topological group H,

(A.4) $\text{Hom}\,\{G, H\}$ has operators Ω.

In particular, if G is discrete or compact,

(A.5) $\text{Char}\, G$ has operators Ω.

Given these interpretations of all our basic groups as groups with operators, we next demonstrate that the various isomorphisms between these groups, as established in Chapters II–IV, are allowable. In particular, an inspection of the construction used to establish the fundamental Theorem 10.1 of Chapter II proves

(A.6) *The isomorphism*

$$\text{Ext}\,\{G, H\} \cong \text{Hom}\,\{R, G\}/\text{Hom}\,\{F \mid R, G\},$$

where $H = F/R$, F free, is allowable.

[31] Practically all the elementary formal facts about groups and homomorphisms apply to operator groups and allowable homomorphisms.

The same conclusion holds for the other isomorphisms stated in that theorem. Also, the isomorphism Ext $\{G, H\} \cong$ Char Hom $\{G, H\}$ established in Theorem 15.1 for compact topological G and discrete H is allowable. The proof of this fact depends essentially on showing that the "trace" used in that theorem has the commutation property,

$$t(\omega\theta, \phi) = t(\theta, \omega\phi), \text{ for any } \theta \in \text{Hom } \{R, G\}, \text{ and } \phi \in \text{Hom } \{G, H\}.$$

The allowability of the other isomorphisms in Chapters II–IV is similarly established. The proofs are closely analogous to the "naturality" proofs of §12, except that here the operators apply to G, while in §12 the operator T applied to H.

Now turn to the homology groups. Let c^q be a chain in the star finite complex K, with coefficients chosen in the group G with operators Ω. For each $\omega \in \Omega$, define

$$\omega(c^q) = \omega(\sum_i g_i\sigma_i^q) = \sum_i (\omega g_i)\sigma_i^q;$$

since the result is a chain, and since the requisite continuity holds, the group $C^q(K, G)$ of q-chains has operators Ω. Moreover, $\omega\partial = \partial\omega$, so that both $Z^q(K, G)$ and $B^q(K, G)$ are allowable subgroups of C^q. Therefore

(A.7) $\qquad H^q(K, G)$ has operators Ω.

The essential tool in establishing the isomorphisms of Chap. V is the Kronecker index $c^q \cdot d^q$ for $d^q \in \mathcal{C}_q(K, I)$, $c^q \in C^q(K, G)$. We verify at once that

(A.8) $$\omega(c^q \cdot d^q) = (\omega c^q) \cdot d^q \qquad (\text{all } \omega \in \Omega).$$

Since the subgroup A^q of Z^q was defined as a certain annihilator under this Kronecker index (see (29.9)), it follows at once that A^q is an allowable subgroup of Z^q. Furthermore, the proof that A^q is a direct factor of C^q depended on a decomposition of $\mathcal{C}_q$ as a direct product $\mathcal{C}_q = \mathcal{Z}_q \times \mathcal{D}_q$, for a suitably chosen group $\mathcal{D}_q$. In the notation of Lemma 16.2, we then had, by means of the Kronecker index (see the proof of Theorem 30.3)

$$C^q \cong \text{Hom } \{\mathcal{C}_q, G\} \cong \text{Hom } \{\mathcal{C}_q, G; \mathcal{D}_q, 0\} \times \text{Hom } \{\mathcal{C}_q, G; \mathcal{Z}_q, 0\}.$$

On the right both factors are allowable subgroups, and the isomorphism to the direct product is allowable;[32] furthermore, the second factor is the one which corresponds to the subgroup A^q of C^q. Therefore C^q has a representation of the form $C^q \mp A^q \times D^q$, where D^q is an allowable subgroup, complementary to A^q. A similar decomposition holds for Z^q and thus for its factor group $H^q = Z^q/B^q$. In terms of the homology subgroup $Q^q = A^q/B^q$ determined by A^q, this proves

(A.9) *The isomorphism* $H^q \cong (H^q/Q^q) \times Q^q$ *is allowable.*

The further analysis of these two factors, as carried out in Chapter V, all depended on the Kronecker index. In view of the property (A.8) of this index,

[32] If A and B are two groups with operators Ω the direct product $A \times B$ has operators Ω defined by $\omega(a, b) = (\omega(a), \omega(b))$ for $\omega \in \Omega$.

and the property (A.6) of the basic group-extension theorem, we have

(A.10) *The isomorphisms*

$$H^q(K, G)/Q^q(K, G) \cong \text{Hom } \{\mathcal{H}_q, G\},$$
$$Q^q(K, G) \cong \text{Ext } \{G, \mathcal{H}_{q+1}\}$$

are allowable, as is the isomorphism $H^q \cong \text{Hom} \times \text{Ext}$, *obtained by combining* (A.9) *and* (A.10).

Similar remarks apply to the representation of the "weak" homology group H^q_w (Theorem 32.2), which is a factor group of H^q by an allowable subgroup. The same holds for the topologized homology group H^q_t (i.e., the isomorphisms of Theorem 34.2 are allowable), for in any topological group G with operators Ω, the continuity of the operators insures that the subgroup $\bar{0} \subset G$ is allowable (recall that $H^q_t = H^q/\bar{0}$).

Turn next to the analysis of the cohomology groups. The groups $\mathcal{C}_q(K, G)$, with G discrete, again have operators in Ω, under the natural definition. As in the case of the homology groups, we have

(A.11) $\mathcal{H}_q(K, G)$ has operators Ω.

The representation of these groups depended on duality; i.e., on the Kronecker index $c^q \cdot d^q$, for $c^q \in Z^q(K, \text{Char } G)$, $d^q \in Z^q(K, G)$. Given the various definitions of the effect of an operator ω, one shows easily that

$$(\omega c^q) \cdot d^q = c^q \cdot (\omega d^q) \qquad \text{(all } \omega \in \Omega).$$

From this formula one may deduce that the well known isomorphism $\mathcal{H}_q(K, G) \cong \text{Char } H^q(K, \text{Char } G)$ is allowable. Thence it follows that the isomorphisms of Theorem 33.1 representing $\mathcal{H}_q$ are allowable.

These considerations yield the following

ADDENDUM TO THE UNIVERSAL COEFFICIENT THEOREM. *If K is any star finite complex, G a group with operators Ω, then the homology groups of K (and, if G is discrete, the finite cohomology groups) with coefficients in G all have operators Ω. All these groups with their operators are determined by the group G (with its operators) and the cohomology groups of the finite integral cocycles of K.*

A similar discussion applies to the results of Chap. VI.

In many important cases the operators form a ring (or even a field). Let us assume then that Ω is a generalized topological ring; that is, a ring which is a generalized topological group under addition and in which the multiplication is continuous.. Then G is called an *Ω-modulus* if G is a generalized topological group with operators Ω (i.e., conditions (i) and (ii) above hold) such that

(iii) $$(\omega_1\omega_2)g = \omega_1(\omega_2 g). \qquad \text{(for } \omega_i \in \Omega, g \in G),$$

(iv) $$(\omega_1 + \omega_2)g = \omega_1 g + \omega_2 g \qquad \text{(for } \omega_i \in \Omega, g \in G).$$

In other words, addition and multiplication of operators are determined in the natural fashion from G.

If the standard coefficient group G is now assumed to be an Ω-modulus, simple arguments will show that all the groups with operators Ω as described above are in fact Ω-moduli. Since the basic isomorphisms are still Ω-allowable, we conclude that the addendum to the universal coefficient group theorem still holds in these circumstances.

It is sometimes convenient to use a set Ω of operators in which only the addition or only the multiplication of operators is defined. More generally, we may consider a space Ω in which only certain sums $\omega_1 + \omega_2$ and products $\omega_1\omega_2$ are defined (and continuous); we then require that conditions (iii) and (iv) above hold only when the terms $\omega_1\omega_2$ or $\omega_1 + \omega_2$ are defined. The derived groups satisfy similar assumptions, and the universal coefficient theorem still holds.

If the coefficient group G is locally compact, one can always take the operators to form a ring, for any such group G has its endomorphism ring Ω_G as a natural ring of operators. Specifically, Ω_G is the additive group Hom $\{G, G\}$ of endomorphisms of G, with its usual topology (§3), and the multiplication $\omega_1\omega_2$ of two endomorphisms is defined by (iii) above. The requisite continuity properties of $\omega_1\omega_2$ and ωg are readily established, in virtue of the local compactness of G. Furthermore, if Ω is any other space of operators on G, each $\omega \in \Omega$ determines uniquely an endomorphism $\bar{\omega} \in \Omega_G$ with $\bar{\omega}g = \omega g$ for each g. The correspondence $\omega \to \bar{\omega}$ is a continuous mapping of Ω into Ω_G which preserves whatever sums and products may be present in Ω (assumed to satisfy (iii) and (iv)). Thus, any group, derived from G, which is an Ω_G-modulus is also a group with operators Ω, and any isomorphism between groups which is Ω_G-allowable is Ω-allowable. This indicates, that, for locally compact groups, one may restrict attention to operators of the ring Ω_G.

The most useful case is that in which the coefficient group is a field F, which is its own ring of operators. In this case all the homology groups, groups of homomorphisms, etc., become F-moduli; that is, vector spaces over F.

All these remarks suggest the following rather negative conclusion: *although in many applications it is convenient to consider a homology theory over coefficients which form more than merely a group, no new topological invariants can be so obtained.*

Appendix B. Solenoids

Here we compute the one-dimensional homology group $H^1(\Sigma, I)$ of regular cycles for the solenoid[33] Σ, or the isomorphic group Ext $\{I, \text{Char } \Sigma\}$ (see (45.4)).

A solenoid is uniquely determined by a Steinitz G-number; that is, by a formal (infinite) product $G = \prod p_i^{e_i}$ of distinct primes with exponents e_i which are non-negative integers or ∞. Any such number G can be represented (in many ways) as a formal product $G = a_1a_2 \cdots a_n \cdots$ of ordinary integers a_i; if G is not an ordinary integer, we can take each $a_i \geqq 2$. Given such a representation of G, take replicas P_n of the additive group P of real numbers modulo 1, and let ϕ_n be the homomorphism which wraps P_n a_{n-1} times around P_{n-1}. The

[33] Solenoids were studied by L. Vietoris, Math. Annalen 97 (1927), p. 459, and more in detail by D. van Dantzig, Fundam. Math. 15 (1930), pp. 102–135. See also L. Pontrjagin [8], p. 171.

P_n then form an inverse system of groups, relative to the homomorphisms $\psi_{n+m,n} = \phi_{n+1} \cdots \phi_{n+m}$, and the solenoid Σ_G is defined as the limit $\Sigma_G = \underleftarrow{\text{Lim}}\, P_n$. Therefore Char $\Sigma_G = \underrightarrow{\text{Lim}}$ Char P_n, where the groups form a direct system under the dual correspondences ϕ_n^*. Here Char P_n is an isomorphic replica I_n of the additive group of integers, and ϕ_n^* maps I_n into I_{n+1} by multiplying each $x \in I_n$ by a_n. Therefore Char $\Sigma_G = \underrightarrow{\text{Lim}}\, I_n$ is a subgroup N_G of the additive group of rational numbers, consisting of all rationals of the form a/d_n, with a an integer and $d_n = a_1 \cdots a_{n-1}$. Alternatively, N_G consists of all rationals r/s with s a "divisor" of G; hence N_G and Σ_G are uniquely determined by G, and are independent of the representation $G = a_1 a_2 \cdots a_n \cdots$.

A Steinitz G-number which is not an ordinary integer also determines a certain topological ring. Set $G = a_1 a_2 \cdots a_n \cdots$, $d_n = a_1 \cdots a_{n-1}$. In the ring I of integers, introduce as neighborhoods of zero the sets (d_n) of all multiples of d_n. Since the intersection of all these (d_n) is the zero element of I, these neighborhoods make I a topological ring. It can be embedded in a unique fashion in a minimal complete topological ring $I_G \supset I$, so that every element of I_G is a limit of a sequence of integers, under the given topology. This is one of the b_ν-adic rings introduced by D. van Dantzig.[34] The additive group of I_G can be alternatively described as a limit of an inverse sequence; specifically, the factor group $I/(d_{n+1})$ has a natural homomorphism into $I/(d_n)$, and the limit group is $I_G \cong \underleftarrow{\text{Lim}}\, I/(d_n)$. In the special case when $G = p^\infty$ is an infinite power of a prime p, I_G is the ordinary ring of p-adic integers.

THEOREM. *If G is any Steinitz G-number which is not an ordinary integer, Σ_G the corresponding solenoid, and I_G the corresponding complete ring containing the ring I of integers, then*

$$\text{(B.1)} \qquad \text{Ext}\,\{I, \text{Char}\,\Sigma_G\} \cong I_G/I.$$

PROOF. As above, Char Σ_G is a group N_G of rationals, generated by the numbers $r_n = 1/d_n$ with relations $a_n r_{n+1} = r_n$. Therefore N_G can be represented as F/R, where F is a free group with generators $z_1, z_2, \cdots$, and R the subgroup with generators $y_n = a_n z_{n+1} - z_n$, $n = 1, 2, \cdots$. By the fundamental theorem on group extensions

$$\text{(B.2)} \qquad \text{Ext}\,\{I, \text{Char}\,\Sigma_G\} \cong \text{Hom}\,\{R, I\}/\text{Hom}\,\{F \mid R, I\}.$$

Let $\theta \in \text{Hom}\,\{R, I\}$ and set

$$x(\theta) = \lim_{n \to \infty} [\theta y_1 + d_2 \theta y_2 + \cdots + d_n \theta y_n].$$

Then $x(\theta)$ is a well-defined element of I_G, and $\theta \to x(\theta)$ is a homomorphic mapping of Hom $\{R, I\}$ into I_G and thus, derivatively, into I_G/I. We assert that the kernel of the latter mapping is Hom $\{F \mid R, I\}$.

Assume first that $\theta \in \text{Hom}\,\{F \mid R, I\}$, and let θ^* be an extension of θ to F. Then

$$\theta(y_1 + d_2 y_2 + \cdots + d_n y_n) = -\theta^* z_1 + d_{n+1} \theta^* z_{n+1},$$

[34] Math. Annalen 107 (1932), pp. 587–626; Compositio Math. 2 (1935), pp. 201–223.

so that the limit $x(\theta)$ is $-\theta^* z_1$, which is an integer in I. Conversely, suppose that $x(\theta) \in I$, and set $x(\theta) = -c_1$. We then have

$$\theta y_1 + d_2\theta y_2 + \cdots + d_n\theta y_n \equiv -c_1 \pmod{d_{n+1}}.$$

By successive applications of this condition we find integers c_n with $\theta y_n = a_n c_{n+1} - c_n$. The homomorphism $\theta^* z_n = c_n$ then provides an extension of θ to F, so that $\theta \in \mathrm{Hom}\,\{F \mid R, I\}$.

Every element in I_G is a limit of integers, hence has the form Lim $[b_1 + d_2 b_2 + \cdots + d_n b_n]$; therefore $\theta \to x(\theta)$ is a mapping onto I_G. We thus have

$$\text{(B.3)} \qquad \mathrm{Hom}\,\{R, I\}/\mathrm{Hom}\,\{F \mid R, I\} \cong I_G/I.$$

The correspondence is topological, as one may readily verify that both (generalized topological) groups carry the trivial topology in which the only open sets are zero and the whole space. Thus (B.2) and (B.3) prove the isomorphism (B.1).

By cardinal number considerations, one shows that the group I_G/I is uncountable, hence not void. The formula (B.1) gives at once all the special properties of the homology group of the solenoid, as found by Steenrod [10] in his partial determination of this group.

THE UNIVERSITY OF MICHIGAN
HARVARD UNIVERSITY

BIBLIOGRAPHY

[1] ALEXANDROFF, P. AND HOPF, H. *Topologie*, vol. I, Berlin, 1935.
[2] BAER, R. *Erweiterungen von Gruppen und ihren Isomorphismen*, Math. Zeit. 38 (1934), pp. 375–416.
[3] ČECH, E. *Les groupes de Betti d'un complexe infini*, Fund. Math. 25 (1935), pp. 33–44.
[4] EILENBERG, S. *Cohomology and continuous mappings*, Ann. of Math. 41 (1940), pp. 231–251.
[5] EILENBERG, S. AND MAC LANE, S. *Infinite Cycles and Homologies*, Proc. Nat. Acad. Sci. U. S. A. 27 (1941), pp. 535–539.
[6] HALL, M. *Group Rings and Extensions, I*, Ann. of Math. 39 (1938), pp. 220–234.
[7] LEFSCHETZ, S. *Algebraic Topology*, Amer. Math. Soc. Colloquium Series, vol. 27, New York, 1942.
[8] PONTRJAGIN, L. *Topological Groups*, Princeton, 1939.
[9] STEENROD, N. E. *Universal Homology Groups*, Amer. Journ. of Math. 58 (1936), pp. 661–701.
[10] ———. *Regular Cycles of Compact Metric Spaces*, Ann. of Math. 41 (1940), pp. 833–851.
[11] TURING, A. M. *The Extensions of a Group*, Compositio Math. 5 (1938), pp. 357–367.
[12] WEIL, A. *L'Integration dans les groupes topologiques et ses applications*, Actualites Sci. et Ind. No. 869, Paris, 1940.
[13] WHITNEY, H. *Tensor Products of Abelian Groups*. Duke Math. Journ. 4 (1938), pp. 495–528.
[14] ———. *On matrices of integers and combinatorial topology*, Duke Math. Journ. 3 (1937), pp. 35–45.
[15] ZASSENHAUS, H. *Lehrbuch der Gruppentheorie*, Hamburg. Math. Einzelschriften, 21, Leipzig, 1937.

GROUPS OF ALGEBRAS OVER AN ALGEBRAIC NUMBER FIELD.*

By S. MacLane and O. F. G. Schilling.

In this note we shall discuss some new aspects of the theory of normal simple algebras over an algebraic number field. The algebras considered are split by a finite normal extension which is the join of two normal subfields. Moreover, the ramifications of the algebras are to be prime to the discriminant of the given normal field. We propose to find conditions insuring that the group of algebras split by the top field be the join of the groups of algebras split by the given component fields. This query leads to certain curious facts concerning the associated Galois groups. One of our results states that the group of algebras for any abelian field is always the join of the groups of algebras split by any set of subfields whose join is the top field. In the general case we have succeeded in finding a partial description of the Galois groups in terms of the condition on the associated algebras. The method of proof of our theorems consists in a reduction of problems on algebras to problems on groups by means of the theory of invariants of algebras.

Let K be a finite normal extension with Galois group $\Gamma = \{\sigma, \cdots\}$ over a finite algebraic field F. Suppose that the field K/F is given as the join $K' \cup K''$ of two normal subfields K'/F and K''/F with the respective Galois groups $\Gamma' = \{\sigma', \cdots\}$ and $\Gamma'' = \{\sigma'', \cdots\}$. Let $\Delta = \{\delta, \cdots\}$ be the Galois group of the intersection $(K' \cap K'')/F$. Let T' and T'' be the homomorphisms mapping the groups Γ' and Γ'' upon Δ. Then [1] the Galois group Γ can be described as the group of all pairs (σ', σ'') such that $\sigma'\mathrm{T}' = \sigma''\mathrm{T}''$. We call Γ the *subdirect product* of the factor groups Γ', Γ''.

In the sequel we are necessarily led to consider a class of fields, in which each field K is the join of two normal subfields K' and K'' such that the following hypothesis (H) holds:

H: *For every prime l and positive integer m such that l^m is the order of an element τ' in Γ' and of an element τ'' in Γ'' there exists at least one element $\sigma = (\sigma', \sigma'')$ in Γ whose components σ', σ'' have orders divisible by l^m.*

Let M be the product of all those finite and infinite prime divisors of F

* Received December 6, 1941; Presented to the Society, December 29-31, 1941.

[1] For an outline of the proof for the following assertion see S. MacLane and O. F. G. Schilling, "A formula for the direct product of crossed product algebras," *Bulletin of the American Mathematical Society*, vol. 48 (1942), pp. 108-114.

which are ramified in K. We shall consider the groups of algebra classes S, S', S'' which are prime to M and split by the normal fields K, K', K'', respectively.[2] The group S contains as subgroups both S', S'' and their join $S' \cup S''$. We prove

THEOREM 1. *Let K, K', K'' be a triple of normal fields with $K = K' \cup K''$. Then $S = S' \cup S''$ if and only if hypothesis (H) holds.*

Proof. We first prove that the condition is sufficient. It is enough to prove the assertion for algebras of prime power degree l^ν. Such algebras we shall term l-primary algebras. For any algebra S is similar [3] to the direct product of algebras S_i, each of which has a degree which is a power of some prime. If the original S is prime to M, it follows that each of the component algebras is also prime to M, by the theory of the local invariants.[4] Hence let S be an algebra whose index is a power of a prime l. Suppose that p is a prime divisor of F which is not ramified in K. Then, for every prime factor P of p in K, the Frobenius symbol $[K/F; P]$ has a unique representation $[K/F; P] = ([K'/F; P'], [K''/F; P''])$ where P', P'' denote the prime divisors induced by P in the subfields K', K'', respectively.[5] Therefore the degree $n(P) = n(p)$ of the local extension K_P/F_p is equal to the l. c. m. of the degrees $n(P')$, $n(P'')$ of $K'_{P'}/F_p$, $K''_{P''}/F_p$, respectively.

Now let $\rho(p)$ be an arbitrary invariant of an l-primary algebra S. Since $\rho(p)$ is a fraction with denominator $n(p)$, we may write

$$(1) \qquad \rho(p) = a(p)/n(p) \equiv a'(p)/n(P') + a''(p)/n(P'') \pmod 1$$

for p is, by assumption on S, prime to M. In general there will be several such representations (1) of $\rho(p)$ as the sum of fractions with denominators $n(P')$, $n(P'')$. We agree to select one such representation for each prime divisor p. By the fundamental relation [6] for the invariants of S we have

$$(2) \qquad \sum_p \rho(p) \equiv \sum_p [a'(p)/n(P')] + \sum_p [a''(p)/n(P'')] \equiv 0 \pmod 1.$$

---

[2] For the definition of these groups of algebras see S. MacLane and O. F. G. Schilling, "Normal algebraic number fields," *Transactions of the American Mathematical Society*, vol. 50 (1941), pp. 295-384, especially pp. 298-303.

[3] A. A. Albert, *Structure of Algebras*, American Mathematical Society Colloquium Publications, vol. 24 (1939), Theorem 18, p. 77.

[4] H. Hasse, "Die Struktur der R. Brauerschen Algebrenklassengruppe über einem algebraischen Zahlkörper," *Mathematische Annalen*, vol. 107 (1933), pp. 731-60. Quoted as HA. See in particular pp. 750, 752.

[5] H. Hasse, "Bericht über neuere Untersuchungen und Probleme der algebraischen Zahlkörper," *Jahresberichte der Deutsche Mathematische Vereinigung*, Ergänzungband 6 (1930), Part II, p. 7. Quoted as HB.

[6] HA. p. 750.

Since all invariants $\rho(p)$ have as denominators powers of l it follows from (2) that

$$(3) \qquad s' = -\sum_p [a''(p)/n(P'')] \text{ and } s'' = -\sum_p [a'(p)/n(P')]$$

have the same denominator l^m. Because there is only a finite number of prime divisors p giving non-zero contributions to s' and s'', it follows that l^m divides the l. c. m. of a finite number of local degrees $n(P')$, $n(P'')$. All prime divisors p under consideration are, by assumption, prime to M. Hence l^m is the order of elements τ' in Γ' and τ'' in Γ''. The elements τ', τ'' may be determined as follows. Pick a prime P'' involved in s', P' involved in s'' such that $n(P')$, $n(P'')$ are divisible by l^m. The associated Frobenius automorphisms σ'_0, σ''_0 are elements in the respective Galois groups. Then suitable powers of σ'_0, σ''_0 are elements τ', τ'' with the required properties.

We next make use of hypothesis (H). There exists at least one element $\sigma = (\sigma', \sigma'')$ in Γ whose components have orders divisible by l^m. The Tschebotareff density theorem [7] asserts that there are infinitely many prime divisors q in F which are prime to M such that

$$[K/F; Q] = \sigma = (\sigma', \sigma'') = ([K'/F; Q'], [K''/F; Q''])$$

where $Q \mid q$ in K and Q induces Q', Q'' in K', K'', respectively. By construction we have

$$l^m \mid n(Q') \qquad \text{and} \qquad l^m \mid n(Q'').$$

Consequently [8] s' and s'' can be realized as invariants of local algebras split by $K'_{Q'}/F_q$ and $K''_{Q''}/F_q$, respectively. Similarly, the fractions $a'(p)/n(P')$, $a''(p)/n(P'')$ are individually invariants of local algebras split by $K'_{P'}/F_p$ and $K''_{P''}/F_p$, respectively. By the definition (3) of s', s'',

$$(4) \qquad \begin{aligned} s' + \sum_p [a'(p)/n(P')] &\equiv 0 \pmod 1, \\ s'' + \sum_p [a''(p)/n(P'')] &\equiv 0 \pmod 1. \end{aligned}$$

Hence [9] there exist algebras S' and S'' split by K', K'', respectively, which have the invariants indicated in (4), where in particular s', s'' are the invariants at q. Relation (2) and definition (3) imply $s'' \equiv s' \pmod 1$. Consequently $S \sim S' \times S''$, for the associated invariants at q are equal, modulo 1.

In order to prove that condition (H) is necessary let us suppose that (H) is false. Then there exists a prime l, an integer m and elements τ' in Γ', τ'' in Γ'' with orders l^m such that for every element $\sigma = (\sigma', \sigma'')$ in Γ either

[7] HB. Part II, p. 133. [8] HA. p. 752. [9] HA. pp. 748, 752.

the order of σ' or that of σ'' is not divisible by l^m. From the elements τ', τ'' in the groups Γ', Γ'' we can then construct elements $\tau_1 = (\tau', \rho'')$ and $\tau_2 = (\rho', \tau'')$ in Γ. We choose ρ'' as any element in Γ'' which is mapped by T'' into the element of Δ which is the image of τ' under T'; similarly we choose ρ'. This construction can be done so that the elements τ_1, τ_2 both have order l^m, using the fact that hypothesis (H) is supposed to be false. By the Tschebotareff density theorem there exist prime divisors p_1 and p_2 of F whose corresponding Artin symbols[10] $(K/F; p_i) = \{\sigma[K/F; P_i]\sigma^{-1}, P_i \mid p_i$ in K, σ in $\Gamma\}$ have

$$\tau_1 \,\varepsilon\, (K/F; p_1) \qquad \text{and} \qquad \tau_2 \,\varepsilon\, (K/F; p_2).$$

This means that for any factor P_1 of p_1 in K the local degree $n(P_1)$ will be exactly the order l^m. We next construct an algebra S over F whose only invariants are

$$\rho(p_1; S) = 1/l^m \qquad \text{and} \qquad \rho(p_2; S) = -1/l^m.$$

Then S is split by K because the local degrees of K at p_1, p_2 are both equal to l^m, which is the local index of the algebra S.

Suppose now that the conclusion of the theorem held. Then $S \sim S' \times S''$, where S' and S'' are algebras split by K' and K'', respectively. We can assume here that the indices of S' and S'' are powers of the prime l, for otherwise we could write[11] $S' = S'_1 \times S'_2$ and $S'' = S''_1 \times S''_2$, where the indices of S'_1 and S''_1 are powers of l, and the indices of S'_2 and S''_2 are prime to l. One then would have the decomposition $S \sim S'_1 \times S''_1$, as asserted.

In the decomposition $S = S' \times S''$, the invariants must satisfy for every p the condition

$$\rho(p, S) \equiv \rho(p, S') + \rho(p, S'') \pmod 1.$$

From this it follows that $i(p, S)$ is the least common multiple of $i(p, S')$ and $i(p, S'')$, where $i(p, S)$ denotes the local index of S at p; that is, the smallest integer such that $i(p, S)\rho(p, S) \equiv 0 \pmod 1$. Since S' and S'' are split by K we have

$$i(p_1, S') \mid l^m \qquad \text{and} \qquad i(p_1, S'') \mid l^m.$$

Moreover, at least one of the indices must equal l^m because the original index $i(p_1, S) = l^m$. But now the Artin symbol $(K/F; p_1)$ is the class of conjugates of τ_1 in the group Γ, so that one may choose a prime factor P_1 of p_1 in K so that $\tau_1 = [K/F; P_1] = ([K'/F; P'_1], [K''/F; P''_1])$, where P'_1, P''_1 are

[10] HB. Part II, p. 6.

[11] By the decomposition theorem, A. A. Albert, *loc. cit.*

the prime multiples of P_1 in K', K'', respectively. We compare this equation with the construction of the element τ_1. We find $[K'/F; P'_1] = \tau'$ has order l^m and $[K''/F; P''_1] = \rho''$ has an order which is not divisible by l^m, for hypothesis (H) is assumed to be false. Therefore the local degree of K''/F at p_1 is not a multiple of l^m. Consequently $i(p_1; S'') = l^m$ is impossible and actually $i(p_1; S'') = l^{m_1}$ with $m_1 < m$. Hence we have

$$(5) \qquad \begin{array}{lll} i(p_1; S') = l^m, & i(p_1; S'') = l^{m_1}, & m_1 < m, \\ i(p_2; S') = l^{m_2}, & i(p_2; S'') = l^m, & m_2 < m. \end{array}$$

Since the sum of the local invariants of S' is 0 (mod 1) we have

$$(6) \qquad \rho(p_1; S') \equiv -\rho(p_2; S') - \sum_{q \neq p_1, p_2} \rho(q; S') \pmod 1.$$

Observing (5) we find that there exists a prime divisor q such that $i(q; S') = l^{m+\nu}$, $\nu \geqq 0$. But the invariant of S at q is equal to 0 (mod 1) by construction. Hence $0 \equiv \rho(q; S) \equiv \rho(q; S') + \rho(q; S'')$ (mod 1) implies $i(q; S'') = l^{m+\nu}$. Therefore the local degress of K', K'' at q are divisible by l^m. Next take a factor Q of q in K. Then $\sigma = [K/F; Q] = ([K'/F; Q'], [K''/F; Q''])$ where the orders of the components are divisible by l^m, Q' and Q'' the prime multiples of Q in K', K'', respectively. This contradicts the second part of the auxiliary assumption that hypothesis (H) is false.

We next discuss pairs of groups Γ', Γ'' satisfying hypothesis (H). Although we have not found a complete characterization of these groups we have analyzed various special cases. In particular, we show by examples that hypothesis (H) does not hold for all pairs of groups.

THEOREM 2. *Let Γ', Γ'' be two groups which have the same homomorphic image Δ. Suppose that $\{l\}$ is the set of all prime factors common to the orders of Γ', Γ''. Then hypothesis (H) holds for Γ', Γ'' if it holds for all pairs of l-Sylow groups Γ'_0, Γ''_0 of Γ', Γ'', respectively, which have the same image in Δ.*

Proof. Let τ', τ'' be elements of Γ', Γ'' which have the same order l^m. We have to prove the existence of an element (σ', σ'') in the subdirect product of Γ' and Γ'' whose components have orders divisible by l^m. Each of the elements τ', τ'' lies in some Sylow group Γ'_1, Γ''_0 of Γ', Γ'', respectively. The maps $\Gamma'_1 T' = \Delta'_1$, $\Gamma''_0 T'' = \Delta''_0$ of the Sylow groups into $\Delta = \Gamma' T' = \Gamma'' T''$ are again[12] *l-Sylow* subgroups of Δ. The two groups Δ'_1, Δ''_0 are conjugate in Δ, thus $\Delta''_0 = \phi \Delta'_1 \phi^{-1}$ for some element ϕ in Δ. Now pick an element ρ in Γ'

[12] See H. Zassenhaus, *Lehrbuch der Gruppentheorie*, Hamb. Math. Einzelschriften (1937), p. 100, Th. 2. Quoted as Z.

with $\rho T' = \phi$. Then $\Gamma'_0 = \rho\Gamma'_1\rho^{-1}$ is an l-Sylow group of Γ' the map of which covers all of Δ''_0. Thus, the groups Γ'_0, Γ''_0 have a common image Δ''_0 in Δ. The pair $(\rho\tau'\rho^{-1}, \tau'')$ has again components whose orders are divisible by l^m. We now use the assumption that hypothesis (H) holds for Γ'_0, Γ''_0. Then there exists an element (σ', σ'') of the subdirect product of Γ'_0, Γ''_0 which has the required properties. Since the latter group lies in the subdirect product of Γ' and Γ'' the hypothesis (H) holds.

We thus reduce the problem of deciding whether a pair of groups satisfies hypothesis (H) to groups of prime power order. For the latter class we prove

LEMMA 1. *If Γ', Γ'' are both regular*[13] *l-groups then hypothesis (H) holds.*

Proof. Suppose that l^{m+1} is the maximal order for the elements of a regular l-group Γ. Let Σ be the set of all elements not of maximal order in Γ. Then for every α in Σ we have $\alpha^{l^m} = 1$, while for the elements γ in the complementary set $\Gamma - \Sigma$

$$\gamma^{l^m} \neq 1 \qquad \text{and} \qquad \gamma^{l^{m+1}} = 1.$$

By a theorem of Hall[14] Σ is a subgroup of Γ. We return now to the groups Γ', Γ'' of the Lemma and consider the respective subgroups Σ', Σ''. By construction the elements of maximal order of Γ', Γ'' are found in the complementary sets $\Gamma' - \Sigma'$, $\Gamma'' - \Sigma''$. The maps $\Sigma'T'$ and $\Sigma''T''$ are subgroups of Δ. Suppose that both are proper subgroups. Then each subgroup $\Sigma'T'$, $\Sigma''T''$ contains at most half the number of elements in Δ. Since both homomorphic maps have the unit element of Δ in common, there must exist at least one element δ in Δ which lies neither in $\Sigma'T'$ nor in $\Sigma''T''$. There exist elements σ' in Γ' and σ'' in Γ'' which are mapped upon δ by the homomorphisms T', T''. These elements do not lie in Σ', Σ'' hence they have maximal order. The element (σ', σ'') of Γ has the required properties.

It remains to consider the case where $\Sigma'T' = \Delta$. Let σ', σ'' be elements in Γ', Γ'' which have maximal orders in the respective groups. Suppose that $\sigma'T' = \delta'$, $\sigma''T'' = \delta''$. By assumption on Σ', there exists then an element τ' in Σ' with $\tau'T' = (\delta')^{-1}\delta''$. By construction, the element $\sigma'\tau'$ has maximal order and $(\sigma'\tau')T' = \delta'(\delta')^{-1}\delta'' = \delta''$. Hence $(\sigma'\tau', \sigma'')$ is an element of Γ which has the desired properties. A similar argument may be applied in case $\Sigma''T'' = \Delta$.

[13] For the definition and properties of regular l-groups see P. Hall, "A contribution to the theory of groups of prime-power order," *Proceedings of the London Mathematical Society*, ser. 2, vol. 36 (1932), pp. 29-95, especially pp. 73-8.

[14] P. Hall, *loc. cit.* Theorem 4.26, p. 76.

Observe that we proved more than is required by the statement of the lemma. We actually proved the existence of an element σ whose components have maximal (possible) orders.

LEMMA 2. *The class of regular l-groups contains the following types of groups:*

(i) *every abelian group of prime power order,*

(ii) *every l-group, for odd prime l, whose commutator group lies in its center.*

Proof. We recall that an l-group of class c is regular whenever[15] $c < l$. Here the class of a group is the length of its lower central series. An abelian group has class 1; this gives case (i) above. In the case (ii) the class of the group is 2, and $2 < l$.

Suppose that Γ is represented as the subdirect product of two factor groups Γ' and Γ'' which satisfy the conditions of Lemma 1, 2 for their respective Sylow groups. We next observe that the direct product of regular groups is again a regular group. Then the Sylow subgroups of Γ satisfy the same conditions, for Γ is contained in the direct product of Γ' and Γ''. Conversely, suppose that the Sylow subgroups of Γ satisfy the above conditions. Let Γ' be any homomorphic image of Γ, say $\Gamma' = \Gamma/\Lambda$. Suppose the commutator groups of Γ, Γ' are Γ_0, Γ'_0, similarly let Z, Z' be the associated centers. Then $\Gamma'_0 = \Gamma_0 \cup \Lambda/\Lambda$ and $Z \cup \Lambda/\Lambda \leqq Z'$. Since $\Gamma_0 \leqq Z$ we have $\Gamma'_0 \leqq Z'$. Hence the validity of the hypotheses of Lemmas 1, 2 for Γ implies their validity for any two factor groups Γ', Γ'' such that Γ is the subdirect product[16] of Γ', Γ''.

We are now in a position to prove

THEOREM 3. *Let K/F be a normal field which is the join of t normal subfields K_i/F, $i = 1, 2, \cdots, t$, then $S = S_1 \cup \cdots \cup S_t$ where the S and S_i, $i = 1, \cdots, t$, are the groups of algebras prime to the module M of K/F and split by K, K_i, respectively, if either*

(i) *K is the direct join of the subfields K_i, or*

(ii) *all Sylow groups of the Galois group of K/F are regular.*

Proof. We shall prove the theorem by induction on the set of subfields K_i. For the first part of the theorem we observe that hypothesis (H) is finally

[15] P. Hall, *loc. cit.*, p. 4a, Definition 2.45 and p. 73, Corollary 4.13.

[16] Observe that the assumptions on the Sylow groups of Γ do not imply that Γ is a subdirect product of suitable factor groups. For an example consider the icosahedral group of $60 = 3 \cdot 4 \cdot 5$ elements. This group satisfies all assumptions on the Sylow groups but is not a subdirect product since it is a simple group.

satisfied. The assertion is true for $t=1$. Suppose it is proved for any subset, say $K_1,\cdots,K_{t-1}$, that an algebra split by $K_1\cup\cdots\cup K_{t-1}$ is the direct product of algebras split by $K_1,\cdots,K_{t-1}$. Then let $K_1\cup\cdots\cup K_{t-1}=K'$ and $K_t=K''$. The Galois groups Γ', Γ'', are homomorphic maps of the group Γ. Thus their Sylow subgroups Γ'_0, Γ''_0 are maps of the Sylow subgroups of Γ. Hence both groups Γ'_0, Γ''_0 are regular groups.[17] Since Theorem 2 combined with Lemmas 1 and 2 and the remark at the beginning of this proof implies the validity of hypothesis (H) for the two cases under consideration, we may apply Theorem 1. Then the algebra S split by K is similar to the direct product $S'\times S''$ where $S'\times K'\sim K'$ and $S''\times K''\sim K''$. Since $S'\sim S_1\times\cdots\times S_{t-1}$, $S''=S_t$ we have $S\sim S_1\times\cdots\times S_t$.

In the preceding arguments the assumption that the algebras concerned are prime to the modulus M cannot be dropped, as we now show by an example in which l is any given prime, F any algebraic number field, K a suitable extension of degree l^2 over F. First choose two distinct rational primes l_1 and l_2 in the arithmetic progression $1+xl$, let p_i be a prime divisor of l_i in F, and let F_{p_i} be the corresponding complete (p-adic) field, for $i=1,2$. Each F_{p_i} has an unramified cyclic extension of degree l; on the other hand, the choice of the l_i insures that each F_{p_i} also has a ramified cyclic extension of the same degree (for example, the extension generated by the l-th root of any prime element). The Grunwald existence theorem[18] then gives two cyclic fields $K^{(i)}$ of prime degree l over F, such that p_i is ramified in $K^{(i)}/F$ and totally inert in $K^{(j)}/F$, for $i\neq j$. However, the join $K=K^{(1)}\cup K^{(2)}$ of these fields has local degree l^2 both at p_1 and p_2, so that there exists a division algebra split by K which has the invariants $1/l^2$, $-1/l^2$ at p_1, p_2, respectively.[19] This algebra is not similar to the direct product of algebras split by $K^{(1)}$ and $K^{(2)}$.

Furthermore, Theorem 3 is not true for arbitrary (non-normal) subfields K', K'', with $K'\cap K''=F$, $K'\cup K''=K$. This may be shown explicitly in the case when F is the rational field, K a field with the symmetric group of order 6, K' and K'' conjugate cubic subfields of K.

The following converse to the preceding theorems holds:

THEOREM 4. *If K/F is normal, and $K'\supset F$, $K''\supset F$ are two subfields*

---

[17] For every homomorphic map of a regular group is regular. See P. Hall, *loc. cit.*, p. 74.

[18] W. Grunwald, "Ein algebraisches Existenztheorem für algebraische Zahlkörper," *Jour. für d. reine und ang. Math.*, vol. 169 (1933), pp. 103-7.

A. A. Albert, "On p-adic fields and rational division algebras," *Annals of Mathematics*, vol. 41 (1940), pp. 674-93; in particular pp. 688-90.

[19] HA. pp. 748, 752.

such that $S = S' \cup S''$, *for the associated group of algebra classes, then* $K = K' \cup K''$.

An indirect proof may be given, using the Tschebotaroff density theorem to construct suitable prime divisors inert in $K' \cup K''$ but not in K. We omit details.

We now wish to indicate the scope of the group theoretical arguments leading to Theorem 3, and in particular, to show that hypothesis (H) is not vacuous. Specifically, we exhibit for each prime l an irregular group for which hypothesis (H) holds and one for which it fails.

LEMMA 3. *Let l be an odd prime. Then the group Γ' with the generators* $Q_1, \cdots, Q_{l-1}$, R *and the relations* $Q_1^{l^2} = Q_2^l = \cdots = Q_{l-1}^l = R^l = 1$, $RQ_1R^{-1} = Q_1Q_2$, $RQ_2R^{-1} = Q_2Q_3, \cdots$, $RQ_{l-2}R^{-1} = Q_{l-2}Q_{l-1}$ *and* $RQ_{l-1}R^{-1} = Q_{l-1}Q_1^{-l}$ *is irregular.*

This group may be conveniently described as a group extension. Let A be the abelian group generated by the elements $Q_1, \cdots, Q_{l-1}$ of orders l^2, $l, \cdots, l$ and set $Q_l = Q_1^{-l}$, $Q_{l+1} = Q_2^{-1} = 1$. Then one may verify that the definition

$$Q_i^{\alpha} = Q_iQ_{i+1}, \qquad i = 1, 2, \cdots, l-1 \tag{1}$$

determines an automorphism α of A. By induction on k one then shows that

$$Q_1^{\alpha^k} = \prod_{i=1}^{k+1} Q_i^{d_i}, \quad \text{where } d_i = \binom{k}{i-1}, \text{ for } k = 1, \cdots, l. \tag{2}$$

In particular, since each binominal coefficient d_i with $k = l$ is divisible by l, one has $Q_1^{\alpha^l} = Q_1$. By (1) and induction on i, $Q_{i+1}^{\alpha^l} = Q_i^{\alpha^l\alpha}Q_i^{-\alpha^l} = Q_{i+1}$; hence α is an automorphism of order l. We may therefore regard Γ' as a group extension [20] of A by a cyclic group of order l, represented by an operator R with $R^l = 1$, $RQ_iR^{-1} = Q_i^{\alpha}$.

To determine the order of elements in Γ', let $NA = A^{1+\alpha+\cdots+\alpha^{l-1}}$ for any element of the group A. By (2) and by properties of binomial coefficients, one shows that $NQ_1 = 1$ and hence by (1) that $NQ_{i+1} = NQ_i^{\alpha}NQ_1^{-1} = 1$. Therefore $NA = 1$ for every A.

Each element of Γ' may be expressed as $B = R^jA$. If $j \not\equiv 0 \pmod{l}$, $B^l = R^{j^l}NA = 1$, so B has order l. The elements of maximal order l^2 in Γ' are thus found in the subgroup A, and are exactly the elements given in the lemma. This implies that Γ' is irregular, for RQ_1 and R^{l-1} both have order l, whereas their product $R^{l-1}RQ_1 = Q_1$ has the larger order l^2. This cannot happen [21] in a regular group.

---

[20] Z., pp. 89-98.

[21] P. Hall, *loc. cit.*, p. 77, Th. 4.28.

From this group Γ' we construct a pair of groups for which hypothesis (H) fails. Let T' be the homomorphism $\Gamma' \to \Gamma'/A$ in which Γ' is mapped on a cyclic group Δ of order l, while all elements of maximal order in Γ' are mapped on the identity of Δ. Let $\Gamma''_1 = \{S\}$ be cyclic of order l^2, and let T''_1 be the homomorphism $\Gamma''_1 \to \Gamma''_1/\{S^l\}$. No element of maximal order l^2 in Γ''_1 is mapped on 1. This shows that hypothesis (H) fails in this case.

We may also use the irregular group Γ' to construct an example in which (H) holds; we let Γ''_2 be an abelian group generated by elements S_1, S_2 with respective orders l^2, l, while T''_2 is the homomorphism $\Gamma''_2 \to \Gamma''_2/\{S_1\} = \Delta$. The maps of elements of maximal orders in Γ', Γ''_2 then have an element of Δ in common; namely, the identity of Δ.

Since l is an odd prime we can realize [22] the above groups Γ', Γ''_1, as Galois groups of fields K'_1, K''_1 over any algebraic number field F. By Theorem 1 we thus have fields K'_1, K''_1 with $K_1 = K'_1 \cup K''_1$, $S > S' \cup S''$.

In the case $l = 2$, exactly similar examples may be constructed. Instead of the group Γ' of Lemma 3 we use the generalized quaternion group (Z., p. 110) of order 2^n, $n \geqq 4$. This group is generated by two elements α, β with $\alpha^{2^{n-1}} = 1$, $\beta^2 = \alpha^{2^{n-2}}$, $\beta\alpha\beta^{-1} = \alpha^{-1}$. The elements of maximal order in this group are the elements α^j with $j \not\equiv 0 \pmod 2$. This group is irregular because $\alpha^2\beta$ and $\beta^{-1}\alpha$ both have order 4, while their product α^3 has order $2^{n-1} > 4$.

HARVARD UNIVERSITY,
UNIVERSITY OF CHICAGO.

---

[22] See for example H. Reichardt, "Konstruktion von Zahlkörpern mit gegebener Galoisgruppe von Primzahlpotenzordnung," *Jour. für d. reine und ang. Math.*, vol. 177 (1937), pp. 1-5.

T. Tannaka, "Ueber die Konstruktion der galoisschen Körper mit vorgegebener p-Gruppe," *Tohoku Math. Journ.*, vol. 43 (1937), pp. 252-260.

Reprinted from the Proceedings of the National Academy of Sciences,
Vol. 29, No. 5, pp. 155–158. May, 1943

RELATIONS BETWEEN HOMOLOGY AND HOMOTOPY GROUPS

By Samuel Eilenberg and Saunders MacLane

Departments of Mathematics, University of Michigan and Harvard University

Communicated April 7, 1943

1. *The Problem.*—Let P be a locally finite simplicial connected polytope. We denote by $\mathfrak{H}^q(P, G)$ the (discrete) qth homology group of the finite chains of P over a (discrete additive) coefficient group G, by $H_q(P, G)$ the (topologized)[1] qth cohomology group of the infinite cochains of P over a (topologized additive) coefficient group G, by $\pi_q(P)$ the qth homotopy group of P relative to a fixed base point $x_0 \in P$, by $\mathfrak{S}^q(P, G)$ the subgroup of $\mathfrak{H}^q(P, G)$ determined by the cycles of P that can be obtained from cycles on the q-sphere S^q by continuous mappings of S^q into P.

H. Hopf[2] has shown that the fundamental group $\pi_1(P)$ determines the group $\mathfrak{H}^2(P, I)/\mathfrak{S}^2(P, I)$ (I = additive group of all integers) and has exhibited a group construction which leads from π_1 to $\mathfrak{H}^2/\mathfrak{S}^2$. He considers a multiplicative discrete group $\pi = F/R$ represented as a factor group of a free group F by a group of relations R and defines

$$\pi_1{}^* = R \cap Com\ F/Com_F\ R, \tag{1.1}$$

where $Com\ F$ stand for the commutator group of F, $Com_F\ R$ is the subgroup of F generated by the elements of the form $xrx^{-1}r^{-1}$, $x \in F$, $r \in R$, while "$\cap$" stands for the intersection.

Hopf shows that the group $\pi_1{}^*$ does not depend upon the particular representation chosen for π and that if $\pi = \pi_1(P)$ then $\pi_1{}^* \cong \mathfrak{H}^2(P, I)/\mathfrak{S}^2(P,I)$.

In a more recent paper[3] Hopf has shown that if $\pi_i(P) = 0$ for all $1 < i < n$ then $\pi_1(P)$ determines the group $\mathfrak{H}^n(P, I)/\mathfrak{S}^n(P, I)$. This proof leads to no method for the algebraic determination of $\mathfrak{H}^n/\mathfrak{S}^n$ by means of π_1.

We outline here a new treatment of the problem which leads to generalizations of Hopf's results and gives intrinsic descriptions for all the groups involved. In particular the group $\pi_1{}^*$ can be defined intrinsically as the character group[4] of the group of central group extensions[5] of the group X (reals reduced mod 1) by the group π:

$$\pi_1{}^* = Char\ Extcent\ \{X, \pi\} \tag{1.2}$$

The groups obtained for the higher cases may rightly be regarded as generalizations of the group of group extensions.

The method used also permits us to generalize and simplify the results of Hopf dealing with the influence of $\pi_1(P)$ upon the intersection theory in P.

The results are formulated for a connected polytope P but are valid for any arcwise connected space, provided suitable singular homologies are used.

2. The Complex $K(\pi)$.—Given a discrete group π written multiplicatively we will consider square matrices $\Delta = \| p_{ij} \|$ with p_{ij} in π satisfying the condition

$$p_{ij}p_{jk} = p_{ik}. \tag{2.1}$$

We will denote by $\Delta^{(i)}$ the matrix obtained from Δ by erasing the ith row and the ith column. The rows and columns will always be numbered starting from 0. The matrices Δ with $q + 1$ rows and columns will be taken as the generators of a free abelian group C^q. We define a homomorphism

$$\alpha : C^q \rightarrow C^{q-1}$$

by putting

$$\alpha\Delta = \sum_{i=0}^{q} (-1)^i \Delta^{(i)}$$

for each generator Δ of C^q. We verify that $\alpha\alpha = 0$. Consequently if we consider each generator Δ of C^q as a q-cell and α as a boundary operation we obtain a closure finite abstract complex $K(\pi)$. The (discrete) qth homology group obtained by using finite chains of $K(\pi)$ over a (discrete) coefficient group G will be written as $\mathfrak{H}^q(\pi, G)$. The (topologized) qth cohomology group obtained using infinite cochains of $K(\pi)$ over a (topologized) coefficient group G will be written as $H_q(\pi, G)$.

3. The Main Theorem.—Let s^q be q-dimensional simplex with vertices $v_0, v_1, \ldots, v_q$ and let $T: s^q \rightarrow P$ be a continuous mapping of s^q into the polytope P such that the vertices v_i are all mapped into the point x_0 in P (here x_0 is the base point for the construction of the fundamental group $\pi_1 = \pi_1(P)$). Each edge v_iv_j in s^q will then determine an element p_{ij} in the group π_1 and the matrix $\| p_{ij} \|$ so obtained will satisfy condition (2.1). Hence the mapping T determines a q-cell of the complex $K(\pi_1)$. This observation leads to a chain transformation

$$\tau : P \rightarrow K(\pi_1(P)) \tag{3.1}$$

A study of this transformation furnishes the following:

THEOREM 1. *If a connected locally finite polytope P has*

$$\pi_i(P) = 0 \quad \text{for} \quad 1 < i < n, \tag{3.2}$$

the following isomorphisms hold:

$$\mathfrak{H}^q(P, G) \cong \mathfrak{H}^q(\pi_1(P), G) \quad \text{for} \quad q < n, \tag{3.3}$$

$$H_q(P, G) \cong H_q(\pi_1(P), G) \quad \text{for} \quad q < n, \tag{3.4}$$

$$\mathfrak{H}^n(P, G)/\mathfrak{S}^n(P, G) \cong \mathfrak{H}^n(\pi_1(P), G). \tag{3.5}$$

4. *Computation of the Groups* $H_q(\pi, G)$.—We will outline a method for computing the groups $\mathfrak{H}^q(\pi, G)$ and $H_q(\pi, G)$ directly in terms of π and G without using the complex $K(\pi)$. It is sufficient to compute the cohomology groups $H_q(\pi, G)$, since the homology-cohomology duality applied to the complex $K(\pi)$ gives

$$\mathfrak{H}^q(\pi, G) \cong \mathit{Char}\ H_q(\pi, \mathit{Char}\ G), \quad G \text{ discrete.}$$

Our calculations are based upon a 1 — 1 correspondence between the matrices Δ with elements in π and the q-tuples $(p_1, \ldots, p_q)$ with $p_i \in \pi$. Given such a q-tuple, the definitions $p_{ii} = 1$, $p_{ij} = p_{i+1} \ldots p_j$ for $i < j$ and $p_{ij} = (p_{ji})^{-1}$ for $j < i$ define a $q + 1$ by $q + 1$ matrix $\Delta = \| p_{ij} \|$ which satisfies (2.1). Conversely, every such matrix Δ can be obtained in this way from exactly one q-tuple. Consequently the group of q-cochains of the complex $K(\pi)$ over the (topologized) group G is nothing but the group $C_q(\pi, G)$ of all functions f of q variables on π to G. The coboundary δf of such a function is a function of $q + 1$ variables defined as follows

$$(\delta f)(p_1, p_2, \ldots, p_{q+1}) = f(p_2, \ldots p_{q+1}) + \sum_{i=1}^{q}(-1)^i f(p_1, \ldots, p_{i-1}, p_i p_{i+1}, p_{i+2}, \ldots, p_{q+1}) + (-1)^{q+1} f(p_1, p_2, \ldots, p_q).$$

The group $Z_q(\pi, G)$ of cocycles consists of the functions f with $\delta f = 0$, while the group $B_q(\pi, G)$ of the cobounding cocycles consists of those functions f of the form $f = \delta g$. The cohomology groups are

$$H_q(\pi, G) = Z_q(\pi, G)/B_q(\pi, G).$$

5. *The Cases* $q = 1, 2, 3$.—The coboundary of a function of one variable $f: \pi \to G$ is

$$(\delta f)(p_1, p_2) = f(p_2) - f(p_1 p_2) + f(p_1). \tag{5.1}$$

Hence $Z_1(\pi, G)$ is composed of the homomorphisms $f: \pi \to G$. Since $B_1(\pi, G) = 0$ we have

$$H_1(\pi, G) = \mathit{Hom}\ \{\pi, G\}.$$

For a function f of two variables the coboundary is

$$(\delta f)(p_1, p_2, p_3) = f(p_2, p_3) - f(p_1 p_2, p_3) + f(p_1, p_2) - f(p_1, p_2). \tag{5.2}$$

The condition that $\delta f = 0$ is exactly the associativity condition for a "central" factor set $f(p_1, p_2)$ of π in G, while the coboundaries δf in (5.1) of functions of one variable are simply the "central" transformation sets. Hence $H_2(\pi, G)$ is the group of factor sets modulo transformation sets, and so is the group of central extensions of G by π,

$$H_2(\pi, G) \cong \mathit{Extcent}\ \{G, \pi\}.$$

For $q = 3$ the coboundary δf of f becomes

$$(\delta f)(p_1, p_2, p_3, p_4) = f(p_2, p_3, p_4) - f(p_1p_2, p_3, p_4) + f(p_1, p_2p_3, p_4) - f(p_1, p_2, p_3p_4) + f(p_1, p_2, p_3). \quad (5.3)$$

This operation and the resulting group $H_3(\pi, G)$ have been defined by O. Teichmüller,[6] in a somewhat more general case (corresponding to group extensions which are not necessarily central).

6. *Higher Dimensions.*—A generalization of the main theorem to higher dimensions can be obtained by constructing for each discrete *abelian* group π and each integer $n > 0$ a complex $K(\pi, n)$ whose cells are suitable systems of elements in π with $n + 1$ indices. Denoting the homology and cohomology groups of $K(\pi, n)$ by $\mathfrak{H}^q(\pi, n, G)$ and $H_q(\pi, n, G)$ we have

Theorem 2.—Given a connected locally finite polytope P such that

$$\pi_i(P) = 0 \quad \text{for} \quad 0 < i < m \quad \text{and for} \quad m < i < n \quad (6.1)$$

the following isomorphisms hold:

$$\mathfrak{H}^q(P, G) \cong \mathfrak{H}^q(\pi_m(P), m, G), \quad q < n, \quad (6.2)$$

$$H_q(P, G) \cong H_q(\pi_m(P), m, G), \quad q < n, \quad (6.3)$$

$$\mathfrak{H}^n(P, G)/\mathfrak{S}^n(P, G) \cong \mathfrak{H}^n(\pi_m(P), m, G). \quad (6.4)$$

[1] The topological groups are understood in a generalized sense, with the Hausdorff separation axiom not postulated.

[2] *Commentarii Math. Helvetici,* **14,** 257–309 (1942).

[3] *Ibid.,* **15,** 27–32 (1942).

[4] The group *Char G* of a given abelian group G is the suitably topologized group of all homomorphisms of the group G into the group X of reals reduced mod 1.

[5] A group E is a central extension of X by H if X is a normal subgroup of E which lies in the center of E and has $E/X = H$. The group of group extensions (including non-central cases) is treated in H. Zassenhaus, *Lehrbuch der Gruppentheorie,* Hamburg Math. Einzelschriften, Leipzig 21, 1937.

[6] O. Teichmüller, "Ueber die sogenannte nichtkommutative Galoische Theorie und die Relation $\xi_{\lambda,\mu\nu}\xi_{\lambda,\mu\nu,\pi}\xi_{\mu,\nu,\pi}^{\lambda} = \xi_{\lambda,\mu,\nu\pi}\xi_{\lambda\mu,\nu,\pi}$," *Deutsche Math.,* **5,** 138–149 (1940).

GENERAL THEORY OF NATURAL EQUIVALENCES

BY

SAMUEL EILENBERG AND SAUNDERS MacLANE

CONTENTS

Introduction. The subject matter of this paper is best explained by an example, such as that of the relation between a vector space L and its "dual"

Presented to the Society, September 8, 1942; received by the editors May 15, 1945.

or "conjugate" space $T(L)$. Let L be a finite-dimensional real vector space, while its conjugate $T(L)$ is, as is customary, the vector space of all real valued linear functions t on L. Since this conjugate $T(L)$ is in its turn a real vector space with the same dimension as L, it is clear that L and $T(L)$ are isomorphic. But such an isomorphism cannot be exhibited until one chooses a definite set of basis vectors for L, and furthermore the isomorphism which results will differ for different choices of this basis.

For the iterated conjugate space $T(T(L))$, on the other hand, it is well known that one can exhibit an isomorphism between L and $T(T(L))$ *without* using any special basis in L. This exhibition of the isomorphism $L \cong T(T(L))$ is "natural" in that it is given *simultaneously* for *all* finite-dimensional vector spaces L.

This simultaneity can be further analyzed. Consider two finite-dimensional vector spaces L_1 and L_2 and a linear transformation λ_1 of L_1 into L_2; in symbols

$$\lambda_1: \quad L_1 \to L_2. \tag{1}$$

This transformation λ_1 induces a corresponding linear transformation of the second conjugate space $T(L_2)$ into the first one, $T(L_1)$. Specifically, since each element t_2 in the conjugate space $T(L_2)$ is itself a mapping, one has two transformations

$$L_1 \xrightarrow{\lambda_1} L_2 \xrightarrow{t_2} R;$$

their product $t_2\lambda_1$ is thus a linear transformation of L_1 into R, hence an element t_1 in the conjugate space $T(L_1)$. We call this correspondence of t_2 to t_1 the mapping $T(\lambda_1)$ *induced* by λ_1; thus $T(\lambda_1)$ is defined by setting $[T(\lambda_1)]t_2 = t_2\lambda_1$, so that

$$T(\lambda_1): \quad T(L_2) \to T(L_1). \tag{2}$$

In particular, this induced transformation $T(\lambda_1)$ is simply the identity when λ_1 is given as the identity transformation of L_1 into L_1. Furthermore the transformation induced by a product of λ's is the product of the separately induced transformations, for if λ_1 maps L_1 into L_2 while λ_2 maps L_2 into L_3, the definition of $T(\lambda)$ shows that

$$T(\lambda_2\lambda_1) = T(\lambda_1)T(\lambda_2).$$

The process of forming the conjugate space thus actually involves two different operations or functions. The first associates with each space L its conjugate space $T(L)$; the second associates with each linear transformation λ between vector spaces its induced linear transformation $T(\lambda)$[1].

(1) The two different functions $T(L)$ and $T(\lambda)$ may be safely denoted by the same letter T because their arguments L and λ are always typographically distinct.

A discussion of the "simultaneous" or "natural" character of the isomorphism $L \cong T(T(L))$ clearly involves a simultaneous consideration of all spaces L and all transformations λ connecting them; this entails a simultaneous consideration of the conjugate spaces $T(L)$ and the induced transformations $T(\lambda)$ connecting them. Both functions $T(L)$ and $T(\lambda)$ are thus involved; we regard them as the component parts of what we call a "functor" T. Since the induced mapping $T(\lambda_1)$ of (2) reverses the direction of the original λ_1 of (1), this functor T will be called "contravariant."

The simultaneous isomorphisms

$$\tau(L): \quad L \rightleftarrows T(T(L))$$

compare two *co*variant functors; the first is the identity functor I, composed of the two functions

$$I(L) = L, \qquad I(\lambda) = \lambda;$$

the second is the iterated conjugate functor T^2, with components

$$T^2(L) = T(T(L)), \qquad T^2(\lambda) = T(T(\lambda)).$$

For each L, $\tau(L)$ is constructed as follows. Each vector $x \in L$ and each functional $t \in T(L)$ determine a real number $t(x)$. If in this expression x is fixed while t varies, we obtain a linear transformation of $T(L)$ into R, hence an element y in the double conjugate space $T^2(L)$. This mapping $\tau(L)$ of x to y may also be defined formally by setting $[[\tau(L)]x]t = t(x)$.

The connections between these isomorphisms $\tau(L)$ and the transformations $\lambda: L_1 \to L_2$ may be displayed thus:

$$\begin{array}{ccc} L_1 & \xrightarrow{\tau(L_1)} & T^2(L_1) \\ {\scriptstyle I(\lambda)}\downarrow & & \downarrow{\scriptstyle T^2(\lambda)} \\ L_2 & \xrightarrow{\tau(L_2)} & T^2(L_2) \end{array}$$

The statement that the two possible paths from L_1 to $T^2(L_2)$ in this diagram are in effect identical is what we shall call the "naturality" or "simultaneity" condition for t; explicitly, it reads

$$\tau(L_2)I(\lambda) = T^2(\lambda)\tau(L_1). \tag{3}$$

This equality can be verified from the above definitions of $t(L)$ and $T(\lambda)$ by straightforward substitution. A function t satisfying this "naturality" condition will be called a "natural equivalence" of the functors I and T^2.

On the other hand, the isomorphism of L to its conjugate space $T(L)$ is a comparison of the covariant functor I with the contravariant functor T. Suppose that we are given simultaneous isomorphisms

$$\sigma(L): \quad L \rightleftarrows T(L)$$

for each L. For each linear transformation $\lambda: L_1 \to L_2$ we then have a diagram

$$\begin{array}{ccc} L_1 & \xrightarrow{\sigma(L_1)} & T(L_1) \\ I(\lambda) \downarrow & & \uparrow T(\lambda) \\ L_2 & \xrightarrow{\sigma(L_2)} & T(L_2) \end{array}$$

The only "naturality" condition read from this diagram is $\sigma(L_1) = T(\lambda)\sigma(L_2)\lambda$. Since $\sigma(L_1)$ is an isomorphism, this condition certainly cannot hold unless λ is an isomorphism of L_1 into L_2. Even in the more restricted case in which $L_2 = L_1 = L$ is a single space, there can be no isomorphism $\sigma: L \to T(L)$ which satisfies this naturality condition $\sigma = T(\lambda)\sigma\lambda$ for every nonsingular linear transformation λ(2). Consequently, with our definition of $T(\lambda)$, there is no "natural" isomorphism between the functors I and T, even in a very restricted special case.

Such a consideration of vector spaces and their linear transformations is but one example of many similar mathematical situations; for instance, we may deal with groups and their homomorphisms, with topological spaces and their continuous mappings, with simplicial complexes and their simplicial transformations, with ordered sets and their order preserving transformations. In order to deal in a general way with such situations, we introduce the concept of a *category*. Thus a category $\mathfrak{A}$ will consist of abstract elements of two types: the objects A (for example, vector spaces, groups) and the mappings α (for example, linear transformations, homomorphisms). For some pairs of mappings in the category there is defined a product (in the examples, the product is the usual composite of two transformations). Certain of these mappings act as identities with respect to this product, and there is a one-to-one correspondence between the objects of the category and these identities. A category is subject to certain simple axioms, so formulated as to include all examples of the character described above.

Some of the mappings α of a category will have a formal inverse mapping in the category; such a mapping α is called an equivalence. In the examples quoted the equivalences turn out to be, respectively, the isomorphisms for vector spaces, the homeomorphisms for topological spaces, the isomorphisms for groups and for complexes, and so on.

Most of the standard constructions of a new mathematical object from given objects (such as the construction of the direct product of two groups,

(2) For suppose σ had this property. Then $(x, y) = [\sigma(x)]y$ is a nonsingular bilinear form (not necessarily symmetric) in the vectors x, y of L, and we would have, for every λ, $(x, y) = [\sigma(x)](y) = [T(\lambda)\sigma\lambda x]y = [\sigma\lambda x]\lambda y = (\lambda x, \lambda y)$, so that the bilinear form is left invariant by every nonsingular linear transformation λ. This is clearly impossible.

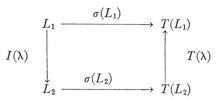

the homology group of a complex, the Galois group of a field) furnish a function $T(A, B, \cdots) = C$ which assigns to given objects $A, B, \cdots$ in definite categories $\mathfrak{A}, \mathfrak{B}, \cdots$ a new object C in a category $\mathfrak{C}$. As in the special case of the conjugate $T(L)$ of a linear space, where there is a corresponding induced mapping $T(\lambda)$, we usually find that mappings $\alpha, \beta, \cdots$ in the categories $\mathfrak{A}, \mathfrak{B}, \cdots$ also induce a definite mapping $T(\alpha, \beta, \cdots) = \gamma$ in the category $\mathfrak{C}$, properly acting on the object $T(A, B, \cdots)$.

These examples suggest the general concept of a functor T on categories $\mathfrak{A}, \mathfrak{B}, \cdots$ to a category $\mathfrak{C}$, defined as an appropriate pair of functions $T(A, B, \cdots)$, $T(\alpha, \beta, \cdots)$. Such a functor may well be covariant in some of its arguments, contravariant in the others. The theory of categories and functors, with a few of the illustrations, constitutes Chapter I.

The natural isomorphism $L \rightarrow T^2(L)$ is but one example of many natural equivalences occurring in mathematics. For instance, the isomorphism of a locally compact abelian group with its twice iterated character group, most of the general isomorphisms in group theory and in the homology theory of complexes and spaces, as well as many equivalences in set theory in general topology satisfy a naturality condition resembling (3). In Chapter II, we provide a general definition of equivalence between functors which includes these cases. A more general notion of a transformation of one functor into another provides a means of comparing functors which may not be equivalent. The general concepts are illustrated by several fairly elementary examples of equivalences and transformations for topological spaces, groups, and Banach spaces.

The third chapter deals especially with groups. In the category of groups the concept of a subgroup establishes a natural partial order for the objects (groups) of the category. For a functor whose values are in the category of groups there is an induced partial order. The formation of a quotient group has as analogue the construction of the quotient functor of a given functor by any normal subfunctor. In the uses of group theory, most groups constructed are obtained as quotient groups of other groups; consequently the operation of building a quotient functor is directly helpful in the representation of such group constructions by functors. The first and second isomorphism theorems of group theory are then formulated for functors; incidentally, this is used to show that these isomorphisms are "natural." The latter part of the chapter establishes the naturality of various known isomorphisms and homomorphisms in group theory[3].

The fourth chapter starts with a discussion of functors on the category of partially ordered sets, and continues with the discussion of limits of direct and inverse systems of groups, which form the chief topic of this chapter.

(3) A brief discussion of this case and of the general theory of functors in the case of groups is given in the authors' note, *Natural isomorphisms in group theory*, Proc. Nat. Acad. Sci. U.S.A. vol. 28 (1942) pp. 537–543.

After suitable categories are introduced, the operations of forming direct and inverse limits of systems of groups are described as functors.

In the fifth chapter we establish the homology and cohomology groups of complexes and spaces as functors and show the naturality of various known isomorphisms of topology, especially those which arise in duality theorems. The treatment of the Čech homology theory utilizes the categories of direct and inverse systems, as discussed in Chapter IV.

The introduction of this study of naturality is justified, in our opinion, both by its technical and by its conceptual advantages.

In the technical sense, it provides the exact hypotheses necessary to apply to both sides of an isomorphism a passage to the limit, in the sense of direct or inverse limits for groups, rings or spaces(4). Indeed, our naturality condition is part of the standard isomorphism condition for two direct or two inverse systems(5).

The study of functors also provides a technical background for the intuitive notion of naturality and makes it possible to verify by straightforward computation the naturality of an isomorphism or of an equivalence in all those cases where it has been intuitively recognized that the isomorphisms are indeed "natural." In many cases (for example, as in the above isomorphism of L to $T(L)$) we can also assert that certain known isomorphisms are in fact "unnatural," relative to the class of mappings considered.

In a metamathematical sense our theory provides general concepts applicable to all branches of abstract mathematics, and so contributes to the current trend towards uniform treatment of different mathematical disciplines. In particular, it provides opportunities for the comparison of constructions and of the isomorphisms occurring in different branches of mathematics; in this way it may occasionally suggest new results by analogy.

The theory also emphasizes that, whenever new abstract objects are constructed in a specified way out of given ones, it is advisable to regard the construction of the corresponding induced mappings on these new objects as an integral part of their definition. The pursuit of this program entails a simultaneous consideration of objects and their mappings (in our terminology, this means the consideration not of individual objects but of categories). This emphasis on the specification of the type of mappings employed gives more insight into the degree of invariance of the various concepts involved. For instance, we show in Chapter III, §16, that the concept of the commutator subgroup of a group is in a sense a more invariant one than that of the center,

(4) Such limiting processes are essential in the transition from the homology theory of complexes to that of spaces. Indeed, the general theory developed here occurred to the authors as a result of the study of the admissibility of such a passage in a relatively involved theorem in homology theory (Eilenberg and MacLane, *Group extensions and homology*, Ann. of Math. vol. 43 (1942) pp. 757–831, especially, p. 777 and p. 815).

(5) H. Freudenthal, *Entwickelung von Räumen und ihren Gruppen*, Compositio Math. vol. 4 (1937) pp. 145–234.

which in its turn is more invariant than the concept of the automorphism group of a group, even though in the classical sense all three concepts are invariant.

The invariant character of a mathematical discipline can be formulated in these terms. Thus, in group theory all the basic constructions can be regarded as the definitions of co- or contravariant functors, so we may formulate the dictum: The subject of group theory is essentially the study of those constructions of groups which behave in a covariant or contravariant manner under induced homomorphisms. More precisely, group theory studies functors defined on well specified categories of groups, with values in another such category.

This may be regarded as a continuation of the Klein Erlanger Programm, in the sense that a geometrical space with its group of transformations is generalized to a category with its algebra of mappings.

Chapter I. Categories and functors

1. **Definition of categories.** These investigations will deal with aggregates such as a class of groups together with a class of homomorphisms, each of which maps one of the groups into another one, or such as a class of topological spaces together with all their continuous mappings, one into another. Consequently we introduce a notion of "category" which will embody the common formal properties of such aggregates.

From the examples "groups plus homomorphisms" or "spaces plus continuous mappings" we are led to the following definition. A *category* $\mathfrak{A} = \{A, \alpha\}$ is an aggregate of abstract elements A (for example, groups), called the *objects* of the category, and abstract elements α (for example, homomorphisms), called *mappings* of the category. Certain pairs of mappings $\alpha_1, \alpha_2 \in \mathfrak{A}$ determine uniquely a product mapping $\alpha = \alpha_2\alpha_1 \in \mathfrak{A}$, subject to the axioms C1, C2, C3 below. Corresponding to each object $A \in \mathfrak{A}$ there is a unique mapping, denoted by e_A or by $e(A)$, and subject to the axioms C4 and C5. The axioms are:

C1. *The triple product $\alpha_3(\alpha_2\alpha_1)$ is defined if and only if $(\alpha_3\alpha_2)\alpha_1$ is defined. When either is defined, the associative law*

$$\alpha_3(\alpha_2\alpha_1) = (\alpha_3\alpha_2)\alpha_1$$

holds. This triple product will be written as $\alpha_3\alpha_2\alpha_1$.

C2. *The triple product $\alpha_3\alpha_2\alpha_1$ is defined whenever both products $\alpha_3\alpha_2$ and $\alpha_2\alpha_1$ are defined.*

Definition. A mapping $e \in \mathfrak{A}$ will be called an *identity* of $\mathfrak{A}$ if and only if the existence of any product $e\alpha$ or βe implies that $e\alpha = \alpha$ and $\beta e = \beta$.

C3. *For each mapping $\alpha \in \mathfrak{A}$ there is at least one identity $e_1 \in \mathfrak{A}$ such that αe_1*

is defined, and at least one identity $e_2 \in \mathfrak{A}$ *such that* $e_2\alpha$ *is defined.*

C4. *The mapping* e_A *corresponding to each object* A *is an identity.*

C5. *For each identity* e *of* $\mathfrak{A}$ *there is a unique object* A *of* $\mathfrak{A}$ *such that* $e_A = e$.

These two axioms assert that the rule $A \to e_A$ provides a one-to-one correspondence between the set of all objects of the category and the set of all its identities. It is thus clear that the objects play a secondary role, and could be entirely omitted from the definition of a category. However, the manipulation of the applications would be slightly less convenient were this done.

LEMMA 1.1. *For each mapping* $\alpha \in \mathfrak{A}$ *there is exactly one object* A_1 *with the product* $\alpha e(A_1)$ *defined, and exactly one* A_2 *with* $e(A_2)\alpha$ *defined.*

The objects A_1, A_2 will be called the *domain* and the *range* of α, respectively. We also say that α acts on A_1 to A_2, and write

$$\alpha: \ A_1 \to A_2 \text{ in } \mathfrak{A}.$$

Proof. Suppose that $\alpha e(A_1)$ and $\alpha e(B_1)$ are both defined. By the properties of an identity, $\alpha e(A_1) = \alpha$, so that axioms C1 and C2 insure that the product $e(A_1)e(B_1)$ is defined. Since both are identities, $e(A_1) = e(A_1)e(B_1) = e(B_1)$, and consequently $A_1 = B_1$. The uniqueness of A_2 is similarly established.

LEMMA 1.2. *The product* $\alpha_2\alpha_1$ *is defined if and only if the range of* α_1 *is the domain of* α_2. *In other words,* $\alpha_2\alpha_1$ *is defined if and only if* $\alpha_1: A_1 \to A_2$ *and* $\alpha_2: A_2 \to A_3$. *In that case* $\alpha_2\alpha_1: A_1 \to A_3$.

Proof. Let $\alpha_1: A_1 \to A_2$. The product $e(A_2)\alpha_1$ is then defined and $e(A_2)\alpha_1 = \alpha_1$. Consequently $\alpha_2\alpha_1$ is defined if and only if $\alpha_2 e(A_2)\alpha_1$ is defined. By axioms C2 and C1 this will hold precisely when $\alpha_2 e(A_2)$ is defined. Consequently $\alpha_2\alpha_1$ is defined if and only if A_2 is the domain of α_2 so that $\alpha_2: A_2 \to A_3$. To prove that $\alpha_2\alpha_1: A_1 \to A_3$ note that since $\alpha_1 e(A_1)$ and $e(A_3)\alpha_2$ are defined the products $(\alpha_2\alpha_1)e(A_1)$ and $e(A_3)(\alpha_2\alpha_1)$ are defined.

LEMMA 1.3. *If* A *is an object,* $e_A: A \to A$.

Proof. If we assume that $e(A): A_1 \to A_2$ then $e(A)e(A_1)$ and $e(A_2)e(A)$ are defined. Since they are all identities it follows that $e(A) = e(A_1) = e(A_2)$ and $A = A_1 = A_2$.

A "left identity" β is a mapping such that $\beta\alpha = \alpha$ whenever $\beta\alpha$ is defined. Axiom C3 shows that every left identity is an identity. Similarly each right identity is an identity. Furthermore, the product ee^1 of two identities is defined if and only if $e = e^1$.

If $\beta\gamma$ is defined and is an identity, β is called a *left inverse* of γ, γ a *right inverse* of β. A mapping α is called an *equivalence* of $\mathfrak{A}$ if it has in $\mathfrak{A}$ at least one left inverse and at least one right inverse.

LEMMA 1.4. *An equivalence α has exactly one left inverse and exactly one right inverse. These inverses are equal, so that the (unique) inverse may be denoted by α^{-1}.*

Proof. It suffices to show that any left inverse β of α equals any right inverse γ. Since $\beta\alpha$ and $\alpha\gamma$ are both defined, $\beta\alpha\gamma$ is defined, by axiom C2. But $\beta\alpha$ and $\alpha\gamma$ are identities, so that $\beta=\beta(\alpha\gamma)=(\beta\alpha)\gamma=\gamma$, as asserted.

For equivalences α, β one easily proves that α^{-1} and $\alpha\beta$ (if defined) are equivalences, and that

$$(\alpha^{-1})^{-1} = \alpha, \qquad (\alpha\beta)^{-1} = \beta^{-1}\alpha^{-1}.$$

Every identity e is an equivalence, with $e^{-1}=e$.

Two objects A_1, A_2 are called *equivalent* if there is an equivalence α such that $\alpha:A_1\rightarrow A_2$. The relation of equivalence between objects is reflexive, symmetric and transitive.

2. **Examples of categories.** In the construction of examples, it is convenient to use the concept of a subcategory. A subaggregate $\mathfrak{A}_0$ of $\mathfrak{A}$ will be called a *subcategory* if the following conditions hold:

1°. *If $\alpha_1, \alpha_2\in\mathfrak{A}_0$ and $\alpha_2\alpha_1$ is defined in $\mathfrak{A}$, then $\alpha_2\alpha_1\in\mathfrak{A}_0$.*
2°. *If $A\in\mathfrak{A}_0$, then $e_A\in\mathfrak{A}_0$.*
3°. *If $\alpha:A_1\rightarrow A_2$ in $\mathfrak{A}$ with $\alpha\in\mathfrak{A}_0$, then $A_1, A_2\in\mathfrak{A}_0$.*

Condition 1° insures that $\mathfrak{A}_0$ is "closed" with respect to multiplication in $\mathfrak{A}$; from conditions 2° and 3° it then follows that $\mathfrak{A}_0$ is itself a category. The intersection of any number of subcategories of $\mathfrak{A}$ is again a subcategory of $\mathfrak{A}$. Note, however, that an equivalence $\alpha\in\mathfrak{A}_0$ of $\mathfrak{A}$ need not remain an equivalence in a subcategory $\mathfrak{A}_0$, because the inverse α^{-1} may not be in $\mathfrak{A}_0$.

For example, if $\mathfrak{A}$ is any category, the aggregate $\mathfrak{A}_e$ of all the objects and all the equivalences of $\mathfrak{A}$ is a subcategory of $\mathfrak{A}$. Also if $\mathfrak{A}$ is a category and S a subclass of its objects, the aggregate $\mathfrak{A}_s$ consisting of all objects of S and all mappings of $\mathfrak{A}$ with both range and domain in S is a subcategory. In fact, every subcategory of $\mathfrak{A}$ can be obtained in two steps: first, form a subcategory $\mathfrak{A}_s$; second, extract from $\mathfrak{A}_s$ a subaggregate, consisting of all the objects of $\mathfrak{A}_s$ and a set of mappings of $\mathfrak{A}_s$ which contains all identities and is closed under multiplication.

The category $\mathfrak{S}$ of all sets has as its objects all sets S(6). A mapping σ of $\mathfrak{S}$ is determined by a pair of sets S_1 and S_2 and a many-one correspondence between S_1 and a subset of S_2, which assigns to each $x\in S_1$ a corresponding element $\sigma x\in S_2$; we then write $\sigma:S_1\rightarrow S_2$. (Note that any deletion of elements from S_1 or S_2 *changes* the mapping σ.) The product of $\sigma_2:S_2^1\rightarrow S_3$ and $\sigma_1:S_1\rightarrow S_2$ is defined if and only if $S_2^1=S_2$; this product then maps S_1 into S_3 by the usual

(6) This category obviously leads to the paradoxes of set theory. A detailed discussion of this aspect of categories appears in §6, below.

composite correspondence $(\sigma_2\sigma_1)x=\sigma_2(\sigma_1 x)$, for each $x\in S_1$(7). The mapping e_S corresponding to the set S is the identity mapping of S onto itself, with $e_S x=x$ for $x\in S$. The axioms C1 through C5 are clearly satisfied. An equivalence $\sigma: S_1\rightarrow S_2$ is simply a one-to-one mapping of S_1 *onto* S_2.

Subcategories of $\mathfrak{S}$ include the category of all finite sets S, with all their mappings as before. For any cardinal number M there are two similar categories, consisting of all sets S of power less than $\mathfrak{m}$ (or, of power less than or equal to $\mathfrak{m}$), together with all their mappings. Subcategories of $\mathfrak{S}$ can also be obtained by restricting the mappings; for instance we may require that each σ is a mapping of S_1 *onto* S_2, or that each σ is a one-to-one mapping of S_1 into a subset of S_2.

The category $\mathfrak{X}$ of all topological spaces has as its objects all topological spaces X and as its mappings all continuous transformations $\xi: X_1\rightarrow X_2$ of a space X_1 into a space X_2. The composition $\xi_2\xi_1$ and the identity e_X are both defined as before. An equivalence in $\mathfrak{X}$ is a homeomorphism (=topological equivalence).

Various subcategories of $\mathfrak{X}$ can again be obtained by restricting the type of topological space to be considered, or by restricting the mappings, say to open mappings or to closed mappings(8).

In particular, $\mathfrak{S}$ can be regarded as a subcategory of $\mathfrak{X}$, namely, as that subcategory consisting of all spaces with a discrete topology.

The category $\mathfrak{G}$ of all topological groups(9) has as its objects all topological groups G and as its mappings γ all those many-one correspondences of a group G_1 into a group G_2 which are homomorphisms(10). The composition and the identities are defined as in $\mathfrak{S}$. An equivalence $\gamma: G_1\rightarrow G_2$ in $\mathfrak{G}$ turns out to be a one-to-one (bicontinuous) isomorphism of G_1 to G_2.

Subcategories of $\mathfrak{G}$ can be obtained by restricting the groups (discrete, abelian, regular, compact, and so on) or by restricting the homomorphisms (open homomorphisms, homomorphisms "onto," and so on).

The category $\mathfrak{B}$ of all Banach spaces is similar; its objects are the Banach spaces B, its mappings all linear transformations β of norm at most 1 of one Banach space into another(11). Its equivalences are the equivalences between two Banach spaces (that is, one-to-one linear transformations which preserve

(7) This formal associative law allows us to write $\sigma_2\sigma_1 x$ without fear of ambiguity. In more complicated formulas, parentheses will be inserted to make the components stand out.

(8) A mapping $\xi: x_1\rightarrow x_2$ is *open* (*closed*) if the image under ξ of every open (closed) subset of X is open (closed) in X_2.

(9) A *topological group* G is a group which is also a topological space in which the group composition and the group inverse are continuous functions (no separation axioms are assumed on the space). If, in addition, G is a Hausdorff space, then all the separation axioms up to and including regularity are satisfied, so that we call G a *regular topological group*.

(10) By a homomorphism we always understand a continuous homomorphism.

(11) For each linear transformation β of the Banach space B_1 into B_2, the norm $\|\beta\|$ is defined as the least upper bound $\|\beta b\|$, for all $b\in B_1$ with $\|b\|=1$.

the norm). The assumption above that the mappings of the category $\mathfrak{B}$ all have norm at most 1 is necessary in order to insure that the equivalences in $\mathfrak{B}$ actually preserve the norm. If one admits arbitrary linear transformations as mappings of the category, one obtains a larger category in which the equivalences are the isomorphisms (that is, one-to-one linear transformations)([12]).

For quick reference, we sometimes describe a category by specifying only the object involved (for example, the category of all discrete groups). In such a case, we imply that the mappings of this category are to be all mappings appropriate to the objects in question (for example, all homomorphisms).

3. **Functors in two arguments.** For simplicity we define only the concept of a functor covariant in one argument and contravariant in another. The generalization to any number of arguments of each type will be immediate.

Let $\mathfrak{A}$, $\mathfrak{B}$, and $\mathfrak{C}$ be three categories. Let $T(A, B)$ be an *object-function* which associates with each pair of objects $A \in \mathfrak{A}$, $B \in \mathfrak{B}$ an object $T(A, B) = C$ in $\mathfrak{C}$, and let $T(\alpha, \beta)$ be a *mapping-function* which associates with each pair of mappings $\alpha \in \mathfrak{A}$, $\beta \in \mathfrak{B}$ a mapping $T(\alpha, \beta) = \gamma \in \mathfrak{C}$. For these functions we formulate certain conditions already indicated in the example in the introduction.

DEFINITION. The object-function $T(A, B)$ and the mapping-function $T(\alpha, \beta)$ form a *functor* T, covariant in $\mathfrak{A}$ and contravariant in $\mathfrak{B}$, with values in $\mathfrak{C}$, if

$$T(e_A, e_B) = e_{T(A,B)}, \tag{3.1}$$

if, whenever $\alpha: A_1 \to A_2$ in $\mathfrak{A}$ and $\beta: B_1 \to B_2$ in $\mathfrak{B}$,

$$T(\alpha, \beta): \quad T(A_1, B_2) \to T(A_2, B_1), \tag{3.2}$$

and if, whenever $\alpha_2\alpha_1 \in \mathfrak{A}$ and $\beta_2\beta_1 \in \mathfrak{B}$,

$$T(\alpha_2\alpha_1, \beta_2\beta_1) = T(\alpha_2, \beta_1)T(\alpha_1, \beta_2). \tag{3.3}$$

Condition (3.2) guarantees the existence of the product of mappings appearing on the right in (3.3).

The formulas (3.2) and (3.3) display the distinction between co- and contravariance. The mapping $T(\alpha, \beta) = \gamma$ induced by α and β acts from $T(A_1, -)$ to $T(A_2, -)$; that is, in the same direction as does α, hence the *co*variance of T in the argument $\mathfrak{A}$. The induced mapping $T(\alpha, \beta)$ at the same time operates in the direction opposite from that of β; thus it is contravariant in $\mathfrak{B}$. Essentially the same shift in direction is indicated by the orders of the factors in formula (3.3) (the covariant α's appear in the same order on both sides; the contravariant β's appear in one order on the left and in the opposite order on the right). With this observation, the requisite formulas for functors in more arguments can be set down.

According to this definition, the functor T is composed of an object func-

([12]) S. Banach, *Théorie des opérations linéaires*, Warsaw, 1932, p. 180.

tion and a mapping function. The latter is the more important of the two; in fact, the condition (3.1) means that it determines the object function and therefore the whole functor, as stated in the following theorem.

THEOREM 3.1. *A function $T(\alpha, \beta)$ which associates to each pair of mappings α and β in the respective categories $\mathfrak{A}$, $\mathfrak{B}$ a mapping $T(\alpha, \beta) = \gamma$ in a third category $\mathfrak{C}$ is the mapping function of a functor T covariant in $\mathfrak{A}$ and contravariant in $\mathfrak{B}$ if and only if the following two conditions hold:*

(i) *$T(e_A, e_B)$ is an identity mapping in $\mathfrak{C}$ for all identities e_A, e_B of $\mathfrak{A}$ and $\mathfrak{B}$.*

(ii) *Whenever $\alpha_2\alpha_1 \in \mathfrak{A}$ and $\beta_2\beta_1 \in \mathfrak{B}$, then $T(\alpha_2, \beta_1)T(\alpha_1, \beta_2)$ is defined and satisfies the equation*

$$T(\alpha_2\alpha_1, \beta_2\beta_1) = T(\alpha_2, \beta_1)T(\alpha_1, \beta_2). \tag{3.4}$$

If $T(\alpha, \beta)$ satisfies (i) *and* (ii), *the corresponding functor T is uniquely determined, with an object function $T(A, B)$ given by the formula*

$$e_{T(A,B)} = T(e_A, e_B). \tag{3.5}$$

Proof. The necessity of (i) and (ii) and the second half of the theorem are obvious.

Conversely, let $T(\alpha, \beta)$ satisfy conditions (i) and (ii). Condition (i) means that an object function $T(A, B)$ can be defined by (3.5). We must show that if $\alpha: A_1 \rightarrow A_2$ and $\beta: B_1 \rightarrow B_2$, then (3.2) holds. Since $e(A_2)\alpha$ and $\beta e(B_1)$ are defined, the product $T(e(A_2), e(B_1))\ T(\alpha, \beta)$ is defined; for similar reasons the product $T(\alpha, \beta)\ T(e(A_1), e(B_2))$ is defined.

In virtue of the definition (3.5), the products

$$e(T(A_2, B_1))T(\alpha, \beta), \qquad T(\alpha, \beta)e(T(A_1, B_2))$$

are defined. This implies (3.2).

In any functor, the replacement of the arguments A, B by equivalent arguments A', B' will replace the value $T(A, B)$ by an equivalent value $T(A', B')$. This fact may be alternatively stated as follows:

THEOREM 3.2. *If T is a functor on $\mathfrak{A}$, $\mathfrak{B}$ to $\mathfrak{C}$, and if $\alpha \in \mathfrak{A}$ and $\beta \in \mathfrak{B}$ are equivalences, then $T(\alpha, \beta)$ is an equivalence in $\mathfrak{C}$, with the inverse $T(\alpha, \beta)^{-1} = T(\alpha^{-1}, \beta^{-1})$.*

For the proof we assume that T is covariant in $\mathfrak{A}$ and contravariant in $\mathfrak{B}$. The products $\alpha\alpha^{-1}$ and $\alpha^{-1}\alpha$ are then identities, and the definition of a functor shows that

$$T(\alpha, \beta)T(\alpha^{-1}, \beta^{-1}) = T(\alpha\alpha^{-1}, \beta^{-1}\beta), \qquad T(\alpha^{-1}, \beta^{-1})T(\alpha, \beta) = T(\alpha^{-1}\alpha, \beta\beta^{-1}).$$

By condition (3.1), the terms on the right are both identities, which means that $T(\alpha^{-1}, \beta^{-1})$ is an inverse for $T(\alpha, \beta)$, as asserted.

4. **Examples of functors.** The same object function may appear in various

functors, as is shown by the following example of one covariant and one contravariant functor both with the same object function. In the category $\mathfrak{S}$ of all sets, the "power" functors P^+ and P^- have the object function

$$P^+(S) = P^-(S) = \text{the set of all subsets of } S.$$

For any many-one correspondence $\sigma: S_1 \to S_2$ the respective mapping functions are defined for any subset $A_1 \subset S_1$ (or $A_2 \subset S_2$) as[13]

$$P^+(\sigma)A_1 = \sigma A_1, \qquad P^-(\sigma)A_2 = \sigma^{-1}A_2.$$

It is immediate that P^+ is a covariant functor and P^- a contravariant one.

The cartesian product $X \times Y$ of two topological spaces is the object function of a functor of two covariant variables X and Y in the category $\mathfrak{X}$ of all topological spaces. For continuous transformations $\xi: X_1 \to X_2$ and $\eta: Y_1 \to Y_2$ the corresponding mapping function $\xi \times \eta$ is defined for any point (x_1, y_1) in the cartesian product $X_1 \times Y_1$ as

$$\xi \times \eta(x_1, y_1) = (\xi x_1, \eta y_1).$$

One verifies that

$$\xi \times \eta: \quad X_1 \times Y_1 \to X_2 \times Y_2,$$

that $\xi \times \eta$ is the identity mapping of $X_1 \times Y_1$ into itself when ξ and η are both identities, and that

$$(\xi_2\xi_1) \times (\eta_2\eta_1) = (\xi_2 \times \eta_2)(\xi_1 \times \eta_1)$$

whenever the products $\xi_2\xi_1$ and $\eta_2\eta_1$ are defined. In virtue of these facts, the functions $X \times Y$ and $\xi \times \eta$ constitute a covariant functor of two variables on the category $\mathfrak{X}$.

The direct product of two groups is treated in exactly similar fashion; it gives a functor with the set function $G \times H$ and the mapping function $\gamma \times \eta$, defined for $\gamma: G_1 \to G_2$ and $\eta: H_1 \to H_2$ exactly as was $\xi \times \eta$. The same applies to the category $\mathfrak{B}$ of Banach spaces, provided one fixes one of the usual possible definite procedures of norming the cartesian product of two Banach spaces.

For a topological space Y and a locally compact (=locally bicompact) Hausdorff space X one may construct the space Y^X of all continuous mappings f of the whole space X into Y ($fx \in Y$ for $x \in X$). A topology is assigned to Y^X as follows. Let C be any compact subset of X, U any open set in Y. Then the set $[C, U]$ of all $f \in Y^X$ with $fC \subset U$ is an open set in Y^X, and the most general open set in Y^X is any union of finite intersections $[C_1, U_1] \cap \cdots \cap [C_n, U_n]$.

This space Y^X may be regarded as the object function of a suitable functor, Map (X, Y). To construct a suitable mapping function, consider any

(13) Here σA_1 is the set of all elements of S_2 of the form σx for $x \in A_1$, while $\sigma^{-1}A_2$ consists of all elements $x \in S_1$ with $\sigma x \in A_2$. When σ is an equivalence, with an inverse τ, $\tau A_2 = \sigma^{-1}A_2$, so that no ambiguity as to the meaning of σ^{-1} can arise.

continuous transformations $\xi: X_1 \to X_2$, $\eta: Y_1 \to Y_2$. For each $f \in Y_1^{X_2}$, one then has mappings acting thus:

$$X_1 \xrightarrow{\xi} X_2 \xrightarrow{f} Y_1 \xrightarrow{\eta} Y_2,$$

so that one may derive a continuous transformation $\eta f \xi$ of $Y_2^{X_1}$. This correspondence $f \to \eta f \xi$ may be shown to be a continuous mapping of $Y_1^{X_2}$ into $Y_2^{X_1}$. Hence we may define object and mapping functions "Map" by setting

$$\text{(4.1)} \qquad \text{Map}\,(X, Y) = Y^X, \qquad [\text{Map}\,(\xi, \eta)]f = \eta f \xi.$$

The construction shows that

$$\text{Map}\,(\xi, \eta): \ \text{Map}\,(X_2, Y_1) \to \text{Map}\,(X_1, Y_2),$$

and hence suggests that this functor is contravariant in X and covariant in Y. One observes at once that Map (ξ, η) is an identity when both ξ and η are identities. Furthermore, if the products $\xi_2\xi_1$ and $\eta_2\eta_1$ are defined, the definition of "Map" gives first,

$$[\text{Map}\,(\xi_2\xi_1, \eta_2\eta_1)]f = \eta_2\eta_1 f \xi_2\xi_1 = \eta_2(\eta_1 f \xi_2)\xi_1,$$

and second,

$$\text{Map}\,(\xi_1, \eta_2)\ \text{Map}\,(\xi_2, \eta_1)f = [\text{Map}\,(\xi_1, \eta_2)]\eta_1 f \xi_2 = \eta_2(\eta_1 f \xi_2)\xi_1.$$

Consequently

$$\text{Map}\,(\xi_2\xi_1, \eta_2\eta_1) = \text{Map}\,(\xi_1, \eta_2)\ \text{Map}\,(\xi_2, \eta_1),$$

which completes the verification that "Map," defined as in (4.1), is a functor on $\mathfrak{X}_{lc}$, $\mathfrak{X}$ to $\mathfrak{X}$, contravariant in the first variable, covariant in the second, where $\mathfrak{X}_{lc}$ denotes the subcategory of $\mathfrak{X}$ defined by the locally compact Hausdorff spaces.

For abelian groups there is a similar functor "Hom." Specifically, let G be a locally compact regular topological group, H a topological abelian group. We construct the set Hom (G, H) of all (continuous) homomorphisms ϕ of G into H. The sum of two such homomorphisms ϕ_1 and ϕ_2 is defined by setting $(\phi_1+\phi_2)g = \phi_1 g + \phi_2 g$, for each $g \in G$[14]; this sum is itself a homomorphism because H is abelian.

Under this addition, Hom (G, H) is an abelian group. It is topologized by the family of neighborhoods $[C, U]$ of zero defined as follows. Given C, any compact subset of G, and U, any open set in H containing the zero of H, $[C; U]$ consists of all $\phi \in$ Hom (G, H) with $\phi C \subset U$. With these definitions, Hom (G, H) is a topological group. If H has a neighborhood of the identity containing no subgroup but the trivial one, one may prove that Hom (G, H) is locally compact.

(14) The group operation in G, H, and so on, will be written as addition.

This function of groups is the object function of a functor "Hom." For given $\gamma: G_1 \to G_2$ and $\eta: H_1 \to H_2$ the mapping function is defined by setting

$$[\text{Hom}(\gamma, \eta)]\phi = \eta\phi\gamma \tag{4.2}$$

for each $\phi \in \text{Hom}(G_2, H_1)$. Formally, this definition is exactly like (4.1). One may show that this definition (4.2) does yield a continuous homomorphism

$$\text{Hom}(\gamma, \eta): \text{Hom}(G_2, H_1) \to \text{Hom}(G_1, H_2).$$

As in the previous case, Hom is a functor with values in the category $\mathfrak{G}_a$ of abelian groups, defined for arguments in two appropriate subcategories of $\mathfrak{G}$, contravariant in the first argument, G, and covariant in the second, H.

For Banach spaces there is a similar functor. If B and C are two Banach spaces, let Lin (B, C) denote the Banach space of all linear transformations λ of B into C, with the usual definition of the norm of the transformation. To describe the corresponding mapping function, consider any linear transformations $\beta: B_1 \to B_2$ and $\gamma: C_1 \to C_2$ with $\|\beta\| \leqq 1$ and $\|\gamma\| \leqq 1$, and set, for each $\lambda \in \text{Lin}(B_2, C_1)$,

$$[\text{Lin}(\beta, \gamma)]\lambda = \gamma\lambda\beta. \tag{4.3}$$

This is in fact a linear transformation

$$\text{Lin}(\beta, \gamma): \text{Lin}(B_2, C_1) \to \text{Lin}(B_1, C_2)$$

of norm at most 1. As in the previous cases, Lin is a functor on $\mathfrak{B}$, $\mathfrak{B}$ to $\mathfrak{B}$, contravariant in its first argument and covariant in the second.

In case C is fixed to be the Banach space R of all real numbers with the absolute value as norm, Lin (B, C) is just the Banach space conjugate to B, in the usual sense. This leads at once to the functor

$$\text{Conj}(B) = \text{Lin}(B, R), \quad \text{Conj}(\beta) = \text{Lin}(\beta, e_R).$$

This is a contravariant functor on $\mathfrak{B}$ to $\mathfrak{B}$.

Another example of a functor on groups is the tensor product $G \circ H$ of two abelian groups. This functor has been discussed in more detail in our Proceedings note cited above.

5. **Slicing of functors.** The last example involved the process of holding one of the arguments of a functor constant. This process occurs elsewhere (for example, in the character group theory, Chapter III below), and falls at once under the following theorem.

THEOREM 5.1. *If T is a functor covariant in $\mathfrak{A}$, contravariant in $\mathfrak{B}$, with values in $\mathfrak{C}$, then for each fixed $B \in \mathfrak{B}$ the definitions*

$$S(A) = T(A, B), \quad S(\alpha) = T(\alpha, e_B)$$

yield a functor S on $\mathfrak{A}$ to $\mathfrak{C}$ with the same variance (in $\mathfrak{A}$) as T.

This "slicing" of a functor may be partially inverted, in that the functor T is determined by its object function and its two "sliced" mapping functions, in the following sense.

THEOREM 5.2. *Let* $\mathfrak{A}$, $\mathfrak{B}$, $\mathfrak{C}$ *be three categories and* $T(A, B)$, $T(\alpha, B)$, $T(A, \beta)$ *three functions such that for each fixed* $B \in \mathfrak{B}$ *the functions* $T(A, B)$, $T(\alpha, B)$ *form a covariant functor on* $\mathfrak{A}$ *to* $\mathfrak{C}$, *while for each* $A \in \mathfrak{A}$ *the functions* $T(A, B)$ *and* $T(A, \beta)$ *give a contravariant functor on* $\mathfrak{B}$ *to* $\mathfrak{C}$. *If in addition for each* $\alpha: A_1 \rightarrow A_2$ *in* $\mathfrak{A}$ *and* $\beta: B_1 \rightarrow B_2$ *in* $\mathfrak{B}$ *we have*

$$T(A_2, \beta)T(\alpha, B_2) = T(\alpha, B_1)T(A_1, \beta), \tag{5.1}$$

then the functions $T(A, B)$ *and*

$$T(\alpha, \beta) = T(\alpha, B_1)T(A_1, \beta) \tag{5.2}$$

form a functor covariant in $\mathfrak{A}$, *contravariant in* $\mathfrak{B}$, *with values in* $\mathfrak{C}$.

Proof. The condition (5.1) merely states the equivalence of the two paths about the following square:

$$\begin{array}{ccc}
T(A_1, B_2) & \xrightarrow{T(\alpha, B_2)} & T(A_2, B_2) \\
{\scriptstyle T(A_1, \beta)} \downarrow & & \downarrow {\scriptstyle T(A_2, \beta)} \\
T(A_1, B_1) & \xrightarrow{T(\alpha, B_1)} & T(A_2, B_1)
\end{array}$$

The result of either path is then taken in (5.2) to define the mapping function, which then certainly satisfies conditions (3.1) and (3.2) of the definition of a functor. The proof of the basic product condition (3.3) is best visualized by writing out a 3×3 array of values $T(A_i, B_j)$.

The significance of this theorem is essentially this: in verifying that given object and mapping functions do yield a functor, one may replace the verification of the product condition (3.3) in two variables by a separate verification, one variable at a time, provided one *also* proves that the order of application of these one-variable mappings can be interchanged (condition (5.1)).

6. **Foundations.** We remarked in §2 that such examples as the "category of *all* sets," the "category of *all* groups" are illegitimate. The difficulties and antinomies here involved are exactly those of ordinary intuitive *Mengenlehre*; no essentially new paradoxes are apparently involved. Any rigorous foundation capable of supporting the ordinary theory of classes would equally well support our theory. Hence we have chosen to adopt the intuitive standpoint, leaving the reader free to insert whatever type of logical foundation (or absence thereof) he may prefer. These ideas will now be illustrated, with particular reference to the category of groups.

It should be observed first that the whole concept of a category is essentially an auxiliary one; our basic concepts are essentially those of a *functor* and of a natural transformation (the latter is defined in the next chapter). The idea of a category is required only by the precept that every function should have a definite class as domain and a definite class as range, for the categories are provided as the domains and ranges of functors. Thus one could drop the category concept altogether and adopt an even more intuitive standpoint, in which a functor such as "Hom" is not defined over the category of "all" groups, but for each particular pair of groups which may be given. The standpoint would suffice for the applications, inasmuch as none of our developments will involve elaborate constructions on the categories themselves.

For a more careful treatment, we may regard a group G as a pair, consisting of a set G_0 and a ternary relation $g \cdot h = k$ on this set, subject to the usual axioms of group theory. This makes explicit the usual tacit assumption that a group is not just the set of its elements (two groups can have the *same* elements, yet different operations). If a pair is constructed in the usual manner as a certain class, this means that each subcategory of the category of "all" groups is a class of pairs; each pair being a class of groups with a class of mappings (binary relations). Any given system of foundations will then legitimize those subcategories which are allowable classes in the system in question.

Perhaps the simplest precise device would be to speak not of *the* category of groups, but of *a* category of groups (meaning, any legitimate such category). A functor such as "Hom" is then a functor which can be defined for any two suitable categories of groups, $\mathfrak{G}$ and $\mathfrak{H}$. Its values lie in a third category of groups, which will in general include groups in neither $\mathfrak{G}$ nor $\mathfrak{H}$. This procedure has the advantage of precision, the disadvantage of a multiplicity of categories and of functors. This multiplicity would be embarrassing in the study of composite functors (§9 below).

One might choose to adopt the (unramified) theory of types as a foundation for the theory of classes. One then can speak of the category $\mathfrak{G}_m$ of all abelian groups of type m. The functor "Hom" could then have both arguments in $\mathfrak{G}_m$, while its values would be in the same category $\mathfrak{G}_{m+k}$ of groups of higher type $m+k$. This procedure affects each functor with the same sort of typical ambiguity adhering to the arithmetical concepts in the Whitehead-Russell development. Isomorphism between groups of different types would have to be considered, as in the simple isomorphism Hom $(\mathfrak{J}, G) \cong G$ (see §10); this would somewhat complicate the natural isomorphisms treated below.

One can also choose a set of axioms for classes as in the Fraenkel-von Neumann-Bernays system. A category is then any (legitimate) class in the sense of this axiomatics. Another device would be that of restricting the cardinal number, considering the category of all denumerable groups, of all groups of cardinal at most the cardinal of the continuum, and so on. The subsequent developments may be suitably interpreted under any one of these viewpoints.

CHAPTER II. NATURAL EQUIVALENCE OF FUNCTORS

7. Transformations of functors. Let T and S be two functors on $\mathfrak{A}$, $\mathfrak{B}$ to $\mathfrak{C}$ which are *concordant*; that is, which have the same variance in $\mathfrak{A}$ and the same variance in $\mathfrak{B}$. To be specific, assume both T and S covariant in $\mathfrak{A}$ and contravariant in $\mathfrak{B}$. Let τ be a function which associates to each pair of objects $A \in \mathfrak{A}$, $B \in \mathfrak{B}$ a mapping $\tau(A, B) = \gamma$ in $\mathfrak{C}$.

DEFINITION. The function τ is a "natural" transformation of the functor T, covariant in $\mathfrak{A}$ and contravariant in $\mathfrak{B}$, into the concordant functor S provided that, for each pair of objects $A \in \mathfrak{A}$, $B \in \mathfrak{B}$,

$$\tau(A, B) : T(A, B) \to S(A, B) \quad \text{in} \quad \mathfrak{C}, \tag{7.1}$$

and provided, whenever $\alpha: A_1 \to A_2$ in $\mathfrak{A}$ and $\beta: B_1 \to B_2$ in $\mathfrak{B}$, that

$$\tau(A_2, B_1) T(\alpha, \beta) = S(\alpha, \beta) \tau(A_1, B_2). \tag{7.2}$$

When these conditions hold, we write

$$\tau : T \to S.$$

If in addition each $\tau(A, B)$ is an equivalence mapping of the category $\mathfrak{C}$, we call τ a *natural equivalence* of T to S (notation: $\tau : T \rightleftarrows S$) and say that the functors T and S are *naturally equivalent*. In this case condition (7.2) can be rewritten as

$$\tau(A_2, B_1) T(\alpha, \beta) [\tau(A_1, B_2)]^{-1} = S(\alpha, \beta). \tag{7.2a}$$

Condition (7.1) of this definition is equivalent to the requirement that both products in (7.2) are always defined. Condition (7.2) is illustrated by the equivalence of the two paths indicated in the following diagram:

$$\begin{array}{ccc} T(A_1, B_2) & \xrightarrow{\quad T(\alpha, \beta) \quad} & T(A_2, B_1) \\ \tau(A_1, B_2) \Big\downarrow & & \Big\downarrow \tau(A_2, B_1) \\ S(A_1, B_2) & \xrightarrow{\quad S(\alpha, \beta) \quad} & S(A_2, B_1) \end{array}$$

Given three concordant functors T, S and R on $\mathfrak{A}$, $\mathfrak{B}$ to $\mathfrak{C}$, with natural transformations $\tau: T \to S$ and $\sigma: S \to R$, the product

$$\rho(A, B) = \sigma(A, B) \tau(A, B)$$

is defined as a mapping in $\mathfrak{C}$, and yields a natural transformation $\rho: T \to R$. If τ and σ are natural equivalences, so is $\rho = \sigma\tau$.

Observe also that if $\tau: T \to S$ is a natural equivalence, then the function τ^{-1} defined by $\tau^{-1}(A, B) = [\tau(A, B)]^{-1}$ is a natural equivalence $\tau^{-1}: S \to T$. Given any functor T on $\mathfrak{A}$, $\mathfrak{B}$ to $\mathfrak{C}$, the function

$$\tau_0(A, B) = e_{T(A,B)}$$

is a natural equivalence $\tau_0 : T \rightleftarrows T$. These remarks imply that the concept of natural equivalence of functors is reflexive, symmetric and transitive.

In demonstrating that a given mapping $\tau(A, B)$ is actually a natural transformation, it suffices to prove the rule (7.2) only in these cases in which all except one of the mappings $\alpha, \beta, \cdots$ is an identity. To state this result it is convenient to introduce a simplified notation for the mapping function when one argument is an identity, by setting

$$T(\alpha, B) = T(\alpha, e_B), \qquad T(A, \beta) = T(e_A, \beta).$$

THEOREM 7.1. *Let T and S be functors covariant in $\mathfrak{A}$ and contravariant in $\mathfrak{B}$, with values in $\mathfrak{C}$, and let τ be a function which associates to each pair of objects $A \in \mathfrak{A}$, $B \in \mathfrak{B}$ a mapping with* (7.1). *A necessary and sufficient condition that τ be a natural transformation $\tau : T \to S$ is that for each mapping $\alpha : A_1 \to A_2$ and each object $B \in \mathfrak{B}$ one has*

$$\tau(A_2, B)T(\alpha, B) = S(\alpha, B)\tau(A_1, B), \tag{7.3}$$

and that, for each $A \in \mathfrak{A}$ and each $\beta : B_1 \to B_2$ one has

$$\tau(A, B_1)T(A, \beta) = S(A, \beta)\tau(A, B_2). \tag{7.4}$$

Proof. The necessity of these conditions is obvious, since they are simply the special cases of (7.2) in which $\beta = e_B$ and $\alpha = e_A$, respectively. The sufficiency can best be illustrated by the following diagram, applying to any mappings $\alpha : A_1 \to A_2$ in $\mathfrak{A}$ and $\beta : B_1 \to B_2$ in $\mathfrak{B}$:

$$\begin{array}{ccccc}
 & T(A_1, B_2) & \xrightarrow{\tau(A_1, B_2)} & S(A_1, B_2) & \\
T(\alpha, B_2) & \downarrow & & \downarrow & S(\alpha, B_2) \\
 & T(A_2, B_2) & \xrightarrow{\tau(A_2, B_2)} & S(A_2, B_2) & \\
T(A_2, \beta) & \downarrow & & \downarrow & S(A_2, \beta) \\
 & T(A_2, B_1) & \xrightarrow{\tau(A_2, B_1)} & S(A_2, B_1) &
\end{array}$$

Condition (7.3) states the equivalence of the results found by following either path around the upper small rectangle, and condition (7.4) makes a similar assertion for the bottom rectangle. Combining these successive equivalences, we have the equivalence of the two paths around the edges of the whole rectangle; this is the requirement (7.2). This argument can be easily set down formally.

8. **Categories of functors.** The functors may be made the objects of a category in which the mappings are natural transformations. Specifically, given three fixed categories $\mathfrak{A}$, $\mathfrak{B}$ and $\mathfrak{C}$, form the category $\mathfrak{T}$ for which the objects are the functors T covariant in $\mathfrak{A}$ and contravariant in $\mathfrak{B}$, with values in $\mathfrak{C}$, and for which the mappings are the natural transformations $\tau: T \rightarrow S$. This requires some caution, because we may have $\tau: T \rightarrow S$ and $\tau: T' \rightarrow S'$ for the same function τ with different functors T, T' (which would have the same object function but different mapping functions). To circumvent this difficulty we define a mapping in the category T to be a triple $[\tau, T, S]$ with $\tau: T \rightarrow S$. The product of mappings $[\tau, T, S]$ and $[\sigma, S', R]$ is defined if and only if $S = S'$; in this case it is

$$[\sigma, S, R][\tau, T, S] = [\sigma\tau, T, R].$$

We verify that the axioms C1–C3 of §1 are satisfied. Furthermore we define, for each functor T,

$$e_T = [\tau_T, T, T], \quad \text{with} \quad \tau_T(A, B) = e_{T(A,B)},$$

and verify the remaining axioms C4, C5. Consequently $\mathfrak{T}$ is a category. In this category it can be proved easily that $[\tau, T, S]$ is an equivalence mapping if and only if $\tau: T \rightleftarrows S$; consequently the concept of the natural equivalence of functors agrees with the concept of equivalence of objects in the category $\mathfrak{T}$ of functors.

This category $\mathfrak{T}$ is useful chiefly in simplifying the statements and proofs of various facts about functors, as will appear subsequently.

9. **Composition of functors.** This process arises by the familiar "function of a function" procedure, in which for the argument of a functor we substitute the value of another functor. For example, let T be a functor on $\mathfrak{A}$, $\mathfrak{B}$ to $\mathfrak{C}$, R a functor on $\mathfrak{C}$, $\mathfrak{D}$ to $\mathfrak{E}$. Then $S = R \otimes (T, I)$, defined by setting

$$S(A, B, D) = R(T(A, B), D), \qquad S(\alpha, \beta, \delta) = R(T(\alpha, \beta), \delta),$$

for objects $A \in \mathfrak{A}$, $B \in \mathfrak{B}$, $D \in \mathfrak{D}$ and mappings $\alpha \in \mathfrak{A}$, $\beta \in \mathfrak{B}$, $\delta \in \mathfrak{D}$, is a functor on $\mathfrak{A}$, $\mathfrak{B}$, $\mathfrak{D}$ to $\mathfrak{E}$. In the argument $\mathfrak{D}$, the variance of S is just the variance of R. The variance of R in $\mathfrak{A}$ (or $\mathfrak{B}$) may be determined by the rule of signs (with + for covariance, − for contravariance): variance of S in $\mathfrak{A}$ = variance of R in $\mathfrak{C} \times$ variance of T in $\mathfrak{A}$.

Composition can also be applied to natural transformations. To simplify the notation, assume that R is a functor in *one* variable, contravariant on $\mathfrak{C}$ to $\mathfrak{E}$, and that T is covariant in $\mathfrak{A}$, contravariant in $\mathfrak{B}$ with values in $\mathfrak{C}$. The composite $R \otimes T$ is then contravariant in $\mathfrak{A}$, covariant in $\mathfrak{B}$. Any pair of natural transformations

$$\rho: R \rightarrow R', \qquad \tau = T \rightarrow T'$$

gives rise to a natural transformation

$$\rho \otimes \tau: R \otimes T' \to R' \otimes T$$

defined by setting

$$\rho \otimes \tau(A, B) = \rho(T(A, B))R(\tau(A, B)).$$

Because ρ is natural, $\rho \otimes \tau$ could equally well be defined as

$$\rho \otimes \tau(A, B) = R'(\tau(A, B))\rho(T'(A, B)).$$

This alternative means that the passage from $R \otimes T'(A, B)$ to $R' \otimes T(A, B)$ can be made either through $R \otimes T(A, B)$ or through $R' \otimes T'(A, B)$, without altering the final result. The resulting *composite transformation* $\rho \otimes \tau$ has all the usual formal properties appropriate to the mapping function of the "functor" $R \otimes T$; specifically,

$$(\rho_2\rho_1) \otimes (\tau_1\tau_2) = (\rho_2 \otimes \tau_2)(\rho_1 \otimes \tau_1),$$

as may be verified by a suitable 3×3 diagram.

These properties show that the functions $R \otimes T$ and $\rho \otimes \tau$ determine a functor C, defined on the categories $\mathfrak{R}$ and $\mathfrak{T}$ of functors, with values in a category $\mathfrak{S}$ of functors, covariant in $\mathfrak{R}$ and contravariant in $\mathfrak{T}$ (because of the contravariance of R). Here $\mathfrak{R}$ is the category of all contravariant functors R on $\mathfrak{C}$ to $\mathfrak{E}$, while $\mathfrak{S}$ and $\mathfrak{T}$ are the categories of all functors S and T, of appropriate variances, respectively. In each case, the mappings of the category of functors are natural transformations, as described in the previous section. To be more explicit, the mapping function $C(\rho, \tau)$ of this functor is not the simple composite $\rho \otimes \tau$, but the triple $[\rho \otimes \tau, R \otimes T', R' \otimes T]$.

Since $\rho \otimes \tau$ is essentially the mapping function of a functor, we know by Theorem 3.2 that if ρ and τ are natural equivalences, then $\rho \otimes \tau$ is an equivalence. Consequently, if the pairs R and R', T and T' are naturally equivalent, so is the pair of composites $R \otimes T$ and $R' \otimes T'$.

It is easy to verify that the composition of functors and of natural transformations is associative, so that symbols like $R \otimes T \otimes S$ may be written without parentheses.

If in the definition of $\rho \otimes \tau$ above it occurs that $T = T'$ and that τ is the identity transformation $T \to T$ we shall write $\rho \otimes T$ instead of $\rho \otimes \tau$. Similarly we shall write $R \otimes \tau$ instead of $\rho \otimes \tau$ in the case when $R = R'$ and ρ is the identity transformation $R \to R$.

10. **Examples of transformations.** The associative and commutative laws for the direct and cartesian products are isomorphisms which can be regarded as equivalences between functors. For example, let X, Y and Z be three topological spaces, and let the homeomorphism

$$(X \times Y) \times Z \cong X \times (Y \times Z) \tag{10.1}$$

be established by the usual correspondence $\tau = \tau(X, Y, Z)$, defined for any point $((x, y), z)$ in the iterated cartesian product $(X\times Y)\times Z$ by

$$\tau(X, Y, Z)((x, y), z) = (x, (y, z)).$$

Each $\tau(X, Y, Z)$ is then an equivalence mapping in the category $\mathfrak{X}$ of spaces. Furthermore each side of (10.1) may be considered as the object function of a covariant functor obtained by composition of the cartesian product functor with itself. The corresponding mapping functions are obtained by the parallel composition as $(\xi\times\eta)\times\zeta$ and $\xi\times(\eta\times\zeta)$. To show that $\tau(X, Y, Z)$ is indeed a natural equivalence, we consider three mappings $\xi: X_1\rightarrow X_2$, $\eta: Y_1\rightarrow Y_2$ and $\zeta: Z_1\rightarrow Z_2$, and show that

$$\tau(X_2, Y_2, Z_2)[(\xi \times \eta) \times \zeta] = [\xi \times (\eta \times \zeta)]\tau(X_1, Y_1, Z_1).$$

This identity may be verified by applying each side to an arbitrary point $((x_1, y_1), z_1)$ in the space $(X_1\times Y_1)\times Z_1$; each transforms it into the point $(\xi x_1, (\eta y_1, \zeta z_1))$ in $X_2\times(Y_2\times Z_2)$.

In similar fashion the homeomorphism $X\times Y\cong Y\times X$ may be interpreted as a natural equivalence, defined as $\tau(X, Y)(x, y)=(y, x)$. In particular, if X, Y and Z are discrete spaces (that is, are simply sets), these remarks show that the associative and commutative laws for the (cardinal) product of two sets are natural equivalences between functors.

For similar reasons, the associative and commutative laws for the direct product of groups are natural equivalences (*or natural isomorphisms*) between functors of groups. The same laws for Banach spaces, with a fixed convention as to the construction of the norm in the cartesian product of two such spaces, are natural equivalences between functors.

If J is the (fixed) additive group of integers, H any topological abelian group, there is an isomorphism

$$\text{Hom}\ (J, H) \cong H \tag{10.2}$$

in which both sides may be regarded as covariant functors of a single argument H. This isomorphism $\tau=\tau(H)$ is defined for any homomorphism $\phi\in\text{Hom}\ (J, H)$ by setting $\tau(H)\phi=\phi(1)\in H$. One observes that $\tau(H)$ is indeed a (bicontinuous) isomorphism, that is, an equivalence in the category of topological abelian groups. That $\tau(H)$ actually is a natural equivalence between functors is shown by proving, for any $\eta: H_1\rightarrow H_2$, that

$$\tau(H_2)\ \text{Hom}\ (e_J, \eta) = \eta\tau(H_1).$$

There is also a second natural equivalence between the functors indicated in (10.2), obtained by setting $\tau'(H)\phi=\phi(-1)$.

With the fixed Banach space R of real numbers there is a similar formula

$$\text{Lin}\ (R, B) \cong B \tag{10.3}$$

for any Banach space B. This gives a natural equivalence $\tau=\tau(B)$ between two covariant functors of one argument in the category $\mathfrak{B}$ of all Banach spaces. Here $\tau(B)$ is defined by setting $\tau(B)l = l(1)$ for each linear transformation $l \in \text{Lin}(R, B)$; another choice of τ would set $\tau(B)l = l(-1)$.

For topological spaces there is a distributive law for the functors "Map" and the direct product functor, which may be written as a natural equivalence

$$\text{Map}(Z, X) \times \text{Map}(Z, Y) \cong \text{Map}(Z, X \times Y) \tag{10.4}$$

between two composite functors, each contravariant in the first argument Z and covariant in the other two arguments X and Y. To define this natural equivalence

$$\tau(X, Y, Z): \text{Map}(Z, X) \times \text{Map}(Z, Y) \rightleftarrows \text{Map}(Z, X \times Y),$$

consider any pair of mappings $f \in \text{Map}(Z, X)$ and $g \in \text{Map}(Z, Y)$ and set, for each $z \in Z$,

$$[\tau(f, g)](z) = (f(z), g(z)).$$

It can be shown that this definition does indeed give the homeomorphism (10.4). It is furthermore natural, which means that, for mappings $\xi: X_1 \to X_2$, $\eta: Y_1 \to Y_2$ and $\zeta: Z_1 \to Z_2$,

$$\tau(X_2, Y_2, Z_1)[\text{Map}(\zeta, \xi) \times \text{Map}(\zeta, \eta)] = \text{Map}(\zeta, \xi \times \eta)\tau(X_1, Y_1, Z_2).$$

The proof of this statement is a straightforward application of the various definitions involved. Both sides are mappings carrying $\text{Map}(Z_2, X_1) \times \text{Map}(Z_2, Y_1)$ into $\text{Map}(Z_1, X_2 \times Y_2)$. They will be equal if they give identical results when applied to an arbitrary element (f_2, g_2) in the first space. These applications give, by the definition of the mapping functions of the functors "Map" and "$\times$," the respective elements

$$\tau(X_2, Y_2, Z_1)(\xi f_2 \zeta, \eta g_2 \zeta), \qquad (\xi \times \eta)\tau(X_1, Y_1, Z_2)(f_2, g_2)\zeta.$$

Both are in $\text{Map}(Z_1, X_2 \times Y_2)$. Applied to an arbitrary $z \in Z_1$, we obtain in both cases, by the definition of τ, the same element $(\xi f_2 \zeta(z), \eta g_2 \zeta(z)) \in X_2 \times Y_2$.

For groups and Banach spaces there are analogous natural equivalences

$$\text{Hom}(G, H) \times \text{Hom}(G, K) \cong \text{Hom}(G, H \times K), \tag{10.5}$$

$$\text{Lin}(B, C) \times \text{Lin}(B, D) \cong \text{Lin}(B, C \times D). \tag{10.6}$$

In each case the equivalence is given by a transformation defined exactly as before. In the formula for Banach spaces we assume that the direct product is normed by the maximum formula. In the case of any other formula for the norm in a direct product, we can assert only that τ is a one-to-one linear transformation of norm one, but not necessarily a transformation preserving the norm. In such a case τ then gives merely a natural transformation of the functor on the left into the functor on the right.

For groups there is another type of distributive law, which is an equivalence transformation,

$$\text{Hom}(G, K) \times \text{Hom}(H, K) \cong \text{Hom}(G \times H, K).$$

The transformation $\tau(G, H, K)$ is defined for each pair $(\phi, \psi) \in \text{Hom}(G, K) \times \text{Hom}(H, K)$ by setting

$$[\tau(G, H, K)(\phi, \psi)](g, h) = \phi g + \psi h$$

for every element (g, h) in the direct product $G \times H$. The properties of τ are proved as before.

It is well known that a function $g(x, y)$ of two variables x and y may be regarded as a function τg of the first variable x for which the values are in turn functions of the second variable y. In other words, τg is defined by

$$[[\tau g](x)](y) = g(x, y).$$

It may be shown that the correspondence $g \rightarrow \tau g$ does establish a homeomorphism between the spaces

$$Z^{X \times Y} \cong (Z^Y)^X,$$

where Z is any topological space and X and Y are locally compact Hausdorff spaces. This is a "natural" homeomorphism, because the correspondence $\tau = \tau(X, Y, Z)$ defined above is actually a natural equivalence

$$\tau(X, Y, Z): \text{Map}(X \times Y, Z) \rightleftarrows \text{Map}(X, \text{Map}(Y, Z))$$

between the two composite functors whose object functions are displayed here.

To prove that τ is natural, we consider any mappings $\xi: X_1 \rightarrow X_2$, $\eta: Y_1 \rightarrow Y_2$, $\zeta: Z_1 \rightarrow Z_2$, and show that

$$(10.7) \quad \tau(X_1, Y_1, Z_2) \text{Map}(\xi \times \eta, \zeta) = \text{Map}(\xi, \text{Map}(\eta, \zeta))\tau(X_2, Y_2, Z_1).$$

Each side of this equation is a mapping which applies to any element $g_2 \in \text{Map}(X_2 \times Y_2, Z_1)$ to give an element of $\text{Map}(X_1, \text{Map}(Y_1, Z_2))$. The resulting elements may be applied to an $x_1 \in X_1$ to give an element of $\text{Map}(Y_1, Z_2)$, which in turn may be applied to any $y_1 \in Y_1$. If each side of (10.7) is applied in this fashion, and simplified by the definitions of τ and of the mapping functions of the functors involved, one obtains in both cases the same element $\zeta g_2(\xi x_1, \eta y_1) \in Z_2$. Hence (10.7) holds, and τ is natural.

Incidentally, the analogous formula for groups uses the tensor product $G \circ H$ of two groups, and gives an equivalence transformation

$$\text{Hom}(G \circ H, K) \cong \text{Hom}(G, \text{Hom}(H, K)).$$

The proof appears in our Proceedings note quoted in the introduction.

Let D be a fixed Banach space, while B and C are two (variable) Banach spaces. To each pair of linear transformations λ and μ, with $\|\lambda\| \leqq 1$ and $\|\mu\| \leqq 1$, and with

$$B \xrightarrow{\lambda} C \xrightarrow{\mu} D,$$

there is associated a composite linear transformation $\mu\lambda$, with $\mu\lambda: B \to D$. Thus there is a correspondence $\tau = \tau(B, C)$ which associates to each $\lambda \in \text{Lin}(B, C)$ a linear transformation $\tau\lambda$ with

$$[\tau\lambda](\mu) = \mu\lambda \in \text{Lin}(B, D).$$

Each $\tau\lambda$ is a linear transformation of Lin (C, D) into Lin (B, D) with norm at most one; consequently τ establishes a correspondence

$$(10.8) \qquad \tau(B, C): \text{Lin}(B, C) \to \text{Lin}(\text{Lin}(C, D), \text{Lin}(B, D)).$$

It can be readily shown that τ itself is a linear transformation, and that $\|\tau(\lambda)\| = \|\lambda\|$, so that τ is an isometric mapping.

This mapping τ actually gives a transformation between the functors in (10.8). If the space D is kept fixed(15), the functions Lin (B, C) and Lin (Lin (C, D), Lin (B, D)) are object functions of functors contravariant in B and covariant in C, with values in the category $\mathfrak{B}$ of Banach spaces. Each $\tau = \tau(B, C)$ is a mapping of this category; thus τ is a natural transformation of the first functor in the second provided that, whenever $\beta: B_1 \to B_2$ and $\gamma: C_1 \to C_2$,

$$(10.9) \qquad \tau(B_1, C_2) \text{ Lin}(\beta, \gamma) = \text{Lin}(\text{Lin}(\gamma, e), \text{Lin}(\beta, e))\tau(B_2, C_1),$$

where $e = e_D$ is the identity mapping of D into itself. Each side of (10.9) is a mapping of Lin (B_2, C_1) into Lin (Lin (C_2, D), Lin (B_1, D)). Apply each side to any $\lambda \in \text{Lin}(B_2, C_1)$, and let the result act on any $\mu \in \text{Lin}(C_2, D)$. On the left side, the result of these applications simplifies as follows (in each step the definition used is cited at the right):

$$\{[\tau(B_1, C_2)] \text{ Lin}(\beta, \gamma)\lambda\}\mu$$
$$= \{[\tau(B_1, C_2)](\gamma\lambda\beta)\}\mu \qquad \text{(Definition of Lin } (\beta, \gamma))$$
$$= \mu\gamma\lambda\beta \qquad \text{(Definition of } \tau(B_1, C_2)).$$

The right side similarly becomes

$$\{\text{Lin}(\text{Lin}(\gamma, e), \text{Lin}(\beta, e))[\tau(B_2, C_1)\lambda]\}\mu$$
$$= \{\text{Lin}(\beta, e)[\tau(B_2, C_1)\lambda] \text{ Lin}(\gamma, e)\}\mu \qquad \text{(Definition of Lin } (—, —))$$
$$= \text{Lin}(\beta, e)\{[\tau(B_2, C_1)\lambda](\mu\gamma)\} \qquad \text{(Definition of Lin } (\gamma, e))$$
$$= \text{Lin}(\beta, e)(\mu\gamma\lambda) \qquad \text{(Definition of } \tau(B_2, C_1))$$
$$= \mu\gamma\lambda\beta \qquad \text{(Definition of Lin } (\beta, e)).$$

(15) We keep the space D fixed because in one of these functors it appears twice, once as a covariant argument and once as a contravariant one.

The identity of these two results shows that τ is indeed a natural transformation of functors.

In the special case when D is the space of real numbers, Lin (C, D) is simply the conjugate space Conj (C). Thus we have the natural transformation

$$(10.10) \qquad \tau(B, C):\text{Lin }(B, C) \rightarrow \text{Lin }(\text{Conj } C, \text{Conj } B).$$

A similar argument for locally compact abelian groups G and H yields a natural transformation

$$(10.11) \qquad \tau(G, H):\text{Hom }(G, H) \rightarrow \text{Hom }(\text{Ch } H, \text{Ch } G).$$

In the theory of character groups it is shown that each $\tau(G, H)$ is an isomorphism, so (10.11) is actually a natural isomorphism. The well known isomorphism between a locally compact abelian group G and its twice iterated character group is also a natural isomorphism

$$\tau(G):G \rightleftarrows \text{Ch }(\text{Ch } G)$$

between functors[16]. The analogous natural transformation

$$\tau(B):B \rightarrow \text{Conj }(\text{Conj } B)$$

for Banach spaces is an equivalence only when B is restricted to the category of reflexive Banach spaces.

11. **Groups as categories.** Any group G may be regarded as a category $\mathfrak{G}_G$ in which there is only one object. This object may either be the set G or, if G is a transformation group, the space on which G acts. The mappings of the category are to be the elements γ of the group G, and the product of two elements in the group is to be their product as mappings in the category. In this category every mapping is an equivalence, and there is only one identity mapping (the unit element of G). A covariant functor T with one argument in $\mathfrak{G}_G$ and with values in (the category of) the group H is just a homomorphic mapping $\eta = T(\gamma)$ of G into H. A natural transformation τ of one such functor T_1 into another one, T_2, is defined by a single element $\tau(G) = \eta_0 \in H$. Since η_0 has an inverse, every natural transformation is automatically an equivalence. The naturality condition (7.2a) for τ becomes simply $\eta_0 T_1(\gamma)\eta_0^{-1} = T_2(\gamma)$. Thus the functors T_1 and T_2 are naturally equivalent if and only if T_1 and T_2, considered as homomorphisms, are conjugate.

Similarly, a contravariant functor T on a group G, considered as a category, is simply a "dual" or "counter" homomorphism $(T(\gamma_2\gamma_1) = T(\gamma_1)T(\gamma_2))$.

A ring R with unity also gives a category, in which the mappings are the elements of R, under the operation of multiplication in R. The unity of the ring is the only identity of the category, and the units of the ring are the equivalences of the category.

(16) The proof of naturality appears in the note quoted in footnote 3.

12. **Construction of functors as transforms.** Under suitable conditions a mapping-function $\tau(A, B)$ acting on a given functor $T(A, B)$ can be used to construct a new functor S such that $\tau: T \to S$. The case in which each τ is an equivalence mapping is the simplest, so will be stated first.

THEOREM 12.1. *Let T be a functor covariant in $\mathfrak{A}$, contravariant in $\mathfrak{B}$, with values in $\mathfrak{C}$. Let S and τ be functions which determine for each pair of objects $A \in \mathfrak{A}$, $B \in \mathfrak{B}$ an object $S(A, B)$ in $\mathfrak{C}$ and an equivalence mapping*

$$\tau(A, B): T(A, B) \to S(A, B) \quad \textit{in} \quad \mathfrak{C}.$$

Then S is the object function of a uniquely determined functor S, concordant with T and such that τ is a natural equivalence $\tau: T \rightleftarrows S$.

Proof. One may readily show that the mapping function appropriate to S is uniquely determined for each $\alpha: A_1 \to A_2$ in $\mathfrak{A}$ and $\beta: B_1 \to B_2$ in $\mathfrak{B}$ by the formula

$$S(\alpha, \beta) = \tau(A_2, B_1) T(\alpha, \beta) [\tau(A_1, B_2)]^{-1}.$$

The companion theorem for the case of a transformation which is not necessarily an equivalence is somewhat more complicated. We first define mappings cancellable from the right. A mapping $\alpha \in \mathfrak{A}$ will be called cancellable from the right if $\beta\alpha = \gamma\alpha$ always implies $\beta = \gamma$. To illustrate, if each "formal" mapping is an actual many-to-one mapping of one set into another, and if the composition of formal mappings is the usual composition of correspondences, it can be shown that every mapping α of one set *onto* another is cancellable from the right.

THEOREM 12.2. *Let T be a functor covariant in $\mathfrak{A}$ and contravariant in $\mathfrak{B}$, with values in $\mathfrak{C}$. Let $S(A, B)$ and $S(\alpha, \beta)$ be two functions on the objects (and mappings) of $\mathfrak{A}$ and $\mathfrak{B}$, for which it is assumed only, when $\alpha: A_1 \to A_2$ in $\mathfrak{A}$ and $\beta: B_1 \to B_2$ in $\mathfrak{B}$, that*

$$S(\alpha, \beta): S(A_1, B_2) \to S(A_2, B_1) \quad \textit{in} \quad \mathfrak{C}.$$

If a function τ on the objects of $\mathfrak{A}$, $\mathfrak{B}$ to the mappings of $\mathfrak{C}$ satisfies the usual conditions for a natural transformation $\tau: T \to S$; namely that

(12.1) $$\tau(A, B): T(A, B) \to S(A, B) \quad \textit{in} \quad \mathfrak{C},$$

(12.2) $$\tau(A_2, B_1) T(\alpha, \beta) = S(\alpha, \beta) \tau(A_1, B_2),$$

and if in addition each $\tau(A, B)$ is cancellable from the right, then the functions $S(\alpha, \beta)$ and $S(A, B)$ form a functor S, concordant with T, and τ is a transformation $\tau: T \to S$.

Proof. We need to show that

(12.3) $$S(e_A, e_B) = e_{S(A,B)},$$

$$(12.4)\qquad S(\alpha_2\alpha_1, \beta_2\beta_1) = S(\alpha_2, \beta_1)S(\alpha_1, \beta_2).$$

Since T is a functor, $T(e_A, e_B)$ is an identity, so that condition (12.2) with $A_1=A_2$, $B_1=B_2$ becomes

$$\tau(A, B) = S(e_A, e_B)\tau(A, B).$$

Because $\tau(A, B)$ is cancellable from the right, it follows that $S(e_A, e_B)$ must be the identity mapping of $S(A, B)$, as desired.

To consider the second condition, let $\alpha_1: A_1 \rightarrow A_2$, $\alpha_2: A_2 \rightarrow A_3$, $\beta_1: B_1 \rightarrow B_2$ and $\beta_2: B_2 \rightarrow B_3$, so that $\alpha_2\alpha_1$ and $\beta_2\beta_1$ are defined. By condition (12.2) and the properties of the functor T,

$$\begin{aligned} S(\alpha_2\alpha_1, \beta_2\beta_1)\tau(A_1, B_3) &= \tau(A_3, B_1)T(\alpha_2\alpha_1, \beta_2\beta_1) \\ &= \tau(A_3, B_1)T(\alpha_2, \beta_1)T(\alpha_1, \beta_2) \\ &= S(\alpha_2, \beta_1)\tau(A_2, B_2)T(\alpha_1, \beta_2) \\ &= S(\alpha_2, \beta_1)S(\alpha_1, \beta_2)\tau(A_1, B_3). \end{aligned}$$

Again because $\tau(A_1, B_3)$ may be cancelled on the right, (12.4) follows.

13. **Combination of the arguments of functors.** For n given categories $\mathfrak{A}_1, \cdots, \mathfrak{A}_n$, the cartesian product category

$$(13.1)\qquad \mathfrak{A} = \prod_i \mathfrak{A}_i = \mathfrak{A}_1 \times \mathfrak{A}_2 \times \cdots \times \mathfrak{A}_n$$

is defined as a category in which the objects are the n-tuples of objects $[A_1, \cdots, A_n]$, with $A_i \in \mathfrak{A}_i$, the mappings are the n-tuples $[\alpha_1, \cdots, \alpha_n]$ of mappings $\alpha_i \in \mathfrak{A}_i$. The product

$$[\alpha_1, \cdots, \alpha_n][\beta_1, \cdots, \beta_n] = [\alpha_1\beta_1, \cdots, \alpha_n\beta_n]$$

is defined if and only if each individual product $\alpha_i\beta_i$ is defined in $\mathfrak{A}_i$, for $i=1, \cdots, n$. The identity corresponding to the object $[A_1, \cdots, A_n]$ in the product category is to be the mapping $[e(A_1), \cdots, e(A_n)]$. The axioms which assert that the product $\mathfrak{A}$ is a category follow at once. The natural correspondence

$$(13.2)\qquad P(A_1, \cdots, A_n) = [A_1, \cdots, A_n],$$

$$(13.3)\qquad P(\alpha_1, \cdots, \alpha_n) = [\alpha_1, \cdots, \alpha_n]$$

is a covariant functor on the n categories $\mathfrak{A}_1, \cdots, \mathfrak{A}_n$ to the product category. Conversely, the correspondences given by "projection" into the ith coordinate,

$$(13.4)\qquad Q_i([A_1, \cdots, A_n]) = A_i, \qquad Q_i([\alpha_1, \cdots, \alpha_n]) = \alpha_i,$$

is a covariant functor in one argument, on $\mathfrak{A}$ to $\mathfrak{A}_i$.

It is now possible to represent a functor covariant in any number of arguments as a functor in one argument. Let T be a functor on the categories $\mathfrak{A}_1, \cdots, \mathfrak{A}_n, \mathfrak{B}$, with the same variance in $\mathfrak{A}_i$ as in $\mathfrak{A}_1$; define a new functor T^* by setting

$$T^*([A_1, \cdots, A_n], B) = T(A_1, \cdots, A_n, B),$$
$$T^*([\alpha_1, \cdots, \alpha_n], \beta) = T(\alpha_1, \cdots, \alpha_n, \beta).$$

This is a functor, since it is a composite of T and the projections Q_i of (13.4); its variance in the first argument is that of T in any A_i. Conversely, each functor S with arguments in $\mathfrak{A}_1 \times \cdots \times \mathfrak{A}_n$ and $\mathfrak{B}$ can be represented as $S = T^*$, for a T with $n+1$ arguments in $\mathfrak{A}_1, \cdots, \mathfrak{A}_n, \mathfrak{B}$, defined by

$$T(A_1, \cdots, A_n, B) = S([A_1, \cdots, A_n], B) = S(P(A_1, \cdots, A_n), B),$$
$$T(\alpha_1, \cdots, \alpha_n, \beta) = S([\alpha_1, \cdots, \alpha_n], \beta) = S(P(\alpha_1, \cdots, \alpha_n), \beta).$$

Again T is a composite functor. These reduction arguments combine to give the following theorem.

THEOREM 13.1. *For given categories $\mathfrak{A}_1, \cdots, \mathfrak{A}_n, \mathfrak{B}_1, \cdots, \mathfrak{B}_m, \mathfrak{C}$, there is a one-to-one correspondence between the functors T covariant in $\mathfrak{A}_1, \cdots, \mathfrak{A}_n$, contravariant in $\mathfrak{B}_1, \cdots, \mathfrak{B}_m$, with values in $\mathfrak{C}$, and the functors S in two arguments, covariant in $\mathfrak{A}_1 \times \cdots \times \mathfrak{A}_n$ and contravariant in $\mathfrak{B}_1 \times \cdots \times \mathfrak{B}_m$, with values in the same category $\mathfrak{C}$. Under this correspondence, equivalent functors T correspond to equivalent functors S, and a natural transformation $\tau: T_1 \rightarrow T_2$ gives rise to a natural transformation $\sigma: S_1 \rightarrow S_2$ between the functors S_1 and S_2 corresponding to T_1 and T_2 respectively.*

By this theorem, all functors can be reduced to functors in two arguments. To carry this reduction further, we introduce the concept of a "dual" category.

Given a category $\mathfrak{A}$, the dual category $\mathfrak{A}^*$ is defined as follows. The objects of $\mathfrak{A}^*$ are those of $\mathfrak{A}$; the mappings α^* of $\mathfrak{A}^*$ are in a one-to-one correspondence $\alpha \rightleftarrows \alpha^*$ with the mappings of $\mathfrak{A}$. If $\alpha: A_1 \rightarrow A_2$ in $\mathfrak{A}$, then $\alpha^*: A_2 \rightarrow A_1$ in $\mathfrak{A}^*$. The composition law is defined by the equation

$$\alpha_2^* \alpha_1^* = (\alpha_1 \alpha_2)^*,$$

if $\alpha_1 \alpha_2$ is defined in $\mathfrak{A}$. We verify that $\mathfrak{A}^*$ is a category and that there are equivalences

$$(\mathfrak{A}^*)^* \cong \mathfrak{A}, \qquad \prod_i \mathfrak{A}_i^* \cong (\prod \mathfrak{A}_i)^*.$$

The mapping

$$D(A) = A, \qquad D(\alpha) = \alpha^*$$

is a contravariant functor on $\mathfrak{A}$ to $\mathfrak{A}^*$, while D^{-1} is contravariant on $\mathfrak{A}^*$ to $\mathfrak{A}$.

Any contravariant functor T on $\mathfrak{A}$ to $\mathfrak{C}$ can be regarded as a covariant functor T^* on $\mathfrak{A}^*$ to $\mathfrak{C}$, and vice versa. Explicitly, T^* is defined as a composite

$$T^*(A) = T(D^{-1}(A)), \qquad T^*(\alpha^*) = T(D^{-1}(\alpha^*)).$$

Hence we obtain the following reduction theorem.

THEOREM 13.2. *Every functor T covariant on $\mathfrak{A}_1, \cdots, \mathfrak{A}_n$ and contravariant on $\mathfrak{B}_1, \cdots, \mathfrak{B}_m$ with values in $\mathfrak{C}$ may be regarded as a covariant functor T' on*

$$\left(\prod_i \mathfrak{A}_i\right) \times \left(\prod_j \mathfrak{B}_j^*\right)$$

with values in $\mathfrak{C}$, and vice versa. Each natural transformation (or equivalence) $\tau: T_1 \rightarrow T_2$ yields a corresponding transformation (or equivalence) $\tau': T_1' \rightarrow T_2'$.

CHAPTER III. FUNCTORS AND GROUPS

14. Subfunctors. This chapter will develop the fashion in which various particular properties of groups are reflected by properties of functors with values in a category of groups. The simplest such case is the fact that subgroups can give rise to "subfunctors." The concept of subfunctor thus developed applies with equal force to functors whose values are in the category of rings, spaces, and so on.

In the category $\mathfrak{G}$ of all topological groups we say that a mapping $\gamma': G_1' \rightarrow G_2'$ is a *submapping* of a mapping $\gamma: G_1 \rightarrow G_2$ (notation: $\gamma' \subset \gamma$) whenever $G_1' \subset G_1$, $G_2' \subset G_2$ and $\gamma'(g_1) = \gamma(g_1)$ for each $g_1 \in G_1'$. Here $G_1' \subset G_1$ means of course that G_1' is a subgroup (not just a subset) of G_1.

Given two concordant functors T' and T on $\mathfrak{A}$ and $\mathfrak{B}$ to $\mathfrak{G}$, we say that T' is a subfunctor of T (notation: $T' \subset T$) provided $T'(A, B) \subset T(A, B)$ for each pair of objects $A \in \mathfrak{A}$, $B \in \mathfrak{B}$ and $T'(\alpha, \beta) \subset T(\alpha, \beta)$ for each pair of mappings $\alpha \in \mathfrak{A}$, $\beta \in \mathfrak{B}$. Clearly $T' \subset T$ and $T \subset T'$ imply $T = T'$; furthermore this inclusion satisfies the transitive law. If T' and T'' are both subfunctors of the same functor T, then in order to prove that $T' \subset T''$ it is sufficient to verify that $T'(A, B) \subset T''(A, B)$ for all A and B.

A subfunctor can be completely determined by giving its object function alone. The requisite properties for this object function may be specified as follows:

THEOREM 14.1. *Let the functor T covariant in $\mathfrak{A}$ and contravariant in $\mathfrak{B}$ have values in the category $\mathfrak{G}$ of groups, while T' is a function which assigns to each pair of objects $A \in \mathfrak{A}$ and $B \in \mathfrak{B}$ a subgroup $T'(A, B)$ of $T(A, B)$. Then T' is the object function of a subfunctor of T if and only if for each $\alpha: A_1 \rightarrow A_2$ in $\mathfrak{A}$ and each $\beta: B_1 \rightarrow B_2$ in $\mathfrak{B}$ the mapping $T(\alpha, \beta)$ carries the subgroup $T'(A_1, B_2)$ into part of $T'(A_2, B_1)$. If T' satisfies this condition, the corresponding mapping function is uniquely determined.*

Proof. The necessity of this condition is immediate. Conversely, to prove the sufficiency, we define for each α and β a homomorphism $T'(\alpha, \beta)$ of $T'(A_1, B_2)$ into $T'(A_2, B_1)$ by setting $T'(\alpha, \beta)g = T(\alpha, \beta)g$, for each $g \in T'(A_1, B_2)$. The fact that T' satisfies the requisite conditions for the mapping function of a functor is then immediate, since T' is obtained by "cutting down" T.

The concept of a subtransformation may also be defined. If T, S, T', S' are concordant functors on $\mathfrak{A}, \mathfrak{B}$ to $\mathfrak{G}$, and if $\tau: T \to S$ and $\tau': T' \to S'$ are natural transformations, we say that τ' is a *subtransformation* of τ (notation: $\tau' \subset \tau$) if $T' \subset T$, $S' \subset S$ and if, for each pair of arguments A, B, $\tau'(A, B)$ is a submapping of $\tau(A, B)$. Any such subtransformation of τ may be obtained by suitably restricting both the domain and the range of τ. Explicitly, let $\tau: T \to S$, let $T' \subset T$ and $S' \subset S$ be such that for each A, B, $\tau(A, B)$ maps the subgroup $T'(A, B)$ of $T(A, B)$ into the subgroup $S'(A, B)$ of $S(A, B)$. If then $\tau'(A, B)$ is defined as the homomorphism $\tau(A, B)$ with its domain restricted to the subgroup $T'(A, B)$ and its range restricted to the subgroup $S'(A, B)$, it follows readily that τ' is indeed a natural transformation $\tau': T' \to S'$.

Let τ be a natural transformation $\tau: T \to S$ of concordant functors T and S on $\mathfrak{A}$ and $\mathfrak{B}$ to the category $\mathfrak{G}$ of groups. If T' is a subfunctor of T, then the map of each $T'(A, B)$ under $\tau(A, B)$ is a subgroup of $S(A, B)$, so that we may define an object function

$$S'(A, B) = \tau(A, B)[T'(A, B)], \qquad A \in \mathfrak{A}, B \in \mathfrak{B}.$$

The naturality condition on τ shows that the function S' satisfies the condition of Theorem 14.1; hence $S' = \tau T'$ gives a subfunctor of S, called the τ-*transform* of T'. Furthermore there is a natural transformation $\tau': T' \to S'$, obtained by restricting τ. In particular, if τ is a natural equivalence, so is τ'.

Conversely, for a given $\tau: T \to S$ let S'' be a subfunctor of S. The inverse image of each subgroup $S''(A, B)$ under the homomorphism $\tau(A, B)$ is then a subgroup of $T(A, B)$, hence gives an object function

$$T''(A, B) = \tau(A, B)^{-1}[S''(A, B)], \qquad A \in \mathfrak{A}, B \in \mathfrak{B}.$$

As before, this is the object function of a subfunctor $T'' \subset T$ which may be called the inverse transform $\tau^{-1}S'' = T''$ of S''. Again, τ may be restricted to give a natural transformation $\tau'': T'' \to S''$. In case each $\tau(A, B)$ is a homomorphism of $T(A, B)$ *onto* $S(A, B)$, we may assert that $\tau(\tau^{-1}S'') = S''$.

Lattice operations on subgroups can be applied to functors. If T' and T'' are two subfunctors of a functor T with values in G, we define their meet $T' \cap T''$ and their join $T' \cup T''$ by giving the object functions,

$$[T' \cap T''](A, B) = T'(A, B) \cap T''(A, B),$$
$$[T' \cup T''](A, B) = T'(A, B) \cup T''(A, B).$$

We verify that the condition of Theorem 14.1 is satisfied here, so that these object functions do uniquely determine corresponding subfunctors of T. Any lattice identity for groups may then be written directly as an identity for the subfunctors of a fixed functor T with values in $\mathfrak{G}$.

15. **Quotient functors.** The operation of forming a quotient group leads to an analogous operation of taking the "quotient functor" of a functor T by a "normal" subfunctor T'. If T is a functor covariant in $\mathfrak{A}$ and contravariant in $\mathfrak{B}$, with values in $\mathfrak{G}$, a *normal subfunctor* T' will mean a subfunctor $T'\subset T$ such that each $T'(A, B)$ is a normal subgroup of $T(A, B)$, while a *closed* subfunctor T' will be one in which each $T'(A, B)$ is a closed subgroup of the topological group $T(A, B)$. If T' is a normal subfunctor of T, the quotient functor $Q=T/T'$ has an object function given as the factor group,

$$Q(A, B) = T(A, B)/T'(A, B).$$

For homomorphisms $\alpha:A_1\rightarrow A_2$ and $\beta:B_1\rightarrow B_2$ the corresponding mapping function $Q(\alpha, \beta)$ is defined for each coset[17] $x+T'(A_1, B_2)$ as

$$Q(\alpha, \beta)[x + T'(A_1, B_2)] = [T(\alpha, \beta)x] + T'(A_2, B_1).$$

We verify at once that Q thus gives a uniquely defined homomorphism,

$$Q(\alpha, \beta):Q(A_1, B_2) \rightarrow Q(A_2, B_1).$$

Before we prove that Q is actually a functor, we introduce for each $A\in\mathfrak{A}$ and $B\in\mathfrak{B}$ the homomorphism

$$\nu(A, B):T(A, B) \rightarrow Q(A, B)$$

defined for each $x\in T(A, B)$ by the formula

$$\nu(A, B)(x) = x + T'(A, B).$$

When $\alpha:A_1\rightarrow A_2$ and $\beta:B_1\rightarrow B_2$ we now show that

$$Q(\alpha, \beta)\nu(A_1, B_2) = \nu(A_2, B_1)T(\alpha, \beta).$$

For, given any $x\in T(A_1, B_2)$, the definitions of ν and Q give at once

$$\begin{aligned} Q(\alpha, \beta)[\nu(A_1, B_2)(x)] &= Q(\alpha, \beta)[x + T'(A_1, B_2)] \\ &= [T(\alpha, \beta)(x)] + T'(A_2, B_1) \\ &= \nu(A_2, B_1)[T(\alpha, \beta)(x)]. \end{aligned}$$

Notice also that $\nu(A, B)$ maps $T(A, B)$ *onto* the factor group $Q(A, B)$, hence is cancellable from the right. Therefore, Theorem 12.2 shows that $Q=T/T'$ is a functor, and that ν is a natural transformation $\nu:T\rightarrow T/T'$. We may call ν *the* natural transformation of T onto T/T'.

In particular, if the functor T has its values in the category of regular topological groups, while T' is a *closed* normal subfunctor of T, the quotient

[17] For convenience in notation we write the group operations (commutative or not) with a plus sign.

functor T/T' has its values in the same category of groups, since a quotient group of a regular topological group by a *closed* subgroup is again regular.

To consider the behavior of quotient functors under natural transformations we first recall some properties of homomorphisms. Let $\alpha: G \rightarrow H$ be a homomorphism of the group G into H, while $\alpha': G' \rightarrow H'$ is a submapping of α, with G' and H' normal subgroups of G and H, respectively, and ν and μ are the natural homomorphisms $\nu: G \rightarrow G/G'$, $\mu: H \rightarrow H/H'$. Then we may define a homomorphism $\beta: G/G' \rightarrow H/H'$ by setting $\beta(x+G') = \alpha x + H'$ for each $x \in G$. This homomorphism is the only mapping of G/G' into H/H' with the property that $\beta\nu = \mu\alpha$, as indicated in the figure

$$\begin{array}{ccc} G & \xrightarrow{\alpha} & H \\ \nu \downarrow & & \downarrow \mu \\ G/G' & \xrightarrow{\beta} & H/H' \end{array}$$

We may write $\beta = \alpha/\alpha'$. The corresponding statement for functors is as follows.

THEOREM 15.1. *Let $\tau: T \rightarrow S$ be a natural transformation between functors with values in $\mathfrak{G}$; and let $\tau': T' \rightarrow S'$ be a subtransformation of τ such that T' and S' are normal subfunctors of T and S, respectively. Then the definition $\rho(A, B) = \tau(A, B)/\tau'(A, B)$ gives a natural transformation $\rho = \tau/\tau'$,*

$$\rho: T/T' \rightarrow S/S'.$$

Furthermore, $\rho\nu = \mu\tau$, where ν is the natural transformation $\nu: T \rightarrow T/T'$ and μ is the natural transformation $\mu: S \rightarrow S/S'$.

Proof. This requires only the verification of the naturality condition for ρ, which follows at once from the relevant definitions.

The "kernel" of a transformation appears as a special case of this theorem. Let $\tau: T \rightarrow S$ be given, and take S' to be the identity-element subfunctor of S; that is, let each $S'(A, B)$ be the subgroup consisting only of the identity (zero) element of $S(A, B)$. Then the inverse transform $T' = \tau^{-1}S'$ is by §14 a (normal) subfunctor of T, and τ may be restricted to give the natural transformation $\tau': T' \rightarrow S'$. We may call T' the kernel functor of the transformation τ. Theorem 15.1 applied in this case shows that there is then a natural transformation $\rho: T/T' \rightarrow S$ such that $\rho = \tau\nu$. Furthermore each $\rho(A, B)$ is a one-to-one mapping of the quotient group $T(A, B)/T'(A, B)$ into $S(A, B)$. If in addition we assume that each $\tau(A, B)$ is an open mapping of $T(A, B)$ *onto* $S(A, B)$, we may conclude, exactly as in group theory, that ρ is a natural equivalence.

16. **Examples of subfunctors.** Many characteristic subgroups of a group

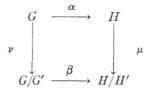

may be written as subfunctors of the identity functor. The (covariant) identity functor I on $\mathfrak{G}$ to $\mathfrak{G}$ is defined by setting

$$I(G) = G, \qquad I(\gamma) = \gamma.$$

Any subfunctor of I is, by Theorem 14.1, determined by an object function

$$T(G) \subset G$$

such that whenever γ maps G_1 homomorphically into G_2, then $\gamma[T(G_1)] \subset T(G_2)$. Furthermore, if each $T(G)$ is a normal subgroup of G, we can form a quotient functor I/T.

For example, the commutator subgroup $C(G)$ of the group G determines in this fashion a normal subfunctor of I. The corresponding quotient functor $(I/C)(G)$ is the functor determining for each G the factor commutator group of G (the group G made abelian).

The center $Z(G)$ does not determine in this fashion a subfunctor of I, because a homomorphism of G_1 *into* G_2 may carry central elements of G_1 into non-central elements of G_2. However, we may choose to restrict the category $\mathfrak{G}$ by using as mappings only homomorphisms of one group *onto* another. For *this* category, Z is a subfunctor of I, and we may form a quotient functor I/Z.

Thus various types of subgroups of G may be classified in terms of the degree of invariance of the "subfunctors" of the identity which they generate. This classification is similar to, but not identical with, the known distinction between normal subgroups, characteristic subgroups, and strictly characteristic subgroups of a single group([18]). The present distinction by functors refers not to the subgroups of an individual group, but to a definition yielding a subgroup for each of the groups in a suitable category. It includes the standard distinction, in the sense that one may consider functors on the category with only one object (a single group G) and with mappings which are the inner automorphisms of G (the subfunctors of I=normal subgroups), the automorphisms of G (subfunctors=characteristic subgroups), or the endomorphisms of G (subfunctors=strictly characteristic subgroups).

Still another example of the degree of invariance is given by the automorphism group $A(G)$ of a group G. This is a functor A defined on the category $\mathfrak{G}$ of groups with the mappings restricted to the isomorphisms $\gamma: G_1 \rightarrow G_2$ of one group onto another. The mapping function $A(\gamma)$ for any automorphism σ_1 of G_1 is then defined by setting

$$[A(\gamma)\sigma_1]g_2 = \gamma\sigma_1\gamma^{-1}g_2, \qquad g_2 \in G_2.$$

The types of invariance for functors on $\mathfrak{G}$ may thus be indicated by a table, showing how the mappings of the category must be restricted in order to make the indicated set function a functor:

([18]) A subgroup S of G is characteristic if $\sigma(S) \subset S$ for every atuomorphism σ of G, and strictly (or "strongly") characteristic if $\sigma(S) \subset S$ for every endomorphism of G.

Functor	*Mappings* $\gamma: G_1 \to G_2$
$C(G)$	Homomorphisms into,
$Z(G)$	Homomorphisms onto,
$A(G)$	Isomorphisms onto.

For the subcategory of $\mathfrak{G}$ consisting of all (additive) abelian groups there are similar subfunctors: 1°. G_0, the set of all elements of finite order in G; 2°. G_m, the set of all elements in G of order dividing the integer m; 3°. mG, the set of all elements of the form mg in G. The corresponding quotient functors will have object functions G/G_0 (the "Betti group" of G), G/G_m, and G/mG (the group G reduced modulo m).

17. **The isomorphism theorems.** The isomorphism theorems of group theory can be formulated for functors; from this it will follow that these isomorphisms between groups are "natural."

The "first isomorphism theorem" asserts that if G has two normal subgroups G_1 and G_2 with $G_2 \subset G_1$, then G_1/G_2 is a normal subgroup of G/G_2, and there is an isomorphism τ of $(G/G_2)/(G_1/G_2)$ to G/G_1. The elements of the first group (in additive notation) are cosets of cosets, of the form $(x+G_2)+G_1/G_2$, and the isomorphism τ is defined as

$$\tau[(x + G_2) + G_1/G_2] = x + G_1. \tag{17.1}$$

This may be stated in terms of functors as follows.

THEOREM 17.1. *Let T_1 and T_2 be two normal subfunctors of a functor T with values in the category of groups. If $T_2 \subset T_1$, then T_1/T_2 is a normal subfunctor of T/T_2 and the functors*

$$T/T_1 \quad \text{and} \quad (T/T_2)/(T_1/T_2) \tag{17.2}$$

are naturally equivalent.

Proof. We assume that the given functor T depends on the usual typical arguments A and B. Since $(T_1/T_2)(A, B)$ is clearly a normal subgroup of $(T/T_2)(A, B)$, a proof that T_1/T_2 is a normal subfunctor of T/T_2 requires only a proof that each $(T_1/T_2)(\alpha, \beta)$, is a submapping of the corresponding $(T/T_2)(\alpha, \beta)$ for any $\alpha: A_1 \to A_2$ and $\beta: B_1 \to B_2$. To show this, apply $(T_1/T_2)\cdot(\alpha, \beta)$ to a typical coset $x+T_2(A_1, B_2)$. Applying the definitions, one has

$$\begin{aligned}(T_1/T_2)(\alpha, \beta)[x + T_2(A_1, B_2)] &= T_1(\alpha, \beta)(x) + T_2(A_2, B_1)\\ &= T(\alpha, \beta)(x) + T_2(A_2, B_1)\\ &= (T/T_2)(\alpha, \beta)[x + T_2(A_1, B_2)],\end{aligned}$$

for $T_1(\alpha, \beta)$ was assumed to be a submapping of $T(\alpha, \beta)$.

The asserted equivalence (17.2) is established by setting, as in (17.1),

$$\tau(A, B)\{[x + T_2(A, B)] + (T_1/T_2)(A, B)\} = x + T_1(A, B).$$

The naturality proof then requires that, for any mappings $\alpha:A_1\to A_2$ and $\beta:B_1\to B_2$,

$$\tau(A_2, B_1)S(\alpha, \beta) = (T/T_1)(\alpha, \beta)\tau(A_1, B_2),$$

where $S=(T/T_2)/(T_1/T_2)$. This equality may be verified mechanically by applying each side to a general element $[x+T_2(A_1, B_2)]+(T_1/T_2)(A_1, B_2)$ in the group $S(A_1, B_2)$.

The theorem may also be stated and proved in the following equivalent form.

THEOREM 17.2. *Let T' and T'' be two normal subfunctors of a functor T with values in the category G of groups. Then $T'\cap T''$ is a normal subfunctor of T' and of T, $T'/T'\cap T''$ is a normal subfunctor of $T/T'\cap T''$, and the functors*

$$(17.3)\qquad T/T' \quad \textit{and} \quad (T/T'\cap T'')/(T'/T'\cap T'')$$

are naturally equivalent.

Proof. Set $T_1=T'$, $T_2=T'\cap T''$.

The second isomorphism theorem for groups is fundamental in the proof of the Jordan-Hölder Theorem. It states that if G has normal subgroups G_1 and G_2, then $G_1\cap G_2$ is a normal subgroup of G_1, G_2 is a normal subgroup of $G_1\cup G_2$, and there is an isomorphism μ of $G_1/G_1\cap G_2$ to $G_1\cup G_2/G_2$. (Because G_1 and G_2 are normal subgroups, the join $G_1\cup G_2$ consists of all "sums" g_1+g_2, for $g_i\in G_i$, so is often written as $G_1\cup G_2=G_1+G_2$.) For any $x\in G_1$, this isomorphism is defined as

$$(17.4)\qquad \mu[x + (G_1\cap G_2)] = x + G_2.$$

The corresponding theorem for functors reads:

THEOREM 17.3. *If T_1, T_2 are normal subfunctors of a functor T with values in G, then $T_1\cap T_2$ is a normal subfunctor of T_1, and T_2 is a normal subfunctor of $T_1\cup T_2$, and the quotient functors*

$$(17.5)\qquad T_1/(T_1\cap T_2) \quad \textit{and} \quad (T_1\cup T_2)/T_2$$

are naturally equivalent.

Proof. It is clear that both quotients in (17.5) are functors. The requisite equivalence $\mu(A, B)$ is given, as in (17.4), by the definition

$$\mu(A, B)[x + (T_1(A, B)\cap T_2(A, B))] = x + T_2(A, B),$$

for any $x\in T_1(A, B)$. The naturality may be verified as before.

From these theorems we may deduce that the first and second isomorphism theorems yield natural isomorphisms between groups in another and more specific way. To this end we introduce an appropriate category $\mathfrak{G}^*$. An object of $\mathfrak{G}^*$ is to be a triple $G^*=[G, G', G'']$ consisting of a group G and two of its normal subgroups. A mapping $\gamma: [G_1, G_1', G_1'']\to[G_2, G_2', G_2'']$ of $\mathfrak{G}^*$ is to be a homomorphism $\gamma: G_1\to G_2$ with the special properties that $\gamma(G_1')\subset G_2'$ and $\gamma(G_1'')\subset G_2''$. It is clear that these definitions do yield a category $\mathfrak{G}^*$. On this category $\mathfrak{G}^*$ we may define three (covariant) functors with values in the category $\mathfrak{G}$ of groups. The first is a "projection" functor,

$$P([G, G', G'']) = G, \qquad P(\gamma) = \gamma;$$

the others are two normal subfunctors of P, which may be specified by their object functions as

$$P'([G, G', G'']) = G', \qquad P''([G, G', G'']) = G''.$$

Consider now the first isomorphism theorem, in the second form,

$$G/G' \cong (G/(G'\cap G''))/(G'/(G'\cap G'')). \tag{17.6}$$

If we set $G^* = [G, G', G'']$, the left side here is a value of the object function of the functor, P/P', and the right side is similarly a value of $(P/P'\cap P'')/(P'/P'\cap P'')$. Theorem 17.2 asserts that these two functors are indeed naturally equivalent. Therefore, the isomorphism (17.6) is itself natural, in that it can be regarded as a natural isomorphism between the object functions of suitable functors on the category $\mathfrak{G}^*$.

The second isomorphism theorem

$$(G'\cup G'')/G'' \cong G'/(G'\cap G'')$$

is natural in a similar sense, for both sides can be regarded as object functions of suitable (covariant) functors on $\mathfrak{G}^*$.

It is clear that this technique of constructing a suitable category $\mathfrak{G}^*$ could be used to establish the naturality of even more complicated "isomorphism" theorems.

18. **Direct products of functors.** We recall that there are essentially two different ways of defining the direct product of two groups G and H. The "external" direct product $G\times H$ is the group of all pairs (g, h) with $g\in G$, $h\in H$, with the usual multiplication. This product $G\times H$ contains a subgroup G', of all pairs $(g, 0)$, which is isomorphic to G, and a subgroup H' isomorphic to H. Alternatively, a group L with subgroups G and H is said to be the "internal" direct product $L=G\mathbf{\times} H$ of its subgroups G and H if $gh=hg$ for every $g\in G$, $h\in H$ and if every element in L can be written uniquely as a product gh with $g\in G$, $h\in H$. The intimate connection between the two types of direct products is provided by the isomorphism $G\times H\cong G\mathbf{\times} H$ and by the equality $G\times H=G'\mathbf{\times} H'$, where $G'\cong G$, $H'\cong H$.

As in §4, the *external* direct product can be regarded as a covariant functor on $\mathfrak{G}$ and $\mathfrak{G}$ to $\mathfrak{G}$, with object function $G\times H$, and mapping function $\gamma\times\eta$, defined as in §4.

Direct products of functors may also be defined, with the same distinction between "external" and "internal" products. We consider throughout functors covariant in a category $\mathfrak{A}$, contravariant in $\mathfrak{B}$, with values in the category $\mathfrak{G}_0$ of discrete groups. If T_1 and T_2 are two such functors, the external direct product is a functor $T_1 \times T_2$ for which the object and mapping functions are respectively

$$(T_1 \times T_2)(A, B) = T_1(A, B) \times T_2(A, B), \tag{18.1}$$

$$(T_1 \times T_2)(\alpha, \beta) = T_1(\alpha, \beta) \times T_2(\alpha, \beta). \tag{18.2}$$

If $T_1'(A, B)$ denotes the set of all pairs $(g, 0)$ in the direct product $T_1(A, B) \times T_2(A, B)$, T_1' is a subfunctor of $T_1 \times T_2$, and the correspondence $g \rightarrow (g, 0)$ provides a natural isomorphism of T_1 to T_1'. Similarly T_2 is naturally isomorphic to a subfunctor T_2' of $T_1 \times T_2$.

On the other hand, let S be a functor on $\mathfrak{A}$, $\mathfrak{B}$ to $\mathfrak{G}_0$ with subfunctors S_1 and S_2. We call S the *internal* direct product $S_1 \mathbf{X} S_2$ if, for each $A \in \mathfrak{A}$ and $B \in \mathfrak{B}$, $S(A, B)$ is the internal direct product $S_1(A, B) \mathbf{X} S_2(A, B)$. From this definition it follows that, whenever $\alpha: A_1 \rightarrow A_2$ and $\beta: B_1 \rightarrow B_2$ are given mappings and $g_i \in S_i(A_1, B_2)$ are given elements $(i = 1, 2)$, then, since $S_i(\alpha, \beta) \subset S(\alpha, \beta)$,

$$S(\alpha, \beta) g_1 g_2 = [S_1(\alpha, \beta) g_1][S_2(\alpha, \beta) g_2].$$

This means that the correspondence τ defined by setting $[\tau(A_1, B_2)](g_1 g_2) = g_2$ is a natural transformation $\tau: S \rightarrow S_2$. Furthermore this transformation is idempotent, for $\tau(A_1, B_2)\tau(A_1, B_2) = \tau(A_1, B_2)$.

The connection between the two definitions is immediate; there is a natural isomorphism of the internal direct product $S_1 \mathbf{X} S_2$ to the external product $S_1 \times S_2$; furthermore any external product $T_1 \times T_2$ is the internal product $T_1' \mathbf{X} T_2'$ of its subfunctors $T_1' \cong T_1$, $T_2' \cong T_2$.

There are in group theory various theorems giving direct product decompositions. These decompositions can now be classified as to "naturality." Consider for example the theorem that every finite abelian group G can be represented as the (internal) direct product of its Sylow subgroups. This decomposition is "natural"; specifically, we may regard the Sylow subgroup $S_p(G)$ (the subgroup consisting of all elements in G of order some power of the prime p) as the object function of a subfunctor S_p of the identity. The theorem in question then asserts in effect that the identity functor I is the internal direct product of (a finite number of) the functors S_p. This representation of the direct factors by functors is the underlying reason for the possibility of extending the decomposition theorem in question to infinite groups in which every element has finite order.

On the other hand consider the theorem which asserts that every finite abelian group is the direct product of cyclic subgroups. It is clear here that the subgroups cannot be given as the values of functors, and we observe that in this case the theorem does not extend to infinite abelian groups.

As another example of non-naturality, consider the theorem which asserts that any abelian group G with a finite number of generators can be represented as a direct product of a free abelian group by the subgroup $T(G)$ of all elements of finite order in G. Let us consider the category $\mathfrak{G}_{af}$ of all discrete abelian groups with a finite number of generators. In this category the "torsion" subgroup $T(G)$ does determine the object function of a subfunctor $T \subset I$. However, there is no such functor giving the complementary direct factor of G.

THEOREM 18.1. *In the category $\mathfrak{G}_{af}$ there is no subfunctor $F \subset I$ such that $I = F \times T$, that is, such that, for all G,*

$$G = F(G) \times T(G). \tag{18.3}$$

Proof. It suffices to consider just one group, such as the group G which is the (external) direct product of the additive group of integers and the additive group of integers mod m, for $m \neq 0$. Then no matter which free subgroup $F(G)$ may be chosen so that (18.3) holds for this G, there clearly is an isomorphism of G to G which does not carry F into itself. Hence F cannot be a functor.

This result could also be formulated in the statement that, for any G with $G \neq T(G) \neq (0)$, there is no decomposition (18.3) with $F(G)$ a (strongly) characteristic subgroup of G. In order to have a situation which cannot be reformulated in this way, consider the closely related (and weaker) group theoretic theorem which asserts that for each G in $\mathfrak{G}_{af}$ there is an isomorphism of $G/T(G)$ into G. This isomorphism cannot be natural.

THEOREM 18.2. *For the category $\mathfrak{G}_{af}$ there is no natural transformation, $\tau: I/T \to I$, which gives for each G an isomorphism $\tau(G)$ of $G/T(G)$ into a subgroup of G.*

This proof will require consideration of an infinite class of groups, such as the groups $G_m = J \times J_{(m)}$ where J is the additive group of integers and $J_{(m)}$ the additive group of integers, modulo m. Suppose that $\tau(G): G/T(G) \to G$ existed. If $\mu(G): G \to G/T(G)$ is the natural transformation of G into $G/T(G)$ the product $\sigma(G) = \tau(G)\mu(G)$ would be a natural transformation of G into G with kernel $T(G)$. For each of the groups G_m with elements $(a, b_{(m)})$ for $a \in J$, $b_{(m)} \in J_{(m)}$, this transformation $\sigma_m = \sigma(G_m)$ must be a homomorphism with kernel $J_{(m)}$, hence must have the form

$$\sigma_m(a, b_{(m)}) = (r_m a, (s_m a)_{(m)}),$$

where r_m and s_m are integers. Now consider the homomorphism $\gamma: G_m \to G_m$ defined by setting $\gamma(a, b_{(m)}) = (0, b_{(m)})$. Since σ_m is natural, we must have $\sigma_m \gamma = \gamma \sigma_m$. Applying this equality to an arbitrary element we conclude that $s_m \equiv 0 \pmod{m}$. Next consider $\delta: G_m \to G_m$ defined by $\delta(a, b_{(m)}) = (0, a_{(m)})$. The condition $\sigma_m\delta = \delta\sigma_m$ here gives $r_m \equiv 0 \pmod{m}$, so that we can write $r_m = mt_m$. Therefore for each m

$$\sigma_m(a, b_{(m)}) = (mt_m a, 0).$$

Now consider two groups G_m, G_n with a homomorphism $\beta: G_m \to G_n$ defined by setting $\beta(a, b_{(m)}) = (a, 0_{(n)})$. The naturality condition $\sigma_n\beta = \beta\sigma_m$ now gives $mt_m = nt_n$. If we hold m fixed and allow n to increase indefinitely, this contradicts the fact that mt_m is a finite integer. The proof is complete.

It may be observed that the use of an infinite number of distinct groups is essential to the proof of this theorem. For any subcategory of $\mathfrak{G}_{af}$ containing only a finite number of groups, Theorem 18.2 would be false, for it would be possible to define a natural transformation $\tau(G)$ by setting $[\tau(G)]g = kg$ for every g, where the integer k is chosen as any multiple of the order of all the subgroups $T(G)$ for G in the given category.

The examples of "non-natural" direct products adduced here are all examples which mathematicians would usually recognize as not in fact natural. What we have done is merely to show that our definition of naturality does indeed properly apply to cases of intuitively clear non-naturality.

19. **Characters**(19). The character group of a group may be regarded as a contravariant functor on the category $\mathfrak{G}_{lca}$ of locally compact regular abelian groups, with values in the same category. Specifically, this functor "Char" may be defined by "slicing" (see §5) the functor Hom of §4 as follows. Let P be the (fixed) topological group of real numbers modulo 1, define "Char" by setting

$$(19.1) \qquad \text{Char}\, G = \text{Hom}\,(G, P), \qquad \text{Char}\, \gamma = \text{Hom}\,(\gamma, e_P).$$

Given $g \in G$ and $\chi \in \text{Char}\, G$ it will be convenient to denote the element $\chi(g)$ of P by (χ, g). Using this terminology and the definition of Hom we obtain for $\gamma: G_1 \to G_2$, $\chi \in \text{Char}\, G_2$ and $g_1 \in G_1$,

$$(19.2) \qquad (\text{Char}\,(\gamma)\chi, g) = (\chi, \gamma g).$$

As mentioned before (§10) the familiar isomorphism Char (Char $G) \cong G$ is a natural equivalence.

The functor "Char" can be compounded with other functors. Let T be any functor covariant in $\mathfrak{A}$, contravariant in $\mathfrak{B}$, with values in $\mathfrak{G}_{lca}$. The composite functor Char T is then defined on the same categories $\mathfrak{A}$ and $\mathfrak{B}$ but is contravariant in $\mathfrak{A}$ and covariant in $\mathfrak{B}$. Let S be any closed subfunctor of T. Then for each pair of objects $A \in \mathfrak{A}$, $B \in \mathfrak{B}$, the closed subgroup $S(A, B) \subset T(A, B)$ determines a corresponding subgroup Annih $S(A, B)$ in Char $T(A, B)$; this annihilator is defined as the set of all those characters $\chi \in \text{Char}\, T(A, B)$ with $(\chi, g) = 0$ for each $g \in S(A, B)$. This leads to a closed

(19) General references: A. Weil, *L'integration dans les groupes topologiques et ses applications*, Paris, 1938, chap. 1; S. Lefschetz, *Algebraic topology*, Amer. Math. Soc. Colloquium Publication, vol. 27, New York, 1942, chap. 2.

subfunctor Annih $(S; T)$ of the functor Char T, determined by the object function

$$[\text{Annih } (S; T)](A, B) = \text{Annih } S(A, B) \text{ in Char } T(A, B).$$

It is well known that

$$\text{Char } [T(A, B)/S(A, B)] = \text{Annih } S(A, B),$$

$$\text{Char } S(A, B) = \text{Char } T(A, B)/\text{Annih } S(A, B).$$

These isomorphisms in fact yield natural equivalences

$$\sigma: \text{Annih } (S; T) \rightleftarrows \text{Char } (T/S), \tag{19.3}$$

$$\tau: \text{Char } T/\text{Annih } (S; T) \rightleftarrows \text{Char } S. \tag{19.4}$$

For example, to prove (19.4) one observes that each $\chi \in \text{Char } T(A, B)$ may be restricted to give a character $\tau_0(A, B)\chi$ of $S(A, B)$ by setting

$$(\tau_0(A, B)\chi, h) = (\chi, h), \qquad h \in S(A, B). \tag{19.5}$$

This gives a homomorphism

$$\tau_0(A, B): \text{Char } T(A, B) \rightarrow \text{Char } S(A, B)$$

with kernel Annih $S(A, B)$. This homomorphism τ_0 will yield the required isomorphism τ of (19.4); by Theorem 15.1 a proof that τ_0 is natural will imply that τ is natural.

To show τ_0 natural, consider any mappings $\alpha: A_1 \rightarrow A_2$ and $\beta: B_1 \rightarrow B_2$ in the argument categories of T. Then $\gamma = T(\alpha, \beta)$ maps $T(A_1, B_2)$ into $T(A_2, B_1)$, while $\delta = S(\alpha, \beta)$ is a submapping of γ. The naturality requirements for τ_0 is

$$(\text{Char } \delta)\tau_0(A_2, B_1) = \tau_0(A_1, B_2) \text{ Char } \gamma. \tag{19.6}$$

Each side is a homomorphism of Char $T(A_2, B_1)$ into Char $S(A_1, B_2)$. If the left-hand side be applied to an element $\chi \in \text{Char } T(A_2, B_1)$, and the resulting character of $S(A_1, B_2)$ is then applied to an element h in the latter group, we obtain

$$(\text{Char } \delta(\tau_0(A_2, B_1)\chi), h) = (\tau_0(A_2, B_1)\chi, \delta h) = (\chi, \delta h)$$

by using the definition (19.2) of Char δ and the definition (19.5) of τ_0. If the right-hand side of (19.6) be similarly applied to χ and then to h, the result is

$$(\tau_0(A_1, B_2)((\text{Char } \gamma)\chi), h) = ((\text{Char } \gamma)\chi, h) = (\chi, \gamma h).$$

Since $\delta \subset \gamma$, these two results are equal, and both τ_0 and τ are therefore natural.

The proof of naturality for (19.3) is analogous.

If R is a closed subfunctor of S which is in turn a closed subfunctor of T, both of these natural isomorphisms may be combined to give a single natural isomorphism

(19.7) $\rho: \mathrm{Char}\,(S/R) \rightleftarrows \mathrm{Annih}\,(S; T)/\mathrm{Annih}\,(R; T).$

Chapter IV. Partially ordered sets and projective limits

20. **Quasi-ordered sets.** The notions of functors and their natural equivalences apply to partially ordered sets, to lattices, and to related mathematical systems. The category $\mathfrak{Q}$ of all quasi-ordered sets(20) has as its objects the quasi-ordered sets P and as its mappings $\pi: P_1 \to P_2$ the order preserving transformations of one quasi-ordered set, P, into another. An equivalence in this category is thus an isomorphism in the sense of order.

An important subcategory of $\mathfrak{Q}$ is the category $\mathfrak{Q}_d$ of all directed sets(21). One may also consider subcategories which are obtained by restricting both the quasi-ordered sets and their mappings. For example, the category of lattices has as objects all those partially ordered sets which are lattices and as mappings those correspondences which preserve both joins and meets. Alternatively, by using these mappings which preserve only joins, or those which preserve only meets, we obtain two other categories of lattices.

The category $\mathfrak{S}$ of sets may be regarded as a subcategory of $\mathfrak{Q}$, if each set S is considered as a (trivially) quasi-ordered set in which $p_1 < p_2$ in S means that $p_1 = p_2$. The category $\mathfrak{W}$ of well-ordered sets is another subcategory of $\mathfrak{Q}$. These categories provide a basis for applying the study of functors to cardinal and ordinal arithmetic. Specifically, the general theory of arithmetic of partially ordered sets, as developed recently by Birkhoff(22), can be viewed as the construction of a large number of functors (cardinal power, ordinal power, and so on) defined on suitable subcategories of $\mathfrak{Q}$, together with a collection of natural equivalences and transformations between these functors(23).

The construction of the category $\mathfrak{Q}$ of all quasi-ordered sets is not the only such interpretation of partial order. It is also possible to regard the elements of a *single* quasi-ordered set P as the objects of a category; with this device, one can represent an inverse or a direct system of groups (or of spaces) as a functor on P.

If a quasi-ordered set P be regarded as a category $\mathfrak{C}_P$, the objects of the category are the elements $p \in P$ and the mappings are the pairs $\pi = (p_2, p_1)$ of elements $p_i \in P$ such that $p_1 < p_2$. To each object p we assign the pair $e_p = (p, p)$ as the corresponding identity mapping, while the product (p_3, p_2') (p_2, p_1) of two mappings of $\mathfrak{C}_P$ is defined if and only if $p_2' = p_2$ and is in this case the mapping (p_3, p_1). The axioms C1 to C5 for a category are readily

(20) A *quasi-ordered set* P is a set of elements $p_1, p_2, \cdots$ with a reflexive and transitive binary relation $p_1 < p_2$ between the elements. If, in addition, the antisymmetric law ($p_1 < p_2$ and $p_2 < p_1$ imply $p_1 = p_2$) holds, P is a *partially ordered set*.

(21) A quasi-ordered set P is *directed* if for each pair of elements $p_1, p_2 \in P$ there exists a $p_3 \in P$ with $p_1 < p_3$, $p_2 < p_3$.

(22) Garrett Birkhoff, *Generalized arithmetic*, Duke Math. J. vol. 9 (1942) pp. 283–302.

(23) Note, however, that the ordinary cardinal sum of two sets A and B does not give rise to a functor, because the definition applies only when the sets A and B are disjoint.

verified, and it develops that the only identities are the pairs (p, p), that the equivalence mappings of $\mathfrak{C}_P$ are the pairs (p_2, p_1) with $p_1 < p_2$ and $p_2 < p_1$ and that any pair (p_2, p_1) with $p_1 < p_2$ is a mapping $(p_2, p_1): p_1 \rightarrow p_2$. It further follows that any two mappings $\pi_1: p_1 \rightarrow p_2$ and $\pi_2: p_1 \rightarrow p_2$ of this category which have the same range and the same domain are necessarily equal. Conversely any given category $\mathfrak{C}$ which has the property that any two mappings π_1 and π_2 of $\mathfrak{C}$ with the same range and the same domain are equal is isomorphic to the category $\mathfrak{C}_P$ for a suitable quasi-ordered set P. In fact, P can be defined to be the set of all objects C of the category $\mathfrak{C}$ with $C_1 < C_2$ if and only if there is in $\mathfrak{C}$ a mapping $\gamma: C_1 \rightarrow C_2$.

Consider now two quasi-ordered sets P and Q, with their corresponding categories $\mathfrak{C}_P$ and $\mathfrak{C}_Q$. A covariant (contravariant) functor T on $\mathfrak{C}_P$ with values in $\mathfrak{C}_Q$ is determined uniquely by an order preserving (reversing) mapping t of P into Q. Specifically, each such correspondence t is the object function $t(p) = q$ of a functor T, for which the corresponding mapping function is defined as $T(p_2, p_1) = (tp_2, tp_1)$ (or, in case t is order-reversing, as (tp_1, tp_2)). Each functor T of one variable can be obtained in this way.

21. **Direct systems as functors.** Let D be a directed set. If for every $d \in D$ a discrete group G_d is defined and for every pair $d_1 < d_2$ in D a homomorphism

$$\phi_{d_2, d_1}: G_{d_1} \rightarrow G_{d_2} \tag{21.1}$$

is given such that $\phi_{d,d}$ is the identity and that

$$\phi_{d_3, d_1} = \phi_{d_3, d_2}\phi_{d_2, d_1} \quad \text{for} \quad d_1 < d_2 < d_3 \tag{21.2}$$

then we say that the groups $\{G_d\}$ and the homomorphisms $\{\phi_{d_2, d_1}\}$ constitute a direct system of groups indexed by D.

Let us now regard the directed set D as a category. For every object $d \in D$ define

$$T(d) = G_d.$$

For every mapping $\delta = (p_2, p_1)$ in D define

$$T(\delta) = T(d_2, d_1) = \phi_{d_2, d_1}.$$

Conditions (21.1) and (21.2) imply that T is a covariant functor on D with values in the category $\mathfrak{G}_0$ of discrete groups. Conversely any such functor gives rise to a unique direct system. Consequently the terms "direct system of groups indexed by the directed set D" and "covariant functor on D to $\mathfrak{G}_0$" may be regarded as synonyms.

With each direct system of groups T there is associated a discrete limit group $G = \mathrm{Lim}_{\rightarrow} T$ defined as follows. The elements of the limit group G are pairs (g, d) for $g \in T(d)$; two elements (g_1, d_1) and (g_2, d_2) are considered equal if and only if there is an index d_3 with $d_1 < d_3$, $d_2 < d_3$ and with $T(d_3, d_1)g_1 = T(d_3, d_2)g_2$. The sum is defined by setting $(g, d) + (g', d) = (g + g', d)$; since the set D is directed, this provides for the addition of any two pairs in G. For a fixed $d \in D$ one may also consider the homomorphisms, called projections, $\lambda(d): T(d) \to G$ defined by setting

$$\lambda(d)g = (g, d) \tag{21.3}$$

for $g \in T(d)$. Clearly

$$\lambda(d_1) = \lambda(d_2)T(d_2, d_1) \quad \text{for} \quad d_1 < d_2. \tag{21.4}$$

To treat this limit group, we enlarge the given directed set D by adjoining one new element ∞, ordered by the specification that $d < \infty$ for each $d \in D$. This enlarged directed set D_∞ also determines a category containing D as a subcategory, with new mappings (∞, d) for each $d \in D$. Let now T be any covariant functor on D to $\mathfrak{G}_0$ (that is, any direct system of groups indexed by D). We define an extension T_∞ of the object function of T by setting

$$T_\infty(\infty) = \operatorname{Lim}_{\to} T = G, \tag{21.5}$$

the limit group of the given directed system T, and we similarly extend the mapping function of T by letting T_∞, for a new mapping (∞, d), be the corresponding projection of $T(d)$ into the limit group

$$T_\infty(\infty, d) = \lambda(d). \tag{21.6}$$

Condition (21.5) implies that T_∞ is indeed a covariant function on D_∞ with values in $\mathfrak{G}_0$. The properties of the limit group may be described in terms of this extended functor T_∞.

THEOREM 21.1. *Let D be a directed set and T a covariant functor on D (regarded as a category) to $\mathfrak{G}_0$. Then the limit group G of the direct system T and the projections of each group $T(d)$ into this limit determine as in* (21.5) *and* (21.6) *an extension of T to a covariant functor T_∞ on D_∞ to $\mathfrak{G}_0$. If S_∞ is any other extension of T to a covariant functor on D_∞ to $\mathfrak{G}_0$, there is a unique natural transformation $\sigma: T_\infty \to S_\infty$ such that each $\sigma(d)$ with $d \neq \infty$ is the identity.*

Proof. We have already seen that T_∞ is a covariant functor on D_∞ to $\mathfrak{G}_0$, extending T. Let now S_∞ be any other functor extending T. Since $S(d_2, d_1) = T(d_2, d_1)$ for $d_2 < d_1$ in D, it follows from the functor condition on S_∞ that

$$S_\infty(\infty, d_2)T(d_2, d_1) = S_\infty(\infty, d_1). \tag{21.7}$$

We define a homomorphism

$$\sigma(\infty): T_\infty(\infty) \to S_\infty(\infty)$$

by setting $\sigma(\infty)(g, d) = S_\infty(\infty, d)g$ for every element $(g, d) \in T_\infty(\infty) = \operatorname{Lim}_{\to} T$. Condition (21.7) implies that $\sigma(\infty)$ is single-valued. If we now set $\sigma(d)$ to be the identity mapping $T_\infty(d) \to S_\infty(d)$ for $d \neq \infty$, we have the desired transformation $\sigma: T_\infty \to S_\infty$.

The extension T_∞ and hence the limit group $G=T_\infty(\infty)$ of the given direct system is completely determined by the property given in the last sentence of the theorem. In fact if T'_∞ is any other extension of T with the same property as T_∞, there will exist transformations $\sigma: T_\infty \to T'_\infty$ and $\sigma': T'_\infty \to T_\infty$. Then $\rho=\sigma'\sigma: T_\infty \to T_\infty$ with $\rho(d)$ the identity whenever $d \neq \infty$. It follows that

$$\rho(\infty)\lambda(d) = \rho(\infty)T(\infty, d) = T(\infty, d)\rho(d) = T(\infty, d) = \lambda(d)$$

and therefore for every (g, d) in G we get

$$\rho(\infty)(g, d) = \rho(\infty)\lambda(d)g = \lambda(d)g = (g, d).$$

Hence $\rho(\infty)$ is the identity and σ is a natural equivalence $\sigma: T_\infty \to T'_\infty$. In this way the limit group of a direct system of groups can be defined up to an isomorphism by means of such extensions of functors. This indicates that the concept (but not necessarily the existence) of direct "limits" could be set up not only for groups, but also for objects of any category.

THEOREM 21.2. *If T_1 and T_2 are two covariant functors on the directed category D with values in $\mathfrak{G}_0$, and τ is a natural transformation $\tau: T_1 \to T_2$, there is only one extension τ_∞ of τ which is a natural transformation $\tau_\infty: T_{1\infty} \to T_{2\infty}$ between the extended functors on D_∞. When τ is a natural equivalence so is τ_∞.*

Proof. The naturality condition for τ, when applied to any mapping (d_2, d_1) with $d_1 < d_2$ in the directed set D reads

$$\tau(d_2)T_1(d_2, d_1) = T_2(d_2, d_1)\tau(d_1). \tag{21.8}$$

Given any element (g_1, d) of the limit group $T_{1\infty}(\infty) = \mathrm{Lim}_{\rightarrow} T_1$ we define

$$\omega(g_1, d) = (\tau(d)g_1, d) \in \mathrm{Lim}_{\rightarrow} T_2 = T_{2\infty}(\infty). \tag{21.9}$$

Condition (21.8) implies that this definition of ω gives a result independent of the special representation (g_1, d) chosen for the limit element. Hence we get a homomorphism

$$\omega: T_{1\infty}(\infty) \to T_{2\infty}(\infty).$$

In virtue of (21.6) and (21.3), the definition (21.9) becomes

$$\omega T_{1\infty}(\infty, d) = T_{2\infty}(\infty, d)\tau(d). \tag{21.10}$$

This means simply that by setting $\tau_\infty(d) = \tau(d)$, $\tau_\infty(\infty) = \omega$ we get an extension of τ which is still natural and which gives a transformation $\tau_\infty: T_{1\infty} \to T_{2\infty}$. Since the naturality condition (21.10) is equivalent with (21.9) which completely determines the value of $\tau_\infty(\infty)$, the requisite uniqueness follows. In particular, if τ is an equivalence, each $\tau(d)$ is an isomorphism "onto," hence it follows that $\omega = \tau_\infty(\infty)$ is also an isomorphism onto, and is an equivalence. This is just a restatement of the known theorem that "isomorphic" direct systems determine isomorphic limit groups.

THEOREM 21.3. *If T is a direct system of groups indexed by a directed set D, while H is a fixed discrete group, regarded as a (constant) covariant functor on D to $\mathfrak{G}_0$, then for each natural transformation $\tau: T \rightarrow H$ there is a unique homomorphism τ_0 of the limit group* $\text{Lim}_{\rightarrow} T$ *into H with the property that $\tau(d) = \tau_0 \lambda(d)$ for each $d \in D$, where $\lambda(d)$ is the projection of $T(d)$ into* $\text{Lim}_{\rightarrow} T$.

Proof. This follows from the preceding theorem and from the remark that H_∞ is also a constant functor on D_∞ to $\mathfrak{G}_0$.

22. **Inverse limits as functors.** Let D be a directed set. If for every $d \in D$ a topological group G_d is defined and for every pair $d_1 < d_2$ in D a homomorphism

$$\phi_{d_2, d_1}: G_{d_2} \rightarrow G_{d_1} \tag{22.1}$$

is given such that $\phi_{d,d}$ is the identity and that

$$\phi_{d_3, d_1} = \phi_{d_2, d_1} \phi_{d_3, d_2} \quad \text{for} \quad d_1 < d_2 < d_3 \tag{22.2}$$

then we say that the groups $\{G_d\}$ and the homomorphisms $\{\phi_{d_2, d_1}\}$ constitute an inverse system of groups indexed by D.

If we now regard D as a category, and define as before

$$T(d) = G_d \tag{22.3}$$

for every object d in D, and

$$T(\delta) = T(d_2, d_1) = \phi_{d_2, d_1} \tag{22.4}$$

for every mapping $\delta = (d_2, d_1)$ in D, it is clear that T is a contravariant functor on D with values in the category $\mathfrak{G}$ of topological groups. Conversely any such functor may be regarded as an inverse system of groups.

With each inverse system of groups T there is associated a limit group $G = \text{Lim}_{\leftarrow} T$ defined as follows. An element of G is a function $g(d)$ which assigns to each element $d \in D$ an element $g(d) \in T(d)$, in such wise that these elements "match" under the mappings; that is, such that $T(d_2, d_1) g(d_2) = g(d_1)$ whenever $d_1 < d_2$. The sum of $g_1 + g_2$ is defined as $(g_1 + g_2)(d) = g_1(d) + g_2(d)$. This limit group G is assigned a topology, in known fashion, by treating G as a subgroup of the direct product of the groups $T(d)$, with the usual direct product topology. For fixed d, the (continuous) projection $\mu(d)$ of the limit group G into $T(d)$ is defined by setting $[\mu(d)]g = g(d)$, for $g \in G$.

Again we may consider the extended category D_∞ and define the extension T_∞ of T by setting

$$T_\infty(\infty) = G, \qquad T_\infty(\infty, d) = \mu(d). \tag{22.5}$$

As before the following theorem can be established:

THEOREM 22.1. *Let D be a directed set and T a contravariant functor on D (regarded as a category) to $\mathfrak{G}$. Then the limit group G of the inverse system T*

and the projections of this limit group into each group $T(d)$ determine as in (22.5) *an extension of T to a contravariant functor T_∞ on D_∞ to $\mathfrak{G}$. If S_∞ is any other extension of T to a contravariant functor on D_∞ to $\mathfrak{G}$, there is a unique natural transformation $\sigma: S_\infty \rightarrow T_\infty$ such that each $\sigma(d)$ with $d \neq \infty$ is the identity.*

As before we can also verify that the second half of the theorem determines the extended functor T_∞ to within a natural equivalence, and therefore it determines the limit group to within an isomorphism.

The following two theorems may also be proved as in the preceding section.

THEOREM 22.2. *If T_1 and T_2 are two contravariant functors on the directed category D with values in $\mathfrak{G}$, and τ is a natural transformation $\tau: T_1 \rightarrow T_2$, there is only one extension τ_∞ of τ which is a natural transformation $\tau_\infty: T_{1\infty} \rightarrow T_{2\infty}$ between the extended functors on D_∞. When τ is a natural equivalence so is τ_∞.*

THEOREM 22.3. *If T is an inverse system of groups indexed by the directed set D, while K is a fixed topological group regarded as a (constant) contravariant functor on D to $\mathfrak{G}$, then for each natural transformation $\tau: T \rightarrow K$ there is a unique homomorphism τ_0: $\text{Lim}_{\leftarrow}\, T \rightarrow K$ such that $\tau_0 = \tau(d)\lambda(d)$ for each $d \in D$.*

The preceding discussion carries over to inverse systems of spaces, by a mere replacement of the category of topological groups $\mathfrak{G}$ by the category of topological spaces $\mathfrak{X}$.

23. **The categories "$\mathfrak{Dir}$" and "$\mathfrak{Inv}$."** The process of forming a direct or inverse limit of a system of groups can be treated as a functor "$\text{Lim}_{\rightarrow}$" or "$\text{Lim}_{\leftarrow}$" which operates on an appropriately defined category. Thus the functor "$\text{Lim}_{\rightarrow}$" will operate on any direct system T defined on any directed set D. Consequently we define a category "$\mathfrak{Dir}$" of directed systems whose objects are such pairs (D, T). Here we may regard D itself as a category and T as a covariant functor on D to $\mathfrak{G}_0$. To introduce the mappings of this category, observe first that each order preserving transformation R of a directed set D_1 into another such set D_2 will give for each direct system T_2 of groups indexed by D_2 an induced direct system indexed by D_1. Specifically, the induced direct system is just the composite $T_2 \otimes R$ of the (covariant) functor R on D_1 to D_2 and the (covariant) functor T_2 on D_2 to $\mathfrak{G}_0$. Given two objects (D_1, T_1) and (D_2, T_2) of $\mathfrak{Dir}$, a mapping

$$(R, \rho): (D_1, T_1) \rightarrow (D_2, T_2)$$

of the category $\mathfrak{Dir}$ is a pair (R, ρ) composed of a covariant functor R on D_1 to D_2 and a natural transformation

$$\rho: T_1 \rightarrow T_2 \otimes R$$

of T_1 into the composite functor $T_2 \otimes R$.

To form the product of two such mappings

(23.1) $(R_1, \rho_1):(D_1, T_1) \to (D_2, T_2), \quad (R_2, \rho_2):(D_2, T_2) \to (D_3, T_3)$

observe first that the functors T_2 and $T_3 \otimes R_2$ on D_2 to $\mathfrak{G}_0$ can be compounded with the functor R_1 on D_1 to D_2, and hence that the given transformation $\rho_2: T_2 \to T_3 \otimes R_2$ can be compounded with the identity transformation of R_1 into itself, just as in §9.

The result is a composite transformation

(23.2) $$\rho_2 \otimes R_1: T_2 \otimes R_1 \to T_3 \otimes R_2 \otimes R_1$$

which assigns to each object $d_1 \in D_1$ the mapping $[\rho_2 \otimes R_1](d_1) = \rho_2(R_1 d_1)$ of $T_2(R_1 d_1)$ into $T_3 \otimes R_2(R_1 d_1)$. The transformations (23.2) and $\rho_1: T_1 \to T_2 \otimes R_1$ yield as in §9 a composite transformation $\rho_2 \otimes R_1 \otimes \rho_1: T_1 \to T_3 \otimes R_2 \otimes R_1$. We may now define the product of two given mappings (23.1) to be

$$(R_2, \rho_2)(R_1, \rho_1) = (R_2 \otimes R_1, \rho_2 \otimes R_1 \otimes \rho_1).$$

With these conventions, we verify that $\mathfrak{Dir}$ is a category. Its identities are the pairs (R, ρ) in which both R and ρ are identities; its equivalences are the pairs (R, ρ) in which R is an isomorphism and ρ a natural equivalence.

The effect of fixing the directed set D in the objects (D, T) of the category $\mathfrak{Dir}$ is to restrict $\mathfrak{Dir}$ to the subcategory which consists of all direct systems of groups indexed by D (that is, the category of all covariant functors on D to $\mathfrak{G}_0$, as defined in §8).

We shall now define $\text{Lim}_{\to}$ as a covariant functor on $\mathfrak{Dir}$ with values in $\mathfrak{G}_0$. For each object (D, T) of $\mathfrak{Dir}$ we define $\text{Lim}_{\to} (D, T)$ to be the group obtained as the direct limit of the direct system of groups T indexed by the directed set D. Given a mapping

(23.3) $$(R, \rho):(D_1, T_1) \to (D_2, T_2) \quad \text{in} \quad \mathfrak{Dir}$$

we define the mapping function of $\text{Lim}_{\to}$,

(23.4) $$\text{Lim}_{\to} (R, \rho): \text{Lim}_{\to} (D_1, T_1) \to \text{Lim}_{\to} (D_2, T_2),$$

as follows. An element in the limit group Lim (D_1, T_1) is a pair (g_1, d_1) with $d_1 \in D_1$, $g_1 \in T_1(d_1)$. For each such element define $\phi(g_1, d_1)$ to be the pair $(\rho(d_1)g_1, Rd_1)$. Since $\rho(d_1)$ maps $T_1(d_1)$ into $T_2(Rd_1)$ we have $\rho(d_1)g_1$ in $T_2(Rd_1)$, so that the resulting pair is indeed in the limit group $\text{Lim}_{\to} (D_2, T_2)$. The mapping ϕ carries equal pairs into equal pairs, and yields the requisite homomorphism (23.4). We verify that $\text{Lim}_{\to}$, defined in this manner, is a covariant functor on $\mathfrak{Dir}$ to $\mathfrak{G}_0$.

Alternatively, the mapping function of this functor "$\text{Lim}_{\to}$" can be obtained by extensions of mappings to the directed sets $D_{1\infty}$, $D_{2\infty}$ (with ∞ added), defined as in §21. Given the mapping (R, ρ) of (23.3), first extend the given objects of $\mathfrak{Dir}$ to obtain new objects $(D_{1\infty}, T_{1\infty})$ and $(D_{2\infty}, T_{2\infty})$. The given functor R on D_1 to D_2 can also be extended by setting $R_\infty(\infty) = \infty$; this gives a functor R_∞ on $D_{1\infty}$ to $D_{2\infty}$. Furthermore, Theorem 21.2 asserts that the transformation $\rho: T_1 \to T_2 \otimes R$ has then a unique extension $\rho_\infty: T_{1\infty} \to T_{2\infty} \otimes R_\infty$. All told, we have a new mapping

$$(R_\infty, \rho_\infty): (D_{1\infty}, T_{1\infty}) \to (D_{2\infty}, T_{2\infty})$$

in $\mathfrak{Dir}$. In particular, when ρ_∞ is applied to the new element ∞ of $D_{1\infty}$, it yields a homomorphism of the limit group of T_1 into the limit group of $T_2 \otimes R$. On the other hand, R determines a homomorphism $R^\#$ of the limit group of $T_2 \otimes R$ into the limit group of T_2; explicitly, for (g_1, d_1) in the first limit group, the image $R^\#(g_1, d_1)$ is the element (g_1, Rd_1) in the second limit group. The requisite mapping function of the functor "$\mathrm{Lim}_\rightarrow$" is now defined by setting

$$\mathrm{Lim}_\rightarrow (R, \rho) = R^\#(\rho_\infty(\infty)).$$

In a similar way we define the category $\mathfrak{Inv}$. The objects of $\mathfrak{Inv}$ are pairs (D, T) where D is a directed set and T is an inverse system of topological groups indexed by D (that is, T is a contravariant functor on D to $\mathfrak{G}$). The mappings in $\mathfrak{Inv}$ are pairs (R, ρ)

$$(R, \rho): (D_1, T_1) \to (D_2, T_2)$$

where R is a covariant functor on D_2 to D_1 (that is, an order preserving transformation of D_2 into D_1) and ρ is a natural transformation of the functors

$$\rho: T_1 \otimes R \to T_2$$

both contravariant on D_2 to $\mathfrak{G}$. The product of two mappings

$$(R_1, \rho_1): (D_1, T_1) \to (D_2, T_2), \qquad (R_2, \rho_2) = (D_2, T_2) \to (D_3, T_3)$$

is defined as

$$(R_2, \rho_2)(R_1, \rho_1) = (R_1 \otimes R_2, \rho_2 \otimes \rho_1 \otimes R_2)$$

where $\rho_1 \otimes R_2$ is the transformation

$$\rho_1 \otimes R_2: T_1 \otimes R_1 \otimes R_2 \to T_2 \otimes R_2$$

induced (as in §9) by

$$\rho_1: T_1 \otimes R_1 \to T_2.$$

With these conventions, we verify that $\mathfrak{Inv}$ is a category.

We shall now define $\mathrm{Lim}_\leftarrow$ as a covariant functor on $\mathfrak{Inv}$ with values in $\mathfrak{G}$. For each object (D, T) in $\mathfrak{Inv}$ we define $\mathrm{Lim}_\leftarrow (D, T)$ to be the inverse limit of the inverse system of groups T indexed by the directed set D. Given a mapping

$$(23.5) \qquad (R, \rho): (D_1, T_1) \to (D_2, T_2) \quad \text{in} \quad \mathfrak{Inv}$$

we define the mapping function of $\mathrm{Lim}_\leftarrow$

$$\text{Lim}_{\leftarrow}(R, \rho):\text{Lim}_{\leftarrow}(D_1, T_1) \to \text{Lim}_{\leftarrow}(D_2, T_2) \tag{23.6}$$

as follows. Each element of $\text{Lim}_{\leftarrow}(D_1, T_1)$ is a function $g(d_1)$ with values $g(d_1) \in T_1(d_1)$, for $d_1 \in D_1$, which match properly under the projections in T_1. Now define a new function h, with

$$h(d_2) = \rho(d_2)g(Rd_2), \qquad d_2 \in D_2;$$

it is easy to verify that h is an element of the limit group Lim (D_2, T_2). The correspondence $g \to h$ is the homomorphism (23.6) required for the definition of the mapping function of $\text{Lim}_{\leftarrow}$. One may verify that this definition does yield a covariant functor $\text{Lim}_{\leftarrow}$ on the category $\mathfrak{Inv}$ to $\mathfrak{G}$.

The mapping function of $\text{Lim}_{\leftarrow}$ may again be obtained by first extending the given mapping (23.5) to

$$(R_\infty, \rho_\infty):(D_{1\infty}, T_{1\infty}) \to (D_{2\infty}, T_{2\infty}) \quad \text{in} \quad \mathfrak{Inv}.$$

In particular, when the extended transformation ρ_∞ is applied to the element ∞ of $D_{1\infty}$, we obtain a homomorphism of the limit group of $T_1 \otimes R$ into the limit group of T_2. On the other hand, the covariant functor R on D_2 to D_1 determines a homomorphism R^* of the limit group of (D_1, T_1) into the limit group of $(D_2, T_1 \otimes R)$; explicitly, for each function $g(d_1)$ in the first limit group, the image $h = R^*g$ in the second limit group is defined by setting $h(d_2) = g(Rd_2)$ for each $d_2 \in D_2$. The mapping function of the functor "$\text{Lim}_{\leftarrow}$" is now $\text{Lim}_{\leftarrow}(R, \rho) = \rho_\infty(\infty)R^*$.

24. **The lifting principle.** Let Q be a functor whose arguments and values are groups, while T is any direct or inverse system of groups. If the object function of Q is applied to each group $T(d)$ of the given system, while the mapping function of Q is applied to each projection $T(d_1, d_2)$ of the given system, we obtain a new system of groups, which may be called $Q \otimes T$. If Q is covariant, T and $Q \otimes T$ are both direct or both inverse, while if Q is contravariant, $Q \otimes T$ is inverse when T is direct, and vice versa.

Actually this new system $Q \otimes T$ is simply the composite of the functor T with the functor Q (see §9). We may regard this composition as a process which "lifts" a functor Q whose arguments and values are groups to a functor Q_L whose arguments and values are direct (or inverse) systems of groups. We may then regard the lifted functor as one acting on the categories $\mathfrak{Dir}$ and $\mathfrak{Inv}$, as the case may be. In every case, the lifted functor has its object and mapping functions given formally by the equations (in the "cross" notation for composites)

$$Q_L(D, T) = (D, Q \otimes T), \qquad Q_L(R, \rho) = (R, Q \otimes \rho). \tag{24.1}$$

This formula includes the following four cases:

(I) *Q covariant on $\mathfrak{G}_0$ to $\mathfrak{G}_0$; Q_L covariant on $\mathfrak{Dir}$ to $\mathfrak{Dir}$.*

(II) *Q contravariant on $\mathfrak{G}_0$ to $\mathfrak{G}$; Q_L contravariant on $\mathfrak{Dir}$ to $\mathfrak{Inv}$.*

(III) *Q covariant on $\mathfrak{G}$ to $\mathfrak{G}$; Q_L covariant on $\mathfrak{Inv}$ to $\mathfrak{Inv}$.*

(IV) *Q contravariant on $\mathfrak{G}$ to $\mathfrak{G}_0$; Q_L contravariant on $\mathfrak{Inv}$ to $\mathfrak{Dir}$.*

For illustration, we discuss case (II), in which Q is given contravariant on $\mathfrak{G}_0$ to $\mathfrak{G}$. The object function of Q_L, as defined in the first equation of (24.1), assigns to each object (D, T) of the category $\mathfrak{Dir}$ a pair $(D, Q\otimes T)$. Since T is covariant on D to $\mathfrak{G}_0$ and Q contravariant on $\mathfrak{G}_0$ to $\mathfrak{G}$, the composite $Q\otimes T$ is contravariant on D to $\mathfrak{G}$, so that $Q\otimes T$ is an inverse system of groups, and the pair $(D, Q\otimes T)$ is an object of $\mathfrak{Inv}$. On the other hand, given a mapping

$$(R, \rho):(D_1, T_1) \rightarrow (D_2, T_2) \quad \text{in} \quad \mathfrak{Dir},$$

with $\rho: T_1 \rightarrow T_2\otimes R$, the composite transformation $Q\otimes\rho$ is obtained by applying the mapping function of Q to each homomorphism $\rho(d_1): T_1(d_1)\rightarrow T_2\otimes R(d_1)$, and this gives a transformation $Q\otimes\rho: Q\otimes T_2\otimes R\rightarrow Q\otimes T_1$. Thus the mapping function of Q_L, as defined in (24.1), does give a mapping $(R, Q\otimes\rho):(D_2, Q\otimes T_2)\rightarrow(D_1, Q\otimes T_1)$ in the category $\mathfrak{Inv}$. We verify that Q_L is a contravariant functor on $\mathfrak{Dir}$ to $\mathfrak{Inv}$.

Any natural transformation $\kappa_1: Q\rightarrow P$ induces a transformation on the lifted functors, $\kappa_L: Q_L\rightarrow P_L$, obtained by composition of the transformation κ with the identity transformation of each T, as

$$\kappa_L(D, T) = (D, \kappa \otimes T).$$

If κ is an equivalence, so is this "lifted" transformation.

Just as in the case of composition, the operation of "lifting" can itself be regarded as a functor "Lift," defined on a suitable category of functors Q. In all four cases (I)-(IV), this functor "Lift" is covariant.

In all these cases the functor Q may originally contain any number of additional variables. The lifted functor Q_L will then involve the same extra variables with the same variance. With proper caution the lifting process may also be applied simultaneously to a functor Q with two variables, both of which are groups.

25. Functors which commute with limits. Certain operations, such as the formation of the character groups of discrete or compact groups, are known to "commute" with the passage to a limit. Using the lifting operation, this can be formulated exactly.

To illustrate, let Q be a covariant functor on $\mathfrak{G}_0$ to $\mathfrak{G}_0$, and Q_L the corresponding covariant lifted functor on $\mathfrak{Dir}$ to $\mathfrak{Dir}$, as in case (I) of §24. Since $\text{Lim}_{\rightarrow}$ is a covariant functor on $\mathfrak{Dir}$ to $\mathfrak{G}_0$, we have two composite functors

$$\text{Lim}_{\rightarrow} \otimes Q_L \quad \text{and} \quad Q \otimes \text{Lim}_{\rightarrow},$$

both covariant on $\mathfrak{Dir}$ to $\mathfrak{G}_0$. There is also an explicit natural transformation

$$\omega_{\text{I}}:\text{Lim}_{\rightarrow} \otimes Q_L \rightarrow Q \otimes \text{Lim}_{\rightarrow}, \tag{25.1}$$

defined as follows. Let the pair (D, T) be a direct system of groups in the category $\mathfrak{Dir}$, and let $\lambda(d)$ be the projection

$$\lambda(d): T(d) \to \mathrm{Lim}_{\to} T, \qquad d \in D.$$

Then, on applying the mapping function of Q to λ, we obtain the natural transformation

$$Q\lambda(d): QT(d) \to Q[\mathrm{Lim}_{\to} T].$$

Theorem 21.3 now gives a homomorphism

$$\omega_{\mathrm{I}}(T): \mathrm{Lim}_{\to} [Q \otimes T] \to Q[\mathrm{Lim}_{\to} T],$$

or, exhibiting D explicitly, a homomorphism

$$\omega_{\mathrm{I}}(D, T): \mathrm{Lim}\, Q_L(D, T) \to Q[\mathrm{Lim}_{\to} (D, T)].$$

We verify that ω_{I}, so defined, satisfies the naturality condition.

Similarly, to treat case (II), consider a contravariant functor Q on $\mathfrak{G}_0$ to $\mathfrak{G}$ and the lifted functor Q_L on $\mathfrak{Dir}$ to $\mathfrak{Inv}$. We then construct an explicit natural transformation

(25.2) $$\omega_{\mathrm{II}}: Q \otimes \mathrm{Lim}_{\to} \to \mathrm{Lim}_{\leftarrow} \otimes Q_L$$

(note the order !), defined as follows. Let the pair (D, T) be in $\mathfrak{Dir}$, and let $\lambda(d)$ be the projection

$$\lambda(d): T(d) \to \mathrm{Lim}_{\to} T, \qquad d \in D.$$

On applying Q, we get

$$Q\lambda(d): Q[\mathrm{Lim}_{\to} T] \to QT(d).$$

The Theorem 22.3 for inverse systems now gives a homomorphism

$$\omega_{\mathrm{II}}(D, T): Q[\mathrm{Lim}_{\to} (D, T)] \to \mathrm{Lim}_{\leftarrow} Q_L(D, T).$$

In the remaining cases (III) and (IV) similar arguments give natural transformations

(25.3) $$\omega_{\mathrm{III}}: Q \otimes \mathrm{Lim}_{\leftarrow} \to \mathrm{Lim}_{\leftarrow} \otimes Q_L,$$

(25.4) $$\omega_{\mathrm{IV}}: \mathrm{Lim}_{\to} \otimes Q_L \to Q \otimes \mathrm{Lim}_{\leftarrow}.$$

DEFINITION. The functor Q defined on groups to groups is said to commute (more precisely to ω-commute) with Lim if the appropriate one of the four natural transformations ω above is an equivalence.

In other words, the proof that a functor Q commutes with Lim requires only the verification that the homomorphisms defined above are isomorphisms. The naturality condition holds in general!

To illustrate these concepts, consider the functor C which assigns to each discrete group G its commutator subgroup $C(G)$, and consider a direct system T of groups, indexed by D. Then the lifted functor Q (case (I) of §24) applied to the pair (D, T) in $\mathfrak{Dir}$ gives a new direct system of groups, still indexed by D, with the groups $T(d)$ of the original system replaced by their commutator subgroups $CT(d)$, and with the projections correspondingly cut down. It may be readily verified that this functor does commute with Lim.

Another functor Q is the subfunctor of the identity which assigns to each discrete abelian group G the subgroup $Q(G)$ consisting of those elements $g \in G$ such that there is for each integer m an $x \in G$ with $mx = g$ (that is, of those elements of G which are divisible by every integer), Q is a covariant functor with arguments and values in the subcategory G_{0a} of discrete abelian groups. The lifted functor Q_L will be covariant, with arguments and values in the subcategory $\mathfrak{Dir}_a$ of $\mathfrak{Dir}$, obtained by restricting attention to abelian groups. This functor Q clearly does not commute with Lim, since one may represent the additive group of rational numbers as a direct limit of cyclic groups Z for which each subgroup $Q(Z)$ is the group consisting of zero alone.

The formation of character groups gives further examples. If we consider the functor Char as a contravariant functor on the category $\mathfrak{G}_{0a}$ of discrete abelian groups to the category $\mathfrak{G}_{ca}$ of compact abelian groups, the lifted functor Char_L will be covariant on the appropriate subcategory of $\mathfrak{Dir}$ to $\mathfrak{Inv}$ as in case (II) of §24. This lifted functor Char_L applied to any direct system (D, T) of discrete abelian groups will yield an inverse system of compact abelian groups, indexed by the same set D. Each group of the inverse system is the character group of the corresponding group of the direct system, and the projections of the inverse system are the induced mappings.

On the other hand, there is a contravariant functor Char on $\mathfrak{G}_{ca}$ to $\mathfrak{G}_{0a}$. In this case the lifted functor Char_L will be contravariant on a suitable subcategory of $\mathfrak{Inv}$ with values in $\mathfrak{Dir}$, just as in case (III) of §24. Both these functors Char commute with Lim.

CHAPTER V. APPLICATIONS TO TOPOLOGY(24)

26. **Complexes.** An abstract complex K (in the sense of W. Mayer) is a collection

$$\{C^q(K)\}, \qquad q = 0, \pm 1, \pm 2, \cdots,$$

of free abelian discrete groups, together with a collection of homomorphisms

$$\partial^q : C^q(K) \to C^{q-1}(K)$$

called boundary homomorphisms, such that

$$\partial^q \partial^{q+1} = 0.$$

By selecting for each of the free groups C^q a fixed basis $\{\sigma_i^q\}$ we obtain a complex which is substantially an abstract complex in the sense of A. W.

(24) General reference: S. Lefschetz, *Algebraic topology*, Amer. Math. Soc. Colloquium Publications, vol. 27, New York, 1942.

Tucker. The σ_i^q will be called q-dimensional cells. The boundary operator ∂ can be written as a finite sum

$$\partial\sigma^q = \sum_{\sigma^{q-1}} [\sigma^q : \sigma^{q-1}]\sigma^{q-1}.$$

The integers $[\sigma^q : \sigma^{q-1}]$ are called incidence numbers, and satisfy the following conditions:

(26.1) *Given* σ^q, $[\sigma^q ; \sigma^{q-1}] \neq 0$ *only for a finite number of* $(q-1)$-*cells* σ^{q-1}.

(26.2) *Given* σ^{q+1} *and* σ^{q-1}, $\sum_{\sigma^q} [\sigma^{q+1} ; \sigma^q][\sigma^q ; \sigma^{q-1}] = 0$.

Condition (26.1) indicates that we are confronted with an abstract complex of the closure finite type. Consequently we shall define (§27) homologies based on finite chains and cohomologies based on infinite cochains.

Our preference for complexes à la W. Mayer is due to the fact that they seem to be best adapted for the exposition of the homology theory in terms of functors.

Given two abstract complexes K_1 and K_2, a chain transformation

$$\kappa : K_1 \to K_2$$

will mean a collection $\kappa = \{\kappa^q\}$ of homomorphisms,

$$\kappa^q : C^q(K_1) \to C^q(K_2),$$

such that

$$\kappa^{q-1}\partial^q = \partial^q \kappa^q.$$

In this way we are led to the category $\mathfrak{K}$ whose objects are the abstract complexes (in the sense of W. Mayer) and whose mappings are the chain transformations with obvious definition of the composition of chain transformations.

The consideration of simplicial complexes and of simplicial transformations leads to a category $\mathfrak{K}_s$. As is well known, every simplicial complex uniquely determines an abstract complex, and every simplicial transformation a chain transformation. This leads to a covariant functor on $\mathfrak{K}_s$ to $\mathfrak{K}$.

27. **Homology and cohomology groups.** For every complex K in the category $\mathfrak{K}$ and every group G in the category $\mathfrak{G}_{0a}$ of discrete abelian groups we define the groups $C^q(K, G)$ of the q-dimensional chains of K over G as the tensor product

$$C^q(K, G) = G \circ C^q(K),$$

that is, $C^q(K, G)$ is the group with the symbols

$$gc^q, \qquad g \in G, \ c^q \in C^q(K)$$

as generators, and

$$(g_1 + g_2)\overset{q}{c} = g_1\overset{q}{c} + g_2\overset{q}{c}, \qquad g(\overset{q}{c_1} + \overset{q}{c_2}) = g\overset{q}{c_1} + g\overset{q}{c_2}$$

as relations.

For every chain transformation $\kappa: K_1 \to K_2$ and for every homomorphism $\gamma: G_1 \to G_2$ we define a homomorphism

$$C^q(\kappa, \gamma): C^q(K_1, G_1) \to C^q(K_2, G_2)$$

by setting

$$C^q(\kappa, \gamma)(g_1\overset{q}{c_1}) = \gamma(g_1)\kappa^q(\overset{q}{c_1})$$

for each generator $g_1c_1^q$ of $C^q(K_1, G_1)$.

These definitions of $C^q(K, G)$ and of $C^q(\kappa, \gamma)$ yield a functor C^q covariant in $\mathfrak{K}$ and in $\mathfrak{G}_{0a}$ with values in $\mathfrak{G}_{0a}$. This functor will be called the q-chain functor.

We define a homomorphism

$$\partial^q(K, G): C^q(K, G) \to C^{q-1}(K, G)$$

by setting

$$\partial^q(K, G)(gc^q) = g\partial c^q$$

for each generator gc^q of $C^q(K, G)$. Thus the boundary operator becomes a natural transformation of the functor C^q into the functor C^{q-1}

$$\partial^q: C^q \to C^{q-1}.$$

The kernel of this transformation will be denoted by Z^q and will be called the q-cycle functor. Its object function is the group $Z^q(K, G)$ of the q-dimensional cycles of the complex K over G.

The image of C^q under the transformation ∂^q is a subfunctor $B^{q-1} = \partial^q(C^q)$ of C^{q-1}. Its object function is the group $B^{q-1}(K, G)$ of the $(q-1)$-dimensional boundaries in K over G.

The fact that $\partial^q\partial^{q+1} = 0$ implies that $B^q(K, G)$ is a subgroup of $Z^q(K, G)$. Consequently B^q is a subfunctor of Z^q. The quotient functor

$$H^q = Z^q/B^q$$

is called the qth homology functor. Its object function associates with each complex K and with each discrete abelian coefficient group G the qth homology group $H^q(K, G)$ of K over G. The functor H^q is covariant in $\mathfrak{K}$ and $\mathfrak{G}_{0a}$ and has values in $\mathfrak{G}_{0a}$.

In order to define the cohomology groups as functors we consider the category $\mathfrak{K}$ as before and the category $\mathfrak{G}_a$ of topological abelian groups. Given a complex K in $\mathfrak{K}$ and a group G in $\mathfrak{G}_a$ we define the group $C^q(K, G)$ of the q-dimensional cochains of K over G as

$$C_q(K, G) = \text{Hom}\,(C^q(K), G).$$

Given a chain transformation $\kappa: K_1 \to K_2$ and a homomorphism $\gamma: G_1 \to G_2$, we define a homomorphism

$$C_q(\kappa, \gamma): C_q(K_2, G_1) \to C_q(K_1, G_2)$$

by associating with each homomorphism $f \in C_q(K_2, G_1)$ the homomorphism $\bar{f} = C_q(\kappa, \gamma)f$, defined as follows:

$$\bar{f}(\overset{q}{c_1}) = \gamma[f(\overset{q}{\kappa}\overset{q}{c_1})], \qquad \overset{q}{c_1} \in \overset{q}{C}(K_1).$$

By comparing this definition with the definition of the functor Hom, we observe that $C_q(\kappa, \gamma)$ is in fact just Hom (κ^q, γ).

The definitions of $C_q(K, G)$ and $C_q(\kappa, \gamma)$ yield a functor C_q contravariant in $\mathfrak{K}$, covariant in $\mathfrak{G}_a$, and with values in $\mathfrak{G}_a$. This functor will be called the qth cochain functor.

The coboundary homomorphism

$$\delta_q(K, G): C_q(K, G) \to C_{q+1}(K, G)$$

is defined by setting, for each cochain $f \in C_q(K, G)$,

$$(\delta_q f)(c^{q+1}) = f(\partial^{q+1} c^{q+1}).$$

This leads to a natural transformation of functors

$$\delta_q: C_q \to C_{q+1}.$$

We may observe that in terms of the functor "Hom" we have $\delta_q(K, G) = \text{Hom}(\partial^{q+1}, e_G)$.

The kernel of the transformation δ_q is denoted by Z_q and is called the q-cocycle functor. The image functor of δ_q is denoted by B_{q+1} and is called the $(q+1)$-coboundary functor. Since $\partial^q \partial^{q+1} = 0$, we may easily deduce that B_q is a subfunctor of Z_q. The quotient-functor

$$H_q = Z_q / B_q$$

is, by definition, the qth cohomology functor. H_q is contravariant in $\mathfrak{K}$, covariant in $\mathfrak{G}_a$, and has values in $\mathfrak{G}_a$. Its object function associates with each complex K and each topological abelian group G the (topological abelian) qth cohomology group $H_q(K, G)$.

The fact that the homology groups are discrete and have discrete coefficient groups, while the cohomology groups are topologized and have topological coefficient groups, is due to the circumstance that the complexes considered are closure finite. In a star finite complex the relation would be reversed.

For "finite" complexes both homology and cohomology groups may be topological. Let $\mathfrak{K}_f$ denote the subcategory of $\mathfrak{K}$ determined by all those complexes K such that all the groups $C^q(K)$ have finite rank. If $K \in \mathfrak{K}_f$ and

G is a topological group, then the group $C^q(K, G) = G \circ C^q(K)$ can be topologized in a natural fashion and consequently $H^q(K, G)$ will be topological. Hence both H^q and H_q may be regarded as functors on $\mathfrak{K}_f$ and $\mathfrak{G}_a$ with values in $\mathfrak{G}_a$. The first one is covariant in both $\mathfrak{K}_f$ and $\mathfrak{G}_a$, while the second one is contravariant in $\mathfrak{K}_f$ and covariant in $\mathfrak{G}_a$.

28. **Duality.** Let G be a discrete abelian group and Char G be its (compact) character group (see §19).

Given a chain

$$c^q \in C^q(K, G)$$

where

$$c^q = \sum_i g_i c_i^q, \qquad g_i \in G, \ c_i^q \in C^q(K),$$

and given a cochain

$$f \in C_q(K, \text{Char } G),$$

we may define the Kronecker index

$$KI(f, c^q) = \sum_i (f(c_i^q), g_i).$$

Since $f(c_i^q)$ is an element of Char G, its application to g_i gives an element of the group P of reals reduced mod 1. The continuity of $KI(f, c^q)$ as a function of f follows from the definition of the topology in Char G and in $C_q(K, \text{Char } G)$.

As a preliminary to the duality theorem, we define an isomorphism

$$\tau^q(K, G): C_q(K, \text{Char } G) \rightleftarrows \text{Char } C^q(K, G), \tag{28.1}$$

by defining for each cochain $f \in C_q(K, \text{Char } G)$ a character

$$\tau^q(K, G)f: C^q(K, G) \to P,$$

as follows:

$$(\tau^q f, c^q) = KI(f, c^q).$$

The fact that $\tau^q(K, G)$ is an isomorphism is a direct consequence of the character theory. In (28.1) both sides should be interpreted as object functions of functors (contravariant in both K and G), suitably compounded from the functors C^q, C_q, and Char. In order to prove that (28.1) is natural, consider

$$\kappa: K_1 \to K_2 \quad \text{in} \quad \mathfrak{K}, \qquad \gamma: G_1 \to G_2 \quad \text{in} \quad \mathfrak{G}_{0a}.$$

We must prove that

$$\tau^q(K_1, G_1)C_q(\kappa, \text{Char } \gamma) = [\text{Char } C^q(\kappa, \gamma)]\tau^q(K_2, G_2). \tag{28.2}$$

If now

$$f \in C_q(K_2, G_2), \qquad c^q \in C^q(K_1, G_1),$$

then the definition of τ^q shows that (28.2) is equivalent to the identity

$$KI(C_q(\kappa, \text{Char } \gamma)f, c^q) = KI(f, C^q(\kappa, \gamma)c^q). \tag{28.3}$$

It will be sufficient to establish (28.3) in the case when c^q is a generator of $C^q(K_1, G_1)$,

$$c^q = g_1 c_1^q, \qquad g_1 \in G_1, \; c_1^q \in C^q(K_2).$$

Using the definition of the terms involved in (28.3) we have on the one hand

$$\begin{aligned} KI(C_q(\kappa, \text{Char } \gamma)f, g_1 c_1^q) &= ([C_q(\kappa, \text{Char } \gamma)f]c_1^q, g_1) \\ &= (\text{Char } \gamma[f(\kappa c_1^q)]g_1) = (f(\kappa c_1^q), \gamma g_1), \end{aligned}$$

and on the other hand

$$KI(f, C^q(\kappa, \gamma)g_1 c_1^q) = KI(f, (\gamma g_1)(\kappa c_1^q)) = (f(\kappa c_1^q), \gamma g_1).$$

This completes the proof of the naturality of (28.1).

Using the well known property of the Kronecker index

$$KI(f, \partial^{q+1} c^{q+1}) = KI(\delta_q f, c^{q+1}),$$

one shows easily that under the isomorphism τ^q of (28.1)

$$\tau^q[Z_q(K, \text{Char } G)] = \text{Annih } B^q(K, G), \quad \tau^q[B_q(K, \text{Char } G)] = \text{Annih } Z^q(K, G),$$

with "Annih" defined as in §19. Both Annih (B^q; C^q) and Annih (Z^q; C^q) are functors covariant in K and G; the latter is a subfunctor of the former, so that τ^q induces a natural isomorphism

$$\sigma^q: Z_q(K, \text{Char } G)/B_q(K, \text{Char } G) \rightleftarrows \text{Annih } B^q(K, G)/\text{Annih } Z^q(K, G).$$

The group on the left is $H_q(K, \text{Char } G)$. The group on the right is, according to (19.7), naturally isomorphic to Char $Z^q(K, G)/B^q(K, G)$. All told we have a natural isomorphism:

$$\rho^q: H_q(K, \text{Char } G) \rightleftarrows \text{Char } H^q(K, G).$$

This is the customary Pontrjagin-type duality between homology and cohomology. Thus we have established the naturality of this duality.

29. **Universal coefficient theorems.** The theorems of this name express the cohomology groups of a complex, for an arbitrary coefficient group, in terms of the integral homology groups and the coefficient group itself. A quite general form of such theorems can be stated in terms of certain groups of group extensions([25]); hence we first show that the basic constructions of group extensions may be regarded as functors.

Let G be a topological abelian group and H a discrete abelian group. A factor set of H in G is a function $f(h, k)$ which assigns to each pair h, k of elements in H an element $f(h, k) \in G$ in such wise that

([25]) S. Eilenberg and S. MacLane, *Group extensions and homology*, Ann. of Math. vol. **43** (1943) pp. 757–831.

$$f(h, k) = f(k, h), \qquad f(h, k) + f(h + k, l) = f(h, k + l) + f(k, l),$$

for all h, k, and l in H. With the natural addition and topology, the set of all factor sets f of H in G constitute a topological abelian group Fact (G, H). If $\gamma: G_1 \to G_2$ and $\eta: H_1 \to H_2$ are homomorphisms, we can define a corresponding mapping

$$\text{Fact } (\gamma, \eta): \text{Fact } (G_1, H_2) \to \text{Fact } (G_2, H_1)$$

by setting

$$[\text{Fact } (\gamma, \eta) f](h_1, k_1) = \gamma f(\eta h_1, \eta k_1)$$

for each factor set f in Fact (G_1, H_2). Thus it appears that Fact is a functor, covariant on the category $\mathfrak{G}_a$ of topological abelian groups and contravariant in the category $\mathfrak{G}_{0a}$ of discrete abelian groups.

Given any function $g(h)$ with values in G, the combination

$$f(h, k) = g(h) + g(k) - g(h + k)$$

is always a factor set; the factor sets of this special form are said to be transformation sets, and the set of all transformation sets is a subgroup Trans (G, H) of the group Fact (G, H). Furthermore, this subgroup is the object function of a subfunctor. The corresponding quotient functor

$$\text{Ext} = \text{Fact}/\text{Trans}$$

is thus covariant in $\mathfrak{G}_a$, contravariant in $\mathfrak{G}_{0a}$, and has values in $\mathfrak{G}_a$. Its object function assigns to the groups G and H the group Ext (G, H) of the so-called abelian group extensions of G by H.

Since $C_q(K, G) = \text{Hom } (C^q(K), G)$ and since $C^q(K, I) = I \circ C^q(K) = C^q(K)$ where I is the additive group of integers, we have

$$C_q(K, G) = \text{Hom } (C^q(K, I), G).$$

We, therefore, may define a subgroup

$$A_q(K, G) = \text{Annih } Z^q(K, I)$$

of $C_q(K, G)$ consisting of all homomorphisms f such that $f(z^q) = 0$ for $z^q \in Z^q(K, I)$. Thus we get a subfunctor A_q of C_q, and one may show that the coboundary functor B_q is a subfunctor of A_q which, in turn, is a subfunctor of the cocycle functor Z_q. Consequently, the quotient functor

$$Q_q = A_q/B_q$$

is a subfunctor of the cohomology functor H_q, and we may consider the quotient functor H_q/Q_q. The functors Q_q and H_q/Q_q have the following object functions

$$Q_q(K, G) = A_q(K, G)/B_q(K, G),$$
$$(H_q/Q_q)(K, G) = H_q(K, G)/Q_q(K, G) \cong Z_q(K, G)/A_q(K, G).$$

The universal coefficient theorem now consists of these three assertions[26]:

(29.1) $Q_q(K, G)$ *is a direct factor of* $H_q(K, G)$.

(29.2) $$Q_q(K, G) \cong \text{Ext}\,(G, H^{q+1}(K, I)).$$

(29.3) $$H_q(K, G)/Q_q(K, G) \cong \text{Hom}\,(H^q(K, I), G).$$

Both the isomorphisms (29.2) and (29.3) can be interpreted as equivalences of functors. The naturality of these equivalences with respect to K has been explicitly verified[27], while the naturality with respect to G can be verified without difficulty. We have not been able to prove and we doubt that the functor Q_q is a direct factor of the functor H_q (see §18).

30. **Čech homology groups.** We shall present now a treatment of the Čech homology theory in terms of functors.

By a covering U of a topological space X we shall understand a finite collection:

$$U = \{A_1, \cdots, A_n\}$$

of open sets whose union is X. The sets A_i may appear with repetitions, and some of them may be empty. If U_1 and U_2 are two such coverings, we write $U_1 < U_2$ whenever U_2 is a refinement of U_1, that is, whenever each set of the covering U_2 is contained in some set of the covering U_1. With this definition the coverings U of X form a directed set which we denote by $C(X)$.

Let $\xi: X_1 \rightarrow X_2$ be a continuous mapping of the space X_1 into the space X_2. Given a covering

$$U = \{A_1, \cdots, A_n\} \in C(X_2),$$

we define

$$C(\xi)U = \{\xi^{-1}(A_1), \cdots, \xi^{-1}(A_n)\} \in C(X_1)$$

and we obtain an order preserving mapping

$$C(\xi): C(X_2) \rightarrow C(X_1).$$

We verify that the functions $C(X)$, $C(\xi)$ define a contravariant functor C on the category $\mathfrak{X}$ of topological spaces to the category $\mathfrak{D}$ of directed sets.

Given a covering U of X we define, in the usual fashion, the nerve $N(U)$ of U. $N(U)$ is a finite simplicial complex; it will be treated, however, as an object of the category K_f of §27.

If two coverings $U_1 < U_2$ of X are given, then we select for each set of the covering U_2 a set of the covering U_1 containing it. This leads to a simplicial mapping of the complex $N(U_2)$ into the complex $N(U_1)$ and therefore gives a chain transformation

(26) Loc. cit. p. 808.
(27) Loc. cit. p. 815.

$$\kappa: N(U_2) \to N(U_1).$$

This transformation κ will be called a projection. The projection κ is not defined uniquely by U_1 and U_2, but it is known that any two projections κ_1 and κ_2 are chain homotopic and consequently the induced homomorphisms

$$H^q(\kappa, e_G): H^q(N(U_2), G) \to H^q(N(U_1), G), \tag{30.1}$$

$$H_q(\kappa, e_G): H_q(N(U_1), G) \to H_q(N(U_2), G) \tag{30.2}$$

of the homology and cohomology groups do not depend upon the particular choice of the projection κ.

Given a topological group G we consider the collection of the homology groups $H^q(N(U), G)$ for $U \in C(X)$. These groups together with the mappings (30.1) form an inverse system of groups defined on the directed set $C(X)$. We denote this inverse system by $\overline{C}^q(X, G)$ and treat it as an object of the category $\mathfrak{Inv}$ (§23).

Similarly, for a discrete G the cohomology groups $H_q(N(U), G)$ together with the mappings (30.2) form a direct system of groups $\overline{C}_q(X, G)$ likewise defined on the directed set $C(X)$. The system $\overline{C}_q(X, G)$ will be treated as an object of the category $\mathfrak{Dir}$.

The functions $\overline{C}^q(X, G)$ and $\overline{C}_q(X, G)$ will be object functions of functors $\overline{C}^q$ and $\overline{C}_q$. In order to complete the definition we shall define the mapping functions $\overline{C}^q(\xi, \gamma)$ and $\overline{C}_q(\xi, \gamma)$ for given mappings

$$\xi: X_1 \to X_2, \qquad \gamma: G_1 \to G_2.$$

We have the order preserving mapping

$$C(\xi): C(X_2) \to C(X_1) \tag{30.3}$$

which with each covering

$$U = \{A_1, \cdots, A_n\} \in C(X_2)$$

associates the covering

$$V = C(\xi)U = \{\xi^{-1}A_1, \cdots, \xi^{-1}A_n\} \in C(X_1).$$

Thus to each set of the covering V corresponds uniquely a set of the covering U; this yields a simplicial mapping

$$\kappa: N(V) \to N(U),$$

which leads to the homomorphisms

$$H^q(\kappa, \gamma): H^q(N(V), G_1) \to H^q(N(U), G_2), \tag{30.4}$$

$$H_q(\kappa, \gamma): H_q(N(U), G_1) \to H_q(N(V), G_2). \tag{30.5}$$

The mappings (30.3)–(30.5) define the transformations

$$\overline{C}^q(\xi, \gamma):\overline{C}^q(X_1, G_1) \to \overline{C}^q(X_2, G_2) \quad \text{in} \quad \mathfrak{Inv},$$

$$\overline{C}_q(\xi, \gamma):\overline{C}_q(X_2, G_1) \to \overline{C}_q(X_1, G_2) \quad \text{in} \quad \mathfrak{Dir}.$$

Hence we see that $\overline{C}^q$ is a functor covariant in $\mathfrak{X}$ and in $\mathfrak{G}_a$ with values in $\mathfrak{Inv}$ while $\overline{C}_q$ is contravariant in $\mathfrak{X}$ covariant in $\mathfrak{G}_{0a}$ and has values in $\mathfrak{Dir}$.

The Čech homology and cohomology functors are now defined as

$$\overline{H}^q = \operatorname{Lim}_{\leftarrow} \overline{C}^q, \qquad \overline{H}_q = \operatorname{Lim}_{\rightarrow} \overline{C}_q.$$

$\overline{H}^q$ is covariant in $\mathfrak{X}$ and $\mathfrak{G}_a$ and has values in $\mathfrak{G}_a$, while $\overline{H}_q$ is contravariant in $\mathfrak{X}$, covariant in $\mathfrak{G}_{0a}$, and has values in $\mathfrak{G}_{0a}$. The object functions $\overline{H}^q(X, G)$ and $\overline{H}_q(X, G)$ are the Čech homology and cohomology groups of the space X with the group G as coefficients.

31. **Miscellaneous remarks.** The process of setting up the various topological invariants as functors will require the construction of many categories. For instance, if we wish to discuss the so-called relative homology theory, we shall need the category $\mathfrak{X}_S$ whose objects are the pairs (X, A), where X is a topological space and A is a subset of X. A mapping

$$\xi:(X, A) \to (Y, B) \quad \text{in} \quad \mathfrak{X}_S$$

is a continuous mapping $\xi:X\to Y$ such that $\xi(A)\subset B$. The category $\mathfrak{X}$ may be regarded as the subcategory of $\mathfrak{X}_S$, determined by the pairs (X, A) with $A=0$.

Another subcategory of $\mathfrak{X}_S$ is the category $\mathfrak{X}_b$ defined by the pairs (X, A) in which the set A consists of a single point, called the base point. This category $\mathfrak{X}_b$ would be used in a functorial treatment of the fundamental group and of the homotopy groups.

Appendix. Representations of categories

The purpose of this appendix is to show that every category is isomorphic with a suitable subcategory of the category of sets $\mathfrak{S}$.

Let $\mathfrak{A}$ be any category. A covariant functor T on $\mathfrak{A}$ with values in $\mathfrak{S}$ will be called a representation of $\mathfrak{A}$ in $\mathfrak{S}$. A representation T will be called faithful if for every two mappings, $\alpha_1, \alpha_2 \in \mathfrak{A}$, we have $T(\alpha_1)=T(\alpha_2)$ only if $\alpha_1=\alpha_2$. This implies a similar proposition for the objects of $\mathfrak{A}$. It is clear that a faithful representation is nothing but an isomorphic mapping of $\mathfrak{A}$ onto some subcategory of $\mathfrak{S}$.

If the functor T on $\mathfrak{A}$ to $\mathfrak{S}$ is contravariant, we shall say that T is a dual representation. T is then obviously a representation of the dual category $\mathfrak{A}^*$, as defined in §13.

Given a mapping $\alpha:A_1\to A_2$ in $\mathfrak{A}$, we shall denote the domain A_1 of α by $d(\alpha)$ and the range A_2 of α by $r(\alpha)$. In this fashion we have

$$\alpha:d(\alpha) \to r(\alpha).$$

Given an object A in $\mathfrak{A}$ we shall denote by $R(A)$ the set of all $\alpha \in \mathfrak{A}$, such that $A = r(\alpha)$. In symbols

(I) $$R(A) = \{\alpha \mid \alpha \in \mathfrak{A}, r(\alpha) = A\}.$$

For every mapping α in $\mathfrak{A}$ we define a mapping

(II) $$R(\alpha): R(d(\alpha)) \rightarrow R(r(\alpha))$$

in the category $\mathfrak{S}$ by setting

(III) $$[R(\alpha)]\xi = \alpha\xi$$

for every $\xi \in R(d(\alpha))$. This mapping is well defined because if $\xi \in R(d(\alpha))$, then $r(\xi) = d(\alpha)$, so that $\alpha\xi$ is defined and $r(\alpha\xi) = r(\alpha)$ which implies $\alpha\xi \in R(r(\alpha))$.

THEOREM. *For every category $\mathfrak{A}$ the pair of functions $R(A)$, $R(\alpha)$, defined above, establishes a faithful representation R of $\mathfrak{A}$ in $\mathfrak{S}$.*

Proof. We first verify that R is a functor. If $\alpha = e_A$ is an identity, then definition (III) implies that $[R(\alpha)]\xi = \xi$, so that $R(\alpha)$ is the identity mapping of $R(A)$ into itself. Thus R satisfies condition (3.1). Condition (3.2) has already been verified. In order to verify (3.3) let us consider the mappings

$$\alpha_1: A_1 \rightarrow A_2, \qquad \alpha_2: A_2 \rightarrow A_3.$$

We have for every $\xi \in R(A_1)$,

$$[R(\alpha_2\alpha_1)]\xi = \alpha_2\alpha_1\xi = [R(\alpha_2)]\alpha_1\xi = [R(\alpha_2)R(\alpha_1)]\xi,$$

so that $R(\alpha_2\alpha_1) = R(\alpha_2)R(\alpha_1)$. This concludes the proof that R is a representation.

In order to show that R is faithful, let us consider two mappings $\alpha_1, \alpha_2 \in \mathfrak{A}$ and let us assume that $R(\alpha_1) = R(\alpha_2)$. It follows from (II) that $R(d(\alpha_1)) = R(d(\alpha_2))$, and, therefore, according to (I), $d(\alpha_1) = d(\alpha_2)$. Consider the identity mapping $e = e_{d(\alpha_1)} = e_{d(\alpha_2)}$. Following (III), we have

$$\alpha_1 = \alpha_1 e = [R(\alpha_1)]e = [R(\alpha_2)]e = \alpha_2 e = \alpha_2,$$

so that $\alpha_1 = \alpha_2$. This concludes the proof of the theorem.

In a similar fashion we could define a faithful dual representation D of $\mathfrak{A}$ by setting

$$D(A) = \{\alpha \mid \alpha \in \mathfrak{A}, d(\alpha) = A\}$$

and

$$[D(\alpha)]\xi = \xi\alpha$$

for every $\xi \in D(r(\alpha))$.

The representations R and D are the analogues of the left and right regular representations in group theory.

We shall conclude with some remarks concerning partial order in categories. Most of the categories which we have considered have an intrinsic partial order. For instance, in the categories $\mathfrak{S}$, $\mathfrak{X}$, and $\mathfrak{G}$ the concepts of subset, subspace, and subgroup furnish a partial order. In view of (I), $A_1 \neq A_2$ implies that $R(A_1)$ and $R(A_2)$ are disjoint, so that the representation R destroys this order completely. The problem of getting "order preserving representations" would require probably a suitable formalization of the concept of a partially ordered category.

As an illustration of the type of arguments which may be involved, let us consider the category $\mathfrak{G}_0$ of discrete groups. With each group G we can associate the set $R_1(G)$ which is the set of elements constituting the group G. With the obvious mapping function, R_1 becomes a covariant functor on $\mathfrak{G}_0$ to $\mathfrak{S}$, that is, R_1 is a representation of $\mathfrak{G}_0$ in $\mathfrak{S}$. This representation is not faithful, since the same set may carry two different group structures. The group structure of G is entirely described by means of a ternary relation $g_1g_2=g$. This ternary relation is nothing but a subset $R_2(G)$ of $R_1(G)\times R_1(G)\times R_1(G)$. All of the axioms of group theory can be formulated in terms of the subset $R_2(G)$. Moreover a homomorphism $\gamma: G_1 \rightarrow G_2$ induces a mapping $R_2(\gamma): R_2(G_1) \rightarrow R_2(G_2)$. Consequently R_2 is a subfunctor of a suitably defined functor $R_1 \times R_1 \times R_1$. The two functors R_1 and R_2 together give a complete description of $\mathfrak{G}_0$, preserving the partial order.

THE UNIVERSITY OF MICHIGAN,
ANN ARBOR, MICH.
HARVARD UNIVERSITY,
CAMBRIDGE, MASS.

Reprinted from the Proceedings of the NATIONAL ACADEMY OF SCIENCES,
Vol. 34, No. 6, pp. 263–267. June, 1948

GROUPS, CATEGORIES AND DUALITY

BY SAUNDERS MACLANE*

DEPARTMENT OF MATHEMATICS, UNIVERSITY OF CHICAGO

Communicated by Marshall Stone, May 1, 1948

It has long been recognized that the theorems of group theory display a certain duality. The concept of a lattice gives a partial expression for this duality, in that some of the theorems about groups which can be formulated in terms of the lattice of subgroups of a group display the customary lattice duality between meet (intersection) and join (union). The duality is not always present, in the sense that the lattice dual of a true theorem on groups need not be true; for example, a Jordan Holder theorem holds for certain ascending well-ordered infinite composition series, but not for the corresponding descending series.[1] Moreover, there are other striking group theoretic situations where a duality is present, but is not readily expressible in lattice-theoretic terms.

As an example, consider the direct product $D = G \times H$ of two groups

G and H, together with its canonical homomorphisms $\gamma(g, h) = g$, $\eta(g, h) = h$ into the given factors G and H. The system $[\gamma:D \to G;\ \eta:D \to H]$ consisting of the direct product together with these homomorphisms is characterized, up to isomorphism, by the following property: given any other such system $[\gamma':D' \to G, \eta':D' \to H]$ for the same groups G and H, there is one and only one homomorphism $\pi:D' \to D$ such that $\gamma' = \gamma\pi$, $\eta' = \eta\pi$. Dually, the free product P of groups G and H is the "most general" group generated by subgroups isomorphic to G and H, respectively. This means that there are canonical homomorphisms $\alpha:G \to P$ and $\beta:H \to P$ of the factors into the corresponding subgroups of P. This system (P, α, β) is characterized by the following property: given any system $[\alpha':G \to P', \beta':H \to P']$ there is one and only one homomorphism $\sigma:P \to P'$ such that $\sigma\alpha = \alpha'$, $\sigma\beta = \beta'$. The theorem that the direct product of any two groups exists is thus dual to the theorem asserting the existence of the free product. The *proofs* of these two theorems are not dual, but the proofs of many other formal properties are dual, as for instance in the case of the associative law $(G \times H) \times K \cong G \times (H \times K)$. For the direct product D, the canonical homomorphisms γ and η are *homomorphisms onto* their respective ranges G and H; in the case of the free product P the canonical homomorphisms α and β are *isomorphisms into P.* The "dual" of a theorem about groups and homomorphisms is to be obtained by inverting the direction of each homomorphism, inverting the order of all products of homomorphisms and replacing homomorphisms onto by isomorphisms into.

For abelian groups the duality is more marked. A free abelian group F can be characterized in terms of homomorphisms of abelian groups by the following property:[2] for any homomorphism $\alpha:F \to A$ and any second homomorphism $\beta:B \to A$ onto the image group A there exists a homomorphism $\gamma:F \to B$ with $\beta\gamma = \alpha$. (The corresponding characterization applies also to free non-abelian groups.) An infinitely divisible abelian group D is one in which there exists for each $d \in D$ and each integer m a solution x of the equation $mx = d$. Any homomorphism of an abelian group A into D can be extended to any abelian group B containing A. This property characterizes the infinitely divisible abelian groups; it may be stated in a form dual to the characteristic property of free groups: given $\alpha:A \to D$ and an isomorphism $\beta:A \to B$ of A into B, there exists a $\gamma:B \to D$ with $\gamma\beta = \alpha$. For an abelian group, free products reduce to direct products. If a factor group of an abelian group is a free group, it is a direct factor. Dually, if a subgroup of an abelian group is infinitely divisible, it is a direct factor.

This duality for abelian groups appears in algebraic topology as a duality between homology and cohomology groups. This phenomenon is especially striking in the axiomatic form of homology theory.[3]

For locally compact topological abelian groups, the duality phenomena can be formulated explicitly by means of character groups;[4] each theorem then gives a dual theorem about the character groups of those groups involved in the original theorem. It is instructive to compare this formulation with the duality of plane projective geometry.[5] A pole-polar reciprocation gives a dual to each projective figure, comparable to the character group of a group. Alternatively, projective geometry has an "axiomatic" or "syntactical" duality: any theorem deducible from the incidence axioms remains true on the interchange of the primitive terms "point" and "line" in the statement of the theorem.

Our objective is a similar formulation of a (partial) axiomatic duality for groups. It clearly must concern the system consisting of all groups and all homomorphisms of one group into another. For certain other investigations of this and similar systems, Eilenberg and the author have introduced the notion of a category.[6] A *category* is a class of "mappings" (say, homomorphisms) in which the product $\alpha\beta$ of certain pairs of mappings α and β is defined. A mapping e is called an *identity* if $\rho\alpha = \alpha$ and $\beta\rho = \beta$ whenever the products in question are defined. These products must satisfy the axioms:

(C-1). *If the products $\gamma\beta$ and $(\gamma\beta)\alpha$ are defined, so is $\beta\alpha$;*

(C-1′). *If the products $\beta\alpha$ and $\gamma(\beta\alpha)$ are defined, so is $\gamma\beta$;*

(C-2). *If the products $\gamma\beta$ and $\beta\alpha$ are defined, so are the products $(\gamma\beta)\alpha$ and $\gamma(\beta\alpha)$, and these products are equal.*

(C-3). *For each γ there is an identity e_D such that γe_D is defined;*

(C-4). *For each γ there is an identity e_R such that $e_R\gamma$ is defined.*

It follows that the identities e_D and e_R are unique; they may be called, respectively, the *domain* and the *range* of the given mapping γ. A mapping θ with a two-sided inverse is an *equivalence.*

These axioms are clearly self dual, and a dual theory of free and direct products may be constructed in any category in which such products exist. These axioms do not, however, suffice to express the duality between "homomorphism onto" and "isomorphism into." These notions can be formulated in terms of subgroups and factor groups; with any subgroup $S \subset G$ we can associate the identity injection $i \cdot S \rightarrow G$ of S into G, and with any normal subgroup N of G we can associate the projection $\tau: G \rightarrow G/N$ mapping each element g of G into its coset gN in the factor group G/N. We propose to axiomatize the dual notions "injection" and "projection."

A *bicategory* is a category with two distinguished classes of mappings, the "injections" and the "projections," subject to the following self dual axioms:

(BC-1). *Every identity is both an injection and a projection;*

(BC-2). *The product of two injections (projections), when defined, is an injection (projection).*

(BC-3). *Every mapping γ can be represented uniquely as a product $\gamma = \kappa\theta\pi$, where π is a projection, θ an equivalence and κ an injection.*

A mapping of the form $\kappa\theta$ is called a mapping within (isomorphism into); one of the form $\theta\pi$ is called a mapping upon (homomorphism onto).

(BC-4). *The product of two mappings within (upon), when defined, is a mapping within (upon).*

(BC-5). *Two injections (projections) with identical domains and identica ranges are identical.*

These concepts suffice to give dual definitions of "subgroups" and "factor groups." Thus e_1 is a "subidentity" of ρ_2 if there exists an injection with domain e_1 and range e_2; this inclusion relation gives a partial order of the identities of a bicategory. We may then define a *lattice-ordered bicategory* as any bicategory in which the subidentities and factor identities of any given identity form a lattice under this partial order.

A group can be interpreted as a lattice-ordered bicategory with an identity; the mappings of the category are all equivalences, and are the elements of the group. A lattice L can be interpreted as a lattice-ordered bicategory in which all mappings are injections: the mappings of the category are the pairs $[a, b]$ with $a \supset b$, and with product $[a, b][b, c] = [a, c]$. Thus the concept "lattice-ordered bicategory" is a common generalization of the notions "group" and "lattice."

We contend that most of the phenomena of universal algebra and of (axiomatic) group duality[7] have appropriate and simple formulations in terms of lattice-ordered bicategories. In particular, for groups, one may use the lattice-ordered bicategory of all homomorphisms of one group into another. In this category we might interpret projection mapping to mean any (canonical) homomorphism $\tau: G \to G/N$ of a group G upon its factor group G/N. For this interpretation the product of two projections is not a projection (axiom BC-2 fails). This axiom might be saved by calling a projection any product of such canonical homomorphisms τ, but in this case the projection factor π of any homomorphism is not unique (axiom BC-3 fails).

This apparent difficulty can be surmounted by an attention to fundamentals. A factor group G/N may be described either as a group in which the *elements* are cosets of N, and the *equality* of elements is the equality of sets, or as a group in which the *elements* are the elements of G, and the "equality" is congruence modulo N. Both approaches are rigorous[8] and can be applied systematically (and with approximately equal inconvenience!) throughout group theory. The difficulties cited disappear when we adopt the second point of view, and regard a group G as a system of elements G with a reflexive symmetric and transitive "equality" relation such that logically identical elements are equal (but not necessarily conversely) and such that products of equal elements are equal.

* John Simon Guggenheim Memorial Fellow.

1 Birkhoff, G., "Lattice Theory," *Am. Math. Soc. Colloq. Pub.*, **25**, 48 (1940).

2 For the case of abelian groups with operators from a group Q, this property is used in Eilenberg, S., and MacLane, S., "Homology Theory of Spaces with Operators II," forthcoming in *Trans. Am. Math. Soc.*

3 Eilenberg, S., and Steenrod, N., Proc. Nat. Acad. Sci., **31**, 117–120 (1945). (The writer has also profited by reading further unpublished work of these authors on this subject.)

4 Pontrjagin, L., *Topological Groups*, Princeton, 1939. Weil, A., *L'Integration dans les groupes topologiques et ses applications*, Paris, 1938.

5 Veblen, O., and Young, J. W., *Projective Geometry*, Boston, 1910.

6 Eilenberg, S., and MacLane, S., Proc. Nat. Acad. Sci., **28**, 537–543 (1942); *Trans. Am. Math. Soc.*, **58**, 231–294 (1945).

7 The formulation with bicategories does not yet indicate the duality between center and factor commutator groups, and similar dual concepts of verbal and marginal subgroups; Hall, P., *J. f. d. reine und angew. Math.*, **182**, 156–157 (1940).

8 A careful treatment, emphasizing the equality approach, appears in the unjustly neglected book by Haupt, O., *Einführung in die Algebra*, Leipzig, 1929.

ANNALS OF MATHEMATICS
Vol. 50, No. 3, July, 1949

COHOMOLOGY THEORY IN ABSTRACT GROUPS. III
Operator Homomorphisms of Kernels

BY SAUNDERS MACLANE[1]

(Received April 1, 1948)

1. Introduction

Let the group Q be represented in any fashion as a quotient group $Q = F/R$ of a free group F; the elements a of F then act as automorphisms $r \to ara^{-1}$ in the kernel R of this representation. On the other hand, let G be an abelian group with the elements x of Q as operators $g \to x \cdot g$; G then also has operators from F, by virtue of the representation $Q = F/R$. An *operator homomorphism* of R in G is a homomorphism $\omega : R \to G$ such that $\omega(ara^{-1}) = a \cdot \omega(r)$ for all $a \in F$ and all $r \in R$. A *crossed homomorphism* of F in G is a one-dimensional cocycle of F in G; i.e., is a function $\tau(a) \in G$, defined for all a in F, such that $\tau(ab) = a \cdot \tau(b) + \tau(a)$. Clearly each such crossed homomorphism, if restricted to the subgroup R, yields an operator homomorphism $\omega : R \to G$. By the group Map $(R, F; G)$ of *operator homomorphism classes* of R into G we understand the group of all operator homomorphisms $\omega : R \to G$, modulo the subgroup of those operator homomorphisms induced by crossed homomorphisms τ of F into G.

THEOREM A. *If R is a normal subgroup of the free group F, and G an abelian group with operators from $Q = F/R$, the group* Map $(R, F; G)$ *of operator homomorphism classes of R into G depends only on the groups Q and G and the operators of Q on G and not on the chosen representation of Q by the free group F.*

This theorem can be stated in a more explicit form.

THEOREM A′. Map $(R, F; G)$ *is isomorphic to the second cohomology group $H^2(Q, G)$ of Q with coefficients in G.*

The second cohomology group $H^2(Q, G)$ is the group of all those group extensions of G by Q which realize the given operators of Q on G, and this theorem formalizes the known representation (cf. [6] and [11] in the bibliography) of these group extensions in terms of selected generators of Q and the group R of all relations on these generators. The formulation given above is essentially that appearing as Theorem 13.1 in the first paper [2] of this series, where it is the zero-dimensional case of a certain "cup product" reduction theorem for the higher cohomology groups of Q.

In the representation $Q = F/R$, the subgroup R of F is a free group, hence is either an infinite cyclic group or a non-abelian free group with center 1. Suppose now more generally that Q is represented as a factor group $Q = E/K$, where the group K has center H and where the factor group E/H is now assumed to be a free group F. The elements e of E again act as automorphisms $k \to eke^{-1}$ on the subgroup K and on its center H. We again denote by Map $(K, F; G)$ the group

[1] John Simon Guggenheim Memorial Fellow.

Reprinted from the Annals of Math. 50 (1949), 736-761, by permission of Princeton University Press.

of all operator homomorphisms of K into G, modulo the subgroup of all those operator homomorphisms which are induced by crossed homomorphisms of F into G. We assert that this group of operator homomorphism classes depends only on H, G, and Q, and explicitly that it can be represented as a direct sum

$$\text{Map}\,(K, F; G) \cong \text{Ophom}\,(H, G) + H^2(Q, G), \tag{1.1}$$

where Ophom (H, G) denotes the group of all operator homomorphisms of H into G.

This generalization of Theorem A' is a direct consequence of that theorem. To show this, denote by M the group of operator homomorphism classes which appears on the left side of (1.1), and by T the subgroup composed of those classes consisting of operator homomorphisms which carry H into 0. Operator homomorphisms of this latter sort can be regarded simply as operator homomorphisms of $R = K/H$ into G; they are to be taken modulo those operator homomorphisms of R into G which can be extended to crossed homomorphisms of F into G. Since this is exactly the situation contemplated in Theorem A', it follows that the subgroup T is isomorphic to $H^2(Q, G)$.

Each operator homomorphism σ of K into G induces an operator homomorphism ρ of the center H of K into G; this correspondence $\sigma \to \rho$ is a homomorphism of M into Ophom (H, G), and its kernel is exactly T. The direct decomposition (1.1) will follow if we can construct a homomorphism of *all* of Ophom (H, G) into M which is a right inverse of this correspondence $\sigma \to \rho$.

Since F is free, one can choose for each a in F a representative $u(a)$ in E such that $u(ab) = u(a)u(b)$. The elements of E can then be written uniquely as products $hu(a)$ for $h \in H$, $a \in F$. Given any operator homomorphism ρ of H into G, we define a function ρ^* on all of E by setting $\rho^*(hu(a)) = \rho(h)$. Since ρ is an operator homomorphism, $\rho(u(a)hu(a)^{-1}) = u(a)\cdot\rho h$; from this one argues that ρ^* is a crossed homomorphism of E. It follows that ρ^*, cut down to the subgroup $K \subset E$, is an operator homomorphism σ on K. Since σ in turn induces on H the originally given ρ, this homomorphism $\rho^* \to \sigma$ is a right inverse of the given homomorphism $\rho \to \sigma$. This completes the proof of (1.1).

Both Theorem A' and its generalization (1.1) treat the operator homomorphisms on the kernel R or K of a representation $Q = F/R$ or $Q = E/K$. Our objective is the generalization of this result to apply to other abstract Q-kernels, in the sense of the second paper [3] of this series. We consider those Q-kernels which have as "graph" (see §3 below) a free group, and prove that the corresponding group of operator homomorphism classes is independent of the choice of this free group, and depends on the group Q, the group G, the center H of the kernel, and on the three dimensional "obstruction" of the kernel. The fashion of this dependence is exactly that studied by Eilenberg and the author in the analysis of the second cohomology group of a space [4, 5]; this allows for certain topological applications of our algebraic theorem (see §4 below).

The theorem of (1.1) deals with the representation $Q = E/K$ and thus with the groups H, K, $F = E/H$, and Q, where $K/H \subset F$ and $F/(K/H) = Q$. Upon replacing the factor groups here by homomorphisms, we have an "exact sequence" of groups and homomorphisms

$$H \to K \to F \to Q.$$

Our main theorem deals directly with such exact sequences. The essential case is precisely that in which the sequence *cannot* be obtained from a group E. It is in these cases that the three dimensional "obstruction" of the kernel K must enter.

CHAPTER I. GROUPS OF OPERATOR HOMOMORPHISMS

2. Formulation of the theorem

Throughout we consider an exact sequence $\mathfrak{S}$ of groups and homomorphisms

$$\mathfrak{S}: H \xrightarrow{\lambda} K \xrightarrow{\mu} F \xrightarrow{\nu} Q; \tag{2.1}$$

the requirement that the sequence be exact means that λ is an isomorphism of H into K, that the image $\lambda H \subset K$ is the kernel of $\mu : K \to F$, that the image $\mu K \subset F$ is the kernel of $\nu : F \to Q$, and that ν is a homomorphism of F *onto* Q. We require also that the elements $a \in F$ serve as operators (automorphisms) $k \to \eta(a) \cdot k$ of K; in other words, we assume that there is given a homomorphism

$$\eta : F \to A(K) \tag{2.2}$$

of F into the group $A(K)$ of automorphisms of K. Furthermore, we assume the following special hypotheses:

(i) For k and k' in K, $\eta(\mu k) \cdot k' = kk'k^{-1}$;

(ii) The image λH contains the center of K.

We note at once certain consequences of these hypotheses.

(iii) The image λH is the center of K.

Indeed, if $k = \lambda h$ for some $h \in H$, then $\mu k = 1$ by the exactness of the sequence; hence, by hypothesis (i), $k' = kk'k^{-1}$ for all k', which shows that k is in the center of K.

(iv) The group Q acts as a group of left operators on the abelian group H in such fashion that $\eta(a) \cdot \lambda(h) = (\nu a) \cdot h$, for $a \in F$, $h \in H$.

The center λH of K is mapped onto itself by all automorphisms of K, hence by the given automorphisms $\eta(a)$. By (i), the automorphisms corresponding to μk leave the center elementwise fixed, so that the effect of $\eta(a)$ on λH depends only on the image $\nu a = x \in Q$. Setting

$$x \cdot h = \lambda^{-1}[\eta(a) \cdot (\lambda h)] \qquad \nu a = x, \tag{2.3}$$

we thus have an automorphism $h \to x \cdot h$ of H for each $x \in Q$, with $x \cdot (y \cdot h) = (xy) \cdot h$. In other words, Q operates on H, as asserted.

(v) For a in F and k in K, $\mu[\eta(a) \cdot k] = a(\mu k)a^{-1}$.

This asserts that μ is an operator homomorphism, if F operates on itself by conjugation. To prove it, we must show that the element

$$b = a(\mu k)a^{-1}\mu[\eta(a)\cdot k]^{-1}$$

of F is 1. On the one hand, $\nu\mu = 1$, by exactness, so that $\nu b = (\nu a)(\nu a)^{-1} = 1$ and b is in the image of μ, as $b = \mu k'$, for some $k' \epsilon K$. On the other hand

$$\eta\mu k' = \eta b = \eta a(\eta\mu k)(\eta a)^{-1}\eta\mu[\eta(a)\cdot k]^{-1}.$$

By hypothesis (i), $\eta\mu k = C[k]$, the inner automorphism induced by k on K. Hence the equation becomes

$$C[k'] = \eta(a)C[k]\eta(a)^{-1}C[\eta(a)\cdot k]^{-1}.$$

But for any automorphism α of K, $\alpha C[k]\alpha^{-1} = C[\alpha\cdot k]$, hence the right hand side is 1. Thus $C[k'] = 1$, k' is the center of K, and by hypothesis (ii) k' is in λH. Therefore $b = \mu k' = 1$, by exactness.

The second group K of the sequence $\mathfrak{S}$ may be regarded in natural fashion as a Q-kernel; as such [3, §7], it determines uniquely a certain three dimensional cohomology class $\{l_3\} \epsilon H^3(Q, H)$. This cohomology class, which may of course be computed directly from the sequences $\mathfrak{S}$ itself (see §12 below) may be called the *obstruction class* of the sequence.

We refer to the sequence $\mathfrak{S}$ and the homomorphism η subject to the hypotheses (i) and (ii) as an exact *sequence with operators* F or as a sequence $(\mathfrak{S}, \eta)$. The abelian group H will be written additively, the other groups of the sequence multiplicatively. R denotes the image $\mu K \subset F$, or equivalently, the kernel of $\nu : F \to Q$.

Throughout we consider also an (additive) abelian group G with operators $g \leftrightarrow x\cdot g$ from Q. The group F then also operates on G, by the convention that $a\cdot g = (\nu a)\cdot g$ for $a \epsilon F$, $g \epsilon G$.

The *operator homomorphisms* of K into G are the functions $\sigma(k) \epsilon G$ defined for $k \epsilon K$ with

$$(2.4) \qquad \sigma(k_1k_2) = \sigma(k_1) + \sigma(k_2) \qquad \sigma(\eta(a)\cdot k) = a\cdot\sigma(k), \qquad a \epsilon F.$$

The sum of two operator homomorphisms σ_1 and σ_2 is an operator homomorphism $(\sigma_1 + \sigma_2)(k) = \sigma_1(k) + \sigma_2(k)$, and under this addition the operator homomorphisms form a group.

The *crossed homomorphisms* of F into G are the functions $\tau(a) \epsilon G$ defined for $a \epsilon F$ with

$$(2.5) \qquad \tau(ab) = \tau(a) + a\cdot\tau(b), \qquad a, b \epsilon F.$$

Each such τ induces a homorphism $\tau\mu : K \to G$; it is an operator homomorphism, for by (v) above and (2.5)

$$\begin{aligned}\tau\mu(\eta(a)\cdot k) &= \tau(a(\mu k)a^{-1}) = a\cdot\tau((\mu k)a^{-1}) + \tau(a)\\ &= a\cdot\tau(\mu k) + a\cdot\tau(a^{-1}) + \tau(a) = a\cdot\tau(\mu k) + \tau(aa^{-1})\\ &= a\cdot\tau\mu(k).\end{aligned}$$

The group of all operator homomorphisms σ of K into G, modulo the subgroup of the operator homomorphisms $\tau\mu$ of K into G which are induced by crossed homomorphisms τ of F, will be called the *group of operator homomorphism classes* of $\mathfrak{S}$ (or of K) into G, and will be designated as

$$\text{Map}\ (\mathfrak{S}, \eta, G). \tag{2.6}$$

Our principal objective is the

THEOREM B. *For an exact sequence* $(\mathfrak{S}, \eta)$ *with operators from a free group* F *the group* Map $(\mathfrak{S}, \eta, G)$ *of operator homomorphism classes of the sequence into a group* G *with operators* Q *depends only on the group* G *(with its operators), the first and last terms* H *and* Q *of the sequence, the fashion in which* Q *operates on* H, *and on the obstruction class* $\{l_3\}$ *of the sequence.*

For given abelian groups G and H with operators Q and for any cocycle l_3 in a given cohomology class $\{l_3\} \in H^3(Q, H)$, Eilenberg and the author ([5], cf. also §8 below) have constructed a certain generalized cohomology group $E^2(Q, H, l_3, G)$ depending on these data. With this construction we have a more explicit form of the theorem:

THEOREM B′. Map $(\mathfrak{S}, \eta, G)$ *is isomorphic to the group* $E^2(Q, H, l_3, G)$, *where* l_3 *is any cocycle in the obstruction class of the sequence* $(\mathfrak{S}, \eta)$.

In the particular case of an exact sequence $\mathfrak{S}$ in which the first group $H = 0$, the obstruction class $\{l_3\}$ is necessarily 1, and the theorem asserts that Map $(\mathfrak{S}, \eta, G)$ depends only on Q and G. In fact the theorem is then identical with Theorem A, provided that the subgroup R of the free group F of Theorem A is not an infinite cyclic group (the proviso is necessary to obtain the hypothesis (ii)).

The Theorem B shows that a certain group depends only upon the constituents H, Q, and $\{l_3\}$ of the given exact sequence $\mathfrak{S}$; it has real content only if there exist many sequences with the same three constituents. This is assured by the following theorem, proved in §12 below.

THEOREM C. *If* H *is an abelian group with operators* Q, $\nu : F \to Q$ *a homomorphism of a free group* F *onto* Q, *such that the kernel* R *of* ν *is not cyclic, and* $\{l_3\}$ *a cohomology class of* $H^3(Q, H)$, *there exists one and (up to equivalence) only one exact sequence* $(\mathfrak{S}, \eta)$ *with operators* F *which has* H *(with the given operators* Q*) as its initial term, has* $\nu : F \to Q$ *as its final homomorphism, and has the given class* $\{l_3\}$ *as its obstruction class.*

This theorem is essentially an interpretation of the one-dimensional cup product reduction theorem of the first paper [2] of this series. This cup product reduction theorem starts from the representation $Q = F/R$ and asserts that every three dimensional cohomology class of $H^3(Q, H)$ can be obtained from (reduced to) a one-dimensional cohomology class of $H^1(Q, \text{Hom}\ (R, G))$, for suitably defined operators of Q on the group Hom (R, G) of all homomorphisms of R into G. It turns out that the cohomology class of $H^1(Q, \text{Hom}\ (R, G))$ suffices to define the sequence $\mathfrak{S}$ and its operators η, and conversely.

3. Kernels and exact sequences

A homomorphism ϕ of a group E onto Q with kernel K induces a homomorphism θ of Q into the group of automorphism classes of K, where $\theta(x)$ for any $x = \phi e$ in Q is determined as the automorphism class of $k \to eke^{-1}$. Extending this notion, a pair (K, θ) may be called (cf. [3]) a *Q-kernel* if K is a group and θ a homomorphism of Q into the group $A(K)/I(K)$ of automorphism classes of K; here $A(K)$ denote the group of automorphisms of K and $I(K)$ the normal subgroup of inner automorphisms. It was observed by Baer [1] that an abstract Q-kernel in this sense need not always be realizable as the kernel of a homomorphism ϕ. An *operator homomorphism* γ of a Q-kernel (K_1, θ_1) into a second Q-kernel (K_2, θ_2) is a homomorphism $\gamma: K_1 \to K_2$ such that for all $x \in Q$ and all automorphisms α_1 in the automorphism class $\theta_1(x)$ there is an automorphism $\alpha_2 \in \theta_2(x)$ such that $\gamma\alpha_1 = \alpha_2\gamma$ (as mappings of K_1 into K_2). An abelian group G with operators Q may be regarded as a Q-kernel, with $\theta(x)\cdot g = x\cdot g$.

With every Q-kernel (K, θ) we can associate its *center* H' and its *graph* Γ. The center is the center of the group K; it is an abelian group with operators Q induced by θ. If we replace H' by an isomorphic replica H, we can say that H has an operator isomorphism $\lambda : H \to K$. The graph Γ is defined (following Baer [1]; cf. also [3, §12]) as the graph of the "many valued" function θ. Thus Γ consists of all those pairs (x, α) of elements $x \in Q$ and automorphisms $\alpha \in A(K)$ for which $\alpha \in \theta(x)$, the multiplication of pairs being that in the direct product $Q \times A(K)$. If $C[k]$ for $k \in K$ denotes the inner automorphism induced by k, then $C[k] \in \theta(1)$ for all k. Hence the correspondences

$$\mu : k \to (1, C[k]), \qquad \nu : (x, \alpha) \to x \tag{3.1}$$

define homomorphisms of the kernel K into its graph Γ and of the graph Γ onto Q, such that the image of K under μ is exactly the kernel of ν. Furthermore, the kernel of μ is the center of K, hence is the image of H under λ. Thus, all told, the kernel leads to the exact sequence

$$H \xrightarrow{\lambda} K \xrightarrow{\mu} \Gamma \xrightarrow{\nu} Q. \tag{3.2}$$

Furthermore, the graph Γ serves as a group of operators on K; indeed, we may define a homomorphism $\eta : \Gamma \to A(K)$ by setting

$$\eta(x, \alpha) = \alpha \qquad x \in Q, \quad \alpha \in \theta(x). \tag{3.3}$$

Since $\eta\mu k = C[k]$, we thus obtain an exact sequence with the properties (i) and (ii) of the previous section.

Now suppose that the abelian group H with its operators Q is given in advance. Let (K, θ) and (K', θ') be two Q-kernels with the same center H, so that operator isomorphisms $\lambda : H \to K$, $\lambda' : H \to K'$ of H into the kernels are given. The kernels are *H-equivalent* if there is an operator isomorphism γ of (K, θ) to (K', θ')

such that $\gamma\lambda = \lambda'$. Similarly, given an exact sequence (3.2) and another exact sequence

$$H \xrightarrow{\lambda'} K' \xrightarrow{\mu'} \Gamma' \xrightarrow{\nu'} Q \tag{3.4}$$

with corresponding associated homomorphisms

$$\eta : \Gamma \to A(K) \qquad \eta' : \Gamma' \to A(K')$$

we say that the two sequences are *H-equivalent* if there are isomorphisms γ of K to K' and ζ of Γ to Γ' such that

$$\gamma\lambda = \lambda', \qquad \zeta\mu = \mu'\gamma, \qquad \nu = \nu'\zeta, \qquad \eta'\zeta p = \gamma\eta(p)\gamma^{-1},$$

for all $p \in \Gamma$.

THEOREM 3.1. *Every Q-kernel* (K, θ) *determines an exact sequence, consisting of the center of* K, K *itself, its graph, and* Q, *which satisfies the hypotheses* (i) *and* (ii) *of* §2, *and this determination provides a one-to-one correspondence between the equivalence classes of Q-kernels with given center H and the equivalence classes of such exact sequences* $\mathfrak{S}$ *with H as initial and Q as final term.*

The first assertion has already been established. Conversely, given an exact sequence

$$H \xrightarrow{\lambda'} K' \xrightarrow{\mu'} F' \xrightarrow{\nu'} Q, \qquad \eta' : F' \to A(K')$$

with properties (i) and (ii), observe that η' must by (i) carry the image $\mu(K')$ into $I(K')$, hence that η' induces a homomorphism θ of Q into $A(K')/I(K')$. By hypothesis (ii), H (with the injection λ') is indeed the center of this kernel (K', θ). Let Γ be the graph of the kernel and define ζ: $F' \to \Gamma$ by setting $\zeta a = (\nu' a, \eta' a)$ for $a \in F'$. It then follows readily that ζ gives an equivalence of the original exact sequence to the sequence derived from the kernel (K', θ). For example, to prove that ζ maps F' onto Γ, take any (x, α) in the graph Γ and select any $a \in F'$ with $\nu' a = x$. Then α and $\eta' a$ both lie in the automorphism class $\theta(x)$, so that there is a $k \in K'$ with $\alpha = C[k]\eta' a$. Then

$$\zeta[(\mu' k)a] = ((\nu'\mu' k)\nu' a, (\eta'\mu' k)\eta' a) = (x, C[k]\eta' a) = (x, \alpha),$$

as desired. The remaining assertions of the theorem are formalities.

It was proved in detail in [3] that every Q-kernel (K, θ) with center H determines a unique 3-dimensional cohomology class $F_3(K, \theta) \in H^3(Q, H)$. In virtue of the above correspondence, we can now assert that each exact sequence $\mathfrak{S}$ with the properties (i), (ii) determines such a class, which we define to be the obstruction class of the sequence.

Our main Theorem B for a sequence $(\mathfrak{S}, \eta)$ would remain valid if the hypothesis (ii) of §2 were replaced by the weaker property (v) of §2. In particular, it has been shown[2] that the obstruction class can be defined for exact sequences with

[2] In unpublished work by Eilenberg and the author.

operators with the properties (i) and (v). These sequences need no longer correspond to kernels with center H. However, if the group F is free, the generalization is slight, for R, as a subgroup of the free group F, is also free. Hence R is either an infinite cyclic group, or a free group with more than one generator. In the latter case R has center 1, K is isomorphic to $(\lambda H) \times R$, and K has λH as center. Thus we obtain the previous hypothesis (ii), except when R is infinite cyclic.

Because of the equivalence asserted in Theorem 3.1, the Theorem B may now be reformulated as a theorem about Q-kernels. It asserts that the group Map $((K, \theta), G)$ of operator homomorphisms σ of a kernel (K, θ) into G, modulo those σ extendable as crossed homomorphisms to the graph of the kernel, depends only on Q, the center, the obstruction, and G—provided the kernel has a free graph.

4. A topological application

Let P be a connected polyhedron, which is not one-dimensional. Its homotopy invariants include its fundamental group $\pi_1 = \pi_1(P)$, its second homotopy group $\pi_2 = \pi_2(P)$, which is an abelian group with operators from π_1, and a certain uniquely defined three dimensional cohomology class $\{k^3\} \in H^3(\pi_1, \pi_2)$. The second homology and cohomology groups of P are determined by these homotopy invariants; specifically [4, 5] if k^3 is any cocycle in the cited class and G any abelian group of coefficients (on which all elements of π_1 act as the identity automorphism), the second cohomology group of P with coefficients in G is

$$H^2(P, G) = E^2(\pi_1, \pi_2, k^3, G). \tag{4.1}$$

Here E^2 denotes the group to be constructed in the proof of Theorem B′.

Now let P^1 be the one-dimensional skeleton of any sufficiently fine triangulation of P; one can then define (cf. [9]) the second relative homotopy group $\pi_2(P, P^1)$, and an exact homotopy sequence

$$\pi_2(P^1) \xrightarrow{i} \pi_2(P) \xrightarrow{j} \pi_2(P, P^1) \xrightarrow{\partial} \pi_1(P^1) \xrightarrow{i'} \pi_1(P). \tag{4.2}$$

Here i and i' are the homomorphisms induced by the identity map of P^1 into P, while j is the homomorphism induced by the identity map of the pair (P, base point) into the pair (P, P^1), and ∂ is the homotopy boundary operator. Since P^1 is one-dimensional, the group $\pi_2(P^1)$ is zero; therefore j is an isomorphism into the relative homotopy group. The group $\pi_1(P^1)$ is known to serve as a group of operators on all the higher groups of the sequence, and in particular on $\pi_2(P, P^1)$; this means that there is a homomorphism

$$\eta : \pi_1(P^1) \to A[\pi_2(P, P^1)].$$

This homomorphism η has the property (i) of §2. To prove the hypothesis (ii) observe that $\pi_1(P^1)$, as the fundamental group of a graph, is a free group.

Since the triangulation of P is sufficiently fine, we can find two triangles with disjoint boundaries. From each triangle we can construct a loop, as an edge-path running (without double points) from the base point to a point on the triangle, thence around the triangle, and thence back to the base point along the original path reversed. These two loops determine elements in the fundamental group $\pi_1(P^1)$, and indeed elements in the subgroup $\partial\pi_2(P, P^1)$. Since the boundaries of the triangles are disjoint, no power of either element can equal a power of the other. The free group $\partial\pi_2(P, P^1)$ therefore is not abelian, hence has center 1. It follows that the image of $\pi_2(P)$ contains the center of $\pi_2(P, P^1)$, as required by our hypothesis (ii).

Theorem 3.1 now implies that $\pi_2(P, P^1)$ is a $\pi_1(P)$-kernel (κ_2, θ_2) with the second homotopy group $\pi_2(P)$ as center and the first homotopy group $\pi_1(P^1)$ of the skeleton as graph. We call (κ_2, θ_2) a *second homotopy kernel* of the polyhedron P.

This second homotopy kernel is clearly not an invariant of P, since it depends strongly on the choice of the skeleton P^1. However, if one computes the obstruction of this kernel, according to the geometric definition of the operators η, and compares the result with the geometric definition of the invariant $\{k^3\}$ of [4], the two prove to be identical. Hence the invariant $\{k^3\}$ of the space is the obstruction of any second homotopy kernel.[3]

The graph $\pi_1(P^1)$ of any second homotopy kernel is free; hence the topological theorem (4.1) may be combined with the fundamental algebraic theorem B′ of this paper to give

THEOREM D. *If* (κ_2, θ_2) *is any second homotopy* π_1*-kernel of the polyhdcron* P, *and* G *any abelian coefficient group, the second cohomology group of* P *is determined as the group of operator homomorphism classes of* (κ_2, θ_2):

$$H^2(P, G) \cong \text{Map}\,((\kappa_2, \theta_2), G).$$

5. Alternative forms of the theorem

In the given sequence $(\mathfrak{S}, \eta)$ we introduce the commutator subgroups $[R, R]$ and $[K, K]$ of R and K, respectively, and the corresponding factor groups

$$K_0 = K/[K, K], \qquad R_0 = R/[R, R], \qquad F_0 = F/[R, R]$$

with $F_0 \supset R_0$. The given homomorphisms $\mu : K \to F$, $\nu : F \to Q$, and $\eta : F \to A(K)$ then induce homomorphisms

$$K_0 \overset{\mu_0}{\to} F_0 \overset{\nu_0}{\to} Q, \qquad \eta_0 : Q \to A(K_0),$$

with $\mu_0(K_0) = R_0 =$ the kernel of ν_0. Hence Q serves as a group of operators for the abelian group K_0.

---

[3] The idea of finding some such geometrico-algebraic interpretation of k^3 is due to J. A. Zilber and D. Gorenstein, who communicated to me such a construction in terms of loops.

Any operator homomorphism σ of K into the abelian group G necessarily carries the commutator subgroup $[K, K]$ into 0, hence induces an operator homomorphism σ_0 of K_0 into G, with

$$\sigma_0(k_0k_0') = \sigma_0(k_0) + \sigma_0(k_0'), \qquad \sigma_0(\eta_0(x)\cdot k_0) = x\cdot\sigma_0(k_0) \tag{5.1}$$

for k_0, k_0' in K_0. Conversely, the induced σ_0 determines σ. Similarly, since F operates on G through $\nu : F \to Q$, the kernel R of ν operates trivially on G, and any crossed homomorphism τ on F acts as an ordinary homomorphism on R, hence carries $[R, R]$ into 0. Thus τ induces (and is determined by) a crossed homomorphism τ_0 of F_0 into G, with

$$\tau_0(a_0a_0') = \tau_0(a_0) + a_0\cdot\tau_0(a_0'), \qquad a_0, a_0' \in F_0. \tag{5.2}$$

Here F_0 operates on the coefficient group G through $\nu_0 : F_0 \to Q$. Finally, each τ_0 induces an operator homomorphism $\tau_0\mu_0 : K_0 \to G$.

The group Map $((\mathfrak{S}, \eta), G)$ of operator homomorphism classes treated in our main theorem can thus alternatively be described as the group of all the operator homomorphisms $\sigma_0 : K_0 \to G$ of (5.1), modulo the subgroup of those induced as $\sigma_0 = \tau_0\mu_0$, for τ_0 as in (5.2). This gives a reformulation of Theorem B analogous to that given for Theorem A in [2, Theorem 13.3].

Hopf has considered, in [7], the representation $Q = F/R$ with a free group F; using the group $[F, R]$ generated by all commutators of the form $ara^{-1}r^{-1}$ for $a \in F$, $r \in R$, and the commutator subgroup $[F, F]$ of F, he proved there that the (abelian) factor group $R \cap [F, F]/[F, R]$ depends only on Q, and is independent of the choice of the representation F/R. This statement, which is in "homology" form, may be deduced from the "cohomology" form of Theorem A by standard arguments on character groups [2, §14]. By the same methods we may deduce the

THEOREM 5.1. *In the exact sequence* $(\mathfrak{S}, \eta)$ *with operators in the free group* F *let* M *denote the subgroup of all elements* $m \in K$ *with* μm *in the commutator group* $[F, F]$, *and* L *the subgroup generated by all elements in* K *of the form* $[\eta(a)\cdot k]k^{-1}$ *for some* $a \in F$, $k \in K$. *Then* $M \supset L$, *and the factor group* M/L *depends only on the first and last terms* H *and* Q *of the sequence, the operators of* Q *on* H, *and the obstruction class* $\{l_3\}$ *of the sequence. If* P *is the additive group of real numbers modulo* 1 (*with simple operators from* Q), *then* M/L *is a character group,*

$$M/L \cong \text{Char } E^2(Q, H, l_3, P).$$

The case $H = 1$, $\{l_3\} = 1$ gives Hopf's theorem just cited, provided the subgroup R in Hopf's theorem is not infinite cyclic.

6. Comparisons of the operator homomorphism groups

In the given sequence $\mathfrak{S}$ we may explicitly display $R = \mu K$, as in

$$H \xrightarrow{\lambda} K \xrightarrow{\mu'} R \xrightarrow{\mu''} F \xrightarrow{\nu} Q,$$

where $\mu'\mu'' = \mu$, μ' is a homomorphism of K onto R and μ'' is the identity injection of R into F. For a given coefficient group G we have to consider the following corresponding types of operator homomorphisms (and crossed homomorphisms) with values in G:

$$\begin{aligned}
\rho \in \text{Ophom}\,(H, G)&: \rho(hh') = \rho(h) + \rho(h'), \qquad \rho(x\cdot h) = x\cdot\rho(h),\\
\sigma \in \text{Ophom}\,(K, G)&: \sigma(kk') = \sigma(k) + \sigma(k'), \qquad \sigma(\eta(a)\cdot k) = a\cdot\sigma(k),\\
\omega \in \text{Ophom}\,(R, G)&: \omega(rr') = \omega(r) + \omega(r'), \qquad \omega(ara^{-1}) = a\cdot\omega(r),\\
\tau \in Z^1\,(F, G)&: \tau(aa') = \tau(a) + a\cdot\tau(a').
\end{aligned}$$

The homomorphisms in the given sequence induce inverse mappings on these groups of operator homomorphisms, as

$$\sigma\lambda \overset{\lambda^*}{\longleftarrow} \sigma, \qquad \omega\mu' \overset{\mu'^*}{\longleftarrow} \omega, \qquad \tau\mu'' \overset{\mu''^*}{\longleftarrow} \tau;$$

thence the sequence

$$\text{Ophom}\,(H, G) \overset{\lambda^*}{\longleftarrow} \text{Ophom}\,(K, G) \overset{\mu'^*}{\longleftarrow} \text{Ophom}\,(R, G) \overset{\mu''^*}{\longleftarrow} Z^1(F, G). \tag{6.1}$$

Here the kernel of λ^* is the image of μ'^*, for if σ restricted to λH is zero, then σ induces and arises from a $\omega : K/\lambda H \cong R \rightarrow G$.

In this notation, the group of operator homomorphism classes of R, as considered in Theorem A, is

$$\text{Map}\,(R, F; G) \cong \text{Ophom}\,(R, G)/\mu'^*Z^1(F, G),$$

while the group of operator homomorphism classes of K, as considered in the main Theorem B, is

$$\text{Map}\,(\mathfrak{S}, \eta, G) \cong \text{Ophom}\,(K, G)/\mu'^*\mu''^*Z^1(F, G).$$

The given mappings λ^* and μ'^* thus induce homomorphisms

$$\text{Ophom}\,(H, G) \overset{\lambda^*}{\longleftarrow} \text{Map}\,(\mathfrak{S}, \eta, G) \overset{\mu'^*}{\longleftarrow} \text{Map}\,(R, F; G). \tag{6.2}$$

The above displays indicate that μ'^* is here an isomorphism into; by the properties of the sequence (6.1), the kernel of λ^* is the image of μ'^*. In view of this formula, our basic group Map $(\mathfrak{S}, \eta, G)$ of operator homomorphism classes may be regarded as a group extension of Map $(R, F; G) \cong H^2(Q, G)$ by a subgroup of Ophom (H, G).

7. The standard sequence

Assuming the truth of Theorem B′, one may derive an explicit algebraic formula for the group E^2 by constructing any convenient exact sequence from the given constituents Q, H, and $l_3 \in Z^3(Q, H)$, and calculating the operator homomorphism groups for the sequence. To this end, we follow the method used in [3, §9].

Let F_1 be the free group generated by symbols $\{x\}$, one for each $x \neq 1$ in Q, $\nu_1 : F_1 \to Q$ the homomorphism determined by the specification that $\nu_1\{x\} = x$. The kernel R_1 of ν_1 is a free group; a set of free generators for R_1 may be obtained as follows from a given set of free generators e of F_1(Schreier [10]; Hurewicz [9]): In each coset x of F_1/R_1 choose a representative element $w(x) \in F_1$ in such fashion that, when $w(x)$ is written as a reduced word in the generators e, each initial section of the word $w(x)$ is also the representative of its coset; for each x and each e the element $w(x)e[w(x\nu_1 e)]^{-1}$ lies in R_1, and the set of all such elements, after deletion of the elements equal to 1, is a set of free generators of R_1. In the present case the representatives $w(x) = \{x\}$ for $x \neq 1$ and $w(1) = 1$ satisfy (trivially) the cited condition; hence R_1 has as free generators the elements

$$\{x, y\} = \{x\}\{y\}\{xy\}^{-1}, \qquad x \neq 1, y \neq 1 \text{ in } Q. \tag{7.1}$$

Let K_1 be the direct product $H \times R_1$, and set $h(x, y) = \{x, y\}$ if $x \neq 1, y \neq 1$, $h(x, y) = 1$ otherwise. For each $x \in Q$ we may define an endomorphism $\eta_1(x)$ of K_1 by the specifications $\eta_1(x)\cdot h = x\cdot h$, for h in H, and

$$\eta_1(x)\cdot\{y, z\} = l_3(x, y, z)h(x, y)h(xy, z)h(x, yz)^{-1} \tag{7.2}$$

for the generators $\{y, z\}$ of R_1. One may then show ([3, (9.4)]) that

$$\eta_1(x)\eta_1(y) = C_1[h(x, y)]\eta_1(xy) \tag{7.3}$$

where $C_1[h]$ denotes conjugation by h in K_1. It follows that $\eta_1(x)$ is an automorphism of K_1, and thus that η_1 induces a homomorphism $\eta_1 : F_1 \to A(K_1)$. By (7.1) and (7.3), $\eta_1\{x, y\} = C_1[h(x, y)] = C_1[\{x, y\}]$, hence hypothesis (i) of §2 holds. If Q has more than two elements, R_1 is a non-abelian free group with center 1, and hypothesis (ii) of §2 follows. By (7.2) and the definition of the obstruction class, it results that the cohomology class of the given cocycle l_3 is the obstruction class of the exact sequence $H \to K_1 \to F_1 \to Q$, with the obvious homomorphisms λ_1, μ_1, and ν_1.

An operator homomorphism $\sigma_1 : K_1 \to G$ is completely determined by its projection $\rho_1 = \lambda_1{}^*\sigma_1 = \sigma_1\lambda_1$, and by the function $p(x, y) = \sigma_1(h(x, y))$. Applying σ_1 to (7.2), and using the fact that σ_1 is an operator homomorphism, we derive for the function p the condition

$$x\cdot p(y, z) = \rho_1 l_3(x, y, z) + p(x, y) + p(xy, z) - p(x, yz);$$

in terms of the coboundary operator δ of [2], this condition may be written as $\rho_1 l_3 = \delta p$.

A crossed character τ_1 of F_1 is completely determined by its values $d(x) = \tau_1\{x\}$ on the generators $\{x\}$ of F_1; these values may be chosen arbitrarily in G. The corresponding induced operator homomorphism $\tau_1\mu_1$ on K_1 may be calculated, for the generators $h(x, y) = \{x\}\{y\}\{xy\}^{-1}$, by applying the defining property of the crossed homomorphism τ_1 to both sides of the equation $h(x, y)\{xy\} = \{x\}\{y\}$; this gives

$$\tau_1\mu_1 h(x, y) + d(xy) = d(x) + x\cdot d(y).$$

Therefore the special operator homomorphism $\tau_1\mu_1$ yields the functions $\rho_1 = 0$ and $p = \delta d$.

Combining these results, we conclude that the group Map $(\mathfrak{S}_1, \eta_1, G)$ for the sequence $\mathfrak{S}_1$ which has been constructed consists of the group of pairs (ρ_1, p) subject to the condition $\rho_1 l_3 = \delta p$, modulo the special pairs $(0, \delta d)$.

It should be noted that this construction requires that the given cocycle l_3 be "normalized," in the sense that

$$l_3(1, y, z) = l_3(x, 1, z) = l_3(x, y, 1) = 1.$$

The function $p(x, y)$ derived above is similarly normalized, in that

$$p(1, y) = p(x, 1) = 0.$$

We can and will assume throughout that all functions defined on groups to groups are normalized in this sense [2, §6].

8. The group construction

Following the indications of §7, we construct a group depending on the constituents Q, H, l_3, and G, where G and H are abelian groups with operators Q and $l_3 \in Z^3(Q, H)$. We consider all pairs (ρ, p) where ρ is an operator homomorphism of H into G, p is a two dimensional normalized cochain of Q over G (i.e., $p(x, y) \in G$) and where $\rho l_3 = \delta p$. These pairs form a group under the addition $(\rho_1, p_1) + (\rho_2, p_2) = (\rho_1 + \rho_2, p_1 + p_2)$. The pairs $(0, \delta d)$, where d is a one-dimensional normalized cochain of Q over G, form a subgroup. The factor group of pairs (ρ, p) modulo pairs $(0, \delta d)$ is the desired group; we denote it as a function of its constituents as

$$E^2(Q, H, l_3, G), \qquad l_3 \in Z^3(Q, H).$$

The correspondence $\chi(\rho, p) = \rho$ defines a homomorphism of E^2 into a subgroup of Ophom (H, G). The kernel of this homomorphism is the group of pairs $(0, p)$ with $\delta p = 0$, modulo the pairs $(0, \delta d)$—in other words, is the group of 2-cocycles of Q modulo the 2-coboundaries, or the group $H^2(Q, G)$. Thus we have, with an identity injection i,

$$\text{(8.1)} \qquad \text{Ophom}\,(H, G) \xleftarrow{\chi} E^2(Q, H, l_3, G) \xleftarrow{i} H^2(Q, G),$$

with kernel χ = image i. In other words, E^2 is a group extension of $H^2(Q, G)$ by a subgroup of Ophom (H, G).

The main theorem can now be refined by the assertion that the group extension E^2 of (8.1) is equivalent to the group Map $(\mathfrak{S}, \eta, G)$ of operator homomorphism classes, also regarded as in (6.2) as a group extension:

THEOREM B″. *For an exact sequence* $(\mathfrak{S}, \eta)$ *with free operators* F *there exist isomorphisms*

$$\text{(8.2)} \qquad \Lambda: \text{Map}\,(\mathfrak{S}, \eta, G) \to E^2(Q, H, l_3, G),$$

$$\text{(8.3)} \qquad \Lambda_0 : \text{Map}\,(R, F; G) \to H^2(Q, G),$$

such that the homomorphisms indicated in (8.1) *and* (6.2) *satisfy*

$$\chi\Lambda = \lambda^*, \qquad \Lambda\mu'^* = i\Lambda_0 . \tag{8.4}$$

The isomorphism Λ_0 here is that used to prove Theorem A′. The proof of the main theorem, in this explicit form, will involve also an isomorphism Λ_1, obtained from the one-dimensional cup product reduction theorem [3]

$$\Lambda_1 : H^1(Q, \text{Hom}(R, H)) \to H^3(Q, H), \tag{8.5}$$

for suitably defined operators of Q on Hom (R, H). Following a suggestion of R. G. Lyndon, we modify the construction of Λ_1 given in [3]. It then appears (§10 below) that each sequence $\mathfrak{S}$ with F free determines uniquely as "deviation class" a corresponding element of the group $H^1(Q, \text{Hom}(R, H))$, and the correspondence Λ_1 of (8.5) is the correspondence of the deviation class of a sequence to its obstruction class. The desired isomorphism Λ is then constructed in §13 by simultaneous treatment of Λ_0 and Λ_1.

Chapter II. Proof of the Main Theorem

9. Proof of Theorem A′

Given $\nu : F \to Q$, onto, with kernel R, choose for each $x \in Q$ a representative $u(x)$ in F with $\nu u(x) = x$, $u(1) = 1$. Then there exists a factor set $f(x, y) \in R$ such that

$$u(x)u(y) = f(x, y)u(xy); \qquad x, y \in Q. \tag{9.1}$$

The associativity condition is

$$[u(x)f(y, z)u(x)^{-1}]f(x, yz) = f(x, y)f(xy, z). \tag{9.2}$$

If $u'(x)$ is a second set of representatives of x in F, then

$$u'(x) = s(x)u(x), \qquad \text{for } s(x) \in R, \tag{9.3}$$

and the corresponding new factor set f' is

$$f'(x, y) = s(x)[u(x)s(y)u(x)^{-1}]f(x, y)s(xy)^{-1}. \tag{9.4}$$

To each operator homomorphism $\omega : R \to G$, we let correspond the 2-cochain $\Lambda_0\omega$ of Q over G defined by

$$(\Lambda_0\omega)(x, y) = \omega(f(x, y)), \qquad x, y \in Q. \tag{9.5}$$

By applying the operator homomorphism ω to the associativity condition (9.2) we conclude that $\Lambda_0\omega$ is a 2-cocycle. If ω is the operator homomorphism induced by a crossed homomorphism τ of F, equation (9.1) shows that $\Lambda_0\omega$ is a coboundary. Hence Λ_0 induces a homomorphism

$$\Lambda_0 : \text{Map}(R, F; G) \to H^2(Q, G).$$

As in [2 §§10, 13], one proves that this correspondence Λ_0 (denoted op. cit. as λ_0, and applied there to $R_0 = R/[R, R]$ rather than to R) is independent of the choice of the factor set f used in its definition, and that it is an isomorphism onto $H^2(Q, G)$. This is the assertion of Theorems A and A'.

For later purposes we reformulate here the proof that Λ_0 is a mapping onto $H^2(Q, G)$. To this end, take any cohomology class in the latter group, and choose a cocycle $g(x, y)$ in this class. For arguments $a, b \in F$ the "lifted" function $g(\nu a, \nu b)$ is then a cocycle of F in G, hence may be used in conjunction with the given operators of F on G to define an extension E of G by F. Indeed, E is the group of elements $gw(a)$, for $g \in G$, $a \in F$, with the multiplication table

$$w(a)w(b) = g(\nu a, \nu b)w(ab), \qquad w(a)g = [(\nu a)\cdot g]w(a),$$
$$[g_1w(a)][g_2w(b)] = [g_1 + \nu(a)\cdot g_2 + g(a, b)]w(ab)$$

and with the homomorphism $gw(a) \to a$ onto F. Since F is a free group, this extension E is trivial [2, Lemma 3.1], so that its cocycle must be the coboundary of some cochain g^* of F in G. To obtain an explicit formula for g^*, we choose a set of free generators $\{e\}$ of the free group F, and define a new set w' of representatives of F in E by the equations

$$(9.6) \qquad w'(e) = w(e), \qquad w'(ab) = w'(a)w'(b).$$

Since $w(a)$ and $w'(a)$ lie in the same coset of E modulo G there is a cochain $g^*(a)$ with values in G such that

$$(9.7) \qquad w'(a) = g^*(a)w(a), \qquad a \in F.$$

One may compute directly that $\delta g^* = -g\nu$. To express g^* directly in terms of g, note that (9.6) and (9.7) give

$$\begin{aligned} g^*(be)w(be) &= w'(be) = w'(b)w'(e) \\ &= [g^*(b) + g(\nu b, \nu e)]w(be), \\ g^*(b)w(b) &= w'(b) = w'(be^{-1})w'(e) \\ &= g^*(be^{-1})w(be^{-1})w(e) = [g^*(be^{-1}) + g(\nu(be^{-1}), \nu e)]w(b). \end{aligned}$$

Comparison of coefficients gives

$$(9.8a) \qquad g^*(be) = g^*(b) + g(\nu b, \nu e),$$

$$(9.8b) \qquad g^*(be^{-1}) = g^*(b) - g(\nu(be^{-1}), \nu e).$$

We can now circumvent the use of the extension E and instead use these equations to define the cochain $g^* \in C^1(F, G)$. Indeed, every element of F can be written uniquely in terms of the generators e as a word which is reduced, in the sense that no formal cancellation of generators and their inverses is possible; any reduced word a of length n can be written as $a = be$ or as $a = be^{-1}$, where e is a generator and b a (reduced) word of length $n - 1$. The equations (9.8a) and (9.8b) with the initial condition $g^*(1) = 0$ then define $g^*(a)$ for all elements a, by induction on the length of a. It then follows that the defining equations (9.8a) and (9.8b) still hold if be or be^{-1}, respectively, is not a reduced word. Using $\delta g = 0$, one can then prove by induction on the length of b that

$$\delta g^*(a, b) = -g(\nu a, \nu b), \qquad a, b \in F.$$

In particular, for $a = r$ in R, $\nu a = 1$, and $g(1, \nu b) = 0$, since g is normalized: hence, since r operates trivially on the coefficient group,

$$g^*(r) + g^*(b) = g^*(rb), \qquad r \in R, b \in F.$$

Thus g^*, applied to elements of R alone, yields a homomorphism $\omega : R \to G$. One verifies that ω is an operator homomorphism, and that $\Lambda_0\omega$ is cohomologous to the original cocycle g, thus proving that Λ_0 is indeed a mapping onto.

10. The deviation of a sequence[4]

In the homomorphism $\mu : K \to F$ of the sequence $\mathfrak{S}$, the image $\mu K = R$, as a subgroup of the free group F, is free; hence K is the direct product of the kernel λH of μ by a group isomorphic to R. In particular, we may choose representatives $v(r) \in K$ with

$$(10.1) \qquad \mu v(r) = r, \qquad v(rr') = v(r)v(r'),$$

thus representing K as the direct product $(\lambda H) \times (vR)$.

Consider now the automorphism $\eta : F \to A(K)$. Since

$$\mu[\eta(u(x)) \cdot v(r)] = u(x)ru(x)^{-1}$$

by (v) of §2, there must exist elements $m(x, r)$ in H such that

$$(10.2) \qquad \eta(u(x)) \cdot v(r) = \lambda m(x, r)v(u(x)ru(x)^{-1}).$$

The function $m(x, r)$ thus obtained serves to determine η completely, for on the one hand $\eta(a) \cdot (\lambda h) = \lambda[(\nu a) \cdot h]$, by the very definition (2.3) of the operators on H, while on the other hand any a in F has the form $su(x)$ for $s \in R$, and thus $\eta(a) = \eta(s)\eta(u(x))$, where $\eta(s)$ is determined by hypothesis (i) of §2 to be the operation of conjugation with $v(s)$. Thus

$$\begin{aligned} \eta(a) \cdot v(r) &= \lambda m(x, r)v(s)v(u(x)ru(x)^{-1})v(s)^{-1} \\ &= \lambda m(x, r)v(su(x)ru(x)^{-1}s^{-1}). \end{aligned}$$

All told, we have the formula

$$(10.3) \qquad \eta(a) \cdot [(\lambda h)(vr)] = \lambda[(\nu a) \cdot h + m(\nu a, r)]v[ara^{-1}].$$

The function m, however, is subject to certain conditions. Since $\eta(u(x))$ is a homomorphism, and since $v(r)v(r') = v(rr')$, one deduces that (recall that the group composition in H is written additively!)

$$(10.4) \qquad m(x, r) + m(x, r') = m(x, rr'), \qquad x \in Q, r, r' \in R.$$

On the other hand $\eta(u(x))\eta(u(y)) = \eta(f(x, y))\eta(u(xy))$. This condition, expressed in terms of m, becomes

$$(10.5) \qquad [x \cdot m(y, r)] + m(x, u(y)ru(y)^{-1}) = m(xy, r), \qquad x, y \in Q, r \in R.$$

We call this function $m(x, r)$, subject to the two conditions (10.4) and (10.5), a *deviation* of the given exact sequence $\mathfrak{S}$.

---

[4] The theorems of this section are special cases of more general results found by Eilenberg and the author in an unpublished study of the relative cohomology groups of a group.

Any such function m with values in H may be interpreted as a cocycle. For fixed x, equation (10.4) asserts that $m(x, -)$ may be regarded as a homomorphism $\theta : R \rightarrow H$, or indeed as a homomorphism θ_0 of $R_0 = R/[R, R]$ into H. The group R_0 is an abelian group with operators from Q, induced by the automorphisms $r \rightarrow u(x)ru(x)^{-1}$ of R; as automorphisms of R_0, they are independent of the choice of the representatives $u(x)$. The group of the homomorphisms θ has both left and right operators from Q, according to the definitions

$$(10.6) \qquad (x\cdot\theta)(r) = x\cdot(\theta r), \qquad (\theta\cdot x)(r) = \theta(u(x)ru(x)^{-1}).$$

The equation (10.5) asserts that m is a one-dimensional cocycle of Q with values in Hom (R, H), provided that the coboundary $\delta' m = 0$ is defined, relative to the two sided operators, as indicated in [2, §5].

The corresponding one dimensional coboundaries are obtained from the homomorphisms $\theta : H \rightarrow R$ by the formula

$$(10.7) \qquad (\delta'\theta)(x, r) = x\cdot\theta(r) - \theta(u(x)ru(x)^{-1}),$$

which is again independent of the choice of the representatives $u(x)$. Indeed, one may check directly that any function $m(x, r) = \delta'\theta(x, r)$ does satisfy the conditions (10.4) and (10.5). The abelian group of all functions $m(x, r)$ subject to these two conditions, modulo the subgroup of the special functions of the form $\delta'\theta(x, r)$ may thus be interpreted as the cohomology group

$$H^1(Q, \text{Hom}\ (R, H))$$

with the two sided operators indicated in (10.6). The coset of a deviation m in this cohomology group is called a *deviation class* of the given exact sequence $\mathfrak{S}$.

THEOREM 10.1. *The deviation class of an exact sequence* $(\mathfrak{S}, \eta)$ *with free operators* F *is a uniquely determined element* $\{m\}$ *of* $H^1(Q, \text{Hom}\ (R, H))$. *By suitable choice of the representatives* v *used in the definition of a deviation (but with an arbitrary choice of the representatives* u*) any element* m *in this deviation class can arise as a deviation of the sequence.*

PROOF. We have to investigate the effect of a change in the choices of the representatives v and of the representatives u. Suppose first that u is unchanged, and that $v'(r)$ is another set of multiplicative representatives of r in K. Then $v'(r) = [\lambda n(r)]v(r)$, where $n(r) \in H$ for $r \in R$. Since v' is multiplicative, n is a homomorphism of R into H. By (10.3)

$$\eta(u(x))\cdot[v'(r)] = \lambda[(x\cdot n(r)) + m(x, r)]v[u(x)ru(x)^{-1}]$$
$$= \lambda[(x\cdot n(r)) + m(x, r)]\lambda[n(u(x)ru(x)^{-1})]^{-1}v'[u(x)ru(x)^{-1}];$$

hence, according to the definition (10.2), the new deviation m' is

$$m'(x, r) = m(x, r) + [x\cdot n(r)] - n(u(x)ru(x)^{-1}).$$

The old deviation has been altered exactly by the coboundary $\delta' n$, in the sense of (10.7). Hence the deviation class of m is unchanged. Furthermore, since an arbitrary homomorphism $n : R \rightarrow H$ may be used in this fashion to define new representatives $v'(r)$ subject to the conditions (10.1), we may alter a given deviation m of the sequence to any cohomologous deviation, as asserted in the theorem.

Suppose, secondly, that the system of representatives u of Q in F is changed to u', as in (9.3). Then, by (10.3)

$$\eta(u'(x))\cdot v(r) = [\lambda m(x, r)]v(u'(x)ru'(x)^{-1}).$$

Hence, if the representatives v are unchanged, the new representatives u' give the same function m as do the old representatives u'. This completes the proof of Theorem 10.1.

The deviation class can conversely be used to construct the exact sequence $\mathfrak{S}$:

THEOREM 10.2. *Given a group H with operators from Q, a homomorphism $\nu : F \rightarrow Q$ of the free group F onto Q, with non-abelian kernel R, and an element $\{m\}$ of $H^1(Q, \text{Hom }(R, H))$, there exists one, and up to equivalence only one, exact sequence $(\mathfrak{S}, \eta)$ with initial term H, final homomorphism ν, and deviation class $\{m\}$.*

PROOF. To establish the uniqueness, consider two such sequences $\mathfrak{S}$ and $\mathfrak{S}^*$; they can differ at most in the groups K and K^*, and in the respective mappings λ and λ^*, μ and μ^*, η and η^*. Choosing fixed representatives u of Q in F, we can by Theorem 10.1 choose representatives v and v^* in such fashion that the two sequences have the same deviation. The direct product $K = \lambda H \times v(R)$ is then mapped isomorphically on $K^* = \lambda^* H \times v^*(R)$ by the correspondence $(\lambda h, v(r)) \leftrightarrow (\lambda^* h, v^*(r))$. This correspondence yields the desired equivalence of the sequences, as one shows with particular reference to the definition of η and η^* by (10.3), with the *same* m in both cases.

To construct a sequence $\mathfrak{S}$ from the given data, choose a cocycle m in the given obstruction class, define K as the direct product $H \times R$, and define $\eta(a)$ as the automorphism

$$(h, r) \rightarrow ((\nu a)\cdot h + m(\nu a, r), ara^{-1})$$

of K, in accord with (10.3). One verifies readily that this does yield a sequence with the desired properties, and with m as a deviation.

11. The modified cup product reduction

Let $u(x)$ be a chosen system of representatives of Q in F, with factor set $f(x, y)$. Each function $m(x, r)$ subject to the conditions (10.4) and (10.5) then determines a function $(\Lambda_1 m)(x, y, z) \in H$ by the formula

(11.1) $$(\Lambda_1 m)(x, y, z) = m(x, f(y, z)) \qquad x, y, z \in Q.$$

This function $\Lambda_1 m$ is a three dimensional cocycle, as one computes by writing out $\delta(\Lambda_1 m)(x, y, z, t)$; by one application of the property (10.5), the result may be expressed as a sum of four terms of the form $m(x, -)$; these terms can then be combined by (10.4) to a single term $m(x, r)$, where r is expressed in terms of f

and is 1 by the associativity condition (9.2) on f. Similarly, one computes that if $m = \delta'\theta$ is a coboundary of the form (10.7), then $\Lambda_1 m$ is a three dimensional coboundary δc, where $c(x, y) = \theta(f(x, y))$. Thus the definition (11.1) induces a homomorphism

$$\Lambda_1 : H^1(\theta, \text{Hom}(R, H)) \rightarrow H^3(Q, H). \tag{11.2}$$

The cup product reduction theorem of [2] uses a similar homomorphism λ_1 between the same groups; the one sided operators used there are equivalent to the two sided operators of Q on Hom (R, H) used here. However, the homomorphism λ_1 there is not the same as the Λ_1 defined above—λ_1 reduces the *first* two arguments x, y to $f(x, y)$, while Λ_1 reduces the *last* two.[5]

One readily proves

LEMMA 11.1. *The homomorphism Λ_1 of (11.2) is independent of the choice of the (representatives u and the) factor set f used in its definition.*

THEOREM 11.2. *Λ_1 is an isomorphism onto H^3 (Q, H).*

We first show that Λ_1 is an isomorphism *into*. Suppose that $\Lambda_1 m_0$ is the coboundary δd of some two dimensional cochain $d \in C^2(Q, H)$. Since any two elements of F may be expressed in terms of the given representatives u as $ru(x)$ and $su(y)$, for r and s in R, x and y in Q, we can define as a 2-cochain q of F over H as

$$q(ru(x), su(y)) = m_0(x, s) - d(x, y). \tag{11.3}$$

Then $\delta q = 0$, for, by the properties (10.4) and (10.5) of m_0,

$$\begin{aligned}\delta q(ru(x), su(y), tu(z)) &= x\cdot m_0(y, t) - x\cdot d(y, z) - m_0(xy, t)\\ &\quad + d(xy, z) + m_0(x, su(y)tu(y)^{-1}f(y, z)) - d(x, yz) - m_0(x, s) + d(x, y)\\ &= m_0(x, f(y, z)) - \delta d(x, y, z) = 0.\end{aligned}$$

Using a set of free generators e of F, we then define a new function $q^* \in C^1(F, G)$ by induction on the length of a reduced word be or be^{-1} as

$$q^*(be) = q^*(b) + q(b, e) \tag{11.4a}$$

$$q^*(be^{-1}) = q^*(b) - q(be^{-1}, e). \tag{11.4b}$$

As in §9 we prove that these formulas still hold if be resp. be^{-1} is not a reduced word, and thence by induction on the length of the word b that

$$\delta q^*(a, b) = -q(a, b), \qquad a, b \in F. \tag{11.5}$$

On the other hand the original definition (11.3) of q shows that $q(r, b) = 0$ for $r \in R$, hence $\delta q^*(r, b) = 0$ and q^*, when restricted to R, yields a homomorphism $\theta : R \rightarrow H$.

---

[5] The modification here, involving the use of Λ_1 and of the two sided operators was suggested in correspondence by R. G. Lyndon.

But, since $\delta q^*(u(x)su(x)^{-1}, u(x)) = 0$,

$$\begin{aligned} -m_0(x, s) &= -q(u(x), s) = \delta q^*(u(x), s) \\ &= x\cdot\theta(s) - q^*(u(x)s) + q^*(u(x)) \\ &= x\cdot\theta(s) - q^*(u(x)su(x)^{-1}u(x)) + q^*(u(x)) \\ &= x\cdot\theta(s) - \theta(u(x)su(x)^{-1}) = \delta'\theta(x, s). \end{aligned}$$

Thus m_0 is a coboundary, and Λ_1 is an isomorphism into.

To show Λ_1 a homomorphism onto, take any 3-cocycle l of Q over H and define the function $l^*(x, a)$ for arguments $x \in Q$ and $a \in F$ by induction on the length of a word $a = be$ or $a = be^{-1}$ as

$$l^*(x, be) = l^*(x, b) + l(x, \nu b, \nu e), \tag{11.6a}$$

$$l^*(x, be^{-1}) = l^*(x, b) - l(x, \nu(be^{-1}), \nu e). \tag{11.6b}$$

We prove again that these equations hold even if the word be or be^{-1} is not reduced; using this and the fact that $\delta l = 0$ we prove by induction on the length of b that

$$l(x, a, b) = x\cdot l^*(\nu a, b) - l^*(x(\nu a), b) + l^*(x, ab) - l^*(x, a). \tag{11.7}$$

Note that the expression on the right would be δl^*, if the first argument of l^* were lifted from Q to the group F. In particular, if $a = r \in R$, then $\nu a = 1$, $l(x, \nu a, \nu b) = 0$, since l is normalized, and (11.7) gives

$$l^*(x, rb) = l^*(x, r) + l^*(x, b). \tag{11.8}$$

Similarly, with $b = r$ and $a = u(y)$ in (11.7) we obtain

$$x\cdot l^*(y, r) - l^*(xy, r) + l^*(x, u(y)r) - l^*(x, u(y)) = 0.$$

Applying (11.8) to the third term, with $u(y)r = [u(y)ru(y)^{-1}]u(y)$, we obtain thence

$$x\cdot l^*(y, r) - l^*(xy, r) + l^*(x, u(y)ru(y)^{-1}) = 0. \tag{11.9}$$

The two results (11.8) and (11.9) assert that $m^*(x, r) = l^*(x, r)$ is a cocycle from $H^1(Q, \mathrm{Hom}\,(R, G))$.

Finally, with $a = u(y)$ and $b = u(z)$ in (11.7) we have

$$\begin{aligned} l(x, y, z) &= x\cdot l^*(y, u(z)) - l^*(xy, u(z)) + l^*(x, f(y, z)u(yz)) \\ &\quad - l^*(x, u(y)) \\ &= x\cdot l^*(y, u(z)) - l^*(xy, u(z)) + l^*(x, f(y, z)) \\ &\quad + l^*(x, u(yz)) - l^*(x, u(y)), \end{aligned}$$

or,

$$l(x, y, z) = \delta d(x, y, z) + (\Lambda_1 m^*)(x, y, z), \tag{11.10}$$

where $d(x, y)$ is defined as $l^*(x, u(y))$. This equation shows that the cohomology class of l is the Λ_1-image of the cohomology class of m^*, and completes the proof of the theorem.

If we start with a function $m(x, r)$, derive the cocycle l as $l = \Lambda_1 m$, and then calculate $m^{\#}$ from l as above, the resulting $m^{\#}$ has $\Lambda_1 m^{\#} \backsim l$. Hence, by the first half of the theorem, m and $m^{\#}$ must be cohomologous in $H^1(Q, \text{Hom}(R, G))$. For later purposes we note a sharper form of this result:

LEMMA 11.3. *If $l(x, y, z) = m(x, f(y, z))$ for fixed representatives $u(x)$ and the corresponding factor set $f(y, z)$, and if $l^*(x, a)$ is defined by* (11.6), *there exists a homorphism θ of R into H and a one-cochain c^1 of Q in H such that*

$$m(x, r) = l^*(x, r) - \delta'\theta(x, r), \tag{11.11}$$

$$\theta(f(x, y)) = \delta c^1(x, y) - l^*(x, u(y)). \tag{11.12}$$

PROOF. Define m_0 and d by

$$m_0(x, r) = m(x, r) - l^*(x, r),$$

$$d(x, y) = l^*(x, u(y)).$$

Then by (11.10) $\delta d = l - \Lambda_1 m^{\#} = \Lambda_1 m - \Lambda_1 m^{\#} = \Lambda_1 m_0$, so that the proof of the first part of the theorem applies. The homomorphism θ there constructed has the property (11.11). Furthermore, setting $a = u(y)$ and $b = u(y)$ in (11.5), and using the definition (11.3) we have

$$x \cdot q^*(u(y)) - q^*(f(x, y)u(xy)) + q^*(u(x)) = d(x, y).$$

Since $\delta q^*(f(x, y), u(xy)) = 0$, the middle term may be expanded. With the 1-cochain $c^1(x) = q^*(u(x))$, this equation thus becomes

$$q^*(f(x, y)) = \delta c^1(x, y) - d(x, y),$$

or

$$\theta(f(x, y)) = \delta c^1(x, y) - l^*(x, u(y)),$$

as in (11.12).

12. Obstruction and deviation

An obstruction cycle l_3 of the given sequence $\mathfrak{S}$ is obtained, according to the definitions of [3] for the equivalent case of a Q-kernel, as follows. For the given factor set $f(x, y)$, choose any elements $h(x, y)$ in K such that $\mu h(x, y) = f(x, y)$. Since μ is an operator homomorphism, as in property (v) of §2, we have $\mu[\eta(u(x)) \cdot h(y, z)] = u(x)f(y, z)u(x)^{-1}$. Hence the associativity condition (9.2) and the exactness of the sequence implies that there exist elements $l_3(x, y, z) \in H$ such that

$$[\eta(u(x)) \cdot h(y, z)]h(x, yz) = [\lambda l_3(x, y, z)]h(x, y)h(xy, z). \tag{12.1}$$

It follows that l_3 is a three dimensional cocycle of Q over H, and that its cohomology class $\{l_3\}$ in $H^3(Q, H)$ is independent of the choice of the factor set f and of the representatives h of this factor set in K [cf. 3, Lemmas 7.1, 7.2, and 7.3]. We call l_3 an *obstruction* of the sequence $(\mathfrak{S}, \eta)$, and its cohomology class the obstruction class of $(\mathfrak{S}, \eta)$.

THEOREM 12.1. *The homomorphism Λ_1 of (11.2) carries the deviation class $\{m\}$ of an exact sequence $(\mathfrak{S}, \eta)$ with free operators F into the obstruction class $\Lambda_1\{m\}$ of this sequence.*

PROOF. Let u and v be the representatives used in the definition of a deviation m of the sequence. In constructing our obstruction we may then choose the elements $h(x, y) \in K$ as $v(f(x, y))$. Then, by the definition (10.2) of m and the associativity condition (9.2) for f,

$$[\eta(u(x))\cdot v(f(y, z))]v(f(x, yz))$$
$$= [\lambda m(x, f(y, z))]v(u(x)f(y, z)u(x)^{-1}f(x, yz))$$
$$= [\lambda m(x, f(y, z))]v(f(x, y))v(f(xy, z)).$$

Hence, by comparison with the definition (12.1) of the obstruction l,

$$l_3(x, y, z) = m(x, f(y, z)) = (\Lambda_1 m)(x, y, z), \tag{12.2}$$

as asserted.

By the argument of the previous section, Λ_1 is an isomorphism onto the cohomology group $H^3(Q, H)$. Theorem 10.2 asserts the existence of a sequence with a given deviation; hence we now deduce as a corollary the existence of a sequence $\mathfrak{S}$ with a given obstruction, as asserted in Theorem C of §2.

13. The main homomorphism

Let σ be an operator homomorphism of K into G, and $h(x, y)$ any representatives in K of the factor set $f(x, y)$. Then the function σh is a 2-cochain $(\sigma h)(x, y) = \sigma(h(x, y))$ of Q over G. Applying the operator homomorphism σ to the definition (12.1) of the obstruction l, we obtain

$$x\cdot\sigma h(y, z) + \sigma h(x, yz) = \sigma\lambda l_3(x, y, z) + \sigma h(x, y) + \sigma h(xy, z),$$

an equation which asserts that $\delta(\sigma h)(x, y, z) = \sigma\lambda l_3(x, y, z)$ and hence that the operator homomorphism $\sigma\lambda = \rho$ induced on H and the cochain $\sigma h = p$ satisfy the condition used in the construction (§8) of the group E^2. Thus we have a homomorphism

$$\Lambda : \text{Ophom}\,(H, G) \to E^2; \qquad \Lambda\sigma = (\sigma\lambda, \sigma h). \tag{13.1}$$

In particular, if $\sigma = \tau\mu$ is induced by a crossed homomorphism μ of F into G, then $\sigma\lambda = 0$ on H, and $\sigma h(x, y) = \tau\mu h(x, y) = \tau f(x, y)$. Applying the crossed homomorphism τ to the definition (9.1) of the factor set, we get

$$\tau u(x) + x\cdot\tau u(y) = \sigma h(x, y) + \tau u(x, y),$$

so that $(\sigma\lambda, \sigma h) = (0, \delta(\tau u))$ is a coboundary. Thus Λ induces a homomorphism

$$\Lambda : \text{Map}\,(\mathfrak{S}, G) \to E^2(Q, H, l_3, G). \tag{13.2}$$

The homomorphism Λ is the one considered in the main theorem B″ of §8. It is also a homomorphism between the two groups of (13.2), regarded as group extensions, in the sense that the desired conditions (8.4) hold; indeed, by the relevant definitions

$$\chi\Lambda\sigma = \chi(\sigma\lambda, \sigma h) = \sigma\lambda = \lambda^*\sigma,$$

hence $\chi\Lambda = \lambda^*$, while for any $\omega \epsilon$ Ophom (R, G),

$$\Lambda\mu'^*\omega = \Lambda\omega\mu' = (\omega\mu'\lambda, \omega\mu'h) = (0, \omega f),$$

$$i\Lambda_0\omega = i\omega f = (0, \omega f),$$

and thus $\Lambda\mu'^* = i\Lambda_0$. This gives (8.4).

The fact that Λ_0 is an isomorphism "into" now yields the same result for Λ. Indeed, letting $\{\sigma\}$ denote the operator homomorphism class of σ in Map (S, η, G), suppose that $\Lambda\{\sigma\} = 0$. Then $0 = \chi\Lambda\{\sigma\} = \lambda^*\sigma$, so that, by §6, σ is induced by an operator homomorphism ω in Ophom (R, G) as $\sigma = \omega\mu' = \mu'^*\omega$. But $i\Lambda_0\{\omega\} = \Lambda\mu'^*\{\omega\} = \Lambda\{\sigma\} = 0$; since $i\Lambda_0$ is an isomorphism into, $\{\omega\} = 0$ and $\{\sigma\} = \{\mu'^*\omega\} = 0$. Hence Λ in (13.2) has kernel 0.

The homomorphism Λ in (13.2) is independent of the choices of f and h made in its definition, in the following sense. First replace u and f by $u'(x) = s(x)u(x)$ and f', as in (9.3) and (9.4). Choose any representatives $k(x) \epsilon K$ for $s(x) \epsilon R$ and define h' by the equation

$$h'(x, y)k(xy) = k(x)[\eta(u(x))\cdot k(y)]h(x, y). \tag{13.3}$$

Then $\mu h' = f'$, by (9.4), and exactly as in [3, Lemma 7.3] it follows that the obstruction cocycle l_3 is unchanged. On the other hand, applying σ to (13.3), one has

$$\sigma h'(x, y) = \delta(\sigma k)(x, y) + \sigma h(x, y);$$

thus σh is altered by the coboundary $\delta(\sigma k)$, and Λ in (13.2) is unchanged.

Secondly, hold u and f fixed and replace the representatives $h(x, y)$ by any other representatives $h'(x, y)$, where

$$h'(x, y) = [\lambda g(x, y)]h(x, y), \qquad g(x, y) \epsilon H.$$

Then as in [3, Lemma 7.2] the obstruction l_3 is changed to the cohomologous cocycle[6]

$$l_3'(x, y, z) = l_3(x, y, z) + \delta g(x, y, z).$$

Now the groups E^2 defined by these cocycles l_3' and l_3 are distinct, but isomorphic. One shows readily [5, §12] that the correspondence

$$\gamma_g(\rho, p) = (\rho, p - \rho g)$$

induces an isomorphism

$$\gamma_g : E^2(Q, H, l_3', G) \cong E^2(Q, H, l_3, G) \tag{13.4}$$

[6] We obtain here $+g$, and not $-g$, as is erroneously indicated in the corresponding (multiplicative) formula, loc. cit.

which is actually an equivalence of the two group extensions in question, in the sense that $\chi' = \chi\gamma_g$, $i = \gamma_g i'$, for the mappings χ and i indicated in (8.1).

We assert that the main homomorphism Λ is changed only by γ_g, in the sense that the new homomorphism Λ' defined by the representatives h' has $\gamma_g\Lambda' = \Lambda$. Indeed,

$$\begin{aligned}\gamma_g\Lambda'\sigma &= \gamma_g(\sigma\lambda, \sigma h') = (\sigma\lambda, \sigma h' - \sigma\lambda g) \\ &= (\sigma\lambda, \sigma h) = \Lambda\sigma,\end{aligned}$$

as asserted.

It remains only to prove that Λ is an isomorphism onto. Since Λ is essentially independent of the various choices, it is convenient in this proof to specialize the choices as follows. We choose fixed representatives u of Q in F, and fixed representatives v of R in K, and set $h = vf$. The deviation m and the obstruction l_3 relative to these choices are then related by $l_3 = \Lambda_1 m$, as in (12.2).

14. The inverse mapping

Take an arbitrary element of the group $E^2(Q, H, l_3, G)$, represented by a pair (ρ, p) with $\rho l_3 = \delta p$, $p(x, y) \in G$. As in the previous inverse constructions, define a new function p^* on F to G by induction on the length of the word be or be^{-1} of F as

(14.1a) $$p^*(be) = p^*(b) + p(\nu b, \nu e),$$

(14.1b) $$p^*(be^{-1}) = p^*(b) - p(\nu(be^{-1}), \nu e),$$

with the initial condition $p^*(1) = 0$. One shows directly that these two equations hold even if be or be^{-1} is not a reduced word.

Now compare p^* with the function l^* constructed from $l = l_3$ by the parallel method of (11.6a) and (11.6b), with the same generators e for the free group F. We assert that

(14.2) $$\delta p^*(a, c) = -p(\nu a, \nu c) + \rho l^*(\nu a, c), \qquad a, c \in F.$$

The proof is by induction on the length of c as a reduced word in the generators e; the result is trivial if $c = 1$ is of the length 0. If $c = be$, expand δp^* and use the definitions (14.1a) and (11.6a)—the first applies to the term $p^*(abe)$ even if abe is not reduced—to obtain the equivalent equation

$$\begin{aligned}&a\cdot p^*(b) + a\cdot p(\nu b, \nu e) - p^*(ab) - p(\nu(ab), \nu e) + p^*(a) \\ &\quad = -p(\nu a, \nu(be)) + \rho l^*(\nu a, b) + \rho l(\nu a, \nu b, \nu e),\end{aligned}$$

or

$$\begin{aligned}&\delta p^*(a, b) + \delta p(\nu a, \nu b, \nu e) \\ &\quad = -p(\nu a, \nu b) + \rho l^*(\nu a, b) + \rho l(\nu a, \nu b, \nu e).\end{aligned}$$

Since $\delta p = \rho l$, the two corresponding terms drop out, and the remaining equation is a consequence of the induction assumption. The case of (14.2) with $c = be^{-1}$ is treated similarly by (14.1b) and (11.6b).

Setting $a = r \in R$ in (14.2) we have $\nu a = 1$, so that the terms in p and l^* vanish by the normalization of these functions, and

$$p^*(rc) = p^*(r) + p^*(c) \qquad r \in R,\ c \in F; \tag{14.3}$$

in particular, p^* induces a homomorphism $R \to G$. Second, set $a = u(x)$, $c = r \in R$ in (14.2); the middle term on the left is then

$$p^*(u(x)r) = p^*(u(x)ru(x)^{-1}u(x)) = p^*(u(x)ru(x)^{-1}) + p^*(u(x))$$

by (14.3), so that the result is

$$x \cdot p^*(r) - p^*(u(x)ru(x)^{-1}) = \rho l^*(x, r); \tag{14.4}$$

the left side is the coboundary $\delta' p^*$ of p^* on R. Finally set $a = u(x)$, $c = u(y)$ in (14.2) to obtain

$$x \cdot p^*(u(y)) - p^*(f(x, y)u(xy)) + p^*(u(x)) = -p(x, y) + \rho l^*(x, u(y)),$$

or, using (14.3),

$$p^*(f(x, y)) = \delta(p^*u)(x, y) + p(x, y) - \rho l^*(x, u(y)). \tag{14.5}$$

Now combine the homomorphism $p^* : R \to G$ with the homomorphism $\theta : R \to H$ of Lemma 11.3 to give the homomorphism $t = p^* - \rho\theta : R \to G$. The coboundary $\delta' t$ is then given directly by combining (14.4) and (11.11) as

$$\delta' t(x, r) = \rho m(x, r). \tag{14.6}$$

In terms of the representatives v the elements of K appear uniquely as $(\lambda h)(vr)$ for $h \in H$, $r \in R$. We define a homomorphism $\sigma : K \to R$ by setting

$$[(\lambda h)(vr)] = \rho(h) + t(r) \qquad h \in H,\ r \in R.$$

To show that it is an operator homomorphism, observe by (10.3) that, for $a \in F$, $\nu a = x \in Q$,

$$\eta(a) \cdot [(\lambda h)(vr)] = \lambda[(x \cdot h) + m(x, r)]v(ara^{-1}).$$

Hence, by the equation (14.6),

$$\begin{aligned} \sigma\{\eta(a) \cdot [(\lambda h)(vr)]\} &= \rho(x \cdot h) + \rho m(x, r) + t(u(x)ru(x)^{-1}) \\ &= \rho(x \cdot h) + x \cdot t(r) \\ &= a \cdot \{\sigma[(\lambda h)(vr)]\}. \end{aligned}$$

The image $\Lambda\sigma$ of this operator homomorphism σ is cohomologous to the given pair (ρ, p). In fact $\Lambda\sigma$ is the pair (ρ, p'), where

$$\begin{aligned} p'(x, y) &= \sigma h(x, y) = \sigma[vf(x, y)] = t(f(x, y)) \\ &= p^*(f(x, y)) - \rho\theta(f(x, y)). \end{aligned}$$

By (14.5) and (11.12) of Lemma 11.3 this becomes

$$\begin{aligned} p'(x, y) &= \delta(p^*u)(x, y) + p(x, y) - \rho l^*(x, u(y)) \\ &\quad - \rho\delta c^1(x, y) + \rho l^*(x, u(y)) \\ &= p(x, y) + \delta b^1(x, y), \end{aligned}$$

where $b^1(x) = p^*u(x) - \rho c^1(x)$ is a one-dimensional cochain. This completes the proof that the image of Λ includes the cohomology class of every pair (ρ, p).

THE UNIVERSITY OF CHICAGO

BIBLIOGRAPHY

[1] BAER, R., *Erweiterung von Gruppen und ihren Isomorphismen*, Math. Zeit. 38 (1934), 375–416.
[2] EILENBERG, S. AND MACLANE, S., *Cohomology theory in abstract groups* I, Ann. of Math. 48 (1947), 51–78.
[3] EILENBERG, S. AND MACLANE, S., *Cohomology theory in abstract groups* II, *Group extensions with a non-Abelian kernel*, Ann. of Math. 48 (1947), 326–341.
[4] EILENBERG, S. AND MACLANE, S., *Determination of the second homology and cohomology groups of a space by means of homotopy invariants*, Proc. Nat. Acad. Sci. 32 (1946), 277–280.
[5] EILENBERG, S. AND MACLANE, S., *Homology of spaces with operators* II, Trans. Amer. Math. Soc. 65 (1949), 49–99.
[6] HALL, M., *Group rings and extensions* I, Ann. of Math. 39 (1938), 220–234.
[7] HOPF, H., *Fundamentalgruppe und zweite Bettische Gruppe*, Comment. Math. Helv. 14 (1942), 257–309.
[8] HUREWICZ, W., *Zu einer Arbeit von O. Schreier*, Abh. Math. Sem. Hansischen Univ. 8 (1930), 307–314.
[9] HUREWICZ, W. AND STEENROD, N. E., *Homotopy relations in fibre spaces*, Proc. Nat. Acad. Sci. 27 (1941), 60–64.
[10] SCHREIER, O., *Die Untergruppen der freien Gruppen*, Abh. Math. Sem. Hansischen Univ. 5 (1927), 161–183.
[11] TURING, A. M., *The extensions of a group*, Compositio Math. 5 (1938), 357–367.

Reprinted from the Proceedings of the NATIONAL ACADEMY OF SCIENCES,
Vol. 36, No. 1, pp. 41–48. January, 1950

ON THE 3-TYPE OF A COMPLEX

BY SAUNDERS MACLANE* AND J. H. C. WHITEHEAD

DEPARTMENT OF MATHEMATICS, THE UNIVERSITY OF CHICAGO, AND MAGDALEN COLLEGE, OXFORD

Communicated November 10, 1949

1. Introduction.—The standard algebraic invariants of a topological space depend only on the homotopy type of the space. This note will deal with part of the converse problem of the determination of the homotopy type by algebraic invariants, and will show in effect that the only one- and two-dimensional invariants which enter are the fundamental group π_1, the second homotopy group π_2, and a certain three-dimensional cohomology class of π_1 in π_2.

As in CH I,[2] we consider connected cell complexes[3] K, and denote by K^n the n-dimensional skeleton of K. We recall that the complexes K and K' are said to be of the same *n-type* if, and only if, there are maps $\phi: K_n \to K'_n$, $\phi': K'_n \to K^n$, and homotopies

$$\phi'\phi|K^{n-1} \simeq i: K^{n-1} \to K^n,$$
$$\phi\phi'|K'^{n-1} \simeq i': K'^{n-1} \to K'^n,$$

where i, i' are the identical maps. In this case we write $\phi: K^n \equiv_{n-1} K'^n$, and we assume that $n > 1$, since any two (connected) complexes are of the same 1-type. Then $\phi: K_n \equiv_{n-1} K'^n$ if, and only if,

$$\phi_r: \pi_r(K^n) \cong \pi_r(K'^n) \qquad (r = 1, \ldots, n-1), \tag{1.1}$$

according to Theorem 2 in CHI, where ϕ_r is the homomorphism induced by ϕ. The classification of complexes according to their 2-type is equivalent, under the correspondence $K \to \pi_1(K)$, to the classification of groups by the relation of isomorphism. The purpose of this note is to define an algebraic equivalent of the 3-type. Since the n-type of K depends only on K^n, we may always replace K by K^3 when we discuss the 3-type. Therefore we assume that any given complex is at most 3-dimensional.

By an *algebraic 3-type* we mean a triple, $T = (\pi_1, \pi_2, \mathbf{k})$, which consists of

(*a*) an arbitrary (multiplicative) group π_1,

(*b*) an additive, Abelian group π_2, which admits π_1 as a group of operators,

(*c*) a 3-dimensional cohomology class $\mathbf{k} \in H^3(\pi_1, \pi_2)$.

The algebraic 3-type of a complex K is the triple $(\pi_1(K), \pi_2(K), \mathbf{k}(K))$ consisting of the fundamental group, the second homotopy group of K (with the usual operators of $\pi_1(K)$ on $\pi_2(K)$), and the "obstruction" invariant $\mathbf{k}^3 = \mathbf{k}(K)$ defined as in CT III[4] (cf. §2 below).

Let $T = (\pi_1, \pi_2, \mathbf{k})$ and $T' = (\pi_1', \pi_2', \mathbf{k}')$ be any algebraic 3-types. By a *homomorphism*

$$\theta = (\theta_1, \theta_2): T \to T'$$

we mean a pair of homomorphisms

$$\theta_1: \pi_1 \to \pi_1', \qquad \theta_2: \pi_2 \to \pi_2' \tag{1.2}$$

such that

$$\theta_2(xa) = (\theta_1 x)\theta_2 a, \qquad x \in \pi_1, a \in \pi_2, \tag{1.3a}$$

$$\theta_2 k(x, y, z) \sim k'(\theta_1 x, \theta_1 y, \theta_1 z), \qquad x, y, z \in \pi_1; \tag{1.3b}$$

where k and k' are (non-homogeneous) cocycles in the classes $\mathbf{k}$, $\mathbf{k}'$, respectively. The homomorphism θ is an isomorphism if, and only if, both θ_1 and θ_2 are isomorphisms; the resulting relation $T \cong T'$ of isomorphism between 3-types is clearly an equivalence relation.

We shall say that a given algebraic 3-type, T, is *realized* by a complex, K, if, and only if, $T \cong T(K)$. Let $K = K^3$ and $K' = K'^3$ be given complexes and $\phi: K \to K'$ a given map. Let $\pi_n = \pi_n(K)$, $\pi_n' = \pi_n(K')$. Then (1.3a) is satisfied by the homomorphisms ϕ_1, ϕ_2, which are induced by ϕ. It follows from the definition of $\mathbf{k}^3$ that (1.3b) is also satisfied. Therefore $\phi: K \to K'$ induces a homomorphism $\phi: T(K) \to T(K')$. If the latter is given, then a map, $K \to K'$, which induces it, will be called a (geometrical) *realization* of $\phi: T(K) \to T(K')$.

The main results are:

THEOREM 1. *Two complexes K and K' are of the same 3-type[5] if, and only if, $T(K) \cong T(K')$.*

THEOREM 2. *Any algebraic 3-type can be realized by some complex.*

THEOREM 3. *For complexes K and K' a given homomorphism $T(K) \to T(K')$ has a geometrical realization, $\phi: K \to K'$, provided that $\dim K \leqq 3$.*

Theorem 1 may be deduced from Theorem 3 and the statement containing equation (1.1).

2. *Crossed Sequences.*—The algebraic constructions relating to 3-types involve certain types of "operator sequences" of groups and homomorphisms. In general, such a sequence consists of additive groups A and B (B abelian, but not necessarily A) which admit the multiplicative groups P and Q, respectively, as groups of left operators, together with the homomorphisms

$$O \to B \xrightarrow{\lambda} A \xrightarrow{\mu} P \xrightarrow{\nu} Q \to 1, \tag{2.1}$$

such that $\mu\lambda(B) = 1$, $\nu\mu(A) = 1$, $\nu(P) = Q$. The homomorphisms λ, μ must be operator homomorphisms, in that

$$\lambda[(\nu p)\cdot b] = p(\lambda b), \qquad p \in P, b \in B; \tag{2.2}$$

$$\mu(pa) = p(\mu a)p^{-1}, \qquad p \in P, a \in A. \tag{2.3}$$

Finally,

$$a + a' - a = (\mu a)a', \qquad a, a' \epsilon A. \tag{2.4}$$

The middle section $\mu: A \to P$, since it is subject to the conditions (2.3) and (2.4), defines A as a crossed (P, μ)-module in the sense of CH II. It follows from equation (2.4), first with $a \epsilon \mu^{-1}(1)$ and then $a' \epsilon \mu^{-1}(1)$ that $\mu^{-1}(1)$ is in the center of A and that μA operates simply on $\mu^{-1}(1)$.

We need only two special types of such operator sequences.

Crossed sequences are operator sequences (2.1) which are exact sequences. The crossed (P, μ) module A of such a sequence determines the crossed sequence up to isomorphism, with $Q \cong P/\mu A$, $B \cong \mu^{-1}(1) \subset A$, and with the operators of Q on B determined by equation (2.2). Hence the theory of crossed modules is equivalent to that of crossed sequences, as developed[6] in CT III.

Each crossed sequence determines an algebraic 3-type in the following way. For each $q \epsilon Q$ select a representative $u(q) \epsilon \nu^{-1}q$ in P, with $u(1) = 1$. Then $u(q)u(q')u(qq')^{-1} = f(q, q')$ lies in $\nu^{-1}(1) = \mu(A)$. Select $a(q, q') \epsilon \mu^{-1}f(q, q')$ with $a(q, 1) = a(1, q') = 0$. Then, for $q, r, s \epsilon Q$,

$$\delta a(q, r, s) = u(q) \cdot a(r, s) + a(q, rs) - a(qr, s) - a(q, r)$$

lies in $\mu^{-1}(1)$. The function k with $k(q, r, s) = \lambda^{-1}[\delta a(q, r, s)]$ is then defined, and is a (non-homogeneous) 3-dimensional cocycle of Q in B. Its cohomology class, $\mathbf{k}$, which is independent[7] of the choice of u and a, is called the *obstruction* of the sequence, and the triple $(Q, B, \mathbf{k})$ is the (unique) algebraic 3-type associated with the sequence.

Homotopy Systems of dimension 3, as defined in CH II, are operator sequences (2.1) in which P is a free group, A a free crossed (P, μ) module, and B a free (abelian) Q-module, and in which $\nu^{-1}\, 1 = \mu A$ (exactness at P). Since $Q \cong P/\mu(A)$, Q need not be given in advance.

Each homotopy system (2.1) determines a certain crossed sequence, as follows. Since $\mu\lambda(B) = 1$, μ induces a homomorphism $\mu': A/\lambda B \to P$. By equation (2.2), the given operators of P on A induce operators of P on $A/\lambda B$. By equation (2.3) $\mu^{-1}(1) \subset A$ is closed under operation by P; since μA operates simply on $\lambda B \subset \mu^{-1}(1)$, operators of $Q \cong P/\mu A$ on $\mu^{-1}(1)/\lambda B$ are induced. Using the identity injection λ', we thus have a crossed sequence

$$0 \to \mu^{-1}(1)/\lambda B \xrightarrow{\lambda'} A/\lambda B \xrightarrow{\mu'} P \xrightarrow{\nu'} Q \to 1. \tag{2.5}$$

We call this the sequence *derived* from the homotopy sequence (2.1).

The geometric applications are as follows. If K is a complex, the sequence of homotopy groups

$$0 \to \pi_2(K) \to \pi_2(K, K^1) \to \pi_1(K^1) \to \pi_1(K) \to 1, \tag{2.6}$$

with the usual mappings and operators, is a crossed sequence. We define $\mathbf{k}(K)$ as the obstruction of this sequence. This agrees with the geometric definition[8] of this invariant, as may be proved using one of the known additivity theorems[9] for relative homotopy groups.

The homotopy system (CH II) of the 3-dimensional cell complex K consists of the homotopy groups $\pi_1(K)$, $\rho_1 = \pi_1(K^1)$ and $\rho_n = \pi_n(K^n, K^{n-1})$ for $n = 2, 3$, together with the usual operators, and the homomorphisms

$$0 \to \rho_3 \xrightarrow{d_3} \rho_2 \xrightarrow{d_2} \rho_1 \xrightarrow{d_1} \pi_1 \to 1, \tag{2.7}$$

where d_1 is the injection homomorphism and $d_n = j_{n-1}\beta_n$, for $n = 2, 3$, is the composite of the boundary and injection homomorphisms

$$\rho_n \xrightarrow{\beta_n} \pi_{n-1}(K_{n-1}) \xrightarrow{j_{n-1}} \rho_{n-1}.$$

The derived sequence

$$0 \to d_2^{-1}(1)/d_3\rho_3 \to \rho_2/d_3\rho_3 \to \rho_1 \to \pi_1 \to 1 \tag{2.8}$$

of this homotopy sequence is isomorphic to the crossed sequence (2.6) of K; in other words $\pi_2(K) \cong d_2^{-1}(1)/d_3\rho_3$ and $\pi_2(K, K^1) \cong \rho_2/d_3\rho_3$, with the mappings and operators corresponding under the isomorphisms. Indeed, the first isomorphism follows from the exactness of the homotopy sequences for the pairs K^2, K^1 and K^3, K^2, together with $\pi_2(K^1) = 0$, $\pi_2(K^3, K^2) = 0$, while the second isomorphism follows from the known exactness of the homotopy sequence

$$\pi_3(K^3, K^2) \to \pi_2(K^2, K^1) \to \pi_2(K^3, K^1)$$

for the triple K^3, K^2, K^1.

3. *Realization of an Algebraic 3-Type.*—Each homotopy system determines a derived crossed sequence and thence an algebraic 3-type. Theorem 2 will be proved by reversing this process. Let $(\pi_1, \pi_2, \mathbf{k})$ be any algebraic 3-type. The group π_1 can be represented as the image of a free group X under a homomorphism ν. By Theorem C in CT III we can construct[10] a crossed sequence

$$0 \to \pi_2 \xrightarrow{\lambda} A \xrightarrow{\mu} X \xrightarrow{\nu} \pi_1 \to 1 \tag{3.1}$$

which realizes the given algebraic 3-type. To construct a corresponding homotopy system with $\rho_1 = X$, take ρ_2, as in §2 of CH II, to be the free crossed (X, d_2)-module with symbolic generators (x, a) for all $x \in X$ and all a in any chosen set of generators of A and with $d_2: \rho_2 \to X$ determined by $d_2(x, a) = x(\mu a)x^{-1}$. By Lemma 2 in CH II an operator homomorphism $\omega: \rho_2 \to A$ onto A is determined by setting $\omega(1, a) = a$. Then $d_2 = \mu\omega$, and, since ω and d_2 are operator homomorphisms, the abelian subgroups $\omega^{-1}(0) \subset d_2^{-1}(1)$ of ρ_2 are invariant under the operators of ρ_1, and the crossed sequence (3.1) is isomorphic (under ω) to the crossed sequence

$$0 \to d_2^{-1}(1)/\omega^{-1}(0) \to \rho_2/\omega^{-1}(0) \to \rho_1 \to \pi_1 \to 1. \qquad (3.2)$$

The abelian group $\omega^{-1}(0)$ admits π_1 as a group of operators according to the rule $(\nu x)b = xb$ for $x \in X$ and $b \in \omega^{-1}(0)$. Hence there is a free π_1-module ρ_3 with an operator homomorphism d_3: $\rho_3 \to \omega^{-1}(0) \subset \rho_2$ onto $\omega^{-1}(0)$. We have thus constructed a homotopy system

$$0 \to \rho_3 \xrightarrow{d_3} \rho_2 \xrightarrow{d_2} \rho_i \xrightarrow{\nu} \pi_i \to 1 \qquad (3.3)$$

for which the derived crossed sequence (3.2) is isomorphic to (3.1).

Theorem 2 of CH II asserts that the homotopy system (3.3) can be realized as the homotopy system of a 3-dimensional complex K. The derived sequence (3.2) is then on the one hand isomorphic to the relative homotopy sequence (2.6) of the complex K, with obstruction the obstruction of the space K, and on the other hand to the given sequence (3.1) with the preassigned obstruction **k**. Hence K realizes the given 3-type, as asserted in Theorem 2.

4. Mappings of Complexes with Operators.—An abstract closure finite cell complex C—that is, a system of free abelian groups C_q and homomorphisms

$$C_0 \xleftarrow{\partial} C_1 \xleftarrow{\partial} C_2 \xleftarrow{\partial} \dots$$

with $\partial\partial = 0$—has free operators in the multiplicative group W if each C_q is a free W-module and each $w \in W$ a chain transformation ($w\partial = \partial w$). Select a preferred W-base for C_0, consisting of certain 0-cells, one of which we call the *special* 0-cell, and define the homomorphism J of C_0 into the group of integers by setting $J(wc_0) = 1$ for each preferred 0-cell c_0. We require that C_0 be augmentable, as[11] in HSO II, p. 54; i.e., that $C_0 \neq 0$ and $J\partial = 0$. Under these conditions we call (W, C) a *complex with free operators.*

A homomorphism (f_0, λ) of one such complex (W, C) into a second (W', C') consists of homomorphisms $f_0{:}\,W \to W'$, $\lambda_q{:}\,C_q \to C_q'$, such that λ is a chain transformation ($\lambda\partial = \partial\lambda$), $\lambda w = (f_0 w)\lambda$ for each $w \in W$, and $\lambda_0 c_0$ is the special 0-cell of C' whenever c_0 is in the preferred W-base of C. Then $J\lambda_0 = J$, and λ is also augmentable, in the sense of HSO II.

Any multiplicative group W determines such a complex K_W, as in HSO II, with q-cells $(w_0, \dots, w_q)$ for $w_i \in W$ and the preferred W-base for C_0 consisting of the special 0-cell (1). Any homomorphism f_0: $W \to W'$ induces a homomorphism $(f_0, f){:}\,K_W \to K_{w}'$ with $f_q{:}\,C_q(K_W) \to C_q(K_{W}')$ determined by the formula

$$f_q(w_0, \ldots, w_q) = (f_0 w_0, \ldots, f_0 w_q), \qquad w_i \in W.$$

For any complex K the abstract cell complex $C(\tilde{K})$, consisting of the chain groups $C_n = H_n(\tilde{K}^n, \tilde{K}^{n-1})$ of the universal covering complex[12] $\tilde{K}$ is a complex with free operators in the group $W \cong \pi_1(K)$ of covering transformations. For the preferred W-base of C_0 select *a* 0-cell over each 0-cell of K, and as the special 0-cell select the 0-cell carried by the base point of $\tilde{K}$. A homomorphism $(f_0, \lambda):C(\tilde{K}) \to C(\tilde{K}')$ is then a chain mapping in the sense of CH II, §9, and in particular $\lambda_0: C_0(\tilde{K}) \to C_0(\tilde{K}')$ can be realized geometrically by a map $K^0 \to e'^0$, where e'^0 is the base point in K' (of course all of $\tilde{K}^0$ need not map into the base point in $\tilde{K}'$).

If the complexes (W, C) and (W', C') are acyclic in dimensions less than q, their integral homology groups $H = H_q(C)$ and $H' = H_q(C')$ in this dimension have operators in W and W', respectively. C (and likewise C') then has an obstruction cohomology class $\mathbf{l} \in H^{q+1}(W, H)$, determined as in HSO II, Theorem 5.1, as the obstruction of any homomorphism of the q-dimensional skeleton of K_W into C. For $q = 2$, the system $(W, H, \mathbf{l})$ determined by C is an algebraic 3-type. For any q, a *homomorphism*

$$(f_0, h): (W, H, \mathbf{l}) \to (W', H', \mathbf{l}') \tag{4.1}$$

will mean a pair of homomorphisms $f_0: W \to W'$ and $h:H \to H'$ which satisfy conditions analogous to equation (1.3).

If C^{q+1} is the $(q + 1)$-dimensional skeleton of C, any homomorphism $(f_0, \lambda): C^{q+1} \to C'^{q+1}$ induces a homomorphism $h:H \to H'$. By the argument of Theorem 5.1 in HSO II, f_0 and h satisfy the analogue of (1.3b); they obviously satisfy (1.3a). Therefore (f_0, h) is a homomorphism of the form (4.1). We then call (f_0, λ) a *combinatorial realization* of (f_0, h).

THEOREM 4. *For complexes C, C' with free operators in W, W', acyclic in dimensions less than $q(q > 0)$, any homomorphism (4.1) has a combinatorial realization $(f_0, \lambda): C^{q+1} \to C'^{q+1}$.*

This theorem is an extension of part of Theorem 7.1 in HSO II, and is established by the argument there (pp. 62, 63) with the following modifications. If $\mathbf{l}$ and $\mathbf{l}'$ are (homogeneous) cocycles in the classes $\mathbf{l}$ and $\mathbf{l}'$, respectively, the analogue of (1.3b) shows that there is a cochain m': $C_q(K_W) \to H'$ such that $h\mathbf{l} = \mathbf{l}f_{q+1} + \delta m'$. Since $C_q(K_w)$ is a free W-module, the homomorphism m' can be lifted to an operator homomorphism $g': C_q(K_W) \to Z_q(C')$, with $\eta' g = m'$. The first equation of (7.5) becomes $\eta' F_q = h\eta E_q$, and the required realization λ is defined by

$$\begin{aligned} \lambda_i &= \alpha_i' f_i \gamma_i \colon C_i \to C_i' \qquad i = 0, \ldots, q-1, \\ \lambda_q &= \alpha_q' f_q \gamma_q + g' \gamma_q + F_q \colon C_q \to C_q'. \end{aligned}$$

The cited calculations then show the existence of a suitable λ_{q+1}.

5. *Proof of Theorem 3.*—For a complex K^3 the associated chain system $C(\tilde{K}^3)$ is acyclic in dimensions less than 2, and the obstruction $\mathbf{k}(K)$, as defined in §2 above, agrees with the obstruction $\mathbf{l}$ of $C(K^3)$. Indeed, choose $u(q) \epsilon \rho_1$ and $a(q, r) \epsilon \rho_2/d_3\rho_3$ as in the definition of $\mathbf{k}$ for the sequence (2.8), and let ω be the natural homomorphism $\omega: \rho_2 \rightarrow \rho_2/d_3\rho_3$. Choose $R^1(q) = u(q)$ and $R^2(q, r) \epsilon \omega^{-1}a(q, r)$. Then $\omega\delta R^2 = \delta a$, whence $\delta R^2(q, r, s) \epsilon \omega^{-1}\mu^{-1}(1) = d_2^{-1}(1)$. Using the operator homomorphisms h_n: $\rho_n \rightarrow C_n$ of §12 of CH II, set $f^n = h_nR^n \epsilon C^n(\pi_1, C_n)$ for $n = 1, 2$, and $f^0 = c^0$, where c^0 is the special 0-cell in C_0. Then it may be verified that $\delta f^i = \partial f^{i+1}$ for $i = 0, 1$; hence, by Theorem 6.3 in HSO II, δf^2 is the obstruction $\mathbf{l}$ of C. By Lemma 5 in CH II, h^2 induces an isomorphism $d_2^{-1}(1)/d_3\rho_3 \cong H$. It follows that the isomorphisms $\pi_1 \cong W$, $d_2^{-1}(1)/d_3\rho_3 \cong H$ carry $\mathbf{k}(K^3)$ into the obstruction $\mathbf{l}$ of C.

Now consider two complexes K^3, K'^3 with their associated chain systems $C = C(\tilde{K}^3)$ and $C' = C(\tilde{K}'^3)$, acyclic in dimensions less than 2. Any homomorphism $\theta: T(K) \rightarrow T(K')$ on the algebraic 3-types of K and K' satisfies the hypotheses of Theorem 4 for $q = 2$, hence has a combinatorial realization $\lambda: C \rightarrow C'$. By Theorem 16 in CH II, λ has a geometrical realization $\phi: K \rightarrow K'$. Because of the natural isomorphisms $H \cong \pi_2(K)$, $H' = \pi_2(K')$, the map ϕ is a geometrical realization of θ, and Theorem 3 is proved.

Theorem 3 can also be proved without the use of chain groups and covering complexes by combining certain theorems of CT III, on the "deviation" of exact sequences, with theorems in CH II on the realizability of homomorphisms of homotopy systems.

6. *A Sufficiency Theorem.*—By a *sufficiency theorem* we mean one which states that certain invariants are sufficiently powerful to insure that, within a definite category, any mapping which induces isomorphisms of these invariants is an equivalence. For instance the theorem quoted in §1, which states that formula (1.1) implies $\phi: K^{n+1} \equiv_n K^{n+1}$, is a sufficiency theorem, within any category of CW-complexes and n-homotopy classes of maps, $K^{n+1} \rightarrow K'^{n+1}$. A realizability theorem, like Theorem 3 or Theorem 4 above, is one which states that a homomorphism of some kind of algebraic invariant can be realized by a mapping of objects in the category.

Let C, C' mean the same as at the beginning of §4, let $W = W'$, and let $C_n = 0$, $C_n' = 0$ if $n > q + 1$. Let $\lambda \simeq_q \mu$ mean the same as $\lambda \simeq \mu$ (dim $\leq q$) in HSO II, where λ is equivariant. We shall write $\lambda: C \equiv_q C'$ if, and only if, there is an equivariant homomorphism $\lambda': C' \rightarrow C$, such that $\lambda'\lambda \simeq_q 1$, $\lambda\lambda' \simeq_q 1$. Then our sufficiency theorem, which is analogous to Theorem 2 in CH I, is

THEOREM 5. *If* $\lambda: C \rightarrow C'$ *induces isomorphisms* $H_n(C) \cong H_n(C')$, *for* $n = 0, \ldots, q$, *then* $\lambda: C \equiv_q C'$.

This follows from the arguments used in a forthcoming paper.[13] It is proved by constructing an "abstract" mapping cylinder of λ and transcribing into algebraic terms the proof of the analogous theorem on CW-complexes.

* This note arose from consultations during the tenure of a John Simon Guggenheim Memorial Fellowship by MacLane.

[2] Whitehead, J. H. C., "Combinatorial Homotopy I and II," *Bull. A.M.S.*, **55**, 214–245 and 453–496 (1949). We refer to these papers as CH I and CH II, respectively.

[3] By a complex we shall mean a connected CW complex, as defined in §5 of CH I. We do not restrict ourselves to finite complexes. A fixed 0-cell $e^0 \epsilon K^0$ will be the base point for all the homotopy groups in K.

[4] MacLane, S., "Cohomology Theory in Abstract Groups III," *Ann. Math.*, **50**, 736–761 (1949), referred to as CT III.

[5] An (unpublished) result like Theorem 1 for the homotopy type was obtained prior to these results by J. A. Zilber.

[6] CT III uses in place of equation (2.4) the stronger hypothesis that λB contains the center of A, but all the relevant developments there apply under the weaker assumption (2.4).

[7] Eilenberg, S., and MacLane, S., "Cohomology Theory in Abstract Groups II," *Ann. Math.*, **48**, 326–341 (1947).

[8] Eilenberg, S., and MacLane, S., "Determination of the Second Homology . . . by Means of Homotopy Invariants," these PROCEEDINGS, **32**, 277–280 (1946).

[9] Blakers, A. L., "Some Relations Between Homology and Homotopy Groups," *Ann. Math.*, **49**, 428–461 (1948), §12.

[10] The hypothesis of Theorem C, requiring that $\nu^{-1}(1)$ not be cyclic, can be readily realized by suitable choice of the free group X, but this hypothesis is not needed here (cf. [6]).

[11] Eilenberg, S., and MacLane, S., "Homology of Spaces with Operators II," *Trans. A.M.S.*, **65**, 49–99 (1949); referred to as HSO II.

[12] $C(\tilde{K})$ here is the $C(K)$ of CH II. Note that $\tilde{K}$ exists and is a *CW* complex by (N) of p. 231 of CH I and that $p^{-1}K^n = \tilde{K}^n$, where p is the projection $p:\tilde{K} \rightarrow K$.

[13] Whitehead, J. H. C., "Simple Homotopy Types." If $W \neq 1$, Theorem 5 follows from (17:3) on p. 155 of S. Lefschetz, *Algebraic Topology*, (New York, 1942) and arguments in §6 of J. H. C. Whitehead, "On Simply Connected 4-Dimensional Polyhedra" (*Comm. Math. Helv.*, **22**, 48–92 (1949)). However this proof cannot be generalized to the case $W \neq 1$.

ACYCLIC MODELS.*

By SAMUEL EILENBERG and SAUNDERS MACLANE.

1. Introduction. There are a number of situations in algebraic topology where one establishes the existence of chain transformations and chain homotopies, dimension by dimension, using the fact that certain homology groups of a local character are zero. Most of the applications can be derived from well known theorems dealing with acyclic carriers. Other investigations (e. g. complexes with operators [2] and homology theory of multiplicative systems [3]) lead to similar proofs in situations no longer covered by theorems on acyclic carriers. The present paper formulates a general theorem, which seems to subsume all the situations of this type hitherto encountered. The theorem is formulated in the language of categories and functors [1].

As applications, we give proofs of the theorems of this type encountered in [2] and [3]. However, the most important application is a new theorem, establishing, by this method, the equivalence of the singular homology theories based respectively on simplexes and on cubes. This result was prompted by recent work of J. P. Serre and H. Cartan.

Another application is included in the paper of Eilenberg-Zilber [4] immediately following.

2. Main definitions and results. Let $\mathcal{A} = (A, \alpha)$ be a category with objects A and maps α. We shall assume given a set $\mathcal{M}$ of objects in $\mathcal{A}$ (called *model objects*).

Let T be any covariant functor on the category $\mathcal{A}$ with values in the category $\mathcal{G}$ of abelian groups. We define a new functor $\tilde{T}$ on $\mathcal{A}$ to $\mathcal{G}$ as follows. For each object $A \varepsilon \mathcal{A}$, the group $T(A)$ is the free abelian group generated by the symbols (ϕ, m) where $\phi: M \to A$ is a map (in $\mathcal{A}$), $M \varepsilon \mathcal{M}$ and $m \varepsilon T(M)$. If $\alpha: A \to B$ in $\mathcal{A}$ then $\tilde{T}(\alpha)$ is defined by $\tilde{T}(\alpha)(\phi, m) = (\alpha\phi, m)$. In addition we define a natural transformation $\Phi: \tilde{T} \to T$ by setting $\Phi(A)(\phi, m) = T(\phi)m$.

Definition. The functor T is said to be *representable* if there is a natural transformation $\Psi: T \to \tilde{T}$ such that the composition $\Phi\Psi: T \to T$ is the identity. Ψ is called a *representation* of Φ.

* Received May 26, 1952.

Reprinted from the Amer. J. of Math. 75 (1953), 189-199, by permission of The Johns Hopkins University Press.

Let $\partial\mathfrak{G}$ denote the category of chain complexes and chain transformations, and let K be a covariant functor on $\mathfrak{A}$ to $\partial\mathfrak{G}$. For each object $A \in \mathfrak{A}$, the functor K then determines a complex $K(A)$ composed of chain groups $K_q(A)$ and boundary homomorphisms $\partial_q : K_q(A) \to K_{q-1}(A)$, with $\partial_{q-1}\partial_q = 0$. The groups $K_q(A)$ yield a functor K_q on $\mathfrak{A}$ to $\mathfrak{G}$ and the boundary operators yield natural transformations $\partial_q : K_q \to K_{q-1}$ with $\partial_{q-1}\partial_q = 0$.

Let K and L be two (covariant) functors on $\mathfrak{A}$ to $\partial\mathfrak{G}$. A *map* $f: K \to L$ is a family of natural transformations $f_q : K_q \to L_q$ such that $\partial_q f_q = f_{q-1}\partial_q$. If f_q is defined and satisfies this equation only for $q \leqq n$, we say that f is a map $K \to L$ in dimensions $\leqq n$.

Let $f, g: K \to L$ be two maps. A *homotopy* $D: f \simeq g$ is a sequence of natural transformations $D_q : K_q \to L_{q+1}$ such that

$$\partial_{q+1}D_q + D_{q-1}\partial_q = g_q - f_q. \tag{2.1}$$

If the maps D_q are defined and satisfy (2.1) only for $q \leqq n$, we say that D is a homotopy in dimensions $\leqq n$.

THEOREM Ia. *Let K and L be covariant functors on $\mathfrak{A}$ with values in $\partial\mathfrak{G}$, and let $f: K \to L$ be a map in dimensions $< q$. If K_q is representable and if $H_{q-1}(L(M)) = 0$ for each model $M \in \mathfrak{M}$, then f admits an extension to a map $K \to L$ in dimensions $\leqq q$.*

Proof. The objective is to define a natural transformation $f_q: K_q \to L_q$ such that $\partial_q f_q = f_{q-1}\partial_q$.

For each $m \in K_q(M)$, $M \in \mathfrak{M}$, we have $f_{q-1}\partial_q m \in L_{q-1}(M)$. Since $\partial_{q-1}f_{q-1}\partial_q = f_{q-1}\partial_{q-1}\partial_q = 0$ it follows that $f_{q-1}\partial_q m$ is a $(q-1)$-dimensional cycle in $L(M)$. Since $H_{q-1}(L(M)) = 0$ we may choose a chain $d(m) \in L_q(M)$ with $\partial_q d(m) = f_{q-1}\partial_q m$.

We now consider the functor K_q associated with K_q and define a natural transformation $\Lambda: K_q \to L_q$ by setting $\Lambda(A)(\phi, m) = L_q(\phi)d(m)$. We have

$$\begin{aligned}\partial_q\Lambda(A)(\phi, m) &= \partial_q L_q(\phi)d(m) = L_{q-1}(\phi)\partial_q d(m)\\ &= L_{q-1}(\phi)f_{q-1}\partial_q m = f_{q-1}\partial_q K_q(\phi)m = f_{q-1}\partial_q\Phi(A)(\phi, m)\end{aligned}$$

where $\Phi: \tilde{K}_q \to K_q$ is defined as above. Thus $\partial_q\Lambda = f_{q-1}\partial_q\Phi$. Now, let $\Psi: K_q \to \tilde{K}_q$ be a representation of K_q, and let $f_q = \Lambda\Psi: K_q \to L_q$. Then $\partial_q f_q = \partial_q\Lambda\Psi = f_{q-1}\partial_q\Phi\Psi = f_{q-1}\partial_q$, as desired.

THEOREM Ib. *Let K and L be covariant functors on $\mathfrak{A}$ with values in $\partial\mathfrak{G}$, let $f, g: K \to L$ be maps, and let $D: f \simeq g$ be a homotopy in dimensions*

$< q$. *If K_q is representable and if $H_q(L(M)) = 0$ for each model $M \varepsilon \mathfrak{M}$, then D admits an extension to a homotopy $f \simeq g$ in dimensions $\leqq q$.*

Proof. The objective is to define a natural transformation $D_q : K_q \to L_{q+1}$ with $\partial_{q+1} D_q + D_{q-1}\partial_q = g_q - f_q$.

For each $m \varepsilon K_q(M)$, $M \varepsilon \mathfrak{M}$, we have $(g_q - f_q - D_{q-1}\partial_q)m \varepsilon L_q(M)$ Since

$$\begin{aligned}\partial(g_q - f_q - D_{q-1}\partial_q) &= \partial_q g_q - \partial_q f_q - (\partial_q D_{q-1})\partial_q \\ &= g_{q-1}\partial_q - f_{q-1}\partial_q - (g_{q-1} - f_{q-1} - D_{q-2}\partial_{q-1})\partial_q = 0,\end{aligned}$$

it follows that $(g_q - f_q - D_{q-1}\partial_q)m$ is a q-cycle in $L(M)$. Since $H_q(L(M)) = 0$ we may choose a chain $e(m) \varepsilon L_{q+1}(M)$ with $\partial_{q+1} e(m) = (g_q - f_q - D_{q-1}\partial_q)m$. We now define $\Gamma : \tilde{K}_q \to L_{q+1}$ by setting $\Gamma(A)(\phi, m) = L_{q+1}(\phi) e(m)$. We have

$$\begin{aligned}\partial_{q+1}\Gamma(A)(\phi, m) &= \partial_{q+1} L_{q+1}(\phi) e(m) = L_q(\phi)\partial_{q+1} e(m) \\ &= L_q(\phi)(g_q - f_q - D_{q-1}\partial_q)m = (g_q - f_q - D_{q-1}\partial_q) K_q(\phi) m \\ &= (g_q - f_q - D_{q-1}\partial_q)\Phi(A)(\phi, m).\end{aligned}$$

Thus $\partial_{q+1}\Gamma = (g_q - f_q - D_{q-1}\partial_q)\Phi$. Now, let $\Psi : K_q \to \tilde{K}_q$ be a representation of K_q, and let $D_q = \Gamma\Psi : K_q \to L_{q+1}$. Then

$$\partial_{q+1} D_q = (g_q - f_q - D_{q-1}\partial_q)\Phi\Psi = g_q - f_q - D_{q-1}\partial_q,$$

as desired.

Theorems Ia and Ib imply

THEOREM II. *Let K and L be covariant functors on $\mathfrak{A}$ with values in $\partial\mathfrak{G}$ and let $f : K \to L$ be a map in dimensions $< q$. If K_n is representable for all $n \geqq q$ and if $H_n(L(M)) = 0$ for all $n \geqq q - 1$ and all $M \varepsilon \mathfrak{M}$, then f admits an extension $f' : K \to L$ (defined in all dimensions). If $f', f'' : K \to L$ are two such extension of f then there is a homotopy $D : f' \simeq f''$ with $D_n = 0$ for all $n < q$.*

3. Groups with operators. Let W be an associative system with a unit element. Each element $w \varepsilon W$ gives rise to a transformation $w : W \to W$ defined by $w(x) = wx$. This gives rise to a category $\mathfrak{A}$ containing one object W and maps $w \varepsilon W$. The object W will be regarded as a model; thus in this case the set $\mathfrak{M}$ contains all the objects of $\mathfrak{A}$.

What is a covariant functor T on $\mathfrak{A}$ with values in $\mathfrak{G}$? It consists of an abelian group $G = T(W)$ and of endomorphisms $T(w) : G \to G$ such that

$T(w_2w_1) = T(w_2)T(w_1)$ and that $T(1) =$ identity. Thus G is an abelian group with W as left operators.

If we now inspect the definition of $\tilde{T}$ we find that $\tilde{G}$ is the free abelian group generated by pairs (w, g), $w \varepsilon W$, $g \varepsilon G$ with operators defined by $w'(w, g) = (w'w, g)$. Thus $\tilde{G}$ is the W-free group with a W-base formed by the elements $g = (1, g)$, $g \varepsilon G$. The map $\Phi : \tilde{G} \to G$ maps the generator g of $\tilde{G}$ into the element g of G. Clearly Φ maps $\tilde{G}$ onto G.

In order that the functor corresponding to G be representable, there must exist a W-map $\Psi : G \to \tilde{G}$ such that $\Phi\Psi =$ identity. Such a Ψ always exists if G is W-free.

These remarks may be more conveniently restated using the algebra Λ of W, i. e. the additive free abelian group generated by the elements $w \varepsilon W$ with a multiplication defined by that of W. Then G becomes a left Λ-module. Natural transformations of functors translate into Λ-homomorphisms. Further, $\tilde{G}$ is the free Λ-module generated by the elements $g \varepsilon G$. The existence of the Λ-homomorphism $\Psi : G \to \tilde{G}$, with $\Phi\Psi =$ identity, is equivalent with the property that G be a projective Λ-module (one of several equivalent definitions: projective = direct summand of a free module).

These remarks and the results of § 2 yield a new proof (in a somewhat more general form) of a basic result concerning complexes with operators [2, § 5].

4. Doubling of subcategories. For the purpose of subsequent applications of the results of § 2 it will be convenient to describe a certain abstract method for constructing categories.

Let $\mathfrak{A}$ be a category and $\mathfrak{B}$ a subcategory of $\mathfrak{A}$. We define a new category $\mathfrak{A}^*$, called the result of *doubling* the subcategory $\mathfrak{B}$, as follows.

Fro each object $B \varepsilon \mathfrak{B}$ we introduce a new object B^*, and for each map $\beta : B \to B'$ in $\mathfrak{B}$ we introduce a new map $\beta^* : B^* \to B'^*$ with $(\beta_2\beta_1)^* = \beta_2{}^*\beta_1{}^*$. These new objects and new maps constitute a category $\mathfrak{B}^*$ isomorphic to $\mathfrak{B}$. The objects in $\mathfrak{A}^*$ are the objects A of $\mathfrak{A}$ and the objects B^* of $\mathfrak{B}^*$. The maps of $\mathfrak{A}^*$ are the maps of $\mathfrak{A}$, the maps of $\mathfrak{B}^*$ (each with the given composition rules), plus new maps $\gamma^\# : B^* \to A$, one for each map $\gamma : B \to A$ in $\mathfrak{A}$ with $B \varepsilon \mathfrak{B}$. The composition rule for these new maps with either previous type are given as follows. If $B' \xrightarrow{\beta} B \xrightarrow{\gamma} A \xrightarrow{\alpha} A'$ with $\beta \varepsilon \mathfrak{B}, \gamma, \alpha \varepsilon \mathfrak{A}$, then $(\gamma\beta)^\# = \gamma^\#\beta^*$, $(\alpha\gamma)^\# = \alpha\gamma^\#$. The axioms for a category are readily verified.

These rules show that for each $\gamma : B \to A$ we have $\gamma^\# = \gamma i_B{}^\#$, where $i_B : B \to B$ is the identity map of B. The map $i_B{}^\# : B^* \to B$ is called the *inclusion* map for B. For $\beta : B' \to B$ we have $\beta^\# = i_B{}^\#\beta^*$. Hence any map in $\mathfrak{A}^*$ is uniquely representable as one of the maps α, β^* or $i_B{}^\#$ or their composites, subject to the rule $i_B{}^\# \beta^* = \beta i_{B'}{}^\#$ for $\beta : B' \to B$ in $\mathfrak{B}$.

In the applications, the set of objects B^* will usually be the set $\mathfrak{M}$ of models for the category $\mathfrak{A}^*$.

5. Homology theories for multiplicative systems. We shall show here how the results of § 2 can be applied to yield the main results of [3]. We shall use, without explanation, the notation and terminology introduced there.

Let F be a free multiplicative system with generators $g_1, \cdots, g_i, \cdots$ and $\mathfrak{M}(F)$ the category of multiplicative systems belonging to F [3, § 2]. Let Φ be an admissible set of endomorphisms of F. The system $\{F, \Phi\}$ is a subcategory of $\mathfrak{M}(F)$ and we denote by $\mathfrak{A}$ the result of doubling up the subcategory $\{F, \Phi\}$ of $\mathfrak{M}(F)$. The new object F^* is chosen as the (only) model.

Let K be a Φ-construction on $\mathfrak{M}(F)$, [3, § 5]. We shall regard K as a covariant functor on $\mathfrak{M}(F)$ with values in $\partial\mathfrak{G}$. We shall show how K can be extended to a functor K^* on $\mathfrak{A}$. We define $K^*(F^*)$ as the Φ-subcomplex $K(F, \Phi)$ of $K(F)$. The map $K^*(i_F{}^\#) : K^*(F^*) \to K^*(F)$ is defined as the inclusion map $K(F, \Phi) \to K(F)$. If $\phi \varepsilon \Phi$ then $K^*(\phi^*) : K^*(F^*) \to K^*(F^*)$ is defined as the endomorphism $K(F, \Phi) \to K(F, \Phi)$ defined by $K(\phi)$. If L is another (augmented) Φ-construction and $f : K \to L$ is an (augmented) map as defined in [3, § 5], then f admits a unique extension $f^* : K^* \to L^*$. Conversely, for each map $f^* : K^* \to L^*$ the restriction $f : K \to L$ is a map in the sense of [3, § 5]. The same applies to homotopies.

We shall now show that each component $K_q{}^*$ of K^* is representable for $q > 0$. (For $q = 0$, $K_0{}^*$ is the augmentation functor, which is not representable). For $M \varepsilon \mathfrak{M}(F)$, each q-dimensional cell in $K(M)$ has a form $[x_1, \cdots, x_r]_t$, and is of type t with entries $x_1, \cdots, x_r \varepsilon M$. Let $\alpha : F \to M$ be the map defined by $\alpha(g_i) = x_i$ for $i = 1, \cdots, r$ and $\alpha(g_i) = 1$ for $i > r$. Then $[g_1, \cdots, g_r]_t$ is a q-cell of $K^*(F^*)$. The mapping

$$[x_1, \cdots, x_r]_t \to (\alpha^\#, [g_1, \cdots, g_r]_t)$$

then defines a representation of K_q. With these preliminaries it is clear that Theorem 6.1 of [3] is a consequence of the results of § 2.

The same remarks apply to the considerations of [2, § 15]. Instead of the single free system F we consider the sequence $\{G_R{}^i\}$ where $G_R{}^i$ is generated by $g_1, \cdots, g_i$ $(i = 0, 1, \cdots)$. In this case, instead of a single model we have a sequence of models.

13

6. Simplicial singular homology. Let X be a topological space. A singular n-simplex T of X is a function $T(\lambda_0, \cdots, \lambda_n) \varepsilon X$ defined for $0 \leqq \lambda_i$, $\lambda_0 + \cdots + \lambda_n = 1$ and continuous in the topology induced by the cartesian product of the variables. The faces $F_i T$ $(i = 0, \cdots, n)$ are $(n-1)$- simplexes defined as

$$(F_i T)(\lambda_0, \cdots, \lambda_{n-1}) = T(\lambda_0, \cdots, \lambda_{i-1}, 0, \lambda_i, \cdots, \lambda_n).$$

Then

$$F_i F_j = F_{j-1} F_i, \qquad i < j. \tag{6.1}$$

We define $S_n(X)$ as the free group generated by all singular n-simplexes in X. Then

$$\partial T = \sum_{i=0}^{n} (-1)^i F_i T$$

is a homomorphism $\partial : S_n(X) \to S_{n-1}(X)$ with $\partial\partial = 0$. This yields a chain complex $S(X)$. It will be convenient to "augment" $S(X)$ by defining $S_{-1}(X)$ to be the group of integers with $\partial T = 1$ for each 0-simplex. Henceforth $S(X)$ will denote the augmented complex.

If $f : X \to Y$ is a continuous map and T is a singular n-simplex in X, then the composition fT is a singular n-simplex in Y. This yields maps $S_n(f) : S_n(X) \to S_n(Y)$ and $S(f) : S(X) \to S(Y)$ and therefore functors S_n and S defined on the category $\mathfrak{a}$ of topological spaces (with continuous maps) and with values in the categories $\mathfrak{G}$ and $\partial\mathfrak{G}$ respectively.

For each n-simplex T $(n \geqq 0)$ in X and each $i = 0, \cdots, n$ we define the $(n+1)$-simplex $D_i T$ in X as follows:

$$(D_i T)(\lambda_0, \cdots, \lambda_{n+1}) = T(\lambda_0, \cdots, \lambda_{i-1}, \lambda_i + \lambda_{i+1}, \lambda_{i+2}, \cdots, \lambda_n).$$

We have the identities

$$D_i D_j = D_{j+1} D_i, \qquad i \leqq j, \tag{6.2}$$

$$F_i D_j = D_{j-1} F_i,\ i < j;\ F_j D_j = F_{j+1} D_j = I;\ F_i D_j = D_j F_{i-1},\ i > j+1. \tag{6.3}$$

These imply that the simplexes $D_n T$ (where $\dim T = n$) form a subcomplex $\bar{D}S(X)$ of $S(X)$. Similarly the simplexes $D_i T$ $(i = 0, 1, \cdots, \dim T)$ form a subcomplex $DS(X)$. We introduce the quotient functors $\bar{S} = S/\bar{D}S$, $S^N = S/DS$. We call $S(X)$ the *singular complex* of X, $S^N(X)$ the singular complex of X *normalized*, $\bar{S}(X)$ the singular complex of X *normalized at the top*. We observe that in dimensions < 1, the three singular complexes of X coincide.

THEOREM III. *Let* $f: S \to S^N$ *be the natural factorization homomorphism. There exists then a map* $g: S^N \to S$ *and homotopies* $H: gf \cong \text{identity}$ $G: fg \cong \text{identity}$. *The same conclusion applies to the natural maps* $f_1: S \to \bar{S}$ *and* $f_2: \bar{S} \to S^N$.

Proof. In the category $\mathcal{A}$ of topological spaces (on which the functors $S, \bar{S}, S^N$ are defined) we consider the set $\mathfrak{M}$ of *models* consisting of all spaces contractible to a point. The theorem now follows from Theorem II and the following two lemmas:

LEMMA 6.1. *For any model* M *we have*

$$H_n(S(M)) = H_n(\bar{S}(M)) = H_n(S^N(M)) = 0.$$

LEMMA 6.2. *The functors* S_n, $\bar{S}_n$, *and* $S_n{}^N$ *are representable for all* n.

Proof of 6.1. Let I denote the unit interval $0 \leqq \lambda \leqq 1$. Since any model M is contractible, there is a homotopy $H: I \times M \to M$ such that $H(0, x) = x$, $H(1, x) = p$ for all $x \in M$, where p is a fixed point of M.

For each n-simplex T ($n \geqq 0$) in M we define the $(n+1)$-simplex hT in M by setting

$$hT(\lambda_0, \cdots, \lambda_{n+1}) = H(\lambda_0, T(\kappa\lambda_1, \cdots, \kappa\lambda_{n+1})), \quad \kappa = 1/(1-\lambda_0), \quad \lambda_0 \neq 1;$$

$$hT(1, 0, \cdots, 0) = p.$$

For $n = -1$ we define $h(1)$ to be the 0-simplex of M located at p. Since $hD_iT = D_{i+1}hT$ we may regard h as a homomorphism

$$S_n(M) \to S_{n+1}(M), \quad \bar{S}_n(M) \to \bar{S}_{n+1}(M), \quad S_n{}^N(M) \to S_{n+1}{}^N(M).$$

For $n > 0$ we have $F_0h = I$, $F_ih = hF_{i-1}$ if $i > 0$. For $n = 0$ we have $F_0h = I$, $F_1h = h(1)$. These relations imply $\partial h + h\partial = I$, thus yielding the conclusion of 6.1.

Proof of 6.2. Let Δ^n be the simplex consisting of all points $(\lambda_0, \cdots, \lambda_n)$, $\lambda_i \geqq 0$, $\Sigma\lambda_i = 1$, with the usual topology. Every singular n-simplex T is then a map $T: \Delta^n \to X$. In particular, the identity map $e_n: \Delta^n \to \Delta^n$ is an n-simplex in $S(\Delta^n)$. The correspondence $T \to (T, e_n)$ yields then a representation of the functor S_n for $n \geqq 0$. For $n = -1$ the proof is trivial.

Next we consider the natural factorization maps $\xi_1: S_n \to \bar{S}_n$, $\xi_2: S_n \to S_n{}^N$. The expressions $\eta_1 T = (I - D_{n-1}F_n)T$, $\eta_2 T = (I - D_0F_1)(I - D_1F_2) \cdots (I - D_{n-1}F_n)T$, satisfy $\eta_1 T = 0$ for $T \in \bar{D}S_{n-1}$ and $\eta_2 T = 0$ for $T \in DS_{n-1}$.

Thus η_1 and η_2 yield maps $\eta_1 : \bar{S}_n \rightarrow S_n$, $\eta_2 : S_n{}^N \rightarrow S_n$. Clearly $\xi_1\eta_1 =$ identity, $\xi_2\eta_2 =$ identity. The representability of $\bar{S}_n$ and $S_n{}^N$ is now a consequence of the following lemma.

LEMMA 6.3. *Let T and T_1 be functors with values in $\mathfrak{G}$ and $\xi : T \rightarrow T_1$, $\eta : T_1 \rightarrow T$ be natural transformations such that $\xi\eta =$ identity. If T is representable then so is T_1.*

Proof. We have the commutative diagram

$$\begin{array}{ccccc} \tilde{T}_1 & \xrightarrow{\tilde{\eta}} & T & \xrightarrow{\tilde{\xi}} & \tilde{T}_1 \\ \downarrow \Phi_1 & & \downarrow \Phi & & \downarrow \Phi_1 \\ T_1 & \xrightarrow[\eta]{} & T & \xrightarrow[\xi]{} & T_1 \end{array}$$

where $\tilde{\eta}(\phi, m) = (\phi, \eta m)$, $\tilde{\xi}(\phi, m) = (\phi, \xi m)$. Let $\Psi : T \rightarrow \tilde{T}$ be a representation of T and define $\Psi_1 : T_1 \rightarrow \tilde{T}_1$ as $\Psi_1 = \tilde{\xi}\Psi\eta$. Then

$$\Phi_1\Psi_1 = \Phi_1\tilde{\xi}\Psi\eta = \xi\Phi\Psi\eta = \xi\eta = \text{identity},$$

as desired.

Remark. Theorem III is known and has been proved [4] for the more general class of complete semi-simplicial complexes. In this more general case the proof still can be carried out using the method of acyclic models.

7. Cubical singular homology. A singular n-cube R in X is a function $R(\mu_1, \cdots, \mu_n) \in X$ defined for $0 \leqq \mu_i \leqq 1$ and continuous in the topology of the cartesian product of the variables. If $n = 0$, then R is interpreted as a single point of X. The *front* and *aft faces* A_iR and B_iR $(i = 1, \cdots, n)$ are defined as $(n-1)$-cubes

$$(A_iR)(\mu_1, \cdots, \mu_{n-1}) = R(\mu_1, \cdots, \mu_{i-1}, 0, \mu_i, \cdots, \mu_{n-1}),$$

$$(B_iR)(\mu_1, \cdots, \mu_{n-1}) = R(\mu_1, \cdots, \mu_{i-1}, 1, \mu_i, \cdots, \mu_{n-1}).$$

Then

$$\begin{array}{lll} & A_iA_j = A_{j-1}A_i, \qquad B_iB_j = B_{j-1}B_i, & \\ (7.1) & & i < j \\ & A_iB_j = B_{j-1}A_i, \qquad B_iA_j = A_{j-1}B_i. & \end{array}$$

As before we regard the singular n-cubes in X as generators of a free group $Q_n(X)$ and introduce the operator

$$\partial R = \sum_{i=1}^{n} (-1)^i (A_i R - B_i R).$$

Then (7.1) implies $\partial\partial = 0$. Thus the groups $Q_n(X)$ and the operator ∂ define a chain complex $Q(X)$. As before we augment $Q(X)$ by setting $Q_{-1}(X) =$ integers, $\partial R = 1$ if $\dim R = 0$. Also as before we convert $Q(X)$ into a functor Q defined on $\mathfrak{A}$ with values $\partial\mathfrak{G}$.

For each n-cube R $(n \geqq 0)$ in X and each $i = 1, \cdots, n+1$ we define the $(n+1)$-cube $E_i R$ of X as follows

$$(E_i R)(\mu_1, \cdots, \mu_{n+1}) = R(\mu_1, \cdots, \mu_{i-1}, \mu_{i+1}, \cdots, \mu_{n+1}).$$

We have the identities

$$E_i E_j = E_{j+1} E_i, \qquad i \leqq j, \tag{7.2}$$

$$\begin{cases} A_i E_j = E_{j-1} A_i, & B_i E_j = E_{j-1} B_i, \qquad i < j; \\ A_j E_j = B_j E_j = I; & \\ A_i E_j = E_j A_{i-1}, & B_i E_j = E_j B_{i-1}, \qquad i > j. \end{cases} \tag{7.3}$$

These imply that the cubes $E_{n+1} R$ (where $\dim R = n$) form a subcomplex $\bar{E}Q(X)$ of $Q(X)$. Similarly the cubes $E_i R$ $(i = 1, \cdots, 1 + \dim R)$ form a subcomplex $EQ(X)$. We introduce the quotient functors $\bar{Q} = Q/\bar{E}Q$, $Q^N = Q/EQ$.

THEOREM IV. *Let $f: \bar{Q} \to Q^N$ be the natural factorization homomorphism. There exist then a map $g: Q^N \to \bar{Q}$ and homotopies $H: gf \simeq$ identity, $G: fg \simeq$ identity.*

The same conclusion does not apply to the maps $Q \to \bar{Q}$ and $Q \to Q^N$.

As in the case of Theorem III, Theorem IV follows from Theorem II and the following two propositions:

PROPOSITION 7.1. *For any model M we have*

$$H_n(\bar{Q}(M)) = H_n(Q^N(M)) = 0.$$

This is not the case for $H_n(Q(M))$.

PROPOSITION 7.2. *The functors Q_n, $\bar{Q}_n$, and $Q_n{}^N$ are representable for all n.*

Proof of 7.1. Using the notation of the proof of 6.1 we define, for each n-cube R of M, an $(n+1)$-cube hR as

$$(hR)(\mu_1, \cdots, \mu_{n+1}) = H(1-\mu_1, R(\mu_2, \cdots, \mu_{n+1})), \qquad n \geqq 0$$

For $n = -1$ we define $h(1)$ to be the 0-cube of M located at p. Since $hE_iR = E_{i+1}hR$ we may regard h as an operator $\bar{Q}_n(M) \to \bar{Q}_{n+1}(M)$, $Q_n{}^N(M) \to Q_{n+1}{}^N(M)$. For $n > 0$ we have

$$A_1 h \,\varepsilon\, \bar{E}Q_{n-1}(M), \qquad B_1 h = I$$

$$A_i h = hA_{i-1}, \qquad B_i h = hB_{i-1} \qquad \text{for } 0 < i.$$

For $n = 0$ we have $A_1 h = h(1)$, $B_1 h = I$. These relations imply $\partial h + h\partial \equiv I$ mod $\bar{E}Q(M)$, thus yielding the conclusion of 7.1. The complex $Q(M)$ without any normalization is *not* acyclic, as can be seen in the case in which M consists of a single point.

Proof of 7.2. The proof of the representability of the functor Q_n is the same as that for S_n with the simplex Δ^n replaced by the cube $\square^n$ given by $(\mu_1, \cdots, \mu_n)$, $0 \leqq \mu_i \leqq 1$.

The representability of $\bar{Q}_n$ and $Q_n{}^N$ is proved in the same way as for $\bar{S}_n$ and $S_n{}^N$ with the expressions η_1 and η_2 replaced by

$$\eta_1 R = (I - E_n A_n)R, \; \eta_2 R = (I - E_1 A_1)(I - E_2 A_2) \cdots (I - E_n A_n)R.$$

Remark. In the simplicial theory the normalizations were essentially a luxury since the functor S already gives the "correct" homology theory. In the cubical theory some normalization is a necessity since the functor Q (without normalization) does not give the "correct" homology groups even in the case of a space consisting of a single point.

8. Comparison of singular and cubical theories. A singular 0-simplex and a singular 0-cube each represent a point of X; thus we are led to identify the functors S_0 and Q_0. Further we identify S_1 with Q_1 by identifying each 1-simplex T with the 1-cube R defined by $R(\mu_1) = T(1-\mu_1, \mu_1)$. These identifications are compatible with the boundary operators $S_1 \to S_0$, $Q_1 \to Q_0$. These identifications induce identifications $\bar{S}_i = \bar{Q}_i$ and $S_i{}^N = Q_i{}^N$ for $i < 2$.

Lemmas 6.1, 6.2, 7.1, 7.2 together with Theorem II yield

THEOREM V. *There exist maps* $f: \bar{S} \to \bar{Q}$, $g: \bar{Q} \to \bar{S}$ *and homotopies* $H: gf \simeq$ *identity*, $G: fg \simeq$ *identity such that* f *and* g *are the identity in dimensions* < 2 *while* H *and* G *are zero in dimensions* < 2. *The same applies to the pair of functors* S^N, Q^N.

We conclude by giving an explicit form (due to H. Cartan) of a map $f:\bar{S}\to\bar{Q}$ which is the identity in dimensions < 2. We define for each n-simplex T

$$(fT)(\mu_1,\cdots,\mu_n)=T(\lambda_0,\cdots,\lambda_n),$$

where

$$\lambda_0=1-\mu_1,\qquad \lambda_1=\mu_1(1-\mu_2),\cdots,$$
$$\lambda_i=\mu_1\cdots\mu_i(1-\mu_{i+1}),\quad 0<i<n,\cdots,$$
$$\lambda_n=\mu_1\cdots\mu_n.$$

Clearly $\lambda_0+\cdots+\lambda_i=1-\mu_1\cdots\mu_{i+1}$, $i<n$. Thus $\Sigma\lambda_i=1$. Further, one easily verifies

$$fF_i=B_{i+1}f,\quad 0\leqq i<n;$$
$$fF_n=A_nf;\quad A_if=EA_iA_nf;\quad fD_n=E_{n+1}f.$$

These formulae imply that f maps $\bar{D}S$ into $\bar{E}Q$ and that $\partial f\equiv f\partial \bmod \bar{E}Q$. Thus f induces a map $f:\bar{S}\to\bar{Q}$ as desired. It should be further noted that $fT=fT'$ implies $T=T'$, so that f actually yields an isomorphic mapping of $\bar{S}$ into $\bar{Q}$.

COLUMBIA UNIVERSITY,
THE UNIVERSITY OF CHICAGO.

BIBLIOGRAPHY.

[1] Samuel Eilenberg and Saunder MacLane, "General theory of natural equivalences," *Transactions of the American Mathematical Society*, vol. 58 (1945), pp. 231-294.

[2] ———, "Homology of spaces with operators, II," *Transactions of the American Mathematical Society*, vol. 65 (1949), pp. 49-99.

[3] ———, "Homology theories for multiplicative systems," *Transactions of the American Mathematical Society*, vol. 71 (1951), pp. 294-330.

[4] Samuel Eilenberg and J. A. Zilber, "Semi-simplicial complexes and singular homology," *Annals of Mathematics*, vol. 51 (1950), pp. 499-513.

[5] ———, "On products of complexes," *American Journal of Mathematics*, vol. 75 (1953), pp. 200-204.

Homologie des anneaux et des modules [1]

par M. Saunders Mac Lane (Chicago) [2]

1. Introduction

Pour plusieurs structures algébriques (groupes, algèbres associatives, algèbres de Lie, groupes abéliens) il existe une théorie de l'homologie et de la cohomologie. Pour chacune de ces structures, cette théorie commence par une étude des extensions de la structure. En effet, si S est un système de la structure donnée, le deuxième (quelquefois, le premier) groupe $H^2(S ; G)$ de cohomologie du système S avec des « coefficients » G représente l'ensemble des classes des extensions, d'une nature à préciser, de G par S.

Par exemple, soit S un groupe multiplicatif, et G un groupe abélien additif. Une extension centrale de G par S est une suite exacte

$$0 \to G \to E \to S \to 0$$

des groupes, de sorte que l'image de G soit contenue dans le centre de E. Pour une telle extension E choisissons pour chaque $x \in S$ un élément $Ux \in E$ dont l'image dans S est x. On a donc $(Ux)(Uy) = f(x, y)U(xy)$ pour des éléments $f(x, y)$ dans l'image de G. La fonction f ainsi définie, appelée « système de facteurs », détermine l'extension E à une équivalence d'extensions près. La loi associative pour E correspond à la condition

$$f(y, z) + f(x, yz) = f(xy, z) + f(x, y) \qquad x, y, z \in S, \quad (1\text{-}1)$$

pour f ; un nouveau choix des représentants Ux donne un nouveau système f' de la forme

$$f'(x, y) = f(x, y) + h(y) - h(xy) + h(x), \qquad (1\text{-}2)$$

pour des éléments $h(x)$ dans G.

(1) This research was supported by the United States Air Force through the office of Scientific Research of the Air Research and Development Command.

(2) Je voudrais remercier M. G. Papy, qui a bien voulu m'indiquer quelques corrections de style à mon texte original.

Pour l'homologie du groupe S on a un complexe standard K(S). Pour la dimension 1, $K_1 = K_1(S)$ est le groupe abélien engendré par les éléments x de S, tandis que $K_2(S)$ est le groupe abélien libre engendré par les couples d'éléments x, y de S. Les fonctions $h(x) \varepsilon G$ et $f(x, y) \varepsilon G$ ci-dessus sont donc des cochaînes de dimensions 1, resp. 2 de K(S). La condition (1-1) signifie que f soit un cocycle, tandis que les cocycles f' et f de (1-2) sont cohomologues. On a donc une correspondance biunivoque entre le groupe $H^2(K(S); G)$ et l'ensemble de classes d'équivalence des extensions centrales de G par S.

Si, de plus, le groupe S opère sur le groupe abélien G, on a un résultat analogue bien connu pour les extensions E qui réalisent ces opérateurs.

Soit maintenant S un groupe abélien. Pour une extension E abélienne, le système f de facteurs doit satisfaire à la condition (1-1) d'associativité et à la condition $f(x, y) = f(y, x)$ de commutativité. Ces deux conditions se traduisent par le fait que f est un cocycle pour un certain complexe A(S) défini pour les groupes abéliens S (cf. [9] et [6, II]) ([3]).

Mais on peut remplacer les lois commutatives et associatives par une seule loi « commutassociative »

$$(x+y)+(z+t)=(x+z)+(y+t)\,, \tag{1-3}$$

jointe à la loi usuelle

$$x+0=x=0+x\,. \tag{1-4}$$

La condition correspondante pour le système f des facteurs signifie que f est un cocycle (cette fois de dimension 1) dans un certain complexe Q(S). Nous donnerons ci-dessous la définition de ce complexe « cubique » Q.

Posons le même problème des extensions pour un anneau Λ. Une extension (spéciale) par Λ est une suite exacte d'anneaux

$$0 \rightarrow \Lambda \xrightarrow{\alpha} E \xrightarrow{\beta} \Lambda \rightarrow 0 \tag{1-5}$$

telle que $\Lambda^2 = 0$ (ce qui veut dire que $ab = 0$ pour toute a, $b \varepsilon \Lambda$). Deux telles extensions E et E' du même anneau Λ par Λ sont équivalentes s'il existe un isomorphisme $\theta : E \rightarrow E'$ d'anneaux tel que $\theta\alpha = \alpha'$, $\beta'\theta = \beta$. Dans chacune de ces extensions l'idéal Λ a la structure d'un Λ-bimodule ; pour un choix arbitraire des éléments $Ux \varepsilon E$ tel que $\beta(Ux) = x \varepsilon \Lambda$, cette structure de bimodule est définie sans ambiguïté par

$$\alpha(xa) = (Ux)(\alpha a), \quad \alpha(ax) = (\alpha a)(Ux), \quad a \varepsilon \Lambda,\ x \varepsilon \Lambda\,.$$

([3]) Les chiffres entre crochets renvoient à la bibliographie à la fin de l'article.

Les équations pour l'addition et la multiplication des représentants Ux ont la forme

$$Ux + Uy = \alpha f(x, y) + U(x+y), \quad (Ux)(Uy) = \alpha g(x, y) + Uxy). \quad (1\text{-}6)$$

L'extension E est déterminée, à une équivalence près, par les deux fonctions $f(x, y) \varepsilon \Lambda$, $g(x, y) \varepsilon \Lambda$. Les quatre identités valables dans un anneau (associativité de la multiplication, distributivité à gauche et à droit, et commutassociativité de l'addition) donnent quatre conditions, faciles à préciser, pour ces fonctions f et g.

PROBLÈME I. — *Construire pour chaque anneau* Λ *un complexe* $R(\Lambda)$ *tel que : 1° les 2-cocycles de* $R(\Lambda)$ *à coefficients dans un* Λ*-bimodule* A *sont les fonctions* f, g *qui répondent aux quatre conditions indiquées plus haut ; 2° avec un nouveau choix des représentants* Ux *le cocycle correspondant à* (f, g) *se change par un cobord, et inversement.*

Le moment décisif de ce problème est la présence simultanée des systèmes de facteurs pour l'addition et la multiplication dans (1-6). C. J. Everett [10] a déjà considéré des extensions d'anneaux plus générales (sans la condition $A^2 = 0$), mais n'a pas considéré des complexes $R(\Lambda)$ correspondants.

Posons le problème analogue pour des modules A et C (à droite) sur un anneau Λ. Une *extension* de A par C est une suite exacte

$$0 \to A \xrightarrow{\alpha} E \xrightarrow{\beta} C \to 0 \quad (1\text{-}7)$$

de Λ-modules à droite. Deux telles extensions (α, E, β) et (α', E', β') sont équivalentes s'il existe un isomorphisme $\theta : E \to E'$ de modules tel que $\theta\alpha = \alpha'$ et $\beta'\theta = \beta$. Pour tout $c \varepsilon C$ choisissons $Uc \varepsilon E$ tel que $\beta(Uc) = c$ et $UO = 0$. Pour $x \varepsilon \Lambda$ et c, $c' \varepsilon C$, il existe des éléments $f(c, c') \varepsilon A$ et $g(c, x) \varepsilon A$ tels que

$$Uc + Uc' = \alpha f(c, c') + U(c + c'), \quad (Uc)x = \alpha g(c, x) + U(cx) \; ; \quad (8\text{-}1)$$

la première équation donne pour E la structure additive, la seconde donne les opérations de Λ à droite pour la structure de Λ-module. Pour les fonctions f et g, on a quatre identités simultanées qui expriment les quatre lois de la structure de module : commutassociativité de l'addition, distributivité à gauche et à droite et associativité des opérateurs de Λ. Ces identités ont été considérées par Bürger [2] ; Lyndon [11] et Whitehead [12] ont donné une généralisation de ces identités, différente de la nôtre.

PROBLÈME II. — *Construire pour chaque* Λ*-module* C *un complexe* $M_\Lambda(C)$ *de* Λ*-modules à droite tel que : 1° les 1-cocycles de ce complexe, avec coefficients dans le* Λ*-module* A, *sont les fonctions* f *et* g *répondant aux quatre identités mentionnées ci-dessus ; 2° avec un changement du choix des représentants* Uc *le cocycle correspondant à* (f, g) *se change par un cobord, et inversement.*

Cette conférence est consacrée à la solution naturelle de ces deux problèmes, en utilisant pour l'addition la construction cubique et pour la multiplication une généralisation convenable du « bar construction » d'Eilenberg-Mac Lane [6, II ; 7]. Notre nouvelle méthode de définition de cette construction, en opérant sur deux catégories abéliennes, donne une méthode universelle pour les constructions standard utilisée en algèbre homologique.

La partie essentielle du deuxième problème est un théorème d'homologie relative pour les groupes de Q, démontré par une méthode d'Eilenberg-Mac Lane pour la construction des homotopies par approximations successives. Il en résulte que le complexe $M_\Lambda(C)$ avec une augmentation convenable $\eta : M_\Lambda(C) \rightarrow C$ est acyclique. Comme $M_\Lambda(C)$ dans chaque dimension est un Λ-module libre, il suit que M fournit une résolution libre et acyclique de C, dans la terminologie de Cartan-Eilenberg [5]. Ces auteurs ont démontré que deux telles résolutions ont mêmes groupes d'homologie ou de cohomologie à coefficients dans un Λ-module donné, et ont défini ainsi les foncteurs Tor_n^Λ et Ext_Λ^n. Donc, pour un Λ-module à gauche G, on a

$$H_n(M_\Lambda(C) \otimes_\Lambda G) = \mathrm{Tor}_n^\Lambda(C, G) \qquad (1\text{-}9)$$

et pour un Λ-module à droite G

$$H^n(\mathrm{Hom}_\Lambda(M_\Lambda(C), G)) = \mathrm{Ext}_\Lambda^n(C, G)\,. \qquad (1\text{-}10)$$

Le complexe M fournit ainsi une définition explicite et fonctorielle des groupes Tor et Ext.

La relation (1-10) pour $n=1$ répond bien au probème II, parce que $\mathrm{Ext}_\Lambda^1(C, G)$ est l'ensemble des classes des extensions de G par C. De plus, nous donnons à la fin quelques relations entre le complexe M et les groupes projectifs d'homotopie des modules définis dans la conférence de M. Eckmann.

2. LA « BAR CONSTRUCTION »

Soit Λ un anneau, toujours muni d'un élément unité

$1 \neq 0$; soit K un Λ-module à droite, toujours unitaire (c'est-à-dire $k1 = k$ pour $k \varepsilon K$). K est *gradué* s'il est somme directe des sous-Λ-modules K_n (n entier) avec $K_n = 0$ pour $n < 0$. Le Λ-module K est *différentiel gradué* si l'on se donne, de plus, un homomorphisme $\partial : K \longrightarrow K$ de Λ-modules tel que

$$\partial^2 = 0 \text{ et } \partial K_n \subset K_{n-1} .$$

Les éléments $k \varepsilon K_n$ sont dits *homogènes* de degré n, et on écrit $dk = n$. L'anneau Λ est lui-même Λ-module (à droite ou à gauche), différentiel gradué, avec la graduation triviale ($\Lambda = \Lambda_0$) et la différentielle triviale ($\partial = 0$).

Pour Λ donné, soit $\mathcal{K} = \mathcal{K}(\Lambda)$ la catégorie dont les objets sont les Λ-modules K différentiels gradués, et dont les applications $f : K \longrightarrow K'$ sont les homomorphismes différentiels homogènes de Λ-modules. Un homomorphisme différentiel *homogène* de degré p (p entier) est un homomorphisme $f : K \longrightarrow K'$ de Λ-modules tel que

$$fK_n \subset K_{n+p} , \quad f\partial = (-1)^p \partial f . \tag{2-1}$$

Un *anneau différentiel gradué* G est un anneau (avec unité), différentiel gradué comme groupe abélien, tel que $G_n G_m \subset G_{n+m}$ et

$$\partial(uv) = (\partial u)v + (-1)^{du} u(\partial v), \quad u, v \varepsilon G, \tag{2-2}$$

(plus exactement, pour $v \varepsilon G$ et u homogène dans G).

Pour G donné, soit $\mathcal{L} = \mathcal{L}(G, \Lambda)$ la catégorie dont les objets sont tous les G-Λ-bimodules différentiels gradués L (la différentielle dans L sera notée ∂_r). Un tel L doit être G-module à gauche et Λ-module différentiel gradué à droite avec les propriétés

$$u(a\lambda) = (ua)\lambda \quad u \varepsilon G, a \varepsilon L, \quad \lambda \varepsilon \Lambda, \tag{2-3}$$

$$\partial_r(ua) = (\partial u)a + (-1)^{du} u \partial_r a , \quad u \varepsilon G, a \varepsilon L . \tag{2-4}$$

Les applications $f : L \longrightarrow L'$ de la catégorie $\mathcal{L}$ sont les homomorphismes de bimodules différentiels gradués, c'est-à-dire les homomorphismes différentiels homogènes de Λ-modules à droite, de degré p, tels que

$$f(ua) = (-1)^{pdu} u(fa), \quad u \varepsilon G, a \varepsilon L . \tag{2-5}$$

L'anneau G est lui-même un objet de de $\mathcal{L}$, et $\mathcal{L}$ est une sous-catégorie de $\mathcal{K}$.

On définit un foncteur covariant $T : \mathcal{K} \longrightarrow \mathcal{L}$ par

$$T(K) = G \otimes K , \quad T(f) = I_G \otimes f , \tag{2-6}$$

le produit tensoriel étant pris sur l'anneau Z des entiers, et I_G désignant l'application identique de G. Dans $G \otimes K$, la graduation, la différentielle ∂_r, les opérateurs de Λ à droite, et les opérateurs de G à gauche sont définis, pour u, $v \varepsilon$G, $h \varepsilon$K et $\lambda \varepsilon \Lambda$, par

$$d(u \otimes k) = du + dk + 1\,, \qquad \text{(notez bien le } +1)$$
$$\partial_r(u \otimes k) = \partial u \otimes k - (-1)^{du} u \otimes \partial_r k, \qquad \text{(notez bien le } -1)$$
$$(u \otimes k)\lambda = u \otimes k\lambda\,, \qquad v(u \otimes k) = vu \otimes k\,.$$

On vérifie que $G \otimes K$ a la structure d'un objet dans $\mathcal{L}$.

On définit un homomorphisme naturel $\tau : K \longrightarrow T(K)$ par

$$\tau(k) = 1 \otimes k, \qquad k \varepsilon K\,. \tag{2-7}$$

On vérifie que $\tau : K \longrightarrow G \otimes K$ est une application de la catégorie $\mathcal{K}$, homogène de degré $+1$; le choix du signe dans la formule pour ∂_r ci-dessus est essentiel pour cela.

Notons que T et τ donnent un plongement universel de tout objet K de $\mathcal{K}$ dans un objet T(K) de $\mathcal{L}$, dans le sens suivant.

Lemme 1. — *Pour chaque application* $f : K \longrightarrow L$, *de la catégorie* $\mathcal{K}$, *d'un objet* K *de* $\mathcal{K}$ *dans un objet* L *de* $\mathcal{L}$ *il existe une application* $f' : T(K) \longrightarrow L$ *unique dans la catégorie* $\mathcal{L}$ *telle que* $f'\tau = f$.

La démonstration est facile ; si f est homogène de degré $p+1$ on pose

$$f'(u \otimes k) = (-1)^{p\,du} u f(k)\ ;$$

f' est alors homogène de degré p.

L'anneau Λ, avec la graduation et la différentielle triviales est un anneau différentiel gradué. Soit $\eta : G \longrightarrow \Lambda$ une homomorphisme d'anneaux qui applique l'élément unité de G sur l'unité de Λ, qui est homogène de degré zéro, et qui commute avec la différentielle. Donc Λ a la structure d'un objet de $\mathcal{L}$, en posant $u\lambda = (\eta u)\lambda$. Inversement si Λ a la structure d'un objet de $\mathcal{L}$, la définition $\eta u = u1$ détermine un homomorphisme $\eta : G \longrightarrow \Lambda$ d'anneaux différentiels gradués.

Pour $\eta : G \longrightarrow \Lambda$ donné, la bar construction $B = B_\Lambda(G, \eta)$ est la G-Λ-bimodule défini par la récurrence

$$B_0 = G \otimes \Lambda\,, \qquad B_{n+1} = T(B_n)\,, \qquad B = \sum_{n=0}^{\infty} B_n\,.$$

Chaque B_n, comme objet de la catégorie $\mathcal{L}$, a une différentielle ∂_r. De (2-7) on tire des applications $\tau : B_n \longrightarrow B_{n+1}$ homogènes de degré $+1$ dans la catégorie $\mathcal{L}$, et donc telles que

$$\partial_r \tau = -\tau \partial_r\,.$$

On veut que B soit acyclique avec τ comme homotopie. Grâce à la propriété universelle de τ, on peut définir des homomorphismes $\partial_s : B_{n+1} \to B_n$ par récurrence, en posant

$$\partial_s(u \otimes \lambda) = 0$$

et

$$\partial_s \tau(u \otimes \lambda) = u \otimes \lambda - 1 \otimes (\eta u)\lambda \quad u \varepsilon G,\ \lambda \varepsilon \Lambda\ ,$$

$$\partial_s \tau b_n = b_n - \tau \partial_s b_n\ , \quad b_n \varepsilon B_n,\ n > 0\ .$$

On en tire $\partial_s^2 = 0$. Les ∂_s sont des applications de degré -1 de la catégorie $\mathcal{L}$, donc $\partial_s \partial_r = -\partial_r \partial_s$ et

$$\partial_s(ub) = (-1)^{du} u \partial_s b\ , \quad u \varepsilon G,\ b \varepsilon B\ .$$

En posant

$$\partial = \partial_r + \partial_s\ ,$$

on a donc $\partial^2 = 0$, ∂ homogène de degré -1, et

$$\begin{aligned} \partial \tau b_n &= b_n - \tau \partial b_n\ , \quad b_n \varepsilon B_n,\ n > 0\ , \\ \partial \tau b_0 &= b_0 - 1 \otimes \eta b_0\ , \quad b_0 \varepsilon B_0\ , \end{aligned} \tag{2-8}$$

où l'application $\eta = \eta_B : B \to \Lambda$ est définie par

$$\eta_B(u \otimes \lambda) = (\eta u)\lambda\ , \quad \eta_B b_n = 0\ ; \quad u \varepsilon G,\ \lambda \varepsilon \Lambda,\ b_n \varepsilon B_n,\ n > 0\ .$$

On a démontré :

Proposition 1. — *Soit Λ un anneau, soit* G *un anneau différentiel gradué, et soit $\eta : G \to \Lambda$ un homomorphisme d'anneaux différentiels gradués. Donc $B = B_\Lambda(G, \eta)$ est un* G-Λ-*bimodule différentiel gradué avec la différentielle ∂_r et aussi avec la différentielle $\partial = \partial_r + \partial_s$. Pour la différentielle ∂, $\eta_B : B \to \Lambda$ est un homomorphisme de bimodules différentiels gradués, homogène de degré zéro, qui induit un isomorphisme pour l'homologie.*

En effet, η_B est une équivalence de chaînes. L'application dans le sens inverse $\zeta : \Lambda \to B$ est $\zeta(\lambda) = 1 \otimes \lambda$, on a $\eta\zeta = I$, et τ donne de l'homotopie de $\zeta\eta$ à l'identité. Autrement dit, B muni de l'augmentation η_B est acyclique, et représente une résolution libre et acyclique de Λ par des G-Λ-bimodules.

Pour des formules explicites, on a

$$B_n(G) = G \otimes \ldots \otimes G \otimes \Lambda$$

avec $n+1$ facteurs G. Les éléments

$$b = t \otimes u_1 \otimes \ldots \otimes u_n \otimes \lambda \text{ de } B_n$$

seront notés

$$b = t\,[u_1 \mid \ldots \mid u_n]\,\lambda\ , \quad t,\ u_i \varepsilon G\ ;\ \lambda \varepsilon \Lambda \tag{2-9}$$

La graduation est donnée par

$$db = dt + n + du_1 + \ldots + du_n ; \tag{2-10}$$

et les différentielles par

$$\partial_r b = \partial t\,[u_1 | \ldots | u_n]\lambda - \sum_{i=1}^{n} (-1)^{\epsilon_{i-1}} t\,[u_1 | \ldots | \partial u_i | \ldots | u_n]\lambda \tag{2-11}$$

$$\begin{aligned}\partial_s b = (-1)^{\epsilon_0} t u_1\,[u_2 | \ldots | u_n]\lambda \\ + \sum_{i=1}^{n} (-1)^{\epsilon_i} t[u_1 | \ldots | u_i u_{i+1} | \ldots | u_n]\lambda \\ + (-1)^{\epsilon_n} t[u_1 | \ldots | u_{n-1}]\eta(u_n)\lambda ,\end{aligned} \tag{2-12}$$

où

$$\epsilon_i = d(t[u_1 | \ldots | u_i]) ,$$

et l'homotopie par

$$\tau b = 1[t | \ldots | u_n]\lambda , \tag{2-13}$$

Les éléments de B_0 sont désignés comme $t[\,]\lambda = t \otimes \lambda$, avec degré dt et différentielle et augmentation données par

$$\partial(t[\,]\lambda) = \partial t[\,]\lambda , \quad \eta_B(t[\,]\lambda) = (\eta t)\lambda , \quad t \varepsilon G, \lambda \varepsilon \Lambda ,$$

On a aussi une bar construction « quotient », $\overline{B}$.

Proposition 2. — *Soient Λ et Λ' des anneaux, G un anneau différentiel gradué, et $\eta : G \longrightarrow \Lambda$, $\eta' : G \longrightarrow \Lambda'$ des homomorphismes d'anneaux différentiels gradués. Donc*

$$\overline{B}(\eta', G, \eta) = \Lambda' \otimes_G B_\Lambda(G, \eta) \tag{2-14}$$

est un Λ'-Λ-bimodule différentiel gradué.

En effet, Λ' devient un G-module différentiel gradué à droite avec les opérateurs $\mu t = \mu \eta'(t)$ pour $\mu \varepsilon \Lambda'$, $t \varepsilon G$. Le produit tensoriel de (2-14) est pris sur G, c'est-à-dire que

$$\mu \otimes tb = \mu t \otimes b , \quad \partial(\mu \otimes b) = \mu \otimes \partial b , \quad (\mu \varepsilon \Lambda', t \varepsilon G, b \varepsilon B)$$

(cas spécial du produit tensoriel $R \otimes_G L$ d'un Λ'-G-bimodule R différentiel gradué par un G-Λ bimodule différentiel gradué L ; cf. § 7).

Pour des formules explicites, on a

$$\overline{B} = \sum \overline{B}_n ,$$

où

$$\overline{B}_n = \Lambda \otimes_Z G \otimes_Z \ldots \otimes_Z G \otimes_Z \Lambda$$

avec n facteurs G. Les éléments

$$b = \mu \otimes u_1 \otimes \ldots \otimes u_n \otimes \lambda$$

de $\overline{B}_n$ seront notés

$$b = \mu[u_1 | \ldots | u_n]\lambda, \quad \mu \varepsilon \Lambda', \ u_i \varepsilon G, \ \lambda \varepsilon \Lambda . \tag{2-9'}$$

La graduation est donnée par

$$db = n + du_1 + \ldots + du_n . \tag{2-10'}$$

Posons

$$\epsilon_i = d(\mu[u_1 | \ldots | u_i]) ;$$

la différentielle est $\partial = \partial_r + \partial_s$, où ∂_r et ∂_s sont donnés par

$$\partial_r b = - \sum_{i=1}^{n} (-1)^{\epsilon_{i-1}} \mu[u_1 | \ldots | \partial u_i | \ldots | u_n]\lambda , \tag{2-11'}$$

$$\begin{aligned} \partial_s b = \mu\eta'(u_1)[u_2 | \ldots | u_n]\lambda \\ + \sum_{i=1}^{n-1} (-1)^{\epsilon_i} \mu[u_1 | \ldots | u_i u_{i+1} | \ldots | u_n]\lambda \\ + (-1)^{\epsilon_n} \mu[u_1 | \ldots | u_{n-1}]\eta(u_n)\lambda . \end{aligned} \tag{2-12'}$$

Les ∂_r et ∂_s étant des homomorphismes de bimodules, il suffit d'ailleurs de donner ces formules pour $\mu = 1 = \lambda$.

Si $\Lambda' = \Lambda$ et $\eta' = \eta$, on pose

$$\overline{B}_\Lambda(G, \eta) = \overline{B}(\eta', G, \eta) .$$

Par un *complexe* C nous entendons une structure (groupe, module, ou bimodule), différentielle graduée ; on définit de la manière usuelle le groupe gradué d'homologie H(C) du complexe C. Si C est un Λ-module à droite, le groupe d'homologie de C à coefficients dans un Λ-module G à gauche est

$$H(C ; G) = H(C \otimes_\Lambda G) ,$$

tandis que le groupe de cohomologie de C à coefficients dans le Λ-module G à droite est

$$H^*(C ; G) = H^*(\mathrm{Hom}(C, G)) ,$$

le groupe de cohomologie du cocomplexe

$$C^* = \sum C_n^* ;$$

où C_n^* est le groupe des Λ-homomorphismes de C_n dans G.

3. La méthode universelle pour les constructions standard

La bar construction quotient $\overline{B}_Z(G, \eta)$ a été introduite par Eilenberg-Mac Lane dans [6, II] (avec une autre graduation) et [8]; les formules (2-9')-(2-12') sont les mêmes que les formules (7-2)-(7-5) de [8].

La bar construction B_Z acyclique (avec une normalisation supplémentaire) est due à Cartan [3] et [4]; en particulier il a utilisé l'homotopie $s=\tau$ pour simplifier la définition de ∂. Il a aussi remplacé l'anneau Z des entiers par un anneau commutatif R quelconque (appelé chez lui Λ; ne confondez pas avec le Λ ci-dessus). Dans ce cas G est une algèbre différentielle graduée sur R et $\eta : G \longrightarrow R$ un homomorphisme d'algèbres.

Ce cas n'est pas compris dans les formules ci-dessus, mais on donne facilement une définition de $B_\Lambda(G, \eta)$ qui comprend les deux cas, en prenant R commutatif, Λ une algèbre sur R, G une algèbre différentielle graduée sur R, et $\eta : G \longrightarrow \Lambda$ un homomorphisme d'algèbres, avec les modifications convenables des catégories $\mathcal{K}$ et $\mathcal{L}$ et du foncteur T.

Cette utilisation des catégories $\mathcal{K}$ et $\mathcal{L}$ avec un foncteur T universel pour les plongements de K dans L donne un schéma qui contient toutes les constructions standard pour l'homologie des systèmes algébriques. La discussion ci-dessus est faite dans le cas le plus compliqué, où les éléments des catégories $\mathcal{K}$ et $\mathcal{L}$ sont des modules différentiels. Par exemple, pour l'homologie d'un groupe Π, on prend pour $\mathcal{K}$ la catégorie des groupes abéliens K, pour $\mathcal{L}$ la catégorie des $Z(\Pi)$-modules à gauche, et pour T le foncteur $T(K)=Z(\Pi)\otimes K$, $Z(\Pi)$ étant l'anneau du groupe Π. On obtient donc un complexe $\overline{B}$ acyclique et un complexe B qui sont les complexes standard pour l'homologie de Π. On peut donner une théorie abstraite de ces constructions, en utilisant pour $\mathcal{K}$ et $\mathcal{L}$ des catégories abéliennes (« exact categories » dans la terminologie de Buchsbaum [1]); nous n'insistons pas sur les détails.

4. La construction cubique

Soit A un groupe abélien additif. Pour $n>0$, soit C_n l'ensemble de tous les sommets e d'un n-cube $I \times \dots \times I$, notés

$$e=(\epsilon_1, \dots, \epsilon_n) \text{ avec } \epsilon_i=0 \text{ ou } 1 .$$

Pour $n=0$, soit C_0 l'ensemble réduit à un seul élément, noté (—). Soit $Q_n'(A)$, pour $n \geqslant 0$, le groupe abélien libre qui a comme générateurs toutes les applications $t: C_n \longrightarrow A$. Soit Q' le groupe gradué

$$Q'(A) = \sum_{n=0}^{\infty} Q_n'(A).$$

Des applications 0_i, $1_i : C_n \longrightarrow C_{n+1}$ sont définies pour $n \geqslant 0$ et $i = 1, \ldots, n+1$ par

$$\mu_i(\epsilon_1, \ldots, \epsilon_n) = (\epsilon_1, \ldots, \epsilon_{i-1}, \mu, \epsilon_i, \ldots, \epsilon_n), \quad \mu = 0, 1.$$

Pour μ, $\nu = 0$, 1 on a les deux relations (équivalentes si $\mu = \nu$)

$$\mu_i \nu_j = \nu_j \mu_{i-1} \qquad i > j, \quad (4\text{-}1)$$
$$\mu_i \nu_j = \nu_{j+1} \mu_i \qquad i \leqq j. \quad (4\text{-}1')$$

Les applications « face » de Q_n', $n \geqslant 1$, sont les 3 n homomorphismes R_i, S_i, $P_i : Q_n'(A) \longrightarrow Q'_{n-1}(A)$ définis pour $i = 1, \ldots, n$ et $e \varepsilon C_n$ par

$$(R_i t) e = t(0_i e), \qquad (S_i t)(e) = t(1_1 e),$$
$$(P_i t) e = t(0_i e) + t(1_i e) \qquad \text{(addition dans A !)}.$$

La différentielle $\partial : Q_n' \longrightarrow Q'_{n-1}$ est définie par

$$\partial t = \sum_{i=1}^{n} (-1)^i (P_i t - R_i t - S_i t) \qquad (\text{addition dans } Q'_{n'-1}). \tag{4-2}$$

En conséquence de (4-1) et (4-1'), on a $\partial\partial = 0$.

Le groupe $Q(A) = \dfrac{Q'(A)}{N_A}$ est défini comme quotient par un certain sous-groupe N_A des « normes » (« tranches » et « diagonales »). Un générateur t de Q_n' s'appelle *tranche* pour $n = 0$ si $t(-) = 0$, et i-*tranche*, pour $n > 0$ et $i = 1, \ldots, n$, lorsque $t(0_i e) = 0$ pour tout $e \varepsilon C_{n-1}$, ou que $t(1_i e) = 0$ pour tout e. Un t s'appelle i-*diagonale* ($n > 1$, $i = 1, \ldots, n-1$) lorsque $\epsilon_i \neq \epsilon_{i+1}$ implique $t(\epsilon_1, \ldots, \epsilon_n) = 0$. Le sous-groupe N_A de Q' engendré par toutes les tranches et toutes les diagonales est gradué et est stable pour la différentielle ; le quotient $Q(A) = \dfrac{Q'(A)}{N_A}$ est donc un groupe différentiel gradué.

Une augmentation $\eta_Q : Q(A) \longrightarrow A$ est définie par $\eta_Q t = 0$ si t est un générateur de degré $n > 0$ et $\eta_Q t = t(-) \varepsilon A$ si t est un générateur de degré $n = 0$.

Proposition 3. — *Pour tout groupe abélien* A, $Q(A)$ *est un groupe abélien différentiel gradué et* $\eta_Q : Q(A) \longrightarrow A$ *un homomorphisme de groupes différentiels gradués (différentielle et graduation triviales pour* A).

Les générateurs de $Q_n'(A)$ peuvent être représentés comme des n-cubes ayant un élément de A attaché à chaque sommet. Pour $n=0$, 1 et 2, on a, par exemple $t \longrightarrow (t(-))$,

$$t \to (t(0), t(1)), \qquad t \to \begin{pmatrix} t(00) & t(01) \\ t(10) & t(11) \end{pmatrix}.$$

Dans cette écriture $\eta(a)=a$ et

$$\partial(a, b)=(a)+(b)-(a+b), \qquad a, b \varepsilon A, \qquad (4\text{-}3)$$

$$\partial \begin{pmatrix} a & b \\ c & d \end{pmatrix} = (a, b)+(c, d)-(a+c, b+d)$$
$$-(a, c)-(b, d)+(a+b, c+d), \qquad (4\text{-}4)$$

$a, b, c, d \varepsilon A$. Les normes sont (0), (0, b), (a, 0), et

$$\begin{pmatrix} 0 & 0 \\ c & d \end{pmatrix}, \begin{pmatrix} a & b \\ 0 & 0 \end{pmatrix}, \begin{pmatrix} 0 & b \\ 0 & d \end{pmatrix}, \begin{pmatrix} a & 0 \\ c & 0 \end{pmatrix}, \begin{pmatrix} a & 0 \\ 0 & d \end{pmatrix}, \qquad (4\text{-}5)$$

(quatre tranches et une diagonale). La formule (4-4) pour la différentielle correspond bien à la loi (1-3) commutassociative, et les normes « tranches » correspondent également à la loi (1-4) pour le zéro. La normalisation (additionnelle) par diagonales n'est pas nécessaire, mais est utile.

Les groupes d'homologie $H_k(Q(A))$ de Q sont isomorphes aux groupes d'homologie $H_{n+k}(K(A, n))$ stables (c'est-à-dire avec $n > k$) des espaces $K(A, n)$ d'Eilenberg-Mac Lane. La construction cubique a été introduite par Eilenberg-Mac Lane [6, I] dans ce but. Cette construction a aussi la propriété d'acyclicité générique ([7], [9]) pour les groupes abéliens, et donne donc la théorie naturelle d'homologie et de cohomologie pour les groupes abéliens.

Dans la formule (4-2) du bord nous avons remplacé le signe $(-1)^{i+n-1}$ de [7] par $(-1)^i$, ce qui est plus commode pour le produit de Dixmier (§ 6).

5. Homologie relative cubique

Pour une somme directe A + C de groupes abéliens, on a [7, théorème 12.1] un isomorphisme canonique (théorème d'additivité)

$$H_n(Q(A+C)) \cong H_n(Q(A)) + H_n(Q(C)). \qquad (5\text{-}1)$$

Comme généralisation nous aurons le

Théorème 1. — *Pour une suite exacte*

$$0 \to A \xrightarrow{\alpha} E \xrightarrow{\beta} C \to 0 \qquad (5\text{-}2)$$

de groupes abéliens, Q(A) *étant identifié par* α *à son image dans* Q(E), *l'application* $\beta_* : \dfrac{Q(E)}{Q(A)} \rightarrow Q(C)$ *induite par* β *est une équivalence des chaînes.*

En effet, il existe des homomorphismes

$$\Gamma : Q(C) \rightarrow \frac{Q(E)}{Q(A)}\,; \qquad D : \frac{Q(E)}{Q(A)} \rightarrow \frac{Q(E)}{Q(A)} \tag{5-3}$$

de groupes gradués, de degrés 0, resp. 1, tels que

$$\begin{aligned} &\partial\Gamma = \Gamma\partial\,, && \beta_*\Gamma = I\,, \\ &\partial D + D\partial = I - \Gamma\beta_*\,, && \beta_* D = 0\,, \end{aligned} \tag{5-4}$$

où les I désignent les applications identiques convenables. En appliquant, à ce résultat, la suite exacte usuelle d'homologie relative d'un complexe Q(E) modulo un sous-complexe Q(A), on déduit la suite exacte

$$\ldots \rightarrow H_n(A\,;\,G) \xrightarrow{\alpha_*} H_n(E\,;\,G) \xrightarrow{\beta_*} H_n(C\,;\,G) \xrightarrow{\partial} H_{n-1}(A\,;\,G) \rightarrow \ldots\,,$$

où G est un groupe abélien et $H_n(Q(A)\,;\,G)$ est noté $H_n(A\,;\,G)$. Pour G = Z (et donc pour tout G), on peut aussi déduire cette suite exacte des calculs [4] de Cartan pour les groupes stables d'Eilenberg-Mac Lane. En effet ([4], exposé 11, théorème 2), tout groupe $H_n(A\,;\,Z)$ est somme directe des exemplaires de $A \otimes Z_p$ et $\mathrm{Tor}(A, Z_p)$ pour des entiers premiers p. La suite ci-dessus devient donc une somme directe des exemplaires de la suite exacte bien connue pour $\otimes$ et Tor.

Les applications Γ et D du théorème 1 seront définies par une méthode d'approximations successives, utilisée par Eilenberg-Mac Lane dans la démonstration ([6, I]; détails non publiés) du fait que Q donne les groupes stables d'Eilenberg-Mac Lane, et aussi dans la démonstration [8, § 12] du théorème de contraction pour la bar construction.

Nous commençons la démonstration avec la définition d'une certaine « homotopie de dédoublement » pour Q(E). Nous identifions A avec le sous-groupe $\alpha A \subset E$, d'après l'application α. Choisissons pour chaque $c \varepsilon C$ un représentant $Uc \varepsilon E$ tel que

$$U0 = 0\,, \qquad \beta(Uc) = c\,, \qquad c \varepsilon C\,.$$

Chaque élément b de E a une représentation $b = a + Uc$ unique, avec $a \varepsilon A$, $c \varepsilon C$. Des applications p_0, $p_1 : E \rightarrow E$ sont définies en posant

$$p_0(a + Uc) = a\,, \qquad p_1(a + Uc) = Uc\,,$$
$$a \varepsilon A\,, \qquad c \varepsilon C\,.$$

On a $p_0 E \subset A$, $p_1 A = 0$ et

$$p_0^2 = p_0 , \qquad p_1^2 = p_1 ,$$
$$p_0 p_1 = 0 = p_1 p_0 , \qquad p_0 + p_1 = I ,$$

I étant l'application identique. On notera que p_0, p_1 ne sont pas des homomorphismes ; on a cependant pour $b \varepsilon E$ et $a \varepsilon A$,

$$p_\epsilon (a+b) = p_\epsilon (a) + p_\epsilon (b), \quad \epsilon = 0, 1 . \tag{5-6}$$

On définit les homomorphismes $p_\epsilon : Q(E) \longrightarrow Q(E)$ de groupes gradués en posant

$$(p_\epsilon t) e = p_\epsilon (te) . \quad \epsilon = 0, 1 .$$

où $e \varepsilon C_n$ et $t : C_n \longrightarrow E$ est un générateur de Q(E). Ces homomorphismes respectent les normes et le sous-complexe Q(A). Ils ne commutent pas avec l'opérateur ∂ ; on a seulement pour les faces R_i et S_i (pas pour P_i)

$$R_i p_\epsilon = p_\epsilon R_i , \quad S_i p_\epsilon = p_\epsilon S_i , \quad \epsilon = 0, 1 ; \ i = 1 , \ldots , n .$$

L'homotopie $V : Q(E) \longrightarrow Q(E)$ de dédoublement est l'homomorphisme de groupes gradués, homogène de degré $+1$, défini par

$$(Vt)(\epsilon , e) = p_\epsilon (te), \quad t : C_n \rightarrow E , \quad e \varepsilon C_n , \quad \epsilon = 0, 1 .$$

Les normes et le sous-groupe Q(A) sont stables pour V. Pour les opérateurs « face », on a

$$R_1 V = p_0 , \qquad S_1 V = p_1 ,$$
$$R_{i+1} V = V R_i , \qquad S_{i+1} V = V S_i ,$$

et $P_1 V = I$, mais $P_{i+1} V \neq V P_i$. En effet, pour tout $t : C_n \longrightarrow E$ les termes

$$\begin{aligned} (P_{i+1} V t)(\epsilon , e) &= p_\epsilon [t(0_i e)] + p_\epsilon [t(1_i e)] , \\ (V P_i t)(\epsilon , e) &= p_\epsilon [t(0_i e) + t(1_i e)] , \quad \epsilon = 0, 1 , \ e \varepsilon C_n , \end{aligned} \tag{5-7}$$

ne sont pas égaux, parce que p_ϵ n'est pas un homomorphisme. En posant

$$J_i t = P_{i+1} V t - V P_i t ,$$
$$J = \sum_{i=1}^{n} (-1)^i J_i : Q_n(E) \rightarrow Q_n(E) ,$$

on vérifie que les normes sont stables pour tout J_i, et on a la formule essentielle pour V

$$-\partial V - V \partial = I - p_0 - p_1 + J . \tag{5-8}$$

Pour le cas spécial d'une somme directe on a

LEMME 2. — *Si* $E = A + C$ *est une somme directe de groupes abéliens, et* p_0 *resp.* $p_1 : Q(E) \rightarrow Q(E)$ *l'endomorphisme induit par la projection de* E *sur* A *resp.* C, *il existe un homomorphisme* $V : Q(E) \rightarrow Q(E)$ *de groupes gradués, homogène de degré* $+1$, *et tel que*

$$-\partial V - V\partial = I - p_0 - p_1 . \qquad (5\text{-}9)$$

En effet, dans ce cas on peut choisir les représentants Uc tels que U et donc p_ϵ sont des homomorphismes ; on a donc $J = 0$ dans (5-8). L'additivité (5-1) de $H(Q(E))$ est une conséquence immédiate de ce lemme. D'ailleurs l'homotopie V est la même, à une inversion des facteurs du cube I^n près, que la homotopie E de dédoublement introduite par Eilenberg-Mac Lane [7, p, 323] pour la démonstration de cette additivité.

Dans le cas général d'une suite (5-2) exacte, on va chasser le terme J de (5-8) par des approximations successives.

LEMME 3. — *Pour une suite exacte* (5-2) *on a des homomorphismes*

$$V\ ,\ p_1\ ,\ J : \frac{Q(E)}{Q(A)} \rightarrow \frac{Q(E)}{Q(A)}$$

de groupes gradués, homogènes de degrés 1, 0, *resp.* 0, *tels que*

$$-\partial V - V\partial = I - p_1 + J , \qquad (5\text{-}10)$$

$$p_1^2 = p_1 , \qquad \beta_* p_1 = \beta_* , \qquad (5\text{-}11)$$

$$p_1 V = 0 , \qquad p_1 J = 0 . \qquad (5\text{-}12)$$

Démonstration. — La première équation résulte de (5-8) en passant au quotient par les sous-groupes $Q(A)$ stable pour V, p_1 et J ; de même pour la deuxième. Quant à la troisième, p_1Vt est une tranche de $Q(E)$, et de même pour $p_1P_{i+1}Vt$ et p_1VP_it et donc pour p_1J_it, en raison de $p_1p_0 = 0$.

LEMME 4. — *Pour un entier* $n \geqslant 0$ *donné, il existe un entier* m *tel que* $J^m t = 0$ *pour chaque* $t \varepsilon Q(E)$ *homogène de degré* n (*en effet, on peut mettre* $m = n + 1$). *Il en résulte la convergence de la série*

$$t + Jt + J^2t + \dots + J^kt + \dots .$$

Pour la démonstration on va utiliser une filtration décroissante

$$Q(E) = Q^{(0)}(E) \supset Q^{(1)}(E) \supset \dots \supset Q^{(k)}(E) \supset \dots$$

du groupe $Q(E)$, en prenant pour $Q^{(k)}(E)$ le sous-groupe engendré par tous les $t : C_n \rightarrow E$ tels que $t(O_j e) \varepsilon A$ pour

$j=1, \ldots, k$. Ces sous-groupes ne sont pas stables pour l'opérateur ∂, mais on a

$$Q_k \cap Q^{(k)} = 0 \pmod{Q(A)}.$$

Pour tout générateur $t : C_n \longrightarrow E$ de Q(E), on va démontrer que

$$t \varepsilon Q^{(k)} \text{ et } i \leqq k \text{ entraîne } J_i t = 0\,, \tag{i}$$

$$t \varepsilon Q^{(k)} \qquad \text{entraîne } J_i t \varepsilon Q^{(k+1)}\,, \tag{ii}$$

$$t \varepsilon Q^{(k)} \qquad \text{entraîne } J t \varepsilon Q^{(k+1)}\,. \tag{iii}$$

Le lemme résultera de (iii). J étant une somme alternée des J_i, (iii) est une conséquence de (ii). De plus, (i) résulte sans peine de (5-6) et (5-7). Il reste (ii) pour $i > k$; il faut démontrer pour $j=0, \ldots, k$ que $P_{i+1}Vt(O_{j+1}e)$ et $VP_it(O_{j+1}e)$ sont dans A pour tout $e \varepsilon C_{n+1}$. Pour $j=0$ c'est immédiat par (5-7) et $p_0 E \subset A$. Pour $j=1, \ldots, k$ et $\epsilon = 0, 1$, on a

$$(VP_i t)(O_{j+1}\epsilon\,, e) = (VP_i t)(\epsilon\,, O_j e) = p_\epsilon\,[t(O_i O_j e) + t(1_i O_j e)].$$

Mais $j \leqslant k < i$, donc

$$O_i O_j = O_j O_{i-1}\,, \quad 1_i O_j = O_j 1_{i-1}$$

et

$$(VP_i t)(O_{j+1}\epsilon\,, e) = p_\epsilon\,[t(O_j O_{i-1} e) + t(O_j 1_{i-1} e)].$$

On a $t \varepsilon Q^{(k)}$, donc $t(O_j —) \varepsilon A$; le terme à droite est donc nul si $\epsilon = 1$ et il appartient à A si $\epsilon = 0$. La démonstration s'achève par un calcul analogue pour $P_{i+1}Vt$.

Lemme 5. — *On a des homomorphismes*

$$\Delta_0,\ \Gamma_0,\ W : Q(C) \rightarrow \frac{Q(E)}{Q(A)}$$

de groupe gradué, homogène de degré 0, 0, *resp.* —1, *tels que*

$$\partial \Gamma_0 - \Gamma_0 \partial = W\,, \tag{5-13}$$

$$\beta_* \Gamma_0 = I\,, \qquad p_1 = \Delta_0 \beta_*\,. \tag{5-14}$$

Démonstration. — On définit Γ_0 en posant

$$(\Gamma_0 t)(e) = U(te), \quad t : C_n \rightarrow C\,, \quad e \varepsilon C_n\,,$$

avec les représentants U déjà choisis. On pose $\Delta_0 = \Gamma_0$; (5-14) est évident, et on utilise (5-13) comme définition de W (en effet,

$$W = \sum_{i=1}^{n} (-1)^i (P_i \Gamma_0 - \Gamma_0 P_i)).$$

La démonstration qu'on va donner maintenant pour le théorème 1 s'appuie seulement sur les propriétés formelles des opérateurs V, J, etc., formulées dans les lemmes 3, 4 et 5. Posant

$$\Gamma_{k+1} = \Gamma_k + (-1)^k \mathrm{VJ}^k \mathrm{W}, \quad k = 0, 1, \dots,$$

on va démontrer par induction que

$$\partial\Gamma_k - \Gamma_k\partial = (-1)^k \mathrm{J}^k \mathrm{W}. \tag{5-15}$$

Pour $k=0$, c'est bien (5-13). Pour k fixe, (5-15) donne $\partial \mathrm{J}^k\mathrm{W} = -\mathrm{J}^k\mathrm{W}\partial$ et donc, en appliquant (5-10),

$$\partial\Gamma_{k+1} - \Gamma_{k+1}\partial = (-1)^k [p_1 \mathrm{J}^k \mathrm{W} - \mathrm{J}^{k+1}\mathrm{W}].$$

Si $k=0$, on déduit de (5-13) et (5-14) que

$$p_1\mathrm{W} = \Delta_0\beta_*\mathrm{W} = \Delta_0 0 = 0.$$

Si $k>0$, on a $p_1\mathrm{J}=0$ par (5-12). Dans tous les cas, il reste seulement $(-1)^{k+1}\mathrm{J}^{k+1}\mathrm{W}$, ce qui achève l'induction. En posant

$$\Gamma = \Gamma_0 + \sum_{k=0}^{\infty} (-1)^k \mathrm{VJ}^k\mathrm{W},$$

on a donc l'application voulue pour le théorème, avec $\partial\Gamma = \Gamma\partial$. On a

$$\beta_*\Gamma = \beta_*\Gamma_0 = \mathrm{I},$$

parce que

$$\beta_*\mathrm{V} = \beta_* p_1 \mathrm{V} = 0.$$

Pour l'homotopie D du théorème, posons $\mathrm{D}_0 = -\mathrm{V}$. On a donc

$$\partial\mathrm{D}_0 + \mathrm{D}_0\partial = \mathrm{I} - \Gamma\beta_* + \mathrm{L}, \tag{5-16}$$

où $\mathrm{L} = \mathrm{J} - p_1 + \Gamma\beta_*$. On a

$$\begin{aligned}\beta_*\mathrm{L} &= \beta_* p_1 \mathrm{J} - \beta_* p_1^2 + \beta_*\Gamma\beta_* = -\beta_* + \beta_*\Gamma\beta_* \\ &= -\beta_* + \beta_* = 0.\end{aligned}$$

Mais $p_1 = \Delta_0\beta_*$, donc $p_1\mathrm{L}=0$. Posant

$$\mathrm{D}_{k+1} = \mathrm{D}_k + (-1)^k \mathrm{VJ}^k \mathrm{L}.$$

on va démontrer par induction que

$$\partial\mathrm{D}_k + \mathrm{D}_k\partial = \mathrm{I} - \Gamma\beta_* + (-1)^k \mathrm{J}^k\mathrm{L}. \tag{5-17}$$

Pour $k=0$, c'est (5-16); pour tout k, (5-17) donne

$$\partial\mathrm{J}^k\mathrm{L} = \mathrm{J}^k\mathrm{L}\partial$$

et donc

$$\partial\mathrm{D}_{k+1} + \mathrm{D}_{k+1}\partial = \mathrm{I} - \Gamma\beta_* + (-1)^k [p_1\mathrm{J}^k\mathrm{L} - \mathrm{J}^{k+1}\mathrm{L}],$$

où $p_1 L = 0$ et $p_1 J = 0$, ce qui complète l'induction. La démonstration s'achève en posant

$$D = D_0 + \sum_{k=0}^{\infty} (-1)^k V J^k L .$$

6. Le complexe standard pour un anneau

Soit Λ un anneau, soit A un Λ-module à droite, et soient $Q(A)$, $Q(\Lambda)$ les complexes cubiques des groupes additifs de A et de Λ. J. Dixmier a signalé (dans des discussions privées) qu'il existe un bon produit $Q(A) \otimes Q(\Lambda) \rightarrow Q(A)$, défini pour les générateurs $t : C_m \rightarrow A$ et $u : C_n \rightarrow \Lambda$ respectifs de $Q(A)$ et de $Q(\Lambda)$ par la formule ($\epsilon_i = 0$ ou 1),

$$(tu)(\epsilon_1, \ldots, \epsilon_{m+n}) = t(\epsilon_1, \ldots, \epsilon_m) u(\epsilon_{m+1}, \ldots, \epsilon_{m+n}), \tag{6-1}$$

la multiplication à droite étant l'action d'un élément de Λ sur un élément de A. Ce produit (6-1) respecte les normes. Pour les faces R_i on a

$$R_i(tu) = (R_i t) u , \qquad 1 \leqq i \leqq m ,$$
$$R_i(tu) = t(R_{i-m} u) , \qquad m < i \leqq m + n ,$$

de même, pour les faces S_i et P_i. On a donc la formule fondamentale

$$\partial(tu) = (\partial t) u + (-1)^{dt} t (\partial u) . \tag{6-2}$$

Tout ceci restant valable si $A = \Lambda$, il en résulte la

Proposition 4. — *Le complexe cubique $Q(\Lambda)$ du groupe additif d'un anneau Λ devient, avec le produit (6-1) de Dixmier, un anneau différentiel gradué, et l'application $\eta_Q : Q(\Lambda) \rightarrow \Lambda$ devient un homomorphisme d'anneaux différentiels gradués. Si de plus, A est un Λ-module à droite, le complexe cubique $Q(A)$ du groupe additif de A devient un module différentiel gradué à droite sur l'anneau différentiel gradué $Q(\Lambda)$.*

L'homologie de l'anneau Λ est maintenant définie par l'homologie du complexe R formé en appliquant la bar construction quotient :

$$R(\Lambda) = B(Q(\Lambda), \eta_Q) . \tag{6-3}$$

Proposition 5. — *Pour chaque anneau Λ, $R(\Lambda)$ est un Λ-Λ-bimodule différentiel gradué.*

Les générateurs de R_n pour les petites dimensions n sont donnés dans le tableau suivant pour $x, y, z, s \varepsilon \Lambda$. Pour simplifier la notation, nous écrivons $[x]$ au lieu de $[(x)]$, etc. Le bord du générateur $\lambda[x]\mu$ pour λ, $\mu\varepsilon\Lambda$ est déterminé par le bord de $[x]$, ∂ étant un homomorphisme de bimodules.

$$
\begin{aligned}
n=0, \quad & [\,] \\
1, \quad & \partial[x]=x[\,]-[\,]x \\
2, \quad & \partial[x, y]=-[x]-[y]+[x+y] \\
& \partial[x \mid y]=x[y]-[xy]+[x]y \\
3, \quad & \partial\begin{bmatrix} x & y \\ z & s \end{bmatrix}=\begin{matrix} -[x, y]-[z, s]+[x+z\,,\; y+s] \\ +[x, z]+[y, s]-[x+y\,,\; z+s] \end{matrix} \\
& \partial[x, y \mid z]=[xz, yz]-[x, y]z-[y \mid z]-[x \mid z] \\
& \qquad\qquad\qquad\qquad\qquad\qquad +[x+y \mid z] \\
& \partial[x \mid y, z]=\begin{matrix} -[xy, xz]+x[y, z] \\ +[x \mid y]+[x \mid z]-[x \mid y+z] \end{matrix} \\
& \partial[x \mid y \mid z]=x[y \mid z]-[xy \mid z]+[x \mid yz]-[x \mid y]z\,.
\end{aligned}
$$

Les normes dans ces dimensions sont engendrées par tous les symboles $[x]$, $[x, y]$, $[x|y]$, etc., où l'un des x, y est nul, et tous les symboles de la liste (4-5) avec a, b, c, $d\varepsilon\Lambda$. Il est évident que ces formules représentent des propriétés combinées de l'addition et de la multiplication.

Revenons au problème 1 de l'introduction. Soit E une extension d'anneau, de la forme (1-5), et soient f et g les deux systèmes de facteurs de (1-6) pour l'addition et la multiplication. En posant

$$F[x, y]=f(x, y)\,, \qquad F[x|y]=-g(x, y) \tag{6-4}$$

(on notera le signe (-1)), on obtient un homomorphisme $F: R_2 \longrightarrow A$ de bimodules. F est donc une cochaîne du complexe $R(\Lambda)$ à coefficients dans le bimodule A. Les quatre types de cellules de dimension trois ci-dessus donnent quatre conditions pour que F soit un cocycle. On vérifie que ces quatre conditions sont les quatre conditions nécessaires pour les extensions. Autrement dit, on a démontré le

Théorème 2. — *Pour un Λ-bimodule* A *donné, les équations* (6-4) *donnent une correspondance biunivoque naturelle entre* $H^2(R(\Lambda); A)$ *et l'ensemble des classes d'équivalence des extensions* (1-5) *d'anneau* (*avec* $A^2=0$) *de* A *par* Λ.

On conjecture que des autres groupes de cohomologie de $R(\Lambda)$ admettent des interprétations analogues.

7. Le complexe standard pour un module

On va utiliser le produit tensoriel usuel $H \otimes_G L$, pris sur un anneau G différentiel gradué, où L est un G-Λ-bimodule différentiel gradué, et H un G-module à droite, différentiel gradué

$$(\text{donc } \partial(hg) = (\partial h)g + (-1)^{dh} h(\partial g) \text{ pour } h \varepsilon \mathrm{H},\ g \varepsilon \mathrm{G}).$$

En effet, $\mathrm{H} \otimes_{\mathrm{G}} \mathrm{L}$ est le Λ-module à droite engendré par les symboles $h \otimes \chi$ ($h \varepsilon \mathrm{H}$, $\chi \varepsilon \mathrm{L}$) additifs en h et χ, et soumis aux identités

$$hg \otimes \chi = h \otimes g\chi, \qquad (h \otimes \chi)\lambda = h \otimes \chi\lambda .$$

La graduation et la différentielle de $\mathrm{H} \otimes_{\mathrm{G}} \mathrm{L}$ sont définies par

$$d(h \otimes \chi) = dh + d\chi \qquad (h, \chi \text{ homogène})$$

$$\partial(h \otimes \chi) = (\partial h) \otimes \chi + (-1)^{dh} h \otimes \partial\chi . \qquad (7\text{-}1)$$

Muni de ces structures, $\mathrm{H} \otimes_{\mathrm{G}} \mathrm{L}$ devient un Λ-module (à droite) différentiel gradué. De plus $\mathrm{H} \otimes_{\mathrm{G}} \mathrm{L}$ est un foncteur covariant de H et L.

Soit maintenant A un Λ-module à droite. Nous définissons le Λ-module différentiel gradué M(A) par

$$\mathrm{M}_\Lambda(\mathrm{A}) = \mathrm{Q}(\mathrm{A}) \otimes_{\mathrm{Q}(\Lambda)} \mathrm{B}(\mathrm{Q}(\Lambda), \eta_{\mathrm{Q}}), \qquad (7\text{-}2)$$

où Q(A) est un $\mathrm{Q}(\Lambda)$-module différentiel gradué à droite en vertu de la proposition 4 et où B est un $\mathrm{Q}(\Lambda)$-Λ-bimodule différentiel gradué en vertu des propositions 1 et 3. De plus on a une augmentation

$$\eta_{\mathrm{M}} = \eta_{\mathrm{Q}} \otimes \eta_{\mathrm{B}} : \mathrm{M}_\Lambda(\mathrm{A}) \to \mathrm{A} \otimes_{\mathrm{Q}(\Lambda)} \Lambda . \qquad (7\text{-}3)$$

En effet $\eta_{\mathrm{B}} : \mathrm{B} \to \Lambda$ est un homomorphisme de bimodules, et $\eta_{\mathrm{Q}} : \mathrm{Q}(\mathrm{A}) \to \mathrm{A}$ un homomorphisme de $\mathrm{Q}(\Lambda)$-modules, Λ étant considéré comme $\mathrm{Q}(\Lambda)$-module avec les opérateurs $au = a(\eta_{\mathrm{Q}}u)$, où $\eta_{\mathrm{Q}} : \mathrm{Q}(\Lambda) \to \Lambda$. On a donc (7-3) avec la simplification

$$\mathrm{A} \otimes_{\mathrm{Q}(\Lambda)} \Lambda = \mathrm{A} \otimes_\Lambda \Lambda = \mathrm{A} .$$

PROPOSITION 6. — *Soit* A *un* Λ*-module à droite. Donc* $\mathrm{M}_\Lambda(\mathrm{A})$ *est un* Λ*-module à droite différentiel gradué, et* $\eta : \mathrm{M}_\Lambda(\mathrm{A}) \to \mathrm{A}$ *un homomorphisme de* Λ*-modules différentiels gradués. Pour chaque degré* n, M_n *est un* Λ*-module à droite libre.*

La décomposition $\partial = \partial_r + \partial_s$ de la différentielle de B induit une décomposition analogue pour la différentielle de M. En effet, notons ∂_r la différentielle de Q(A) et de $\mathrm{Q}(\Lambda)$, et posons $\partial_s = 0$ dans ces deux complexes. La définition (7-1) donne alors deux différentielles ∂_r et ∂_s dans le produit tensoriel M de (7-2) et on a $\partial = \partial_r + \partial_s$ dans M. Les éléments

$$m = t \otimes_{\mathrm{Q}(\Lambda)} 1[u_1 | \ldots | u_n]\lambda$$

de M seront notés

$$m = t[u_1 | \ldots | u_n]\lambda \qquad t \varepsilon \mathrm{Q}(\mathrm{A}),\ u_i \varepsilon \mathrm{Q}(\Lambda),\ \lambda \varepsilon \Lambda .$$

comme pour $b \varepsilon B_n$ dans (2-9), avec la seule différence qu'ici on a $t \varepsilon Q(A)$. Les formules (2-10), (2-11) et (2-12) pour db, $\partial_r b$ et $\partial_s b$ restent valables si b est remplacé par m.

Pour les petits degrés n, les générateurs de M_n comme Λ-module libre sont donnés dans le tableau suivant, où a, b, c, $d \varepsilon A$ et x, $y \varepsilon \Lambda$.

$0,\ \eta(a) = a$

$1,\ \partial(a, b) = (a) + (b) - (a + b)$

$\quad \partial(a)[x] = (ax) - (a)x$

$2,\ \partial\begin{pmatrix} a & b \\ c & d \end{pmatrix} = \text{formule de } (4\text{-}4)$

$$\partial(a, b)[x] = (a)[x] + (b)[x] - (a + b)[x] - (ax, bx) + (a, b)x$$

$$\partial(a)[x, y] = (ax, ay) - (a)[x] - (a)[y] + (a)[x + y]$$

$$\partial(a)[x \mid y] = (ax)[y] - (a)[xy] + (a)[xy] .$$

Les normes sont les symboles (a), (a, b), $(a)[x]$, etc., comprenant un argument a, b, x, ... nul, et les symboles de (4-5).

Revenons au problème 2 de l'introduction ; soit E l'extension (1-7) des modules et soient f et g les deux systèmes (1-8) correspondants de facteurs. En posant

$$F(c, c') = f(c, c'), \quad F((c)[x]) = -g(c, x) \tag{7-4}$$

on obtient un homomorphisme $F : M_1(C) \longrightarrow A$ de Λ-modules, donc une cochaîne de degré 1 du complexe $M_\Lambda(C)$, à coefficients dans A. En appliquant les formules explicites dérivées plus haut, on a le

THÉORÈME 3. — *Pour des Λ-modules à droite* A *et* C, *les formules* (7-4) *donnent une correspondance biunivoque entre* $H^1(M_\Lambda(C); A)$ *et l'ensemble des classes d'équivalence des extensions* (1-7) *de modules de* A *par* C.

Remarque 1. — En se servant de l'homomorphisme trivial $O : Q(\Lambda) \longrightarrow Z$ on a la bar construction $\overline{B}(O, Q(\Lambda), \eta_Q)$ de (2-14) qui est un Λ-module à droite gradué muni des différentielles ∂, ∂_r et ∂_s. L'application

$$t[u_1|\ldots|u_n]\lambda \longrightarrow t \otimes [u_1|\ldots|u_n]\lambda$$

pour $t \varepsilon Q(A)$, $u_i \varepsilon Q(\Lambda)$, $\lambda \varepsilon \Lambda$ donne évidemment un isomorphisme

$$M_\Lambda(A) \cong Q(A) \otimes_Z \overline{B}(0, Q(\Lambda), \eta_Q) \tag{7-5}$$

pour la structure de Λ-module à droite gradué, permis pour la différentielle ∂_r — mais non pour ∂_s ou ∂, à cause du premier terme de la formule (2-12).

Remarque 2. — La filtration croissante introduite dans [8, p. 100] par Eilenberg-Mac Lane est définie pour les complexes B et M en posant

$$B^{(k)} = \sum_{n=0}^{k} B_n(Q(\Lambda), \eta_Q),$$
$$M^{(k)} = Q(A) \otimes_{Q(\Lambda)} B^{(k)} .$$

En effet, un élément de $B^{(k)}$ contient dans son symbole au plus $k-1$ symboles « | ». Les $B^{(k)}$ et $M^{(k)}$ sont stables, et on a de plus

$$\partial_s B^{(k)} \subset B^{(k-1)}, \quad \partial_s M^{(k)} \subset M^{(k-1)} . \tag{7-6}$$

8. Homologie relative pour modules

On va démontrer des théorèmes analogues aux résultats du paragraphe 5.

Théorème 4. — *Pour une somme directe* $A+C$ *de* Λ*-modules à droite* A *et* C, *pour chaque* Λ*-module* G *à gauche, et pour chaque entier* $n \geqslant 0$, *on a un isomorphisme naturel*

$$H_n(M_\Lambda(A+C); G) \cong H_n(M_\Lambda(A); G) + H_n(M_\Lambda(C); G) . \tag{8-1}$$

Le résultat analogue pour les groupes de cohomologie est aussi vrai.

Théorème 5. — *Pour une suite exacte*

$$0 \to A \xrightarrow{\alpha} E \xrightarrow{\beta} C \to 0 \tag{8-2}$$

de Λ*-modules à droite,* $M_\Lambda(A)$ *étant identifié par* α *à son image dans* $M_\Lambda(E)$, *l'application* $\beta_* : \dfrac{M_\Lambda(E)}{M_\Lambda(A)} \to M_\Lambda(C)$ *induite par* β *est une équivalence de chaînes.*

En effet, il existe des homomorphismes

$$\varnothing : M_\Lambda(C) \to \frac{M_\Lambda(E)}{M_\Lambda(A)}, \qquad T : \frac{M_\Lambda(E)}{M_\Lambda(A)} \to \frac{M_\Lambda(E)}{M_\Lambda(A)} \tag{8-3}$$

de Λ-modules gradués, homogènes de degré 0 resp. 1, tels que

$$\partial\varnothing = \varnothing\partial , \qquad \beta_*\varnothing = I ,$$
$$\partial T + T\partial = I - \varnothing\beta_* . \qquad \beta_* T = 0 ,$$

où les symboles I désignent les applications identiques convenables.

Commençons les démonstrations en prenant une suite exacte (8-2) et les applications V, p et J, définies pour Q(E) au paragraphe 5. Ecrivons M_Λ (E) sous la forme (7-5). On a des homomorphismes de Λ-modules

$$V' = V \otimes \bar{I} , \quad p' = p \otimes \bar{I} , \quad J' = J \otimes \bar{I} : M_\Lambda(E) \to M_\Lambda(E),$$

où $\bar{I}$ désigne l'application identique de $\vec{B}$. Ce sont des applications homogènes de degré 1, 0, resp. 0, qui respectent la filtration (7-6) de M. Comme (7-5) est un isomorphisme pour la différentielle ∂_r, la propriété essentielle (5-8) de V donne

$$-\partial_r V' - V'\partial_r = I - p_0' - p_1' + J' .$$

Or on a $\partial = \partial_r + \partial_s$; posant

$$Y_0 = -\partial_s V' - V'\partial_s , \qquad Y = J' + Y_0 , \tag{8-6}$$

il vient

$$-\partial V' - V'\partial = I - p_0' - p_1' + Y . \tag{8-7}$$

Prenons l'hypothèse du théorème 4, avec $E = A + C$. En choisissant U comme homomorphisme de Λ-modules, on a $J = 0$ d'après le lemme 2 ; les $p_\epsilon : Q(E) \to Q(E)$ sont aussi des homomorphismes de modules. Pour le produit de Dixmier des générateurs t et u de $Q(E)$ resp. $Q(\Lambda)$, on a donc

$$p_\epsilon (tu) = (p_\epsilon t)u .$$

La définition de V donne alors $V(tu) = (Vt)u$. En appliquant la formule explicite (2-12) pour ∂_s, on a $\partial_s V' = -V'\partial_s$ et donc $Y_0 = 0$. Il en résulte (8-7) avec $Y = 0$, ce qui donne l'analogue du lemme 2 et démontre le théorème 4.

Reprenons le cas général du théorème 5.

Lemme 3'. — *Pour une suite exacte* (8-2) *on a des homomorphismes*

$$V', p_1', Y : \frac{M_\Lambda(E)}{M_\Lambda(A)} \to \frac{M_\Lambda(E)}{M_\Lambda(A)}$$

de Λ-modules à droite homogènes de degrés 1, 0, *resp.* 0 *et tels que*

$$-\partial V' - V'\partial = I - p' + Y , \tag{8-8}$$

$$p_1'^2 = p_1' , \qquad \beta_* p_1' = \beta_* , \tag{8-9}$$

$$p_1' V' = 0 , \qquad p_1' Y = 0 . \tag{8-10}$$

Toutes ces équations, sauf la dernière, résultent de (8-7) et des équations correspondantes du lemme 3 du paragraphe 5, compte tenu du fait que l'on a $M(\beta) = Q(\beta) \otimes \bar{I}$ pour les applications $\beta_*' = M(\beta)$ et $\beta_* = Q(\beta)$ induites. Quant à la dernière équation de (8-10) on a déjà $\beta_* V = \beta_* p_1 V = 0$ d'après (5-12). Mais $\partial_s \beta_*' = \beta_*' \partial_s$ et donc

$$\beta_*' Y_0 = -\beta_*'(\partial_s V' + V'\partial_s) = -\partial_s \beta_*' V' - \beta_*' V'\partial_s = 0 .$$

Or, d'après (5-14), $p_1 = \Delta_0 \beta_*$, d'où $p_1' Y_0 = 0$, ce qui donne (8-10).

LEMME 4'. — *Pour un entier donné* $n \geqslant 0$, *il existe un entier* m *tel que* $Y^m h = 0$ *pout tout* $h \varepsilon M_\Lambda(E)$ *homogène de degré* n.

Démonstration. — Dans la filtration de M on a

$$Y_0 M^{(k)} \subset M^{(k-1)}$$

d'après (7-6). On peut donc choisir $m = n^2 + 2n + 1$; dans chaque terme de l'expression de $Y^m = (J' + Y_0)^m$ on a ou bien $n+1$ termes Y_0 et donc un terme de filtration négative, ou bien $n+1$ termes J' adjacents et donc zéro par le lemme 4.

LEMME 5'. — *On a des homomorphismes*

$$\Delta_0', \ \varnothing_0, \ X : M_\Lambda(C) \rightarrow \frac{M_\Lambda(E)}{M_\Lambda(A)}$$

des groupes gradués, homogènes de degrés 0, 0, *resp.* —1, *tels que*

$$\partial\varnothing_0 - \varnothing_0\partial = X, \tag{8-11}$$

$$\beta_*\varnothing_0 = I, \qquad p_1' = \Delta_0'\beta_*, \tag{8-12}$$

Démonstration. — Posant $\Delta_0' = \Gamma_0 \otimes I$, on a $p_1' = \Delta_0'\beta_*$. En vertu du théorème 1, on a une application $\Gamma : Q(C) \rightarrow Q(E)$ de degré zéro telle que $\partial_r\Gamma = \Gamma\partial_r$ (mod $Q(A)$) ; posons

$$\phi_0 = \Gamma \otimes \overline{I};$$

on a donc $\beta_* \phi_0 = I$. On utilise (8-11) comme définition de X ; en effet $X = \partial_s \phi_0 - \phi_0 \partial_s$.

Le reste de la démonstration du théorème 5 repose seulement sur les lemmes 3', 4' et 5', et est tout à fait analogue à la démonstration correspondante du théorème 1.

9. ACYCLICITÉ DE LA CONSTRUCTION M

Par des méthodes connues on établit maintenant le

THÉORÈME 6. — *Pour tout* Λ*-module à droite* C *et pour tout entier* $n > 0$ *on a*

$$H_n(M_\Lambda(C)) = 0, \tag{9-1}$$

tandis que pour $n = 0$ *l'augmentation* η_M *induit un isomorphisme de* Λ*-modules*

$$\eta_* : H_0(M_\Lambda(C)) \cong C. \tag{9-2}$$

Commençons la démonstration avec (9-2). L'application η donne un homomorphisme $\eta_* : H_0 \rightarrow C$. Pour chaque $c \varepsilon C$ soit $\psi c \varepsilon H_0$ la classe d'homologie du cycle (c) de $M_\Lambda(C)$. Les formules du paragraphe 7 pour les bords de dimension zéro montrent que $\psi : C \rightarrow H_0$ est un homomorphisme de Λ-modules. Evidemment ψ est l'inverse de η_*, ce qui donne (9-2)

Pour un Λ-module libre F, on a $H_n(M_\Lambda(F))=0$, pour pour $n>0$. En effet, tout élément de M étant exprimé par un nombre fini d'éléments de F, il suffit de démontrer l'affirmation pour un F à un nombre fini de générateurs. En raison de l'additivité de H_n, du théorème 4, il suffit de prendre $F=\Lambda$. Dans ce cas $M_\Lambda(\Lambda)$ est identique au complexe $B_\Lambda(Q(\Lambda), \eta)$ qui est acyclique en vertu de la proposition 1.

Pour un C arbitraire, démontrons (9-1) par induction sur n. On peut toujours présenter C comme quotient $\frac{F}{A}$ d'un module libre F, ce qui donne la suite exacte

$$0 \to A \to F \to C \to 0 .$$

Pour $n=1$ on a, comme conséquence du théorème 5, la suite exacte d'homologie relative (avec $H(A)=H(M_\Lambda(A))$, etc.),

$$\to H_1(F) \to H_1(C) \to H_0(A) \to H_0(F) \to H_0(C) \to 0$$

On a $H_1(F)=0$; par les isomorphismes (9-2) on peut remplacer cette suite par une autre suite exacte

$$0 \to H_1(C) \to A \to F \to C \to 0 .$$

Les trois derniers termes étant déjà exacts, on a $H_1(C)=0$.

Pour $n>1$ on a la suite exacte

$$H_n(F) \to H_n(C) \to H_{n-1}(A) .$$

On a démontré ci-dessus que $H_n(F)=0$ et on a $H_{n-1}(A)=0$ en vertu de l'hypothèse de l'induction. On a donc $H_n(C)=0$. C. Q. F. D.

On a aussi une relation avec les groupes projectifs d'homotopie des modules définis par M. Eckmann dans ce même colloque. En effet, désignons par $Z_n=ker(M_n \to M_{n-1})$ le n^{me} groupe des cycles de M (C). Comme M est Λ-libre (donc projectif) et acyclique, on a $M_0=C$, $\frac{M_n}{Z_n}=Z_{n-1}$, $Z_n=\Omega Z_{n-1}$, et donc $Z_n=\Omega^{n+1}C$ dans l'écriture d'Eckmann. Donc son groupe d'homotopie projectif a la forme

$$\underline{\Pi}_{n+1}(A, C) = \underline{\Pi}(A, \Omega^{n+1}C) = \frac{\operatorname{Hom}(A, Z_n)}{\partial \operatorname{Hom}(A, M_{n+1})},$$

qui est le n^{me} groupe d'homologie du complexe

$$\operatorname{Hom}(A, M_\Lambda(C)) .$$

On a donc démontré

$$\underline{\Pi}_{n+1}(A, C) = H_n(\operatorname{Hom}(A, M_\Lambda(C))).$$

Université de Chicago.

NATURAL ASSOCIATIVITY AND COMMUTATIVITY

by Saunders Mac Lane[1]

1. **Introduction.** The usual associative law $a(bc) = (ab)c$ is known to imply the "general associative law," which states that any two iterated products of the same factors in the same order are equal, irrespective of the arrangement of parentheses. Here we are concerned with an associativity given by an isomorphism a: $A(BC) \cong (AB)C$; more exactly, with the case where the product AB is a covariant functor of its arguments A and B, while associativity is an isomorphism a natural in its arguments A, B, and C. The general associative law again shows that any two iterated products F and F' of the n arguments $A_1, \ldots, A_n$ are naturally isomorphic, under a natural isomorphism $F \cong F'$ given by "iteration" of a. We then ask: what conditions must be placed upon a if there is to be just *one* such isomorphism $F \cong F'$ for each pair F, F'? This question arises in categorical algebra, as do the corresponding questions for a natural commutative law $AB \cong BA$ and for an identity element K for the multiplication AB, with $KA \cong A$ natural. Here we present answers to each of these questions. The first question (associativity alone) has already been treated by Stasheff [10] in connection with homotopy associative H-spaces. After these lectures formulating our answers had been delivered, we found that the same questions had been answered by Epstein and formulated in a privately circulated preprint of his study [3] of Steenrod operations in Abelian categories. His results are certainly independent of and probably prior to ours.

For categories we employ the usual terminology (see for example, [7], Chapters I and IX). A *category* $\mathscr{C}$ is a class of "objects" $A, B, C, \ldots$ together with a family of disjoint sets $\hom(A,B) = \hom_{\mathscr{C}}(A,B)$, one for each ordered pair of objects. When $f \in \hom(A,B)$ we also write $f\colon A \to B$ and we call f a *morphism* of $\mathscr{C}$ with *domain* A and *range* B. A *composite* gf is defined whenever $f\colon A \to B$ and $g\colon B \to C$ are morphisms with range(f) = domain(g): it is a morphism $gf\colon A \to C$. There are two axioms: the triple composite is associate

[1] This research was supported in part by the Air Office of Scientific Research, with a grant No. AF-AFOSR-62-170.

Editor's Note: This paper was presented in three lectures in Anderson Hall, Rice University, 23, 24, 26 September 1963. Mr. Mac Lane is the Max Mason Distinguished Service Professor of Mathematics at the University of Chicago.

(whenever defined); to each object B there is a morphism $1_B: B \to B$ which acts as a left and a right identity under composition.

If $\mathscr{B}$ and $\mathscr{C}$ are categories, a (covariant) *functor* $F: \mathscr{B} \to \mathscr{C}$ consists of an *object function* and a *mapping function*. The object function assigns to each $B \in \mathscr{B}$ an object $F(B)$ in $\mathscr{C}$: the mapping function assigns to each morphism $f: B \to B'$ in $\mathscr{B}$ a morphism $F(f): F(B) \to F(B')$ in $\mathscr{C}$ in such a way that

$$F(1_B) = 1_{FB}, \qquad F(gf) = (Fg)(Ff),$$

the latter whenever the composite gf is defined.

If $F, G: \mathscr{B} \to \mathscr{C}$ are functors, a *natural transformation* $t: F \to G$ is a function t which assigns to each object B of $\mathscr{B}$ a morphism

$$t(B): F(B) \to G(B) \quad \text{in} \quad \mathscr{C}$$

such that for every morphism $f: B \to B'$ in $\mathscr{B}$ the diagram

$$\begin{array}{ccc} F(B) & \xrightarrow{F(f)} & F(B') \\ \downarrow{\scriptstyle t(B)} & & \downarrow{\scriptstyle t(B')} \\ G(B) & \xrightarrow{G(f)} & G(B') \end{array} \qquad \text{in } \mathscr{C}$$

is commutative. Moreover, t is a *natural equivalence* (or a natural isomorphism) if each $t(B)$ has in $\mathscr{C}$ a two-sided inverse. It follows that this inverse is a natural transformation $t^{-1}: G \to F$.

If $\mathscr{C}$ and $\mathscr{C}'$ are categories, their (cartesian) product $\mathscr{C} \times \mathscr{C}'$ has as objects the pairs (C, C') of objects and as morphism the pairs

$$(f, f'): (C, C') \to (D, D')$$

of morphisms $f: C \to D$ and $f': C' \to D'$, with the evident composition and identities. A functor $F: \mathscr{C} \times \mathscr{C}' \to \mathscr{D}$ is called a *bifunctor* on the categories $\mathscr{C}$ and $\mathscr{C}'$, and similarly for functors of more arguments and for their natural transformations. In particular, $\mathscr{C}^n$ will denote the n-fold product of the category $\mathscr{C}$ with itself, so that a functor $F: \mathscr{C}^n \to \mathscr{D}$ is a covariant functor on n arguments in $\mathscr{C}$ with values in $\mathscr{D}$. We say that F has *multiplicity* n.

2. Categories with a multiplication. By a category with a multiplication we mean a category $\mathscr{C}$ together with a covariant bifunctor on $\mathscr{C}$ to $\mathscr{C}$. This bifunctor will be denoted by the symbol $\otimes$, written between its arguments. Thus the statement that $\otimes$ is a bifunctor means:

(i) Each pair A, B of objects of $\mathscr{C}$ yields an object $A \otimes B$ of $\mathscr{C}$;

(ii) Each pair $f: A \to A'$ and $g: B \to B'$ of morphisms of $\mathscr{C}$ yields

$$f \otimes g : A \otimes B \to A' \otimes B', \tag{2.1}$$

a morphism of $\mathscr{C}$;

(iii) Identity morphisms $1_A : A \to A$ and $1_B : B \to B$ of $\mathscr{C}$ yield

$$1_A \otimes 1_B = 1_{A \otimes B} : A \otimes B \to A \otimes B, \tag{2.2}$$

the identity morphism of $A \otimes B$;

(iv) If the composites $f'f$ and $g'g$ are defined in $\mathscr{C}$, then

$$(f'f) \otimes (g'g) = (f' \otimes g')(f \otimes g). \tag{2.3}$$

Examples abound. For instance, in a category $\mathscr{C}$ with finite products the product is a multiplication. In detail, a diagram

$$A \xleftarrow{p} P \xrightarrow{q} B \qquad \text{in } \mathscr{C} \tag{2.4}$$

(with fixed ends A and B) is called a *product diagram* if to each diagram

$$A \xleftarrow{f} C \xrightarrow{g} B \qquad \text{in } \mathscr{C}$$

with the same ends there exists a unique morphism $h : C \to P$ such that $f = ph$ and $g = qh$. The category has *finite products* if there is such a product diagram for each pair of objects A and B of $\mathscr{C}$. When this is the case, the middle object P of the product diagram is uniquely determined, up to equivalence, by A and B. Choosing one $P = A \times B$ for each A and B yields a bifunctor $\times$ on $\mathscr{C}$, hence a multiplication. This includes examples such as the cartesian product of sets or of topological spaces and the direct product of groups (in the category of all groups, with morphisms all group homomorphisms).

The dual notion is that of a coproduct. In a category $\mathscr{C}$, a diagram

$$A \xrightarrow{i} D \xleftarrow{j} B$$

is called a *coproduct diagram* if to any diagram

$$A \xrightarrow{f} C \xleftarrow{g} B$$

with the same ends there is a unique $h : D \to C$ such that $f = hi$ and $g = hj$. The category $\mathscr{C}$ has *finite coproducts* if there is such a coproduct diagram for each pair of objects A and B. When this is the case, the middle object D in the coproduct diagram is uniquely determined, up to equivalence, by A and B, and a choice $D = A * B$ provides a bifunctor $*$ which is a multiplication for $\mathscr{C}$. Examples are the free product of two groups (in the category of groups) or the "wedge" of two spaces (in the category of topological spaces with a selected base point, with morphisms continuous maps carrying base point to base point).

More important for our applications are the categories with a multiplication given by one of the usual tensor products. For example, in the category $\mathcal{M}_F$ of all vector spaces $V, W, \cdots$ over a fixed field F, the usual tensor product $V \otimes W$ of two vector spaces is a multiplication. The same holds for the category $\mathcal{M}_K$ of all modules over a fixed commutative ring K. Similarly, let $DG(\mathcal{M}_K)$ be the category of all differential graded K-modules (i.e., of all chain complexes of K-modules). The usual tensor product of differential graded modules (defined, say, as in [7], Chap. VI. 7) is a multiplication in this category.

If $\mathscr{C}$ is a category, a subcategory $\mathscr{C}'$ (with the *same* objects as $\mathscr{C}$) is given by a subclass of the class of all morphisms of $\mathscr{C}$, such that this subclass contains every identity morphism of $\mathscr{C}$ and with each pair of morphisms their composite (whenever defined in $\mathscr{C}$). If $\mathscr{C}$ is a category with a multiplication $\otimes$, a *multiplicative subcategory* is a subcategory $\mathscr{C}'$ which contains with any two morphisms f and g the morphism $f \otimes g$; then $\mathscr{C}'$ is itself a category with a multiplication. Given any set S of morphisms of $\mathscr{C}$, we may speak of the multiplicative subcategory *generated* by the morphisms of S; it is defined to be the intersection of all multiplicative subcategories of $\mathscr{C}$ which contain S. The process of generation can be described more explicitly. By an *expansion* of a morphism f we mean a morphism such as $f \otimes 1_A$, $1 \otimes (f \otimes 1)$, $[1 \otimes (f \otimes 1)] \otimes 1$, etc. More formally, the set of *expansions* of f is the smallest set of morphisms of $\mathscr{C}$ which contains f and with any morphism e all morphisms $e \otimes 1_C$ and $1_C \otimes e$ for any object C of the category.

THEOREM 2.1. *Let S be a set of morphisms in a category $\mathscr{C}$ with a multiplication. The multiplicative subcategory of $\mathscr{C}$ generated by S consists of all identity morphisms of $\mathscr{C}$ and of all composites of expansions of morphisms of S.*

For a proof, we must show that the indicated set of morphisms is closed under composition and under $\otimes$-multiplication. For composition, this is immediate. For $\otimes$-multiplication, use (2.3), which states that a $\otimes$-product of two composites can be rewritten as a composite of two tensor products. By iterated application of this result, the problem is reduced to the tensor product of two expansions, say $f \otimes g$. But $f = f1$ and $g = 1'g$, where 1 and $1'$ are suitable identity morphisms, so that (2.3) yields

$$f \otimes g = (f1) \otimes (1'g) = (f \otimes 1')(1 \otimes g); \tag{2.4}$$

this states that the $\otimes$-product of f and g can be rewritten as a composite of expansions of f and g. This completes the proof.

For any category $\mathscr{C}$, let $\mathrm{Fct}(\mathscr{C})$ be the category whose objects are all functors $F:\mathscr{C}^n \to \mathscr{C}$, of any multiplicity n, and whose morphisms are all natural transformations $t: F \to F'$ between two functors of the same multiplicity. The composite of t with $t': F' \to F''$ is defined in the expected way, for any arguments $A_1, \cdots, A_n$, as

$$(2.5) \qquad (t't)(A_1,\cdots,A_n) = t'(A_1,\cdots,A_n)\, t(A_1,\cdots,A_n).$$

If $\mathscr{C}$ has a $\otimes$-multiplication, so does $\mathrm{Fct}(\mathscr{C})$; to functors $F:\mathscr{C}^n \to \mathscr{C}$ and $G:\mathscr{C}^m \to \mathscr{C}$ construct the functor $F \otimes G:\mathscr{C}^{n+m} \to \mathscr{C}$ with object function defined for any arguments $C_1, \cdots, C_{n+m}$ by

$$(2.6) \qquad (F \otimes G)(C_1,\cdots,C_{n+m}) = F(C_1,\cdots,C_n) \otimes G(C_{n+1},\cdots,C_{n+m}),$$

and with the corresponding definition for the mapping function. To natural transformations $t: F \to F'$ and $u: G \to G'$ construct the natural transformation $t \otimes u: F \otimes G \to F' \otimes G'$ defined for arguments C_i by

$$(2.7) \qquad (t \otimes u)(C_1,\cdots,C_{n+m}) = t(C_1,\cdots,C_n) \otimes u(C_{n+1},\cdots,C_{n+m}).$$

These definitions give $\mathrm{Fct}(\mathscr{C})$ a $\otimes$-multiplication.

An *iterate* of the functor $\otimes:\mathscr{C}^2 \to \mathscr{C}$ will mean any functor formed by repeated applications of $\otimes$-multiplication. More exactly, the set of iterates of $\otimes$ is the smallest set of functors $F:\mathscr{C}^n \to \mathscr{C}$ which contains the identity functor $1:\mathscr{C} \to \mathscr{C}$ and with any two functors F and G the functor $F \otimes G$. By $\mathrm{It}_\otimes(\mathscr{C})$ we denote the category whose objects are all iterates of $\otimes$ and whose morphisms are all natural transformations between such iterates. Then $\mathrm{It}_\otimes(\mathscr{C})$ is a category with a $\otimes$-multiplication.

3. Higher Associativity Laws. Let $\mathscr{C}$ be a category with a multiplication $\otimes$ and with a natural transformation.

$$(3.1) \qquad a = a(A,B,C):\ A \otimes (B \otimes C) \to (A \otimes B) \otimes C$$

such that each $a(A,B,C)$ has a two-sided inverse $a^{-1}(A,B,C)$ in $\mathscr{C}$. Call a the *associativity isomorphism*. If the functors F, G, and H are three iterates of $\otimes$, of multiplicities n, m, and k, respectively, the natural isomorphism

$$(3.2) \qquad a = a(F,G,H):\ F \otimes (G \otimes H) \to (F \otimes G) \otimes H:\ \mathscr{C}^{n+m+k} \to \mathscr{C}$$

given for arguments C_i as

$$a(F(C_1,\cdots,C_n),\ G(C_{n+1},\cdots,C_{n+m}),\ H(C_{m+n+1},\cdots,C_{m+n+k}))$$

will be called an *instance* of a; define instances of a^{-1} similarly.

An *iterate* of a will be any morphism in the multiplicative subcategory of $\mathrm{It}_\otimes(\mathscr{C})$ generated by all the instances of a and of a^{-1}. By Theorem 2.1, any iterate of a may be written as a composite of expansions of instances of a. It follows that each iterate of a has a two-sided inverse.

Now call the associativity isomorphism a *coherent* if to each pair $F, F' : \mathscr{C}^p \to \mathscr{C}$ of iterates of the functor $\otimes$ there is at most one iterate of a which is a natural isomorphism $t : F \cong F'$. Thus coherence requires that any two formally different iterates between the same pair of functors be equal. Coherence can also be defined in terms of diagrams: it requires that any diagram with vertices iterates of $\otimes$ and edges expansions of instances of a be commutative. This means, for example, that the diagram

$$
(3.5)\qquad
\begin{array}{ccccc}
A \otimes (B \otimes (C \otimes D)) & \xrightarrow{a_1} & (A \otimes B) \otimes (C \otimes D) & \xrightarrow{a_2} & ((A \otimes B) \otimes C) \otimes D \\
\downarrow{\scriptstyle 1 \otimes a_3} & & & & \uparrow{\scriptstyle a_5 \otimes 1} \\
A \otimes ((B \otimes C) \otimes D) & & \xrightarrow{\qquad a_4 \qquad} & & (A \otimes (B \otimes C)) \otimes D
\end{array}
$$

must be commutative. Here each a_i is an evident instance of a; thus

$$a_1 = a(A, B, C \otimes D), \quad a_2 = a(A \otimes B, C, D), \quad a_3 = a(B, C, D)$$

and so on (and similarly in subsequent diagrams, where we will omit the subscripts i in a_i and the corresponding specification as to which instance is involved). Now the one condition (3.5) suffices to insure coherence:

THEOREM 3.1. *In a category $\mathscr{C}$ with a multiplication $\otimes$, an associativity isomorphism a is coherent if and only if the pentagonal diagram* (3.5) *is commutative for every quadruple A, B, C, D of objects.*

The proof will be by induction on a suitably defined *rank* ρ for the iterates of $\otimes$. The rank of the identity functor is defined to be 0, while for functors $F : \mathscr{C}^m \to \mathscr{C}$ and $G : \mathscr{C}^n \to \mathscr{C}$ the rank of $F \otimes G$ is defined in terms of the ranks of F and G as

$$p(F \otimes G) = \rho(F) + \rho(G) + n - 1 .$$

The presence of $n-1$ in this formula insures that a functor F has rank zero precisely then when it is expressed by a formula in which all pairs of parentheses start "in front."

Now consider all iterates F of fixed multiplicity n; these are given exactly by the different arrangements of parentheses in an n-fold product. Draw the graph with vertices all these iterates and with edges all expansions a_i, $1 \otimes a_i$, $1 \otimes (a_i \otimes 1)$, $(1 \otimes a_i) \otimes 1, \cdots$ of instances of a. For $n = 4$, the graph is exactly the figure displayed in (3.5). In general, any path along successive edges in this graph from vertex F to vertex G represents a natural isomorphism $t : F \cong G$; namely, that isomorphism given as a composite of the instances on its edges.

If the edges of a path, taken in order, involve only instances of a (and none of a^{-1}), call the path *directed*. Let $H_{(n)}:\mathscr{C}^n \to \mathscr{C}$ be that iterate of $\otimes$ which has all pairs of parentheses starting in front, so that $H_{(2)} = \otimes$, $H_{(n+1)} = H_{(n)} \otimes 1$. There is a directed path from any F to $H_{(n)}$; indeed, we may choose such a path in a canonical way, say by moving an outermost parenthesis toward the front. This proves, for every F and G, that there is at least one natural isomorphism $t: F \cong G$; observe that this proof is really just the known proof of the "general associative law" from $a(bc) = (ab)c$.

The essential point is that remaining: A proof that any two paths from F to G yield the same natural isomorphism $F \cong G$. Along an arbitrary path from F to G, join each "vertex" F_i to the "bottom" vertex $H_{(n)}$ by the canonical directed path. A glance at the diagram

$$\begin{array}{ccccccccc} F & \to & F_1 & \leftarrow & F_2 & \to & F_3 & \leftarrow & G \\ \downarrow & & \downarrow & & \downarrow & & \downarrow & & \downarrow \\ H_{(n)} & = & H_{(n)} & = & H_{(n)} & = & H_{(n)} & = & H_{(n)} \end{array}$$

indicates that it will suffice to show that any two *directed* paths from an F_i to $H_{(n)}$ will yield the same isomorphism. This we prove by induction on the rank of $F = F_i$, it being immediate for rank 0. Suppose it true for all F_i of smaller rank, and consider two different directed paths starting at F with the two expansions e and f, as in the following figure:

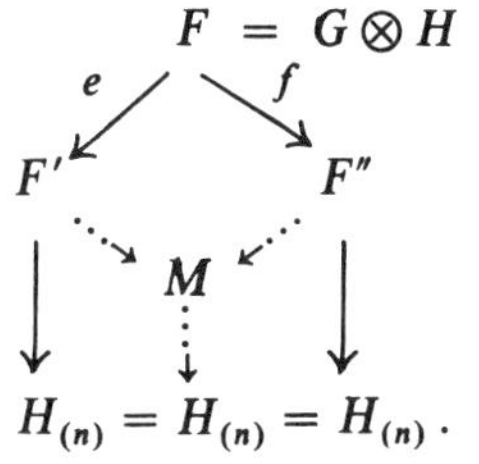

Both e and f decrease the rank. Hence it will suffice to show that one can "rejoin" e and f by directed paths to some common vertex M in such a way that the diamond from F to M is commutative. If $e = f$, take $F' = M = F''$. If $e \neq f$, the functor F, as an iterate of $\otimes$, can be expressed uniquely as $F = G \otimes H$. Now the edge e represents an expanded instance of a; it has one of three forms:

e acts "inside" G; that is, $e = e' \otimes 1_H$ for some e';

e acts "inside" H; that is, $e = 1_G \otimes e''$ for some e'';

e is an instance of a; that is, $e = a(G, K, L)$ and $H = K \otimes L$.

For f there are the same three choices.

Now compare e with f. If both act inside the same G, use the induction assumption on G. If e acts inside G and f inside H, say as $f = 1_G \otimes f''$, use the diamond

$$\begin{array}{ccccc} & & F = G \otimes H & & \\ & {\scriptstyle e = e' \otimes 1} \swarrow & & \searrow {\scriptstyle f = 1 \otimes f''} & \\ G' \otimes H & & & & G \otimes H' \\ & {\scriptstyle 1' \otimes f''} \searrow & & \swarrow {\scriptstyle e' \otimes 1'} & \\ & & G' \otimes H' & & \end{array}$$

which commutes in view of (2.4). There remains the case when e, say, is an instance of a, so that F has the form

$$F = G \otimes H = G \otimes (K \otimes L).$$

Since $e \neq f$, f must act inside G or inside H. If f acts inside G as $f' : G \to G'$, use the diamond

$$\begin{array}{ccccc} & & G \otimes (K \otimes L) & & \\ & {\scriptstyle a} \swarrow & & \searrow {\scriptstyle f' \otimes 1} & \\ (G \otimes K) \otimes L & & & & G' \otimes (K \otimes L) \\ & {\scriptstyle (f' \otimes 1) \otimes 1} \searrow & & \swarrow {\scriptstyle a} & \\ & & (G \otimes K) \otimes L & & \end{array}$$

This commutes because a is natural. We are left with f acting inside $H = K \otimes L$ as $f = 1_G \otimes f''$. If it is actually inside K or inside L, we are again done, by naturality. If it is inside neither, f'' must be an instance of a, say with $L = P \otimes Q$ and

$$f'' = a : K \otimes (P \otimes Q) \to (K \otimes P) \otimes Q.$$

Here the diamond starts

$$\begin{array}{ccccc} & & G \otimes (K \otimes (P \otimes Q)) & & \\ & {\scriptstyle a} \swarrow & & \searrow {\scriptstyle f = 1 \otimes a} & \\ (G \otimes K) \otimes (P \otimes Q) & & & & G \otimes ((K \otimes P) \otimes Q); \end{array}$$

it may be completed by the commutative pentagon of our condition (3.5). This is the final case of the proof. Observe that the diagrams used have been quite analogous to those appearing in one of the familiar proofs of the Jordan-Hölder theorem for groups (cf. [9], 295).

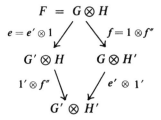

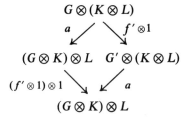

For $n = 5$ the proof may be visualized in the following graph, where AB is short for $A \otimes B$ and $A \cdot BC$ is short for $A(BC)$:

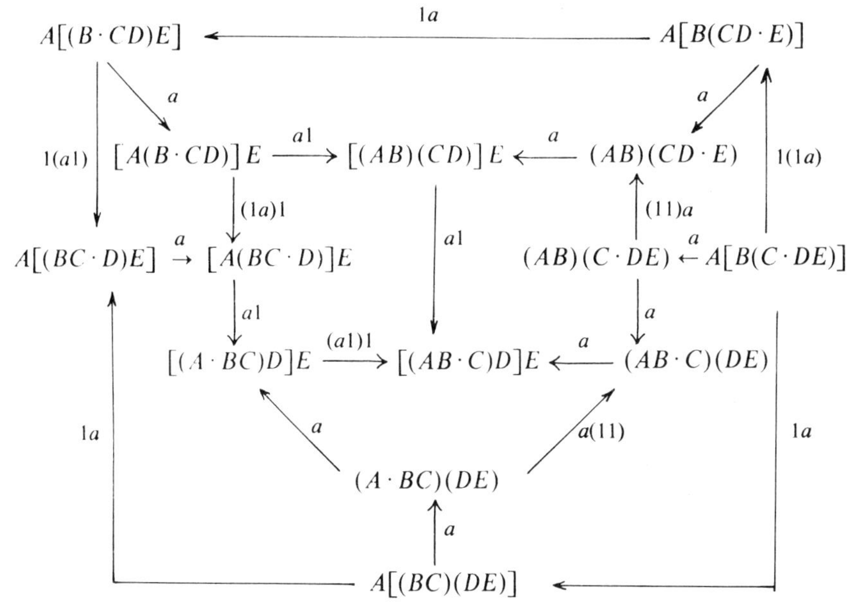

Each complementary region for the graph is either a pentagon (an instance of [3.5]) or a square (an instance of naturality). The whole graph may be regarded as the skeleton of a 3-cell; the regions are its faces and correspond to products $ABCDE$ with two pairs of parentheses omitted, while the edges correspond to products with one pair of parentheses omitted. Stasheff [10] has shown that the corresponding graph for every n gives an $(n-2)$-cell.

This coherence result applies in the example of categories with multiplications which we have cited in §2. In the cases of cartesian or tensor products the usual rules

$$(a,(b,c)) \to ((a,b),c), \quad a \otimes (b \otimes c) \to (a \otimes b) \otimes c$$

yield associativity isomorphisms which evidently satisfy the pentagon condition, hence are coherent. However, not *every* associativity isomorphism need satisfy the pentagon condition. A counter example (suggested by Fred Richman) is the isomorphism

$$a \otimes (b \otimes c) \longrightarrow -(a \otimes b) \otimes c$$

in the category of all K-modules.

4. Higher Commutativity Laws. Let $\mathscr{C}$ be a category with a $\otimes$-multiplication and with a natural homomorphism

$$c = c(A,B)\colon\ A \otimes B \to B \otimes A \tag{4.1}$$

such that

$$c(B,A)\,c(A,B) = 1\colon\ A \otimes B \to A \otimes B \tag{4.2}$$

for every A and B. This insures that $c(B,A)$ is a two-sided inverse for $c(A,B)$. Call c the *commutativity isomorphism*, and assume that $\mathscr{C}$ also has an associativity isomorphism a. Our problem is to describe the "coherence" property for $\mathscr{C}$ under iterates of a, a^{-1}, and c. To this end, we must arrange to permute the arguments of our functor.

To each functor F of multiplicity n and each permutation α in the symmetric group $S(n)$ on n letters we construct a new functor αF of the same multiplicity with the object function

$$(\alpha F)(A_1,\cdots,A_n) = F(A_{\alpha 1},\cdots,A_{\alpha n}), \qquad A_i \in \mathscr{C}$$

and the evident corresponding mapping function. Similarly, if $t\colon F \to F'$ is a natural transformation between two functors of the same multiplicity n, define $\alpha t\colon \alpha F \to \alpha F'$ by

$$(\alpha t)(A_1,\cdots,A_n) = t(A_{\alpha 1},\cdots,A_{\alpha n}), \qquad A_i \in \mathscr{C}.$$

If β is also in $S(n)$, then $(\beta\alpha)F = \beta(\alpha F)$ and $(\beta\alpha)t = \beta(\alpha t)$. Moreover, if the composite transformation $t't$ is defined, then

$$\alpha(t't) = (\alpha t')(\alpha t). \tag{4.3}$$

Again, consider natural transformations $t\colon F \to F'$ and $u\colon G \to G'$ of multiplicities n and m, respectively, so that $t \otimes u$ is defined. Let $\alpha \in S'(n)$ and $\gamma \in S(m)$. Define $\alpha \times \gamma$ to be that permutation of $S(n+m)$ which acts on the first n letters as does α and on the remaining m letters, in order, as does γ. Then the definitions show that $\alpha F \otimes \gamma G = (\alpha \times \gamma)(F \otimes G)$ and that

$$\alpha t \otimes \gamma u = (\alpha \times \gamma)(t \otimes u)\colon\ \alpha F \otimes \gamma G \to \alpha F' \otimes \gamma G'. \tag{4.4}$$

A *permuted iterate* ("pit" for short) of $\otimes$ will be any functor of the form αF, for $\alpha \in S(n)$ and F an iterate of $\otimes$ of multiplicity n. An *instance* of c is (as before), any natural transformation $c(F,G)$, where the functors F and G are any iterates of $\otimes$.

DEFINITION. The isomorphisms a and c are *coherent* if every diagram of the following form is commutative: Vertices, permuted instances of $\otimes$; edges, permuted expansions of instances of a, a^{-1}, and c.

For example, a special case of this coherence is the commutativity of the following "hexagonal" diagram

$$\begin{array}{ccccc} A\otimes(B\otimes C) & \xrightarrow{a} & (A\otimes B)\otimes C & \xrightarrow{c} & C\otimes(A\otimes B) \\ \downarrow{\scriptstyle 1\otimes c} & & & & \downarrow{\scriptstyle a} \\ A\otimes(C\otimes B) & \xrightarrow{a} & (A\otimes C)\otimes B & \xrightarrow{c\otimes 1} & (C\otimes A)\otimes B. \end{array} \tag{4.5}$$

Coherence can also be described in more categorical terms. Let $\mathrm{Pit}_{\otimes}(\mathscr{C})$ be the functor category with objects all permuted iterates $H:\mathscr{C}^n \to \mathscr{C}$ of the functor $\otimes$ and morphisms all natural transformations t between such functors H. Then Pit is a category with a multiplication—and with an action of the groups $S(n)$. Call a subcategory of Pit *symmetric* if it contains with each $t:H \to H'$ of multiplicity n all αt, for $\alpha \in S(n)$. Consider the symmetric multiplicative subcategory of $\mathrm{Pit}_{\otimes}(\mathscr{C})$ generated by all instances of a, a^{-1}, and c. Then the isomorphisms a and c are coherent if and only if this subcategory contains, for objects H and H', at most one morphism $t:H \to H'$. The equivalence of this description of coherence to the previous definition follows from

THEOREM 4.1. *The symmetric multiplicative subcategory of* $\mathrm{Pit}_{\otimes}(\mathscr{C})$ *generated by any set S of morphisms has as its morphisms all composites of permutations of expansions of morphisms of S.*

Proof. The indicated set of morphisms is clearly closed under composition; we must also show it closed under $\otimes$-multiplication and under permutation. The latter follows by (4.3), the former by the previous argument for Theorem 2.1 together with the observation that (4.4) turns permutations followed by $\otimes$ into $\otimes$ followed by a suitable permutation.

In the case at hand, when S consists of instances of a, a^{-1}, and c, this theorem shows that every morphism in the corresponding subcategory has a two-sided inverse. In this case we will show that this subcategory contains with any two objects H and H' at least one morphism $t:H \to H'$. Moreover, the pentagonal and hexagonal conditions suffice for coherence:

THEOREM 4.2. *In a category with a $\otimes$-multiplication, the associativity and commutativity isomorphisms a and c will be coherent if and only if they satisfy the commutativity conditions given by the pentagon of* (3.5) *and the hexagon* (4.5).

The result may be restated by saying that three conditions suffice for coherence: the condition $c^2 = 1$ of (4.2) on c alone, the pentagon condition on a alone, and the hexagon condition on a and c together.

For the proof, consider all the permuted iterates of $\otimes$ of some multiplicity n. Write each as αF, for α in $S(n)$ and F an iterate of $\otimes$, and arrange these αF as "vertices" in $n!$ boxes corresponding to the $n!$ choices of α. Consider the graph formed by the functors as vertices and with edges

$$\square \rightarrow \square \leftarrow \square \rightarrow \cdots \leftarrow \square \qquad (n!\text{ boxes})$$

all permuted expansions of instances of a and c. Each directed path in this graph corresponds to a (composite) natural transformation between the functors represented by the vertices at the ends of the path. We need show only that each closed path in this graph yields the identity natural transformation. By the pentagon condition and the previous treatment for associativity, we know that the result holds for a closed path staying inside any *one* box. Hence we may concentrate on the portions of the path between the boxes, filling in the path within each box as may be convenient.

The vertices are functors of n arguments, say $A_1, \ldots, A_n$. In the box corresponding to the permutation α, the arguments appear in the order $A_{\alpha 1}, \cdots, A_{\alpha n}$. Each edge between boxes comes from a permuted and expanded instance of the commutativity isomorphism c. Thus the edge $1_{A_2} \otimes c(A_3 \otimes A_5, A_4 \otimes A_1)$ must start at the permutation (23541) in the association $2[(35)(41)]$; it must then exchange the block 35 with the block 41. In general, any such edge will interchange two successive blocks of letters in $A_{\alpha 1}, \cdots, A_{\alpha n}$. But now consider the hexagon condition (4.5). In the top row the instance $c(A \otimes B, C)$ interchanges the block AB with the single letter C; the hexagon condition states that this interchange may be replaced by two instances of c which interchange single letters with C. Repeated such replacement using instances of the hexagon shows that any interchange of successive blocks may be replaced by interchanges of successive letters, say A_i and A_{i+1}. This means that any closed path can be replaced by one in which each edge between boxes corresponds to the application of one of the transpositions $\sigma_i = (i, i+1)$, $i = 1, \cdots, n-1$, of successive letters.

Now the symmetric group $S(n)$ is generated by the transpositions σ_i; hence there exists at least one path between any two vertices (and hence at least one natural isomorphism generated by a and c between any two permuted iterates of $\otimes$ of the same multiplicity). More important, any closed path will correspond to a relation between these generators σ_i of $S(n)$. Now all relations can be represented as products of conjugates of suitable "defining relations," and the closed paths can be reduced correspondingly. Hence our proof will be complete if we show that each defining relation in these generators gives a closed path which corresponds to a natural transformation equal to the identity. A classical result ([8], [2]) asserts that a set of defining relations on the generators $\sigma_1, \cdots, \sigma_{n-1}$ of $S(n)$ is

$$\begin{array}{ll} \sigma_i^2 = 1, & i = 1, \cdots, n-1; \\ (\sigma_i \sigma_{i+1})^3 = 1, & i = 1, \cdots, n-2; \\ \sigma_i \sigma_j = \sigma_j \sigma_i, & 1 \leqq i < j-1 \leqq n-2. \end{array}$$

For the first relation $\sigma_i^2 = 1$ we quote the assumed property $c^2 = 1$ of (4.2). For the third relation it suffices to observe the diagram

$$\begin{array}{ccc} (A \otimes B) \otimes (C \otimes D)) & \xrightarrow{1 \otimes c'} & (A \otimes B) \otimes (D \otimes C) \\ \downarrow{\scriptstyle c \otimes 1} & & \downarrow{\scriptstyle c \otimes 1} \\ (B \otimes A) \otimes (C \otimes D) & \xrightarrow{1 \otimes c'} & (B \otimes A) \otimes (D \otimes C) \end{array},$$

which commutes in view of the property (2.3) of $\otimes$-multiplication. Finally, consider $(\sigma_1 \sigma_2)^3 = \sigma_1 \sigma_2 \sigma_1 \sigma_2 \sigma_1 \sigma_2$. To draw the corresponding closed path, we must after each transposition σ_i insert an associativity so that the next transposition can indeed be accomplished by the commutativity c. We get the 12-sided polygon written below (with $\otimes$ omitted)

$$\begin{array}{ccccccc} A(BC) & \overset{1 \cdot c}{—\!—} & A(CB) & \rightarrow & (AC)B & \overset{c \cdot 1}{—\!—} & (CA)B \\ \downarrow & & & & & & \uparrow \\ (AB)C & \cdots & \cdots & \cdots & \cdots & \cdots & C(AB) \\ {\scriptstyle c \cdot 1} \Big| & & & & & & \Big| {\scriptstyle 1 \cdot c} \\ (BA)C & \cdots & \cdots & \cdots & \cdots & \cdots & C(BA) \\ \uparrow & & & & & & \downarrow \\ B(AC) & \overset{1 \cdot c}{—\!—} & B(CA) & \rightarrow & (BC)A & \overset{c \cdot 1}{—\!—} & (CB)A \end{array}$$

(the headless arrows are expanded instances of c; those with heads, of a). We must show this diagram commutative. Insert the two dotted horizontal lines, each an instance of c. The middle rectangle commutes because this c is natural, while the top and bottom are hexagons, both instances of our basic hexagon (4.5). The proof is complete.

This theorem (like the previous one) depends upon a suitable description of the set of all natural transformations $t: F \rightarrow F'$ "generated" by the given isomorphism a, a^{-1}, and c. We have described this set in two ways; first (for the proof) as all compositions of permutations of expansions of instances of a, a^{-1}, and c, and second (more conceptually) as the morphisms of the symmetric multiplicative subcategory generated by these same instances. There is a third description of this set of transformations; namely, as the smallest set of natural transformations containing a, a^{-1}, and c and closed under composition, tensor product, permutation, and substitution. Here substitution is the process which leads from a to one of its instances. More generally, following Godement [4], we define substitution as follows. Let $t: F \to F'$ be a natural transformation between functors of multiplicity n. Let $G_1, \cdots, G_n$ be n functors (permuted iterates of $\otimes$) of multiplicities $p_1, \cdots, p_n$. Then *substitution* of the G_i in t yields the natural transformation

$$t*(G_1, \cdots, G_n);\ F(G_1, \cdots, G_n) \to F'(G_1, \cdots, G_n)$$

between functors of multiplicity $p_1 + \cdots + p_n$, and defined for arguments $C_{ij}, \cdots$, by,

$$t(\cdots C_{ij}, \cdots) = t(G_1(C_{11}, \cdots, C_{1p_1}), \cdots, G_n(C_{n1}, \cdots, C_{np_n})).$$

5. Higher Identity Laws. Let $\mathscr{C}$ be a category with a $\otimes$-multiplication and a distinguished object K, called the *ground object*. (This notation is intended to suggest the case when K is a commutative ring and $\mathscr{C}$ the category of all K-modules, so that K is a left and right identity for the usual tensor product).

Assume that there is a natural transformation

$$e = e(A):\ K \otimes A \to A \tag{5.1}$$

such that each $e(A)$ has a two-sided inverse in $\mathscr{C}$; call e a (left) *identity isomorphism*. An *instance* of e is a natural transformation $e(F)$, where F is any iterate of $\otimes$. If $\mathscr{C}$ also has associativity and commutativity isomorphisms a and c, we again pose the question of coherence, where coherence means that commutativity holds in every diagram with edges permuted expanded instances of a, a^{-1}, c, e, and e^{-1}. In more detail, the set of functors at issue is the smallest set of functors $\mathscr{C}^n \to \mathscr{C}$ which contains both the identity functor and the functor $K \otimes A$ of one variable and which is closed under permutation and tensor multiplication of functors. In the category with objects all such functors we consider the symmetric multiplicative subcategory generated by all instances of a, a^{-1}, c, e, and e^{-1}. Just as in Theorem 4.1, this subcategory consists of all composites of permutations of expansions of such instances. Therefore, a, c, and e are coherent if the subcategory contains to any two of its objects at most one morphism $t: H \to H'$ (it follows readily that it always contains at least one such morphism).

One special case of this coherence condition is

$$e(K) = e(K)\,c(K,K)\colon\ K \otimes K \to K\,; \tag{5.2}$$

two others are the commutativity of the diagrams (with AB short for $A \otimes B$)

$$\begin{array}{ccc} K(BC) & \xrightarrow{a} & (KB)C \\ \big\downarrow e & & \big\downarrow e1 \\ BC & = & BC \end{array} \qquad\qquad \begin{array}{ccc} A(KC) & \xrightarrow{a} & (AK)C \\ \big\downarrow 1e & & \big\downarrow c1 \\ AC & \xleftarrow{e1} & (KA)C\,. \end{array} \tag{5.3}$$

THEOREM 5.1. *In a category with a $\otimes$-multiplication and a ground object K, the associativity, commutativity, and identity isomorphisms a, c, and e will be coherent if and only if they satisfy the commutativity conditions given by the three diagrams* (5.2) *and* (5.3), *the pentagon* (3.5), *the hexagon* (4.5), *and the condition $c^2 = 1$ of* (4.2).

All these necessary conditions hold in the examples of §2 for the usual choices of a, c, and e.

For the proof it is convenient to introduce the "right identity" isomorphism

$$e'(A) = e(A)\,c(A,K)\colon\ A \otimes K \to K \otimes A \to A\,. \tag{5.4}$$

Now consider the graph with vertices all functors F at issue, edges all permutations of expansions of instances of a, a^{-1}, c, e, and e^{-1}. In this graph any two vertices can be joined by at least one path, for suitable instances of e and e' can be used to successively remove all factors K. We must prove that any two paths from F to F' correspond to the same natural isomorphism $F \cong F'$. By applying e or e' to F', we can assume that F' has no factors K. The problem will be reduced to the previous theorem if we show that any path is equivalent to a path which starts by first removing all factors K, and if we show that two different factors K can be removed in either order. This last result is an immediate consequence of (5.2) and naturality. It remains only to show, in effect, that an application of e or of e' after an application of a or of c can be replaced by an application of e or e' first. In many cases this is a consequence of naturality. In other cases, when c is applied first, this is a consequence of the definition (5.4) of e' in terms of e. There remains the case when $a\colon A(BC) \to (AB)C$ is followed by e or e'; this case happens only when one of A, B, or C is K. The result in the first two cases is given by the two cases of commutativity of (5.3). The third case (when $C = K$) requires that the outside rectangle in the following diagram be commutative

$$
\begin{array}{ccccccc}
A(BK) & & & \xrightarrow{\quad\quad a \quad\quad} & & & (AB)K \\
\Big\downarrow {\scriptstyle 1c} & & & & & & \Big\downarrow {\scriptstyle c} \\
A(KB) & \overset{a}{\cdots\rightarrow} & (AK)B & \overset{c1}{\cdots\cdots} & (KA)B & \overset{a}{\leftarrow\cdots} & K(AB) \\
\Big\downarrow {\scriptstyle 1e} & & & & \vdots\, {\scriptstyle e1} & & \Big\downarrow {\scriptstyle e} \\
AB & = & = & = & AB & = & AB.
\end{array}
$$

Fill in the inside dotted arrows as indicated. The top rectangle is then an instance of the hexagon (4.5), while the two bottom squares are the two assumed conditions (5.3). With this argument, the theorem is reduced to the previous cases.

This theorem refers to functors $F:\mathscr{C}^n \to \mathscr{C}$ of *positive* multiplicity n. If we include also the constant K as a functor of zero multiplicity, the condition (5.3) must be replaced by the condition that $c(K,K)=1$.

If commutativity is absent, we must assume two natural isomorphisms

$$
(5.5) \qquad e(A): K \otimes A \to A, \qquad e'(A): A \otimes K \to A.
$$

In this case, replace (5.2) by

$$
(5.6) \qquad e(K) = e'(K):\ K \otimes K \to K
$$

and consider also the commutativities

$$
(5.7) \qquad
\begin{array}{ccc}
A(KC) & \xrightarrow{1e} & AC \\
\Big\downarrow {\scriptstyle a} & & \| \\
(AK)C & \xrightarrow{e'1} & AC
\end{array}
\quad , \qquad
\begin{array}{ccc}
A(BK) & \xrightarrow{1e'} & AB \\
\Big\downarrow {\scriptstyle a} & & \| \\
(AB)K & \xrightarrow{e'} & AB
\end{array}
$$

THEOREM 5.2. *In a category with a $\otimes$-multiplication and a ground object K, the associativity and identity isomorphisms a, e, and e' will be coherent if and only if they satisfy the following five commutativity conditions: The condition* (5.6), *the pentagon condition* (3.5), *the first condition of* (5.3), *and the two conditions* (5.7).

This can be proved by reducing any path to one which first removes all K's and then applying Theorem 3.1, all much as in the proof of the previous theorem.

6. Tensored categories. By a *bicategory* we mean a sextuple

$$(\mathscr{C}, \otimes, K, a, c, e),$$

where $\mathscr{C}$ is a category with a multiplication $\otimes$, K is a selected object of $\mathscr{C}$, and a, c, and e are associativity, commutativity, and identity isomorphisms which are coherent; that is, which satisfy the diagrammatic conditions listed in Theorem 5.1. This coherence allows us to "identify" $A \otimes (B \otimes C)$ with $(A \otimes B) \otimes C$, $A \otimes B$ with $B \otimes A$, and $K \otimes A$ with A according to the given isomorphisms a, c, and e and in the fashion familiar, say, for modules. For example, any category with products and with a terminal object T can be regarded as a bicategory (an object T is *terminal* in $\mathscr{C}$ if to each object C of $\mathscr{C}$ there is a unique morphism $C \to T$ of $\mathscr{C}$). For, take the product $\times$ as the multiplication $\otimes$ in $\mathscr{C}$; this product is naturally associative and commutative, while T satisfies $T \times A \cong A$, and these isomorphisms are readily seen to be coherent.

Bicategories have been introduced independently by several authors. They are in Bénabou [1], with a different but equivalent definition of "coherence," but without any finite list of conditions sufficient for the coherence. In [6] the bicategories are introduced for several purposes: to formulate the notion of a category with a hom functor to some other category, to give a general theory of algebras, and to treat categories of operators and higher homotopies. We will describe briefly each of these objectives.

First, for an arbitrary category $\mathscr{C}$, hom is a bifunctor

$$\hom : \mathscr{C}^{op} \times \mathscr{C} \to \mathscr{S}$$

where $\mathscr{C}^{op}$ is the "opposite" or "dual" of the category $\mathscr{C}$, and $\mathscr{S}$ is the category of sets. Here the composition of homomorphisms is a map of (the product of) sets

$$\hom(B,C) \times \hom(A,B) \to \hom(A,C). \tag{6.1}$$

In an additive category (defined as usual; see for example [7]), hom is a functor to the category of abelian groups, and composition is a morphism

$$\hom(B,C) \otimes \hom(A,B) \to \hom(A,C) \tag{6.2}$$

of abelian groups (this is the condition which states that composition and addition of morphisms satisfy the distributive law). In a differential category, as used by Eilenberg-Moore (in unpublished work), hom is a functor to a category of differential graded modules, and composition is a morphism of such modules, much as in (6.2) except that now $\otimes$ denotes the tensor product of such modules. In all three of these cases we have to do with a category whose hom-functor has values in a bicategory (e.g., sets with $\otimes = \times$ or abelian groups with the usual $\otimes$-multiplication) and whose composition is a morphism as in (6.2).

Second, define a *tensored category* (cf. [6]) to be a bicategory $\mathscr{C}$ in which

$\mathscr{C}$ is abelian and the functor $\otimes$ is right exact in each of its variables separately. An *algebra* in $\mathscr{C}$ may thus be defined to be an object Λ of $\mathscr{C}$ together with two morphisms

$$p = p_\Lambda : \Lambda \otimes \Lambda \to \Lambda, \quad u = u_\Lambda : K \to \Lambda$$

of $\mathscr{C}$ which represents the usual "product" and "identity element" of Λ. Associativity and the other axioms for an algebra can now be expressed via diagrams in the tensored category $\mathscr{C}$, and the usual formal properties of Λ-modules and tensor products of Λ-modules and of algebras can be developed (cf. [6]) so as to include all the familiar cases (rings, graded algebras, bigraded algebras, differential graded algebras, and the like).

Finally, consider a bicategory $\mathscr{C}$ whose objects are generated under $\otimes$-multiplication by a single object B. We can then take the powers of B, in canonical form, to be

$$B^0 = K, \; B^1 = B, \quad B^{n+1} = B^n \otimes B$$

By coherence, every other object of $\mathscr{C}$ will have a unique isomorphism to some one B^n. Hence we may without loss restrict the objects of the category to the powers B^n; indeed, we may say that the objects of the category are just the natural numbers n. However, each permutation α in $S(n)$ induces by coherence a unique morphism $B^n \to B^n$. Thus the bicategory with one generator B may be presented as follows. It is a category $\mathscr{H}$ with objects the natural numbers $\{0, 1, 2, \cdots\}$. For each n, the symmetric group $S(n)$ is given as a subgroup of the group of all invertible elements in $\hom(n, n)$; in particular, the identity permutation in $S(n)$ is the identity morphism $n \to n$. There is also a given bifunctor $\otimes$: $\mathscr{H} \times \mathscr{H} \to \mathscr{H}$ with object function

$$m \otimes n = m + n$$

and with mapping function which assigns to $f: m \to n$ and $g: m' \to n'$ a morphism $f \otimes g: m + m' \to n + n'$ satisfying the usual conditions (2.2) and (2.3) for a functor. These structures $S(n)$ and $\otimes$ on $\mathscr{H}$ satisfy the following three axioms. The associative law for $\otimes$ asserts that

$$f \otimes (g \otimes h) = (f \otimes g) \otimes h;$$

the permutation law requires that $\alpha \in S(n)$ and $\gamma \in S(m)$ give

$$\alpha \otimes \gamma = \alpha \times \gamma \in S(n + m)$$

for $\alpha \times \gamma$ defined as in (4.4). Finally, for any m and m', let $\tau_{m,m'}$ be that permutation in $S(m + m')$ which interchanges the first block of m letters and the second block of m' letters. For any $f: m \to n$ and $f': m' \to n'$ in $\mathscr{H}$, the third axiom requires that

$$\tau_{(n,n')}(f \otimes f') = (f' \otimes f)\tau_{(m,m')}.$$

A category with these structures satisfying these axioms is called a PROP (short for product and permutation category). Categories of this type have arisen in current studies by J. F. Adams and this author on higher homotopies for cohomology operations. Similar types of categories have arisen in Lawvere's studies in functorial semantics [5]. A number of further results on such categories will be available: for example, it can be demonstrated that there exists a "free" PROP on a given set $g_i: m_i \to n_i$ of morphisms as free generators. And, as noted above, each bicategory generated by a single object can be represented as a PROP.

To summarize: in a category, the functor hom is part of the formal structure: a *tensored category* formalizes both of the basic functors hom and $\otimes$ of homological algebra.

REFERENCES

1. Bénabou, Jean, *Categories avec multiplication*, Comptes Rendue Acad. Sci. Paris *256* (1963), 1887–1890.
2. Burnside, W., *Theory of Groups of Finite Order*, Dover, 1955, Note C.
3. Epstein, D. B. A., *Steenrod operations in abelian categories*, forthcoming.
4. Godement, R., *Théorie des Faisceaux*, Hermann, Paris, 1958.
5. Lawvere, W. V., *Functorial Semantics*, Proc. Nat. Acad. Sci. *50* (1963), 869-872.
6. Mac Lane, S., *Categorical Algebra*, Colloquium Lectures, Am. Math. Soc., 1963.
7. Mac Lane, S., *Homology*, Springer and Academic Press, Heidelberg and New York, 1963.
8. Moore, E. H., *Concerning the abstract group of order k!... isomorphic with the symmetric ... substitution group on k letters*, Proc. London Math. Soc. (Ser. 1) *28* (1897), 357–366.
9. Mostow, G. D., Sampson, J. H., and Meyer, J.-P., *Fundamental Structures of Algebra*, McGraw Hill, New York, 1963.
10. Stasheff, J. D., *Homotopy associativity of H-spaces*, I, Trans. Am. Math. Soc. *108* (1963), 275-292.

Note added April 4, 1979

The definition of coherence used in this paper involves a subtle error. Coherence for associativity a is defined by saying that all diagrams of the following class commute: Vertices interates of the binary tensor product $\otimes$, edges expansions of instances of the associativity a. Now in a particular category, it might happen that two formally different iterates of $\otimes$ were in fact equal; if so, the proof of coherence (Theorem 3.1) is ambiguous. This trouble was pointed out to me by G. M. Kelly. It can be repaired (as in our joint paper 109 below) by insisting that the diagrams involved have vertices which are formal, not actual, iterates.

THE MILGRAM BAR CONSTRUCTION AS A TENSOR PRODUCT OF FUNCTORS

BY

SAUNDERS MAC LANE

1. Introduction. The classifying space for a topological group can be constructed "algebraically" or "geometrically". For graded differential modules, Eilenberg-Mac Lane [7] used a strictly algebraic bar construction, and Cartan [1] pointed out that this bar construction could be handled effectively by algebraic analogues of classifying space and fiber bundle techniques. Several authors re-translated these algebraic devices back into geometry (Dold-Lashof [4], Rothenberg-Steenrod [14]. Finally in [13] Milgram described for a topological monoid (= associative H-space with identity) a classifying space which had direct ties to the corresponding algebraic bar construction. Steenrod in [15] has developed further properties of Milgram's classifying space. Here we shall show how these properties can be conveniently reformulated using certain recently developed categorical techniques, notably the construction of the tensor product of two functors. The possibility of this reformulation was recognized by both Allen Clark [2] and the author; the author has also much profited from conversations with Milgram on this topic.

Our treatment will begin with a description of some functorial methods. We then indicate that these methods may illuminate the geometrical facts by giving a simple proof of the Hausdorff character of the classifying space, and by using the triangulation of prisms in proving that this construction preserves products.

2. Coends. A category $\mathcal{C}$ is cocomplete if every functor $T: J \to \mathcal{C}$ from a small category J to $\mathcal{C}$ has a colimit $\underrightarrow{\mathrm{Lim}}\ T$ in $\mathcal{C}$. We recall that such a colimit is an object $A = \underrightarrow{\mathrm{Lim}}\ T$ in $\mathcal{C}$, regarded as a constant functor $A: J \to \mathcal{C}$, together with a natural transformation $\lambda: T \to A$ which is universal among natural transformations from T to constant functors. If J is discrete (= all arrows in T are identities = J is a set), a colimit is just a coproduct $\coprod Ti$ of the objects for $i \in J$. If J is the category with just two non-identity arrows $\rightrightarrows$, a colimit is a coequalizer. A category $\mathcal{C}$ which has all small coproducts and all coequalizers (of pairs $\rightrightarrows$) is necessarily cocomplete.

Let $H(j, k)$ be a functor of two variables $j, k \in J$, contravariant in the first variable and covariant in the second; we write $H: J^{op} \times J \to \mathcal{C}$, where J^{op} denotes the category "opposite" or "dual" to J. It has recently turned out to be very useful to define a supernatural transformation $\beta: H \to B$, with B an object of $\mathcal{C}$, to be a family of arrows $\beta_j: H(j, j) \to B$ of $\mathcal{C}$, one for each object j, such that the square

$$(1) \qquad \begin{array}{ccc} H(k,j) & \xrightarrow{H(1,t)} & H(k,k) \\ {\scriptstyle H(t,1)}\downarrow & & \downarrow{\scriptstyle \beta_k} \\ H(j,j) & \xrightarrow[\beta_j]{} & B \end{array}$$

commutes for every arrow $t: j \to k$ of J. (In [12], Mac Lane called β a diagonal spread; Eilenberg-Kelly [6] called β an extraordinary natural transformation and considered more general cases.) There is a familiar example of the notion dual to supernatural. If $*$ denotes the one point set and J is any category, the family of arrows $\beta_j: * \to \hom(j, j)$ which send the one point to the various identity arrows $j \to j$ define a dually supernatural transformation.

A coend of a functor $H: J^{op} \times J \to \mathcal{C}$ is an object E of $\mathcal{C}$ together with a supernatural $\kappa: H \to E$ which is universal among supernatural transformations from H. Thus κ is supernatural, and if $\beta: H \to B$ is any supernatural transformation, there is a unique arrow $f: E \to B$ with $\beta_j = f\kappa_j$ for all j. As for any universal, a coend of H is unique, up to isomorphism, if it exists. Following Yoneda [18] we use an integral notation for the coend

$$B = \int H = \int^j H(j, j) = \text{the coend of } H: J^{op} \times J \to \mathcal{C}.$$

As the square (1) indicates, a coend is a special sort of colimit, taken not over J but over a suitable category constructed from J (and indeed this construction goes back to Kan). Put differently, if J is small and $\mathcal{C}$ cocomplete, every functor H has a coend, which may be calculated as the coequalizer of a pair of arrows h_1 and h_2,

(2) $$\coprod_j H(k, j) \underset{h_2}{\overset{h_1}{\rightrightarrows}} \coprod_j H(j, j) \longrightarrow \int^j H(j, j),$$

where h_1 and h_2 are defined on the components of the coproduct by $H(1, t)$ and $H(t, 1)$. Here the first coproduct is taken over all arrows t of J. It is clearly sufficient to take this coproduct over arrows t which (with the identities) generate J under composition.

Though a coend is just a special kind of colimit, it appears to deserve a special name and notation because Day and Kelly [3] and Benabou (unpublished) have shown that "coend" does not reduce to colimit for relative categories (categories with hom-sets in some "closed" category like the category of abelian groups). Starting there, Dubuc in [5] has shown the further effective use of the concept of coend in treating closed categories and categories relative to closed categories.

3. Tensor Products of Functors. Let $\mathcal{X}$ be a cocomplete category with a functor $\otimes: \mathcal{X} \times \mathcal{X} \to \mathcal{X}$; often this "tensor product" will be associative up to a coherent natural isomorphism and will have an identity, up to coherent isomorphism, but we do not require these assumptions. Then two functors $F: J^{op} \to \mathcal{X}$ and $G: J \to \mathcal{X}$ of opposite variance have a tensor product which is an object

(3) $$F \otimes_J G = \int^j (F_j) \otimes (G_j)$$

of $\mathcal{X}$. In view of the description given above for the coend, this object may also be described as the coproduct $\coprod_j (F_j) \otimes (G_j)$ modulo suitable identification (one for each arrow $j \to k$ of J). If J is a ring, regarded as a preadditive category with one object, while $\mathcal{X}$ is the category $\mathbf{Ab}$ of abelian groups with the usual tensor product $\otimes = \otimes_Z$, then a contravariant additive functor $F: J^{op} \to \mathcal{X}$ is just a right J-module, while a covariant additive $G: J \to \mathcal{X}$ is a left J-module. The tensor product $F \otimes_J G$ of these functors is then exactly the usual tensor product of modules: There is just one object j in J, so the coproduct $\coprod_j$ is just the abelian group $F \otimes G$ (generated by the usual elements $f \otimes g$ for $f \in F$, $g \in G$); the arrows $t: j \to j$ are just the elements of the ring J, and the identifications a[re] the usual ones $ft \otimes g = f \otimes tg$.

The interested reader may verify that many of the formal identities for tensor products of modules carry over to tensor products of functors.

4. Compactly Generated Spaces. In our case $\mathcal{X}$ will be the category $\mathcal{C}$[$\mathcal{G}$] of compactly generated Hausdorff topological spaces. This category has been carefully examined by Steenrod in [15]. Recall that a Hausdorff space Y is compactly generated when each subset that intersects every compact set of Y in a closed set is itself a closed set, and that a morphism between compactly generat[ed] spaces $Y \to Y'$ is just a continuous map between the associated topological spac[es]. This last property states that $\mathcal{CG}$ is a full subcategory of $\mathbf{Haus}$, the category of Hausdorff spaces and continuous maps between them; we write the inclusion functor as $i: \mathcal{CG} \to \mathbf{Haus}$. Steenrod in [15] constructs to each Hausdorff space X a compactly generated space $k(X)$ and a continuous map $\varepsilon_X: ikX \to X$ (in fact, is the identity function) which is universal from compactly generated spaces to X

i.e., that every continuous $f: iY \to X$ with compactly generated Y factors uniquely as $f = \varepsilon_X f'$ (in fact f' is the same function as f). This statement (without the rest of Steenrod's Theorem 3.2) proves that k is a functor $\mathbf{Haus} \to \mathcal{CG}$ right adjoint to the inclusion functor i and hence that $\mathcal{CG}$ is a coreflective subcategory of $\mathbf{Haus}$, with k the coreflection.

Now any right adjoint preserves products, when they exist. The usual cartesian product topology gives a product $X_1 \times_C X_2$ on $\mathbf{Haus}$; hence any two spaces Y_1, Y_2 in $\mathcal{CG}$ have a product there given as $Y_1 \times Y_2 = k(iY_1 \times_C iY_2)$, with the evident projections on Y_1 and Y_2; we note that this categorical product is the usual cartesian product set with a topology which is not always the usual cartesian product topology. With this product, $\times : \mathcal{CG} \times \mathcal{CG} \to \mathcal{CG}$ is a tensor product, in the sense of §3.

5. <u>The Simplicial Category</u>. Regard each finite ordinal number n as the ordered set $\{0, 1, \ldots, n-1\}$ of all smaller ordinals, and let Δ be the category whose objects are all finite ordinals and whose arrows $f: n \to m$ are all weakly monotone (i.e., order-preserving) functions. The arrows of Δ thus include the usual "face" and "degeneracy" operations d_i and s_i,

$$0 \longrightarrow 1 \rightrightarrows 2 \rightrightarrows\!\!\!\!\rightarrow 3, \ldots, \quad 1 \longleftarrow 2 \leftleftarrows 3 \ \ldots$$

Explicitly, $d_i: n \to n+1$ for $i = 0, \ldots, n$ is that monotone injection with image $\{0, \ldots, \hat{i}, \ldots, n\}$, and $s_i: n+1 \to n$ is that monotone surjection with $s_i(i) = s_i(i+1)$ for $i = 0, \ldots, n-1$. These arrows satisfy the familiar identities and generate under composition the whole category Δ. If Δ^+ is the full subcategory omitting the object 0, a functor $(\Delta^+)^{op} \to \mathcal{C}$ is thus a simplicial object in the category $\mathcal{C}$ in the usual sense, except that our dimension n is <u>one greater</u> than the usual geometric dimension. In particular, a functor on $(\Delta^+)^{op}$ to the category of sets is a simplicial set (= complete semisimplicial complex), while a functor on

$\mathbf{\Delta}^{op}$ to sets is an augmented simplicial set.

We systematically include 0 as an object of $\mathbf{\Delta}$ because it will enable us to wholly dispense with the familiar but troublesome manipulation of identities on faces and degeneracies.

Now $\mathbf{\Delta}$ can be interpreted topologically as the category of "standard simplices". This amounts to defining the functor $\sigma_o : \mathbf{\Delta} \to \mathcal{CG}$ which sends the ordinal n to the standard $(n-1)$-dimensional simplex σ_{n-1} (and 0 to the empty set). With Milgram we regard σ_{n-1} as the set of real numbers $t_1, \ldots, t_{n-1}$ with $0 \le t_1 \le \ldots \le t_{n-1} \le 1$ with the Euclidean topology, while $\sigma_o(d_i) = d_i$ and $\sigma_o(s_i) = s_i$ are the usual face and degeneracy maps

(4)
$$\begin{aligned} d_i(t_1, \ldots, t_{n-1}) &= (0, t_1, \ldots, t_{n-1}), & i &= 0, \\ &= (t_1, \ldots, t_i, t_i, \ldots, t_{n-1}), & 0 &< i < n, \\ &= (t_1, \ldots, t_{n-1}, 1), & i &= n, \end{aligned}$$

(5)
$$s_i(t_1, \ldots, t_n) = (t_1, \ldots, \hat{t}_{i+1}, \ldots, t_n), \qquad 0 \le i < n.$$

6. The Milgram Bar Construction. By a $\mathcal{CG}$-monoid G we mean a CG-space G with a continuous multiplication $G \times G \to G$ (written as a product, $(x, y) \longmapsto xy$) which is associative and has a (two-sided) unit e; thus G is just like an associative H-space with unit except that the topology of G must be $\mathcal{CG}$ and the multiplication must be continuous in the compactly generated topology on the product $G \times G$. A left action of G is a pair (ν, A) where A is in $\mathcal{CG}$ and $\nu : G \times A \to A$ in $\mathcal{CG}$ is associative (in the usual way) and makes the unit e act as the identity. All possible left actions (ν, A) of G, with the evident morphisms, form a category $\mathbf{Mod}_G$, and $(\nu, A) \mapsto A$ is a faithful functor to $\mathcal{CG}$. Moreover, each power $G^n = G \times \ldots \times G$ carries an evident left action of G (multiply in the left-most factor). In particular, the one point space $G^o = *$

carries a (trivial) left-action.

For each G we can now define a functor

$$G^{\bullet}: \mathbf{\Delta}^{op} \longrightarrow \mathbf{Mod}_G$$

which sends n to the n-fold product G^n with the left-action just described, and $d_i: n \to n+1$, $s_i: n+1 \to n$ to the $\mathbf{CG}$-maps $d^i: G^{n+1} \to G^n$ and $s^i: G^n \to G^{n+1}$ defined on (n+1)-tuples $(x_o, \ldots, x_n) \in G^{n+1}$ by

$$\begin{aligned}(6)\qquad d^i(x_o, \ldots, x_n) &= (x_o, \ldots, x_{i-1}, x_i x_{i+1}, \ldots, x_n)\ , & 0 \le i < n\\ &= (x_o, \ldots, x_{n-1})\ , & i = n\end{aligned}$$

$$(7)\qquad s^i(x_o, \ldots, x_{n-1}) = (x_o, \ldots, x_i, e, x_{i+1}, \ldots, x_{n-1})\ , \quad 0 \le i < n.$$

These maps all respect the action of G (on the first factor x_o); they satisfy (the duals of) the face and degeneracy relations on d_i and s_i, so do define the requisite contravariant functor. These maps d_i and s_i are, of course, the familiar ones used to describe the standard resolution used in defining the cohomology of a group G, and these identities are well known from that case.

For the cohomology of groups one can also "divide out" the left-action of G by applying the functor $\mathbf{Z} \otimes_G -$ to this resolution (with trivial right action of the group G on $\mathbf{Z}$). In the present case there is a similar functor $* \times_G -$, where the one point set $*$ has trivial right-action by G. Applied to $G^{\bullet}$, it gives a functor

$$G^{\#}: \mathbf{\Delta}^{op} \longrightarrow \mathbf{CG}$$

which we may describe directly as follows: $G^{\#}$ sends both 0 and 1 to the one point space $*$, and each $n > 1$ to the (n-1)-fold product space G^{n-1} with face and degeneracy maps $d^i: G^n \to G^{n-1}$, $s^i: G^{n-1} \to G^n$ defined by

$$\begin{aligned}(8)\qquad d^i(x_1, \ldots, x_n) &= (x_2, \ldots, x_n)\ , & i = 0\ ,\\ &= (x_1, \ldots, x_i x_{i+1}, \ldots, x_n)\ , & 0 < i < n,\\ &= (x_1, \ldots, x_{n-1})\ , & i = n\ .\end{aligned}$$

(9) $\quad s^i(x_1,\ldots,x_{n-1}) = (x_1,\ldots,x_i,e,x_{i+1},\ldots,x_{n-1})$, $0 \le i < n$;

these formulas apply for $n \geq 1$; for $n = 0$, $d^o: * \to *$ is the only possible function.

Both functors $G^\bullet$ and $G^\#$ can be defined conceptually (no d_i, s_i) using the fact that the monoid $2 \to 1 \leftarrow 0$ in $\mathbf{\Delta}$ is universal.

If we regard the first functor $G^\bullet$ as a functor to $\mathcal{CG}$ (compose with the forgetful functor $\mathcal{M}od_G \to \mathcal{CG}$), there is an evident natural transformation

$$p: G^\bullet \to G^\#,$$

with $p_n(x_o,x_1,\ldots,x_n) = (x_o,\ldots,x_n)$; this amounts to "dividing out" the left-action of G.

The Milgram bar construction now assigns to each $\mathcal{CG}$-monoid G two topological spaces, a "total space" E(G) and a "base space" B(G),

(10) $$E(G) = G^\bullet \times_{\mathbf{\Delta}} \sigma_o = \int^n G^n \times \sigma_{n-1} ,$$

(11) $$B(G) = G^\# \times_{\mathbf{\Delta}} \sigma_\bullet = \int^n G^{n-1} \times \sigma_{n-1} ,$$

moreover, the natural transformation p defined above gives a continuous map $p_G: E(G) \to B(G)$.

In order that the coends used above should exist, we must work in a cocomplete category. So let us calculate the coends above in $\mathcal{T}op$, and so regard E(G) and B(G) for the moment as topological spaces.

This construction is identical, up to language, to that given by Milgram in [13]. For example, the description of E(G) as a coend shows that E(G) is the disjoint union $\coprod_n (G^n \times \sigma_{n-1})$ with certain identifications. Since $\sigma_o(0)$ is the empty set, this is a union for $n > 0$; this amounts to saying that the tensor products could just as well have been taken over the category $\mathbf{\Delta}^+$. The identifications are

(12) $\quad (d^i\xi, \tau) = (\xi, d_i\tau) \qquad \xi \in G^{n+1}$, $\tau \in \sigma_{n-1}$,

(13) $\quad (s^i\xi, \tau) = (\xi, s_i\tau) \qquad \xi \in G^n$, $\tau \in \sigma_n$;

these, with the explicit formulas above for d's and s's , are exactly Milgram's identifications.

We next show that E(G) is a Hausdorff space in two ways -- by direct examination (§ 7) and by categorical devices (§ 8).

7. The Separation Axiom Explicitly.

THEOREM. For each $\mathcal{CG}$-monoid G, the spaces E(G) and B(G) are compactly generated.

Proof. The space E(G) is defined by an identificatiom map

$$\coprod_n G^n \times \sigma_{n-1} \longrightarrow E(G).$$

Now by hypothesis each product $G^n \times \sigma_{n-1}$ is completely generated, hence so is the coproduct (disjoint union) of these spaces. Hence it will be enough to show that E(G) is Hausdorff, since Steenrod ([15], Lemma 2.6) proves that an identification map (= a proclusion) $X \to Y$ with X in $\mathcal{CG}$ and Y Hausdorff has Y in $\mathcal{CG}$.

Now write a point of $G^n \times \sigma_{n-1}$ as $(x_o, x_1, \ldots, x_{n-1}; t_1, \ldots, t_{n-1})$ where the $x_i \in G$ and the t_i are real numbers with $0 \leq t_1 \leq \ldots \leq t_{n-1}$. The identifications above may be used to (possibly) shorten the presentation of the corresponding point of E(G): If $t_1 = 0$, then $(t_1, \ldots, t_{n-1}) = d_o(t_2, \ldots, t_{n-1})$, so

$$(x_o, x_1, \ldots, x_{n-1}; 0, t_2, \ldots, t_{n-1}) = (x_o x_1, x_2, \ldots, x_{n-1}; t_2, \ldots, t_{n-1}).$$

Similarly, if $t_i = t_{i+1}$ we may replace x_i and x_{i+1} by their product and drop t_{i+1}, while if $t_{n-1} = 1$ we may drop both x_{n-1} and t_{n-1}. Again, if one of $x_1, \ldots, x_{n-1}$ is the unit e of G, we may shorten the presentation. As a result, every point of E(G) is equal to a point in normal form

$\eta = (y_o, y_1, \ldots, y_{m-1}; s_1, \ldots, s_1, \ldots, x_{m-1})$, with $y_1 \neq e, \ldots, y_{m-1} \neq e$ and $0 < s_1 < \ldots < s_{m-1} < 1$. As Steenrod proves ([16], Corollary 5.4) the normal form is unique. This can also be observed by giving a direct process for getting the normal form in one step: Take $y_o = x_o x_1 \cdots x_i$ where i is 0 or the last index with $t_i = 0$, take $y_1 = x_{i+1} \dots x_j$ and $s_1 = t_{i+1} = \dots = t_j$, where $t_j < t_{j+1}$ unless this $y_1 = e$, in which case we drop this y_1 and s_1 and take y_1 to be the product of the next (and largest) string of x's for which the corresponding string of t's is equal, and continue in this way.

This presentation allows us to define certain elementary neighborhoods of a point η in normal form $\eta = (y_o, y_1, \dots, y_{m-1}; s_1, \dots, s_{m-1})$. In G, choose open neighborhoods

$$U_o, U_1, \dots, U_{m-1}, U_m \quad \text{of} \quad y_o, y_1, \dots, y_{m-1}, \text{ and } e,$$

respectively, with U_k and U_m disjoint, for every $k \neq m$; in the unit interval, choose disjoint open neighborhoods

$$K_o, K_1, \dots, K_{m-1}, K_m \quad \text{of} \quad 0, s_1, \dots, s_{m-1}, \text{ and } 1 .$$

Then for each n define an open set $V_n \subset G^n \times \sigma_{n-1}$, as follows: A point $\xi = (x_o, x_1, \dots, x_{n-1}; t_1, \dots, t_{n-1})$ lies in V_n if (i) each t_i lies in some K_j; (ii) there is at least one t_i in each of $K_1, \dots, K_{m-1}$ (and hence $n \geq m$); (iii) if $t_1, \dots, t_j$ is the list (possibly empty) of all the t's in K_o, then the product $xx_1 \dots x_j$ is in U_o; (iv) if $t_i, \dots, t_\ell$ is the list of all the t's in K_k, then the product $x_i \dots x_\ell$ is in U_k, for each $k = 1, \dots, m$. Since the product map $\mu : G \times G \to G$ is continuous, this description does give an open set V_n in $G^n \times \sigma_{n-1}$. Moreover, any identification $\xi = \xi'$ in the tensor product clearly carries a ξ in V_n onto a ξ' in $V_{n'}$; in particular, the provision that $U_k \cap U_m = \emptyset$ for $k \neq m$ insures that no identification can drop all the t's lying in K_k. Hence the union of all these V_n for various n provides an open set in $\coprod G^n \times \sigma_{n-1}$ which defines an open set, call it V, in $E(G)$. This V is the desired "elementary" neighborhood of the given point η in normal form.

Now take two distinct points η and η', both in normal form. We can then separate them by elementary neighborhoods V and V' of the above special form.

Thus, if $\eta' = (y'_0, \ldots, y'_{n-1}; s'_1, \ldots, s'_{n-1})$ is such that there is some s'_j equal to none of the $s_1, \ldots, s_{m-1}$ in η, we choose the K_i above so that none of them contains s', and then choose a suitable neighborhood K'_j of s'_j; an easy argument shows $V \cap V' = \emptyset$. On the other hand if $\eta' = (y'_0, \ldots, y'_{m-1}; s_1, \ldots, s_m)$ with the same s_i as for η, we can appeal to the fact that G is Hausdorff to choose strings of neighborhoods U_i, U'_i so that $V \cap V' = \emptyset$. This completes the proof of the theorem for E(G). The proof for B(G) is analogous.

In the treatment by Steenrod [16], the Hausdorff property is obtained by making the additional assumption that G and its unit e form an NDR (a neighborhood deformation retract; see [15]). The above argument avoids the assumption, essentially by showing that the identifications made in forming E(G) involve only very "nicely" situated face and degeneracy maps on the standard simplices.

8. The Separation Axiom Categorically. One may also construct E(G) and B(G) directly in the category $\mathcal{CG}$ of compactly generated spaces in virtue of the following result:

THEOREM. The category $\mathcal{CG}$ is cocomplete.

Proof. Since the coproduct of compactly generated spaces is again such, it will suffice to construct coequalizers in $\mathcal{CG}$. The construction depends on the adjoints to inclusion functors i and i':

$$\mathcal{CG} \underset{i}{\overset{k}{\rightleftarrows}} \mathbf{Haus} \underset{h}{\overset{i'}{\rightleftarrows}} \mathbf{Top} \;;$$

here k is the right adjoint used by Steenrod, which turns each Hausdorff space X into a compactly generated space kX (same points, more closed sets), while h is the left adjoint to i', which sends each topological space Y to its "largest Hausdorff quotient" h(Y). The existence of h follows readily from Freyd's adjoint functor theorem [8]. Since h is a left adjoint, it preserves all colimits which exist, and in particular preserves coequalizers.

The coequalizer of a pair of maps $X \rightrightarrows Y$ in $\mathcal{CH}$ may now be calculated as follows. First take the coequalizer in $\mathcal{T}op$ and apply the left adjoint h to get a coequalizer $p: Y \to C$ in $\mathcal{H}aus$. Form kC.

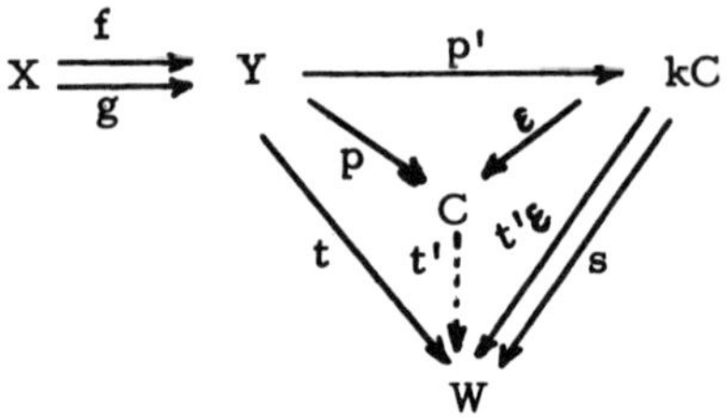

Since $\varepsilon: kC \to C$ is universal, there is a unique $p': Y \to kC$ with $\varepsilon p' = p$; we claim that p' is the desired coequalizer in $\mathcal{CH}$. Clearly $p'f = p'g$. If $t: Y \to W$ is any arrow in $\mathcal{CH}$ with $tf = tg$, then $t = t'p$ for some t', because p is a coequalizer in $\mathcal{H}aus$ and hence $t = (t'\varepsilon)p'$. If also $t = sp'$ for some other arrow s in $\mathcal{CH}$, then $t'\varepsilon = s$ on the level of sets (because p' is surjective); hence $t'\varepsilon$ is unique with $(t'\varepsilon)p' = t$, and $p': Y \to kC$ is the desired coequalizer. This proves the theorem.

For each $\mathcal{CH}$-monoid G we can now calculate the spaces E(G) and B(G) directly as coends in $\mathcal{CH}$. This will give all requisite properties of the classifying space B(G) directly, without recourse to the detailed separation argument of the last paragraph. That argument now serves only to show that, given the $\mathcal{CH}$ spaces $G^n \times \sigma_{n-1}$, we get the <u>same</u> coends

$$E(G) = \int^n G^n \times \sigma_{n-1}, \quad B(G) = \int^n G^{n-1} \times \sigma_{n-1}$$

calculated in $\mathcal{T}op$ or in $\mathcal{CH}$ -- and this information does not seem essential.

9. <u>Acyclicity and other Properties</u>. The essential property of the space E(G) is its acyclicity, which may be established by constructing a suitable contraction of E(G) to a point. This contraction is in fact just the tensor product of two familiar contractions: The contraction h which carries the face $d_0\sigma_{n_1}$ of the

n-simplex to the opposite vertex, and the standard contraction h' of the (algebraic) bar construction. To state this in detail, we use the <u>translation</u> functor $T: \mathbf{\Delta} \to \mathbf{\Delta}$ which carries each ordinal n to $n+1$ and each monotone $f: n \to m$ to $Tf: n+1 \to m+1$; here $(Tf)(0) = 0$ and $(Tf)(i) = 1 + fi$ for $i \neq 0$. Now, if t is any real number, set $\beta(t) = t$ or 1 according as $t \leq 1$ or $t > 1$. Then for I the unit interval define for each ordinal n a continuous map $h_n : I \times \sigma_{n-1} \to \sigma_n$ by (for $s \in I$, $0 \leq t_1 \leq \ldots \leq t_{n-1} \leq 1$)

$$h_n(s, t_1, \ldots, t_{n-1}) = (s, \beta(s + t_1), \ldots, \beta(s + t_n)) .$$

Then $h_n(0, --): \sigma_{n-1} \to \sigma_n$ is the face operator d_o and $h_n(1, --)$ maps σ_{n-1} onto one point -- the point $(1, 1, \ldots, 1)$; geometrically, h_n is the obvious linear map deforming $d_o \sigma_n = \sigma_{n-1}$ through σ_n to the opposite vertex. Moreover, one verifies readily that h is in fact a natural transformation of functors

$$h: I \times \sigma_\bullet \to \sigma_\bullet T, \qquad h_n(0, n) = d_o .$$

(This can be checked readily by replacing the $t_1, \ldots, t_{n-1}$ by the usual barycentric coordinates in a simplex.)

On the other hand, for a $\mathbf{CH}$-monoid G, there is a natural transformation

$$h': G^\bullet \to GT ,$$

defined for elements $x_o, \ldots, x_{n-1}$ of G by

$$h'_n(x_o, \ldots, x_{n-1}) = (e, x_o, \ldots, x_{n-1}) \in G^{n+1} .$$

This is not a G-map, but a continuous map $G^n \to G^{n+1}$, and is indeed just the standard contracting homotopy of the algebraic bar construction; one verifies that h' is natural and that $d_o h'_n$ is the identity. In order to combine these two we first construct the diagram

$$\begin{array}{ccc} G^{Tn} \times \sigma_{T(n-1)} & \longrightarrow & \int^n G^{Tn} \times \sigma_{T(n-1)} \\ \| & & \downarrow g \\ G^{n+1} \times \sigma_n & \longrightarrow & \int^n G^n \times \sigma_{n-1} \end{array}$$

where the horizontal maps are the (universal) supernatural transformations for the

the two coends displayed. By universality, there exists then a (continuous) map g, as displayed, vertically, on the right.

Since h' and h are both natural transformations, we have a continuous map $\chi = \int^n h' \times h$ as in the diagram

$$\begin{array}{ccccc} I \times E(G) & & \xrightarrow{\quad \bar{h} \quad} & & E(G) \\ \| & & & & \| \\ \int^n G^n \times (I \times \sigma_{n-1}) & \xrightarrow{\chi} & \int^n G^{Tn} \times \sigma_{T(n-1)} & \xrightarrow{g} & \int^n G^n \times \sigma_{n-1} . \end{array}$$

The composite $\bar{h}$ is the desired contraction of E(G); $\bar{h}(1, --)$ maps E(G) to a point, while $\bar{h}(0, --)$ is the identity in virtue of the two equations $h(0, --) = d_o$, and $d_o h'$ = identity, as noted above.

This argument shows that Milgram's contraction map, like his classifying space, is a tensor product.

Other properties of E(G) and B(G), as previously established by Milgram and Steenrod, can now be put in this language. For example, one may prove readily that the projection $E(G) \to B(G)$ is an identification map (a proclusion in the terminology of Steenrod); indeed, it is the identification map obtained by "dividing out" the action of G.

10. <u>The Product Theorem.</u> Steenrod, going beyond Milgram, proved in [16] a product formula

THEOREM. For two $\mathcal{CH}$-monoids G and H there is a homeomorphism

$$E(G) \times E(H) \cong E(G \times H)$$

which is also a homomorphism for the action of $G \times H$ on both sides.

We now show how this follows readily from the tensor product formalism. First, the product $G \times H$ in $\mathcal{CH}$ has two projections $p: G \times H \to G$ and $q: G \times H \to H$ which give a diagram

(14) $$E(G) \longleftarrow E(G \times H) \longrightarrow E(H).$$

LEMMA. In sets, the diagram (14) is a product diagram.

Proof. Represent an element of $E(G)$ by a pair $(\xi, u) \in G^n \times \sigma_{n-1}$, where $\xi = (x_0, \dots, x_{n-1})$ and $u = (t_1, \dots, t_{n-1}) \in \sigma_{n-1}$. The identification rules above, by inserting more units e in ξ, allow us to replace u by any longer string involving at least the same t's. Therefore points $(\xi, u) \in E(G)$ and $(\eta, v) \in E(H)$ can be written in the form $(\xi, u) \equiv (\xi', w)$ and $(\eta, v) \equiv (\eta', w)$ with the same string w. This shows that every pair of points in $E(G)$ and $E(H)$ comes by projections from a single point $((\xi', \eta'), w)$ in $E(G \times H)$. By putting the latter point into normal form (see §7 above) its uniqueness readily follows. This proves that we have a product in (14).

Now return to the theorem. Since $E(G) \times E(H)$ and $E(G \times H)$ are both products (as sets) of $E(G)$ and $E(H)$ there are unique functions

$$E(G) \times E(H) \underset{\psi}{\overset{\varphi}{\rightleftarrows}} E(G \times H)$$

which commute with the projections to $E(G)$ and to $E(H)$. Since $E(G) \times E(H)$ is a product in $\mathcal{CH}$, the function ψ is continuous. Since both composites $\psi\varphi$ and $\varphi\psi$ are identities, it remains only to show φ continuous. By a standard result, it will suffice to prove continuity on $G^{n+1} \times \sigma_n \times H^{m+1} \times \sigma_m$, regarded as embedded in $E(G) \times E(H)$. Now $\sigma_n \times \sigma_m$ is a prism (product of simplices) and has a standard triangulation into simplices, with one simplex of dimension $n+m$ for each shuffle κ of n letters through m letters (see Mac Lane [12], page 243). Specifically, the point $(u, v) \in \sigma_n \times \sigma_m$, with $u = (t_1, \dots, t_n)$ and $v = (s_1, \dots, s_m)$, belongs to the shuffle κ if κ puts the real numbers $t_1, \dots, t_n, s_1, \dots, s_m$ into non-decreasing order $w = (r_1, \dots, r_{n+m})$. Moreover, this shuffle κ can then be represented by the monotonic surjections $f: n+m+1 \to n+1$ and $g: n+m+1 \to n+1$ in the category Δ so that, for the maps $\sigma_f: \sigma_{n+m} \to \sigma_n$ and $\sigma_g: \sigma_{n+m} \to \sigma_n$ corresponding to f and g under the functor $\sigma_\bullet$ are

$$\sigma_f(r_1, \ldots, r_{n+m}) = (t_1, \ldots, t_n), \quad \sigma_g(r_1, \ldots, r_{n+m}) = (s_1, \ldots, s_m).$$

But this means that

$$(\xi, u) = (\xi, \sigma_f w) = (G^f \xi, w),$$

$$(\eta, v) = (\eta, \sigma_g w) = (G^g \eta, w).$$

Therefore, by the lemma above, the map φ when restricted to this particular $(n+m)$-simplex is

$$(\xi, u), (\eta, v) \longmapsto (G^g \xi \times G^g \eta, w),$$

and this map is clearly continuous. Since φ is continuous on each simplex $G^{n+1} \times H^{n+1} \times (\sigma_{n+m})$ of maximum dimension, in the subdivision it is continuous. Q.E.D.

The shuffles here used are exactly those involved in the usual Eilenberg-Zilber theorem. The proof of this theorem (for simplicial complexes) uses only the formal properties of the shuffles, and not the triangulation which they provide for the prism $\sigma_n \times \sigma_m$. The present proof really uses that triangulation. Moreover, Steenrod's proof of this result also depends essentially on the same triangulation.

Steenrod also proves that $E(G) \to B(G)$ is a quasifibration, as well as other results when G is a CW-complex. We have not tried to present these in functorial form.

Bibliography

[1] Cartan, H. Sur les groupes d'Eilenberg-Mac Lane $H(\pi, n)$: I, II, Proc. Nat. Acad. Sci. USA 40 (1954), 467-471, 704-707.

[2] Clark, Allen, Categorical Constructions in Topology, (Preprint, 1969).

[3] Day, B.J. and Kelly, G.M. Enriched Functor Categories, Reports of the Midwest Category Seminar III, Berlin, Heidelberg, New York (Springer) 1969, pp. 178-191.

[4] Dold, A. and Lashof, R. Principal Quasifibrations and Fiber Homotopy Equivalence of Bundles, Ill. J. Math 3 (1959), 285-305.

[5] Dubuc, E. Kan Extensions in Enriched Category Theory. Forthcoming in Springer Lecture Notes Series.

[6] Eilenberg, S. and Kelly, G.M. A Generalization of the Functorial Calculus, Journal of Algebra 1 (1964), 397-402.

[7] Eilenberg, S. and Mac Lane, S. On the Groups $H(\pi, n)$, I, II. Ann. of Math. 58 (1953), 55-106 and 60 (1954), 49-139.

[8] Freyd, P. Abelian Categories, An Introduction to the Theory of Functors, New York (Harper and Row), 1963.

[9] Kan, D.M. Adjoint Functors. Trans. Amer. Math. Soc. 87 (1958), 294-329.

[10] __________, Functors Involving C.S.S. Complexes, Trans. Amer. Math. Soc. 87 (1958), 330-346.

[11] Mac Lane, Saunders. Categorical Algebra, Bull. Amer. Math. Soc. 71 (1965), 40-106.

[12] __________, Homology, Heidelberg and New York (Springer), 1963.

[13] Milgram, R.J. The Bar Construction and Abelian H-Spaces. Ill. J. Math. 11 (1967), 242-250.

[14] Rothenberg, M. and Steenrod, N.E. The Cohomology of Classifying Spaces and H-Spaces. Bull. Amer. Math. Soc. 71 (1965), 872-875.

[15] Steenrod, N.E. A Convenient Category of Topological Spaces, Mich. Math. J. 14 (1967), 132-152.

[16] __________, Milgram's Classifying Space of a Topological Group. Topology, 7 (1968), 319-368.

[17] Ulmer, F. Representable Functors with Values in Arbitrary Categories. Journal of Algebra 8 (1968), 96-129.

[18] Yoneda, N. On Ext and Exact Sequences. Jour. Fac. Sci., Univ. Tokyo 8 (1960), 507-576.

Added in Proof:

Stasheff in his theses (Homotopy Associativity of H-Spaces I & II, Trans. AMS 108 (1963)) points out that his construction of $XP(\infty)$ reduces to

$$\cup \Delta^n \times X^n / \sim$$

if X is a monoid and is homeomorphic to a reduced version of the Dold Lashof B_X (p. 289-290).

In addition, Stasheff and Milgram point out the correspondence

$$\overline{B}(C_*(X)) \to C_*(B_X) ,$$

which for Milgram is an isomorphism (using cellular theory in the rare cases X has a cellular multiplication) and in general is a homotopy equivalence.

HAMILTONIAN MECHANICS AND GEOMETRY

SAUNDERS MAC LANE, University of Chicago

1. Introduction. Let me begin by thanking the organizers of the Chauvenet Symposium for the opportunity to talk at the dedication of Chauvenet Hall at the U. S. Naval Academy. It is a pleasure to honor the long-tradition of devotion to Mathematics at the Academy, and to recognize again the impetus which William Chauvenet gave, in his teaching and writing, to the importance of effective exposition. Mathematical ideas do not live fully till they are presented clearly, and we never quite achieve that ultimate clarity. Just as each generation of historians must analyse the past again, so in the exact sciences we must in each period take up the renewed struggle to present as clearly as we can the underlying ideas of mathematics.

The thesis of this article is that classical mechanics and recent conceptual methods in abstract mathematics have a lot to do with each other. Many abstract ideas of pure geometry originally arose in the study of mechanics; for example, the cotangent bundle of a differentiable manifold appeared first as the phase space of Hamiltonian mechanics. Many cumbersome developments in the standard treatments of mechanics can be simplified and better understood when formulated with modern conceptual tools, as in the well-known case of the use of the "universal" definition of tensor products of vector spaces to simplify some of the notational excesses of tensor analysis as traditionally used in relativity theory. This article will develop a few more such cases. It has been stimulated by the pioneering work of Mackey [**12**], Smale [**13**], Abraham [**1**], Sternberg [**14**], and many others.

2. Quantities. Newton's Laws give differential equations for the trajectory of a mass point moving in Euclidean 3-space or along some surface or curve in that space. When there are several such points (say k of them) each moving in 3-space, they can be regarded as a single point in $3k$-space. Hence we consider Newton's Laws for a point moving in *any* "configuration space" M. On M we

Reprinted from the Amer. Math. Monthly 77 (1970) 570-586, by permission of the Mathematics Association of America.

are to deal with various measurable physical "quantities" such as the distance of a point from the origin, or the potential energy of a point of given mass, or the x and y coordinates of a point in the plane, or the latitude and longitude of a point on an equatorial belt U of a sphere S^2. Each such quantity is a continuous real-valued function g defined on all the points of M or of some "piece" U of M. These quantities, taken together, have a number of formal properties: To define a quantity g on the union $U' \cup U''$ of two pieces, it suffices to define quantities g' on U' and g'' on U'', so that they agree on the intersection $U' \cap U''$; if g and h are quantities defined on U, so is any smooth function $f(g, h)$ of g and h. These properties (the list is not complete) which describe quantities on configuration space have a direct mathematical interpretation. Take configuration space to be a differentiable manifold (of class C^∞), the pieces U of the space M to be open sets, and the quantities defined on U to be those continuous functions $g: U \to \mathbf{R}$ to the real numbers $\mathbf{R}$ which are smooth functions (of class C^∞). We recall that such a manifold can be defined as a topological space together with a suitable function S which assigns to each open set U of the space the set $S(U)$ of those continuous functions $g: U \to R$ which are the smooth functions on U. The axioms on S include those suggested above; for example, the description of a quantity g on $U' \cup U''$ in terms of its restrictions g' and g'' to U' and U'' is the essential element in the assertion that S is a "sheaf". Using S, one also describes the smooth maps of M to another manifold N: A function $f: M \to N$ is continuous if the inverse image $f^{-1}V$ of any open set V of N is open in M; it is smooth if for each smooth $h: V \to \mathbf{R}$ the composite hf is smooth (i.e., $hf \in S(f^{-1}V)$).

The axioms for a manifold M also describe the dimension n of M in terms of the existence of coordinates. Early in mechanics it was noted that problems are often best treated not with the original (rectangular) coordinates $x^1, \cdots, x^n$, but in terms of other "curvilinear" coordinates, say $q^1, \cdots, q^n$. These appear in the axiom on M which requires for each point $a \in M$ a neighborhood U (a "coordinate neighborhood") and n smooth functions, $q^1, \cdots, q^n: U \to \mathbf{R}$ such that the correspondence $p \mapsto q^1(p), \cdots, q^n(p)$ is a diffeomorphism (one-one, onto, and smooth in both directions) of U to an open set in $\mathbf{R}^n$. Here the smooth functions f on $\mathbf{R}^n$ are to be exactly the functions $f(x^1, \cdots, x^n)$ of class C^∞ (continuous derivatives of all orders). Hence the axiom states that the smooth functions g on a coordinate neighborhood U are precisely the C^∞ functions $g = f(q^1, \cdots, q^n)$ of the n coordinates. Many basic properties of manifolds and mechanical systems can be understood and expressed more clearly when they are formulated in a fashion independent of any particular choice of coordinates. Thus at the very beginning we speak of a quantity g or of a map $g: U \to \mathbf{R}$, not of a smooth function $g(q^1, \cdots, q^n)$ of some coordinates.

This review points up the fact that the physicists' use of "quantities" anticipates the mathematicians' use of smooth function, is a starting point for the currently active subject of sheaf theory **[15]**, and matches the modern emphasis on coordinate-free presentation of geometry. Often a book on physics will contrast a mathematical presentation of an idea (using coordinates) with a physical presentation (not using coordinates); in such cases the contrast is really between two mathematical presentations, one with coordinates and one coordinate-free.

3. Velocity and tangent vectors. With Newton's Laws we may write down the usual equations of motion for our point of mass m moving on the configuration space M. These equations involve both velocity and acceleration, and so are second order differential equations; initial conditions will be given by specifying the initial position (by coordinates $q^1, \cdots, q^n$) and initial velocity (by the corresponding coordinates $v^1, \cdots, v^n$) (the coordinate v^i is that often written as $\dot{q}^i$, where the dot suggests "time derivative"). Taken all together, these $2n$ coordinates are not coordinates on M, but coordinates on another smooth manifold of twice the dimension, the *tangent bundle* $T_\bullet M$ of M. This bundle may be described in physical terms as the manifold of all positions-and-velocities on M, or in pictorial terms as the manifold of all tangent arrows attached to points a of M, or more formally as the manifold $T_\bullet M$ whose points are all ordered pairs (a, u) where a is a point of M and u a tangent vector at that point. To complete its description as a manifold, one then specifies that the smooth functions are (locally) the smooth functions of the $2n$ coordinates $q^1, \cdots, q^n, v^1, \cdots, v^n$, taken in a coordinate neighborhood of $T_\bullet M$ which is composed of a coordinate neighborhood U on M together with *all* tangent vectors at all of its points. With this description of the tangent bundle $T_\bullet M$ we note that the function indicated by $(a, u) \mapsto a$ is a smooth map $\pi_\bullet : T_\bullet M \to M$, called the *projection* of the tangent bundle. Under this projection the inverse image of each point $a \in M$ is the tangent space $T_a M$, which consists of all vectors tangent to M at a. This is not only a submanifold, but a linear vector space, and the full linear group (of nonsingular $n \times n$ matrices) acts on this space. These data, present and suitably interrelated in the tangent bundle, are essential elements in the recent treatment of many geometric ideas by vector bundles, K-theory, fiber bundles, and fiber spaces (see Husemoller [7]): They were really discovered and used a long time ago in mechanics.

4. Tangent and cotangent bundles. But we are ahead of our story; to complete the description of the tangent bundle we need an invariant description of the tangent spaces $T_a M$. Sometimes tangent vectors (or vector fields) are described as operators on smooth functions; we want a more symmetric treatment. Recall [9] that every finite dimensional vector space W determines in conceptual fashion another vector space of the same dimension, the *dual space* W^* consisting of all linear functions $W \to \boldsymbol{R}$. If $W = T_a M$ is a tangent space, this dual is called the *cotangent space* $T^a M$ to the manifold M at a. We describe them together: Every path determines a tangent vector, every quantity a cotangent vector. In detail, let $g: M \to \boldsymbol{R}$ be a quantity defined on M (or at least in some neighborhood of the chosen point $a \in M$) and let h be a smooth *path* on M passing through the point a. If we use the coordinate t (for "time") on the real numbers (a manifold) $\boldsymbol{R}$, this path can be represented as a smooth map $h: \boldsymbol{R} \to M$ sending the origin $t=0$ to the chosen point $a\in M$. Thus we have two smooth maps

$$R \xrightarrow{h} M \xrightarrow{g} R;$$

their composite, differentiated at $t=0$, yields a real number

$$\langle g, h\rangle_a = \left[\frac{d}{dt}(g \circ h)\right]_{t=0}.$$

If h' is a second path through a we say that h is *tangent* to h' at a (in symbols, $h\equiv_a h'$) if and only if $\langle g, h\rangle_a = \langle g, h'\rangle_a$ for all quantities g. The *tangent vector* $\tau_a h$ to h at a is then defined to be just the equivalence class consisting of all h' with $h\equiv_a h'$. Dually, two quantities g and g' are *cotangent* at a (in symbols $g\equiv_a g'$) if and only if $\langle g, h\rangle_a = \langle g', h\rangle_a$ for all smooth paths h through a; then the differential (cotangent vector) $d_a g$ to g at a is just the equivalence class of all these g'. Let $T^a M$ be the set of all cotangent vectors at a. Now all quantities $f, g, \cdots$ defined near a form a real vector space under the operations of addition and multiplication by real scalars: These operations, transferred to equivalence classes $d_a f$, $d_a g$, make $T^a M$ a real vector space. Moreover, the formula above for $\langle g, h\rangle_a$ defines a pairing (or function of two variables) which assigns to a cotangent vector $d_a g$ and a tangent vector $\tau_a h$ a real number,

$$(d_a g, \tau_a h) \mapsto \langle d_a g, \tau_a h\rangle = \langle g, h\rangle_a.$$

Addition of tangent vectors is now defined so as to make this function bilinear. This makes the set $T_a M$ of all tangent vectors to M at a into a real vector space, and the pairing makes $T^a M$ its dual space; indeed each $d_a g$ is the linear function $\langle d_a g, -\rangle : T_a M \to R$.

We emphasize that this conceptual description of tangent and cotangent vectors does match the intuitive picture of these vectors: A tangent vector to a path at a is an arrow measuring the velocity along the path; a cotangent vector is a set of level lines (contour lines straightened up) given by a quantity g defined near a, and the pairing of the cotangent vector with the tangent vector gives the rate of change of the level-lined quantity along the path.

This invariant description also agrees with the description of tangent vectors in terms of local coordinates $q^1, \cdots, q^n$ on M: The pairing is given by the familiar formula

$$\langle d_a g, \tau_a h\rangle = \sum_{i=1}^{n}\left(\frac{\partial g}{\partial q^i}\right)_a\left(\frac{dq^i}{dt}\right)_0, \qquad \frac{d}{dt} \quad \text{on} \quad R,$$

which gives the derivative of a composite function. In writing this formula we are dealing with n quantities $q^i: M\to R$, the coordinates on M, and a fixed map $h: R\to M$. By this map h the quantities q^i become quantities $q^i h: R\to R$ on the 1-dimensional manifold R, and $(dq^i/dt)_0$ in the formula refers to the derivatives of those latter composite quantities with respect to the coordinate t on R. Thus the tangent vector $\tau_a h$ has the n coordinates

$$v^1 = \left(\frac{dq^1h}{dt}\right)_0, \cdots, v^n = \left(\frac{dq^nh}{dt}\right)_0,$$

the tangent vector space T_aM is n-dimensional, and the whole tangent bundle $T_\bullet M$ can be made a differentiable manifold in exactly one way so that the $2n$ local coordinates are $q^1, \cdots, q^n, v^1, \cdots, v^n$. At the same time in the cotangent vector space T^aM each vector d_af has the n coordinates

$$p_1 = \left(\frac{\partial g}{\partial q^1}\right)_a, \cdots, p_n = \left(\frac{\partial g}{\partial q^n}\right)_a,$$

and T^aM is n-dimensional. The cotangent bundle $T^\bullet M$ has points all pairs (a, d_af), and can be made a differentiable manifold in exactly one way so that the $2n$ quantities $q^1, \cdots, q^n, p_1, \cdots, p_n$ become local coordinates.

There is an evident function which sends every point (a, d_af) on the cotangent bundle to the original point a on configuration space. This is a smooth map $\pi^\bullet$, called the *projection* of the cotangent bundle, displayed as

$$\pi^\bullet : T^\bullet M \to M \text{ by } (a, d_af) \mapsto a.$$

In coordinates, the point $(q^1, \cdots, q^n, p_1, \cdots, p_n)$ projects to $(q^1, \cdots, q^n)$.

5. Differentials and differential forms. If g is a quantity defined on an open set U of M, it determines at each point a of U a cotangent vector d_ag. This determination is really a function from U to the cotangent bundle. We call this function the *differential* of g: It is a smooth map

$$dg: U \to T^\bullet M \quad \text{by} \quad a \mapsto (a, d_ag).$$

This map is a *cross section* of the cotangent bundle in the sense that the composite $\pi^\bullet \circ dg: U \to U$ is the identity. This cross section dg is just the usual "differential" of the function g.

In general *any* cross section θ of the cotangent bundle $T^\bullet M$ over U is called a 1-*form* on U. In case U is a coordinate neighborhood with coordinates $q^1, \cdots, q^n$, these 1-forms θ are precisely the things which can be written as expressions

$$\theta = f_1dq^1 + \cdots + f_ndq^n$$

in terms of the 1-forms dq^i (the differentials of the coordinates) and arbitrary smooth functions f_i on U. In this expression the formal operations of addition (of two 1-forms) and multiplication (of a 1-form by a quantity) have the natural invariant description: Since cotangent vectors at a can be added, we can add two 1-forms θ, ψ by adding their values $\theta a + \psi a$ in each T^aM; since cotangent vectors can be multiplied by scalars, we can multiply θ by a quantity f by the rule $(f\theta)a = (fa)(\theta a)$. Under these operations the 1-forms on U constitute a module over the ring $S(U)$ of quantities on U. Observe by the way that this (classical? modern?) concept of module dominates many recent developments in algebra **[9]** and homological algebra **[10]**.

The systematic use of cotangent bundles, 1-forms, and the like is a major tool in geometry (see, for examples, Flanders [4]). In a moment we shall see that the cotangent bundle is simply the physicists' phase space. First we pause to consider the dual concepts. A cross section X of the tangent bundle $T_{\bullet}M$ over U is called a *vector field* on U because it assigns, in a smooth way, a tangent vector X_a at a to each $a \in U$. Moreover, a vector field X operates on quantities, assigning to each quantity g on U the quantity Xg defined at each point a by $(Xg)(a) = \langle d_a g, X_a \rangle$; this is called the *Lie derivative* of g along the field X. For example, in a coordinate neighborhood we can take at each point a the path $h_{1,a}$ "along the q^1 axis"; this is the path specified in coordinates as $h_{1,a}(t) = (q^1a+t, q^2a, \cdots, q^na)$. Then $a \mapsto \tau_a h_{1a}$ is a vector field, the field of "unit" tangent vectors along the q^1 axis. This vector field is usually written as $X_1 = \partial/\partial q^1$, because the coordinate formula above for the basic pairing shows that its action on a function g is given by $X_1 g = \partial g/\partial q^1$. In the coordinate neighborhood every vector field X can be written uniquely in the form

$$X = f_1 \frac{\partial}{\partial q^1} + \cdots + f_n \frac{\partial}{\partial q^n},$$

where the f_i are quantities in U. In this formula, the addition and scalar multiplication by quantities f are defined for vector fields just as for forms: Under these operations the vector fields on U form a module over the ring $S(U)$ of quantities on U.

6. Lagrange's equations. We return to mechanics, where Newton's Law gives the equations of motion for a point in configuration space in the familiar form

$$m_i(d^2x^i/dt^2) = F_i, \qquad i = 1, \cdots, n,$$

where the $x^1, \cdots, x^n$ are rectangular coordinates and F_i is the ith component of the force. These equations may be rewritten in more invariant form using the quantities which represent potential and kinetic energy. The *kinetic energy* T, in the familiar rectangular coordinates, has an expression such as

$$T = (1/2)\, m_1(v^1)^2 + \cdots + (1/2)\, m_n(v^n)^2 = (1/2)\, m_1(\dot{q}^1)^2 + \cdots + (1/2)\, m_n(\dot{q}^n)^2.$$

In other coordinates there will be different formulas, but *any* such T depends on the coordinates in the tangent bundle, so T is a quantity $T: T_{\bullet}M \to R$ on that bundle. (The letter T is overused here, thanks to separate traditions, in which plain T is a quantity (kinetic energy) and dotted T a functor, the tangent bundle functor.) In a conservative system, the ith component F_i of the force on the particle can be expressed as the quantity

$$F_i = -\frac{\partial V}{\partial x^i},$$

where V is the potential energy. This V, originally a quantity $V: M \to \mathbf{R}$ on the configuration space M, is also a quantity on the tangent bundle $T.M$, namely the composite function $T.M \to M \to \mathbf{R}$. Therefore the difference

$$L = T - V: T.M \to \mathbf{R}$$

is also a quantity on $T.M$. It is called the *Lagrangian*. Now in the rectangular coordinates above,

$$m_i \frac{d^2x^i}{dt^2} = \frac{d}{dt}\left(\frac{\partial T}{\partial \dot{x}^i}\right) = \frac{d}{dt}\left(\frac{\partial L}{\partial \dot{x}^i}\right), \qquad F_i = -\frac{\partial V}{\partial x^i} = \frac{\partial L}{\partial x^i},$$

so the familiar equations of motion can be written as

$$\frac{d}{dt}\left(\frac{\partial L}{\partial \dot{x}^i}\right) - \frac{\partial L}{\partial x^i} = 0, \qquad i = 1, \cdots, n.$$

Now transform to *any* other local coordinates $q^1, \cdots, q^n$; a calculation shows that these equations again take the form

$$\frac{d}{dt}\left(\frac{\partial L}{\partial \dot{q}^i}\right) - \frac{\partial L}{\partial q^i} = 0, \qquad i = 1, \cdots, n.$$

The *Lagrange equations* have the *same* form in *any* local coordinates; we omit the classical explanation: These are the equations for the extremal curves (trajectories) which minimize a suitable integral.

7. The Legendre transformation. Now we shift from the tangent bundle to the cotangent bundle. This shift to "phase space" is accomplished by a "Legendre transformation" which uses the quantity L to carry tangent spaces to cotangent spaces. This transformation, which sometimes appears adventitious when presented via coordinates, can be described invariantly, following an observant suggestion of Sternberg [**14**]; we start with the case of one tangent space (at one point $a \in M$) and get the following situation:

THEOREM. *If W is a real vector space, regarded as a differentiable manifold, then at each point $u \in W$ the tangent and cotangent spaces have canonical isomorphisms $T_u W \cong W$, $T^u W \cong W^*$. Any quantity L on W determines by $u \mapsto d_u L$ a smooth mapping $\mathfrak{L}_L: W \to W^*$, called the Legendre transformation for L. In particular, if L is a positive definite quadratic form on W, making W an inner product space, then $\mathfrak{L}_L$ is the standard natural isomorphism which identifies the inner product space W with its dual. If $\mathfrak{L}_L$ is invertible, then the inverse $(\mathfrak{L}_L)^{-1}$ is also a Legendre transformation for the quantity H on W^* which is defined in terms of (linear) coordinates $v^1, \cdots, v^n$ on W and the dual coordinates $p_1, \cdots, p_n$ on W^* as*

(1) $$H = v^1 p_1 + \cdots + v^n p_n - L.$$

Proof. Each vector $w \in W$ determines a path $h_w: \mathbf{R} \to W$ at $u \in W$ by $t \mapsto u + tw$; this is just the "radial" path from the point u along the vector w. The correspondence $w \mapsto \tau_u h_w$ gives the desired isomorphism $W \cong T_u W$; its dual is $T^u W \cong W^*$.

The mapping $\mathcal{L}_L: W\to W^*$ comes from the very definition of the differential; if we describe differentials by coordinates as above, the ith coordinate of $\mathcal{L}_L u$, in symbols $p_i(\mathcal{L}_L u)$, is just the value $(\partial L/\partial V^i)_u$ of the ith partial derivative of L at this point u. Thus we can write (with $v^i = \dot{q}^i$)

$$(2) \qquad p_i \circ \mathcal{L}_L = \partial L/\partial v^i = \partial L/\partial \dot{q}^i, \qquad i = 1, \cdots, n.$$

These equalities (between quantities $W\to \mathbf{R}$) describe the map $\mathcal{L}_L: W\to W^*$ completely in terms of its composites with coordinates on W^*. These equations are often written just as $p_i = \partial L/\partial \dot{q}^i$. This coordinate description of $\mathcal{L}$ is to be understood as an equation between quantities on W; we emphasize that it is just a coordinate version of the simple conceptual assignment $u \mapsto d_u L \in W^*$; in words, send each point of u to the value at u of the differential dL, where the space $T^u W$ of differentials is identified with W^* by the canonical isomorphism we have described.

Next suppose that L is a quadratic form. This is the important case for our purposes, in which W is the tangent space $T_a M$ to a configuration space M at some point a, while $L = T - V$ is the classical Lagrangian. Restricted to one tangent space $T_a M$ this L is (up to a constant) just the familiar positive definite form $\frac{1}{2}\sum m_i v_i^2$ for kinetic energy. This also provides an interpretation for the coordinates p_i, which we introduced just as the coordinates on W^* dual to the given v^i; with this L, $p_i = \partial L/\partial v^i$ is just $p_i = m_i v^i$, the ith-component of the *momentum*. For this reason the $p_1, \cdots, p_n$ are called the *momentum coordinates* in W^*.

Now the statement that the function L is a "quadratic form" is usually explained by saying that L has a quadratic expression in some coordinate system, but it can be also explained in invariant fashion: L is a function $L: W\to \mathbf{R}$ with $L(bu) = b^2 L(u)$ for any scalar b and such that the expression

$$(u, u') = (1/2)\,[L(u + u') - L(u) - L(u')]$$

is a bilinear function—that is, a function linear in each of the vectors u and $u' \in W$. (In other words, a function L is quadratic when its deviation $L(u+u') - L(u) - L(u')$ from additivity is bilinear.) Now the formula above gives a real inner product (u, u') for any two vectors of W; this inner product is symmetric in u and u' by definition, and is positive definite by the assumption on L. Hence W with L is an inner product space (a Euclidean vector space = a finite dimensional Hilbert space) and L is given in terms of the inner product as $L(u) = (u, u)$.

We return to the proof of the theorem. Along the standard path h_w which we introduced above we have

$$L(u + tw) = (u + tw, u + tw) = (u, u) + 2t(u, w) + t^2(w, w),$$

since the inner product is bilinear. Differentiating at $t = 0$ gives

$$\langle d_u L, w\rangle = 2(u, w).$$

In other words (up to the factor 2), $d_u L$ is that linear form on W given by the function $(u, -): W \to W$. This correspondence $u \mapsto d_u L$ is thus precisely the familiar map $u \mapsto (u, -)$ sending W to W^*, for W an inner product space.

The map $\mathfrak{L}_L: W \to W^*$ will be invertible (locally) provided the familiar Jacobian matrix $\|\partial^2 L/\partial v^i \partial v^j\|$ is nonzero at every point of W. When it is invertible, the equation

$$H = \sum_{i=1}^{n} v^i p_i - L \qquad \text{on } W^* \text{ along } W \xleftarrow{\mathfrak{L}_L^{-1}} W^*$$

does define a (smooth) quantity H on W^*. Specifically, in this equation the coordinates p_i on W^* are given as quantities on W^*, while the coordinates v^i and L, originally given as quantities on W, become quantities on W^* by composition with $\mathfrak{L}_L^{-1}$. The equation is thus short for the more explicit equation

$$H = \sum_i (v^i \mathfrak{L}_L^{-1}) p_i - L \mathfrak{L}_L^{-1}$$

(and we emphasize this point because it explains a general method of reading an equation in quantities defined on different spaces in terms of quantities *on* one space *along* a given diagram $W \leftarrow W^*$ of arrows between those spaces). From this formula we may now calculate the Legendre transformation $\mathfrak{L}_H: W^* \to W$. The coordinate formula for $\mathfrak{L}$ (above) is $v^j \mathfrak{L}_H = \partial H/\partial p_j$, with partial derivative calculated on W^*. By the usual rules for the derivative of a product

$$\frac{\partial H}{\partial p_j} = v^j + \sum_i \frac{\partial v^i}{\partial p_j} \cdot p_i - \sum_i \frac{\partial L}{\partial v^i} \frac{\partial v^i}{\partial p_j} = v^j;$$

the two sums cancel because $p_i \circ \mathfrak{L}_L$ is $\partial L/\partial v^i$, by the definition of $\mathfrak{L}_L$. This formula proves that $v^j \mathfrak{L}_H \mathfrak{L}_L = v^j$, so $\mathfrak{L}_H \mathfrak{L}_L = 1$ and $\mathfrak{L}_H$ is indeed (locally) the inverse of $\mathfrak{L}_L$. This completes the proof.

The formula for H may be stated in invariant terms, for any $y \in W^*$, as

$$Hy = \langle y, \mathfrak{L}_L^{-1} y \rangle - L \mathfrak{L}_L^{-1} y.$$

Here $\mathfrak{L}_L^{-1} y \in W$ and $\langle , \rangle$ denotes the pairing $W^* \times W \to \boldsymbol{R}$ which makes W^* the dual of W. With coordinates $p_1, \cdots, p_n$ for y and dual coordinates $v^1, \cdots, v^n$ for $\mathfrak{L}_L^{-1} y$, the value of this pairing is exactly the sum $\sum p_i v^i$ of the previous formula. The above calculation that $\mathfrak{L}_H$ is $\mathfrak{L}_L^{-1}$ may also be put in invariant form, but the translation to this form appears to be cumbersome.

The notation for this theorem has been chosen so as to fit the natural extension from one tangent space to the whole tangent bundle of any manifold M. Let L be a quantity on $T_\bullet M$, and identify each $T_u(T_a M)$ with $T_a M$ and $T^u(T_a M)$ with $(T_a M)^* = T^a M$, as above. A point of $T_\bullet M$ is then a pair (a, u) for $u \in T_a M$, and the quantity L determines, by restriction to each tangent space, a quantity $L | T_a M$ on that space. Then we define

$$\mathfrak{L}_L: T_\bullet M \longrightarrow T^\bullet M, \qquad (a, u) \mapsto (a, d_u(L | T_a M)),$$

a smooth mapping of the tangent bundle $T_{\bullet}M$ to the cotangent bundle $T^{\bullet}M$. This mapping is sometimes called the *fiber derivative* of L (see Abraham [1]) because it arises by taking the differential of L restricted to each fiber. If we put in the projections of these two bundles, the diagram

$$\begin{array}{ccc} T_{\bullet}M & \xrightarrow{\mathcal{L}_L} & T^{\bullet}M \\ \downarrow{\scriptstyle \pi_{\bullet}} & & \downarrow{\scriptstyle \pi^{\bullet}} \\ M & = & M \end{array}$$

commutes; one says that the Legendre transformation carries fibers (tangent spaces) to fibers (cotangent spaces). In coordinates, the formulas for $\mathcal{L}_L$ are just those already given in (2), plus the equations $q^i\mathcal{L}_L = q^i$ on the position coordinates. The case where L restricted to each fiber is a positive definite quadratic form is still applicable as above. Indeed, a manifold M with a smooth function assigning to each tangent space a quadratic form there is just a manifold with a *Riemann metric*. In other words, the kinetic energy in a classical mechanical system may be regarded as a Riemann metric on the configuration space.

8. Hamilton's Equations. This Legendre transformation $\mathcal{L}$ is now applied to the trajectories of the given mechanical system. These trajectories are originally the paths $h: \mathbf{R} \to M$ on configuration space which satisfy the Lagrange equations. Now at each point $h(t)$ of M on such a path h the path itself determines a tangent vector $\tau_{h(t)}h$ to M, and the map $t \mapsto (h(t), \tau_{h(t)}h) \in T_{\bullet}M$ "lifts" the path h to a path $\tilde{h}: \mathbf{R} \to T_{\bullet}M$ on the tangent bundle; this is just the path giving position *and* velocity at any time $t \in \mathbf{R}$. The image under $\mathcal{L}$ of these lifted trajectories will be the trajectories in $T^{\bullet}M$ of the mechanical system. A straightforward calculation from the Lagrange equations shows that these trajectories $\tilde{h}$ on $T^{\bullet}M$ are the solutions of the systems of equations

$$\frac{dq^i}{dt} = \frac{\partial H}{\partial p_i}, \qquad \frac{dp_i}{dt} = -\frac{\partial H}{\partial q^i}, \qquad i = 1, \cdots, n,$$

on $\mathbf{R}$, evaluated along $\tilde{h}: \mathbf{R} \to T^{\bullet}M$. Here "evaluated along $\tilde{h}$" means (as in a previous case) that each of the q^i, the p_i, and the partial derivatives of H, all originally quantities on $T^{\bullet}M$, are to be regarded as quantities on $\mathbf{R}$ (that is, as functions of t) by composition with $\tilde{h}$.

These $2n$ equations are *Hamilton's equations*. They describe the trajectories on *phase space* $T^{\bullet}M$ by a system of first order differential equations. These equations depend on the quantity H on $T^{\bullet}M$; this quantity is the *Hamiltonian* defined above in terms of the Lagrangian. The formulation of these equations also seems to depend on the fact that phase space is the manifold of "positions and momenta" (mathematically, that phase space is a cotangent bundle). This means that we can get these equations in this form for any system of $2n$ local coordinates on $T^{\bullet}M$, *provided* these coordinates are split into two sets: position coordinates $q^1, \cdots, q^n$ and the corresponding momentum coordinates $p_1, \cdots, p_n$. Any position coordinates will do—provided that they are matched with the corresponding momentum coordinates. There is another mystery beyond the use of such pairs of coordinates. A system of first order differential equations on a manifold T^*M is usually given by a vector field X on T^*M; the solutions of the equations may then be described as those paths on the manifold which thread through the vector field, with tangent vectors at each point b exactly that specified as Xb by the vector field at that point. But this system of equations is determined by the quantity H, and what a quantity H gives naturally is a covector field dH, not a vector field.

So we reconsider the properties of phase space. The formula $H = \sum q^i p_i - L$ used to define H suggests the expression

$$\theta = \sum p_i dq^i.$$

Now p_i and q^i are quantities on T^*M, so each dq^i is a 1-form on T^*M, and so is each scalar multiple $p_i dq^i$ and hence θ itself. Now if we make any change in the local coordinates q^i and the corresponding change in the momenta p_i, a calculation shows that this 1-form θ is unchanged. Indeed, this 1-form can be described in invariant fashion, using only the fact that T^*M is a cotangent bundle. As a 1-form on T^*M, θ must be a cross section of the cotangent bundle of T^*M; that is, a smooth map $T^*M \to T^*(T^*M)$ which is a right inverse of the projection $\pi^{**}: T^*(T^*M) \to T^*M$. So take any point $b = (a, d_a g)$ on T^*M and define θ as the map

$$b = (a, d_a g) \mapsto ((a, d_a g), d_b(g\pi^*)).$$

Here $\pi^*: T^*M \to M$ is the projection of the given tangent bundle, so $g\pi^*$ is a quantity on T^*M, and the expression on the right in this formula is a point of $T^*(T^*M)$ with first component $(a, d_a g)$ the given point b. Thus this formula does define a 1-form θ'. Moreover, $\theta' = \theta$: For $\theta = \sum p_i dq^i: T^*M \to T^*(T^*M)$ is a linear combination of the 1-forms dq^i, so is given by the formula (definition of the linear operations $\sum_i$)

$$\theta b = (b, \sum p_i(b) d_b q^i).$$

Now we can write $b = (a, d_a g) = (a, \sum k_i d_a q^i)$ with n scalars k_i; these scalars being exactly the p_i coordinates $k_i = p_i b$. Therefore

$$\begin{aligned} \theta(b) = \theta(a, d_a g) &= (b, \sum k_i d_b q^i) \\ &= (b, \sum k_i d_b(q^i \pi^*)) \\ &= \theta'(a, \sum k_i d_a q^i) = \theta'(a, d_a g), \end{aligned}$$

so $\theta = \theta'$, as claimed. On *every* cotangent bundle there is an invariant 1-form θ.

9. Forms and symplectic manifolds. On any manifold S we now consider 1-forms θ and 2-forms ω. In terms of local coordinates $x^1, x^2, \cdots$ these are expressions (in quantities f_i, g_i) like

$$\theta = f_1 dx^1 + f_2 dx^2 + \cdots, \quad \omega = g_{12} dx^1 \wedge dx^2 + g_{13} dx^1 \wedge dx^3 + \cdots;$$

and any 1-form θ determines a 2-form $d\theta$ by formal application of the differential d as

$$d\theta = \left(\frac{\partial f_2}{\partial x^1} - \frac{\partial f_1}{\partial x^2}\right) dx^1 \wedge dx^2 + \cdots.$$

The conceptual presentation uses the exterior algebra of the cotangent spaces $W^* = T^b S$ of S (see, for example [9]). First, the exterior square $W^* \wedge W^*$ is the vector space generated by all formal products $e \wedge e'$ of the vectors $e, e' \in W^*$, where the exterior product $\wedge$ is required to be bilinear and alternating ($e \wedge e' = -e' \wedge e$, hence $e \wedge e = 0$). If W has a basis of n vectors e, then $W^* \wedge W^*$ has a basis of $n(n-1)/2$ vectors $e^i \wedge e^j$ for $i < j$. Over the manifold S we can construct the exterior bundle $T^* S \wedge T^* S$ whose fiber over each point b is the vector space $T^b S \wedge T^b S$. A 2-form ω on S is then defined to be a smooth cross section of this bundle $T^* S \wedge T^* S \to S$. Since the dx^i for local coordinates x^i form a module basis of the 1-forms, the exterior algebra above shows that the $dx^i \wedge dx^j$ for $i < j$ form a module basis for the 2-forms. This means exactly that any 2-form can be written uniquely in the form $\sum g_{ij} dx^i \wedge dx^j$ displayed above.

In particular, on any cotangent bundle $S = T^* M$, our invariant 1-form $\theta = \sum p_i dq^i$ yields by exterior differentiation a 2-form $\omega = d\theta$ with

$$\omega = \sum_{i=1}^{n} dp_i \wedge dq^i.$$

It is a closed 2-form ($d\omega = 0$ because $d\omega = dd\theta$) and is of maximal rank. The latter means that its exterior power $\omega^n = \omega \wedge \cdots \wedge \omega$ to n factors is nonzero; indeed $\omega^n = \pm k dp_1 \wedge \cdots \wedge dp_n \wedge dq^1 \wedge \cdots \wedge dq^n$ is (up to a factor k) just the familiar "element of $2n$-dimensional volume" on $T^* M$.

Now an even dimensional manifold S equipped with a specified 2-form ω, with $d\omega = 0$, of maximal rank is called a *symplectic manifold*. We have just seen that every cotangent bundle (every phase space $T^* M$) is a symplectic manifold. One reason for naming such manifolds is precisely that the Hamilton equation above can be written down directly for any quantity H on *any* symplectic manifold. This can be seen by choices of coordinates: Given any 2-form ω, with $d\omega = 0$ and of maximum rank, a theorem due to Darboux states that in some neighborhood of each point one can successively choose coordinates $q^1, \cdots, q^n$ and $p_1, \cdots, p_n$ so that ω takes on the standard form $\omega = \sum dp_i dq^i$. With these coordinates we can write down Hamilton's equations. Calculation then shows that a different choice of coordinates with the same properties yields the same trajectories for Hamilton's equation.

There is a better and more conceptual proof which uses the 2-form ω directly. If W^* is the dual of the vector space W, then each $e \in W^*$ is a linear function

$W \to \mathbf{R}$. Because of this, we shall show that each element $t \in W^* \wedge W^*$ yields a linear function

$$t^\flat : W \longrightarrow W^*.$$

Since any t is a linear combination of vectors $e \wedge e'$ for e and $e' \in W^*$, it will suffice to take $t = e \wedge e'$ (and use the corresponding sums of functions $t^\flat$). Thus we define $(e \wedge e')^\flat$ on each $w \in W$ to be an element of W^*; that is, to be the linear function of $w' \in W^*$ given by the formula

$$[(e \wedge e')^\flat w](w') = e(w)e'(w'),$$

where $e(w)$ and $e'(w')$ are real numbers, since e, $e' : W \to \mathbf{R}$. Now consider the 2-form ω on S; at each point $b \in S$, $\omega_b \in T^b S \wedge T^b S$ gives a map $T_b S \to T^b S$ of tangent space to cotangent space. Put together at all points, this is a smooth map

$$\begin{array}{ccc} T_\bullet S & \xrightarrow{\ \omega^\flat\ } & T^\bullet S \\ \downarrow & & \downarrow\uparrow \; dH \\ S & = & S \end{array}$$

of the tangent bundle to the cotangent bundle. When ω is of maximal rank on S, this map is one-one and onto, so its inverse is a smooth map of the cotangent bundle back to the tangent bundle. In particular, given any 1-form on S, such as the differential dH of the Hamiltonian, this map gives a vector field $(\omega^\flat)^{-1} \circ dH$ on S. This is the vector field which is used in Hamilton's equations; therefore these equations arise directly from the quantity H and the basic 2-form ω on the symplectic manifold S.

10. Canonical transformations and generating functions. Specific problems in mechanics are often treated by picking coordinates convenient to the problem in hand; in the symplectic formulation of mechanics, this means picking on S new local coordinates, conventionally written $Q^1, \cdots, Q^n, P_1, \cdots, P_n$, in such a way that the 2-form ω is still expressed as $\omega = \sum dP_i \wedge dQ^i$. Classically, this is called a *canonical transformation* from the canonical coordinates p_i, q^i to the new canonical coordinates P_i, Q^i. These new coordinates need no longer have a direct physical interpretation by momentum and position.

Certain "generating functions" F are classically used to give such canonical transformations. The traditional presentation ([6], [16]) goes about as follows. Take a smooth "generating function" of $2n$ coordinates $q^1, \cdots, q^n, P_1, \cdots, P_n$, and assume the determinant

$$\det \left\| \frac{\partial^2 F}{\partial q^i \partial P_j} \right\| \neq 0$$

everywhere. Now the definitions

$$q^i = q^i, \qquad p_i = \partial F / \partial q^i, \qquad i = 1, \cdots, n,$$

give $2n$ functions which we regard as coordinates, while the parallel definitions

$$Q^i = \partial F/\partial P_i, \qquad P_i = P_i, \qquad i = , \cdots , n$$

again give $2n$ coordinates. The determinant condition implies that we can solve for the $Q^1, \cdots, Q^n, P_1, \cdots, P_n$ as functions of the $q^1, \cdots, q^n, p_1, \cdots, p_n$ or conversely. A calculation (to be done below) shows that this transformation carries the form $\sum dP_i \wedge dQ^i$ to the form $\sum dp_i \wedge dq^i$. Thus F has "generated" a transformation from one set of canonical coordinates to another.

This situation could also be described as follows. Let S be a manifold of even dimension $2n$ which is given as a product $S = S_1 \times S_2$ of two n-dimensional manifolds S_1 and S_2, and let F be a quantity on S. If α is any form on S (a 2-form, a 1-form, or a 0-form; that is, a quantity such as F), then the exterior derivative $d\alpha$ can be written in a natural way as a sum $d\alpha = d_1\alpha + d_2\alpha$, where d_1 is the differential along S_1 and d_2 that along S_2. In particular, d_1F is a 1-form on S, and $d_2(d_1F) = d_1(d_2F)$ is a 2-form on S. Our basic assumption now reads: F is a quantity on a manifold $S = S_1 \times S_2$ such that the 2-form $\omega = d_1d_2F$ has maximal rank. This assumption means exactly that S with the 2-form ω is a symplectic manifold. Indeed $d\omega = 0$ becomes

$$d\omega = d(d_1d_2\, F) = (d_1 + d_2)(d_1d_2F) = d_1^2d_2F + d_1d_2^2F = 0.$$

Choose any local coordinates $q^1, \cdots, q^n$ on S_1 and $P_1, \cdots, P_n$ on S_2. Then $q^1, \cdots, q^n, P_1, \cdots, P_n$ are $2n$ local coordinates on $S = S_1 \times S_2$, and

$$d_1F = \frac{\partial F}{\partial q^1}\, dq^1 + \cdots + \frac{\partial F}{\partial q^n}\, dq^n,$$

hence, applying d_2 in the evident way,

$$\omega = d_2d_1F = \sum_{i=1}^{n}\sum_{j=1}^{n} \frac{\partial^2 F}{\partial q^i \partial P_j}\, dq^i \wedge dP_j.$$

This is the formula by which the quantity F defines a 2-form. Calculating the exterior product $\omega \wedge \cdots \wedge \omega$, to n factors, the condition that it is nonzero becomes exactly the condition that the determinant above is nonzero at all points of S.

From the data $S = S_1 \times S_2$ and F we have thus defined a symplectic manifold $(S, \omega = d_1d_2F)$. In this manifold (and in these coordinates) the basic 2-form does *not* have the usual canonical form. But there are other local coordinates on S; for example, the determinant condition above implies that the quantities $q^1, \cdots, q^n, p_1, \cdots, p_n$ with $p_i = \partial F/\partial q^i$ are also local coordinates. In these coordinates we may calculate the 2-form $\omega_0 = \sum dp_i \wedge dq^i$ by the usual formula

$$\omega_0 = \sum_i dp_i \wedge dq^i = \sum_i \sum_j \frac{\partial^2 F}{\partial p_j \partial q^i} dp_j \wedge dq^i + \sum_i \sum_j \frac{\partial^2 F}{\partial q^j \partial q^i}\, dq^j \wedge dq^i.$$

The first double summation is $\omega = d_1d_2F$, while the second contains terms with

$dq^i \wedge dq^i = 0$, and terms with $dq^j \wedge dq^i = -dq^i \wedge dq^j$ and $i \neq j$, which cancel in pairs; therefore $\omega_0 = \omega$. In the coordinates q^i, p_i, ω becomes canonical. An exactly analogous calculation shows that ω is canonical in the coordinates P_i, $Q^i = \partial F/\partial P_i$. These two calculations together give the calculations previously suggested which show that the transformation from the coordinates q^i, p_i to Q^i, P_i takes the canonical 2-form to itself. But the calculation is best understood by going through the form ω expressed in terms of the intermediate (or original) coordinates $q^1, \cdots, q^n, P_1, \cdots, P_n$.

The same intermediate step can be exhibited in other classical cases (see [6], [16]) of generating functions for canonical transformations. The presentation of these ideas in classical books is sometimes impoverished because the notion of a differential form (other than a 1-form) is wholly avoided. The general methods of E. Cartan for invariant integrals use the calculus of differential forms extensively and go far beyond what we have presented here.

We have considered only changes of coordinates. Dually, there are symplectic transformations between two symplectic manifolds (S, ω) and (S', ω') of the same dimension. These are the smooth maps $f: S \to S'$ which carry the form ω' on S' back to the form ω on S (in the natural sense in which forms go backwards).

11. Hamilton-Jacobi Equations. Many problems of classical mechanics can be solved by first obtaining solutions of a certain first order partial differential equation, the Hamilton-Jacobi equation. We sketch the situation very briefly. Given the Hamiltonian $H: T^*M \to \boldsymbol{R}$, each quantity S on M determines a 1-form $dS: M \to T^*M$ which is a cross section of the cotangent bundle. The composite of these two functions is a quantity on M; setting it equal to 0 gives an equation

$$H \circ dS = 0 \qquad \text{on } M$$

called the *Hamilton-Jacobi* partial differential equation. If we write the Hamiltonian H as a function $H(q^1, \cdots, q^n, p_1, \cdots, p_n)$ of local coordinates and the differential $dS: M \to T^*M$ as the map $a \mapsto (a, d_aS)$, where a has the coordinates $q^1, \cdots, q^n$, this equation appears as

$$H\left(q^1, \cdots, q^n, \frac{\partial S}{\partial q^1}, \cdots, \frac{\partial S}{\partial q^n}\right) = 0;$$

it is a partial differential equation of the first order in S.

On the other hand, for any path $h: \boldsymbol{R} \to M$, giving the quantity S on M allows us to lift this path to $\bar{h} = dS \circ h: \boldsymbol{R} \to T^*M$, a path on the cotangent bundle. The role of the Hamilton-Jacobi partial differential equations may then be formulated in the following theorem (which I take from some lecture notes of George Mackey, who treated the more general case of a "time dependent" Hamiltonian):

THEOREM. *Let S be a solution of the Hamilton-Jacobi partial differential equations for a Hamiltonian quantity H on a cotangent bundle T^*M. Then, if a path h on M satisfies*

$$\frac{dq^i}{dt} = \frac{\partial H}{\partial p_i}, \qquad i = 1, \cdots, n \quad \text{on } \mathbf{R} \text{ along } h \text{ and } \tilde{h},$$

(the first n Hamilton equations) it also satisfies the second n equations

$$\frac{dp_i}{dt} = -\frac{\partial H}{\partial q^i}, \qquad i = 1, \cdots, n \text{ on } \mathbf{R} \text{ along } h \text{ and } \tilde{h}.$$

Conversely, if S is a quantity on M which has this property for all h, then S satisfies the Hamilton-Jacobi partial differential equation.

There are many further connections with the theory of characteristics of first order partial differential equations. This beautiful theory—well presented, say, in [3] and [5], is normally treated analytically, with heavy use of coordinates. It should be possible and useful to have a coordinate-free and more geometrical presentation for this and similar "analytic" theories.

12. Summary. With these lines we have just begun the exploration of the relation between ideas arising in classical mechanics and conceptual methods of modern differential geometry. Much more could be said in differential geometry (see [2], [14]) or in the translation of the classical presentation of mechanics ([6], [16], [18], etc.) into geometrical language. Some portions of this task are carried out at an elementary level in [8] and [11], and at a more advanced level in [1], [12], [17]. We have not touched on the active research in the related topics of structural stability and qualitative dynamics. We have emphasized the idea that quantities are functions and that equations hold on a manifold along a diagram of functions. More use of categorical ideas may be indicated. There has been a suggestion (F. W. Lawvere) that "categorical dynamics" will replace the category of differentiable manifolds by other more flexible categories, axiomatically characterized so as to bring out the tangent bundle and infinitesimal constructions. Here, as throughout mathematics, conceptual methods should penetrate deeper to give clearer understanding.

References

1. Ralph Abraham, Foundations of Mechanics, Benjamin, New York, 1967.

2. R. L. Bishop and S. I. Goldberg, Tensor Analysis on Manifolds, Macmillan, New York, 1968.

3. Constantin Caratheodory, Calculus of Variations and Partial Differential Equations of the First Order (Translated from the German), Holden-Day, San Francisco, 1965.

4. Harley Flanders, Differential Forms, with Applications to the Physical Sciences, Academic Press, New York, 1963.

5. Philipp Frank and Richard von Mises, Die Differentialgleichungen der Mechanik und Physik, 2nd edition, Vieweg, Braunschweig, 1930 and 1935.

6. Herbert Goldstein, Classical Mechanics, Addison-Wesley, Cambridge, Mass., 1951.

7. Dale Husemoller, Fibre Bundles, McGraw-Hill, New York, 1966.

8. L. H. Loomis and S. Sternberg, Advanced Calculus, Addison-Wesley, Reading, Mass., 1968.

9. Saunders Mac Lane and Garrett Birkhoff, Algebra, Macmillan, New York, 1967.

10. Saunders Mac Lane, Homology, Springer, Heidelberg and New York, 1963.

11. ———, Geometrical Mechanics, Parts I and II. Lecture notes by Raphael Zahler, assisted by George Angwin, William Corvette, David Golber, Paul Palmquist, Reinhard Schultz, Martha Smith, and Stanley Weiss. Department of Mathematics, The University of Chicago, 1968.

12. George W. Mackey, The Mathematical Foundations of Quantum Mechanics, Benjamin, New York, 1963.

13. S. Smale, Differentiable dynamic systems, Bull. Amer. Math. Soc., 73 (1967) 747–817.

14. Shlomo Sternberg, Lectures on Differential Geometry, Prentice-Hall, Englewood Cliffs, N. J., 1964.

15. Richard G. Swan, The Theory of Sheaves, The University of Chicago Press, 1964.

16. E. T. Whittaker, A Treatise on the Analytical Dynamics of Particles and Rigid Bodies, 4th edition, Cambridge University Press, 1959.

17. R. Hermann, Differential Geometry and the Calculus of Variations, Academic Press, New York, 1968.

18. L. A. Pars, A Treatise on Analytical Dynamics, Wiley, New York, 1965.

JOURNAL OF PURE AND APPLIED ALGEBRA – Volume 1, No. 1 (1971) pp. 97–140.

COHERENCE IN CLOSED CATEGORIES

G.M. KELLY
The University of New South Wales, Australia

and

S. MACLANE
The University of Chicago, Ill., USA

Received 20 February 1970

§ 1. Introduction

For the purposes of this paper we understand by a *closed category* the following collection of data:

(i) a category $\underline{V}$;

(ii) functors $\otimes$: $\underline{V} \times \underline{V} \to \underline{V}$ and $[\ ,\ ]$: $\underline{V}^{\mathrm{op}} \times \underline{V} \to \underline{V}$;

(iii) an object I of $\underline{V}$;

(iv) natural isomorphisms

$$a = a_{ABC}\colon (A \otimes B) \otimes C \to A \otimes (B \otimes C)\ ,$$

$$b = b_A : A \otimes I \to A\ ,$$

$$c = c_{AB}\colon A \otimes B \to B \otimes A\ ;$$

(v) natural transformations (in the generalized sense of [1])

$$d = d_{AB}\colon A \to [B, A \otimes B]\ ,$$

$$e = e_{AB}\colon [A, B] \otimes A \to B\ .$$

The axioms to be satisfied by these data are that, for all $A, B, C, D \in \underline{V}$, the following diagrams should commute:

Reprinted from the J. Pure Appl. Algebra 1 (1971) 97-140, by permission of North-Holland Publishing Company.

C1

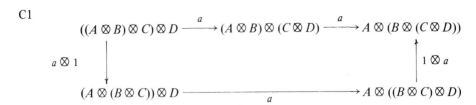

$$\begin{array}{ccccc}
((A\otimes B)\otimes C)\otimes D & \xrightarrow{a} & (A\otimes B)\otimes(C\otimes D) & \xrightarrow{a} & A\otimes(B\otimes(C\otimes D)) \\
\downarrow{\scriptstyle a\otimes 1} & & & & \uparrow{\scriptstyle 1\otimes a} \\
(A\otimes(B\otimes C))\otimes D & & \xrightarrow[a]{\qquad\qquad} & & A\otimes((B\otimes C)\otimes D)
\end{array}$$

C2

$$\begin{array}{ccc}
(A\otimes B)\otimes I & \xrightarrow{a} & A\otimes(B\otimes I) \\
{\scriptstyle b}\searrow & & \swarrow{\scriptstyle 1\otimes b} \\
& A\otimes B &
\end{array}$$

C3

$$\begin{array}{ccc}
A\otimes B & \xrightarrow{c} & B\otimes A \\
& {\scriptstyle 1}\searrow & \downarrow{\scriptstyle c} \\
& & A\otimes B
\end{array}$$

C4

$$\begin{array}{ccccc}
(A\otimes B)\otimes C & \xrightarrow{a} & A\otimes(B\otimes C) & \xrightarrow{c} & (B\otimes C)\otimes A \\
\downarrow{\scriptstyle c\otimes 1} & & & & \downarrow{\scriptstyle a} \\
(B\otimes A)\otimes C & \xrightarrow[a]{} & B\otimes(A\otimes C) & \xrightarrow[1\otimes c]{} & B\otimes(C\otimes A)
\end{array}$$

C5

$$\begin{array}{ccc}
[B,A] & \xrightarrow{d} & [B,[B,A]\otimes B] \\
& {\scriptstyle 1}\searrow & \downarrow{\scriptstyle [1,e]} \\
& & [B,A]
\end{array}$$

C6

$$\begin{array}{ccc}
A\otimes B & \xrightarrow{d\otimes 1} & [B,A\otimes B]\otimes B \\
& {\scriptstyle 1}\searrow & \downarrow{\scriptstyle e} \\
& & A\otimes B
\end{array}$$

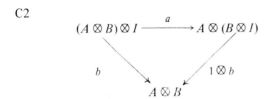

Such a closed category, which we denote by the single letter $\underline{V}$, is not essentially different from what was called in [2] a "symmetric monoidal closed category". In particular we have a natural isomorphism

$$\pi : \underline{V}(A \otimes B, C) \to \underline{V}(A, [B, C])$$

where $\pi(f)$ is the composite

$$(1.1) \qquad A \xrightarrow[d]{} [B, A \otimes B] \xrightarrow[[1,f]]{} [B, C]$$

and $\pi^{-1}(g)$ is the composite

$$(1.2) \qquad A \otimes B \xrightarrow[g \otimes 1]{} [B, C] \otimes B \xrightarrow[e]{} C\,;$$

indeed the commutativity of C5 and C6 is exactly the condition that the natural transformations π and π^{-1} defined by (1.1) and (1.2) should be mutually inverse.

If we omit [,], d, and e from the data and C5 and C6 from the axioms, we obtain the description of what we shall call a *monoidal category*. (This was called a "symmetric monoidal category" in [2], but we shall consider no other kind.) The axioms C1–C4 are exactly (see [9] and [5]) what is needed to ensure that the natural isomorphisms a, b, c are *coherent* in the sense of [9]. Roughly speaking, this means that any diagram will commute if (as in the diagrams C1–C4) each arrow is a natural isomorphism manufactured from $1, a, b, c, a^{-1}, b^{-1}, c^{-1}$ by taking repeated $\otimes$-products. Another example of such a diagram would be

$$\begin{array}{ccccc}
(I \otimes A) \otimes B & & \xrightarrow{\quad a \quad} & & I \otimes (A \otimes B) \\
{\scriptstyle c \otimes 1}\downarrow & & & & \downarrow{\scriptstyle c} \\
(A \otimes I) \otimes B & \xrightarrow[b \otimes 1]{} & A \otimes B & \xrightarrow[b^{-1}]{} & (A \otimes B) \otimes I
\end{array}$$

Note that coherence asserts equality of *natural transformations,* and not of morphisms in $\underline{V}$ except insofar as these are components of natural transformations; thus it does not assert that $c : A \otimes A \to A \otimes A$ and $1 : A \otimes A \to A \otimes A$ coincide, these being components of quite different natural transformations $c : A \otimes B \to B \otimes A$ and $1 : A \otimes B \to A \otimes B$.

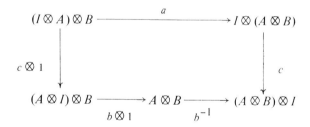

The question naturally arises whether the analogous coherence result holds for a *closed* category: does a diagram commute if each arrow is a natural transformation manufactured from $1, a, b, c, a^{-1}, b^{-1}, c^{-1}, d, e$ by the use of $\otimes$ and [,]? Evidence that *something* of this kind is true was provided by the partial results in this direction due to Epstein [3] (cf. also MacDonald [8]), and by the mass of diagrams proved to be commutative in [2]. Nevertheless the answer to the question as asked is negative. Write $k_A : A \to [[A, I], I]$ for the natural transformation given by the composite

$$A \xrightarrow[d]{} [[A,I], A \otimes [A,I]] \xrightarrow[[1,c]]{} [[A,I], [A,I] \otimes A] \xrightarrow[[1,e]]{} [[A,I],I] \;;$$

then it is easy to see that the diagram

$$\begin{array}{ccc} [A,I] & \xrightarrow{k_{[A,I]}} & [[[A,I],I],I] \\ & {\searrow}^{1} & \downarrow{\scriptstyle [k_A,1]} \\ & & [A,I] \end{array} \qquad (1.3)$$

commutes; however the diagram

$$\begin{array}{ccc} [[[A,I],I],I] & \xrightarrow{[k_A,1]} & [A,I] \\ & {\searrow}^{1} & \downarrow{\scriptstyle k_{[A,I]}} \\ & & [[[A,I],I],I] \end{array} \qquad (1.4)$$

does not commute in general. For if (1.4) commuted as well as (1.3), $k_{[A,\,I]}$ would be an isomorphism; but this is not so when $\underline{V}$ is the category of real vector spaces with the usual $\otimes$ and $[\;,\;]$, in which case k_A is the usual embedding of a vector space into its double dual.

It is the chief purpose of the present paper to show that we do get a coherence result of the desired kind provided that we impose a restriction on the functors which form the "vertices" of the diagram: in the formation of these functors we must never write $[T, S]$ where S (like I in the example above) is a *constant* functor, unless T too is a constant functor. Both of the diagrams (1.3) and (1.4), then, escape this modified coherence result, as they stand; but (1.3) can equally well be written with a variable B in place of I, and then the result applies. Diagram (1.4) ceases to make sense, *as a diagram of natural transformations,* if we replace I by a variable B, and in fact as we have seen does not commute in general; but we could replace A in (1.4) by the constant I, and then our result applies and in this special case (1.4) commutes. The full statement of our results is given in § 2 below.

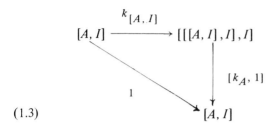

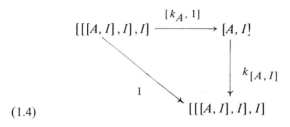

The method we have used is inspired by the work of Lambek [6, 7], who deals with a similar problem for a structure closely related to a closed category but differing from it in certain essential ways. (In particular, Lambek's structures lack the "symmetry isomorphism" c, and this does seem to make an essential difference.) From his work we have learnt the possibility of replacing *composition* of morphisms in a closed category by other processes of combination more adapted to proofs by induction. By his own account, Lambek himself came to recognize this possibility by generalizing the work of Gentzen, whose scheme for eliminating the "cut" in certain logical systems (see [4]) is essentially a special case of the above elimination of composition. An essential step in our §6 below, the proof that what we there call "constructible morphisms" are closed under composition, does not yield to a direct inductive argument – one must go round about and prove instead our Proposition 6.4; and this trick too we learnt from Lambek's work. These essential insights leave us heavily in Lambek's debt. For the rest, however, our results differ considerably from those of Lambek, being expressed in the context of the generalized natural transformations introduced in [1], with which the reader is supposed to be familiar.

§ 2. Statement of results

For a particular closed category $\underline{V}$, the functors T, S: $\underline{V} \times \underline{V}^{op} \times \underline{V} \to \underline{V}$, given by $T(A, B, C) = A \otimes [B, C]$ and $S(A, B, C) = [[A, B], C]$, might fortuitously coincide; they do so, in fact, if $\underline{V}$ is the category with one object and one morphism (and also in less trival cases). Since for our inductive arguments it is essential that each such functor be assigned a *rank*, and since the above two functors are to have different ranks, it is clear that rank should be an attribute not of the functor as such but of its formal expression. We proceed to introduce these formal expressions under the name of *shapes*.

We define shapes, without reference to any particular closed category, by the following inductive rules:

S1 I is a shape.

S2 1 is a shape.

S3 If T and S are shapes there is a shape $T \otimes S$.

S4 If T and S are shapes there is a shape $[T, S]$.

Shapes, therefore, are formal expressions involving I, $1, \otimes$, and $[\ ,\ ]$, with parentheses where necessary; for instance $[1, I] \otimes [1, (1 \otimes I) \otimes 1]$ is a shape.

We define a *variable-set* to be a totally-ordered finite set X, provided with a function called *variance* from X to the two-element set {covariant, contravariant}. Define the *ordinal sum* $X \hat{+} Y$ of two variable-sets X and Y to be the disjoint union $X + Y$ of X and Y, so ordered that X and Y retain their orders and that every $x \in X$ precedes every $y \in Y$, and with the variance of $t \in X \hat{+} Y$ being its variance in X or in Y as the case may be. Define the *twisted sum* $X \tilde{+} Y$ to be the same totally-ordered set as $X \hat{+} Y$, but with the variance of $t \in X \tilde{+} Y$ being its variance in Y when $t \in Y$ and the opposite of its variance in X when $t \in X$.

With each shape T is associated a variable-set $v(T)$ called the *set of variables* of T. This is defined inductively by the rules:

V1 $v(I)$ is the empty set.

V2 $v(1)$ is a chosen one-element set $\{*\}$, with $*$ covariant.

V3 $v(T \otimes S) = v(T) \hat{+} v(S)$.

V4 $v([T, S]) = v(T) \tilde{+} v(S)$.

In many contexts it is convenient to suppose that $v(T)$, if it has n elements, is actually the set $\{1, 2, ..., n\}$. We can accomplish this under the above conventions if we take $\{*\}$ to be $\{1\}$, and if we agree that the disjoint union of $\{1, ..., n\}$ and $\{1, ..., m\}$ is $\{1, ..., n+m\}$ with the given sets embedded as the complementary sets $\{1, ..., n\}$ and $\{n+1, ..., n+m\}$. We also, however, want to speak of $v(T)$ and $v(S)$ as being disjoint complementary subsets of $v(T) + v(S)$; the reader will recognize that we are then speaking of the *images* of $v(T)$ and $v(S)$ in $v(T) + v(S)$.

If T and S are shapes we define a *graph* ξ: $T \to S$ to be a fixed-point-free involution on the disjoint union $v(T) + v(S)$, with the property that mates under ξ have opposite variances in the *twisted* sum $v(T) \tilde{+} v(S)$. Given graphs ξ: $T \to S$ and η: $S \to R$, we define a composite graph $\eta\xi$: $T \to R$ as follows: different elements

$x, y \in v(T) + v(R)$ are mates under $\eta\xi$ if and only if there is a sequence $x = t_0, t_1, \ldots, t_r = y$, with each $t_i \in v(T) + v(S) + v(R)$, such that, for each i, t_{i-1} and t_i are mates either under ξ or under η. Then $\eta\xi$ is indeed a graph, and this law of composition is associative. Moreover for each shape T there is an evident identity graph $1: T \to T$, so that shapes and graphs form a category $\underline{G}$.

Two graphs $\xi: T \to S$, $\eta: S \to R$ are said to be *compatible* if there is no sequence $t_1, t_2, \ldots, t_{2r}$ $(r \geq 1)$ of elements of $v(S)$ such that t_{2i-1} and t_{2i} are mates under ξ for $1 \leq i \leq r$, t_{2i} and t_{2i+1} are mates under η for $1 \leq i \leq r-1$, and t_{2r} and t_1 are mates under η.

The definitions of composition of graphs and of compatibility of graphs become more perspicuous if we consider a graph $\xi: T \to S$ to be a graph in the literal sense, with the disjoint union $v(T) + v(S)$ as its vertex-set, and with one edge (or *linkage*) joining each pair of mates under ξ. The linkages in the graph $\eta\xi$ are then what we get by following alternately the linkages of ξ and of η, ignoring any closed loops that may arise; and ξ and η are compatible when in fact no closed loops do arise. All this is treated in detail in [1].

If $\xi: T \to T'$ and $\eta: S \to S'$ are graphs, we can define a graph $\xi \otimes \eta: T \otimes S \to T' \otimes S'$ by taking the linkages in $v(T) + v(S) + v(T') + v(S')$ to be those of ξ together with those of η. Similarly we can define a graph $[\xi, \eta]: [T', S] \to [T, S']$. It is easy to verify that $\otimes$ and $[\ ,\ ]$ are thereby made into functors $\underline{G} \times \underline{G} \to \underline{G}$ and $\underline{G}^{\mathrm{op}} \times \underline{G} \to \underline{G}$ respectively.

For any shapes T, S, R there are evident graphs

$$\alpha: (T \otimes S) \otimes R \to T \otimes (S \otimes R),$$

$$\beta: T \otimes I \to T,$$

$$\gamma: T \otimes S \to S \otimes T,$$

$$\delta: T \to [S, T \otimes S],$$

$$\epsilon: [T, S] \otimes T \to S;$$

it is easy to verify that these are natural transformations (natural isomorphisms in the case of α, β, γ), and that $\underline{G}$ becomes a closed category if we take these as its a, b, c, d, e.

Now let $\underline{V}$ be any closed category. For each shape T, with $v(T) = \{i_1, \ldots, i_n\}$ say (as an *ordered* set), we define a functor $|T|: \underline{V}_{i_1} \times \underline{V}_{i_2} \times \ldots \times \underline{V}_{i_n} \to \underline{V}$, where $\underline{V}_{i_r}$ is $\underline{V}$ or $\underline{V}^{\mathrm{op}}$ according as i_r is covariant or contravariant in $v(T)$. (If $v(T)$ is empty then $n = 0$ and we understand $\underline{V}_{i_1} \times \ldots \times \underline{V}_{i_n}$ to mean the unit category $\underline{I}$ with one object and one morphism.) The inductive definition of $|T|$ is the following:

F1 $|I|$ is the constant functor $I: \underline{I} \to \underline{V}$.

F2 $|1|$ is the identity functor $1: \underline{V} \to \underline{V}$.

F3 $|T \otimes S|$ is the composite functor

$$\underline{V}_{i_1} \times \ldots \times \underline{V}_{i_n} \times \underline{V}_{j_1} \times \ldots \times \underline{V}_{j_m} \xrightarrow{|T| \times |S|} \underline{V} \times \underline{V} \xrightarrow[\otimes]{} \underline{V},$$

where $v(T) = \{i_1, ..., i_n\}$ and $v(S) = \{j_1, ..., j_m\}$.

F4 $|[T, S]|$ is the composite functor

$$\underline{V}^{\text{op}}_{i_1} \times ... \times \underline{V}^{\text{op}}_{i_n} \times \underline{V}_{j_1} \times ... \times \underline{V}_{j_m} \xrightarrow[|T|^{\text{op}} \times |S|]{} \underline{V}^{\text{op}} \times \underline{V} \xrightarrow[[\,,\,]]{} \underline{V},$$

where $v(T)$ and $v(S)$ are as in F3.

Let T and S be shapes, with $v(T) = \{i_1, ..., i_n\}$ and $v(S) = \{j_1, ..., j_m\}$. Then, as in [1], a *natural transformation* $f\colon |T| \to |S|$ consists of a graph $\xi\colon T \to S$, called the *graph* Γf of f, and morphisms

$$f(A_{i_1}, ..., A_{i_n}, A_{j_1}, ..., A_{j_m})\colon |T|(A_{i_1}, ..., A_{i_n}) \to |S|(A_{j_1}, ..., A_{j_m})$$

of $\underline{V}$, called the *components* of f; here $A_x = A_y$ whenever x and y are mates under ξ, and for each such pair of mates there is a *naturality condition* to be satisfied by these components. (In practice one suppresses, in writing the components of f, one of each pair of equal arguments in $f(A_{i_1}, ..., A_{i_n}, A_{j_1}, ..., A_{j_m})$, and one often writes the remaining arguments as subscripts. Thus one writes $e_{AB}\colon [A, B] \otimes A \to B$ or $e(A, B)$, and not $e(A, B, A, B)$.) If $g\colon |S| \to |R|$ is another natural transformation, of graph $\eta\colon S \to R$, and if η and ξ are *compatible*, we can define as in [1] a composite natural transformation $gf\colon |T| \to |R|$ of graph $\eta\xi\colon T \to R$; the component

$$(gf)(A_{i_1}, ..., A_{i_n}, A_{k_1}, ..., A_{k_l})\colon |T|(A_{i_1}, ..., A_{i_n}) \to |R|(A_{k_1}, ..., A_{k_l})$$

of gf is the composite of the components

$$f(A_{i_1}, ..., A_{i_n}, A_{j_1}, ..., A_{j_m})$$

and

$$g(A_{j_1}, ..., A_{j_m}, A_{k_1}, ..., A_{k_l}),$$

where $A_x = A_y$ if x and y are either mates under ξ or mates under η. In fact we can in the present circumstances define the composite gf, of graph $\eta\xi$, even when η and ξ are not compatible, for we have here a recourse not available in the more general situation of [1]: we define the components of gf just as above, setting $A_{j_r} = I$ for any $j_r \in v(S)$ which occurs in one of the closed loops. That the composite so formed is still natural is clear, as we have merely modified f and g by specializing some of the arguments before composing them as in [1]. This law of composition is associative, and there is an evident identity natural transformation $1\colon |T| \to |T|$ of graph $1\colon T \to T$.

We can define, therefore, a new category $\underline{N}(\underline{V})$ depending upon $\underline{V}$. The objects of $\underline{N}(\underline{V})$, like those of $\underline{G}$, are to be all the shapes; a morphism $f\colon T \to S$ in $\underline{N}(\underline{V})$ is to be a natural transformation $f\colon |T| \to |S|$, which we shall often call "a natural transformation $f\colon T \to S$"; and composition in $\underline{N}(\underline{V})$ is to be the above composition of natural transformations. We can call $\underline{N}(\underline{V})$ "the category of shapes and natural transformations for $\underline{V}$"; and we shall often abbreviate $\underline{N}(\underline{V})$ to $\underline{N}$ when $\underline{V}$ is clear from the context. There is an evident functor $\Gamma: \underline{N} \to \underline{G}$ which is the identity on objects and which takes each natural transformation f to its graph Γf.

From natural transformations $f: T \to T'$ and $g: S \to S'$ of graphs ξ and η we get a natural transformation $f \otimes g: T \otimes S \to T' \otimes S'$ of graph $\xi \otimes \eta$ by taking the components of $f \otimes g$ to be the $\otimes$-products of the components of f and those of g. Similarly we get a natural transformation $[f, g]: [T', S] \to [T, S']$ of graph $[\xi, \eta]$. It is easy to verify that $\otimes$ and $[\ ,\ ]$ are thereby made into functors $\underline{N} \times \underline{N} \to \underline{N}$ and $\underline{N}^{\mathrm{op}} \times \underline{N} \to \underline{N}$; clearly $\Gamma: \underline{N} \to \underline{G}$ commutes with $\otimes$ and $[\ ,\ ]$.

For any shapes T, S, R we get a natural transformation $a_{TSR}: (T \otimes S) \otimes R \to T \otimes (S \otimes R)$ of graph $\alpha_{TSR}: (T \otimes S) \otimes R \to T \otimes (S \otimes R)$ by taking the component

$$a_{TSR}(A_{i_1}, \ldots, A_{i_n}, A_{j_1}, \ldots, A_{j_m}, A_{k_1}, \ldots, A_{k_l})$$

of a_{TSR} to be the component

$$a(|T|(A_{i_1}, \ldots, A_{i_n}), |S|(A_{j_1}, \ldots, A_{j_m}), |R|(A_{k_1}, \ldots, A_{k_l}))$$

of a. Then it follows easily that the morphism a_{TSR} of $\underline{N}$ is the (T, S, R)-component of a natural isomorphism between the functors $(-\otimes-)\otimes-$ and $-\otimes(-\otimes-)$ of $\underline{N} \times \underline{N} \times \underline{N}$ into $\underline{N}$. This natural isomorphism we again call a, and we often write $a: (T \otimes S) \otimes R \to T \otimes (S \otimes R)$, abbreviating as usual a_{TSR} to a. In the same way we define natural isomorphisms $b: T \otimes I \to T$, $c: T \otimes S \to S \otimes T$ and natural transformations $d: T \to [S, T \otimes S]$, $e: [T, S] \otimes T \to S$, of respective graphs $\beta, \gamma, \delta, \epsilon$; and we verify that a, b, c, d, e give to $\underline{N}$ the structure of a closed category.

We now have closed categories $\underline{N} = \underline{N}(\underline{V})$ and $\underline{G}$, and a functor $\Gamma: \underline{N} \to \underline{G}$ which is the identity on objects, which commutes with $\otimes$ and $[\ ,\ ]$, and which sends a, b, c, d, e to $\alpha, \beta, \gamma, \delta, \epsilon$. In order to make statements that will embrace at once the closed categories $\underline{N}$ and $\underline{G}$, we shall suppose, throughout this paper, that $\underline{H}$ is some closed category with the same objects as $\underline{G}$, and that $\Gamma: \underline{H} \to \underline{G}$ is a functor which is the identity on objects, which commutes with $\otimes$ and $[\ ,\ ]$, and which sends a, b, c, d, e to $\alpha, \beta, \gamma, \delta, \epsilon$. The cases of interest are that where $\underline{H} = \underline{N}(\underline{V})$ and Γ is as above, and that where $\underline{H} = \underline{G}$ and $\Gamma = 1$.

Given any such $\underline{H}$, we define a subcategory of $\underline{H}$, whose objects are all shapes, and whose morphisms shall be called the *allowable* morphisms of $\underline{H}$. These are to be the smallest class of morphisms of $\underline{H}$ satisfying the following five conditions (in which T, S, R, ... denote arbitrary shapes):

AM1. For any T, S, R each of the following morphisms is in the class:

$$1: T \to T$$
$$a: (T \otimes S) \otimes R \to T \otimes (S \otimes R),$$
$$a^{-1}: T \otimes (S \otimes R) \to (T \otimes S) \otimes R,$$
$$b: T \otimes I \to T,$$
$$b^{-1}: T \to T \otimes I,$$
$$c: T \otimes S \to S \otimes T.$$

AM2. For any T, S each of the following morphisms is in the class:

$$d\colon\ T \to [S, T \otimes S]\ ,$$
$$e\colon\ [T, S] \otimes T \to S.$$

AM3. If $f\colon T \to T'$ and $g\colon S \to S'$ are in the class so is $f \otimes g\colon T \otimes S \to T' \otimes S'$.

AM4. If $f\colon T \to T'$ and $g\colon S \to S'$ are in the class so is $[f, g]\colon [T', S] \to [T, S']$.

AM5. If $f\colon T \to S$ and $g\colon S \to R$ are in the class so is $gf\colon T \to R$.

The allowable morphisms of $\underline{G}$ are called the *allowable graphs*, and those of $\underline{N}(\underline{V})$ are called the *allowable natural transformations* (for $\underline{V}$). It is evident that the functor $\Gamma\colon \underline{N} \to \underline{G}$ takes allowable natural transformations to allowable graphs, since those natural transformations $f \in \underline{N}$ for which Γf is allowable clearly satisfy AM1–AM5.

The first two of our principal results deal with the case $\underline{H} = \underline{G}$, and are:

Theorem 2.1. *There is an algorithm for deciding whether a graph $\xi\colon T \to S$ is allowable.*

Theorem 2.2. *If the graphs $\xi\colon T \to S$, $\eta\colon S \to R$ are allowable, they are compatible.*

The proofs will be given in §7 and in §6 respectively.

Since we shall be interested only in *allowable* natural transformations, we see from Theorem 2.2 that there was no real need to introduce the composition of incompatible ones; it was merely a convenience so that $\underline{N}$ could be described as a category. Our third principal result is:

Theorem 2.3. *Let $\underline{V}$ be any closed category. If $\xi\colon T \to S$ is an allowable graph, there is in $\underline{N}(\underline{V})$ at least one allowable natural transformation $f\colon T \to S$ of graph ξ.*

Proof. Those allowable graphs ξ which are images under Γ of allowable natural transformations satisfy AM1–AM5, and therefore constitute the totality of allowable graphs.

For our final main result we pick out a subset of the shapes called the *proper* shapes. Call a shape T *constant* if its set of variables $v(T)$ is empty. Then the proper shapes are defined inductively by:

PS1 I is a proper shape.

PS2 1 is a proper shape.

PS3 If T and S are proper shapes so is $T \otimes S$.

PS4 If T and S are proper shapes so is $[T, S]$, *unless* S is constant and T is not constant.

Our final principal result then is:

Theorem 2.4. *Let $\underline{V}$ be any closed category, and let $f, f': T \to S$ be two allowable natural transformations in $\underline{N}(\underline{V})$ with the same graph $\xi = \Gamma f = \Gamma f'$. Then, provided that the shapes T and S are proper, we have $f = f'$.*

The proof will be given in §7.

§3. The monoidal case

We are going to build on the known coherence theorem for the monoidal case, proved in [9] and simplified a little in [5]. The purpose of this section is to re-state this result in terms entirely analogous to those used in §2 above, so that it is easily available for our use.

We have seen that we get the description of a monoidal category from that of a closed category by omitting the data [,], d, e and the axioms C5 and C6. All the concepts introduced in §2 have analogues in the monoidal case, as follows.

The shapes we need here are those defined by the inductive rules S1, S2, S3 of §2, omitting S4; we call these the *integral shapes* (for it is reasonable to think of $\otimes$ as a kind of multiplication, and of [,] as a kind of division). For integral T the rules V1, V2, V3 suffice to describe the set of variables $v(T)$; clearly each element of $v(T)$ is covariant. Because of this, a pair of mates under a graph $\xi: T \to S$, where T and S are integral, consists of an element of $v(T)$ and an element of $v(S)$; thus we may identify the graph ξ with the corresponding bijection of $v(T)$ onto $v(S)$. It is especially for integral T (where there are no complications of variance) that it is convenient to identify $v(T)$, when it has n elements, with the ordered set $\{1, 2, ..., n\}$; and we shall do so freely. The integral shapes and the graphs connecting them form a full subcategory $\underline{G}_0$ of $\underline{G}$; we can look upon $\otimes$ as a functor $\underline{G}_0 \otimes \underline{G}_0 \to \underline{G}_0$, and the graphs $\alpha: (T \otimes S) \otimes R \to T \otimes (S \otimes R)$, $\beta: T \otimes I \to T$, $\gamma: T \otimes S \to S \otimes T$ turn $\underline{G}_0$ into a monoidal category.

If $\underline{V}$ is any monoidal category, each integral shape T determines a functor $|T|: \underline{V} \times ... \times \underline{V} \to \underline{V}$ by the rules F1, F2, F3 of §2. Since we can again speak of a natural transformation $f: |T| \to |S|$ of graph $\xi: T \to S$, we have a category $\underline{N}_0(\underline{V})$, whose objects are the integral shapes and whose morphisms $f: T \to S$ are the natural transformations $f: |T| \to |S|$ of arbitrary graph. Like the category $\underline{N}(\underline{V})$ of §2, $\underline{N}_0(\underline{V})$ becomes a monoidal category with the obvious definitions of $f \otimes g$ and of a, b, c; and there is a functor $\Gamma: \underline{N}_0(\underline{V}) \to \underline{G}_0$ which is the identity on objects and which sends each natural transformation to its graph. The functor Γ commutes with $\otimes$ and sends a, b, c, to α, β, γ.

In this monoidal case we shall need to compare the $\underline{N}_0(\underline{V})$'s for different monoidal categories $\underline{V}$. If $\underline{V}$ and $\underline{V}'$ are monoidal categories, a *strict monoidal functor* $\Delta: \underline{V} \to \underline{V}'$ shall mean a functor that commutes with $\otimes$ and for which $\Delta a = a'$, $\Delta b = b'$, and $\Delta c = c'$ (where, for example, this last assertion means that $\Delta c_{A,B} = c'_{\Delta A, \Delta B}$). In particular, $\Gamma: \underline{N}_0(\underline{V}) \to \underline{G}_0$ is a strict monoidal functor. So is the $\Gamma: \underline{N} = \underline{N}(\underline{V}) \to \underline{G}$ of §2 when $\underline{V}$ is closed; and this latter clearly induces a strict monoidal functor $\underline{N}_0(\Gamma): \underline{N}_0(\underline{N}) \to \underline{N}_0(\underline{G})$, which is the identity on objects and which sends the natural transformation f: T → S to the natural transformation whose components are the images under $\Gamma: \underline{N} \to \underline{G}$ of those of f. It is further clear that the composite of $\Gamma: \underline{N}_0(\underline{G}) \to \underline{G}_0$ with $\underline{N}_0(\Gamma): \underline{N}_0(\underline{N}) \to \underline{N}_0(\underline{G})$ is $\Gamma: \underline{N}_0(\underline{N}) \to \underline{G}_0$.

For any monoidal category $\underline{V}$ we define the *central* morphisms of $\underline{V}$ to be the smallest class of morphisms of $\underline{V}$ satisfying the conditions AM1, AM3, and AM5 of

§ 2, where $T, S, R, \ldots$ now denote arbitrary objects of $\underline{V}$; since the isomorphisms of $\underline{V}$ satisfy AM1, AM3 and AM5, every central morphism is an isomorphism. These central morphisms constitute a subcategory Cent $\underline{V}$ of $\underline{V}$ with the same objects as $\underline{V}$; clearly Cent $\underline{V}$ is itself a monoidal category, and the inclusion Cent $\underline{V} \to \underline{V}$ is a strict monoidal functor. It is clear that any strict monoidal functor $\Delta : \underline{V} \to \underline{V}'$ carries central morphisms of $\underline{V}$ into central morphisms of $\underline{V}'$.

The analogue of Theorem 2.2 for the monoidal case is trivially true, for any graphs $\xi\colon T \to S$ and $\eta\colon S \to R$ are clearly compatible when T, S and R are integral. The analogues of Theorems 2.1, 2.3 and 2.4 are contained in the following result, which expresses essentially what was proved in [9]:

Theorem 3.1. *Let $\underline{V}$ be any monoidal category. If T and S are integral shapes, then any graph $\xi\colon T \to S$ is central in $\underline{G}_0$, and there is in* Cent $\underline{N}_0(\underline{V})$ *one and only one natural transformation $f\colon T \to S$ of graph ξ.*

In other words we have Cent $\underline{N}_0(\underline{V}) \cong$ Cent $\underline{G}_0 = \underline{G}_0$. We shall write $|\xi|_{\underline{V}}\colon T \to S$ for the unique morphism of Cent $\underline{N}_0(\underline{V})$ with $\Gamma|\xi|_{\underline{V}} = \xi$. It is immediate that $|\eta\xi|_{\underline{V}} = |\eta|_{\underline{V}}|\xi|_{\underline{V}}$, and that $|\xi \otimes \eta|_{\underline{V}} = |\xi|_{\underline{V}} \otimes |\eta|_{\underline{V}}$; and further that $|\alpha|_{\underline{V}} = a$, $|\beta|_{\underline{V}} = b$, $|\gamma|_{\underline{V}} = c$. Moreover, for a closed category $\underline{V}$, it is clear that $\underline{N}_0(\Gamma)\colon \underline{N}_0(\underline{N}) \to \underline{N}_0(\underline{G})$, where again $\underline{N}$ denotes $\underline{N}(\underline{V})$, carries $|\xi|_{\underline{N}}$ to $|\xi|_{\underline{G}}$.

§4. Central morphisms in $\underline{N}(\underline{V})$ and in $\underline{G}$

This section will use Theorem 3.1 to handle, for a closed category $\underline{V}$, that part of the coherence problem involving only a, b and c. In other words, we shall deal here with the *central* morphisms of $\underline{N}(\underline{V})$. These, like the composite

$$(T \otimes [S, R]) \otimes I \xrightarrow[a]{} T \otimes ([S, R] \otimes I) \xrightarrow[1 \otimes b]{} T \otimes [S, R] \ ,$$

involve in general non-integral shapes. We bring them within the ambit of Theorem 3.1 by showing that the central morphisms of $\underline{N}(\underline{V})$ and of $\underline{G}$ admit an alternative description: they arise from the morphisms of $\underline{G}_0$ by the substitution of "$\otimes$-irreducible" or "prime" shapes for the variables.

We suppose then that $\underline{V}$ is a closed category, and as in §2 we use $\underline{H}$ to denote either $\underline{N}(\underline{V})$ or $\underline{G}$, with $\Gamma: \underline{H} \to \underline{G}$ sending f to its graph in the first case and being the identity in the second case. Since $\underline{H}$, being a closed category, is a monoidal category, we can speak as in §3 of the central morphisms of $\underline{H}$; it is immediate from the definition of these that they are a subset of the allowable morphisms of $\underline{H}$. Since $\Gamma: \underline{N}(\underline{V}) \to \underline{G}$ is a strict monoidal functor, it takes a central morphism of $\underline{N}(\underline{V})$ (which we shall call a *central natural transformation*) to a central morphism of $\underline{G}$ (which we shall call a *central graph*). As $\underline{V}$ will be fixed, we shall abbreviate $\underline{N}(\underline{V})$ to $\underline{N}$.

If P is any integral shape we have as in §3, since $\underline{H}$ is a monoidal category, a functor $|P|: \underline{H} \times ... \times \underline{H} \to \underline{H}$. Thus for arbitrary shapes $X_1, ..., X_n$ (where n is the number of elements of $v(P)$) we get a shape $|P|(X_1, ..., X_n)$, and for arbitrary morphisms $f_i: X_i \to X_i'$ in $\underline{H}$ we get a morphism $|P|(f_1, ..., f_n): |P|(X_1, ..., X_n) \to |P|(X_1', ..., X_n')$ in $\underline{H}$. It is evident that $|P|(X_1, ..., X_n)$ is the same shape whether we take $\underline{H}$ to be $\underline{N}$ or $\underline{G}$, and that $\Gamma(|P|(f_1, ..., f_n)) = |P|(\Gamma f_1, ..., \Gamma f_n)$.

Now let P, Q be integral shapes. A graph $\xi: P \to Q$ may be identified with a bijection of $v(P)$ onto $v(Q)$ and hence, if $v(P)$ and $v(Q)$ have n elements, with a permutation ξ of $\{1, ..., n\}$. As in §3 we have a unique $|\xi|_{\underline{H}}: P \to Q$ in Cent $\underline{N}_0(\underline{H})$ of graph ξ. We can write its typical component as

$$(4.1) \qquad |\xi|_{\underline{H}}(X_1, ..., X_n): |P|(X_{\xi 1}, ..., X_{\xi n}) \to |Q|(X_1, ..., X_n) \ ;$$

it is a morphism of $\underline{H}$.

Proposition 4.1. *For any graph $\xi: P \to Q$ between integral shapes P, Q and for any shapes $X_1, ..., X_n$ the morphism* (4.1) *of $\underline{H}$ is central.*

Proof. Consider the family of all those graphs ξ in $\underline{G}_0$ for which (4.1) is indeed central in $\underline{H}$ for all $X_1, ..., X_n$; it suffices to show that this family satisfies AM1, AM3 and AM5, for then it contains Cent $\underline{G}_0$ which, by Theorem 3.1, is all of $\underline{G}_0$. Now this family satisfies AM1 because $|\alpha|_{\underline{H}} = a$, etc.; it satisfies AM3 because the components of $|\xi \otimes \eta|_{\underline{H}} = |\xi|_{\underline{H}} \otimes |\eta|_{\underline{H}}$ are the tensor products of the components of $|\xi|_{\underline{H}}$ and those of $|\eta|_{\underline{H}}$; and it satisfies AM5 because the components of $|\eta\xi|_{\underline{H}} = |\eta|_{\underline{H}}|\xi|_{\underline{H}}$ are the composites of certain components of $|\eta|_{\underline{H}}$ and of $|\xi|_{\underline{H}}$.

Since, as we saw in §3, $\underline{N}_0(\Gamma)$ takes $|\xi|_{\underline{N}}$ to $|\xi|_{\underline{G}}$, it follows from the definition in §3 of $\underline{N}_0(\Gamma)$ that

$$\Gamma(|\xi|_{\underline{N}}(X_1, \ldots, X_n)) = |\xi|_{\underline{G}}(X_1, \ldots, X_n)\,. \tag{4.2}$$

It is easy to calculate $|\xi|_{\underline{G}}(X_1, \ldots, X_n)$. First, it is clear by induction that the variable-set $v(|P|(X_1, \ldots, X_n))$ is $v(X_1) \hat{+} \ldots \hat{+} v(X_n)$.

Proposition 4.2. *The graph*

$$|\xi|_{\underline{G}}(X_1, \ldots, X_n)\colon |P|(X_{\xi 1}, \ldots, X_{\xi n}) \to |Q|(X_1, \ldots, X_n)$$

is the involution on the set

$$v(X_{\xi 1}) + \ldots + v(X_{\xi n}) + v(X_1) + \ldots + v(X_n)$$

corresponding to the evident bijection induced by ξ *of*

$$v(X_{\xi 1}) + \ldots + v(X_{\xi n}) \quad \textit{with} \quad v(X_1) + \ldots + v(X_n)\,.$$

Proof. Again it suffices to show that the family of those ξ in $\underline{G}_0$ for which this is true satisfies AM1, AM3, and AM5; the verifications are immediate.

We now proceed to show that all the central morphisms of $\underline{H}$ are obtainable in the form (4.1). Define the *prime* shapes to be the shape 1 and all shapes of the form $[T, S]$. It follows easily from the inductive definition of shapes that any shape T can be expressed *uniquely* in the form $T = |P|(X_1, \ldots, X_n)$ where P is an integral shape and $X_1, \ldots, X_n$ are prime shapes. We call this the *prime factorization* of T, and call $X_1, \ldots, X_n$ the list of *prime factors* of T. Note that n may be 0, so that this list may be empty; namely when T is a constant integral shape. In general, if T is an integral shape, its prime factorization is $T = |T|(1, 1, \ldots, 1)$. Observe that if the prime factorizations of T and S are $T = |P|(X_1, \ldots, X_n)$ and $S = |Q|(Y_1, \ldots, Y_m)$, then that of $T \otimes S$ is $|P \otimes Q|(X_1, \ldots, X_n, Y_1, \ldots, Y_m)$.

Proposition 4.3. *Let* $f\colon T \to S$ *be a central morphism of* $\underline{H}$, *and let the prime factorizations of* T *and of* S *be* $T = |P|(X_1, \ldots, X_n)$ *and* $S = |Q|(Y_1, \ldots, Y_m)$. *Then* $m = n$, *and there is a permutation* ξ *of* $\{1, \ldots, n\}$ *such that* $X_i = Y_{\xi i}$ *for each i and such that* $f = |\xi|_{\underline{H}}(Y_1, \ldots, Y_n)\colon |P|(Y_{\xi 1}, \ldots, Y_{\xi n}) \to |Q|(Y_1, \ldots, Y_n)$.

Proof. Consider the family of all those morphisms of $\underline{H}$ that *are* of the form

$$|\eta|_{\underline{H}}(Z_1, \ldots, Z_r)\colon |J|(Z_{\eta 1}, \ldots, Z_{\eta r}) \to |K|(Z_1, \ldots, Z_r)$$

for integral shapes J, K and prime shapes Z_i. It suffices to show that this family satisfies AM1, AM3 and AM5, and therefore contains all central morphisms of $\underline{H}$. When we advert to the relation between the prime factorization of $T \otimes S$ and those of T and of S, the verifications are immediate from the facts that $|\alpha|_{\underline{H}} = a$, etc., $|\eta \otimes \zeta|_{\underline{H}} = |\eta|_{\underline{H}} \otimes |\zeta|_{\underline{H}}$, and $|\zeta\eta|_{\underline{H}} = |\zeta|_{\underline{H}} |\eta|_{\underline{H}}$.

Since a shape T is integral exactly when its prime factors are all 1, and is constant exactly when its prime factors are all constant, we have

Corollary 4.4. *If $f: T \to S$ is a central morphism in $\underline{H}$ and if either one of T, S is integral (resp. constant), so is the other.*

In view of Proposition 4.2 we also have

Corollary 4.5. *If $\phi: T \to S$ is a central graph, each pair of mates under ϕ consists of an element of $v(T)$ and an element of $v(S)$.*

Returning to Proposition 4.3, we may observe that the permutation ξ therein is not in general uniquely determined by f. For instance if T and S are both $[I, I] \otimes [I, I]$, so that $P = Q = 1 \otimes 1$ and $X_1 = Y_1 = X_2 = Y_2 = [I, I]$, then it will follow from Proposition 4.8 below that $|\xi|_{\underline{H}}(Y_1, Y_2)$ is $1: T \to S$ for both permutations ξ of $\{1, 2\}$. However:

Proposition 4.6. *If in Proposition 4.3 permutations ξ and ξ' both satisfy the stated conditions, we have $\xi' = \lambda\xi$ where λ is a permutation of $\{1, ..., n\}$ for which $\lambda i \neq i$ implies that Y_i and $Y_{\lambda i}$ are equal constant shapes.*

Proof. Since $X_{\xi^{-1} i} = Y_i$ and also $X_{\xi^{-1} i} = Y_{\xi'\xi^{-1} i} = Y_{\lambda i}$ we have $Y_i = Y_{\lambda i}$. Since $|\xi|_{\underline{H}}(Y_1, ..., Y_n) = |\xi'|_{\underline{H}}(Y_1, ..., Y_n)$ we have by (4.2) that $|\xi|_{\underline{G}}(Y_1, ..., Y_n) = |\xi'|_{\underline{G}}(Y_1, ..., Y_n)$, and we conclude from Proposition 4.2 that $\xi' j = \xi j$ unless $v(Y_{\xi j})$ is empty; that is, $\lambda i = i$ unless Y_i is constant.

For the desired main result of this section, we need to show that the permutations λ of the type described in Proposition 4.6 are *exactly* those for which $|\lambda|_{\underline{H}}(Y_1, ..., Y_n) = 1$. First we prove:

Lemma 4.7. *If T is a constant shape there is an isomorphism $k_T: T \to I$ in $\underline{H}$ which, together with its inverse, is allowable.*

Proof. From the natural isomorphism

$$\underline{H}(A, I) \xrightarrow[\underline{H}(b,1)]{} \underline{H}(A \otimes I, I) \xrightarrow[\pi]{} \underline{H}(A, [I, I])$$

we deduce, by the Yoneda Lemma, the existence of an isomorphism $h: [I, I] \to I$.

Using (1.1) and (1.2) we find that h and h^{-1} are the respective composites

$$[I,I] \xrightarrow[b^{-1}]{} [I,I]\otimes I \xrightarrow[e]{} I, \qquad I \xrightarrow[d]{} [I, I\otimes I] \xrightarrow[[1,b]]{} [I,I],$$

so that both are allowable. We now define k_T inductively for constant shapes T by setting $k_I = 1$, by taking $k_{T\otimes S}$ to be the composite

$$T\otimes S \xrightarrow[k_T\otimes k_S]{} I\otimes I \xrightarrow[b]{} I,$$

and by taking $k_{[T,S]}$ to be the composite

$$[T,S] \xrightarrow[[k_T^{-1}, k_S]]{} [I,I] \xrightarrow[h]{} I.$$

Proposition 4.8. *Let Q be an integral shape, Let $Y_1, \ldots, Y_n$ be prime shapes, and let λ be a permutation of $\{1, \ldots, n\}$ for which $\lambda i \neq i$ implies that Y_i and $Y_{\lambda i}$ are equal constant shapes. Then $|\lambda|_{\underline{H}}(Y_1, \ldots, Y_n) = 1\colon |Q|(Y_{\lambda 1}, \ldots, Y_{\lambda n}) \to |Q|(Y_1, \ldots, Y_n)$.*

Proof. We can express λ as a product of transpositions; since $|\mu\nu|_{\underline{H}} = |\mu|_{\underline{H}}|\nu|_{\underline{H}}$ we may suppose that λ is such a transposition. Replacing λ by a suitable conjugate $\mu\lambda\mu^{-1}$, we may suppose that λ is the transposition interchanging 1 and 2 and leaving fixed $3, \ldots, n$, while $Y_1 = Y_2$ are equal constant shapes. In $\underline{G}_0$, Q is isomorphic to $(1\otimes 1)\otimes R$ for some integral R, and since $|\mu\otimes 1|_{\underline{H}} = |\mu|_{\underline{H}}\otimes 1$ we may suppose that Q is in fact the shape $1\otimes 1$. But then $|\lambda|_{\underline{H}} = c$, and it remains to prove that $c_{YY} = 1\colon Y\otimes Y \to Y\otimes Y$ if Y is a constant shape. Using the isomorphism k_Y of Lemma 4.7 we have by the naturality of c a commutative diagram

$$\begin{array}{ccc} Y\otimes Y & \xrightarrow{c_{YY}} & Y\otimes Y \\ {\scriptstyle k_Y\otimes k_Y}\downarrow & & \downarrow{\scriptstyle k_Y\otimes k_Y} \\ I\otimes I & \xrightarrow[c_{II}]{} & I\otimes I \end{array};$$

since $c_{II} = 1$ by Theorem 3.1, it follows that $c_{YY} = 1$.

Theorem 4.9. *Let $\underline{V}$ be a closed category. Then each central graph $\phi\colon T\to S$ in $\underline{G}$ is Γf for a unique central natural transformation $f\colon T\to S$ in $\underline{N}(\underline{V})$.*

Proof. Let $T = |P|(X_1, \ldots, X_n)$, $S = |Q|(Y_1, \ldots, Y_m)$ be the prime factorizations. Applying Proposition 4.3 with $\underline{H} = \underline{G}$, we conclude that $m = n$, and that for some permutation ξ we have $X_i = Y_{\xi i}$ and $\phi = |\xi|_{\underline{G}}(Y_1, \ldots, Y_n)$. Setting $f = |\xi|_{\underline{N}}(Y_1, \ldots, Y_n)$, which is central by Proposition 4.1, we see by (4.2) that $\Gamma f = \phi$, thus proving the existence of f.

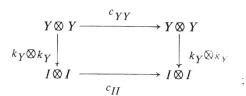

To prove the uniqueness of f, let $f' : T \to S$ be another central natural transformation with $\Gamma f' = \phi$. Applying Proposition 4.3 with $\underline{H} = \underline{N}$, we conclude that $f' = |\xi'|_{\underline{N}}(Y_1, \ldots, Y_n)$ for some permutation ξ' with $X_i = Y_{\xi' i}$. Now (4.2) gives $\phi = \Gamma f' = |\xi'|_{\underline{G}}(Y_1, \ldots, Y_n)$; and Proposition 4.6 with $H = G$ shows that $\lambda = \xi'\xi^{-1}$ has the properties described therein. We conclude from Proposition 4.8 with $\underline{H} = \underline{N}$ that $|\lambda|_{\underline{N}}(Y_1, \ldots, Y_n) = 1$. Thus, since $|\lambda\xi|_{\underline{N}} = |\lambda|_{\underline{N}}\, |\xi|_{\underline{N}}$, we have $|\xi'|_{\underline{N}}(Y_1, \ldots, Y_n) = |\xi|_{\underline{N}}(Y_1, \ldots, Y_n)$, or $f' = f$.

We conclude this section with two useful propositions that could in fact have been proved immediately after Proposition 4.3. In the situation of that proposition, we may call ξ the *association of the prime factors of T and of S*, and then call $Y_{\xi i}$ *the prime factor of S associated, via f, with the prime factor X_i of T*. This language is a little imprecise, because of the non-uniqueness of ξ; we sometimes have a *choice of association*. The statements of the results below allow for this choice.

Proposition 4.10. *Let $f: A \otimes B \to C \otimes D$ be a central morphism of $\underline{H}$. For some choice of association, let each prime factor of A, considered as a prime factor of $A \otimes B$, be associated via f with a prime factor of C. Then there are a shape E and central morphisms $g: A \otimes E \to C$ and $h: B \to E \otimes D$ such that f is the composite*

$$A \otimes B \xrightarrow{1 \otimes h} A \otimes (E \otimes D) \xrightarrow{a^{-1}} (A \otimes E) \xrightarrow{g \otimes 1} C \otimes D.$$

Proof. Let the prime factorizations be $A = |P|(X_1, \ldots, X_n)$, $B = |Q|(Y_1, \ldots, Y_m)$, $C = |R|(Z_1, \ldots, Z_l)$, $D = |S|(V_1, \ldots, V_k)$. We must have $n + m = l + k$, and the hypothesis of the proposition means that $f = |\xi|_{\underline{H}}(Z_1, \ldots, Z_l, V_1, \ldots, V_k)$ for some permutation ξ of $\{1, \ldots, n{+}m\}$ that maps the subset $\{1, \ldots, n\}$ into the subset $\{1, \ldots, l\}$. Let $j_1, \ldots, j_{l-n}$ be those elements of $\{1, \ldots, l\}$, in ascending order, that are not in the image under ξ of $\{1, \ldots, n\}$. Set $E = (Z_{j_1} \otimes Z_{j_2}) \otimes \ldots \otimes Z_{j_{l-n}}$; any way of inserting parentheses will do. Let ρ be the permutation of $\{1, \ldots, l\}$ given by $\rho i = \xi i$ for $i \leq n$, $\rho(n + i) = j_i$ for $i \leq l - n$. Let $\bar{\rho}$ be the permutation of $\{1, \ldots, n + m\}$ which is equal to ρ on $\{1, \ldots, l\}$ and which is the identity on $\{l + 1, \ldots, n + m\}$. Let $\bar{\sigma}$ be the permutation $\bar{\rho}^{-1}\xi$ of $\{1, \ldots, n + m\}$; clearly $\bar{\sigma}$ is the identity on $\{1, \ldots, n\}$. Let σ be the permutation of $\{1, \ldots, m\}$ given by $\sigma i = \bar{\sigma}(n + i) - n$. Define g as $|\rho|_{\underline{H}}(Z_1, \ldots, Z_l)$ and h as $|\sigma|_{\underline{H}}(Z_{j_1}, \ldots, Z_{j_{l-n}}, V_1, \ldots, V_k)$. Then $1 \otimes h = |\bar{\sigma}|_{\underline{H}}(X_1, \ldots, X_n, Z_{j_1}, \ldots, Z_{j_{l-n}}, V_1, \ldots, V_k)$; $a^{-1} = |\alpha^{-1}|_{\underline{H}}(X_1, \ldots, X_n, Z_{j_1}, \ldots, Z_{j_{l-n}}, V_1, \ldots, V_k) = |1|_{\underline{H}}(X_1, \ldots, V_k)$; and $g \otimes 1 = |\bar{\rho}|_{\underline{H}}(Z_1, \ldots, Z_l, V_1, \ldots, V_k)$. It follows that $(g \otimes 1)\, a^{-1}(1 \otimes h) = |\rho\sigma|_{\underline{H}}(Z_1, \ldots, Z_l, V_1, \ldots, V_k) = |\xi|_{\underline{H}}(Z_1, \ldots, V_k) = f$, as required.

Proposition 4.11. *Let $f: [P, Q] \otimes B \to [R, S] \otimes D$ be a central morphism of $\underline{H}$, and for some choice of association let the prime factor $[P, Q]$ of $[P, Q] \otimes B$ be asso-*

ciated via f with the prime factor $[R, S]$ *of* $[R, S] \otimes D$. *Then* $P = R$, $Q = S$, *and there is a central morphism* $k: B \to D$ *such that* $f = 1 \otimes k: [P, Q] \otimes B \to [P, Q] \otimes D$.

Proof. Since associated prime factors must be equal, we have $P = R$ and $Q = S$. We apply Proposition 4.10 with $A = [P, Q]$ and $C = [R, S]$; in this case $E = I$, and g is clearly $b: [P, Q] \otimes I \to [P, Q]$. Writing k for the composite

$$B \xrightarrow[h]{} I \otimes D \xrightarrow[c]{} D \otimes I \xrightarrow[b]{} D,$$

it follows from Theorem 3.1 that $f = 1 \otimes k$.

§5. Processes of construction

In the next section we shall show that the allowable natural transformations and the allowable graphs can be classified by a numerical *rank*, and that those of higher rank can be built up from those of lower rank, modulo central ones, by the use of three simple processes now to be described.

We consider a closed category $\underline{H}$, which will in our applications be either $\underline{G}$ or $\underline{N}(\underline{V})$. The first process of construction is the formation of the tensor product $f \otimes g : A \otimes B \to C \otimes D$ of two given morphisms $f: A \to C$ and $g : B \to D$. Observe that

$$(5.1) \qquad hf \otimes kg = (h \otimes k)(f \otimes g)$$

whenever hf and gk are defined. The second process of construction is the formation of the morphism $\pi(f): A \to [B, C]$ as in §1 from a given morphism $f: A \otimes B \to C$. Since π is natural we have commutativity in

$$(5.2) \qquad \begin{array}{ccc} A' & & \\ {\scriptstyle g}\downarrow & \searrow^{\pi(f(g \otimes 1))} & \\ A & \xrightarrow[\pi(f)]{} & [B, C] \end{array} ,$$

where $f(g \otimes 1)$ is the obvious composite $A' \otimes B \to A \otimes B \to C$. The third process of construction begins with morphisms $f: A \to B$ and $g : C \otimes D \to E$ and produces the composite

$$(5.3) \qquad ([B, C] \otimes A) \otimes D \xrightarrow[(1 \otimes f) \otimes 1]{} ([B, C] \otimes B) \otimes D \xrightarrow[e \otimes 1]{} C \otimes D \xrightarrow[g]{} E .$$

Rather than introduce a special symbol for the composite (5.3), we find it convenient to denote the composite

$$(5.4) \qquad [B, C] \otimes A \xrightarrow[1 \otimes f]{} [B, C] \otimes B \xrightarrow[e]{} C$$

by $\langle f \rangle : [B, C] \otimes A \to C$, so that (5.3) may be written as $g(\langle f \rangle \otimes 1)$. The symbol $\langle f \rangle$ is of course ambiguous, inasmuch as the value of C must be understood from the context. It is clear that $\langle\ \rangle$ is natural, in the sense that, for

$$A' \xrightarrow[u]{} A \xrightarrow[f]{} B \xrightarrow[v]{} B' , \quad C' \xrightarrow[w]{} C$$

we have commutativity in

$$
(5.5)\qquad
\begin{array}{ccc}
[B,C]\otimes A & \xrightarrow{\langle f\rangle} & C \\
{\scriptstyle [v,w]\otimes u}\uparrow & & \uparrow{\scriptstyle w} \\
[B',C']\otimes A' & \xrightarrow[\langle vfu\rangle]{} & C'
\end{array}
$$

We need the following two lemmas giving connections between these processes.

Lemma 5.1. *Let* $f\colon A\otimes B\to C$ *and* $g\colon D\to I$. *Then the image under* π *of the composite*

$$(A\otimes D)\otimes B \xrightarrow[u]{} (A\otimes B)\otimes D \xrightarrow[f\otimes g]{} C\otimes I \xrightarrow[b]{} C\ ,$$

where u *is the evident central morphism* $a^{-1}(1\otimes c)a$, *is the composite*

$$A\otimes D \xrightarrow[\pi(f)\otimes g]{} [B,C]\otimes I \xrightarrow[b]{} [B,C]\ .$$

Proof. Set $\pi(f)=h$, so that $f=e(h\otimes 1)$ by the definition of π. Then

$$b(f\otimes g)u = b(e(h\otimes 1)\otimes g)u = b(e\otimes 1)((h\times 1)\otimes g)u\ ,$$

which is $ebu((h\otimes g)\otimes 1)$ by the naturality of b and of u. The bu in this last expression is, by Theorem 3.1, the unique central morphism $b\otimes 1\colon ([B,C]\otimes I)\otimes B\to [B,C]\otimes B$. Thus, by the definition of π again,

$$b(f\otimes g)u = e(b\otimes 1)((h\otimes g)\otimes 1) = \pi^{-1}(b(h\otimes g))\ .$$

Lemma 5.2. *For* $f\colon A\to B$ *and* $g\colon C\otimes B\to D$, *we have*

$$g(1\otimes f) = \langle f\rangle(\pi(g)\otimes 1)\colon C\otimes A\to D\ .$$

Proof. By (5.1) and the definition of π,

$$\langle f\rangle(\pi(g)\otimes 1) = e(1\otimes f)(\pi(g)\otimes 1)) = e(\pi(g)\otimes 1)(1\otimes f) = g(1\otimes f)\ .$$

The remainder of this section concerns compatibility of graphs, for the closed category $\underline{G}$. We mostly omit the proofs, which are entirely evident but tedious to put into words.

Lemma 5.3. *Let $\xi : Q \to R$, $\eta : R \to S$, $\zeta : S \to T$ be graphs in $\underline{G}$; then the following assertions are equivalent:*

(i) *ζ is compatible with η and $\zeta\eta$ is compatible with ξ;*

(ii) *η is compatible with ξ and ζ is compatible with $\eta\xi$.*

When the assertions of Lemma 5.3 are true, we say that the three graphs ξ, η, ζ are compatible. The concept clearly extends to any number of graphs $\xi_i : T_{i-1} \to T_i$.

Lemma 5.4. (a) *Graphs $\xi : R \to S$ and $\eta : S \to T$ are compatible if either is central.*

(b) *If two graphs $\xi : R \to S$ and $\eta : S \to T$ are compatible while the graphs $\rho : R' \to R$, $\sigma : S \to S'$, $\tau : T \to T'$ are central, then the graphs $\sigma\xi\rho : R' \to S'$ and $\tau\eta\sigma^{-1} : S' \to T'$ are compatible.*

Proof. Immediate from Corollary 4.5.

Lemma 5.5. *In the situation of (5.1) above, if $\underline{H} = \underline{G}$, $h \otimes k$ is compatible with $f \otimes g$ if and only if h is compatible with f and k compatible with g.*

Lemma 5.6. *In the situation of (5.2) above, if $\underline{H} = \underline{G}$, then $\pi(f)$ is compatible with g if and only if f is compatible with $g \otimes 1$.*

Lemma 5.7. *Let $f: A \to B$, $g: C \otimes D \to E$, $k: E \otimes F \to G$ in $\underline{G}$. If k is compatible with $g \otimes 1 : (C \otimes D) \otimes F \to E \otimes F$, it is compatible with $(g \otimes 1)(((\langle f \rangle \otimes 1) \otimes 1)$: $(([B, C] \otimes A) \otimes D) \otimes F \to E \otimes F$.*

Lemma 5.8. *Let $f: A \to B$, $g: C \otimes D \to E$, $u: A' \to A$, $v: D' \to D$ in $\underline{G}$. If f is compatible with u and if g is compatible with $1 \otimes v$ then $g(\langle f \rangle \otimes 1): ([B, C] \otimes A) \otimes D \to E$ is compatible with $(1 \otimes u) \otimes v: ([B, C] \otimes A') \otimes D' \to ([B, C] \otimes A) \otimes D$.*

Lemma 5.9. *Let $\underline{H} = \underline{G}$, let f and g be as in Lemma 5.2, and let $h: D \otimes E \to F$. Then if $h: D \otimes E \to F$, $g \otimes 1: (C \otimes B) \otimes E \to D \otimes E$, and $(1 \otimes f) \otimes 1$: $(C \otimes A) \otimes E \to (C \otimes B) \otimes E$ are compatible, so are $h(\langle f \rangle \otimes 1): ([B, D] \otimes A) \otimes E \to F$ and $(\pi(g) \otimes 1) \otimes 1: (C \otimes A) \otimes E \to ([B, D] \otimes A) \otimes E$.*

§6. Constructibility of allowable morphisms

We place ourselves once again in the general situation $\Gamma: \underline{H} \to \underline{G}$ envisaged in §2 and §4; we recall that the cases of interest are $\underline{H} = \underline{N}(\underline{V})$ and $\underline{H} = \underline{G}$. The object of this section is to show that the allowable morphisms may be built up, modulo central morphisms, by the three processes described in §5. It is convenient to introduce the temporary name of *constructible* morphisms for those allowable morphisms that can be so built up; our aim is then to show that all allowable morphisms are constructible. We also give in this section the proof of Theorem 2.2.

We therefore define the *constructible* morphisms of $\underline{H}$ to be the smallest class of morphisms of $\underline{H}$ satisfying the following five conditions:

CM1 Every central morphism is in the class.

CM2 If $f: T \to S$ is in the class and if $u: T' \to T$ and $v: S \to S'$ are central then $vfu: T' \to S'$ is in the class.

CM3 If $f: A \to C$ and $g: B \to D$ are in the class so is $f \otimes g: A \otimes B \to C \otimes D$.

CM4 If $f: A \otimes B \to C$ is in the class so is $\pi(f): A \to [B, C]$.

CM5 If $f: A \to B$ and $g: C \otimes D \to E$ are in the class so is $g(\langle f \rangle \otimes 1)$: $([B, C] \otimes A) \otimes D \to E$.

In view of the definitions (1.1) of π and (5.4) of $\langle\ \rangle$, it is evident that the allowable morphisms satisfy CM1–CM5, so that the constructible morphisms are a subclass of the allowable ones.

We call an allowable morphism $f: T \to S$ in $\underline{H}$ *trivial* if both T and S are constant integral shapes.

Lemma 6.1. *A trivial constructible morphism in* $\underline{H}$ *is central.*

Proof. Consider the subclass of the constructible morphisms consisting of the following morphisms $f: T \to S$: if T and S are both constant integral shapes, f is to be central; otherwise, f is to be constructible. This subclass clearly satisfies CM1, CM4 and CM5. It satisfies CM2 because, by Corollary 4.4, if T' and S' are constant integral shapes so are T and S; and then vfu is central if f is. It satisfies CM3 because if $A \otimes B$ and $C \otimes D$ are constant integral shapes so are A, B, C and D; and then $f \otimes g$ is central if f and g are. Hence this subclass contains all constructible morphisms.

Proposition 6.2. *For each constructible* $h: T \to S$ *in* $\underline{H}$, *at least one of the following is true:*

(i) *h is central*

(ii) *h is of the form*

$$T \xrightarrow[x]{} A \otimes B \xrightarrow[f \otimes g]{} C \otimes D \xrightarrow[y]{} S$$

where f and g are constructible and non-trivial, and x and y are central.

(iii) *h is of the form*

$$T \xrightarrow[\pi(f)]{} [B, C] \xrightarrow[y]{} S$$

where f is constructible and y is central.

(iv) *h is of the form*

$$T \xrightarrow[x]{} ([B, C] \otimes A) \otimes D \xrightarrow[\langle f\rangle \otimes 1]{} C \otimes D \xrightarrow[g]{} S$$

where g and f are constructible and x is central.

Proof. Consider those constructible morphisms that *are* of one of the above forms (i)–(iv); we show that this class satisfies CM1–CM5 and therefore consists of all constructible morphisms. That it satisfies CM1, CM4 and CM5 is clear.

To see that CM2 is satisfied, let $u: T' \to T$ and $v: S \to S'$ be central. Then if h is central, so is vhu. If h is as in (ii) above, vhu is $(vy)(f \otimes g)(xu)$, which is of the same form. If h is as in (iii) above, vhu is $(vy)\,\pi(f(u \otimes 1))$, which is of the same form, $f(u \otimes 1)$ being constructible by CM2 since $u \otimes 1$ is central. If h is as in (iv) above, vhu is $(vg)(\langle f\rangle \otimes 1)(xu)$, which is of the same form, vg being constructible by CM2.

That CM3 is satisfied is clear unless f or g is trivial. If g is trivial it is central by Lemma 6.1. In this case B, I, D are constant integral shapes, and the empty graphs $B \to I$ and $I \to D$ give, by Proposition 4.1, central morphisms $u: B \to I$ and $v: I \to D$ in $\underline{H}$. Then by Theorem 4.9 we have $g = vu$. It follows at once from the naturality of b that $f \otimes g$ is then the composite

$$A \otimes B \xrightarrow[1 \otimes u]{} A \otimes I \xrightarrow[b]{} A \xrightarrow[f]{} C \xrightarrow[b^{-1}]{} C \otimes I \xrightarrow[1 \otimes v]{} C \otimes D;$$

since $(1 \otimes v)b^{-1}$ and $b(1 \otimes u)$ are central, and since CM2 is satisfied, this lies in the class because f does. Finally if f is trivial then, by the naturality of c, $f \otimes g$ is the composite

$$A \otimes B \xrightarrow[c]{} B \otimes A \xrightarrow[g \otimes f]{} D \otimes C \xrightarrow[c]{} C \otimes D,$$

which is in the class since CM2 is satisfied and since, by what we have just proved, $g \otimes f$ is in the class.

Remark. For brevity, morphisms h of the forms (ii), (iii), (iv) of Proposition 6.2 will be said to be respectively *of type* $\otimes$, *of type* π, and *of type* $\langle\ \rangle$.

For the purposes of our inductive proofs we introduce for each shape T a non-negative integer $r(T)$ called its *rank*, defined by the following inductive rules:

R1 $\quad r(I) = 0$.

R2 $r(1) = 1$.

R3 $r(T \otimes S) = r(T) + r(S)$.

R4 $r([T, S]) = r(T) + r(S) + 1$.

Note that $r(T) = 0$ if and only if T is a constant integral shape.

Lemma 6.3. *If $f: T \to S$ is central then $r(T) = r(S)$.*

Proof. Those central morphisms for which this is true clearly satisfy AM1, AM3 and AM5, and therefore constitute the totality of central morphisms.

The non-trivial step in the proof that all the allowable morphisms are constructible is the proof that the constructible morphisms are closed under composition. In fact, because of the exigencies of the inductive argument, we prove the variant of closure-under-composition given in Proposition 6.4 below. Moreover, because the same inductive argument applies, we prove at the same time the corresponding fact about compatibility, which will lead to a proof of Theorem 2.2.

Proposition 6.4. *If the morphisms $h: T \to S$ and $k: S \otimes U \to V$ of $\underline{H}$ are constructible, so is the composite morphism*

$$T \otimes U \xrightarrow[h \otimes 1]{} S \otimes U \xrightarrow[k]{} V .$$

Moreover, if $\underline{H} = \underline{G}$, the graphs k and $h \otimes 1$ are compatible.

Proof. The proof is by a double induction; we suppose the results to be true for all pairs of constructible morphisms $h': T' \to S'$ and $k': S' \otimes U' \to V'$ for which $r(T') + r(S') + r(U') + r(V') < r(T) + r(S) + r(U) + r(V)$; we also suppose them to be true for any pair h', k' for which $r(T') + r(S') + r(U') + r(V') = r(T) + r(S) + r(U) + r(V)$, provided that $r(T') + r(S') < r(T) + r(S)$.

By Proposition 6.2, each of h and k is central, or of type $\otimes$, or of type π, or of type $\langle\ \rangle$; we distinguish cases accordingly. We shall use Lemma 5.4, the Axiom CM2, and Lemma 6.3 freely without further explicit mention to "ignore" or to "absorb" central morphisms wherever convenient.

Case 1: either h or k is central. If h is central, so is $h \otimes 1$; the results follow from CM2 and from Lemma 5.4.

Case 2: h is of type $\langle\ \rangle$. Let h be $g(\langle f \rangle \otimes 1)x$ as in Proposition 6.2 (iv). Then the desired composite is

$$k(h \otimes 1) = k(g \otimes 1)\,((\langle f \rangle \otimes 1) \otimes 1)\,(x \otimes 1)$$

$$= (k(g \otimes 1)a^{-1})(\langle f \rangle \otimes 1)(a(x \otimes 1)) ,$$

which is again of type $\langle\ \rangle$, provided only that $k(g\otimes 1)a^{-1}$ is constructible. Since a^{-1} is central, we need the constructibility of the composite

$$(C\otimes D)\otimes U \xrightarrow[g\otimes 1]{} S\otimes U \xrightarrow[k]{} V\ ;$$

this follows by the inductive hypothesis since T, whose rank is equal by Lemma 6.3 to that of $([B,C]\otimes A)\otimes D$, has been replaced by $C\otimes D$, clearly of lower rank.

The same induction shows that k and $g\otimes 1$ are compatible; so by Lemma 5.7 and Lemma 5.4, k is compatible with $(g\otimes 1)((\langle f\rangle\otimes 1)\otimes 1)(x\otimes 1) = h\otimes 1$.

Case 3: h is of type $\otimes$. Let h be $y(f\otimes g)x$ as in Proposition 6.2 (ii). We are to consider the composite $k(h\otimes 1)$; without loss of generality we may suppose that $x = 1$ and absorb $y\otimes 1$ into k. Then we have

$$k(h\otimes 1) = k((f\otimes g)\otimes 1) = ka^{-1}(f\otimes(g\otimes 1))a$$

$$= ka^{-1}(f\otimes 1)(1\otimes(g\otimes 1))a\ ,$$

so that finally $k(h\otimes 1)$ is the composite

$$(6.1)\qquad (A\otimes B)\otimes U \xrightarrow[wa]{} B\otimes(A\otimes U) \xrightarrow[g\otimes 1]{} D\otimes(A\otimes U) \xrightarrow[ka^{-1}(f\otimes 1)w^{-1}]{} V\ ,$$

where the w's stand for two instances of the central morphism aca^{-1}. Now the composite

$$A\otimes(D\otimes U) \xrightarrow[f\otimes 1]{} C\otimes(D\otimes U) \xrightarrow[ka^{-1}]{} V$$

is constructible by the induction hypothesis, because U has been replaced by $D\otimes U$ (*this* is the reason for formulating Proposition 6.4 for $k(h\otimes 1)$ instead of just kh) and we have

$$r(T)+r(S)+r(U)+r(V) = r(A\otimes B)+r(C\otimes D)+r(U)+r(V)$$

$$> r(A)+r(C)+r(D\otimes U)+r(V)$$

unless $r(B) = 0$; in the latter case we get equality, but then $r(D) > 0$ since g is non-trivial, and

$$r(T) + r(S) = r(A \otimes B) + r(C \otimes D) > r(A) + r(C),$$

so that the second half of the induction hypothesis applies. Since w^{-1} is central, $k' = ka^{-1}(f \otimes 1)w^{-1}$ is constructible; and since wa is central, (6.1) will be constructible if the composite $k'(g \otimes 1)$ is. But now the induction hypothesis shows that $k'(g \otimes 1)$ is indeed constructible, by essentially the same calculation with ranks as above, with g replacing f.

The same inductions show that ka^{-1} is compatible with $f \otimes 1$, so that k is compatible with $a^{-1}(f \otimes 1)w^{-1}$; and that k' is compatible with $g \otimes 1$, and hence with $(g \otimes 1)wa$. Therefore, by Lemma 5.3, k is compatible with $a^{-1}(f \otimes 1)w^{-1}(g \otimes 1)wa = h \otimes 1$.

Case 4: k is of type π. Thus $k = y\pi(f)$ for central y and constructible f. We can take $y = 1$, so that $k = \pi(f)\colon S \otimes U \to V = [B, C]$ for some constructible $f\colon (S \otimes U) \otimes B \to C$. By (5.2) we have

$$k(h \otimes 1) = \pi(f)(h \otimes 1) = \pi(f((h \otimes 1) \otimes 1));$$

this is constructible if $f((h \otimes 1) \otimes 1)$ is, and hence, if $f((h \otimes 1) \otimes 1)a^{-1} = fa^{-1}(h \otimes 1)$ is. This last is the composite

$$T \otimes (U \otimes B) \xrightarrow[h \otimes 1]{} S \otimes (U \otimes B) \xrightarrow[fa^{-1}]{} C,$$

which is constructible by induction, since $r(U \otimes B) + r(C) < r(U) + r([B, C])$.

The same induction proves fa^{-1} compatible with $h \otimes 1$, hence f with $a^{-1}(h \otimes 1)a = (h \otimes 1) \otimes 1$; and thence, by Lemma 5.6, $\pi(f)$ with $h \otimes 1$.

Case 5: h is of type π *and k of type* $\otimes$. There are central morphisms x, y and z such that $h = z\pi(m)$ for some constructible $m\colon T \otimes P \to Q$ and $k = y(f \otimes g)x$ for some constructible and non-trivial f and g. We may take $y = 1$ and absorb $z \otimes 1$ into x, so that the composite $k(h \otimes 1)$ to be considered has the form

$$T \otimes U \xrightarrow[h \otimes 1]{} [P, Q] \otimes U \xrightarrow[x]{} A \otimes B \xrightarrow[f \otimes g]{} C \otimes D = V.$$

Interchanging A and B if necessary, we can assume that the central morphism x associates $[P, Q]$ with a prime factor of A. Then Proposition 4.10 gives a shape R such that x has the form of a composite

$$[P, Q] \otimes U \xrightarrow[1 \otimes s]{} [P, Q] \otimes (R \otimes B) \xrightarrow[a^{-1}]{} ([P, Q] \otimes R) \otimes B \xrightarrow[t \otimes 1]{} A \otimes B,$$

for suitable central s and t. Since $(1 \otimes s)(h \otimes 1) = (h \otimes 1)(1 \otimes s)$ we can drop s and write $U = R \otimes B$, while t can be absorbed into f. The composite $k(h \otimes 1)$ to be considered now has the form

$$T \otimes (R \otimes B) \xrightarrow[h \otimes 1]{} [P, Q] \otimes (R \otimes B) \xrightarrow[a^{-1}]{} ([P, Q] \otimes R) \otimes B \xrightarrow[f \otimes g]{} C \otimes D .$$

This may be rewritten as

$$(f \otimes g)a^{-1}(h \otimes 1) = (f \otimes g)((h \otimes 1) \otimes 1)\, a^{-1} = (f(h \otimes 1) \otimes g)\, a^{-1} ,$$

which will be constructible by CM2 and CM3 if the composite

$$T \otimes R \xrightarrow[h \otimes 1]{} [P, Q] \otimes R \xrightarrow[f]{} C$$

is. That this is indeed so follows by induction, since g is non-trivial and therefore $r(U) + r(V) = r(R \otimes B) + r(C \otimes D) > r(R) + r(C)$.

The induction argument also shows that f is compatible with $h \otimes 1$, whence, by Lemmas 5.5 and 5.4, $f \otimes g$ is compatible with $((h \otimes 1) \otimes 1)a^{-1} = a^{-1}(h \otimes 1)$; finally, by Lemma 5.4 again, $k = (f \otimes g)\, a^{-1}$ is compatible with $h \otimes 1$.

Case 6: h is of type π and k is of type $\langle\ \rangle$. Thus there are central morphisms z and x such that $h = z\pi(m)$ and $k = g(\langle f \rangle \otimes 1)x$ for constructible morphisms

$$m \colon T \otimes P \to Q , \quad f \colon A \to B , \quad g \colon C \otimes D \to V .$$

By absorbing $z \otimes 1$ in x, we may suppose that $S = [P, Q]$ and that $z = 1$. Then the composite to be considered has the form

$$T \otimes U \xrightarrow[\pi(m) \otimes 1]{} [P, Q] \otimes U \xrightarrow[x]{} ([B, C] \otimes A) \otimes D \xrightarrow[g(\langle f \rangle \otimes 1)]{} V .$$

We distinguish three subcases, according as $[P, Q]$ is associated via the central morphism x with $[B, C]$, with a prime factor of A, or with a prime factor of D. (We recall that these possibilities need not be mutually exclusive, if $[P, Q]$ is constant.)

Subcase 1: $[P, Q]$ is associated with $[B, C]$. By Proposition 4.11, $P = B$, $Q = C$, and x is the composite

$$[B, C] \otimes U \xrightarrow[1 \otimes s]{} [B, C] \otimes (A \otimes D) \xrightarrow[a^{-1}]{} ([B, C] \otimes A) \otimes D$$

for a suitable central s. Since $(1 \otimes s)(h \otimes 1) = (h \otimes 1)(1 \otimes s)$ we may, arguing as in Case 5, suppose that $U = A \otimes D$ and $s = 1$. The composite to be considered then has the form

$$T \otimes (A \otimes D) \xrightarrow[\pi(m)\otimes 1]{} [B, C] \otimes (A \otimes D) \xrightarrow[a^{-1}]{} ([B, C] \otimes A) \otimes D \xrightarrow[g(\langle f \rangle \otimes 1)]{} V.$$

Now $a^{-1}(\pi(m) \otimes 1) = ((\pi(m) \otimes 1) \otimes 1)\, a^{-1}$ by naturality, while by Lemma 5.2 we have $\langle f \rangle\, (\pi(m) \otimes 1) = m(1 \otimes f)$. The composite thus becomes

$$k(h \otimes 1) = g(m(1 \otimes f) \otimes 1)\, a^{-1} = g(mc(f \otimes 1)\, c \otimes 1)\, a^{-1}\,.$$

This formula involves two successive composites, first a composite h',

$$A \otimes T \xrightarrow[f \otimes 1]{} B \otimes T \xrightarrow[mc]{} C = Q$$

and second the composite

$$(T \otimes A) \otimes D \xrightarrow[h'c \otimes 1]{} C \otimes D \xrightarrow[g]{} V\,.$$

Both are of the form considered in our induction. The induction assumption does apply to both because the original rank, with $U = A \otimes D$ and $S = [P, Q] = [B, C]$, is

$$rT + rS + rU + rV = rT + rV + rB + rC + rA + rD + 1\,,$$

and this clearly exceeds either of the ranks $rA + rB + rT + rC$ or $rT + rA + rD + rC + rV$ involved in the two composites above.

The same induction shows that g is compatible with $h'c \otimes 1$ and mc is compatible with $f \otimes 1$; so that by Lemmas 5.3, 5.4 and 5.5 $g(m \otimes 1)$ is compatible with $(1 \otimes f) \otimes 1$. It follows from Lemma 5.9 and Lemma 5.4 that $g(\langle f \rangle \otimes 1)$ is compatible with $((\pi(m) \otimes 1) \otimes 1)\, a^{-1} = a^{-1}(\pi(m) \otimes 1)$, so that $k = g(\langle f \rangle \otimes 1)\, a^{-1}$ is compatible with $h \otimes 1 = \pi(m) \otimes 1$, as required.

Subcase 2: $[P, Q] = S$ is associated with a prime factor of A. By Proposition 4.10, there is a shape R such that $a(c \otimes 1)x$ is the composite

$$S \otimes U \xrightarrow[1 \otimes s]{} S \otimes (R \otimes ([B, C] \otimes D))$$

$$\xrightarrow[a^{-1}]{} (S \otimes R) \otimes ([B, C] \otimes D) \xrightarrow[t \otimes 1]{} A \otimes ([B, C] \otimes D)$$

for suitable central s and t. By the naturality of a and c, therefore, x is the composite

$$S \otimes U \xrightarrow[1 \otimes s]{} S \otimes (R \otimes ([B, C] \otimes D))$$

$$\xrightarrow[w]{} ([B, C] \otimes (S \otimes R)) \otimes D \xrightarrow[(1 \otimes t) \otimes 1]{} ([B, C] \otimes A) \otimes D,$$

where w is the central natural transformation $w = (c \otimes 1)a^{-1}a^{-1}$. Once again, since $(1 \otimes s)(h \otimes 1) = (h \otimes 1)(1 \otimes s)$, we may suppose that $s = 1$ and $U = R \otimes ([B, C] \otimes D)$. Moreover, since $\langle f \rangle (1 \otimes t) = \langle ft \rangle$ by (5.5) (the naturality of $\langle\ \rangle$), we may absorb t in f and hence suppose that $A = S \otimes R$ and $t = 1$. The desired composite $k(h \otimes 1)$ thus has the form

$$g(\langle f \rangle \otimes 1)\, w(h \otimes 1) = g(\langle f \rangle \otimes 1)((1 \otimes (h \otimes 1)) \otimes 1)\, w$$

$$= g(\langle f(h \otimes 1) \rangle \otimes 1) w,$$

by the naturality of w and of $\langle\ \rangle$. It thus suffices by CM5 to prove the composite

$$T \otimes R \xrightarrow[h \otimes 1]{} S \otimes R \xrightarrow[f]{} B$$

constructible. But this is of the form considered in the induction, and since $r(U) > r(R) + r(B)$ the inductive hypothesis applies.

By the same inductive argument f is compatible with $h \otimes 1$; by Lemmas 5.8 and 5.4, therefore, $g(\langle f \rangle \otimes 1)$ is compatible with $((1 \otimes (h \otimes 1)) \otimes 1)w = w(h \otimes 1)$; finally, by Lemma 5.4 again, $k = g(\langle f \rangle \otimes 1)\, w$ is compatible with $h \otimes 1$.

Subcase 3: $[P, Q] = S$ *is associated with a prime factor of* D. By Proposition 4.10, there is a shape R such that cx is the composite

$$S \otimes U \xrightarrow[1 \otimes s]{} S \otimes (R \otimes ([B, C] \otimes A))$$

$$\xrightarrow[a^{-1}]{} (S \otimes R) \otimes ([B, C] \otimes A) \xrightarrow[t \otimes 1]{} D \otimes ([B, C] \otimes A)$$

for suitable central s and t. By the naturality of c, therefore, x is the composite

$$S \otimes U \xrightarrow[1 \otimes s]{} S \otimes (R \otimes ([B, C] \otimes A))$$

$$\xrightarrow[u]{} ([B, C] \otimes A) \otimes (S \otimes R) \xrightarrow[1 \otimes t]{} ([B, C] \otimes A) \otimes D\,,$$

where u is the central natural transformation $u = ca^{-1}$. Since $(1 \otimes s)(h \otimes 1) = (h \otimes 1)(1 \otimes s)$ we may again suppose that $s = 1$, so that $U = R \otimes ([B, C] \otimes A)$. Since $(\langle f \rangle \otimes 1)(1 \otimes t) = (1 \otimes t)(\langle f \rangle \otimes 1)$, we may absorb $1 \otimes t$ in g and hence suppose that $t = 1$ and $D = S \otimes R$. The desired composite $k(h \otimes 1)$ is then

$$g(\langle f \rangle \otimes 1)\, u(h \otimes 1) = g(\langle f \rangle \otimes 1)(1 \otimes (h \otimes 1))\, u$$

$$= g(1 \otimes (h \otimes 1))(\langle f \rangle \otimes 1)\, u$$

$$= gu(h \otimes 1)u^{-1}(\langle f \rangle \otimes 1)\, u\,,$$

using (5.1) and the naturality of u. It thus suffices by CM5 to prove the constructibility of $gu(h \otimes 1)u^{-1}$, and therefore of $gu(h \otimes 1)$. This is the composite

$$T \otimes (R \otimes C) \xrightarrow[h \otimes 1]{} S \otimes (R \otimes C) \xrightarrow[gu]{} V,$$

which is constructible by the inductive hypothesis since $r(R \otimes C) < r(U)$.

By the same inductive argument gu is compatible with $h \otimes 1$, so that g is compatible with $u(h \otimes 1)u^{-1} = 1 \otimes (h \otimes 1)$. By Lemmas 5.8 and 5.4, therefore, $g(\langle f \rangle \otimes 1)$ is compatible with $(1 \otimes (h \otimes 1))u = u(h \otimes 1)$; so that finally $k = g(\langle f \rangle \otimes 1)u$ is compatible with $h \otimes 1$.

This concludes the proof of Proposition 6.4.

Theorem 6.5. *The constructible morphisms of $\underline{H}$ are exactly the allowable ones.*

Proof. We have already observed that the allowable morphisms clearly satisfy CM1–CM5, so that every constructible morphism is allowable. It remains to show that the constructible morphisms satisfy AM1–AM5.

They satisfy AM1 because $1, a, b, c, a^{-1}, b^{-1}$ are central. As for AM2, $d: T \to [S, T \otimes S]$ is $\pi(1)$ where $1: T \otimes S \to T \otimes S$, so that d is constructible by CM4; and $e: [T, S] \otimes T \to S$ is $\langle 1 \rangle$, which by the naturality of b is the composite

$$[T, S] \otimes T \xrightarrow[b^{-1}]{} ([T, S] \otimes T) \otimes I \xrightarrow[\langle 1 \rangle \otimes 1]{} S \otimes I \xrightarrow[b]{} S,$$

so that e is constructible by CM2 and CM5. AM3 is trivially satisfied, as it coincides with CM3. In AM4, let $f: T \to T'$ and $g: S \to S'$ be constructible. Then the composite

$$[T', S] \otimes T \xrightarrow[b^{-1}]{} ([T', S] \otimes T) \otimes I \xrightarrow[\langle f \rangle \otimes 1]{} S \otimes I \xrightarrow[gb]{} S'$$

is constructible by CM2 and CM5; but this composite is $g\langle f \rangle$ by the naturality of b. It follows from CM4 that $\pi(g\langle f \rangle)$ is constructible; but $\pi(g\langle f \rangle) = \pi(ge(1 \otimes f))$ is equal by the naturality of π to $[g, f]\ \pi(e) = [g, f]\ 1 = [g, f]$. Thus AM4 is satisfied.

There remains AM5. Let $f: T \to S$ and $g: S \to R$ be constructible. Then the composite

(6.2) $$S \otimes I \xrightarrow[b]{} S \xrightarrow[g]{} R$$

is constructible by CM2, whence the composite

(6.3) $$T \otimes I \xrightarrow[f \otimes 1]{} S \otimes I \xrightarrow[gb]{} R$$

is constructible by Proposition 6.4; by CM2 again, the composite of (6.3) with $b^{-1}: T \to T \otimes I$ is also constructible, and by the naturality of b this composite is gf.

Proof of Theorem 2.2. Let $f: T \to S$, $g: S \to R$ be allowable graphs in $\underline{G}$. By Theorem 6.5, they are constructible. The composite (6.2) is then also constructible, so that in (6.3) gb is compatible with $f \otimes 1$ by Proposition 6.4. We conclude from Lemma 5.4 that g is compatible with $b(f \otimes 1)b^{-1} = f$.

§7. Proofs of Theorem 2.1 and Theorem 2.4

We still use $\underline{H}$ to denote $\underline{N}(\underline{V})$ or $\underline{G}$, with $\Gamma: \underline{H} \to \underline{G}$ as before. For the purposes of this section we need a slight refinement of Proposition 6.2. Let us call an integral shape P *reduced* if it is either the shape I or else is constructed by the rules S2 and S3 alone; that is, it contains no I's unless it reduces to I alone. By *an iterated tensor product* of shapes $X_1, \ldots, X_n$ we mean $|P|(X_1, \ldots, X_n)$ for any reduced integral shape P with $v(P) = \{1, \ldots, n\}$; if $n = 0$ it is just I. Let us call an arbitrary shape T *reduced* if, in its prime factorization $T = |P|(X_1, \ldots, X_n)$, the integral shape P is reduced. (This is a consistent use of language since the prime factorization of the integral shape P is $|P|(1, 1, \ldots, 1)$.)

Lemma 7.1. *Given any shape T we can find a reduced shape T' and a central isomorphism $z: T \to T'$ in $\underline{H}$.*

Proof. Let the prime factorization of T be $|P|(X_1, \ldots, X_n)$. Let P' be a reduced integral shape with $v(P') = v(P) = \{1, \ldots, n\}$, and let $\xi: P \to P'$ be the graph corresponding to the identity permutation of $\{1, \ldots, n\}$. Set $T' = |P'|(X_1, \ldots, X_n)$ and set $z = |\xi|_{\underline{H}}(X_1, \ldots, X_n)$.

Lemma 7.2. *In Proposition 6.2 we can suppose that the shapes A, B, C, D in (ii) and the shapes A, D in (iv) are reduced.*

Proof. In case (ii), replace A, B, C, D by reduced isomorphs as in Lemma 7.1, absorbing the central isomorphims thereby introduced into x and y; similarly for case (iv).

We define the *rank* $r(h)$ of a morphism $h: T \to S$ in $\underline{H}$ to be the sum $r(T) + r(S)$ of the ranks of T and of S. If h is allowable, which by Theorem 6.5 is the same thing as constructible, Proposition 6.2 asserts that h has one of the four following forms:

$$(7.1) \qquad h = x\,, \quad h = y(f \otimes g)x\,, \quad h = y\pi(f)\,, \quad h = g(\langle f \rangle \otimes 1)x\,,$$

where x and y are central, f and g are allowable, and moreover in the $y(f \otimes g)x$ case neither f nor g is trivial. The basis of our inductive arguments is the obvious fact that in each case we have $r(f) < r(h)$ and (where applicable) $r(g) < r(h)$. In using the forms (7.1) we shall always suppose that the reductions of Lemma 7.2 have been carried out.

Proof of Theorem 2.1. We are to construct an algorithm for deciding whether a graph $h: T \to S$ in $\underline{G}$ is allowable. We suppose inductively that we possess such an algorithm for all smaller values, if any, of $r(h)$. Since finding the prime factorizations of T and of S is algorithmic, Propositions 4.3 and 4.2 enable us to decide whether h is central. It remains to test whether h is of one of the remaining types in (7.1), which we again refer to as type $\otimes$, type π, and type $\langle\ \rangle$.

To test whether h is of type $\otimes$, with the notation as in Proposition 6.2 and with A, B, C, D reduced, first observe that, by Proposition 4.3, the prime factors of A and of B must together make up those of T; so that there are only a finite number of possibilities for A and for B to be tried (because A and B are reduced!). Similarly there are only a finite number of possibilities for C and for D; and for a given choice of A, B, C, D there are only a finite number of possibilities for x and for y. When these choices are all made, the graph $y^{-1}hx^{-1}$ is either not of the form $f \otimes g$, or else is of this form for a unique f and g. Since $r(f) < r(h)$ and $r(g) < r(h)$, we can now test f and g for allowability.

Entirely similar procedures allow us to test whether h is of type π or of type $\langle\ \rangle$, so that we have the desired algorithm.

Remark. There are non-allowable graphs in $\underline{G}$; the unique graph $[1, 1] \to I$ is one such.

Before proving Theorem 2.4 we establish some facts about proper shapes, as defined in §2. Observe that every constant shape is proper; that if $[T, S]$ is proper then T and S are proper; and that $T \otimes S$ is proper if and only if T and S are proper – whence T is proper if and only if each of its prime factors is proper.

Lemma 7.3. *If $h: T \to S$ is a central morphism in $\underline{H}$ and if either T or S is proper so is the other.*

Proof. By Proposition 4.3, T and S have the same prime factors.

Proposition 7.4. *Let $h: T \to S$ be allowable in $\underline{H}$, with the shape S constant and the shape T proper. Then the shape T is constant.*

Proof. Suppose inductively that it is so for all smaller values, if any, of $r(h)$, and consider h of one of the four possible types in (7.1). If h is central the result is immediate by Corollary 4.4. By this same Corollary 4.4, together with Lemma 7.3 and Lemma 6.3, we may ignore central factors x and y in the other types in (7.1). If $h = f \otimes g: A \otimes B \to C \otimes D$ then C and D are constant because S is and A and B are proper because T is, so that by induction A and B are constant, whence T is constant. If $h = \pi(f): T \to [B, C]$ then B and C are constant because S is, so that $T \otimes B$ is proper because T is, and then T is constant by the inductive hypothesis applied to $f: T \otimes B \to C$. Finally, if $h = g(\langle f \rangle \otimes 1): ([B, C] \otimes A) \otimes D \to S$, then the inductive hypothesis applied to $g: C \otimes D \to S$ shows that C and D are constant. Since T is proper so is $[B, C]$, whence B is constant. Finally the inductive hypothesis applied to $f: A \to B$ shows that A is constant, so that T is constant.

We next show how to eliminate constant prime factors from a shape T.

Lemma 7.5. *Given a shape T we can find a shape S with $r(S) \leq r(T)$ and an allowable isomorphism $f\colon T \to S$ in $\underline{H}$ with allowable inverse such that*

(a) *S is reduced;*

(b) *S has no constant prime factors, its prime factors being precisely the non-constant ones of T;*

(c) *if T is proper, so is S;*

(d) *there is a constant shape R and a central isomorphism $T \to S \otimes R$ with the same graph as f.*

Proof. Let S be any iterated tensor product of the non-constant prime factors of T, and R any iterated tensor product of the constant prime factors of T. There is an evident central isomorphism $T \to S \otimes R$, and f is the composite of this with

$$S \otimes R \xrightarrow[1 \otimes k_R]{} S \otimes I \xrightarrow[h]{} S\,,$$

where k_R is the isomorphism of Lemma 4.7.

Proposition 7.6. *Let $h\colon P \otimes Q \to M \otimes N$ be an allowable morphism in $\underline{H}$, where P, Q, M, N are proper shapes. Suppose that the graph Γh is of the form $\xi \otimes \eta$ for graphs $\xi: P \to M$ and $\eta: Q \to N$. Then there are allowable morphisms $p: P \to M$, $q: Q \to N$ such that $h = p \otimes q$, $\Gamma p = \xi$, and $\Gamma q = \eta$.*

Proof. Suppose inductively that it is so for all smaller values, if any, of $r(h)$. By Lemma 7.5 we may without loss of generality suppose each of P, Q, M, N to be reduced and to have only non-constant prime factors.

If h is central, so is $\Gamma h = \xi \otimes \eta$. From Propositions 4.3 and 4.2, it is clear that ξ and η are then central. By Theorem 4.9 there are central $p: P \to M$, $q: Q \to N$ with $\Gamma p = \xi$ and $\Gamma q = \eta$; then $\Gamma h = \Gamma(p \otimes q)$, so that by Theorem 4.9 again we have $h = p \otimes q$.

If h is of type $\otimes$, say h is the composite

$$P \otimes Q \xrightarrow[x]{} A \otimes B \xrightarrow[f \otimes g]{} C \otimes D \xrightarrow[y]{} M \otimes N\,;$$

let an iterated tensor product, in the order in which they occur in P, of those prime factors of P that are associated via x with a prime factor of A [resp. B] be X [resp. Y]; similarly let an iterated $\otimes$-product of those prime factors of Q associated via x with a prime factor of A [resp. B] be U [resp. V]. In the same way let X', Y', U', V' be iterated $\otimes$-products of the prime factors "common" to M and C, M and D, N and C, N and D respectively. Define a graph $\rho: X \to X'$ as the restriction of Γh to $v(X) + v(X')$; this is indeed a graph because Γh is of the form $\xi \otimes \eta$. Define similarly graphs $\sigma: Y \to Y'$, $\tau: U \to U'$, $\kappa: V \to V'$. The graphs of the allowable morphisms

$$(7.2)\qquad X \otimes U \longrightarrow A \xrightarrow[f]{} C \longrightarrow X' \otimes U' ,$$

$$(7.3)\qquad Y \otimes V \longrightarrow B \xrightarrow[g]{} D \longrightarrow Y' \otimes V' ,$$

where the unnamed arrows denote the obvious central morphisms, are respectively $\rho \otimes \tau$ and $\sigma \otimes \kappa$. By the inductive hypothesis we conclude that (7.2) and (7.3) are respectively $r \otimes t$ and $s \otimes k$ for allowable morphisms r, t, s, k with the respective graphs $\rho, \tau, \sigma, \kappa$. Define p and q to be the composites

$$P \longrightarrow X \otimes Y \xrightarrow[r \otimes s]{} X' \otimes Y' \longrightarrow M ,$$

$$Q \longrightarrow U \otimes V \xrightarrow[t \otimes k]{} U' \otimes V' \longrightarrow N ,$$

where once again the unnamed arrows denote the obvious central morphisms. That $h = p \otimes q$ is then immediate from Theorem 4.9, while evidently $\Gamma p = \xi$ and $\Gamma q = \eta$.

If h is of type π, say h is the composite

$$P \otimes Q \xrightarrow[\pi(f)]{} [B, C] \xrightarrow[y]{} M \otimes N ,$$

then by Proposition 4.3 either $M = [B, C]$ and $N = I$ or else $N = [B, C]$ and $M = I$; by replacing h by chc if necessary we may suppose the former to be the case. Then h is the composite

$$P \otimes Q \xrightarrow[\pi(f)]{} [B, C] \xrightarrow[b^{-1}]{} [B, C] \otimes I .$$

Since $\Gamma h = \xi \otimes \eta$ it follows from Lemma 5.1 that the graph of the composite

$$(P \otimes B) \otimes Q \xrightarrow[u]{} (P \otimes Q) \otimes B \xrightarrow[f]{} C \xrightarrow[b^{-1}]{} C \otimes I ,$$

where u is the evident central morphism, is $\pi^{-1}(\xi) \otimes \eta$. By induction, therefore, $b^{-1} fu$ is $r \otimes q$ for allowable $r : P \otimes B \to C$ with graph $\pi^{-1}(\xi)$ and $q : Q \to I$ with graph η. Set $p = \pi(r)$; then $\Gamma p = \xi$ and $h = p \otimes q$ by another application of Lemma 5.1.

If h is of type $\langle\ \rangle$, with the notation of Proposition 6.2, we may (replacing h by chc if necessary) suppose that $[B, C]$ is associated via x with a prime factor of P. Let an iterated $\otimes$-product of those prime factors of A associated via x with a prime factor of P [resp. Q] be X [resp. Y]. The mate under Γh of an element of $v(Y)$ is in $v(A) + v(B)$ by the form $g(\langle f \rangle \otimes 1)x$ of h, but is in $v(Q) + v(N)$ by the hypothesis that $\Gamma h = \xi \otimes \eta$; it must therefore be in $v(Y)$. Thus the graph of the composite

$$(7.4)\qquad X \otimes Y \longrightarrow A \xrightarrow[f]{} B \xrightarrow[b^{-1}]{} B \otimes I ,$$

where the unnamed arrow is the obvious central morphism, is of the form $\rho \otimes \sigma$ for graphs $\rho : X \to B$ and $\sigma : Y \to I$. By the inductive hypothesis, (7.4) is $r \otimes s$ for allowable $r : X \to B$, $s : Y \to I$. It then follows from Proposition 7.4 that Y is constant; since none of the prime factors of Q is constant, this means that $Y = I$ and that all the prime factors of A are therefore associated via x with prime factors of P.

It follows then from Proposition 4.10 that there are a shape R and central morphisms y and z such that x is the composite

$$P \otimes Q \xrightarrow[y \otimes 1]{} (([B, C] \otimes A) \otimes R) \otimes Q \xrightarrow[a]{} ([B, C] \otimes A) \otimes (R \otimes Q)$$

$$\xrightarrow[1 \otimes z]{} ([B, C] \otimes A) \otimes D.$$

It is clear that the graph of the composite

$$(C \otimes R) \otimes Q \xrightarrow[a]{} C \otimes (R \otimes Q) \xrightarrow[1 \otimes z]{} C \otimes D \xrightarrow[g]{} M \otimes N$$

is $\zeta \otimes \eta$, where $\zeta : C \otimes R \to M$ is the restriction of ξ to $v(C) + v(R) + v(M)$. By induction, $g(1 \otimes z)a$ is $r \otimes q$ for allowable $r : C \otimes R \to M$ and $q : Q \to N$ with the appropriate graphs. Setting p equal to the composite

$$P \xrightarrow[y]{} ([B, C] \otimes A) \otimes R \xrightarrow[r(\langle f \rangle \otimes 1)]{} M,$$

we have $p \otimes q = (r \otimes q)((\langle f \rangle \otimes 1) \otimes 1)(y \otimes 1) = g(1 \otimes z)a((\langle f \rangle \otimes 1) \otimes 1)(y \otimes 1)$; by the naturality of a this is $g(1 \otimes z)(\langle f \rangle \otimes 1)a(y \otimes 1) = g(\langle f \rangle \otimes 1)(1 \otimes z)a(y \otimes 1) = g(\langle f \rangle \otimes 1)x = h$.

This completes the proof of Proposition 7.6.

Proposition 7.7. *Let $f : A \otimes B \to C$ be an allowable morphism in $\underline{H}$, where A, B, C are proper shapes. Suppose that the mate under Γf of each element of $v(B)$ is again in $v(B)$. Then B is constant.*

Proof. The composite

$$A \otimes B \xrightarrow[f]{} C \xrightarrow[b^{-1}]{} C \otimes I$$

has a graph of the form $\xi \otimes \eta$; therefore by Proposition 7.6 there is an allowable morphism $q : B \to I$. It follows from Proposition 7.4 that B is constant.

Proposition 7.8. *Let $h : ([Q, M] \otimes P) \otimes N \to S$ be an allowable morphism between proper shapes in $\underline{H}$, with $[Q, M]$ not constant. Suppose that the graph Γh is of the form $\eta(\langle \xi \rangle \otimes 1)$ for graphs $\xi : P \to Q$, $\eta : M \otimes N \to S$. Suppose finally that ξ cannot be written in the form*

$$P \xrightarrow[\omega]{} ([F, G] \otimes E) \otimes H \xrightarrow[\rho(\langle \sigma \rangle \otimes 1)]{} Q \tag{7.5}$$

for any graphs ω, ρ, σ with ω central. Then there are allowable morphisms $p : P \to Q$, $q : M \otimes N \to S$ such that $h = q(\langle p \rangle \otimes 1)$, $\Gamma p = \xi$ and $\Gamma q = \eta$.

Proof. Suppose inductively that it is so for all smaller values, if any, of $r(h)$. Use Lemma 7.5 to replace P, N, S by reduced shapes which have no constant prime factors; we must show that in doing so we lose no generality. It follows from (5.5) that doing so makes no difference to the expressibility Γh in the form $\eta(\langle \xi \rangle \otimes 1)$ or the expressibility of h in the form $q(\langle p \rangle \otimes 1)$ for allowable p and q. We must show that it makes no difference to the expressibility of ξ in the form (7.5); but this is a very easy deduction from Lemma 7.5 (d).

Note that once we have $h = q(\langle p \rangle \otimes 1)$, it is automatic that $\Gamma p = \xi$ and $\Gamma q = \eta$.

Suppose that h is central. By Corollary 4.5, the mate under Γh of an element of $v(P)$ or of $v(Q)$ is then an element of $v(S)$. On the other hand, by the form $\eta(\langle \xi \rangle \otimes 1)$ of Γh the mate of an element of $v(P)$ is an element of $v(Q)$, and conversely. It follows that P and Q are both constant, so that by the reduction above we must have $P = I$. Lemma 4.7 now gives an allowable $p = k_Q^{-1} : I \to Q$. By the naturality of e,

$$\langle p \rangle = e(1 \otimes p) = e([p, 1] \otimes 1) : [Q, M] \otimes I \longrightarrow [I, M] \otimes I \longrightarrow M .$$

On the other hand, $[A, [I, M]] \cong [A \otimes I, M] \cong [A, M]$ in any closed category, so the Yoneda Lemma provides an isomorphism $[1, b]d : M \cong [I, M]$ with inverse eb^{-1}. Therefore $[1, b]de = b : [I, M] \otimes I \cong [I, M]$, so that from the display above

$$[1, b]d\langle p \rangle = b([p, 1] \otimes 1) : [Q, M] \otimes I \longrightarrow [I, M] .$$

Since the right-hand morphism is an isomorphism we have constructed the following factorization of the identity

$$1 = t\langle p \rangle : [Q, M] \otimes I \longrightarrow [Q, M] \otimes I ,$$

with an allowable $t = ([p^{-1}, 1] \otimes 1) b^{-1} [1, b] d$. The originally given allowable morphism h can now be factored as

$$h = h1 = h(t \otimes 1)(\langle p \rangle \otimes 1) = q(\langle p \rangle \otimes 1)$$

with q allowable, as required.

If h is of the form $y(f \otimes g)x$ for $f : A \to C, g : B \to D$, we may without loss of generality suppose that $[Q, M]$ is associated via x with a prime factor of A. Let an iterated $\otimes$-product of those prime factors of P associated via x with a prime factor of A [resp. B] be X [resp. Y]. The mate under Γh of an element of $v(Y)$ is in $v(B) + v(D)$ by the form $y(f \otimes g)x$ of h, but is in $v(Q) + v(P)$ by the hypothesis that $\Gamma h = \eta(\langle \xi \rangle \otimes 1)$; it must therefore be in $v(Y)$. It now follows from Proposition

7.7 that Y is constant; since none of the prime factors of P is constant, this means that $Y = I$ and that all the prime factors of P are associated via x with prime factors of A.

It follows then from Proposition 4.10 that there are a shape R and central morphisms s and t such that x is the composite

$$([Q, M] \otimes P) \otimes N \xrightarrow[1 \otimes t]{} ([Q, M] \otimes P) \otimes (R \otimes B)$$

$$\xrightarrow[a^{-1}]{} (([Q, M] \otimes P) \otimes R) \otimes B \xrightarrow[s \otimes 1]{} A \otimes B .$$

It is clear that the graph of the composite

$$([Q, M] \otimes P) \otimes R \xrightarrow[s]{} A \xrightarrow[f]{} C$$

is $\zeta(\langle\zeta\rangle \otimes 1)$, where $\zeta : M \otimes R \to C$ is the restriction of η to $v(M) + v(R) + v(C)$. It follows by induction that fs is $r(\langle p\rangle \otimes 1)$ for allowable $p: P \to Q$ and $r: M \otimes R \to C$. Then

$$h = y(f \otimes g)x = y(f \otimes g)(s \otimes 1)a^{-1}(1 \otimes t)$$

$$= y(fs \otimes g)a^{-1}(1 \otimes t) = y(r \otimes g)((\langle p\rangle \otimes 1) \otimes 1)a^{-1}(1 \otimes t)$$

$$= y(r \otimes g)a^{-1}(\langle p\rangle \otimes 1)(1 \otimes t) = y(r \otimes g)a^{-1}(1 \otimes t)(\langle p\rangle \otimes 1)$$

is of the required form, with $q = y(r \otimes g)a^{-1}(1 \otimes t)$.

If h is of the form $y\pi(f)$ for some $y: [B, C] \to S$, we must have $[B, C] = S$ and $y = 1$. Then $\Gamma h = \eta(\langle\xi\rangle \otimes 1)$, $\pi^{-1}h = f$ and the naturality of π^{-1} shows that $\Gamma f = \pi^{-1}\eta((\langle\xi\rangle \otimes 1) \otimes 1)$. If we rewrite this as

$$\Gamma(fa^{-1}) = (\Gamma f)\alpha^{-1} = (\pi^{-1}\eta)\alpha^{-1}(\langle\xi\rangle \otimes 1): ([Q, M] \otimes P) \otimes (N \otimes B) \to C,$$

we can apply the induction assumption to fa^{-1} to get $fa^{-1} = r(\langle p\rangle \otimes 1)$ for allowable $p: P \to Q$ and $r: M \otimes (N \otimes B) \to C$. Then $f = ra\ ((\langle p\rangle \otimes 1) \otimes 1)$, so by the naturality of π we get

$$h = \pi f = \pi(ra^{-1})(\langle p\rangle \otimes 1),$$

which is in the desired form.

In the final case where h is of the form

$$([Q, M] \otimes P) \otimes N \xrightarrow[x]{} ([B, C] \otimes A) \otimes D \xrightarrow[\langle f\rangle \otimes 1]{} C \otimes D \xrightarrow[g]{} S,$$

we distinguish cases according as $[Q, M]$ is associated via x with (i) $[B, C]$; (ii) a prime factor of A; or (iii) a prime factor of D.

Case (i). By Proposition 4.11, $B = Q$, $C = M$, and x is the composite

$$([Q, M] \otimes P) \otimes N \xrightarrow{a} [Q, M] \otimes (P \otimes N)$$

$$\xrightarrow{1 \otimes y} [Q, M] \otimes (A \otimes D) \xrightarrow{a^{-1}} ([Q, M] \otimes A) \otimes D$$

for some central y. Let X be an iterated $\otimes$-product of those prime factors of P associated via y with a prime factor of D. The mate under Γh of an element of $v(X)$ is in $v(M) + v(D) + v(S)$ by the form $g(\langle f \rangle \otimes 1)x$ of h, but is in $v(P) + v(Q)$ by the hypothesis that $\Gamma h = \eta(\langle \xi \rangle \otimes 1)$; it must therefore be in $v(X)$. Then X is constant by Proposition 7.7, and since P has no constant prime factors, this means that all the prime factors of P are associated via y with prime factors of A. A similar argument shows that all the prime factors of N are associated via y with prime factors of D. We may therefore, absorbing central morphisms into g and f where necessary, suppose without loss of generality that $A = P$, $D = Q$, and $x = 1$. Then $h = q(\langle p \rangle \otimes 1)$ with $q = g$ and $p = f$.

Case (ii). $[Q, M]$ is associated via x with a prime factor of A. Suppose if possible that $[B, C]$ were associated via x with a prime factor of P. Let X be an iterated $\otimes$-product of all those prime factors of $([Q, M] \otimes P) \otimes N$ that either are prime factors of P or else are associated via x with prime factors of A. The mate under Γh of an element of $v(A)$ is in $v(A) + v(B)$ by the form $g(\langle f \rangle \otimes 1)x$ of h, while the mate under Γh of an element of $v(P)$ is in $v(P) + v(Q)$ by the hypothesis that $\Gamma h = \eta(\langle \xi \rangle \otimes 1)$; thus the mate under Γh of an element of $v(X)$ is again in $v(X)$. It follows from Proposition 7.7 that X is constant, which contradicts the hypothesis that $[Q, M]$ is not constant.

Thus no prime factor of P is associated via x with $[B, C]$. Let Y be an iterated $\otimes$-product of those prime factors of P associated via x with prime factors of D. The mate under Γh of an element of $v(Y)$ is in $v(C) + v(D) + v(S)$ by the form $g(\langle f \rangle \otimes 1)x$ of h, but is in $v(P) + v(Q)$ by the hypothesis that $\Gamma h = \eta(\langle \xi \rangle \otimes 1)$; it must therefore be in $v(Y)$. Then Y is constant by Proposition 7.7, and since P has no constant prime factors this means that every prime factor of P is associated via x with a prime factor of A.

Then by Proposition 4.10 there are a shape R and central morphisms t and s such that $a(c \otimes 1)x$ is the composite

$$([Q, M] \otimes P) \otimes N \xrightarrow{1 \otimes s} ([Q, M] \otimes P) \otimes (R \otimes ([B, C] \otimes D))$$

$$\xrightarrow{a^{-1}} (([Q, M] \otimes P) \otimes R) \otimes ([B, C] \otimes D)$$

$$\xrightarrow{t \otimes 1} A \otimes ([B, C] \otimes D);$$

thus by naturality x is the composite

$$([Q,M]\otimes P)\otimes N \xrightarrow[1\otimes n]{} ([Q,M]\otimes P)\otimes (([B,C]\otimes R)\otimes D)$$

$$\xrightarrow[v]{} ([B,C]\otimes (([Q,M]\otimes P)\otimes R))\otimes D$$

$$\xrightarrow[(1\otimes t)\otimes 1]{} ([B,C]\otimes A)\otimes D\,,$$

where n is the central morphism $(c\otimes 1)a^{-1}s$ and v is the evident central morphism. It is clear that the graph of the composite

$$([Q,M]\otimes P)\otimes R \xrightarrow[t]{} A \xrightarrow[f]{} B$$

is $\zeta(\langle\xi\rangle\otimes 1)$, where $\zeta\colon M\otimes R\to B$ is the restriction of η to $v(M)+v(R)+v(B)$. So by induction ft is $r(\langle p\rangle\otimes 1)$ for allowable $p\colon P\to Q$ and $r\colon M\otimes R\to B$. Setting q equal to the composite

$$M\otimes N \xrightarrow[1\otimes n]{} M\otimes(([B,C]\otimes R)\otimes D)$$

$$\xrightarrow[v]{} ([B,C]\otimes (M\otimes R))\otimes D \xrightarrow[g(\langle r\rangle\otimes 1)]{} S\,,$$

we have $q(\langle p\rangle\otimes 1) = g(\langle r\rangle\otimes 1)v(1\otimes n)(\langle p\rangle\otimes 1) = g(\langle r\rangle\otimes 1)\,v(\langle p\rangle\otimes 1)(1\otimes n)$; by the naturality of v this is $g(\langle r\rangle\otimes 1)((1\otimes(\langle p\rangle\otimes 1))\otimes 1)\,v(1\otimes n)$. Using (5.5), $\langle r\rangle(1\otimes(\langle p\rangle\otimes 1)) = \langle r(\langle p\rangle\otimes 1)\rangle$; which is $\langle ft\rangle$, or $\langle f\rangle(1\otimes t)$ by (5.5) again. Thus finally,

$$q(\langle p\rangle\otimes 1) = g(\langle f\rangle\otimes 1)((1\otimes t)\otimes 1)\,v(1\otimes n) = g(\langle f\rangle\otimes 1)x = h\,.$$

Case (iii). $[Q,M]$ is associated via x with a prime factor of D. Suppose if possible that $[B,C]$ were associated via x with a prime factor of P. Let X be an iterated $\otimes$-product of those prime factors of A that are associated via x with prime factors of N. The mate under Γh of an element of $v(X)$ is in $v(A)+v(B)$ by the form $g(\langle f\rangle\otimes 1)x$ of h, but is in $v(M)+v(N)+v(S)$ by the hypothesis that $\Gamma h = \eta(\langle\xi\rangle\otimes 1)$; it must therefore be in $v(X)$. Then X is constant by Proposition 7.7, and since N has no constant prime factors we conclude that every prime factor of A is associated via x with a prime factor of P. This implies that ξ is of the form

$$P \xrightarrow[\omega]{} ([B,C]\otimes A)\otimes H \xrightarrow[\rho(\langle\sigma\rangle\otimes 1)]{} Q$$

for some integral ω, which is excluded by hypothesis.

Thus no prime factor of P is associated via x with $[B,C]$. Let Y be an iterated $\otimes$-product of those prime factors of P associated via x with prime factors of A. The mate under Γh of an element of $v(Y)$ is in $v(A)+v(B)$ by the form $g(\langle f\rangle\otimes 1)x$ of h, but is in $v(P) + v(Q)$ by the hypothesis that $\Gamma h = \eta(\langle\xi\rangle \otimes 1)$; it must therefore be in $v(Y)$. It follows from Proposition 7.7 that Y is constant, and since P has no constant prime factors this means that all the prime factors of P are associated via x with prime factors of D.

For brevity, let us write

$$M' = [Q, M] \otimes P, \qquad C' = [B, C] \otimes A$$

so that $\langle f\rangle : C' \to C$. In the central morphism $x : M' \otimes N \to C' \otimes D$, we now know that all the prime factors of M' are associated with prime factors of D. Apply Proposition 4.10 to this situation; it gives a shape R, a central morphism $N \to C' \otimes R$ (which without loss we can take to be the identity) and a central morphism $t : M' \otimes R \to D$, so that $x = (1 \otimes t)w$, as in the first row of the following diagram, in which w is the evident central morphism:

$$\begin{array}{ccccccc}
M' \otimes (C' \otimes R) & \xrightarrow{\;w\;} & C' \otimes (M' \otimes R) & \xrightarrow{\;1 \otimes t\;} & C' \otimes D & & \\
\Big\downarrow{\scriptstyle 1 \otimes (\langle f\rangle \otimes 1)} & & \Big\downarrow{\scriptstyle \langle f\rangle \otimes 1} & & \Big\downarrow{\scriptstyle \langle f\rangle \otimes 1} & & \\
M' \otimes (C \otimes R) & \xrightarrow[\;w\;]{} & C \otimes (M' \otimes R) & \xrightarrow[\;1 \otimes t\;]{} & C \otimes D & \xrightarrow[\;g\;]{} & S\,.
\end{array}$$

The diagram evidently commutes. Now the hypothesis $\Gamma h = \eta(\langle\xi\rangle \otimes 1)$ for η: $M \otimes N \to S$ clearly means that the graph of the composite $g(1 \otimes t)w$ is $\zeta(\langle\xi\rangle \otimes 1)$, where ζ: $M \otimes (C \otimes R) \to S$ is the restriction of η to $vM + vC + vR + vS$. By induction, $g(1 \otimes t)w$ is $r(\langle p\rangle \otimes 1)$ for allowable $p : P \to Q$ and $r : M \otimes (C \otimes R) \to S$. Therefore, since $\langle p\rangle$: $M' \to M$,

$$h = g(1 \otimes t)w(1 \otimes (\langle f\rangle \otimes 1)) = r(\langle p\rangle \otimes 1)(1 \otimes (\langle f\rangle \otimes 1))\quad.$$

$$= r(1 \otimes (\langle f\rangle \otimes 1))(\langle p\rangle \otimes 1)$$

has the requisite form $q(\langle p\rangle \otimes 1)$ for $q = r(1 \otimes (\langle f\rangle \otimes 1))$.

This concludes the proof of Proposition 7.8.

Proof of Theorem 2.4. Let T, S be proper shapes and let h, h': $T \to S$ be allowable natural transformations in $\underline{N}(\underline{V})$ with $\Gamma h = \Gamma h'$; we are to prove that $h = h'$. Suppose inductively that it is so for all smaller values (if any) of $r(h)$; note that $r(h) = r(T) + r(S) = r(h')$. By Lemma 7.5, we may suppose that none of the prime factors of T or of S is constant.

If both h and h' are central we have $h = h'$ by Theorem 4.9. So we may suppose that h is of one of the other forms (7.1).

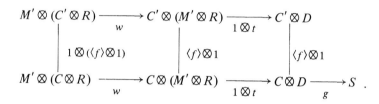

If h is of the form $y\pi(f)$, we have $y^{-1}h = \pi(f)$ where f is allowable. But then $y^{-1}h' = \pi(f')$, where $f' = \pi^{-1}(y^{-1}h')$ is also allowable by (1.2). Since h and h' have the same graph, so do f and f'; hence $f' = f$ by the inductive hypothesis, whence $h' = h$.

If h is of the form $y(f \otimes g)x$, we have $y^{-1}hx^{-1} = f \otimes g: A \otimes B \to C \otimes D$. Then $\Gamma(y^{-1}h'x^{-1}) = \Gamma(y^{-1}hx^{-1}) = \Gamma f \otimes \Gamma g$; so that by Proposition 7.6 $y^{-1}h'x^{-1} = f' \otimes g'$ for allowable $f': A \to C$ and $g': B \to D$ with $\Gamma f = \Gamma f'$ and $\Gamma g = \Gamma g'$, whence $f' = f$ and $g' = g$ by the inductive hypothesis. Hence $h' = h$.

There remains the case where h is of the form $g(\langle f \rangle \otimes 1)x$. Then it may be the case that the graph Γf of f is of the form

$$A \xrightarrow[\omega]{} ([F, G] \otimes E) \otimes H \xrightarrow[\rho(\langle \sigma \rangle \otimes 1)]{} B$$

for some central ω and some ρ, σ; in this case we have $\Gamma h = \Gamma g(\langle \Gamma f \rangle \otimes 1)\Gamma x$. Here, since $\langle\ \rangle$ is natural, as in (5.5),

$$\langle \Gamma f \rangle = \langle \rho(\langle \sigma \rangle \otimes 1)\,\omega \rangle = \langle \rho \rangle (1 \otimes (\langle \sigma \rangle \otimes 1))(1 \otimes \omega)\,.$$

It now follows easily that $\Gamma h = \tau(\langle \sigma \rangle \otimes 1)\psi$ for some τ and for some central ψ. Perhaps σ is of the form

$$E \xrightarrow[\phi]{} ([X, Y] \otimes Z) \otimes W \xrightarrow[\kappa(\langle \lambda \rangle \otimes 1)]{} F$$

for some central ϕ and some κ, λ; but E has strictly fewer prime factors than A, since $[F, G]$ is a prime factor of A but not of E; Z has strictly fewer prime factors than E; and so on. Thus this process terminates, and ultimately we have an expression for Γh of the form

$$T \xrightarrow[\mu]{} ([Q, M] \otimes P) \otimes N \xrightarrow[\eta(\langle \xi \rangle \otimes 1)]{} S$$

where μ is central and ξ is not of the form (7.5). Moreover $[Q, M]$ is not constant since T has no constant prime factors. By Theorem 4.9 there is a central natural transformation $y: T \to ([Q, M] \otimes P) \otimes N$ with $\Gamma y = \mu$. From Proposition 7.8 applied to hy^{-1} and $h'y^{-1}$ we conclude that $hy^{-1} = q(\langle p \rangle \otimes 1)$ and $h'y^{-1} = q'(\langle p' \rangle \otimes 1)$ for allowable $p, p': P \to Q$ and allowable $q, q': M \otimes N \to S$ with $\Gamma p = \Gamma p'$ and $\Gamma q = \Gamma q'$. It follows from the inductive hypothesis that $p = p'$ and $q = q'$, so that $h = h'$.

This completes the proof of Theorem 2.4.

References

[1] S.Eilenberg and G.M.Kelly, A generalization of the functorial calculus, J. Algebra 3 (1966) 366–375.

[2] S.Eilenberg and G.M.Kelly, Closed categories, in: Proc. Conf. on Categorical Algebra, La Jolla, 1965 (Springer-Verlag, 1966) pp. 421–562.

[3] D.B.A.Epstein, Functors between tensored categories, Invent. Math. 1 (1966) 221–228.

[4] G.Gentzen, Untersuchungen über das logische Schliessen I, II, Math. Z. 39 (1934–1935) 176–210 and 405–431.

[5] G.M.Kelly, On MacLane's conditions for coherence of natural associativities, commutativities, etc., J. Algebra 4 (1964) 397–402.

[6] J.Lambek, Deductive systems and categories I. Syntactic calculus and residuated categories, Math. Systems Theory 2 (1968) 287–318.

[7] J.Lambek, Deductive systems and categories II. Standard constructions and closed categories, in: Lecture Notes in Mathematics 86 (Springer-Verlag, 1969) pp. 76–122.

[8] J.L.MacDonald, Coherence of adjoints, associativities, and identities, Arch. Math. 19 (1968) 398–401.

[9] S.MacLane, Natural associativity and commutativity, Rice University Studies 49 (1963) 28–46.

Saunders Mac Lane as a Shaper of Mathematics and Mathematicians

Roger Lyndon

A technical appreciation of Saunders Mac Lane's mathematical work is beyond my competence, for, although I was his student, the body of my own work has had very little overlap with his. I emphasize this, for I am not alone. Mac Lane's former students have gone on to work in a remarkable diversity of fields, quite other than those of his own primary activity. This, I believe, attests to Mac Lane's breadth of interest and knowledge, and to his enthusiasm. But, even more, it is evidence that the strength of his teaching, his example, and his influence generally, lies more in imparting a way of looking at mathematics, a feeling for it, a sense of values, than in indoctrination in any special topic or point of view.

These almost intangible things are hard to spell out, but I will begin by trying to give some hint of them from my own recollections. I excuse myself for talking about myself on the grounds that any attempt to get at what constitutes effective teaching and inspiration should be of general interest. As I go back and look at Mac Lane through the eyes of a not very mature graduate student, the first word that comes to my mind is 'stern', which I then hastily temper with 'patient', 'kind', and 'humorous'. Objectively, perhaps, 'patience' should come first, for I was a student during the years of the Second World War, when there was little thought beyond tomorrow and everyone's mind was on other things, not always of cosmic importance. I worked very slowly—or intermittently—without any plan beyond random exploration, and with very little foundation, having read almost nothing. I would bring Mac Lane my few results at intervals as rare as I could manage, and, if I chanced to see him coming down the street in the times between I would fling myself behind a hedge and lie still till he had passed.

Through all this he patiently expressed enthusiasm for my meager results, suggesting ramifications, insisting on strict rigor and clarity, and, above all, demanding elegance. Thanks to this, and to the fact that he had pushed me into virgin territory abounding in unplucked plums, I finally accumulated a thesis—one day a theorem emerged that could be used as justification for all my erratic wanderings. It is perhaps worthwhile to note here also a 'negative' service. For over a year, at the beginning of my thesis work, I had 'my own topic'. I had become obsessed with the problem of the set theoretic representation of abstract relation algebras, a slippery subject fraught with algebraic, logical, and set theoretic subtleties that have barely been resolved many years later. Surely not suitable for a thesis, and Mac Lane somehow managed to pry me loose from that.

Today, after having directed theses myself, when I look back I find it hard to say how much my thesis was my own work. Surely I discovered all my theorems for myself, but it may well be that Mac Lane knew very well what theorems would be there and headed me in the right direction to find them. On balance, if I found the theorems, he found their significance, and I think that, while watching closely over me and protecting me from utter folly, he taught me independence. This, as we all know, is beyond value, and I have tried to pass it on to my own students; it is the good fortune of my stronger students, and the bane of others, that I am almost incapable of setting a 'problem', but only of pointing out a topic that would be 'worth looking at'. It is largely the same with my own research.

I cannot end these subjective recollections without adding that Mac Lane, like a good parent, gave me security. Although I often felt guilty of not working hard enough, and imagined his veiled wrath, he never caused me to doubt his faith in me, or to lose my own faith in myself and in the importance and excitement of the ideas I was exploring. Although I was rather reserved and aloof, I was always put at ease by Mac Lane's warmth and enthusiasm, and by his sometimes wry humor. When my degree was finally complete, he invited me for a weekend in Connecticut, to learn to play beach croquet and to address him, thenceforth, as Saunders.

We shall now proceed to generalize the above remarks. In my judgement it is not a mere quirk of fate that Mac Lane began his mathematical career in logic. Attention to symbolism, to notation, and to obtaining just the natural expression of an idea at the appropriate level of abstraction,—these have paid off well in all his work, most conspicuously in his development with Sammy Eilenberg (Polish, ergo logician) first of abstract cohomology theory and then of category theory. One popular caricature sees the classical analyst, for example, with an almost mystical feeling for functions as Platonic entities, that he can translate only with great pain into the language of epsilons and deltas needed to convince others of his vision. By contrast, we see the algebraist, or even more the logician, so entranced with what is before him on the page, with formal concepts and formal proofs, that he will lose sight of what he is talking about, and of the questions and applications that led him to them. But not all abstract algebra is general nonsense and corollaries of definitions, as the most superficial inspection of the proof of any major theorem in the area will demonstrate. Here comes into play the formal 'poetry' of mathematics, the skillful and graceful marshalling of pages of symbols, so that the thread of meaning shines through successive lines of calculation. To be fair to our analyst, there is as much to be said on the other side, for a 'logic' that deals with concepts or intuitions that do not naturally reduce themselves to linear strings of letters from a finite alphabet. Some such nebulous vision must have sustained the group-cohomologists before the revelation of the $K(\Pi, n)$.

Mathematics, like other things, is half sweat and half inspiration. Students, in my experience, tend to divide all too sharply into those who like to multiply matrices and those who like to chase diagrams (in fact, rather similar activities), and this same division sometimes persists in fratricidal strife between full professors. Of course, both activities are needed. Experimentation and concrete computation are needed to test the edge of mathematics and provide substantial content. But without the unification provided by abstract theory mathematics would long ago have degenerated into little more than an encyclopedic cookbook. I believe that most of those who have most shaped the development of mathematics have combined these two tastes, and I would certainly count Mac Lane as pre-eminent among these.

It was a few months ago that I most recently saw Saunders. On that occasion he gave a lecture on the history of the cohomology of groups. His talk was not one of those collections of reminiscences that establish myths, but rather a balanced analysis of the various sources from which the theory sprang, and of the forces, including persons and accidents, that shaped its development. Such studies in the 'history of ideas', attributing events neither to capricious gods nor to inexorable forces, but taking into account persons and circumstances, and the critical role of seemingly technical details, are the kind of history from which we can learn and profit. And this especially if the history is recorded, fairly, by one of the central figures.

If Saunders were still my supervisor, I should have written here far less. But I have meandered, in search of a few hints towards the appreciation of Saunders Mac Lane as a great living mathematician.

The Early Work of Saunders Mac Lane on Valuations and Fields

Irving Kaplansky

1. Introduction

In an essay such as this a highly personal touch is perhaps excusable. So let me turn the clock back to the autumn of 1939 when I arrived at Harvard, a hopeful but almost hopelessly naive graduate student. Something over a year earlier I had heard Saunders Mac Lane lecture at an algebra conference at the University of Chicago; that was enough to convince me that I liked his style of mathematics. The valuation period of his work was coming to an end, but luckily there was still enough left for me. Ever since I have had an unshakeable fondness for valuation rings; my thoughts keep coming back to them. The infatuation is contagious; some of my students were infected and even transmitted the infection to succeeding generations.

So this essay is a welcome opportunity to take a fresh look at Mac Lane's work on valuations and related topics in the theory of fields. If one puts aside a paper [1] that resulted from his 1929 summer job at General Electric, and his early flirtation with logic, this is his first body of mathematics.

2. Extensions of Valuations to a Polynomial Ring

Mac Lane's interest in valuations can be traced quite directly to the influence of Øystein Ore. In 1929–1930, as an undergraduate at Yale, he was a student in two of Ore's algebra courses. In 1933–1934, when he returned to Yale as a Sterling Fellow, the contact was renewed and reinforced.

Ore and others had studied two related problems: finding explicitly the factorization of an ordinary prime into prime ideals in an algebraic number field, and giving "higher" irreducibility criteria of the Eisenstein type. The methods used were in essence valuation-theoretic. Mac Lane perceived that these questions could be unified and the results made definitive by studying the following problem: given a field K carrying a discrete rank one valuation V, find all extensions of V to the field $K(x)$ of rational functions in an indeterminate x. Having posed the problem, he proceeded to solve it in [10]. The solution goes as follows. First extend V by giving Vx a preassigned value and letting nature take its course. Proceed if necessary to jack up the valuation by increasing it on a suitable "key" polynomial, again letting nature take its course. Iterate the process, perhaps infinitely often. All valuations of $K(x)$ which extend V are obtainable in this fashion.

The application to factorization into prime ideals appeared in [7], the application to irreducibility in [17]. Incidentally, I believe that this study of irreducibility criteria stands as definitive to this day.

I shift the scene to the work of Zariski. Early in his program of bringing abstract algebra to bear on algebraic geometry he gave a new proof for the resolution of the singularities of a surface [X]. Subsequently this was generalized and polished by Abhyankar [I].

There was a remarkable overlap! To explain this I shall give some detail on the relevant theorems, as formulated by Abhyankar. Let R be a local ring (commutative, Noetherian, with unique maximal ideal M). The dimension d of R is the height of M (the height of a prime ideal P being the maximal length of a chain of prime ideals descending from P). Let n be the smallest number of elements that can generate M. The principal ideal theorem asserts $d \leqslant n$. If equality holds, R is said to be *regular*. Regular local rings enjoy many pleasant properties of which I mention only one: there are no divisors of 0.

Now let R be a two-dimensional regular local ring with quotient field K. Let x and y be a minimal generating set for the maximal ideal of R. Form the ring $T = R[y/x]$. In passing to T the local property is lost; indeed T has an infinite number of maximal ideals of height 1 and also an infinite number of height 2. Form a ring U by localizing T at a maximal ideal of height 2. It turns out that U is again a two-dimensional regular local ring. Let us call U a *quadratic transform* of T. The process of forming a quadratic transform can be iterated. Theorem 1: Any two-dimensional regular local ring lying between R and K can be reached by a finite sequence of quadratic transforms. What happens if we repeat indefinitely the formation of a quadratic transform? Theorem 2: The result is a valuation ring. Theorem 3: Any valuation ring between R and K is obtained in the fashion just mentioned or by localizing at a height 1 prime the result of a finite sequence of quadratic transforms.

In Mac Lane's case the pertinent two-dimensional regular local ring is the localization $V[x]_{(\pi, x)}$, where V is a discrete valuation ring with prime π and x is an indeterminate. In Zariski's case it is the local ring at a nonsingular point of a surface. (Remark: this is by no means the whole story on resolving the singularities of a surface, but only the local part, so to speak.)

The connection between these two enterprises apparently went unnoticed for over thirty years. Then it was set forth in some detail in Judy Sally's 1971 Chicago thesis; the published paper [IX] has a brief sketch at the end. By a remarkable coincidence van der Put, at very nearly the same time, mentioned the connection in his Zentralblatt (v. 198, p. 366) review of [IV]. One might even see a triple coincidence in the publication in 1972 of [XI]; on pages 307–312 there is an incisive account by Hironaka of the development of the ideas that culminated in his successful resolution of singularities in any dimension in characteristic 0.

There was no lack of communication between Saunders Mac Lane and Oscar Zariski. Here is a little personal recollection dating (I think) from 1940 or 1941. Zariski was expected to visit Harvard. To prepare the prospective audience, Mac Lane asked me to give a background talk on valuations. I don't know whether the audience was thereby adequately prepared (I doubt it) but it certainly helped me gather my thoughts.

It seems fair to ascribe to Mac Lane and Zariski independently the discovery of this body of theorems, with Mac Lane having a slight priority. This calls for a glance at the relevant dates. I note with surprise that there is no date of receipt for [4]. But the missing date should presumably precede that of [5]: Oct. 24, 1935. The date of receipt of [X] is Feb. 15, 1939. Incidentally, the two years studying algebra mentioned on page xi of [XI] may well have been 1935-7, with ideas involved in [X] being worked out soon afterwards.

3. *The Structure Theorem for Unramified Complete Discrete Valuation Rings*

Starting with [18] there was a shift to a different topic: the uniqueness of the extensions of valuations. In [18] the context was valuations of higher rank. In due course there followed [22], deservedly a classic, on the rank one case of major importance: unequal characteristic and unramified.

The early literature on valuations emphasized a valuation function on a field. Subsequently there was a large scale shift to the valuation ring as the important thing. That is the language I shall choose in stating the theorem under discussion. I shall even take the space to define a *discrete valuation ring* (DVR) as a principal ideal domain R with exactly one prime π (up to units). R is a *complete* DVR if it is complete relative to the topology given by the powers of π. The field $k=R/(\pi)$ is the *residue class field*. In the characteristic unequal case R has characteristic 0 and k has characteristic p; R is *unramified* if $(p)=(\pi)$. Assume all this. Then the only visible invariant for R is the field k. The big theorem says that indeed no other invariant is needed and so one can write $R=W(k)$. Here W stands for Witt.

The first substantial paper on these matters was by Hasse and Schmidt [III]. But this paper had a flaw. In the opening lines of [22], Mac Lane politely says that the Hasse-Schmidt paper "unfortunately uses an elaborate and unproven lemma" and adds in a footnote that he will discuss in a forthcoming paper "difficulties present in the proof of such a lemma". We could all learn a lesson in tact from this.

Later Witt and Teichmüller gave a correct proof, but it was not easily accessible.

Mac Lane hit the problem hard on three fronts. 1. As I just hinted, he showed in [20] that the offending lemma was actually false. 2. In collaboration with Schmidt [34] he proved a weakened lemma which is good enough for the purpose. 3. In addition, he gave a short, elegant account of his own [22].

Here is a sketchy summary of [22]. One needs to know that $W(k)$ exists. It turns out that this is a special case of an easy general proposition about the existence of suitable extensions of a valuation ring that match a prescribed extension of its residue class field. The uniqueness of $W(k)$ is harder. But if one is content just to prove uniqueness, one need not write down the ring operations explicitly. Mac Lane's uniqueness proof is short and readable. To be sure, this didn't bury Witt vectors forever (for Witt vectors are indeed the explicit construction that got bypassed); they keep reappearing. You can find Witt's own second thoughts on the subject as an exercise in Lang's textbook [VII, Ex. 21 on p. 233].

4. *Separability*

I have short-changed the reader by failing to state the main theorem of [22] in its full force. In its stronger form it was a "relative" uniqueness theorem. In other words, Mac Lane proved the uniqueness of an appropriate kind of extension of a valuation from a given field to a larger field. In the "absolute" case described in the preceding section the bottom field is the field of p-adic numbers. Now it turned out that a certain hypothesis was needed on the induced extension of residue class fields; in a followup paper [28] he showed that the condition in question was indispensable.

He called the condition "preservation of p-independence". In due course the powers that be decreed this was *the* relation between a field of characteristic p and a subfield that deserves the name "separability". (This designation does have the advantage of being easily converted into an adjective.) Of course, for algebraic extensions it coincides with the well-established notion of separability. Among other pleasant properties, separability is automatic when the bottom field is perfect, and this is the reason there was no need to mention it in § 3, where the bottom field was the integers mod p.

The emergence of separability as a standard concept was perhaps the single most important thing that came out of Mac Lane's study of fields of characteristic p. It has added importance as an early significant example of linear disjointness.

5. *F. K. Schmidt and Towers of Fields*

The collaboration with F. K. Schmidt recorded in [34] came about as follows. When Mac Lane discovered the error in [III] he corresponded with Schmidt about it. Later, there was an opportunity to discuss it in person, when Schmidt visited the United States as an emissary of Springer and the Zentralblatt, in connection with the American plans to establish Mathematical Reviews. In these conversations a joint paper was projected. But when the time came to write it, World War II had broken communication. So Mac Lane prepared the announcement [34] on his own responsibility. After the war they got together again, but by then interest in towers of fields had waned. A full manuscript does exist. I take it that the situation is that these delicate, intricate results, along with the related published ones in [20], [21], and [32], await the appearance of a new application. (Remark: The subsequent paper or papers promised by the title of [21] never got beyond the stage of preliminary notes.)

Since the late F. K. Schmidt looms so large in this story I shall take the opportunity to make a belated apology to the mathematical public. The introduction to [V] should have recorded the fact that Krull [VI, pages 163 and 191] said that Schmidt had extensive unpublished results concerning maximal valuation rings. When I wrote my thesis it was impossible to check this out. I unfortunately let the matter slide for a long time, but I did finally write Schmidt on Jan. 8, 1971. There was a prompt and cordial reply dated Feb. 15, 1971, but unfortunately all it said was that he had quite forgotten and had no notes on the subject.

6. *Miscellaneous*

To conclude this essay I offer brief remarks on two more papers and a more extended comment on a third.

(a) Is the phrase "equation without affect" used in [9] becoming obsolete? At any rate, it means that the Galois group is the full symmetric group. Valuation-theoretic methods are used in [9] to construct such equations.

(b) The paper [14] studies certain lattices of subfields of a given field. They rarely satisfy the modular law. But enough modularity survives to prove the invariance of things like the number of elements in a transcendence basis or a p-basis. This got successfully formulated as a piece of abstract lattice theory.

(c) The paper [24] was written, in part, to clarify a theorem published by Gleyzal [II]. I would say that the main result of [24] is the following (although it does not quite appear explicitly). Let K be a field maximal with respect to a valuation. Assume that the value group is divisible and that the residue class field is algebraically closed. Then K is algebraically closed.

I find it interesting that a generation later [VIII] Nagata needed and proved the companion theorem obtained by replacing "algebraically closed" by "real closed".

Several years ago a related observation arose during a conversation with my colleague M. P. Murthy. Assume that the residue class field has characteristic 0. Then the assumption of maximality is unnecessarily strong; the Hensel lemma will do (observe that validity of the Hensel lemma is certainly necessary). I leave the proof as an exercise (and I suspect it can be found somewhere in the literature). So, for instance, in the rank one case completeness will do in place of maximality.

This fails if the residue class field has characteristic p; note that in characteristic 2 the formal power series

$$x = t^{-1/2} + t^{-1/4} + t^{-1/8} + \cdots$$

satisfies $x^2 + x = t^{-1}$.

References

I. S. Abhyankar, On the valuations centered in a local domain, *Amer. J. of Math.* 78(1956), 321–348.

II. A. Gleyzal, Transfinite numbers, *Proc. Nat. Acad. Sci. USA* 23(1937), 581–587.

III. H. Hasse and F. K. Schmidt, Die Struktur diskret bewerteter Körper, *J. für Math.* 170(1934), 4–63.

IV. H. Inoue, On valuations of polynomial rings of many variables. I, *J. Fac. Sci. Hokkaido Univ., Ser I* 21(1970), 46–70. II, ibid., 248–297. III, *Hokkaido Math. J.* 3(1974), 35–64. (These papers develop the program of extending Mac Lane's work to more than one variable.)

V. I. Kaplansky, Maximal fields with valuations, *Duke Math. J.* 9(1942), 303–321.

VI. W. Krull, Allgemeine Bewertungstheorie, *J. für Math.* 167(1932), 160–196.

VII. S. Lang, *Algebra*, Corrected reprinting, Addison-Wesley, 1971.

VIII. M. Nagata, Some remarks on ordered fields, *Jap. J. of Math.*, new Series, 1(1975), 1–4.

IX. J. Sally, Regular overrings of regular local rings, *Trans. Amer. Math. Soc.* 171(1972), 291–300.

X. O. Zariski, The reduction of the singularities of an algebraic surface, *Annals of Math.* 40(1939), 639–689.

XI. ____, *Collected Papers*, Vol. I, MIT Press, 1972.

Some Remarks on the Interface of Algebra and Geometry

Samuel Eilenberg

In my contribution to this volume honoring Saunders Mac Lane, it is natural and expected that I should write about some aspects of our long collaboration. At the same time I should try to avoid saying things that are already published [see 114, 116], and also observe an agreement, made long ago with Saunders, not to disclose who had which idea and did what.

One of the many qualities that I admire in Saunders is his capacity for orderly and intelligent computing. Not just ordinary 2-3 page calculations where the result is known in advance, but real computational fishing expeditions running to 30-40 pages. I admire this talent all the more since my own capacities in this direction are nil; a computation running more than five lines usually defeats me.

These computations were usually homological in nature. Consequently signs abounded and were a real nuisance. After some experience, it was decided in the first approximation to abandon all signs and to pretend that the universe has characteristic two. We used an unproved metatheorem that if a computation works mod 2, then the signs can be fixed later so that it continues to be valid. The principle never failed, but the metatheorem is still unproved!

Our work on the complexes $K(\Pi,n)$, and their cohomology, started early in 1943 and continued through 1953. It involved an enormous amount of preliminary computations, most of which remained buried in voluminous manuscripts (sometimes called "sketches") that never saw daylight.

I believe that it might be of some interest to look into the reasons why this more or less had to be so.

Shortly after our first preliminary note [45] concerning $K(\Pi,n)$ appeared in 1943, I received a letter from Heinz Hopf. He said that the result for $n=1$ did not surprise him because he systematically looked at the fundamental group π_1 of a space X as a group of transformations on the universal covering of X. This made our result for $n=1$ plausible and more or less expected. In fact Hopf himself had proved something quite similar and quite independently. However, he saw no analogue for $n>1$ and thus he found the result very surprising and pointing to something new.

Saunders and I were as much in the dark as Hopf was. The situation remained unchanged when the full paper [47] was published in 1945.

While the case $n=1$ (i.e. the "ordinary" cohomology theory of groups) was the subject of much activity following 1945, the case of $n>1$ was not mentioned in print until 1950 [60,61,65]. In the middle of page 444 of [60] we say

"This note will state some of the results of a systematic study of the groups of $K(\Pi,n)$ by purely algebraic methods".

The term "algebraic methods" in this case was a euphemism for "blood, sweat, and tears" or "brutal and voluminous computations". Thanks to the skill with which the computations were conducted, and to the unlimited faith that the subject was worth pursuing, we managed to extract something fairly pretty. It was a "cubical" substitute for the groups $H^q(\Pi,n)$ which lead to a Freudenthal type suspension theorem. Fortunately the full details never had to be published, because we managed to bypass the heaviest parts. Some of the bypassing was done in the next note [61]. Here two new notions made their first appearance: the "bar construction" and "generic acylicity". Both of these conceptual algebraic constructs grew out of similar and fairly elaborate exploratory computations. While the bar construction has been much used and is quite well understood today, the notion of generic acyclicity [66, 76] has remained undeservedly unexplored.

While all this was going on, other people in another country were learning about fibrations and spectral sequences. Serre's thesis (and history) were being written. Then around 1951 somebody (probably Henri Cartan) observed that if a contractible space E is fibered then the base is a $K(\Pi,n)$ space if and only if the fiber is a $K(\Pi,n-1)$ space. This neatly generalizes what Hopf understood in 1943 for $n=1$. The remark above, combined with the theory of spectral sequences, suggested algebraic procedures which enormously advanced the understanding of the groups $H(\Pi,n)$, even though it did not lead to an absolutely complete solution of the problem.

This development shows once more how a geometric notion (like a fibration) can successfully intervene in a problem which a priori looked entirely algebraic.

Both approaches have some merits and some demerits. The method of spectral sequences, by its nature, tends to give a solution "in pieces", usually without sufficient information about how the pieces are to be put together. The more direct computational method tends to give, in those cases where it succeeded [72], complete results including naturality.

There are reasons to believe that a marriage of the two methods would be a success. Indeed, spectral sequences can be used to prove that given morphisms are isomorphisms. This in fact becomes an iterated use of the "five lemma". Thus once a complete closed natural formulation is guessed (by preliminary computations or inspiration or both), one can use the spectral sequence technique to show that it works.

Saunders Mac Lane and Category Theory

Max Kelly

1. *Introduction*

When Irving Kaplansky invited me to contribute an essay to this volume, I decided to write mainly about Saunders Mac Lane's work on category theory; partly because this has been his chief interest during the years of our friendship, and partly because Sammy Eilenberg is clearly better placed to comment on an earlier period.

Re-reading Mac Lane's articles on the subject soon revealed a difficulty: we already have, in his Retiring Presidential Address [115]* and in his 1965 Bulletin survey of categorical algebra [93], his own succinct, skilful, and enthusiastic accounts of the high points of his interests; what could I usefully add to these?

It seemed to me that his own writings play down the extent of his influence on others—primarily as a creator, but also as a developer, as an expositor, and as an organizer; and that I might bring this out by discussing his influence on me. Doing so would also allow me to look back over the ideas he had introduced, in collaboration or alone; and to point up the later developments, as I knew them, of his germinal notions.

Such an essay necessarily includes more about myself than is decent. Although exercised by this danger, I have thought it worth accepting in the interests of a personal testimony to my friend.

2. *Categories as a Language*

Mac Lane's influence on my generation first made itself felt, long before our meeting, by providing the language in which, as graduate students at Cambridge in the years 1953 to 1956, we learnt algebraic topology. I mean, of course, the basic notions of category, functor, and natural transformation, as developed and illustrated in the original Eilenberg–Mac Lane papers on category theory, [42] and [48].

Major advances, once made, seem so inevitable that a younger generation, brought up familiar with these ideas, may not realize how great their impact was.

*These references in Roman type are to the bibliography, in this volume, of Mac Lane's works; those in Italic type are to the bibliography at the end of this essay.

Two undergraduate courses in homology theory, one at Sydney in 1950 and one at Cambridge in early 1953, had set me reading parts of Seifert–Threlfall and dipping into the books of Lefschetz; and I had found it hard to orient myself in the subject. When I learnt the categorical language, probably in late 1953, it gave a remarkably vigorous boost to my understanding, and quickly put an end to many difficulties.

Why did I learn so late a language available since 1945? I am sure it wasn't used in my last undergraduate year 1952/53, in spite of courses on Pontryagin duality and on group representations; unless it was introduced lightly in that homology theory course by Wylie. It must have been familiar, as a research tool, to the senior people in algebraic topology; but I suspect that it was popularized by the 1952 book [*27*] of Eilenberg-Steenrod. At any rate it became from that time the normal, and indeed the only imaginable, medium of intercourse for us and our teachers—Wylie, Hilton, Rees, Atiyah, Zeeman, James, and Adams, at Cambridge; and Henry Whitehead and Barratt, at Oxford. Some of the above consciously enriched its vocabulary and syntax; for us students, it was the vehicle for even our earliest papers.

Of course I was conscious of Mac Lane's strong influence on the *content* of our studies, as distinct from the language; his name was prominent in our courses and our seminars, and I observe that my note [*46*] on the Künneth theorem refers to [72] and [84]. However this enters the field I am leaving to Eilenberg.

Category theory, at any stage of its development, may doubtless be called a language; but the title of this section is meant to suggest the period during which the *popularized* version scarcely contained more than the three basic notions and a few facts about finite direct sums in additive categories. For most of us this stage lasted until the late fifties or early sixties, although it does not properly represent even the original [48], much less Mac Lane's 1950 paper [64]; the time-lag is seen again. It was characteristic of this period to treat all the objects of a given category on the same footing, not distinguishing one from another by internal properties, unless perhaps to distinguish the zero object in an additive category.

3. *Categories with Individuality*

Even during the period just mentioned, and within its terms, one's attitude could so change that a category, from having been a rather structureless domain or codomain for a functor, might become something more tangible and individual. This clearly happened independently to many users of the language. In this highly personal section I discuss my own conversion, once again influenced by Mac Lane.

The first edition of Hilton-Wylie [*40*], which I was using as a text at Sydney in 1963, falsely asserted that the cohomology ring of a product was determined by those of the factors. Of course one would know the cohomology ring of the product if for each factor, with free abelian chain complex A, one knew the homotopy class $d{:}A \to A \otimes A$ of the chain map representing the diagonal. In passing to the cohomology ring $H^*(A)$ with any given coefficient ring, one lost too much information to reconstruct d; and I wanted to know just what invariants of d, at the homology level, precisely served to determine it, to within a suitable isomorphism.

I soon realized that I didn't know what I meant by the question. The "suitable isomorphism" involved seeing d as an object in an appropriate category; but what was "a complete set of invariants"? Luckily I recalled the Mac Lane-Whitehead paper [63] on the 3-type of a complex. Although barely using the word "category", they in effect said that TA was a complete set of invariants of A if $T: \mathcal{A} \to \mathcal{B}$ was a functor, full on morphisms and essentially surjective on objects, that reflected isomorphisms. Various "Whitehead theorems" were classical examples, and their **k**-invariant was a new one.

I looked at such functors in [*47*]. The essential surjectivity of T meant that only its kernel need be examined. In the additive case, the result was that T reflects isomorphisms exactly when this kernel $\mathcal{K}$ is contained in the "Jacobson radical" of $\mathcal{A}$, which is surely the case if $\mathcal{K}$ is nilpotent. In one classical Whitehead result, where $\mathcal{A}$ is free abelian chain complexes, where $\mathcal{B}$ is graded abelian groups, and where T is homology, we have in fact $\mathcal{K}^2 = 0$ since $\mathrm{Ext}^2 = 0$.

I used this in [*48*] to look at invariants of a chain map $f: A \to B$ and of a chain endomorphism $f: A \to A$, as preliminaries to the harder case of $d: A \to A \otimes A$. In fact I never finished that case, partly because it was too hard, partly because my interests were drawn elsewhere, and partly because I learnt of an answer in the literature [*78*], in terms of the cohomology rings with coefficients $\mathbf{Z}$ and $\mathbf{Z}/p$, and of the Bockstein boundaries. However this was not really the invariant kind of answer I was seeking; and I did go far enough to see that such an answer would need the analogue of Mac Lane's triple torsion products [84].

To finish briefly with this story: I used the results of [*48*] in [*49*] to show that a complete system of invariants of a short exact sequence of free abelian chain complexes was given precisely by the long exact homology sequence, although without determining the nilpotence class of the kernel. This was later [*50*] shown to be 2, or more generally $p+1$ for complexes over a ring of global dimension p. Finally Street, in his thesis ([*86*] and [*87*]), transformed these questions out of sight. With powerful new techniques that rendered mine obsolete, he gave the most general Künneth-theorem-and-splitting result I have seen, and in particular found a complete set of invariants for a finitely-filtered free abelian chain complex: it is something *richer* than the spectral sequence.

I had now reached an important stage: I was about to meet Mac Lane; and in [*49*], to deal with the homotopy classes, I had introduced the notion of differential graded category (under the name "complex category"), bringing me close to current interests both of Mac Lane and of Eilenberg.

4. Categories and Universal Properties

We now turn the clock back. If Mac Lane were asked today what single most important insight had been made possible (in its appropriate generality) by category theory, he would say, I think, the notion of universal property, in its various manifestations: limits, universal arrows, representable functors, adjoint functors. This is a much deeper use of categories than those we have been considering; far from the objects being on the same footing, *one* is distinguished as the possessor of the universal property, and the definition of this property involves its relation with *all* the others.

Although this use was popularized much later, its beginnings go back to the original [48], which considered limits for directed systems. (In view of this I find rather inconsistent the authors' assertion in the first paragraph on p. 247 of [48]:

that a category *is*, after all, only a domain or a codomain, whose objects are never related as a totality to one another.)

I note that even here the authors considered "colimit", for a cocomplete category $\mathcal{A}$, as a functor colim: $D\mathcal{A} \to \mathcal{A}$ defined on the category $D\mathcal{A}$ of all diagrams in $\mathcal{A}$; except of course that our modern $D\mathcal{A}$ has all small categories as diagram-types, while they had only directed sets. Still, it is interesting to see, so early on, essentially 2-categorical constructions such as $D\mathcal{A}$. This construction has been much pursued. Kock [*61*] expressed, as a quotient of D, the monad on **Cat** whose algebras are cocomplete categories. Guitart studied D in [*39*], and in subsequent unpublished work has shown that the algebras for the monad D on **Cat** are 2-categories admitting all lax colimits. If it seems odd that the algebras are categories with an *extra structure* that is complete, recall that the algebras for the covariant power-set monad on **Set** are complete ordered sets. I tried to understand better the *clubs* of § 9 below by exhibiting them in [*58*] as monoids for a monoidal structure $\circ$ on a mild variant of **DCat**, while Wells [*98*] quite independently put forward the same $\mathcal{A} \circ \mathcal{B}$ as the proper and functorial generalization of wreath products. The whole matter is of course closely related to the Grothendieck construction [*38*].

The first big step, however, in the study of properties internal to a category was Mac Lane's 1950 paper [64], first announced in the 1948 note [55]. Here we find free and divisible abelian groups characterized dually as the projective and injective objects, with the recognition that Ext can be calculated equally from injectives as from projectives; a rather strong notion of generator (called "integral object") which is unique when it exists; the characterization of products and coproducts by their universal properties; a discussion of additive categories, with the *equational* characterization of the biproduct in these, and with the recognition that the additive structure of the category is unique and is recoverable from the biproducts; the notion (under the name of "bicategory structure") of what some would now call a proper factorization system; and finally, under the name "abelian bicategory", the modern concept of abelian category, although one required to have both a generator and a cogenerator in his strong sense.

A few comments. It had always been a nuisance in algebraic topology that the topological product of CW-complexes was not a CW-complex; Dowker [*22*] had written about the problem. I first learnt from Atiyah in 1962 Mac Lane's categorical notion of product, that put the CW and topological products each in its place, removing much unnecessary confusion or complication. Note the time-lag again: I had left Cambridge at the end of 1956.

The observations about additive categories put matrices in their right place and, once joined to the notion of splitting idempotents, simplified the exposition of Wedderburn theory, while showing that Krull-Schmidt applied in any additive category if the summands had local endomorphism rings.

As Mac Lane has often remarked, the notion of universal property took a long time to gel. Samuel [*82*] had the idea in a very general form in 1948, but without the categorical language it needed. Not only is the Yoneda Lemma lacking in [48] and [64], but one looks in vain for the set-valued hom-functor, although surprisingly there is no lack of internal ones. We learnt around 1954 in Cambridge that the Eilenberg-Mac Lane spaces were universal for cohomology operations, as the spheres were for homotopy ones; but representable functors were not mentioned as such. Mac Lane says, in §13 of [116], that until the late fifties, he and Eilenberg doubted the value of further serious research on category theory.

When he returned to it in the early sixties, much had changed. Yoneda had written [*99*], [*100*], and [*101*]; Kan had adjoint functors [*45*], while Grothendieck [*35*], following Buchsbaum [*14*], had the definitive form of abelian categories, as well as the notion of representable functor ([*36*) and [*37*]). The work of Gabriel [*32*] and Freyd [*30*], although not yet published, was in the air.

5. *Factorization Systems*

Having been myself involved, I should like to trace briefly the later development of one new notion introduced by Mac Lane in [64]: that of "bicategory". The word itself is out of date, being now appropriated to a quite different concept due to Bénabou [*8*]; I shall instead say "factorization system".

Monomorphisms and epimorphisms are now always defined as left cancellable and right cancellable maps. In respectable categories every map factorizes as an epimorphism followed by a monomorphism, and in some important cases this factorization is unique to within isomorphism: in a topos (such as **Set**), or in an abelian category (cf. Mac Lane [64] Thm 21.1), or in the categories of groups or of compact hausdorff spaces; but in general it is not so.

We can often restore canonical factorization (to within isomorphism) by cutting down to a subclass $\mathcal{M}$ of the monos, containing the isomorphisms and closed under composition, and a similar subclass $\mathcal{E}$ of the epis, chosen to match with $\mathcal{M}$. This is then a *factorization system*. In the category of hausdorff spaces, where the monos are the injective maps and the epis are those with dense image, three possibilities for $\mathcal{M}$ are the injections, the subspace inclusions, and the closed subspace inclusions; the corresponding possibilities for $\mathcal{E}$ are the quotient maps, the surjections, and the maps with dense image. If we define *subobject of* A, relative to such a system, to be an isomorphism class of maps in $\mathcal{M}$ with codomain A, we get in good cases a complete lattice sub A of such subobjects; while a map $f: A \to B$ induces direct and inverse image functions $f_*: \text{sub } A \to \text{sub } B$ and $f^*: \text{sub } B \to \text{sub } A$ with the desired adjunction property; and dually for quotient objects.

All of this Mac Lane set out in [64]. Yet his system was complicated by requiring a *canonical* representative in $\mathcal{M}$ of each subobject, with the dual for quotient objects, and requiring strong closure properties for these canonical choices. These requirements, suggested by such concrete categories as groups, are now seen as contrary to the categorical spirit; and Isbell in [*42*] presented Mac Lane's notion freed of this rigidity. The German school of categorical topologists has made much use of factorizations; many references will be found in [*102*].

I went on in [*53*] to investigate conditions under which the *regular* epimorphisms (that is, the coequalizers, analogues of "quotient maps by a congruence") form a suitable $\mathcal{E}$; it suffices that every pullback of a regular epi be an epi, and in the additive case this is also necessary. Barr (cf. [*2*]) has emphasized the importance of *regular* categories: those finitely-complete ones in which every pullback of a regular epi is itself a regular epi.

The question of the amalgamation property for monos is closely related; see Ringel [*81*] and Tholen [*93*].

The co-well-poweredness so often useful in applying Freyd's adjoint functor theorem may fail for all epis, but hold for those in $\mathcal{E}$. Such uses led Kennison [*60*] to consider generalized factorization systems where, while the $\mathcal{E}$'s were still epis, the $\mathcal{M}$'s need not be monos. Freyd-Kelly [*31*] went all the way, relaxing $\mathcal{E}$'s as well as $\mathcal{M}$'s, and called the original factorization systems the *proper* ones. It turned out that important mathematical questions depended on whether, for given classes $\mathcal{E}$ and $\mathcal{M}$, every map in fact admitted an $(\mathcal{E}, \mathcal{M})$-factorization. A necessary condition was that $\mathcal{E}$ and $\mathcal{M}$ be related by a unique-diagonal-fill-in Galois connexion; this defines a *pre-factorization* system [*31*]. Earlier, Ringel ([*79*],[*80*]) had studied a similar connexion without uniqueness of the diagonal, but closely related. The latest existence results I know of for factorizations are those of Bousfield [*13*].

6. *Additive Relations*

In a category $\mathcal{C}$ with finite limits a *relation* from A to B is defined as a subobject (usually in the $\mathcal{M}=$all monos sense) of the product $A \times B$. Composing relations involves taking images (to get a subobject again), and this composition is associative only if images are well behaved: the precise condition is that $\mathcal{C}$ be *regular* in Barr's sense above. Thus for a regular $\mathcal{C}$ the objects and relations form a category $\mathcal{R}$, in which we embed $\mathcal{C}$ by treating a map as a relation. In fact $\mathcal{R}$ is more than a category, for the hom-set $\mathcal{R}(A,B)$ has an order, compatible with composition, which makes it a lattice, and moreover every relation $r\colon A \to B$ determines an opposite relation $r^{\circ}\colon B \to A$.

In his 1961 paper [86], Mac Lane considers the case of an abelian category $\mathcal{C}$. (By the way, there is nothing "additive" about his relations except the fact that his category $\mathcal{C}$ is additive: they are exactly the relations in the above sense.) He establishes nine properties that hold in $\mathcal{R}$, connecting composition, the lattice structures, and the operation $(\ )^{\circ}$; and then forgets $\mathcal{C}$, considering merely an $\mathcal{R}$ with these nine properties as axioms. He derives a host of further consequences, in which the symmetric idempotents play a large role, and shows that there is only one candidate for $\mathcal{C}$. He mentions the possibility of a representation theorem, but does not establish one.

This was not followed up at the time; perhaps some of the axioms are not very transparent. The calculus of relations is convenient in homology theory for things like connecting homomorphisms and especially spectral sequences, and was so used by Puppe (who had the idea independently) and by Mac Lane (cf. the treatment of the Lyndon spectral sequence in [88]); but these applications did not demand great depth. In a different direction, the notion of relation was generalized by Bénabou's notion of distributor [*9*].

Relations became important again with the discovery of elementary topoi by Lawvere-Tierney (cf. [*71*]); an elementary topos is definable as a finitely-complete category in which the relations from A to B are representable as maps from A to a certain $\mathcal{P}B$. Such topoi have an intimate relation to higher-order logic; and the construction of free topoi is usefully carried out in steps, corresponding to weaker parts of logic and hence to weaker kinds of category: one important stage is that of regular categories.

A new approach to this has recently been given by Freyd (so far, only in lectures) in his notion of *allegory*. This is an object like $\mathcal{R}$ above, satisfying some evident formal axioms and the less-evident axiom $rs \cap t \subset (r \cap ts^{o})s$. Here Freyd gets representation theorems by formally splitting the idempotents or the symmetric idempotents, and using embedding theorems of Barr [*2*]; later adding more axioms to $\mathcal{R}$ to build up to richer logical theories. It seems that the approach suggested by Mac Lane, long dormant, is very much alive 17 years later.

In a similar vein, Succi Cruciani [*91*] has given the precise axioms on $\mathcal{R}$ in order that it should be the relations of a regular category $\mathcal{Q}$, and determined those 2-functors $\mathcal{R} \to \mathcal{R}'$ that come from exact functors $\mathcal{Q} \to \mathcal{Q}'$; using this to construct (following an idea of Joyal) the reflexion of regular categories into exact ones (again in the sense of Barr [*2*]).

7. *Monoidal and Closed Categories*

In the northern Fall of 1963 I went on leave from Sydney to Tulane, where the ever-kind Gail Young made it possible for me to attend the advertised lectures by Eilenberg, at Las Cruces after Christmas, on differential graded categories. These turned out to be the same things as the "complex categories" I was using in [*49*]. Sammy graciously encouraged me to seek an extra year's leave from Sydney, and asked Alex Heller to arrange me a job at Urbana for 1964/65.

A few weeks later, having given a short talk at the A.M.S. meeting in Miami, I was charmed when a member of the audience came up, introduced himself as Saunders Mac Lane, and invited me to visit Chicago. For me it was indeed a vintage year: for I also met Peter Freyd at the same meeting.

Mac Lane had for some time been interested in categories supplied with a tensor product, meaning thereby a bifunctor which, to within isomorphism, was associative and commutative and admitted an identity object; he had worked out in [90] the conditions for "coherence" of the isomorphisms involved (see §9 below). He called them "bicategories" in [90] and [93], continuing his run of bad luck with this term; today they are often called *symmetric monoidal* (s.m.) categories. (These, as well as the nonsymmetric ones, had also been considered by Bénabou in [*5*] and [*6*], who had however no finite list of conditions for coherence.)

Mac Lane used such categories to give a unified treatment of relative homological algebra in [93] (which, although published in 1965, represented basically his Colloquium Lectures at Boulder in 1963). He based the account on the concept of a *monoid* (or *algebra*) in such a monoidal category; later Barr [*1*] would show its close relation to the new *monad* (or *triple*) cohomology of Barr-Beck [*4*]. Note that the s.m. categories Mac Lane uses here are required to be abelian ones in which each $-\otimes A$ is right exact; in fact $-\otimes A$ in all the examples has a right adjoint, and these categories are *closed* (see below).

His next examples in [93] of s.m. categories are the PROPs and PACTs to which I devote §10 below. He ends by pointing out a third possible use: one might perhaps unify such notions as additive category and differential graded category by seeing them as "enriched" categories with hom-functors taking values in an s.m. category $\mathcal{V}$.

During 1964 some progress was made in this last direction by me [*52*], and independently by Linton [*74*], who however considered only the case where the underlying functor $\mathcal{V} \to \mathbf{Set}$ was faithful; and Eilenberg too was thinking about it. In the good cases, the category $\mathcal{V}$ had itself a $\mathcal{V}$-valued, and thus an internal, Hom: which turned out to be a right adjoint for $\otimes$, just as for abelian groups. Such categories came to be called *closed*. (It is interesting to note that many of the prime examples occurred right back in the original [48], so that internal Hom's were noticed long before **Set**-valued ones. The examples of natural transformations in §10 of [48] are virtually all canonical maps in closed categories.)

At Eilenberg's invitation, he and I worked together on this during 1964/65, producing an encyclopaedic but flawed article [*26*]. One of its major faults was an over-concentration on the vanishingly-rare case where the internal Hom exists but $\otimes$ does not; the other was a fastidiousness that, at the cost of equationality of the theory, wanted an underlying functor $\mathcal{V} \to \mathbf{Set}$ that did its job on the nose, instead of introducing messy isomorphisms as did the unpretentious $\mathcal{V}(I, -)$. It is also unfortunate that we overlooked, until [*26*] was nearly in print, the independent and largely contemporaneous work [*7*] of Bénabou.

Further developments of these ideas can be found in several papers, individual and joint, of Day and Kelly ([*54*],[*20*],[*17*],[*18*],[*19*]), and in works of Bunge [*15*], Dubuc [*23*], Kock [*62*], [*63*], and Lawvere [*72*].

8. Mac Lane as Expositor and Stimulator

Back in Australia from mid-1965, I didn't see much of Mac Lane for several years; we met a few times during my short leave to New York in early 1968. Involved with my first students Ross Street and Brian Day, I felt his influence and support in another way.

First, as an expositor. Street's subject was homological (cf. §3 above); we used on the one hand Cartan-Eilenberg [*16*], together with Eilenberg's Las Cruces lectures on his work with Moore; and on the other, the very different and complementary "Homology" [88] of Mac Lane, combined with his survey [93]. I have ever since been impressed by Mac Lane's capacity for pithy expression of what is central. Other notable and more recent examples are the "categories work" text [111], the Retiring Presidential Address [115], and the historical surveys [104], [116], [117], [118].

At this time Mac Lane too was back at homology, using functor-categories, in his paper [96] with Dold and Oberst, to simplify acyclic models. For further developments along this line, see Ulmer ([*94*],[*95*],[*96*]).

The other thing of great value we owed to Mac Lane was the Midwest Category Seminar. In a subject developing so fast, and one whose practitioners seemed amazingly reluctant to publish, information had been hard to come by even in North America, being verbally communicated within a coterie; how should we keep in touch from Australia? In the Midwest Category Seminar Mac Lane provided, in North America, a focal point, a forum, with encouragement and stimulation; and by publishing its Reports in *Lecture Notes in Mathematics*—a lead later followed by others in category theory—made awareness possible in distant places. Day and I gratefully accepted his offer to publish there our own papers of this period; and its importance for such communication declined only with the launching of the *Journal of Pure and Applied Algebra*.

In the (northern) Winter of 1969 Mac Lane came, under "Fulbright" auspices (now technically the Australian-American Educational Foundation), to the Summer Research Institute of the Australian Mathematical Society, held in Canberra. In a course of 14 lectures, he aroused great and lasting enthusiasm for the subject of category theory; so much so that Fulbright have since sent two other ambassadors on longer visits: Peter Freyd in 1971 and John Gray in 1975.

What is more, Mac Lane and I began then to work together.

9. *Coherence*

Mac Lane's coherence result [90] of §7 above considered the natural isomorphisms built up from the basic ones $(A\otimes B)\otimes C\cong A\otimes(B\otimes C)$, $A\otimes B\cong B\otimes A$, and $A\otimes I\cong A$, using composition, the tensor product $\otimes$, and the unit object I; and showed that all diagrams involving these commute if a certain finite list (reduced in [*51*] to four) did so. The associativity part had been independently discovered, a little earlier and in a different context and language, by Stasheff [*85*]; using this, Epstein found results identical to Mac Lane's at almost the same time.

Now Epstein [*28*], completed in respect of the unit object by MacDonald [*75*], had a further coherence result involving Hom as well as $\otimes$, and involving isomorphisms $\mathrm{Hom}(A\otimes B,C)\cong\mathrm{Hom}(A,\mathrm{Hom}(B,C))$ and $A\cong\mathrm{Hom}(I,A)$ besides those above. These seemed to envisage a *closed* category $\mathcal{V}$, but on closer inspection do not. The existence of such isomorphisms does *not* make the s.m. category $\mathcal{V}$ closed; and the proper context for these results is a category $\mathcal{V}'$ on which the s.m. category $\mathcal{V}$ *acts* via $\otimes'\colon\mathcal{V}\times\mathcal{V}'\to\mathcal{V}'$; we are merely dealing with the special case $\mathcal{V}'=\mathcal{V}^{op}$, $\otimes'=\mathrm{Hom}^{op}$.

Yet Eilenberg and I had proved in [*26*] the commutativity of dozens of diagrams for a closed category. Their legs were not in general isomorphisms, as in the above results, but were always natural transformations in our generalized sense [*25*]. They could be built up from the basic natural isomorphisms above, along with the natural transformations giving the unit $A\to\mathrm{Hom}(B,A\otimes B)$ and the counit $\mathrm{Hom}(A,B)\otimes A\to B$ of the $\otimes$-Hom adjunction. We conjectured by 1968 that *all* such diagrams commuted, but got nowhere near a proof.

Mac Lane and I attacked this problem, at his suggestion, in Canberra; but nothing seemed to work. How could one do induction when the middle term S of a composite $R\to S\to T$ might be arbitrarily more complicated than R and T?

We recalled that Lambek had faced a similar difficulty in [*64*], on which he had lectured at Chicago in April 1968, in studying an apparently different problem (that of free structures) for a somewhat different system (a non-symmetric $\otimes$, a left Hom and a right Hom$'$, and no I). He had overcome the difficulty by adapting the cut-elimination results in proof-theory of Gentzen [*33*]. One result in [*64*] even looked tantalizingly like what we wanted, but the proof was obscure. So we imitated the techniques, without clearly seeing the connexion; and within a few days came near to our conjectured result—except for one wholly intractible case in the proof.

That night, depressed, I drank beer with our German visitor Otto Kegel, explaining our problem to a group-theorist. Doing so brought light: what fools we were! There were obvious counter-examples to the conjecture, involving double duals. Twenty-four hours later, we had proved a modified form of the conjecture, which appeared as [109].

There have been many coherence results since. Mac Lane's student Voreadou [*97*] completed our work above by delineating (a most difficult problem) the *precise* class of diagrams that commute in general; his protégé Laplaza studied categories with $\otimes$ where $(A\otimes B)\otimes C \to A\otimes(B\otimes C)$ is not invertible [*65*], categories with two s.m. structures $\otimes$ and $\oplus$ and a distributivity of $\otimes$ over $\oplus$ ([*66*],[*67*]), and "closed categories" [*68*] with a Hom but no $\otimes$ (again a difficult problem, and a challenge to coherence techniques, even though examples in nature are rare). Mac Lane and I extended the results of [109] to enriched situations in [110], and my student Lewis, generalizing Epstein [*28*], considered morphisms of closed categories [*73*]. A coherence result with a rather different flavour, in the context of 2-categories, is given by Gray [*34*].

It soon became clear that, apart from solving such problems, we needed new insights even to pose them exactly. Categories with some structure, and the strict maps between them (which are those appropriate here), form a category **Str** with a forgetful functor to **Cat**; in reasonable cases the latter has a left adjoint $\mathfrak{F}$, in strictly equational ones it is monadic (and we may use $\mathfrak{F}$ to denote the monad), and in certain purely-covariant ones $\mathfrak{F}$ is a 2-monad. The general coherence problem consists in finding $\mathfrak{F}(\mathcal{Q})$ for a category $\mathcal{Q}$; for this is clearly the free category on a certain evident graph, modulo "the diagrams which commute".

In a restricted but important class of cases, this can be done independently of a particular $\mathcal{Q}$; for in these it can be seen *a priori* that $\mathfrak{F}(\mathcal{Q})$ has the form $\mathcal{K}\circ\mathcal{Q}$, where this $\circ$ is a generalization of that on **DCat** mentioned in §4 above. The monoid $\mathcal{K}$ here is what I called a *club* in [*55*], [*56*], [*57*]; its objects are formal functors and its maps are formal natural transformations; this explains why the coherence problems of [90] and [109] and several others above could be formulated in natural-transformation terms, and makes clear *a posteriori* the exact relation with Lambek. Clubs are used further in [*59*], leading to Blackwell's thesis [*10*]; but my attempt in [*58*] to understand them better, at least in the covariant case, is not really satisfactory. It has recently appeared that they admit an absolute definition as monads $\mathfrak{F}$ on **Cat** with a special property, which in particular ensures that $\mathbf{Cat}/\mathfrak{F}1$ is monoidal.

Finally I note (as has been impressed on me by both Lawvere and Joyal) that the theory of categories-with-structures is too narrow if restricted to structures monadic over **Cat** (which topoi are not). We need a 2-dimensional analogue of lim-theories, rather than equational theories; and the ideal setting has not yet been found. The reader will observe the increasing importance of Ehresmann's 1963 notion [*24*] of 2-*category*.

10. PROPs *and* PACTs

Mac Lane reports in [93] on work he was doing with Adams, stimulated by a desire to study higher homotopies.

Write **P** for the symmetric monoidal category whose object-set is the natural numbers, whose morphisms are the permutations (so that $\mathbf{P}(m,n)$ is empty for $m\neq n$), and whose tensor product is + with unit object 0. For a good closed category $\mathcal{V}$, define a $\mathcal{V}$-PROP to be a symmetric monoidal $\mathcal{V}$-category $\mathcal{H}$, together with a strict functor $\mathbf{P}\to\mathcal{H}$ of symmetric monoidal categories that is the identity on objects. Taking $\mathcal{V}=\mathbf{Set}$ gives the Mac Lane notion of PROP in [93]; taking $\mathcal{V}$ = differential graded K-modules gives that of PACT. The *operads* of May [*76*] are special $\mathcal{V}$-PROPs in the case $\mathcal{V}$ = compactly generated spaces.

There is a close analogy with the *theories* of Lawvere [*69*]. There $\mathbf{P}$ is replaced by the skeleton $\mathbf{S}$ of finite sets, the symmetric monoidal structure on $\mathcal{H}$ is given by the categorical cartesian product, and $\mathcal{V}$ (classically **Set**) is either cartesian closed, or is at best one of the "suitable" closed categories of Borceux-Day [*12*].

Take $\mathcal{V}=\mathbf{Set}$ for the moment. A Lawvere theory $\mathcal{H}$ has models in any category $\mathcal{C}$ with finite products. Analogously a PROP $\mathcal{H}$ has models in any *symmetric monoidal* category $\mathcal{C}$; the difference is that, if A^n in such a category means *tensor* power $A\otimes A\otimes\ldots\otimes A$, there are no *canonical* maps $A^f\colon A^n\to A^m$ except when $m=n$ and $f\colon m\to n$ is a permutation. The kinds of thing that are models for a PROP are algebras ($=\otimes$-monoids), coalgebras, and Hopf algebras. These ideas are much used in the thesis [*29*] of Barr's student Fox, generalizing Barr's extensions [*3*] of Sweedler's results [*92*].

May's operads are rather special PROPs (for the appropriate $\mathcal{V}$), in which the spaces $\mathcal{H}(m,n)$ have the form $\mathcal{H}(1,n)^m$, and are hence determined by the functor $H\colon\mathbf{P}\to\mathcal{V}$ sending n to $Hn=\mathcal{H}(1,n)$. In fact they may be described (unpublished) as monoids for a suitable non-symmetric monoidal closed structure on the functor-category $[\mathbf{P},\mathcal{V}]$; this works for any $\mathcal{V}$, and when $\mathcal{V}=\mathbf{Cat}$ there is a close relation to clubs. A corresponding result for theories is noted by Johnstone-Wraith [*44*] in the topos case, with an analogue in the Borceux-Day situation [*12*].

The results on higher homotopies expressed by Mac Lane in Theorem 25.1 of [93] have been carried further: in the language of PROPs by Boardman and Vogt [*11*]; in that of operads by May [*76*]; and in terms of geometric realizations of topological categories by Segal [*83*], [*84*].

The matter of classifying spaces for compactly generated monoids is closely related. In [105] Mac Lane exhibits Milgram's construction [*77*] in terms of the tensor product of functors, thereby achieving a considerable simplification.

The results in Diaconescu's thesis [*21*], on tensor products of internal functors for topoi, were shown by Johnstone [*43*] to hold more generally for exact categories; in fact with a little care they hold for compactly generated spaces. In this language, the functor called σ_0 on p. 140 of [105], considered as an internal functor, is flat; so that tensoring with it is left exact. This flatness, proved by a shuffle argument, provides another approach to the product theorem in §10 of [105].

11. Foundations

Like a parent concerned for his child's legitimacy, Mac Lane has returned time and again to the question of foundations. The Gödel-Bernays system of sets and classes, used in [48], was adequate only while categories remained no more than a purely basic language. Mac Lane himself, probably before anyone else, noted four major difficulties in his 1959 paper [87]. Leaving aside the least pressing of these —the category of all categories—he showed how the others could be overcome in practice by restriction to small subcategories that were "adequate". (Although

Isbell in his papers [*41*], [*42*] on adequacy—later called density by Ulmer—does not mention Mac Lane, the Isbell and Mac Lane concepts are so allied that I suspect some kind of informal connexion.)

As the importance of functor categories became clearer with the work of Grothendieck, Kan, and Freyd, as well as his own [96], Mac Lane came back to the question many times ([99], [102], [103], [108]). If we have to accept Grothendieck universes, let us try [103] to get away with only one. But perhaps the answer lay in Lawvere's elementary theory of categories [*70*].

This line has been most actively pursued by Street ([*88*], [*89*], [*90*], the last in collaboration with Walters). For him, the category of categories being described by first-order axioms, a "category of sets" is merely a chosen internal full subcategory $\mathbb{S}$ with suitable first-order properties. Such a choice determines notions of smallness and local-smallness. A recent theorem, conjectured by Street and proved by Freyd, asserts that the functor-category $[A, \mathbb{S}]$ cannot be locally small if A is locally small, unless A itself is (skeletally) small. This is further evidence that the consideration of categories of different sizes is inescapable in mathematics, even if we begin with a first-order theory of categories.

This is a fitting place to remark that the first-order aspect, emphasized with deep insight by Lawvere, and leading to the 1970 notion (cf. [*71*]) of elementary topos, with its close relations both to algebraic geometry and to logic, has given rise to what are today the most penetrating and rapidly-developing applications of category theory. Mac Lane's 1974 survey article [113] on topoi speaks of "exciting prospects"; it seems that he was right, and that the consequences of an idea, adumbrated in 1942 and published in 1945, are still far from being exhausted.

12. Conclusion

After writing this last heading, I see that no real conclusion is called for. No man could so stimulate others unless, alongside an incisive intellect, he was possessed of enthusiasm and warmth, a deep interest in his fellow man, and a sympathy the more real for being unsentimental. Those who proudly call themselves his friends know of these things: others will infer them in reading these essays.

In any case, what place is there for a conclusion when writing about a person still, as ever, thrusting vigorously ahead?

Bibliography

[*1*] M. Barr, Cohomology in tensored categories, *Proc. Conf. on Categorical Algebra* (*La Jolla* 1965), Springer-Verlag 1966, 344–354.

[*2*] ____, Exact categories, *Lecture Notes in Math.* 236 (1971), 1–120.

[*3*] ____, Coalgebras over a commutative ring, *Jour. Algebra* 32 (1974), 600–610.

[*4*] M. Barr and J. Beck, Homology and standard constructions, *Lecture Notes in Math.* 80 (1969), 245–335.

[*5*] J. Bénabou, Catégories avec multiplication, *C. R. Acad. Sci. Paris* 256 (1963), 1887–1890.

[6] ____, Algèbre élémentaire dans les catégories avec multiplication, *C. R. Acad. Sci. Paris* 258 (1964), 771–774.

[7] ____, Catégories relatives, *C. R. Acad. Sci. Paris* 260 (1965), 3824–3827.

[8] ____, Introduction to bicategories, *Lecture Notes in Math.* 47 (1967), 1–77.

[9] ____, Les distributeurs, *Séminaires de Math. Pure, Univ. Catholique de Louvain*, No. 33, Vander, Louvain, 1973.

[10] R. Blackwell, Some existence theorems in the theory of doctrines (Thesis, Sydney 1976).

[11] J. M. Boardman and R. M. Vogt, *Homotopy Invariant Algebraic Structures on Topological Spaces*, = *Lecture Notes in Math.* 347 (1973).

[12] F. Borceux and B. J. Day, Universal algebra in a closed category, *Jour. Pure Applied Algebra* (to appear).

[13] A. K. Bousfield, Constructions of factorization systems in categories, *Jour. Pure Applied Algebra* 9 (1977), 207–220.

[14] D. A. Buchsbaum, Exact categories and duality, *Trans. Amer. Math Soc.* 80 (1955), 1–34.

[15] M. C. Bunge, Relative functor categories and categories of algebras, *Jour. Algebra* 11 (1969), 64–101.

[16] H. Cartan and S. Eilenberg, *Homological Algebra*, Princeton Univ. Press, New Jersey 1956.

[17] B. J. Day, On closed categories of functors, *Lecture Notes in Math.* 137 (1970), 1–38.

[18] ____, A reflection theorem for closed categories, *Jour. Pure Applied Algebra* 2 (1972), 1–11.

[19] ____, On closed categories of functors II, *Lecture Notes in Math.* 420 (1974), 20–54.

[20] B. J. Day and G. M. Kelly, Enriched functor categories, *Lecture Notes in Math.* 106 (1969), 178–191.

[21] R. Diaconescu, Change of base for toposes with generators, *Jour. Pure Applied Algebra* 6 (1975), 191–218.

[22] C. H. Dowker, Topology of metric complexes, *Amer. Jour. Math.* 74 (1952), 555–577.

[23] E. J. Dubuc, *Kan Extensions in Enriched Category Theory*, = *Lecture Notes in Math.* 145 (1970).

[24] C. Ehresmann, Catégories structurées, *Ann. Sci. Ecole Norm. Sup.* 80 (1963), 349–425.

[25] S. Eilenberg and G. M. Kelly, A generalization of the functorial calculus, *Jour. Algebra* 3 (1966), 366–375.

[26] ____, Closed categories, *Proc. Conf. on Categorical Algebra* (*La Jolla* 1965), Springer-Verlag 1966, 421–562.

[27] S. Eilenberg and N. Steenrod, *Foundations of Algebraic Topology*, Princeton Univ. Press, New Jersey 1952.

[28] D. B. A. Epstein, Functors between tensored categories, *Invent. Math.* 1 (1966), 221–228.

[29] T. F. Fox, Universal coalgebras (Thesis, McGill Univ., 1976).

[30] P. J. Freyd, *Abelian Categories: An Introduction to the Theory of Functors*, Harper and Row, New York, 1964.

[31] P. J. Freyd and G. M. Kelly, Categories of continuous functors I, *Jour. Pure Applied Algebra* 2 (1972), 169–191.

[32] P. Gabriel, Des catégories abéliennes, *Bull. Soc. Math. France* 90 (1962), 323–448.

[33] G. Gentzen, Untersuchungen über das logische Schliessen I, II, *Math. Zeit.* 39 (1934–1935), 176–210 and 405–431.

[34] J. W. Gray, Coherence for the tensor product of 2-categories, and braid groups; in *Algebra, Topology, and Category Theory: A Collection of Papers in Honor of Samuel Eilenberg*, Academic Press, New York, 1976; 63–76.

[35] A. Grothendieck, Sur quelques points d'algèbre homologique, *Tohoku Math. J.* 9 (1957), 119–221.

[36] ____, Techniques de descente et théorèmes d'existence en géométrie algébrique II, *Sém. Bourbaki* 12 (1959/1960), exp. 195; Secrétariat Mathématique, Paris, 1961.

[37] ____, Techniques de construction en géométrie analytique IV: Formalisme général des foncteurs représentables, *Sém. Cartan* 13 (1960/1961), exp. 11; Secrétariat Mathématique, Paris, 1962.

[38] ____, Catégories fibrées et descente, *Sém. de Géométrie Algébrique de l'I.H.E.S.*, Paris 1961.

[39] R. Guitart, Sur le foncteur diagramme, *Cahiers de Top. et Géom. Diff.* 14 (1973), 181–182.

[40] P. J. Hilton and S. Wylie, *Homology Theory*, Cambridge Univ. Press 1960.

[41] J. R. Isbell, Adequate subcategories, *Illinois J. Math.* 4 (1960), 541–552.

[42] ____, Subobjects, adequacy, completeness and categories of algebras, *Rozprawy Math.* 36 (1964) 3–33.

[43] P. T. Johnstone, Some aspects of internal category theory in an elementary topos (Thesis, Univ. of Cambridge, 1974).

[44] P. T. Johnstone and G. C. Wraith, Algebraic theories in toposes, *Lecture Notes in Math.* 661 (1978), 141–242.

[45] D. M. Kan, Adjoint functors, *Trans. Amer. Math. Soc.* 87 (1958), 294–329.

[46] G. M. Kelly, Observations on the Künneth theorem, *Proc. Cambridge Phil. Soc.* 59 (1963), 575–587.

[47] ____, On the radical of a category, *Jour. Austral. Math. Soc.* 4 (1964), 299–307.

[48] ____, Complete functors in homology: I. Chain maps and endomorphisms, *Proc. Cambridge Phil. Soc.* 60 (1964), 721–735.

[49] ____, Complete functors in homology: II. The exact homology sequence, *Proc. Cambridge Phil. Soc.* 60 (1964), 737–749.

[50] ____, Chain maps inducing zero homology maps, *Proc. Cambridge Phil. Soc.* 61 (1965), 847–854.

[51] ____, On Mac Lane's conditions for coherence of natural associativities, commutativities, etc., *Jour. Algebra* 1 (1964), 397–402.

[52] ____, Tensor products in categories, *Jour. Algebra* 2 (1965), 15–37.

[53] ____, Monomorphisms, epimorphisms, and pull-backs, *Jour. Austral. Math. Soc.* 9 (1969), 124–142.

[54] ____, Adjunction for enriched categories, *Lecture Notes in Math.* 106 (1969), 166–177.

[55] ____, Many-variable functorial calculus I, *Lecture Notes in Math.* 281 (1972), 66–105.

[56] ____, An abstract approach to coherence, *Lecture Notes in Math.* 281 (1972), 106–147.

[57] ____, A cut-elimination theorem, *Lecture Notes in Math.* 281 (1972), 196–213.

[58] ____, On clubs and doctrines, *Lecture Notes in Math.* 420 (1974), 181–256.

[59] ____, Coherence theorems for lax algebras and for distributive laws, *Lecture Notes in Math.* 420 (1974), 281–375.

[60] J. F. Kennison, Full reflective subcategories and generalized covering spaces, *Illinois J. Math.* 12 (1968), 353–365.

[61] A. Kock, Limit monads in categories, *Aarhus Universitet Mathematisk Institut Preprint Series* 1967/68 No. 6.

[62] ____, On double dualization monads, *Math Scand.* 27 (1970), 151–165.

[63] ____, Strong functors and monoidal monads, *Archiv Math.* 23 (1972), 113–120.

[64] J. Lambek, Deductive systems and categories I. Syntactic calculus and residuated categories, *Math. Systems Theory* 2 (1968), 287–318.

[65] M. L. Laplaza, Coherence for associativity not an isomorphism, *Jour. Pure Applied Algebra* 2 (1972), 107–120.

[66] ____, Coherence for distributivity, *Lecture Notes in Math.* 281 (1972), 29–65.

[67] ____, A new result of coherence for distributivity, *Lecture Notes in Math.* 281 (1972), 214–235.

[68] ____, Coherence in non-monoidal closed categories, *Trans. Amer. Math. Soc.* 230 (1977), 293–311.

[69] F. W. Lawvere, Functorial semantics of algebraic theories, *Proc. Nat. Acad. Sci. U.S.A* 50 (1963), 869–873 (Fuller account in Thesis of same title, Columbia Univ., 1963).

[70] ____, The category of categories as a foundation for mathematics, *Proc. Conf. on Categorical Algebra* (*La Jolla* 1965), Springer-Verlag 1966, 1–20.

[71] ____,Ed. *Toposes, Algebraic Geometry and Logic*, = *Lecture Notes in Math.*274 (1972).

[72] ____, Metric spaces, generalized logic, and closed categories, *Rend. del Sem. Mat. è Fis. di Milano* 43 (1973), 135–166.

[73] G. Lewis, Coherence for a closed functor, *Lecture Notes in Math.* 281 (1972), 148–195.

[74] F. E. J. Linton, Autonomous categories and duality of functors, *Jour. Algebra* 2 (1965), 315–349.

[75] J. L. MacDonald, Coherence of adjoints, associativities, and identities, *Archiv Math.* 19 (1968), 398–401.

[76] J. P. May, *The Geometry of Iterated Loop Spaces*, = *Lecture Notes in Math.* 271 (1972).

[77] R. J. Milgram, The bar construction and abelian *H*-spaces, *Illinois J. Math.* 11 (1967), 242–250.

[78] F. P. Palermo, The cohomology ring of product complexes, *Trans. Amer. Math. Soc.* 86 (1957), 174–196.

[79] C. M. Ringel, Diagonalisierungspaare I, *Math. Zeit.* 117 (1970), 249–266.

[80] ____, Diagonalisierungspaare II, *Math. Zeit.* 122 (1971), 10–32.

[81] ____, The intrinsic property of amalgamations, *Jour. Pure Applied Algebra* 2 (1972), 341–342.

[82] P. Samuel, On universal mappings and free topological groups, *Bull. Amer. Math. Soc.* 54 (1948), 591–598.

[83] G. B. Segal, Classifying spaces and spectral sequences, *Publ. Math. Inst. des Hautes Études Scient. Paris* 34 (1968), 105–112.

[84] ____, Categories and cohomology theories, *Topology* 13 (1974), 293–312.

[85] J. D. Stasheff, Homotopy-associativity of *H*-spaces I, *Trans. Amer. Math. Soc.* 108 (1963), 275–292.

[86] R. H. Street, Projective diagrams of interlocking sequences, *Illinois J. Math.* 15 (1971), 429–441.

[87] ____, Homotopy classification of filtered complexes, *Jour. Austral. Math. Soc.* 15 (1973), 298–318.

[88] ____, Fibrations and Yoneda's lemma in a 2-category, *Lecture Notes in Math.* 420 (1974), 104–133.

[89] ____, Elementary cosmoi I, *Lecture Notes in Math.* 420 (1974), 134–180.

[90] R. H. Street and R. F. C. Walters, Yoneda structures on 2-categories, *Jour. Algebra* 50 (1978), 350–379.

[91] R. Succi Cruciani, La teoria delle relazioni nello studio di categorie regolari e di categorie esatte, *Riv. Mat. Univ. Parma* (4) 1 (1975), 143–158.

[92] M. E. Sweedler, *Hopf Algebras*, Benjamin, New York, 1969.

[93] W. Tholen, Amalgamations in categories, *Algebra Universalis*, (to appear).

[94] F. Ulmer, On cotriples and André (co)homology, their relationship with classical homological algebra, *Lecture Notes in Math.* 80 (1969), 376–398.

[95] ____, Acyclic models and Kan extensions, *Lecture Notes in Math.* 86 (1969), 181–204.

[*96*] ——, Kan extensions, cotriples and André (co)homology, *Lecture Notes in Math.* 92 (1969), 278–308.

[*97*] R. Voreadou, Coherence and non-commutative diagrams in closed categories, *Memoirs Amer. Math. Soc.* Vol. 9, No. 182, 1977.

[*98*] C. Wells, Wreath product decomposition of categories, preprint (Case Western Reserve Univ. Ohio U.S.A.), 1976.

[*99*] N. Yoneda, On the homology theory of modules, *J. Fac. Sci. Tokyo Sec I.* 7 (1954), 193–227.

[*100*] ——, Note on products in Ext, *Proc. Amer. Math. Soc.* 9 (1958), 873–875.

[*101*] ——, On ext and exact sequences, *J. Fac. Sci. Tokyo Sec I*, 8 (1960), 507–576.

[*102*] Fernuniversität Hagen, Fachbereich Mathematik, Kategorienseminar Nr. 1, 1976; Seminarberichte Nr. 2, 1976; Nr. 3, 1977.

Bibliography of the Publications of Saunders Mac Lane

Papers 38, 41, 42, 43, 45, 47, 48, 49, 50, 51, 52, 53, 57, 59, 60, 61, 65, 66, 68, 69, 70, 72, 73, and 76 are joint with Samuel Eilenberg and are marked with an asterisk.

Papers 23, 26, 29, 36, 39, 40, and 44 are joint with O. F. G. Schilling and are marked with a double asterisk.

1930

[1] The effect of end losses on the characteristics of filaments of tungsten and other materials, *Physical Review* **35**(1930), 478–503 (with Irving Langmuir and Katharine B. Blodgett).

1934

[2] *Abgekürzte Beweise im Logikkalkul*, Thesis, Göttingen, 1934, 61 pp.

1935

[3] A logical analysis of mathematical structure, *Monist* **45**(1935), 118–130.

[4] Abstract absolute values which give new irreducibility criteria, *Proc. Nat. Acad. Sci. USA* **21**(1935), 472–474.

[5] The ideal-decomposition of rational primes in terms of absolute values, *Proc. Nat. Acad. Sci. USA* **21**(1935), 663–667.

[6] Some unique separation theorems for graphs, *Amer. J. of Math.* **57**(1935), 805–820.

1936

[7] A construction for prime ideals as absolute values of an algebraic field, *Duke Math. J.* **2**(1936), 492–510.

[8] Some interpretations of abstract linear dependence in terms of projective geometry, *Amer. J. of Math.* **58**(1936), 236–240.

[9] Note on some equations without affect, *Bull. Amer. Math. Soc.* **42**(1936), 731–736.

[10] A construction for absolute values in polynomial rings, *Trans. Amer. Math. Soc.* **40**(1936), 363–395.

1937

[11] A combinatorial condition for planar graphs, *Fundamenta Math.* **28**(1937), 22–32.

[12] A structural characterization of planar combinatorial graphs, *Duke Math. J.* **3**(1937), 460–472.

[13] Planar graphs whose homeomorphisms can all be extended for any mapping on the sphere, *Amer. J. of Math.* **59**(1937), 823–832 (with V. W. Adkisson).

1938

[14] A lattice formulation for transcendence degrees and *p*-bases, *Duke Math. J.* **4**(1938), 455–468.

[15] Review of *The Logical Syntax of Language* by Rudolf Carnap, *Bull. Amer. Math. Soc.* **44**(1938), 171–176 (the review is entitled "Carnap on logical syntax").

[16] Fixed points and the extension of the homeomorphisms of a planar graph, *Amer. J. of Math.* **60**(1938), 611–639 (with V. W. Adkisson).

[17] The Schönemann-Eisenstein irreducibility criteria in terms of prime ideals, *Trans. Amer. Math. Soc.* **43**(1938), 226–239.

[18] The uniqueness of the power series representation of certain fields with valuations, *Ann. of Math.* **39**(1938), 370–382.

1939

[19] Symbolic logic, *Amer. Math. Monthly* **46**(1939), 289–294.

[20] Steinitz field towers for modular fields, *Trans. Amer. Math. Soc.* **46**(1939), 23–45.

[21] Modular fields, I. Separating transcendence bases, *Duke Math. J.* **5**(1939), 372–393.

[22] Subfields and automorphism groups of *p*-adic fields, *Annals of Math.* **40**(1939), 423–442.

[23]** Infinite number fields with Noether ideal theories, *Amer. J. of Math.* **61**(1939), 771–782.

[24] The universality of formal power series fields, *Bull. Amer. Math. Soc.* **45**(1939), 888–890.

[25] Some recent advances in algebra, *American Math. Monthly* **46**(1939), 3–19. Reprinted as pp. 9–34 of *Studies in Modern Algebra*, edited by A. A. Albert, Math. Assoc. of Amer. Studies in Mathematics vol. 2, 1963.

[26]** Zero-dimensional branches of rank one on algebraic varieties, *Ann. of Math.* **40**(1939), 507–520.

1940

[27] Modular fields, *Amer. Math. Monthly* **47**(1940), 259–274.

[28] Note on the relative structure of p-adic fields, *Ann. of Math.* **41**(1940), 751–753.

[29]** Normal algebraic number fields, *Proc. Nat. Acad. Sci. USA* **26**(1940), 122–126.

[30] *Algebraic Functions*, planographed, Edwards Brothers, Ann Arbor, Mich. 1940, 62 pp.

[31] Extending maps of plane Peano continua, *Duke Math. J.* **6**(1940), 216–228 (with V. W. Adkisson).

[32] The minimum number of generators for inseparable algebraic extensions, *Bull. Amer. Math. Soc.* **46**(1940), 182–186 (with M. F. Becker).

[33] Extensions of homeomorphisms on the sphere, pages 223–236 of *Lectures in Topology*, Conference at the University of Michigan in 1940, Univ. of Mich. Press, 1941 (with V. W. Adkisson).

1941

[34] The generation of inseparable fields, *Proc. Nat. Acad. Sci. USA* **27**(1941), 583–587 (with F. K. Schmidt).

[35] Factor sets of a group in its abstract unit group, *Trans. Amer. Math. Soc.* **50**(1941), 385–406 (with A. H. Clifford).

[36]** Normal algebraic number fields, *Trans. Amer. Math. Soc.* **50**(1941), 295–384.

[37] *A Survey of Modern Algebra*, Macmillan, New York, 1941, 2nd ed. 1953, 3rd ed. 1965, 4th ed. 1977 (with Garrett Birkhoff). A shorter version appeared under the title *A Brief Survey of Modern Algebra* in 1953, 2nd ed. 1965.

[38]* Infinite cycles and homologies, *Proc. Nat. Acad. Sci. USA* **27**(1941), 535–539.

1942

[39] A formula for the direct product of crossed product algebras, *Bull. Amer. Math. Soc.* **48**(1942), 108–114.

[40]** A general Kummer theory for function fields, *Duke Math. J.* **9**(1942), 125–167.

[41]* Group extensions and homology, *Ann. of Math.* **43**(1942), 757–831.

[42]* Natural isomorphisms in group theory, *Proc. Nat. Acad. Sci. USA* **28**(1942), 537–543. (See Math. Reviews 40 no. 7360 for a reprinting with an added note in Portuguese by A. V. Ferreira.)

[43]* On homology groups of infinite complexes and compacta, Appendix A, pp. 344–349, in *Algebraic Topology* by Solomon Lefschetz, Amer. Math. Soc. Coll. Publ. vol. 27, 1942.

1943

[44]** Groups of algebras over an algebraic number field, *Amer. J. of Math.* **65**(1943), 299–308.

[45]* Relations between homology and homotopy groups, *Proc. Nat. Acad. Sci. USA* **29**(1943), 155–158.

[46] A conjecture of Ore on chains in partially ordered sets, *Bull. Amer. Math. Soc.* **49**(1943), 567–568.

1945

[47]* Relations between homology and homotopy groups of spaces, *Ann. of Math.* **46**(1945), 480–509.

[48]* General theory of natural equivalences, *Trans. Amer. Math. Soc.* **58**(1945), 231–294.

1946

[49]* Determination of the second homology and cohomology groups of a space by means of homotopy invariants, *Proc. Nat. Acad. Sci. USA* **32**(1946), 277–280.

1947

[50]* Cohomology theory in abstract groups, I. *Ann. of Math.* **48**(1947), 51–78.

[51]* Cohomology theory in abstract groups, II. Group extensions with a non-Abelian kernel, *Ann. of Math.* **48**(1947), 326–341.

[52]* Algebraic cohomology groups and loops, *Duke Math J.* **14**(1947), 435–463.

1948

[53]* Cohomology and Galois theory, I. Normality of algebras and Teichmüller's cocycle, *Trans. Amer. Math. Soc.* **64**(1948), 1–20.

[54] Symmetry of algebras over a number field, *Bull. Amer. Math. Soc.* **54**(1948), 328–333.

[55] Groups, categories and duality, *Proc. Nat. Acad. Sci. USA* **34**(1948), 263–267.

[56] A nonassociative method for associative algebras, *Bull. Amer. Math. Soc.* **54**(1948), 897–902.

1949

[57]* Homology of spaces with operators, II. *Trans. Amer. Math. Soc.* **65**(1949), 49–99.

[58] Cohomology theory in abstract groups, III. Operator homomorphisms of kernels, *Ann. of Math.* **50**(1949), 736–761.

1950

[59]* Relations between homology and homotopy groups of spaces, II. *Ann. of Math.* **51**(1950), 514–533.

[60]* Cohomology theory of abelian groups and homotopy theory, I. *Proc. Nat. Acad. Sci. USA* **36**(1950), 443–447.

[61]* Ibid. II. *Proc. Nat. Acad. Sci. USA* **36**(1950), 657–663.

[62] Cohomology theory of abelian groups, *Proc. Int. Congress of Math.*, Cambridge, Mass., 1950, vol. 2, pp. 8–14.

[63] On the 3-type of a complex, *Proc. Nat. Acad. Sci. USA* **36**(1950), 41–48 (with J. H. C. Whitehead).

[64] Duality for groups, *Bull. Amer. Math. Soc.* **56**(1950), 485–516.

1951

[65]* Cohomology theory of Abelian groups and homotopy theory, III. *Proc. Nat. Acad. Sci. USA* **37**(1951), 307–310.

[66]* Homology theories for multiplicative systems, *Trans. Amer. Math. Soc.* **71**(1951), 294–330.

[67] Gilbert Ames Bliss (biographical memoir), *Year Book of the American Philosophical Society for 1951*, pp. 288–291.

1952

[68]* Cohomology groups of abelian groups and homotopy theory, IV. *Proc. Nat. Acad. Sci. USA* **38**(1952), 325–329.

1953

[69]* Acyclic models, *Amer. J. of Math.* **75**(1953), 189–199.

[70]* On the groups $H(\Pi, n)$, I. *Ann. of Math.* **58**(1953), 55–106.

1954

[71] The homology products in $K(\Pi, n)$, *Proc. Amer. Math. Soc.* **5**(1954), 642–651.

[72]* On the groups $H(\Pi, n)$, II. Methods of computation, *Ann. of Math.* **60**(1954), 49–139.

[73]* Ibid. III. Operations and obstructions, *Ann. of Math.* **60**(1954), 513–557.

[74] Curso de Topologia Geral, *Notas de Matematica* no. 11, Instituto de Matematica Pura e Aplicada, Rio de Janeiro, 1954. (This is a Portuguese translation by Joviano C. Valadares of lecture notes issued by the University of Chicago.)

[75] Of course and courses, *Amer. Math. Monthly* **61**(1954), 151–157.

1955

[76]* On the homology theory of abelian groups, *Can. J. of Math.* **7**(1955), 43–53.

[77] Slide and torsion products for modules, *Univ. e Politec. Torino Rend. Sem. Mat.* **15**(1955–56), 281–309.

1956

[78] Homologie des anneaux et des modules, *Colloque de topologie algébrique*, Louvain, 1956, pp. 55–80.

1957

[79] Algebra, *Twenty-third Year Book*, *Natl. Council of Teachers of Math.*, 1957, pp. 100–144.

1958

[80] Extensions and obstructions for rings, *Ill. J. of Math.* **2**(1958), 316–345.

[81] A proof of the subgroup theorem for free products, *Mathematika* **5**(1958), 13–19.

1959

[82] Metric postulates for plane geometry, *Amer. Math. Monthly* **66**(1959), 543–555.

1960

[83] Group extensions by primary abelian groups, *Trans. Amer. Math. Soc.* **95**(1960), 1–16.

[84] Triple torsion products and multiple Künneth formulas, *Math. Ann.* **140**(1960), 51–64.

1961

[85] The cohomology theory of a pair of groups, *Ill. J. of Math.* **5**(1961), 45–60 (with Franklin Haimo).

[86] An algebra of additive relations, *Proc. Nat. Acad. Sci. USA* **47**(1961), 1043–1051.

[87] Locally small categories and the foundations of set theory, *Proc. of* 1959 *Sympos. Foundations of Math.*, Warsaw, 1961, pp. 25–43.

1963

[88] *Homology*, Grundlehren vol. 114, Springer, 1963.

[89] Some additional advances in algebra, pp. 35–58 of *Studies in Modern*

Algebra, edited by A. A. Albert, Math. Assoc. of America Studies in Mathematics vol. 2, 1963.

[90] Natural associativity and commutativity, *Rice Univ. Studies* **49**(1963), no. 4, 28–46.

1964

[91] Oswald Veblen (biographical memoir), *Nat. Acad. of Sci.*, 1964, pp. 325–341.

1965

[92] Leadership and quality in science, Basic research and national goals, Report to committee on science and astronautics, U.S. House of Representatives, *Nat. Acad. of Sci.*, pp. 189–202, March, 1965.

[93] Categorical algebra, *Bull. Amer. Math. Soc.* **71**(1965), 40–106.

[94] Preliminary meeting on college level mathematics education, *Amer. Math. Monthly* **72**(1965), 174–175. (This is a report to the National Science Foundation of a meeting held under the auspices of the U.S. Japan Program on Scientific Cooperation.)

1967

[95] *Algebra*, Macmillan, 1967. (A translation into French by J. Weil was published in two volumes by Gauthier-Villars, 1970, 1971.) (With Garrett Birkhoff). Second edition, 1979.

[96] Projective classes and acyclic models, pp. 78–91 of *Reports of the Midwest Category Seminar*, Springer Lecture Notes no. 47, 1967 (with A. Dold and U. Oberst).

[97] Mappings as a basic mathematical concept, pp. 200–218 in *Journeys in Science*, Twelfth AFOSR Science Seminar, edited by David L. Arm, Univ. of New Mexico, Albuquerque, 1967.

[98] The future role of the federal government in mathematics, *Amer. Math. Monthly*, pp. 92–100 of the Fiftieth Anniversary Issue, Part II of Jan., 1967, vol. 74.

1968

[99] Logic and foundations of mathematics, pp. 286–294 in *Contemporary Logic, a Survey*, edited by Raymond Klibansky, Florence, La Nuova, Italia Editrise, 1968.

[100] Geometrical Mechanics, Parts I and II, *Lecture Notes*, Univ. of Chicago, 1968.

1969

[101] Possible programs for categorists, pp. 123–131 of *Category Theory, Homology Theory and their Applications*. I, Proc. of 1968 Battelle Conf., Springer Lecture Notes no. 86, 1969.

[102] Foundations for categories and sets, pp. 146–164 of Ibid. II, Springer Lecture Notes no. 92, 1969.
[103] One universe as a foundation for category theory, pp. 192–200 of *Reports of the Midwest Category Seminar*. III, Springer Lecture Notes no. 106, 1969.

1970

[104] The influence of M. H. Stone on the origins of category theory, pp. 228–241 of *Functional Analysis and Related Fields*, edited by Felix E. Browder, Springer, 1970.
[105] The Milgram bar construction as a tensor product of functors, pp. 135–152 of *The Steenrod Algebra and its Applications*, Proc. of a 1970 conference honoring N. E. Steenrod's 60th birthday, Springer Lecture Notes no. 168, 1970.
[106] Coherence and canonical maps, pp. 231–242 of *Symposia Mathematica*, vol. IV, Azzoguidi, Bologna, 1970 (conference held March 25, 1969).
[107] Hamiltonian mechanics and geometry, *Amer. Math. Monthly* **77**(1970), 570–586.

1971

[108] Categorical algebra and set-theoretic foundations, pp. 231–240 of *Axiomatic Set Theory*, Proc. Symp. Pure Math. vol. 13, Part I, Amer. Math. Soc. 1971.
[109] Coherence in closed categories, *J. Pure Appl. Algebra* **1**(1971), 97–140. (With G. M. Kelly.)

1972

[110] Closed coherence for a natural transformation, pp. 1–28 of *Coherence in Categories*, Springer Lecture Notes no. 281, 1972 (with G. M. Kelly).
[111] *Categories for the Working Mathematician*, Springer, 1972.

1973

[112] Edwin Bidwell Wilson (biographical memoir), *Nat. Acad. of Sci.*, 1973, pp. 285–320 (with Jerome Hunsaker).

1975

[113] Sets, topoi, and internal logic in categories, pp. 119–134 in *Logic Colloquium* '73 (Bristol, 1973), Studies in Logic and the Foundations of Mathematics, vol. 80, North-Holland, 1975.

1976

[114] The work of Samuel Eilenberg in topology, pp. 133–144 of *Algebra, Topology and Category Theory* (a collection of papers in honor of Samuel Eilenberg), Academic Press, 1976.

[115] Topology and logic as a source of algebra, Retiring Presidential Address, *Bull. Amer. Math. Soc.* **82**(1976), 1–40.

1978

[116] Origins of the cohomology of groups, *L'Enseignement Math.* **24**(1978), 1–29.

1979

[117] Bertrand Russell's contribution to mathematics: A retrospective view, *to appear*.

[118] History of abstract algebra: Origin, rise, and decline, *to appear*.

List of Ph.D. Theses Written Under the Supervision of Saunders Mac Lane

1. Kaplansky, Irving, *Maximal fields with valuations*, Harvard, 1941.
2. Putnam, Alfred L., *Integral domains complete in a valuation*, Harvard, 1942.
3. Phelps, C. R., *A homogeneous algebra with limited associativity*, Harvard, 1942.
4. Lyndon, Roger, *The cohomology theory of group extensions*, Harvard, 1946.
5. Moyls, B. N., *Extensions of valuations with prescribed value groups and residue class fields*, Harvard, 1947.
6. Carter, W. C., Jr., *On the cohomology theory of fields*, Harvard, 1947.
7. Novosad, Robert S., *Relations between homotopy and homology groups*, Chicago, 1952.
8. Hughart, Stanley P., *Representations for dicategories*, Cal. Tech., 1954.
9. Nunke, Ronald J., *Modules of extensions over Dedekind rings*, Chicago, 1955.
10. Kruse, Arthur H., *Introduction to the theory of block assemblages and developments in the theory of retraction*, Chicago, 1956.
11. Howard, William A., *k-fold recursion and well-ordering*, Chicago, 1956.
12. Nerode, Anil R., *Composita, equations, and recursive definitions*, Chicago, 1956.
13. Ballard, William R., *Cohomology theory in fields*, Chicago, 1957.
14. Halpern, Edward, *On the structure of hyperalgebras*, Chicago, 1957.
15. Thompson, John G., *A proof that finite groups with a fixed-point-free automorphism of prime order are nilpotent*, Chicago, 1959.
16. Yao, Joseph, *Moore-Cartan theorems and Leray-Serre theorem*, Chicago, 1960.
17. Szczarba, Robert, *The homology of twisted Cartesian products*, Chicago, 1960.
18. Liulevicius, Arunas, *The factorization of cyclic reduced powers by secondary cohomology operations*, Chicago, 1960.
19. Kristensen, Leif, *On the cohomology of two-stage Postnikov systems*, Chicago, 1961.
20. Morley, Michael, *Categoricity in power*, Chicago, 1962.
21. Hungerford, Thomas W., *Bockstein spectra*, Chicago, 1963.
22. Solovay, Robert, *A functorial form of the differentiable Riemann-Roch theorem*, Chicago, 1964.
23. Kuo, T. C., *Universal objects for spectral sequences*, Chicago, 1964.

24. Mountjoy, Robert H., *Abelian varieties attached to representations of discontinuous groups*, Chicago, 1964.
25. Zvengrowski, Peter, *Vector fields and vector products*, Chicago, 1965.
26. MacDonald, John, *Group derived functors and relative representability*, Chicago, 1965.
27. Schafer, James, *On the homology ring of an abelian group*, Chicago, 1965.
28. Shafer, David, *The Hamel basis theorem and the countable axiom of choice; an exercise in the method of Paul Cohen*, Chicago, 1966.
29. Palmquist, Paul, *The double category of adjoint squares*, Chicago, 1969.
30. Stauffer, Howard, *Completions of categories, satellites, and derived functors*, Berkeley, 1969.
31. Dubuc, Eduardo, *V-completion by V-monads through the use of Kan extensions*, Chicago, 1969.
32. Eisenbud, David, *Torsion modules over Dedekind prime rings*, Chicago, 1970.
33. Voreadou, Rodiani, *A coherence theorem for closed categories*, Chicago, 1972.
34. Hamsher, Ross, *Eilenberg-Mac Lane algebras and their computation; an invariant description of* $H(\Pi, 1)$, Chicago, 1973
35. Decker, Gerald John, *The integral homology algebra of an Eilenberg–Mac Lane space*, Chicago, 1974.
36. Ginali, Susanna M., *Iterative algebraic theories, infinite trees and program schemata*, Chicago, 1976.